Fascinating Life Sciences

This interdisciplinary series brings together the most essential and captivating topics in the life sciences. They range from the plant sciences to zoology, from the microbiome to macrobiome, and from basic biology to biotechnology. The series not only highlights fascinating research; it also discusses major challenges associated with the life sciences and related disciplines and outlines future research directions. Individual volumes provide in-depth information, are richly illustrated with photographs, illustrations, and maps, and feature suggestions for further reading or glossaries where appropriate.

Interested researchers in all areas of the life sciences, as well as biology enthusiasts, will find the series' interdisciplinary focus and highly readable volumes especially appealing.

Vincent L. Bels • Anthony P. Russell
Editors

Convergent Evolution

Animal Form and Function

Springer

Editors
Vincent L. Bels
Institute of Systematics,
Evolution, Biodiversity, ISYEB – UMR
7205 – CNRS/MNHN/EPHE/UA
National Museum of Natural History
Sorbonne University
Paris, France

Anthony P. Russell
Department of Biological Sciences
University of Calgary
Calgary, AB, Canada

ISSN 2509-6745 ISSN 2509-6753 (electronic)
Fascinating Life Sciences
ISBN 978-3-031-11440-3 ISBN 978-3-031-11441-0 (eBook)
https://doi.org/10.1007/978-3-031-11441-0

Front cover photos: *Paraxerus cepapi* © Grégoire Boulinguez-Ambroise, *Loxodonta africana* © Grégoire Boulinguez-Ambroise, *Ateles geoffroyi* © Florence Kirchhoff, *Ailurus fulgens* © Florence Kirchhoff.

This Springer imprint is published by the registered company Springer Nature Switzerland AG
The registered company address is: Gewerbestrasse 11, 6330 Cham, Switzerland

The editors wish to dedicate this volume to the memory of Professor David Wake (July 8, 1936–April 29, 2021) who was a source of inspiration in initiating this volume on convergent evolution.

Preface

Convergence is a concept of paramount importance for explaining similarities, across all length scales (from macroscopic to the nanometer level) between organisms whose similarities cannot be attributed to the sharing of features due to common ancestry. The investigation of evolutionary convergence is conducted in a variety of ways and employs various methodological, technical, and conceptual approaches.

The idea for this book, which focuses on convergence of functional approaches to environmental challenges, was spurred by the plethora of studies that explore this phenomenon from the perspective of pattern, process, or both. Our aim is not to provide a catalogue of the thousands of examples of biological convergence but to provide thoroughly considered examples at several levels within the taxonomic hierarchy of animals, both extant and extinct. Since the 1980s, functional convergence has been a major focus of investigators seeking to understand how organisms respond to their invasion of new habitats or to changing environmental conditions over evolutionarily long spans of time.

We are extremely grateful to the many authors who have contributed to this volume. They have provided fascinating insights into the extent of convergence between distantly related organisms and the role that phylogenetic history plays in the structural responses to major challenges imposed by the environment. We are also greatly appreciative of the selfless contributions made by reviewers of each chapter who devoted their time and energy to improving the chapters. We are particularly grateful to P. Aerts, A. Bauer, Ch. Boehmer, C. D'Août, Ph. Cox, P. Flammang, P. Gignac, S. Gorb, E. Heiss, O. Lambert, D. Schmitt, M. Segal, R. Tague, C. Turcotte, S. Van Wassenberg, and A. Werth.

Paris, France — Vincent L. Bels
Calgary, AB, Cananda — Anthony P. Russell

Contents

Chapter 1
The Concept of Convergent Evolution and Its Relationship to the Understanding of Form and Function

Vincent L. Bels and Anthony P. Russell

Within the field of evolutionary biology, divergence and convergence are two major phenomena that have helped shape the diversity and disparity of the Earth's biota throughout the history of life. Exploration of them has contributed to the interpretation of dissimilarities (divergence) and similarities (convergence) in organismic form, function and behaviour at various hierarchical levels and how they favour, in some fashion, the emergence of optimal traits via natural and/or sexual selection across the full spectrum of occupied environments.

Convergent evolution is one of the most commonly invoked evolutionary processes in macro- and micro-evolutionary studies for explaining the repeated appearance of organismic forms and traits. Agrawal (2017) emphasizes that "*The search for convergent evolution and its causes is one way to make sense of the wonderfully bewildering biological diversity on our planet*". Various definitions of convergence have been advanced, but all attempt to explain the resemblance between traits resulting from the influence of proximal and ultimate factors impinging upon organisms in their habitats. Stated most concisely, convergence is "*the evolution of the same or very similar traits independently in different lineages of organisms*" (McGhee, 2011). At the organismal level this results in phenotypic similarity (Mahler et al., 2017) either in particular suites of features or in whole body form (Arbuckle & Speed, 2016; Arbuckle et al., 2014; Almeida et al., 2021). Historically this similarity, initially established with regard to the form of organisms, became further refined through the description of anatomical traits that related morphological

V. L. Bels (✉)
Institute of Systematics, Evolution, Biodiversity, ISYEB – UMR 7205 – CNRS/MNHN/EPHE/UA, National Museum of Natural History, Sorbonne University, Paris, France
e-mail: vincent.bels@mnhn.fr

A. P. Russell
Department of Biological Sciences, University of Calgary, Calgary, AB, Canada

V. L. Bels, A. P. Russell (eds.), *Convergent Evolution*, Fascinating Life Sciences,
https://doi.org/10.1007/978-3-031-11441-0_1

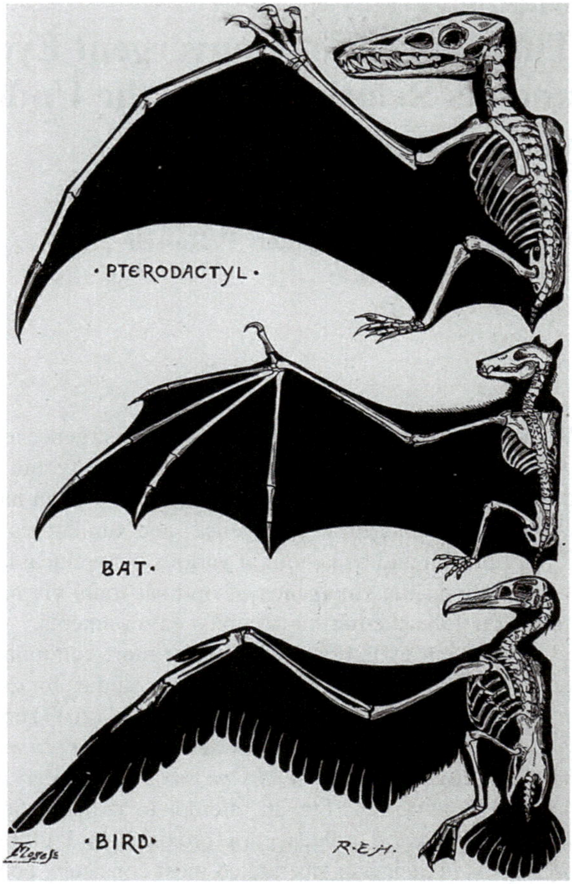

Fig. 1.1 A classical historical view of how homologous morphological traits of the forelimbs of distantly related vertebrates have evolved in a similar fashion in response to similar functional demands (powered flight) acting as selective forces. (**a**) Pterosaur, (**b**) Bat, (**c**) Bird (source: Wikipedia.org/wiki/Convergent_evolution#/media/File:Homology.jpg)

form to function in different groups of organisms, a classical case in point being modifications relating to powered flight (Fig. 1.1). The literature is rich in such examples of evolutionary convergence (and parallelism) that are based upon comparative studies at different scales of the classificatory hierarchy.

In this volume we explore the idea of convergent evolution across a broad range of phylogenetic scales, focusing upon the ways in which environmental challenges drive organisms towards similar outcomes through the evolution of complex form-function systemic adaptations (e.g., Shadwick, 2005; Quinn, 2020; Ballell et al., 2019; Gómez & Lois-Milevicich, 2021; Higham et al., 2021). The contemporary study of convergent evolution, as it relates to form and function, employs a variety of approaches, allowing form to be investigated at various levels, from the overall shape of the structures to their constituent elements. The application of novel technological imaging approaches (such as 2D and 3D geometric morphometrics and digital measurement; 3D tomography; surface scanning; digital reconstruction; finite element analysis) has become increasingly prevalent (Brainerd & Camp, 2019; Rayfield, 2019). These investigative tools have become important for addressing "*big evolutionary questions*" (Agrawal, 2017) and have enabled demonstration of the complexity of the traits involved in the realization of perceived function.

1.1 Recognition of the Phenomenon of Convergent Evolution

What ultimately became recognized as convergent evolution had its beginnings in the investigation and interpretation of morphology. Although no longer restricted to morphology, the issues raised during the initial investigation of what came to be recognized as convergent pathways in the history of life still influence the understanding and discussion of this concept.

The study of morphology experienced fundamental changes throughout the nineteenth century, in parallel with the emergence of biology as a unified discipline devoted to the scholarly exploration of life. Three major phases of research on animal morphology occurred throughout the nineteenth century (Russell, 1916). In the early years, during which Wolfgang Goethe characterized the study of organismal form as morphology (Nyhart, 1995; Singer, 1959), a transcendental approach was carried over from the previous century. In the second half of the nineteenth century this gradually gave way to an evolutionary approach (led by scholars such as Ernst Haeckel and Carl Gegenbaur), which emphasized phylogenetic trees and employed the methods of comparative anatomy to try to explain form (Nyhart, 1995). This, in turn, was superseded in the latter part of the century by a causal approach to morphology that integrated experimental embryology (as practiced by Wilhelm Roux). Collectively these changes characterized a transition from "structural" to "functional" morphology as the century progressed (Singer, 1959).

These shifts in the study of animal form resulted in the replacement of the idea that purpose was sufficient for explaining both organic structure and the succession of life through geological time (commonly held by morphologists at the beginning of the nineteenth century; Ospovat, 1978) by a nascent non-teleological interpretation of form by its end. Even so, morphology as a whole was not particularly successful in the assimilation of evolutionary ideas into its *modus operandi* (Ghiselin, 1980). This was so because the evolutionary approach to morphology, embarked upon as Darwinian reasoning gained increasing acceptance, brought with it fundamental questions about structural resemblances as indicators of evolutionary relationship (Bowler, 1996). Morphology played a prominent role in attempts to reconstruct patterns of historical connectivity (Bowler, 1996). Such investigations soon revealed that structural resemblances could not be universally explained as resulting from common descent and were thus not always reliable indicators of phylogenetic affinity. Instead, it became increasingly evident that structural resemblances could be independently acquired on different branches of the emerging tree of life. For morphologists the possibility of the same structure evolving independently in more than one group became problematic because, if evolution was to be a guide to classification, then shared characteristics should be a consequence of common ancestry (Bowler, 1996). It gradually became realized that when two or more originally very distinct forms came to occupy the same general environment they could, independently, acquire similar structural characteristics (Fig. 1.1). As a result, what became recognized as convergent evolution was identified as a topic worthy of investigation in its own right.

In the 1840s Richard Owen advanced the concepts of homology and analogy for, respectively, structural similarities fundamental to idealistic archetypes and superficial adaptive modifications (Bowler, 1996). Darwin, however, intimated that adaptive specialization was an inextricable consequence of natural selection and that, therefore, homology and analogy had to be assessed in a genealogical, rather than idealistic, framework. Lankester (1870) addressed the multiple origins of seemingly shared characteristics from a Darwinian perspective and coined the term homoplasy for adaptive analogies that were not due to common descent. Osborn (1902) subsequently advocated that convergence was the result of similar adaptation to particular environmental conditions.

Even so, morphologists were generally reluctant to acknowledge convergence as a major feature of evolution. Indeed, Willey (1911) regarded it as being essentially a hindrance to the identification of "genuine" homologies and a detraction from the reconstruction of the tree of life, which he saw as the primary goal of evolutionary morphology. Morphologists of the time lacked techniques that were able to unequivocally distinguish homologies from homoplasies (Bowler, 1996). Ultimately, they had to grapple with the competing ideas of whether evolution was driven by functional adaptation or by formal constraints dependent upon developmental attributes innate to the organism. By the 1940s, however, "The Modern Synthesis" (Huxley, 1942) enabled accommodation of the demands of both form and function. Interpretations of convergence were thus able to be framed within this unifying philosophy, with common descent implying that evolutionary novelties have arisen only once and that convergence results when putative evolutionary novelties are interpreted to have had more than one origin, reflecting similar adaptive responses to particular environmental challenges. Increasingly detailed morphological analyses have shown that seemingly identical structures often exhibit lineage-specific characteristics that confirm their homoplastic nature (e.g. Bergmann & Morinaga, 2018).

1.2 Approaches to the Study of Convergent Evolution in Contemporary Biology

Various terms have been employed in attempts to define and explain convergent evolution. Prominent among these are homology and homoplasy (see above), which Ochoa and Rasskin-Gutman (2015) regard, along with homogeny and parallelism, as evolutionary developmental (evo-devo) mechanisms acting along an evolutionary continuum. Hall (2007) and Wake (2013) conceptualize these various mechanisms as those relating to deep or more recent shared ancestry and that are based upon shared cellular mechanisms, processes, genes, and genetic pathways and networks.

Arendt and Reznick (2008) opine that distinguishing between evolutionary convergence and parallelism is problematic because these terms infer mechanisms that are seemingly dependent on the phylogenetic level at which the comparison is undertaken, as well as upon underlying genetic mechanisms that largely remain uninvestigated. They suggest that convergence be invoked for all evolutionary hypotheses that relate the independent acquisition of phenotypic form: "*Biologists often distinguish 'convergent' from 'parallel' evolution. This distinction usually assumes that when a given phenotype evolves, the underlying genetic mechanisms are different in distantly related species (convergent) but similar in closely related species (parallel). However, several examples show that the same phenotype might evolve among populations within a species by changes in different genes. Conversely, similar phenotypes might evolve in distantly related species by changes in the same gene. We thus argue that the distinction between 'convergent' and 'parallel' evolution is a false dichotomy, at best representing ends of a continuum. We can simplify our vocabulary; all instances of the independent evolution of a given phenotype can be described with a single term - convergent.*" Contrastingly De Lisle and Bolnick (2020) retain the two terms that Arendt and Reznick (2008) regard as representing a false dichotomy and argue that "*...conserved directional selection across lineages could result in parallelism without convergence. Alternatively, adaptation towards a shared optimum by lineages with unique evolutionary histories, and thus unique ancestral positions in trait space, could result in convergence without parallelism. Yet, parallel evolutionary processes can lead to divergence if, for example, one lineage evolves faster along a shared trajectory. Thus, separating parallelism and convergence may often be necessary to link evolutionary pattern and process*".

McGhee (2018), seemingly in at least partial agreement with Arendt and Reznick (2008), provides another viewpoint on the employment of terminology for distinguishing between the outcomes of evo-devo processes that operate along an evolutionary continuum. He suggests broadening the scope of the term "convergence" by recognizing three categories: allo-convergent evolution to account for the independent evolution of the same or very similar traits from different precursors in different lineages (= the conventional representation of convergence); iso-convergent evolution to account for the independent evolution of the same or very similar traits from the same precursor in different lineages (= the classical view of parallelism—see, for example, Riedel et al., 2021); and retro-convergent evolution to account for the independent re-evolution of the same or very similar ancestral trait in different lineages (= the classical view of reversal). Whether such revised terminology will become more universally applied in future remains to be seen. Based upon developmental processes, Hall (2013) regards homology and parallelism as concepts associated with features that arise from similar developmental processes (thus incorporating iso-convergent evolution as proposed by McGhee, 2018); reversals as potentially, but not necessarily arising from different developmental processes (thus incorporating [depending upon the developmental evidence that applies] retro-convergent, iso-convergent or allo-convergent evolution as proposed by McGhee, 2018); and convergence being dependent upon features arising from different developmental processes and characterized by homoplasies [thus incorporating allo-convergent evolution as proposed by McGhee (2018)].

It is evident from the foregoing that different authors regard convergence in various ways, leading to inconsistency and potential confusion in the application of the concept and the sorts of phenomena to which the term "convergence" should be applied. In this volume we deal essentially with what McGhee (2018) regards as "allo-convergent evolution" and use "convergence" in this sense, as advocated by Hall (2013) and Ochoa and Rasskin-Gutman (2015). Convergence, regarded this way, can result from the operation of various evolutionary processes (adaptation, constraint, development) or from purely stochastic events (Losos, 2011; Stayton, 2015a; Agrawal, 2017; Speed & Arbuckle, 2017) and appears to be widespread throughout the history of life. Two approaches have been taken to analyse the occurrence of the repetition of similar phenotypic traits of organisms at all levels: "pattern-based" and "process-based" (Stayton, 2015a). Pattern-based explanations rely upon the estimation of convergent evolution by comparing "terminal" traits without invoking any particular underlying evolutionary process, the observed pattern potentially resulting from shared or different evolutionary processes or simply from stochastic events. Process-based explanations, in contrast, incorporate adaptive and developmental factors (Stayton, 2015a), with convergence regarded as having arisen through shared evolutionary processes, such as natural selection, in response to similar internal or external environmental challenges. Interpretations of the convergence of form and function are almost exclusively process-based.

Recently, increasingly quantitative explorations of various traits, assessed in a phylogenetic context, have been employed to identify convergent patterns at various levels within the hierarchy of organismic structure and behaviour (for example Morris et al., 2018; De Lisle & Bolnick, 2020). In such instances convergence is viewed as a deterministic process that drives patterns that can be quantified and compared (De Lisle & Bolnick, 2020), in contrast to contingency that underpins less predictable, and thus less repeatable, outcomes. The potential for convergence in quantifiable traits, however, is dependent upon their own evolutionary history that is itself subject, at any time, to some degree of contingency (Blount et al., 2018; Powell, 2020).

During the last 20 years conceptual and methodological (such as the application of phylogenetic algorithms) analyses have enabled the interpretation of whether traits recorded in various lineages are attributable to the operation of selective pressures or simply the result of stochastic events (whereby random evolutionary changes result in descendant species becoming more similar to one another than were their ancestors) (Losos, 2011). Wake (2013) indicated that studying convergent evolution may provide insights into both the proximate and ultimate mechanisms that generate diversity and can be informative concerning the extent to which the evolutionary process is both repeatable and predictable (Mahler et al., 2013). Employment of robust phylogenetic hypotheses coupled with more stringent analyses of morphological data (Assis et al., 2011), together with the application of molecular, genomic and developmental techniques, will enable a more integrated investigation of the mechanisms through which similarity arises (Wake et al., 2011; Hall, 2013). Sackton and Clark (2019) note that widespread gene sequencing fosters the investigation of how convergence evolves at the molecular level and indicates that there is increasing evidence of correlation between morphological and genetic information with regard to trait convergence. Changes in regulatory genes, such as homeobox (Hox) genes, that govern the expression of traits make it likely that some traits will appear repeatedly in closely- as well as distantly-related taxa, resulting in convergence. Speed and Arbuckle (2017) and Lamichhaney et al. (2019) suggest that new genomic data sets will enable detailed and tractable analysis of the genetic underpinnings of convergent phenotypes. The examination of genomic data relating to morphological, behavioural, physiological and developmental traits holds great potential for revealing a mechanistic understanding of convergent phenotypes, especially if taxa expressing intermediary conditions along evolutionary continua are incorporated (rather than comparing only binary endpoints) (Lamichhaney et al., 2019).

The phylogenetic history and relationships of particular organismic groups influence the actual expression of their adaptive complexes, so in this volume we also examine the potential constraints that lead to differences in morphological expression in the face of similar environmental challenges. In recent years comparative phylogenetic analyses of sets of structural traits have led to the recognition of what has been called "imperfect convergence" (Bergmann & Morinaga, 2018) or "incomplete convergence" (Grossnickle et al., 2020). For example, Bergmann and Morinaga (2018) considered the convergence upon snake-like body form in six squamate clades and noted that each lineage has, from its different structural starting point, evolved a slightly different expression of it, accompanied by convergent changes in associated anatomical systems. Bergmann and Morinaga (2018) emphasized the important role that historical contingency plays in the attainment of a particular morphological form (Dolezal & Lipps, 2019). Such so-called imperfect or incomplete convergence is what might be expected, because lineages may be anticipated to differ in their phenotypic responses to similar selective demands (Collar et al., 2014) due to constraints imposed by the ancestral forms from which the convergent pathways begin. Different approaches, leading to different morphological combinations that generate essentially the same functional outcomes (Thompson et al., 2017; Russell & Gamble, 2019), are underpinned by genetic (Hu et al., 2017), developmental (Gutierrez et al., 2019; Tokita et al., 2012), phylogenetic, morphological and physical factors that affect the modularity of the convergent structures (Fernández et al., 2020; Grossnickle et al., 2020; Quinn, 2020).

Stayton (2015b) notes that although pattern-based and process-based definitions can both legitimately be applied to biological phenomena, most measures of convergence that are based upon the quantification of patterns do not demonstrate the action of any particular process. Furthermore, he notes that when the distinction between modalities for interpreting convergence is not explicitly stated, assumptions about the potential underlying process are often made, thereby obfuscating alternative explanations. As Mahler et al. (2017) state, convergence itself is not indicative of any particular evolutionary process and only through careful analysis can a particular process be invoked. Indeed, Powell and Mariscal (2015) note that it is difficult to demonstrate non-convergence because many of the statements advocating it as being responsible for observed pattern are too vague. Accordingly, it is advocated that robust evidence be presented for cases to qualify as "true" cases of convergence.

1.3 Convergence of Form and Function

Formally, the concept of convergence is based upon analogous traits in relation to their adaptive value, wherein analogy relates to similarity of function in structures with different origins. In many instances analogous structures tend to become similar in appearance, which leads to the invocation of the concept of convergence of form. In similar fashion, analogous convergence may also occur for physiological processes (for example, Schott et al., 2019) and behaviour. Extension of the concept of convergence into studies in the realm of the "-omics" has assisted in our understanding of the underlying processes that drive convergence. Waters and McCulloch (2021) note that genomic analyses have led to the recognition of convergence as the major driver of repetitive evolution above the species level, whereas repeated sorting of standing variation is the main impetus for repeated shifts within species. The latter is exemplified by the alternation of direction of change in Darwin's finches on the Galapagos Islands, as documented by Grant et al. (2004). Over a period of several decades they noted that species lacking postmating barriers to gene exchange can alternate between what manifests as convergence and divergence as environmental conditions oscillate.

Many papers that deal with convergence, whether focusing on pattern or process, allude to function and thus address functional convergence, with function playing a key role in explanations of the similarity of phenotypic traits (Ochoa & Rasskin-Gutman, 2015). Indeed, Powell and Mariscal (2015) and Kuhn et al. (2020) define convergence as being an evolutionary phenomenon that results in the independent origin of both form and function. The incorporation of function into the concept of convergence, as applied to comparative studies, has fostered the generation of hypotheses about how organisms may respond similarly to selective pressures when under the influence of potential constraints (Casinos & Gasc, 2002).

Investigations of functional convergence generally focus on aspects of form that relate to recurrent "everyday behaviour" (e.g., foraging, feeding, ventilation, locomotion, adhesion) that influences the survival of the organisms through the maintenance of homeostasis, and on aspects of social behaviour (e.g., communication) that involve intra- and interspecific interactions between individuals (e.g., territoriality, social systems). Various examples that attempt to assess convergence in different disciplines show that the concept of function is, however, complex because it encapsulates various factors that might drive evolutionary convergence. Thus, the assessment of "functional convergence" may be conducted to explain similarities of any phenotypic traits across the organizational and structural spectrum from molecules to behaviour. All such behaviours are under complex sensory-motor control governed by neuro-hormonal mechanisms. With regard to this, Fischer et al. (2019) state: "*A major strength of evolutionary comparisons is the ability to determine whether shared genomic and/or neural mechanisms are associated with similar behaviours across species or if there are many alternative mechanistic 'solutions' that can produce similar behaviours.*".

Regardless of whether or not any particular evolutionary process is invoked to account for the convergence advocated, the literature is replete with accounts of similarity attributed to convergence. For almost all animal lineages quantitative analyses of phenotypic traits, particularly those relating to morphology, reveal multiple instances of convergence (Fig. 1.2).

Powell and Mariscal (2015) note that the strongest examples of convergence are those representing adaptive complexes that perform the same function and that incorporate several hierarchical layers of underlying form (Speed & Arbuckle, 2017). Such assessments stem from the recognition that certain design "problems" are pervasive in the history of life and can only be "solved" by a limited number of structurally specific outcomes. These may be regarded as being ecomechanical challenges (Higham et al., 2021) that are based upon "simple" physical laws and are perceived to play a key role in impacting performance. Such challenges are seen to be agents of selection but are modulated in different lineages by the internal constraints imposed by phylogenetic history.

Given that when colonizing new environments organisms are challenged to adapt, both morphologically and behaviourally, to survive and thrive (Kowalko et al., 2013), it is evident that such challenges initiate responses governed by physical and mechanical demands (Higham et al., 2021). These demands potentially drive morphological systems (and the behaviours responsible for their operation) toward similar endpoints. Such responses lead to ecomorphological patterns (*sensu* Norton & Brainerd, 1993) related to those environmental challenges (Toussaint et al., 2019).

Comparative studies related to such challenges tend to reveal that a relatively limited number of morphological attributes account for the similarities between organisms. For example, many recent studies of birds, fishes and marine vertebrates, at various taxonomic levels, tend to confirm the notion that form-function associations will reflect convergence towards predictable trait combinations, suggesting that

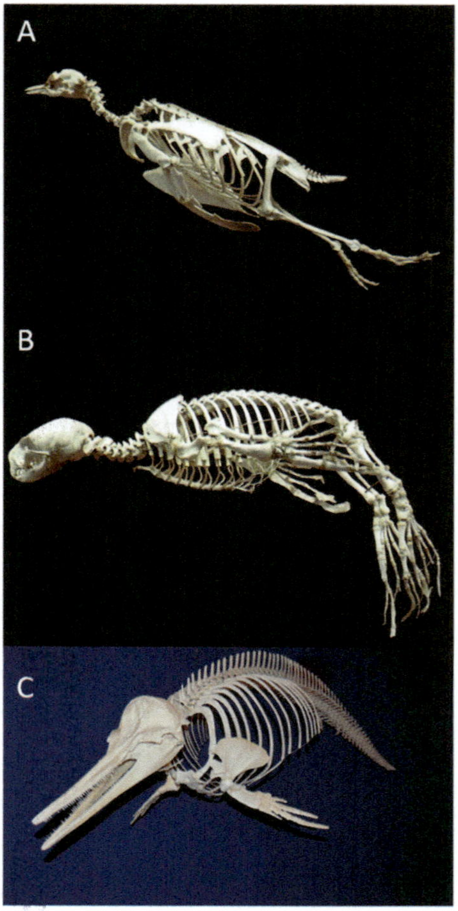

Fig. 1.2 Examples of convergent morphology in two distantly related vertebrate classes (birds and mammals) in relation to modifications for swimming in the marine environment. (**a**) penguin, (**b**) pinniped, (**c**) whale (Photographs courtesy of Eric Pellé, Muséum national d'Histoire naturelle, Paris, France)

morphological variation is channeled into a limited set of possibilities (Pigot et al., 2020) that lead to similar structural adaptations (Donley et al., 2004; Shadwick, 2005). Repeated adaptive radiations within the same group frequently produce convergent forms (Mahler et al., 2013). Although these are often encountered in geographic isolation, it is becoming evident that such convergent patterns may also occur in sympatry in different clades within radiations (Muschick et al., 2012; Llaurens et al., 2021). This occurs when the number of species exceeds the number of available niches. In such instances, sister taxa tend to deviate from one another in morphospace (Muschick et al., 2012) and come to resemble species in other lineages.

The way in which any particular functional response under investigation is related to convergent evolution is dependent upon the discipline or domain within which comparative studies are conducted. Behavioural traits have been the focus of study for many interested in responses to functional demands that lead to convergence.

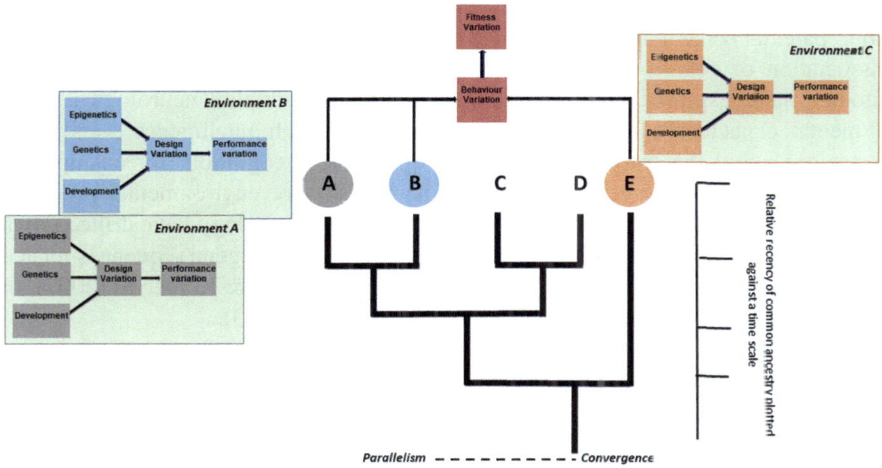

Fig. 1.3 Theoretical framework based on Arnold's (1983) paradigm of support for evolutionary convergence. Each highlighted species A, B, E, is exposed to similar environmental challenges and responds to them. At one level the different environments impose, overall, essentially similar challenges (e.g., aquatic environment), but at other levels may have more particular properties (e.g., surface of the substrate, narrow branches in the arboreal habitat). These responses constitute the links between organismal properties (e.g., design variation and performance variation) resulting from genetic, epigenetic, and/or developmental processes, or from stochastic events (Losos, 2011). These links result in similar responses (behavioural or functional variation) that enhance their fitness in the environment. A, B, phylogenetically closely-related species, E, phylogenetically distantly-related species relative to A and B

Arnold's (1983) paradigm (Fig. 1.3) has been adopted by many when assessing anatomical and functional morphological (including paleontological) approaches to questions relating to design (*sensu* Lauder, 1996) and fitness (e.g., Garland Jr & Losos, 1994; Reilly & Wainwright, 1994; Johnson et al., 2008; Lailvaux & Husak, 2014; Irschick & Higham, 2016; Binning et al., 2017). Within this framework particular behaviours or functions (constituting "performance") are viewed as resulting from one or several traits being used by an individual throughout its life in response to environmental challenges. Characterization of such performances is employed to determine the potential for convergence in distantly-related organisms. Such performances are defined as "*...the ability of an animal to conduct a key task...*" (Irschick & Higham, 2016). Many approaches to the study of functional convergence in the domains of ethology (including neuroethology) and behavioural ecology have focused on the mechanisms of neuro-motor and physiological (e.g., hormonal) control of morphological structures and the manner in which these influence various responses, including actions/movements, complex sensory-motor coordination and control, cognition and memory, and thus performance. For example, several studies have demonstrated that convergence has probably played a major role in the evolution of the anatomy and functioning of the nervous system and the behaviours that it controls. Consideration of convergence has also focused upon the expression of increasingly complex behaviours or abilities (such as, social behaviour, parental behaviour, nest-building), as illustrated by many studies devoted to the association of cognitive abilities, brain structure, and behaviour. Some of these studies have explored the convergent evolution of complex neurological traits ("mental capacities", *sensu* Roth, 2015) such as "high intelligence", and raise questions about the convergent evolution of advanced cognition (for example, in corvid birds, apes and cetaceans) under shared selective environmental pressures (Seed et al., 2009). Some approaches to convergence in ethology have drilled down to the genomic level, investigating the recurrent evolution of gene sequences related to echolocation patterns in different lineages of bats (Jones & Holderied, 2007), or more broadly between bats and cetaceans (Parker et al., 2013).

1.4 The Objectives and Structure of this Volume

McGhee (2011) notes that, in reality, convergent evolution (except in cases in which resemblance in form results from purely stochastic processes) is grounded in the interplay between form and function, a recognition that is well-supported by the literature in general. Indeed, an understanding of the interplay between form, performance(s), and behaviour(s) in the context of determining function is essential for understanding how organisms survive, persist in their ecological environment, and reproduce (Lauder, 1996). In this book we provide examples of how functional considerations are necessary for understanding the integrative evolution of complex adaptations (*sensu* Frazzetta, 1975) in organisms faced with similar environmental challenges. The physical challenges imposed by the environment and the ecological factors that drive convergence (Donley et al., 2004; Shadwick, 2005; Higham et al., 2021) are emphasized in each chapter.

As noted above, and as argued by McGhee (2018), establishing the boundaries for what constitutes convergent evolution, as opposed to parallelism and reversal, is conceived of differently by different authors. The contributors to this volume (who report on taxa and aspects of functional morphology that fall within their areas of expertise) were invited to select and document cases of evolution of form and function that they perceived to be driven by environmental abiotic and/or biotic challenges and that have resulted in convergence. The chapters comprising this book show that examples have been selected across a broad phylogenetic spectrum, representing different levels in the classificatory hierarchy. At the least inclusive level purported convergence among lineages within a single order (Chaps. 2–4) form the basis for investigation. In contrast, at the most inclusive level convergence of structure and function are contemplated across entire Classes (Chaps. 11, 12, 14, and 15) or even Phyla (Chaps. 13 and 16).

The sequence of chapters in this book reflects the phylogenetic breadth of the taxa selected for study, from least to most inclusive, rather than the clustering of chapters by perceived similarity of the phenotypic features or biomechanical challenges being considered. All chapters stress the need for integrative approaches for the elucidation of both pattern and process as they relate to convergence. Employment of multi-pronged approaches to the study of convergence (Mahler et al., 2017) characterizes the included contributions. We adopt the tenet of Mahler et al. (2017) that evolutionary pattern and process cannot be decoupled in comparative studies of convergence, even though such a decoupling could, in theory, guard against adaptationist bias.

Chapter 17 focuses on how links between convergent evolution and bioinspiration can be forged. Such a chapter has been included because we believe that the study of convergence can be instrumental in refining approaches to biomimetic applications of biological knowledge.

In Chaps. 2–4, convergent patterns are explored at the ordinal level within the confines of a single taxon (from Insecta [Arthropoda] to Mammalia) in relation to eco-mechanical challenges (*sensu* Higham et al., 2021). Nel and Piney (Chap. 2) discuss the highly complex morphological structure of wing venation in representatives of the superorder Odonatoptera (Insecta), emphasizing convergence in both extant and extinct representatives. Gomes-Rodriguez et al. (Chap. 3) demonstrate for rodents (Mammalia) the impact of the fossorial habitat on digging abilities as they relate to behavioural, functional, and morphological convergence. Examination of traits related to the long-distance communication songs of mysticetes (baleen whales) by Park et al. (Chap. 4) exemplifies the difficulties associated with approaching the question of convergence of sensory function and discusses the problems related to the demonstration of potentially convergent patterns in the context of our limited understanding of hearing in the largest creatures in the oceans.

Chapters 5–10 explore the question of convergent evolution between more distantly-related taxa within groups of organisms with relatively similar “Bauplans”. Pallandre et al. (Chap. 5) address the question of convergence of the articulation between the hind limbs and the axial skeleton within the Carnivora and contrast this with the situation in ungulates (Mammalia). They compare the form and function of the sacroiliac joint in relation to forces generated during predatory bouts that are associated with different behaviours and strategies in felids and canids, thereby exploring the functional morphology of this arthrological complex beyond that related to routine locomotor associations.

Three chapters (Chaps. 6–8) investigate convergence of approaches to aquatic feeding within different Orders of vertebrates. A salient point of these chapters is the application of the comparative approach to what are considered homologous and convergent traits within these Orders. Heiss and Lemell (Chap. 6) approach the question of convergent evolution in extant lissamphibians, a group generally characterized by an aquatic larval stage. In many species, however, there is a secondarily reduced, free-living larval stage that undergoes direct development as an adaptation to terrestrial life. After metamorphosis, some lissamphibians become terrestrial, others adopt a semiterrestrial/semiaquatic lifestyle, and others remain fully aquatic. The authors summarize the form and function of the trophic structures in these vertebrates to reveal how morphological features have evolved independently in several lineages to increase the efficiency of suction feeding, one of the major modes of aquatic feeding strongly influenced by hydrodynamic constraints. Regarding the

Sauropsida, Heiss et al. (Chap. 7) compare the spectrum of convergently-evolved responses to aquatic feeding, again centring their arguments on the hydrodynamic constraints imposed by the physical properties of water. Werth and Marshall (Chap. 8) demonstrate the large number of instances of convergence of the trophic system in relation to secondarily aquatic feeding in mammals, exploring both form (e.g., jaws, dentition, musculature, overall shape of the head and mouth) and function (e.g., methods for separating food from water, neural and behavioural adaptations for locating and capturing prey).

Chapters 9 and 10 deal with the convergence of adhesive mechanisms in terrestrial habitats in two distantly related Orders. Russell and Garner (Chap. 9) examine convergence in the form and function of the adhesive system within the Squamata by comparing geckos and anoles. Subsequent to considering the purely physical aspects that govern the functioning of dry, fibrillar adhesive devices at the microscopic and nanoscopic levels, they examine the spectrum of morphological configurations of the adhesive system that have evolved within the Gekkota and assess which of these configurations most closely resembles that exhibited by anoles. Büscher and Gorb (Chap. 10) examine patterns of convergence of the adhesive system among insects. They document adaptive solutions at various scales from similar attachment organs in different lineages to different organ properties in the same lineage, drawing attention to the biomimetic potential that these diverse form-function mechanisms present. Both of these chapters demonstrate the complexity required for organisms to attach to and release from the substrata they exploit during locomotion and document the physical and organismic drivers of convergence of the design of fibrillar adhesive systems.

Chapters 11–16 explore functional convergence across even broader scales, comparing structural and functional patterns between Classes within a Phylum (Chaps. 11, 12, 14 and 15) and even across Phyla (Chaps. 13 and 16). Chapters 11–15 are concerned with aspects of locomotion whereas Chap. 16 examines attachment mechanisms in aquatic animals.

In Chaps. 11 (Young) and 12 (Pouydebat et al.) the authors explore convergently-evolved responses to the biomechanical problems of moving on narrow supports, emphasizing both morphological traits (e.g., small body size, enhanced mobility of appendicular joints, grasping extremities, and long and mobile tails) and behaviourally-based functional attributes (such as speed reduction, modulation of the duty factor, a switch to gaits facilitating continuous contact with the substrate, exaggerated limb joint excursions during the support phase). In these chapters, comparative analyses of both form and function reveal convergences in relation to locomotion on narrow substrates. Young (Chap. 11) identifies widespread convergence of locomotor morphology and behaviour among arboreal lissamphibians, lizards, and metatherian/eutherian mammals that has arisen in response to a common set of physical challenges imposed by the complexity (such as compliance, narrow perch diameter, 3D orientation of the substrates) of these habitats. Animals living in such habitats must be able to stabilize their body and manipulate their substrate to move efficiently. This is particularly evident for narrow supports, as is also demonstrated in Chap. 12 (Pouydebat et al.)

In a complimentary contribution Khandelwal et al. (Chap. 13) discuss the evolutionary trajectories relating to morphology and behaviour leading to convergence of gliding abilities in animals, focusing upon selective pressures that have been brought to bear on morphological and behavioural traits that permit distantly-related taxa to generate aerodynamic forces to exploit and control glide paths and to land successfully at the termination of the flight. Much of this chapter concerns vertebrates that live in the complex arboreal habitats considered in Chaps. 11 and 12, but also includes consideration of those animals that glide in air at the air-water interface in marine situations. The latter extends beyond vertebrate examples to include cephalopd mollusks.

Druelle et al. (Chap. 14) approach the question of evolutionary convergence as it relates to bipedal locomotion in terrestrial vertebrates. By considering the various form and function relationships within the distantly-related vertebrate taxa that exhibit bipedal locomotion, whether this be employed occasionally, habitually or obligately, they argue that tetrapods practicing this kind of locomotor behaviour converge upon the same functional anatomical outcomes (such as cyclic limb loading being higher when compared with quadrupedal locomotion). For habitual and obligate bipeds, the evolutionary pathways taken to achieve these capabilities have resulted in structurally and functionally different ways of overcoming the physical challenges involved, the differences being attributable to the differences in anatomical configurations from which the trend to bipedalism began.

Chapters 15 and 16 explore convergent patterns relating to two quite different behaviours that favour survival in aquatic (marine) habitats. Fish (Chap. 15) examines convergence of morphological and physiological traits relating to the locomotor performance of three distantly-related taxa (sharks, ichthyosaurs, and dolphins), a comparative scenario recognized as "*the quintessential example of evolutionary convergence*". Convergence in these vertebrates has been governed by the physical parameters of the fluid in which they live, driving morphological (such as body surface attributes, form of the appendages), physiological (body temperature), mechanical (such as frequency of oscillation of the thrust-generating hydrofoil, development of lift-based thrust, enhancement of speed), and behavioural (stability, manoeuverability, swimming) traits. These have resulted in independent evolutionary trajectories leading to great similarity in form of these high-performance marine predators and a remarkable amount of homoplasy. Delroisse et al. (Chap. 16) discuss the attachment strategies (glue-like bioadhesive secretions and pressure-driven suction) adopted for attachment.to the substrate by aquatic animals. They highlight convergence in underwater attachment mechanisms across a huge array of Phyla, examining these mechanisms across a hierarchy of length-scales (organism, organ, microscopic and molecular), and provide a synthetic overview of how organisms attach to various substrata. This broadly comparative approach permits the authors to posit that, at the molecular level of adhesion, the global amino acid composition of bioadhesives shows homologous characteristics.

In Chap. 17 Broeckhoven and du Plessis highlight the central role that the concept of convergent evolution can play in the quest for using biodiversity as a key source of bioinspiration. They provide valuable insights into methodological and experimental studies of biological systems at various scales to obtain a comprehensive understanding of biological structures, performance, and behaviour. Such knowledge can be used to inspire future directions for seeking innovative solutions to problems associated with human activities. Multiple repetition of functional assemblages, although not identically constructed, should provide biomimeticists with insights about what is collectively necessary and sufficient to carry out identified tasks and about how phylogenetic constraints and contingency have led to variation superimposed upon the fundamental configurational/operational commonality.

Acknowledgments We are highly grateful to Alex Werth for his comments on this introduction.

References

Agrawal, A. A. (2017). Toward a predictive framework for convergent evolution: Integrating natural history, genetic mechanisms, and consequences for the diversity of life. *The American Naturalist, 190*(S1), 1–12.

Almeida, A. L., Francoy, T. M., Àlvarez-Presas, M., & Carbayo, F. (2021). Convergent evolution: A new subfamily for bipaliin-like Chilean land planarians (platyhelminthes [sic]). *Zoologica Scripta, 50*, 500–508.

Arbuckle, K., & Speed, M. P. (2016). Analysing convergent evolution: A practical guide to methods. In P. Pontarotti (Ed.), *Evolutionary biology* (pp. 23–36). Springer International Publishing.

Arbuckle, K., Bennett, C. M., & Speed, M. P. (2014). A simple measure of the strength of convergent evolution. *Methods in Ecology and Evolution, 5*, 685–693.

Arendt, J., & Reznick, D. (2008). Convergence and parallelism reconsidered: What have we learned about the genetics of adaptation? *Trends in Ecology and Evolution, 23*(1), 26–32.

Arnold, S. J. (1983). Morphology, performance and fitness. *American Zoologist, 23*(2), 347–361.

Assis, L. C. S., de Carvalho, M. R., & Wheeler, Q. D. (2011). Homoplasy: From detecting pattern to determining process in evolution, but with a secondary role for morphology? *Zootaxa, 2984*, 67–68.

Ballell, A., Moon, B. C., Porro, L. B., Benton, M. J., & Rayfield, E. J. (2019). Convergence and functional evolution of longirostry in crocodylomorphs. *Palaeontology, 62*(6), 867–887.

Bergmann, P. J., & Morinaga, G. (2018). The convergent evolution of snake-like forms by divergent evolutionary pathways in squamate reptiles. *Evolution, 73*, 481–496.

Binning, S. A., Shaw, A. K., & Roche, D. G. (2017). Parasites and host performance: Incorporating infection into our understanding of animal movement. *Integrative and Comparative Biology, 57*(2), 267–280.

Blount, Z. D., Lenski, R. E., & Losos, J. B. (2018). Contingency and determinism in evolution: Replaying life's tape. *Science, 362*, eaam5979.

Bowler, P. J. (1996). *Life's splendid drama: Evolutionary biology and the reconstruction of life's ancestry* (pp. 1860–1940). The University of Chicago Press. xiii + 1—525.

Brainerd, E. L., & Camp, A. L. (2019). Functional morphology of vertebrate feeding systems: New insights. In V. L. Bels & I. Wishaw (Eds.), *Feeding in vertebrates: Evolution, morphology, behavior, biomechanics* (pp. 21–44). Springer.

Casinos, A., & Gasc, J.-P. (2002). Contraints, size and shape. In P. Aerts, K. D'Août, A. Herrel, & R. Van Damme (Eds.), *Topics in functional and ecological vertebrate morphology. A tribute to Frits de Vree* (pp. 325–335). Shaker Publishing.

Collar, D. C., Reece, J. S., Alfaro, M. E., Wainwright, P. C., & Mehta, R. S. (2014). Imperfect morphological convergence: Variable changes in cranial structures underlie transitions to durophagy in moray eels. *The American Naturalist, 183*(6), 168–184

De Lisle, S. P., & Bolnick, D. I. (2020). A multivariate view of parallel evolution. *Evolution, 74*, 1466–1481.

Dolezal, M., & Lipps, J. (2019). Digest: How many ways to make a snake? Evidence for historical contingency of the convergence of squamate reptiles. *Evolution, 73*, 626–627.

Donley, J. M., Sepulveda, C. A., Konstantinidis, P., Gemballa, S., & Shadwick, R. E. (2004). Convergent evolution in mechanical design of lamnid sharks and tunas. *Nature, 249*, 61–65.

Fernández, M. S., Vlachos, E., Buono, M. R., Alzugaray, L., Campos, L., Sterli, J., Herrera, Y., & Paolucci, F. (2020). Fingers zipped up or baby mittens? Two main tetrapod strategies to return to the sea. *Biology Letters, 16*, 20200281. https://doi.org/10.1098/rsbl.2020.0281

Fischer, E. K., Nowicki, J. P., & O'Connell, L. A. (2019). Evolution of affiliation: Patterns of convergence from genomes to behaviour. *Philosophical Transactions of the Royal Society B, 374*(1777), 20180242. https://doi.org/10.1098/rstb.2018.0242

Frazzetta, T. (1975). *Complex adaptations in evolving populations*. Sinauer Associates Inc.

Garland, T., Jr., & Losos, J. B. (1994). Ecological morphology of locomotor performance in squamate reptiles. In P. Wainwright & S. Reilly (Eds.), *Ecological morphology: Integrative organismal biology* (pp. 240–302). The University of Chicago Press.

Ghiselin, M. T. (1980). The failure of morphology to assimilate Darwinism. In E. Mayr & W. B. Provine (Eds.), *The evolutionary synthesis* (pp. 180–193). Harvard University Press.

Gómez, R. O., & Lois-Milevicich, J. (2021). Why the long beak? Phylogeny, convergence, feeding ecology, and evolutionary allometry shaped the skull of the Giant Cowbird *Molothrus oryzivorus* (Icteridae). *Journal of Morphology, 282*(11), 1587–1603. https://doi.org/10.1002/jmor.21408/

Grant, P. R., Grant, B. R., Markert, J. A., Keller, L. F., & Petren, K. (2004). Convergent evolution of Darwin's finches caused by introgressive hybridization and selection. *Evolution, 58*(7), 1588–1599.

Grossnickle, D. M., Chen, M., Wauer, J. G. A., Pevsner, S. K., Weaver, L. N., Meng, Q.-J., Liu, D., Zhang, Y. G., & Luo, Z. X. (2020). Incomplete convergence of gliding mammal skeletons. *Evolution, 74*, 2662–2680.

Gutierrez, H. L., Tsutsumi, R., Moore, T. Y., & Cooper, K. L. (2019). Convergent metatarsal fusion in jerboas and chickens is mediated by similarities and differences in the patterns of osteoblast and osteoclast activities. *Evolution and Development, 21*, 320–329.

Hall, B. K. (2007). Homoplasy and homology: Dichotomy or continuum? *Journal of Human Evolution, 52*(5), 473–479.

Hall, B. K. (2013). Homology, homoplasy, novelty, and behavior. *Developmental Psychobiology, 55*(1), 4–12.

Higham, T. E., Ferry, R. A., Schmitz, L., Irschick, D. J., Starko, S., Anderson, P. S. L., Bergmann, P. J., Jamniczky, H. A., Monteiro, L. R., Navon, D., Messier, J., Carrington, E., Farina, S. C., Feilich, K. L., Hernandez, L. P., Johnson, M. A., Kawano, S. M., Law, C. J., Longo, S. J., Martin, C. H., Martone, P. T., Rico-Guevara, A., Santana, S. E., & Niklas, K. J. (2021). Linking ecomechanical models and functional traits to understand phenotypic diversity. *Trends in Ecology and Evolution, 2866*. https://doi.org/10.1016/j.tree.2021.05.009

Hu, Y., Wu, Q., Ma, S., Ma, T., Shan, L., Wang, X., & Wei, F. (2017). Comparative genomics reveals convergent evolution between the bamboo-eating giant and red pandas. *Proceedings of the National Academy of Sciences, 114*(5), 1081–1086.

Huxley, J. (1942). *Evolution: The modern synthesis* (p. 645). George Allen and Unwin Ltd.

Irschick, D. J., & Higham, T. E. (2016). *Animal athletes: An ecological and evolutionary approach*. Oxford University Press.

Johnson, J. B., Burt, D. B., & DeWitt, T. J. (2008). Form, function, and fitness: Pathways to survival. *Evolution, 62*(5), 1243–1251.

Jones, G., & Holderied, M. W. (2007). Bat echolocation calls: Adaptation and convergent evolution. *Proceedings of the Royal Society B, 274*, 905–912.

Kowalko, J. E., Rohner, N., Linden, T. A., Rompani, S. B., Warren, W. C., Borowsky, R., & Yoshizawa, M. (2013). Convergence in feeding posture occurs through different genetic loci in independently evolved cave populations of *Astyanax mexicanus*. *Proceedings of the National Academy of Sciences, 110*(42), 16933–16938.

Kuhn, A., Skipwith, P., & Overcast, I. (2020). Digest: An emerging model system for understanding ecomorphological convergence. *Evolution, 74*, 696–697.

Lailvaux, S. P., & Husak, J. F. (2014). The life history of whole-organism performance. *The Quarterly Review of Biology, 89*(4), 285–318.

Lamichhaney, S., Card, D. C., Grayson, P., Tonini, J. F. R., Bravo, G. A., Näpflin, K., Termignoni-Garcia, F., Torres, C., Burbrink, F., Clarke, J. A., Sackton, T. B., & Edwards, S. V. (2019). Integrating natural history collections and comparative genomics to study the genetic architecture of convergent evolution. *Philosophical Transactions of the Royal Society B, 374*, 20180248. https://doi.org/10.1098/rstb.2018.0248

Lankester, E. R. (1870). On the use of the term homology in modern zoology and the distinction between homogenetic and homoplastic agreements. *Annals and Magazine of Natural History*, 4th Series, *6*, 34–43.

Lauder, G. V. (1996). The argument from design. In R. Rose & G. V. Lauder (Eds.), *Adaptation* (pp. 55–91). Academic.

Llaurens, V., Le Poul, Y., Puissant, A., & Blandin, P. (2021). Convergence in sympatry: Evolution of blue-banded wing pattern in *Morpho* butterflies. *Journal of Evolutionary Biology, 34*, 284–295.

Losos, J. B. (2011). Convergence, adaptation, and constraint. *Evolution, 65*, 1827–1840.

Mahler, L. D., Ingram, T., Revell, L. J., & Losos, J. B. (2013). Exceptional convergence on the macroevolutionary landscape in island lizard radiations. *Science, 341*, 292–295.

Mahler, D. L., Weber, M. G., Wagner, C. E., & Ingram, T. (2017). Pattern and process in the comparative study of convergent evolution. *The American Naturalist, 190*(S1), 513–528.

McGhee, G. R. (2011). *Convergent evolution: Limited forms most beautiful*. MIT Press.

McGhee, G. (2018). Limits on the possible forms of stone tools: A perspective from convergent biological evolution. In M. J. O'Brien, B. Buchanan, & M. I. Eren (Eds.), *Convergent evolution in stone-tool technology* (Vol. 22, pp. 23–46). MIT Press.

Morris, P. J. R., Cobb, S. N. F., & Cox, P. G. (2018). Convergent evolution in the Euarchontoglires. *Biology Letters, 14*, 20180366. https://doi.org/10.1098/rsbl.2018.0366

Muschick, M., Indermauer, A., & Salzburger, W. (2012). Convergent evolution within an adaptive radiation of cichlid fishes. *Current Biology, 22*, 2362–2368.

Norton, S. F., & Brainerd, E. L. (1993). Convergence in the feeding mechanics of ecomorphologically similar species in the Centrarchidae and Cichlidae. *Journal of Experimental Biology, 176*(1), 11–29.

Nyhart, L. K. (1995). *Biology takes form. Animal morphology and the German Universities* (pp. 1800–1900). The University of Chicago Press. xiii + 1-414.

Ochoa, C., & Rasskin-Gutman, D. (2015). Evo–devo mechanisms underlying the continuum between homology and homoplasy. *Journal of Experimental Zoology Part B: Molecular and Developmental Evolution, 324*(2), 91–103.

Osborn, H. F. (1902). Homoplasy as a law of latent or potential homology. *The American Naturalist, 36*, 259–271.

Ospovat, D. (1978). Perfect adaptation and teleological explanation: Approaches to the problem of the history of life in the mid-nineteenth century. In W. Coleman & C. Limoges (Eds.), *Studies in the history of biology* (Vol. 2, pp. 33–56). The Johns Hopkins University Press.

Parker, J., Tsagkogeorga, G., Cotton, J. A., Liu, Y., Provero, P., Stupka, E., & Rossiter, S. J. (2013). Genome-wide signatures of convergent evolution in echolocating mammals. *Nature, 502*, 228–231.

Pigot, A. L., Sheard, C., Miller, E. T., Bregman, T. P., Freeman, B. G., Roll, U., & Tobias, J. A. (2020). Macroevolutionary convergence connects morphological form to ecological function in birds. *Nature Ecology and Evolution, 4*(2), 230–239.

Powell, R. (2020). *Contingency and convergence: Toward a cosmic biology of body and mind*. MIT Press.

Powell, R., & Mariscal, C. (2015). Convergent evolution as natural experiment: The tape of life reconsidered. *Interface Focus, 5*, 20150040. https://doi.org/10.1098/rsfs.2015.0040

Quinn, B. L. (2020). Digest: Incomplete convergence drives form-function relationships in gliders. *Evolution, 74*, 246–247.

Rayfield, E. J. (2019). What does musculoskeletal mechanics tell us about evolution of form and function in vertebrates? In V. L. Bels & I. Wishaw (Eds.), *Feeding in vertebrates: Evolution, morphology, behavior, biomechanics* (pp. 45–70). Springer.

Reilly, S. M., & Wainwright, P. C. (1994). Conclusion: Ecological morphology and the power of integration. In P. Wainwright & S. Reilly (Eds.), *Ecological morphology: Integrative organismal biology* (pp. 339–354). The University of Chicago Press.

Riedel, J., Zozaya, S. M., Hoskin, C. J., & Schwarzkopf, L. (2021). Parallel evolution of toepads in rock-dwelling lineages of a terrestrial gecko (Gekkota: Gekkonidae: *Heteronotia binoei*) *Zoological Journal of the Linnean Society, 167*. https://doi.org/10.1093/zoolinnean/zlaa167

Roth, G. (2015). Convergent evolution of complex brains and high intelligence. *Philosophical Transactions of the Royal Society B, 370*, 20150049. https://doi.org/10.1098/rstb.2015.0049

Russell, E. S. (1916). *Form and function. A contribution to the history of animal morphology*. John Murray (Publishers) Ltd.. ix + 383 + [1] pp.

Russell, A. P., & Gamble, T. (2019). Evolution of the gekkotan adhesive system: Does digit anatomy point to one or more origins? *Integrative and Comparative Biology, 59*, 131–147.

Sackton, T. B., & Clark, N. (2019). Convergent evolution in the genomics era: New insights and directions. *Philosophical Transactions of the Royal Society B, 374*, 20190102. https://doi.org/10.1098/rstb.2019.0102

Schott, R. K., Bhattacharyya, N., & Chang, B. S. W. (2019). Evolutionary signatures of photoreceptor transmutation in geckos reveal potential adaptation and convergence with snakes. *Evolution, 73*, 1958–1971.

Seed, A., Emery, N., & Clayton, N. (2009). Intelligence in corvids and apes: A case of convergent evolution? *Ethology, 115*, 401–420.

Shadwick, R. (2005). How tunas and lamnid sharks swim: An evolutionary convergence. *American Scientist, 93*, 524–531.

Singer, C. (1959). *A short history of scientific ideas to 1900*. Oxford University Press, xviii + 525 pp.

Speed, M. P., & Arbuckle, K. (2017). Quantification provides a conceptual basis for convergent evolution. *Biological Reviews of the Cambridge Philosophical Society, 92*, 815–829.

Stayton, C. T. (2015a). What does convergent evolution mean? The interpretation of convergence and its implications in the search for limits to evolution. *Interface Focus, 5*, 20150039. https://doi.org/10.1098/rsfs.2015.0039

Stayton, C. T. (2015b). The definition, recognition, and interpretation of convergent evolution, and two new measures for quantifying and assessing the significance of convergence. *Evolution, 69*(8), 2140–2153.

Thompson, C. J., Ahmed, N. I., Veen, T., Peichel, C. L., Hendry, A. P., Bolnick, D. I., & Stuart, Y. E. (2017). Many-to-one form-to-function mapping weakens parallel morphological evolution. *Evolution, 71*, 2738–2749.

Tokita, M., Chaeychomsri, W., & Siruntawineti, J. (2012). Developmental basis of toothlessness in turtles: Insight into convergent evolution of vertebrate morphology. *Evolution, 67*, 260–273.

Toussaint, E. F. A., Bybee, S. M., Erickson, R., & Condamine, F. L. (2019). Forest giants on different evolutionary branches: Ecological convergence in helicopter damselflies. *Evolution, 73*, 1045–1054.

Wake, D. B. (2013). The enigmatic history of the European, Asian and American plethodontid salamanders. *Amphibia-Reptilia, 34*(3), 323–336.

Wake, D. B., Wake, M. H., & Specht, C. D. (2011). Homoplasy: From detecting pattern to determining process and mechanism of evolution. *Science, 331*(6020), 1032–1035.

Waters, J. M., & McCulloch, G. A. (2021). Reinventing the wheel? Reassessing the roles of gene flow, sorting and convergence in repeated evolution. *Molecular Ecology, 30*(17), 4162–4172.

Willey, A. (1911). *Convergence in evolution*. John Murray.

Chapter 2
Odonatopteran Approaches to the Challenges of Flight: Convergence of Responses Subject to a Common Set of Morphological Constraints

André Nel and Bertrand Piney

Abstract Effective flight capacity is a crucial survival attribute of volant animals. Several vertebrate clades have acquired gliding capabilities and at least three of them independently acquired powered flight. Contrastingly, wings were probably acquired only once by pterygotan insects. Despite this, insects have developed a great variety of structural approaches that have diversified their collective flight capacity. Flight was a key contributor to their diversification during the Late Carboniferous (at least 330 Ma), and flying insects have remained the most diverse animal clade since then. Among pterygotans, representatives of the superorder Odonatoptera, which includes the extant Odonata, have developed impressive performance associated with the highly complex morphological structure of their wing venation. Some venation patterns, such as the nodus, discoidal complex, and arculus, were acquired only once, whereas others have been convergently acquired several times. One example of a pattern acquired more than once is the sclerotized pterostigma, convergently appearing in the Permian Protanisoptera and its sister group, the Discoidalia, these comprising the modern Odonata. All odonatopterans with broad wings were confronted by a major problem, that of 'how to strengthen the basal third of the wing' to prevent it from breaking longitudinally. At least eight different convergent 'solutions' have been 'adopted' that have resulted in the incorporation of structures oriented perpendicular to the main axis of the wing. Additionally, several clades within the Odonatoptera have convergently developed petiolated wings, adapted for flying in cluttered environments. The width and length of the petiole can vary greatly, with the most impressive ones being those of the 'giant' Permian-Triassic Triadophlebiomorpha. This great morphological disparity represents 'variations on a theme' of the already complex wing venation established by the first Carboniferous odonatopterans. It is possible that some of the 'solutions'

A. Nel (✉) · B. Piney
Institut de Systématique, Évolution, Biodiversité (ISYEB) Muséum National d'Histoire Naturelle, CNRS, Sorbonne Université, EPHE, Université des Antilles, Paris, France

3, rue de la Frégate, Thouaré sur Loire, France
e-mail: anel@mnhn.fr

V. L. Bels, A. P. Russell (eds.), *Convergent Evolution*, Fascinating Life Sciences,
https://doi.org/10.1007/978-3-031-11441-0_2

arrived at by extinct clades allowed for performance that was more effective than that of modern odonatans. Many of these groups flourished and co-existed with the ancestors of modern taxa for millions of years. Extant odonatans have been adopted as models for the bio-mimetic development of small drones. The wing patterns of extinct clades should also be investigated for their potential for bio-mimetic inspiration and application.

Keywords Arculus · Crossveins · Discoidal complex · Hawker · Nodus · Percher · Petiole · Pterostigma

2.1 Introduction

Flight is a spectacular and effective tool that promotes survival. It allows for escape from potential predators, hunting, searching for new resources and sexual partners, and colonization of new habitats. All clades that have developed the capacity for powered flight have diversified extensively. Among the vertebrates, birds are the most impressive example, with more than 10,000 extant species, nearly double the number of species of extant mammals. Bats constitute the second largest order of mammals, after rodents. This phenomenon of extensive diversity is especially so for the Pterygota (winged insects) (Engel et al., 2013), which is the most diverse animal clade on Earth. Flying insects had already achieved an impressive level of diversity by the late Carboniferous.

Due to the physical and mechanical constraints related to lift and air turbulence, along with the optimization of flight performance associated with effective prey capture and/or predator escape, the acquisition of the ability to fly has resulted in major evolutionary convergences between the different groups of flying animals. Such constraints have been extensively investigated for vertebrates (e.g., Palmer & Dyke, 2012; Amador et al., 2020; Williams et al., 2020) but much less so for insects, even though they were the first animals to have achieved powered flight. This lack of intensive study is due, in part, to the extraordinary diversity of insects, extant and fossil, and their small size. Experimental studies have been conducted for only a relatively few groups, such as some flies, butterflies, and beetles (e.g., Bomphrey et al., 2009; Phan et al., 2019).

Among insects the members of the superorder Odonatoptera (including the extant dragonflies and damselflies of the order Odonata) have developed unique flight performance capabilities, such as the ability to hover, fly backwards, achieve very rapid accelerations, and very fast changes of direction (Salami et al., 2019). Some species, such as *Pantala flavescens*, the 'globe-trotter' dragonfly, are capable of migrating over considerable distances and have thus reached all intertropical oceanic islands. Others, such as the Pseudostigmatidae (Zygoptera), hover to capture prey or lay eggs in the cavities of bromeliad leaves. The Odonatoptera were the first aerial predators and held this role exclusively for at least 70 million years, prior to the appearance of the first small Middle Permian vertebrate gliders (Steyer, 2009), and the later flying vertebrates of the Triassic. From the beginning of the late

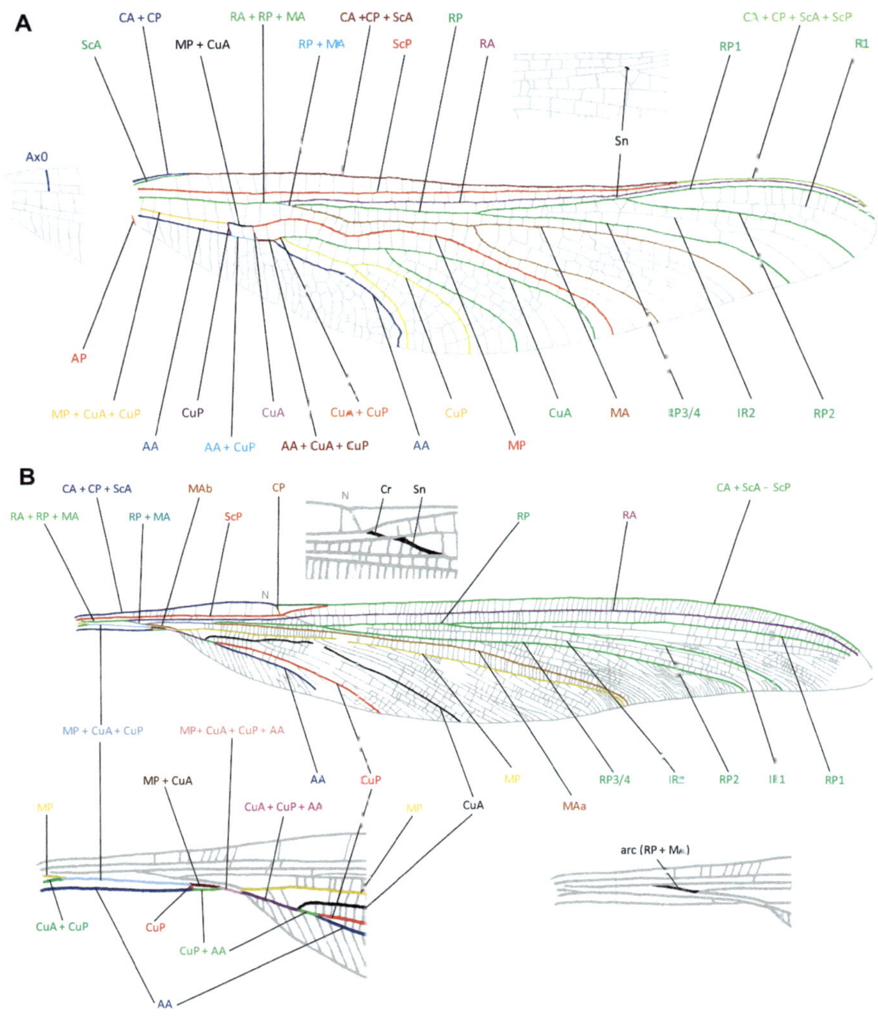

Fig. 2.1 Different types of wing among the Odonatoptera. (**a**) Hawker type with large and broad wings (Meganeuridae); (**b**) percher type with narrower, petiolated wings (Triadophlebiomorpha). Abbreviations and color legend are provided in the Materials and Methods section

Carboniferous two main types of 'dragonflies' co-existed: the large 'hawkers' that hunt while patrolling and hovering over open water, and the 'perchers' that wait in ambush, perched on a branch or twig, and launch themselves at any prey item that passes nearby, landing again on their support to eat it (Fig. 2.1). The Odonatoptera have developed a unique thoracic shape, modified into a greatly reduced prothorax and a very large pterothorax carrying the legs such that they project beneath the mouthparts and serve to capture, stab, and hold the prey. Their wing muscles are also different from those of other insects, being sub-vertically-oriented (Goldsworthy &

Wheeler, 1989: Fig. 2.1), thus permitting independent flapping of all four wings and the twisting of each wing during flight (bi-motor flight) (Hatch, 1966; Brodsky, 1994).

The Odonatoptera is one of the rare insect groups, along with the Asilidae (Diptera) and Myrmeleontoidea (in particular the Ascalaphidae) (Neuroptera), that capture their prey during flight.

Carboniferous and Permian 'hawkers' were represented by the Meganeuridae, among which are found the largest-winged insects, some attaining wing spans of over 70 cm, but also including much smaller species approximating the size of a modern dragonfly (Nel et al., 2009). It is possible that the 'giants' preyed on the smaller 'hawkers' (Nel et al., 2018). The 'perchers' were small 'damselflies' comparable in size to the modern Zygoptera (Nel et al., 2019).

2.2 Material and Methods

Wing venation terminology for the Odonatoptera follows that of Riek and Kukalová-Peck (1984), modified by Nel et al. (1993) and Jacquelin et al. (2018), and the phylogeny of the entire superorder is based mainly on that of Bechly (1996).

Wing venation abbreviations (see Fig. 2.1) (each vein and its branches is represented by a different color in all figures illustrating wing structure): AA/AP anal vein (anterior/posterior); Ax0, Ax1, Ax2, primary antenodal crossveins; arc arculus (transverse common base of RP + MA from R + MA); Bqs, specialized crossveins situated in the triangular zone between IR2, RP and RP2; C, costal vein (in fact a complex vein formed by the fusion of the Costa and the subcostal anterior); Cr nodal crossvein, a reinforced crossvein situated in the Nodus, below the apex of ScP; CuA/CuP, cubital vein (anterior/posterior); CuAa and CuAb, first distal branches of CuA; IR1 and IR2, supplementary convex longitudinal veins in the radial area, emerging from the stem of RP; MA/MP, median vein (anterior/posterior); MAa, anterior branch of MA; MAb, posterior branch of MA, forming the distal side of the discoidal cell; Mspl, supplementary median vein between MAa and MP; N nodus, apex of vein ScP ending in the costal vein C; 'O' oblique vein, an oblique crossvein situated between RP2 and IR2 not far from the base of RP2 (two can be present in some families); Pt, pterostigma; RA/RP, radial vein (anterior/posterior) (RP is divided into three main branches RP1, RP2 and RP3/4 in the Odonatoptera); Rspl, supplementary radial vein between RP2 and IR2; ScA/ScP, subcostal vein; Sn subnodus, a specialized reinforced crossvein more or less aligned with Cr, between RA and RP; Tp, trigonal planate, a specialized secondary longitudinal vein emerging from MAb in the area between MAa and MP.

2.3 Specialized Morphological Structures of Modern Odonatan Wings

Modern odonatans have a very complex wing venation with several structures that are absolutely unique among modern insects and crucial for their flight capacities and performance (Rajabi & Gorb, 2020; Chitsaz et al., 2020), viz.

- the nodus (N—Fig. 2.1b), located near the middle of the anterior margin of the wing, formed by the fusion of the posterior subcostal vein (ScP—Fig. 2.1b) and the anterior margin or costal vein C, with a small portion of the latter not sclerotinized and thus able to fold. The nodal structure is reinforced posteriorly by two well-aligned crossveins: the nodal crossvein (Cr) and the subnodus (Sn).
- The pterostigma (Pt, Fig. 2.4), located near the distal end of the anterior margin of the wing, is a more or less large, bubble-shaped cell with reinforced and colored walls. Its exact function is unclear but it seems to serve as a reinforcement for the apical part of the wing, a structure that reduces vibrations within the wing, a visual reference point for the insect to know where its wings are (!), and as a sexual signal (different colorations depending on the sex in some species).
- The discoidal complex (Figs. 2.3 and 2.4). This structure is located in the middle part of the basal quarter of the wing. It is more or less complex depending on the clade. It is composed of a single large cell with reinforced sides in the Zygoptera, and is divided by a crossvein into a hypertriangle and a discoidal triangle in the Anisoptera (Fig. 2.4). It is supported by a more or less complex subdiscoidal space.
- The arculus (Fig. 2.1b). This is a structure crossing the wing antero-basally to the discoidal complex and strengthening the wing. It is composed of two parts in the Odonatoptera, anteriorly by RP + MA separating from RA and (in the Odonata) by a crossvein between RP + MA and MP + CuA.

2.4 Convergences in Wing Structures

Although the arculus was already present in the wings of the earliest known Carboniferous Odonatoptera, the other structures listed above did not exist in the Meganeuridae, giant ‘hawkers’ of the Paleozoic Era (Fig. 2.1a). Their absence does not seem to have handicapped them with regard to being the dominant aerial predators of the more than 50 Ma period between the late Carboniferous and the middle Permian. Regardless, they became extinct around the Late Permian.

2.4.1 Nodus

A more or less specialized nodus was present in the 'perchers' of the Paleozoic. This structure appears to have developed quite 'gradually', with an unaligned and sparsely thickened 'nodal crossing' and subnodus and no well-defined fold zone on the costal vein in the Lapeyriidae (Fig. 2.2a). It is better defined in the Protanisoptera (Fig. 2.2b) and is comparable in structure to the modern nodus in the Triadophlebiomorpha and Protozygoptera (Figs. 2.1b and Fig. 2.3a). It even regressed in some Mesozoic Protozygoptera (Fig. 2.3b) and some Cenozoic

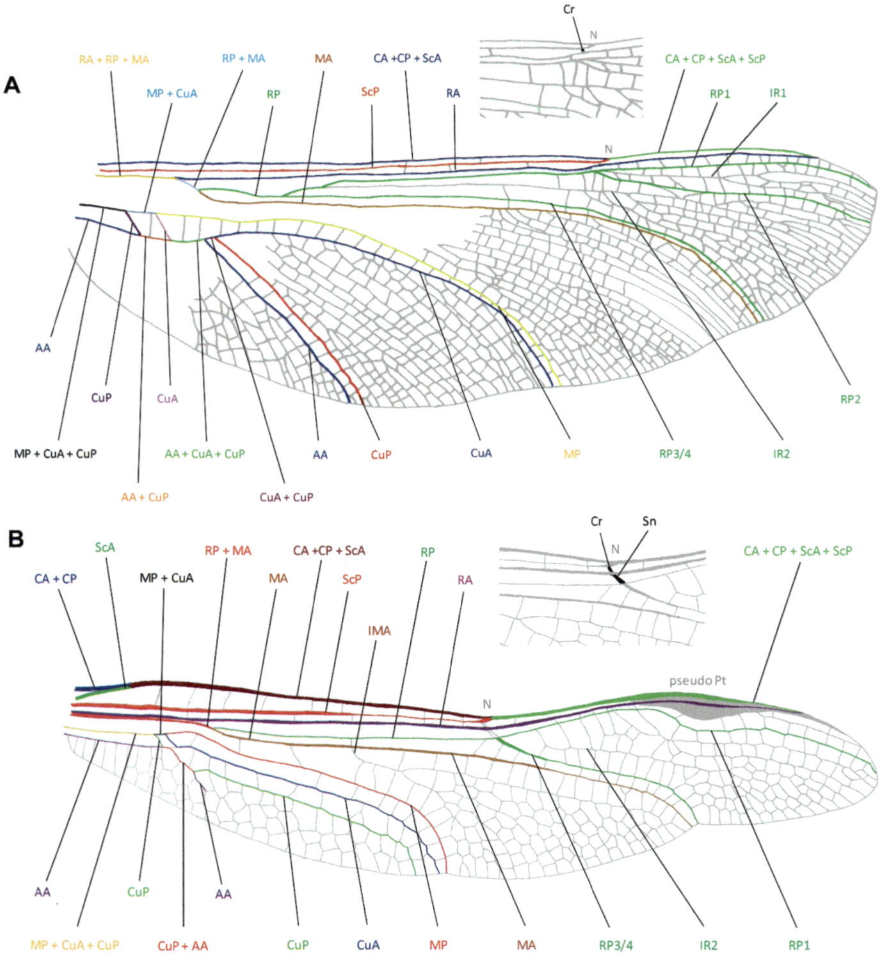

Fig. 2.2 Different types of nodus among Permian hawkers (inset above each wing). (**a**) Lapeyriidae with less well-organized nodal structures; (**b**) Protanisoptera, with well-aligned nodal crossing veins and a subnodus; note the pseudo-pterostigma (pseudoPt)

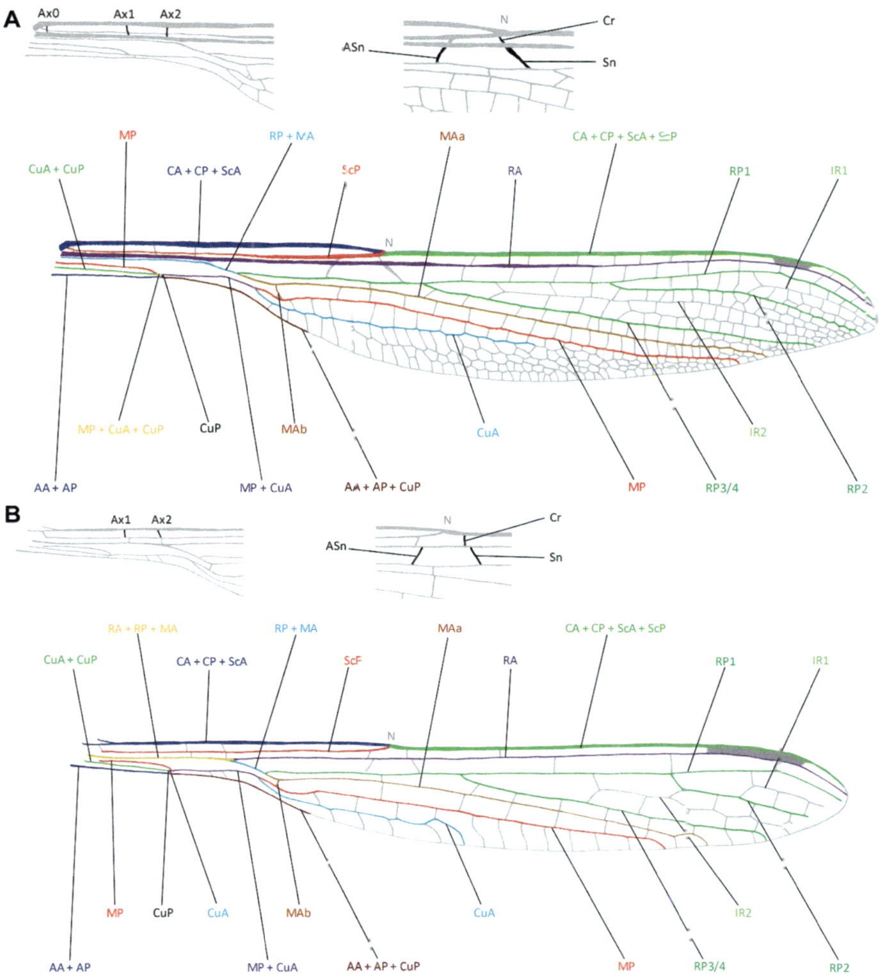

Fig. 2.3 Different types of nodus among the Permian Protozygoptera. (**a**) *Permolestes* sp. with well organized nodal structures; (**b**) *Kennedya* sp. exhibiting reduced nodal structures. Insets correspond to the details of the structure of the wing base and nodus

Zygoptera (Sieblosiidae), in which the subnodus has an 'abnormal' direction and becomes significantly thinner.

The nodus was acquired only once, although it has undergone some variation in its structure. The Odonatoptera that possess the nodus (the Nodialata) probably had a considerable evolutionary advantage because this structure allows significant twists of the costal margin of the wing and therefore better control of forces tending to break the wing. Odonata carry specialized spine-shaped sensilla which, by touching the veins during the twisting of the wing, inform the insect that it is approaching the point of rupture of the membrane or the vein. Being able to better fold the wing is, therefore, an evident advantage. However, the nodus-free Meganeuridae coexisted with nodialatans from the Upper Carboniferous to the Upper Permian. It appears that the appearance of the first gliding vertebrates brought about the demise of the former group.

2.4.2 Pterostigma

The Carboniferous odonatopterans lacked a pterostigma. Two types of pterostigma have been acquired through evolutionary convergence, by the Protanisoptera on one hand and their sister group the Discoidalia (a clade comprising all the dragonflies known from the Triassic and more recent periods) on the other. The Protanisoptera had a pseudo-pterostigma, viz. a sclerotinized structure covering the costo-apical part of the wing, extending from the costal vein to the posterior radial vein RP1 and including a portion of the anterior radial vein RA (Fig. 2.2b). The Discoidalia have a true pterostigma located between C and RA (Fig. 2.3) that is still present in the modern Odonata. This, therefore, represents a case of evolutionary convergence between two similar, but non-homologous, structures, most likely fulfilling the same functions. Pseudo-pterostigma and pterostigma were present in contemporary taxa during the Middle Permian. The pterostigma was 'lost' in parallel by several clades: the Triassic Triadophlebiomorpha had no pterostigma (Fig. 2.1) while some Permian representatives had one. Some Cretaceous Odonata: Anisoptera (Aeschnidiidae) and some extant Zygoptera (Calopteygoidea) have no pterostigma. These insects have large wings with significant multiplications of small cells. Different 'stages' of the regression of the pterostigma are clearly evident among the Aeschniididae, within which some species had a true pterostigma, others had a weakly sclerotinized one traversed by crossveins (Fig. 2.4a), and others had none (Fleck & Nel, 2003). The independent loss of the pterostigma in several clades suggests that this structure is not 'absolutely essential' for flight.

2.4.3 Discoidal Complex

The discoidal complex was not present in Paleozoic odonatopterans. A first crossvein, stronger than the others and running between the median anterior vein (MA) and the median posterior vein (MP), is present in the Discoidalia. It defines a rudimentary discoidal cell that is still open basally (Fig. 2.3). This cell became closed in some Protozygoptera and its sister group, the Panodonata, and formed a discoidal cell of 'modern' type. Note that some modern Zygoptera (e.g., Hemiphlebiidae, sister group of Lestoidea), have lost this closure of the discoidal cell.

The division of the discoidal cell by a specialized crossvein, into a triangle and a hypertriangle arose in the Epiproctophora, the stem group of the modern Anisoptera (Figs. 2.4 and 2.5). It was first acquired on the hind wings and then 'passed' to the

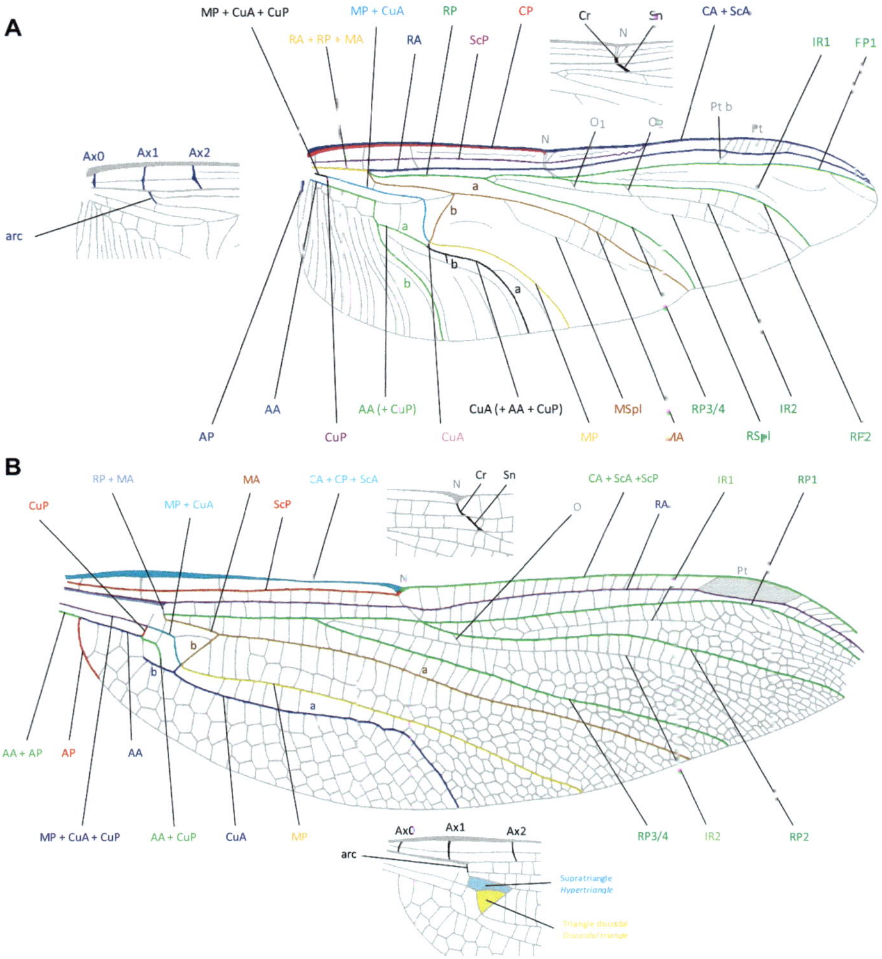

Fig. 2.4 Epiproctophoran types of wings. (**a**) Aeschnidiidae with pterostigma crossed by weak veinlets; (**b**) Heterophlebiidae with division of the discoidal cell into a hypertriangle and a discoidal triangle. Insets: wing base and nodus

forewings (Nel et al., 1993). It seems that this subdivision allows a reinforcement of the basal part of the wing and provides greater resistance to longitudinal breakage of the wing. It is crucial for the hind wings of epiproctophorans that experience greater air pressure due to the larger area of the basal part of the wing.

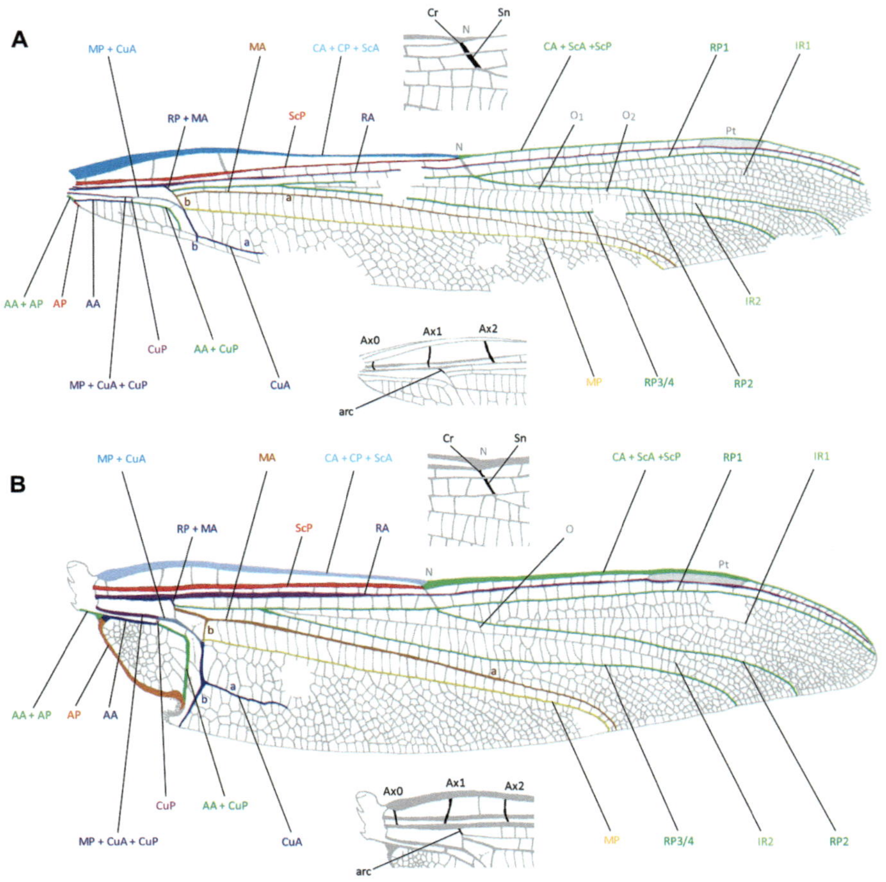

Fig. 2.5 Epiproctophora Campterophlebiidae. (**a**) Forewing; (**b**) hind wing. Insets: wing base and nodus

2.4.4 *Arculus*

The concept of the arculus does not correspond to homologous structures in all insects (Wootton, 1979; Wootton & Ennos, 1989). It is defined as a transverse reinforcing structure located in the basal quarter of the wing, allowing for twists of the wing while avoiding longitudinal fracture of the wing, thus having a similar function to the discoidal complex of the Odonata. If an arculus is present in almost all insects, the odonatopteran type is constituted by the RP + MA vein diverging from the RA: the plecopteran and holometabolan type is constituted by vein M, a crossvein m-cua and the cubitus anterior CuA; the orthopteran type by the vein M + CuA and an anterior branch of the cubitus posterior CuP; and the acercarian type by M + CuA and a modified crossvein cua-cup between CuA and CuP (Nel et al., 2012a, b). These different structures that have the same function represent several cases of evolutionary convergence that 'make it possible to solve' the problem of the longitudinal folding of the base of the wing. The basal closure of the discoidal cell in the Odonata strengthens the arculus, which becomes a rigid structure that crosses the mid part of the wing. The Odonatoptera with a basally open discoidal cell have a 'gap' between RP + MA and MP.

2.4.5 Transverse Reinforcing Structures of the Basal Halves of Wings

The Odonatoptera differ greatly from all other insects with large wings in not having an anal fan that can be folded when at rest. On the contrary, their wings have a rigid cubito-anal area. It appears that the problem of longitudinal folding of the wing in its basal half is crucial for the Odonatoptera because all taxa with wide wings have, more or less, developed transverse rigid structures in this part of the wing. These structures have similar functions but are not homologous.

In the case of the Paleozoic Meganeuridae, Lapeyriidae, and Protanisoptera, the transverse structures are the CuA and CuP veins and their re-emerging distal parts from the common stem, the AA+CuA + CuP stem, in all four wings (Figs. 2.1a and 2.2). In the case of the Mesozoic Epiproctophora: Campterophlebiidae, it is the CuA vein, after its separation from the MP vein, and its CuAb branch, that runs parallel to AA+CuP (Fig. 2.5). In most males of the Epiproctophora there is an additional basal anal triangle supported by a strong posterior branch of the AA, reinforcing the wing but also having a function of guidance of the female abdomen during mating. In the Zygoptera: Calopterygoidea, it is the CuA vein and only its posterior branch CuAb (Fig. 2.6a). In the Aeschnidiidae (in all four wings) it is the transverse discoidal triangle and a series of posterior branches of the anterior anal vein AA (Fig. 2.4a). In the other Anisoptera, the 'solution' is provided by a posterior branch of AA and CuA, with its CuAb branch posteriorly closing a large 'anal loop' that may or may not be subdivided into small cells (Fig. 2.6b). This structure corresponds to a modification of the 'campterophlebiid-type'. The shape of the anal loop varies a great deal by family and genus. It is rudimentary in the Heterophlebioidea, the sister group of the Anisoptera (Fig. 2.4b), small in modern and fossil Petalurida and Gomphida, more or less developed in the Aeshnoptera (Fig. 2.7a), and becomes more and more complex within the Libelluloidea, to form a 'boot' with a well-defined Cuspl midrib ('foot-shaped anal loop') in most Libellulidae (Fig. 2.6b). On the other hand, the Libellulidae has developed a transverse discoidal triangle in the forewings, very similar to that of the Aeschnidiidae, to the point that it is difficult to distinguish the forewings in these two clades, which are however 'phylogenetically' very distant. The Stenophlebioptera, another clade of the Epiproctophora, has developed, in all four wings, a transverse discoidal triangle, similar to, but not homologous with, those of the Aeschnidiidae and the Anisoptera (Figs. 2.4a and

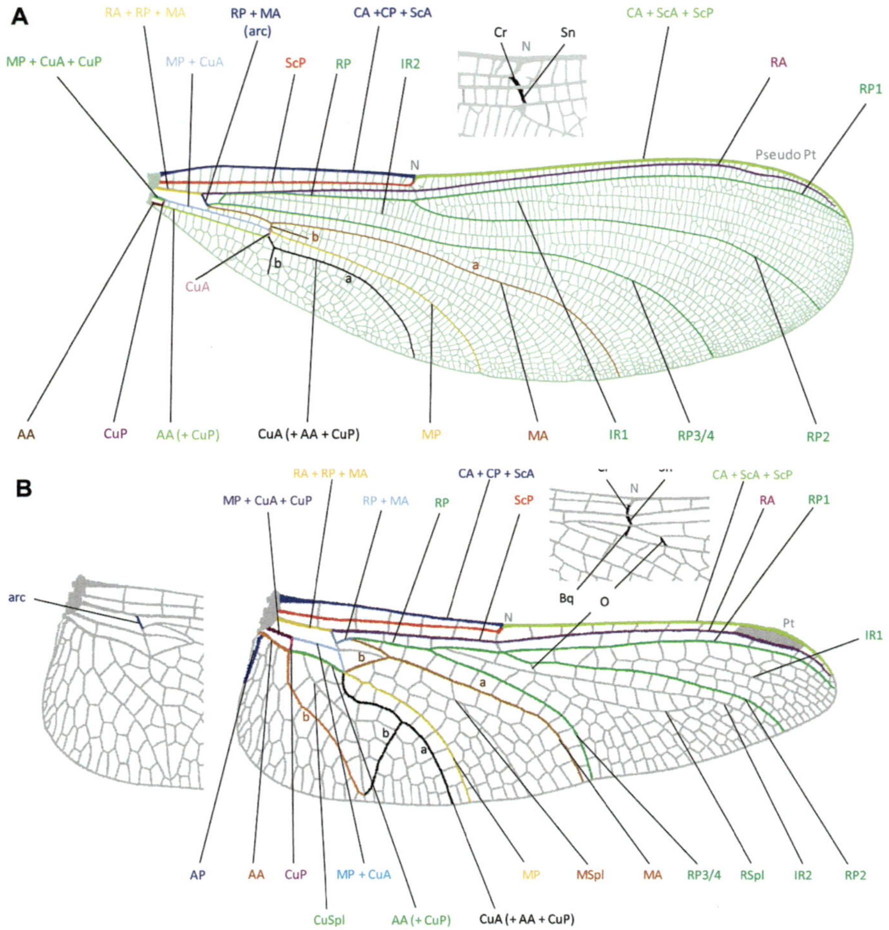

Fig. 2.6 (**a**) Zygoptera Calopterygidae; (**b**) Anisoptera Libellulidae showing the complex boot-shaped anal loop, limited by AAb, CuA and CuAb, and reinforced by a midrib. Insets: wing base and nodus

2.7b). It is indeed less well-separated from the hypertriangle than it is in the latter groups.

Odonatopterans with broad wings, therefore, exhibit at least eight convergent 'solutions' of the incorporation of a transverse reinforcing structure into the basal half of the hind wing, and sometimes also in the forewing. These reinforcing structures allow for wider wings and therefore greater lift. Only the Libellulidae, with the broadest basal part of the hind wings with a large anal loop, is able to migrate over very long distances (Suarez-Tovar & Sarmiento, 2016).

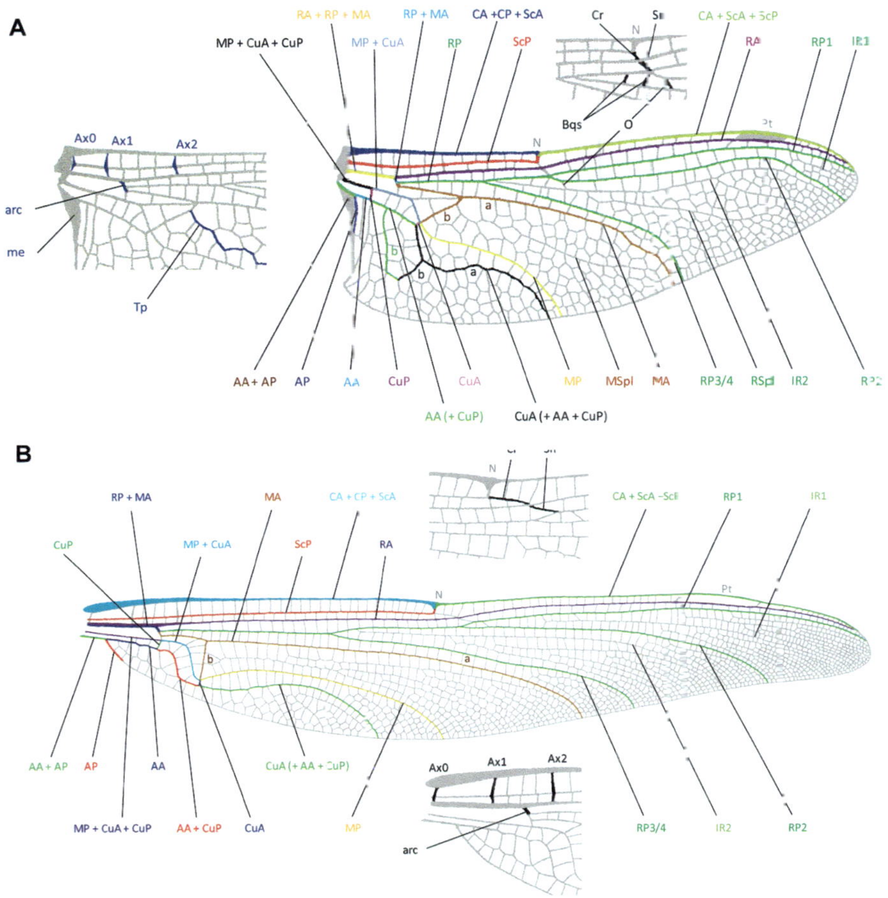

Fig. 2.7 (**a**) Anisoptera Aeshnidae, showing the small anal loop; (**b**) Epiproctophora Stenophlebiidae, showing the transverse discoidal cell. Insets: wing base and nodus

2.4.6 Wing Petiolation

Many odonatopterans have developed another type of wing, one that is basally narrow and petiolate. This wing form is found in distantly-related clades from the Late Carboniferous onward. It is generally present in taxa of relatively small-bodied forms, but some 'giants' with a very long petiole and narrow wings existed during the Permian-Triassic, such as the Triadophlebiomorpha, with wing spans two to three times greater than those of the largest modern dragonflies (Fig. 2.1a).

Petiolation is achieved by the narrowing of the spaces between the AA and posterior anal (AP) veins (Fig. 2.1). Petioles attain various levels of expression (relative length and width of the structure) in modern Zygoptera (except the Calopterygoidea), and Protozygoptera (Fig. 2.3) (Nel et al., 2012a, b). Some

Triadophlebiomorpha and Coenagrionoidea have extremely narrow petioles, with the fusion of the M + Cu, AA and AP veins. Some authors have hypothesized that the Odonata ancestrally possessed petiolate wings in their morphological 'ground plan' (Fraser, 1957), but it seems that the phenomenon of petiolation was developed, to varying degrees, by evolutionary convergence independently in the Triadophlebiomorpha, Protozygoptera, Zygoptera (except Calopterygoidea), and some large clades within the Epiproctophora (in particular the Stenophlebioptera). The modern stalked-wing taxa are either 'perchers' or small forms that are capable of flying in environments uncluttered by vegetation (such as in the vicinity of small streams; among forest undergrowth). Small protozygopterans probably lived in similar environments. Unfortunately, it is not possible to establish with certainty the environments occupied by the 'giant' Triadophlebiomorpha.

Other insects have also developed stalked wings, in particular some Palaeozoic Megasecoptera (Pecharová & Prokop, 2018), the Mecoptera: Bittacidae, Diptera: Tipuloidea, and also some Asilidae and Bombylidae. Their flight performances can vary greatly. The Tipuloidea are rather slow and relatively poor flyers, whereas the Asilidae are formidable predators that fly very rapidly. The aerodynamic impact of wing petiolation is still poorly understood (Phillips et al., 2017).

2.5 Conclusion

Even though the impact of wing morphology on the flight capabilities of extant odonatans is becoming more extensively and better understood (Rajabi & Gorb, 2020), studies of the wing mechanics of extinct clades are essentially lacking. The latter have wing shapes and venations that are very different from those of the extant taxa (Wootton et al., 1998). As these often existed for many tens of millions of years, it is evident that their wings allowed them to fly satisfactorily for hunting, to escape from predators, and for reproductive purposes. It is even possible that some of them, such as the Mesozoic widespread, common and diverse aeschnidiids were as, if not more, efficient, in flight than are the modern Odonata. These fossil species may well be employed as a source for bio-inspiration for minute drones in the near future.

Acknowledgements We sincerely thank Dr. Günter Bechly, Dr. Jakub Prokop, and Dr. Anthony P. Russell for their helpful remarks on the first version of the paper.

References

Amador, L. I., Almeida, F. C., & Giannini, N. P. (2020). Evolution of traditional aerodynamic variables in bats (Mammalia: Chiroptera) within a comprehensive phylogenetic framework. *Journal of Mammalian Evolution, 27*, 549–561.

Bechly, G. (1996) Morphologische Untersuchungen am Flügelgeäder der rezenten Libellen und deren Stammgruppenvertreter (Insecta; Pterygota; Odonata), unter besonderer Berücksichtigung

der phylogenetischen Systematik und des Grundplanes der Odonata. *Petalura*, Special Volume 2, 402 p.
Bomphrey, R. J., Walker, S. M., & Taylor, G. K. (2009). The typical flight performance of blowflies: Measuring the normal performance envelope of *Calliphora vicina* using a novel corner-cube arena. *PLoS One, 4*(11), e7852.
Brodsky, A. K. (1994). The evolution of insect flight. Oxford University Press, i-xiii + 1-229.
Chitsaz, N., Marian, R., Chitsaz, A., & Chahl, J. S. (2020). Parametric and statistical study of the wing geometry of 75 species of Odonata. *Applied Sciences, 10*(15) (5389), 1–17.
Engel, M. S., Davis, S. R., & Prokop, J. (2013). Insect wings: The evolutionary development of Nature's first flyers. In: A. Minelli, G. Boxshall, & G. Fusco (Eds.), *Arthropod biology and evolution: Molecules, development, morphology* (pp. 269–298). Springer, ix + 532 pp.
Fleck, G., & Nel, A. (2003). Revision of the Mesozoic family Aeschnidiidae (Odonata: Anisoptera). *Zoologica, 153*, 1–180.
Fraser, F. C. (1957). A reclassification of the order Odonata, based on some new interpretations of the venation of the dragonfly wing. *Handbook of the Royal Zoological Society of New South Wales, 12*, 1–133.
Goldsworthy, G. J., & Wheeler, C. H. (1989). Insect flight. CRC Press, Taylor & Francis, i-vii + 1-371.
Hatch, G. (1966). Structures and mechanics of the dragonfly pterothorax. *Annals of the Entomological Society of America, 59*, 702–714.
Jacquelin, L., Desutter-Grandcolas, L., Chintauan-Marquier, I., Boistel, R. Zheng, D., Prokop, J., & Nel, A. (2018). New insights on basivenal sclerites using 3D tools and homology of wing veins in Odonatoptera (Insecta). *Scientific Reports, 8*, 1–7.
Nel, A., Martínez-Delclòs, X., Paicheler, J.-C., & Henrotay, M. (1993). Les 'Anisozygoptera' fossiles. Phylogénie et classification (Odonata). *Martinia, Numéro Hors Série, 3*, 1–311.
Nel, A., Fleck, G., Garrouste, R., Gand, G., Lapeyrie, J., Bybee, S. M., & Prokop, J. (2009). Revision of Permo-Carboniferous griffenflies (Insecta: Odonatoptera: Meganisoptera) based upon new species and redescription of selected poorly known taxa from Eurasia. *Palaeontographica (A), 289*, 89–121.
Nel, A., Bechly, G., Prokop, J., Béthoux, O., & Fleck, G. (2012a). Systematics and evolution of Paleozoic and Mesozoic damselfly-like Odonatoptera of the 'Protozygopteran' grade. *Journal of Paleontology, 86*, 81–104.
Nel, A., Prokop, J., Nel, P., Grandcolas, P., Huang, D.-y., Roques, P., Guilbert, E., Dostál, O., & Szwedo, J. (2012b). Traits and evolution of wing venation pattern in paraneopteran insects. *Journal of Morphology, 273*, 480–506.
Nel, A., Prokop, J., Pecharová, M., Engel, M. S., & Garrouste, R. (2018). Palaeozoic giant dragonflies were hawker predators. *Scientific Reports, 8*, 1–5.
Nel, A., Roques, P., Prokop, J., & Garrouste, R. (2019). A new, extraordinary 'damselfly-like' Odonatoptera from the Pennsylvanian of the Avion locality in Pas-de-Calais, France (Insecta: 'Exopterygota'). *Alcheringa, 43*, 241–245.
Palmer, C., & Dyke, G. (2012). Constraints on the wing morphology of pterosaurs. *Proceedings of the Royal Society, (B) Biological Sciences, 279*, 1218–1224.
Pecharová, M., & Prokop, J. (2018). The morphology of mouthparts, wings and genitalia of Paleozoic insect families Protohymenidae and Scytohymenidae reveals new details and supposed function. *Arthropod Structure and Development, 47*, 117–129.
Phan, H. V., Truong, Q. T., & Park, H. C. (2019). Extremely large sweep amplitude enables high wing loading in giant hovering insects. *Bioinspiration and Biomimetics, 14*(6). https://doi.org/10.1088/1748-3190/ab3d55
Phillips, N., Knowles, K., & Bomphrey, R. J. (2017). Petiolate wings: Effects on the leading-edge vortex in flapping flight. *Interface Focus, 7*(1), 20160084.
Rajabi, H., & Gorb, S. N. (2020). How do dragonfly wings work? A brief guide to functional roles of wing structural components. *International Journal of Odonatology, 23*, 23–30.

Riek, E. F., & Kukalová-Peck, J. (1984). A new interpretation of dragonfly wing venation based upon Early Carboniferous fossils from Argentina (Insecta: Odonatoidea) and basic characters states in pterygote wings. *Canadian Journal of Zoology, 62*, 1150–1166.

Salami, E., Ward, T. A., Montazer, E., & Ghazali, N. N. N. (2019). A review of aerodynamic studies on dragonfly flight. *Proceedings of the Institution of Mechanical Engineers, Part C, Journal of Mechanical Engineering Science, 233*, 6519–6537.

Steyer, J. S. (2009) *La terre avant les dinosaures*. Editions Belin – Pour la Science, 205 p.

Suarez-Tovar, C. M., & Sarmiento, C. E. (2016). Beyond the wing planform: Morphological differentiation between migratory and non-migratory dragonfly species. *Journal of Evolutionary Biology, 29*, 690–703.

Williams, H. J., Shepard, E. L. C., Holton, M. D., Alarcón, P. A. E., Wilson, R. P., & Lambertucci, S. A. (2020). Physical limits of flight performance in the heaviest soaring bird. *Proceedings of the National Academy of Sciences of the United States of America, 117*, 1–7.

Wootton, R. J. (1979). Function, homology and terminology in insect wing. *Systematic Entomology, 4*, 81–93.

Wootton, R. J., & Ennos, A. R. (1989). The implications of function on the origin and homologies of the dipterous wing. *Systematic Entomology, 14*, 507–520.

Wootton, R. J., Kukalová-Peck, J., Newman, J. S., & Muzon, J. (1998). Smart engineering in the Mid-Carboniferous: How well could Paleozoic dragonflies fly? *Science, 282*, 749–751.

Chapter 3
Digging Up Convergence in Fossorial Rodents: Insights into Burrowing Activity and Morpho-Functional Specializations of the Masticatory Apparatus

Helder Gomes Rodrigues, Radim Šumbera, Lionel Hautier, and Anthony Herrel

Abstract Fossorial habits are tightly related to digging abilities in vertebrates and the most extreme fossorial specialization is being restricted to conducting the entire life underground. Many mammals, especially rodents, show behavioural, morphological and physiological adaptations to fossorial life, mainly for gaining access to sources of food and escaping predators and extreme climatic fluctuations. Adaptations to fossorial life are found in more than ten families of extant and extinct rodents, on most continents. Examples are Eurasian mole voles (Cricetidae), African mole-rats (Bathyergidae) and root-rats, Asian zokors and bamboo rats (Spalacidae), North American pocket gophers (Geomyidae) and mountain beavers (Aplodontidae), and South American tuco-tucos (Ctenomyidae) and cururos (Octodontidae). The constraints imposed by digging and living underground have led to strong behavioural and morphological convergences, notably involving the functioning of the rodent masticatory apparatus. Whereas most mammals use their claws for digging, rodents are unique in that some species use their ever-growing incisors for this purpose, with most subterranean species having become chisel-tooth

H. G. Rodrigues (✉)
Centre de Recherche en Paléontologie - Paris (CR2P), UMR CNRS 7207, CP38, Muséum National d'Histoire Naturelle, Sorbonne Université, Paris, France
e-mail: helder.gomes-rodrigues@mnhn.fr

R. Šumbera
Faculty of Science, University of South Bohemia, České Budějovice, Czech Republic
e-mail: sumbera@prf.jcu.cz

L. Hautier
Institut des Sciences de l'Evolution, cc64, Université de Montpellier, CNRS, IRD, EPHE, Montpellier, France
e-mail: Lionel.Hautier@umontpellier.fr

A. Herrel
Mécanismes Adaptatifs et Évolution (MECADEV), UMR 7179, CNRS, Funevol Team, Muséum National d'Histoire Naturelle, Paris Cedex 5, France
e-mail: anthony.herrel@mnhn.fr

V. L. Bels, A. P. Russell (eds.), *Convergent Evolution*, Fascinating Life Sciences,
https://doi.org/10.1007/978-3-031-11441-0_3

diggers. Here, we review examples of convergence found in the main morphological and functional components of the rodent digging apparatus in relation to burrowing activity. We first present the different modes of digging in rodents, focusing on the chisel-tooth digging mechanisms and their associated burrowing behaviours. Following this, several morphological specializations of the skull and the main jaw adductor muscles are described in relation to their associated contribution to biting efficiency. Specialized incisors allow subterranean rodents to dig in hard soil and to consume hard subterranean parts of plants, and their morphological and structural characteristics are considered in the last part of this chapter. Data on incisor bite force of fossorial rodents are also compiled to highlight the enhanced efficacy of the masticatory apparatus of chisel-tooth digging species. Despite the different cranial and muscular morphotypes in rodents, we underscore the fact that multiple modifications of the different components of the masticatory apparatus have led to similar overall morphologies and functions, overcoming phylogenetic inheritance. This remarkable example of convergence needs further scrutiny at both the micro- and macroevolutionary level to more fully understand how different rodent families evolved to deal with such external constraints.

Keywords Chisel-tooth digging activity · Cranial specializations · Masticatory muscles · Incisor procumbency · Bite force

3.1 Introduction

Fossorial mammals (that is, mammals adapted for digging) spend most of their life burrowing and transporting excavated soil. Although burrowing is energetically very costly (see Zelová et al., 2010 for an overview), many vertebrate taxa have adapted to the subterranean environment as it provides microclimatic stability, a relatively stable food supply, and a low risk of predation (Nevo, 1979, 1999; Burda et al., 2007). Many lineages of mammaliaforms and mammals, especially rodents, have independently colonised this environment at different times since the Mesozoic (e.g. Nevo, 1979; Cook et al., 2000; Luo & Wible, 2005). Life in subterranean burrow systems and the need to dig for large part of the day have stimulated the independent, but repeated evolution of many genetic, morphological, behavioural, and physiological adaptations (e.g. Nevo, 1979; Lacey et al., 2000; Partha et al., 2017). These specializations make underground mammals one of the best animal models for studying convergent evolution, rivalling traditional textbook examples such as aquatic or flying vertebrates.

Among mammals, rodents show numerous examples of convergent evolution toward fossorial life (e.g. Ellerman, 1956; Nevo, 1979; Stein, 2000). Specializations to both fossorial life and life in burrows have been observed in more than ten families of extant (e.g. Aplodontidae, Bathyergidae, Cricetidae, Ctenomyidae, Echimyidae, Geomyidae, Muridae, Octodontidae Sciuridae, Spalacidae) and extinct rodents (e.g. Cylindrodontidae, Mylagaulidae, Tsaganomyidae; see Fig. 3.1). Due to the independent origin and different timelines of colonisation of subsurface

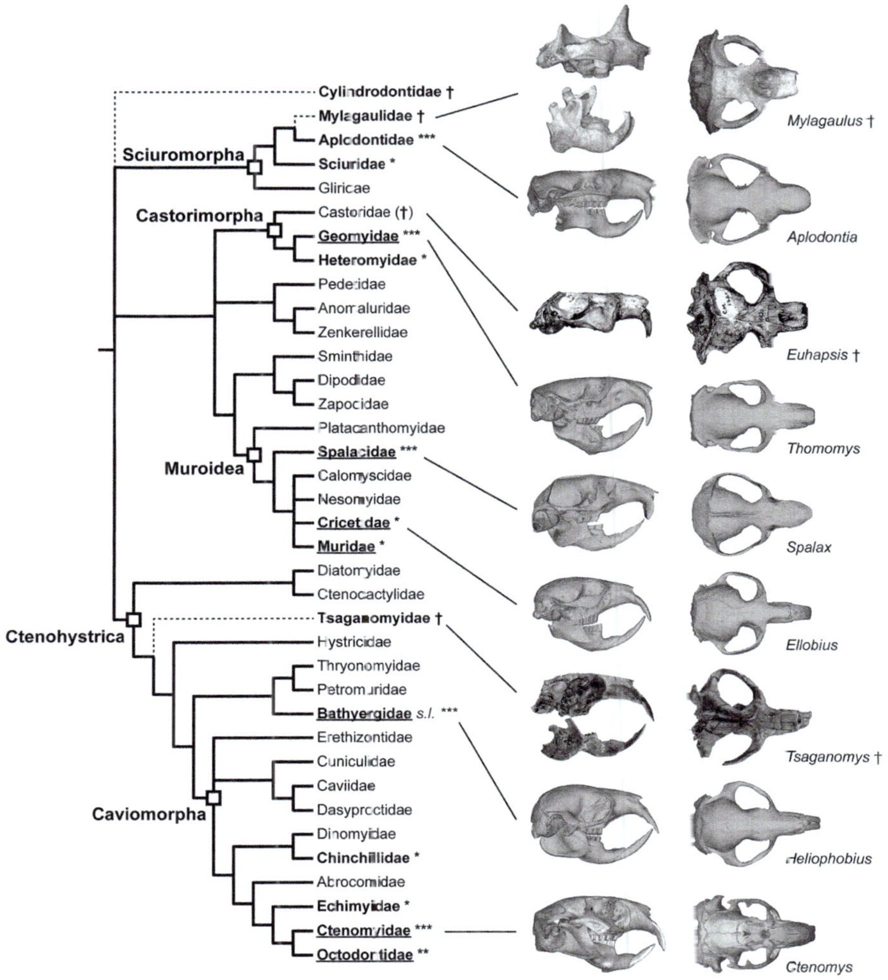

Fig. 3.1 Phylogeny of rodents (d'Elia et al., 2019) showing the main extant fossorial groups (in bold) and their associated convergent skull morphologies (in lateral and dorsal views). The number of asterisks indicates the proportion of fossorial species in each family (*< 30%; 30% < ** < 70%; 70% < ***). Scratch digging behaviour is present in all fossorial families, and underlined names indicate the presence of chisel-tooth digging species. Skull imaging results from X-ray microtomography (*Aplodontia rufa*, MNHN.ZM.MO1981–683; *Thomomys bottae*, BMNH.98.3.1.131; *Spalax* sp., BMNH.10.3.12.10; *Ellobius talpinus*, BMNH.34.2.11.30; *Heliophobius argenteocinereus*, ID13; imaging of *Ctenomys* species is modified from Korbin et al., (2020), except for *Mylagaulus cornusaulax* (modified from Czaplewski, 2012), *Euhapsis platyceps* [modified from Samuels and van Valkenburgh (2009)], and *Tsaganomys altaicus* [modified from Wang (2001)]. †, extinct taxa

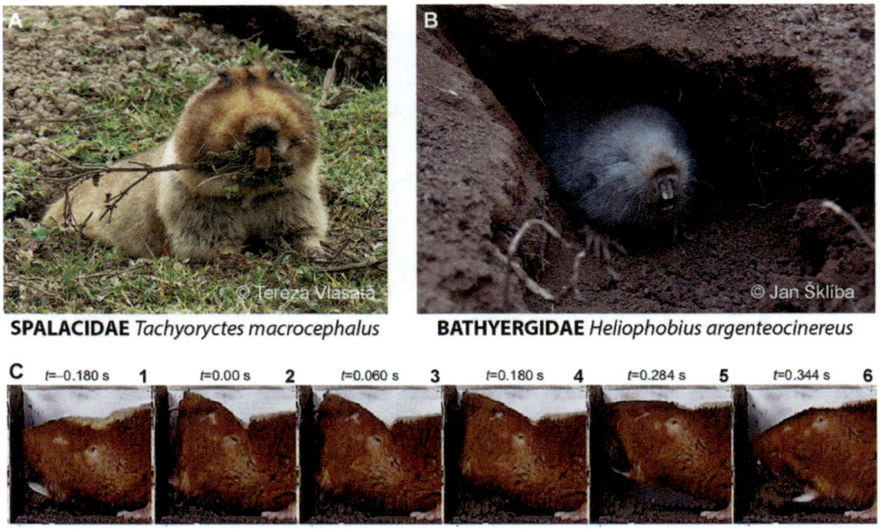

Fig. 3.2 (**a**, **b**) Examples of chisel-tooth digging rodents in the vicinity of their burrow. The protruding incisors can be observed. (**c**) Kinematics of chisel tooth digging in *Fukomys micklemi* (Bathyergidae; modified from Van Wassenbergh et al., 2017). Video frames show: 1. the start of nose-up cranial rotation, 2. reaching maximum gape, 3. initial soil penetration by the incisors, 4. mouth closing mainly by the lifting of the lower incisors, 5. nose-down cranial rotation bringing the grasped soil parcel down, and 6. release of the soil

environments, together with particular ecological conditions of habitats occupied by particular taxa, different degrees of adaptation to subterranean life are observed among fossorial rodents (Fig. 3.2a, b). Fossorial species feeding regularly on the surface can be distinguished from highly specialized fossorial forms that are adapted exclusively for life below ground and categorized as "subterranean" (e.g. Nevo, 1999; Begall et al., 2007; Šumbera, 2019). The most iconic cases of convergent adaptations to subterranean conditions are observed in the naked mole-rat, *Heterocephalus* (Heterocephalidae *sensu* Patterson and Upham, 2014, considered in the present study as included in the Bathyergidae *s.l.*), and the blind mole-rat, *Spalax/Nannospalax* (Spalacidae), both of which have evolved tolerance to hypoxia, exhibit extensive longevity (i.e. live more than 20 years), and resistance to cancer (e.g. Kim et al., 2011; Manov et al., 2013; Fang et al., 2014; Kirby et al., 2018).

Fossorial rodents also display a high number of morphological convergences, primarily those related to digging (Ellerman, 1956; Hildebrand, 1985) but also secondarily to life underground (e.g. fusiform body shape, reduced eyes and pinnae; Nevo, 1979; Stein, 2000). Two main ways of breaking and loosening soil have evolved, relying upon two types of digging tools. Scratch diggers alternate flexion and extension of their forelimbs endowed with enlarged claws, whereas chisel-tooth diggers are characterized by an increase in incisor procumbency, which is suggested to be combined with specific head movements in a few cases (Hildebrand, 1985; Laville et al., 1989; Stein, 2000). The degree of adaptation to a subterranean life and

Table 3.1 Ecological and geographical information for fossorial rodents that mainly use incisors for digging

Family	Subfamily	Genus or species	Modes of digging	Geographic range
Geomyidae	Geomyinae	*Thomomys*	Chisel-tooth/scratch	North America
Spalacidae	Spalacinae	*Spalax*	Chisel-tooth (with head)	Eurasia
	Spalacinae	*Nannospalax*	Chisel-tooth (with head)	Eurasia, Africa
	Rhizomyinae	*Cannomys*	Chisel-tooth, scratch	Asia
	Rhizomyinae	*Rhizomys*	Chisel-tooth, scratch	Asia
	Rhizomyinae	*Tachyoryctes*	Chisel-tooth, scratch	Africa
Cricetidae	Arvicolinae	*Ellobius*	Chisel-tooth (with head?)	Eurasia
	Arvicolinae	*Arvicola scherman*	Chisel-tooth, scratch	Europe
Muridae	Murinae	*Bandicota bengalensis*	Chisel-tooth	Asia
	Murinae	*Nesokia indica*	Chisel-tooth	Africa, Asia
Bathyergidae *s.l.*	Bathyerginae	*Fukomys*	Chisel-tooth	Africa
	Bathyerginae	*Cryptomys*	Chisel-tooth	Africa
	Bathyerginae	*Georychus*	Chisel-tooth	Africa
	Bathyerginae	*Heliophobius*	Chisel-tooth	Africa
	Heterocephalinae	*Heterocephalus*	Chisel-tooth	Africa
Ctenomyidae	Ctenomyinae	*Ctenomys*	Chisel-tooth/scratch	South America
Octodontidae	Octodontinae	*Spalacopus*	Chisel-tooth	South America

the frequency and length of surface forays thus depend on digging mode and performance. These different degrees of specialization provide striking cases of morphological convergence in the limbs and skull (Agrawal, 1967; Nevo, 1979; Stein, 2000), which remain to be more fully explored from both morphological and functional viewpoints.

The aim of the present chapter is to highlight the main adaptations of skull morphology and muscular anatomy in relation to chisel-tooth digging in rodents (Table 3.1). In order to better define fossorial adaptations of the masticatory apparatus, we review the morphological convergences in light of functional aspects and structural constraints, as previously suggested (e.g. Agrawal, 1967; Cook et al., 2000; Lessa, 2000). We first detail the burrowing activity of the most fossorial rodents and compare it to the morphological specializations of the masticatory apparatus. Then, we focus on the characteristics of the main digging tool, the incisors, and the associated bite force, for which new data are provided. This study allows us to link digging behaviours with the morphology and function of the masticatory apparatus for tracing the steps leading to convergent evolution for fossorial life during the evolutionary history of rodents.

3.2 A Brief Overview of Burrowing in Rodents

3.2.1 Extensive Burrowing Activity

Burrowing is a crucial activity for fossorial mammals for finding food, selecting a stable microclimate, locating sexual partners, or simply dispersing (Nevo, 1979). The biomass of food in the form of the subterranean parts of plants (e.g. bulbs, roots, and tubers) is generally less than that of plant parts above the surface. Consequently, solitary species that feed exclusively below ground need to build very large burrow systems of several tens or even hundred metres (e.g. the bathyergid *Heliophobius* and *Bathyergus*; the spalacids *Tachyoryctes* and *Spalax*; Jarvis & Sale, 1971; Cuthbert, 1975; Heth, 1989; Šumbera et al., 2003). The Cape dune mole-rat *Bathyergus suillus* is able to move 13.5 kg of sand in less than one hour and excavate up to five metres of burrows per day with its claws (Cuthbert, 1975). In social species such as the octodontid cururos *Spalacopus cyanus* (Begall & Gallardo, 2000) and chisel-tooth digging bathyergids, such as the naked mole-rat *Heterocephalus glaber* (Brett, 1991), Ansell's mole-rat *Fukomys anselli* (Šklíba et al., 2012), the giant mole-rat *Fukomys mechowii* (Šumbera et al., 2012), and the Damaraland mole-rat *Fukomys damarensis* (Jarvis et al., 1998), burrows can reach several hundred metres or even kilometres in length. Building such extensive burrow systems imposes strong selective pressures on digging efficiency. The silvery mole-rat *Heliophobius argenteocinereus* (Bathyergidae) is able to dig about one metre of burrow per day during the dry season when the soil is as hard as concrete (Šklíba et al., 2009). Based on the weight of material deposited in mounds and burrow diameter data it was estimated that a family of 87 individuals of the naked mole-rat excavated between 2.3 and 2.9 km of burrows in 2 years (Brett, 1991). Similarly, a family of 16 individuals of the Damaraland mole-rat, together weighing 2.2 kg, excavated and moved 2.6 tonnes of soil in less than 2 months (Jarvis et al., 1998). In fact, these values are probably considerable underestimates since excavated soil can also be deposited below ground, especially during the dry season. Burrow systems of fossorial rodents, especially those of subterranean species, do not constitute stable or rigid structures, but are instead dynamic, with new burrows being continuously opened and old ones filled in, especially at the periphery of the burrow system (Jarvis et al., 1998; Šumbera et al., 2003; Šklíba et al., 2009).

Burrowing through substrate requires the application of large forces to the soil, so that soil characteristics and food distribution are the main determinants of burrowing success. Apart from the geo-mechanical quality of the soil, the body mass of the burrow inhabitants also seems to influence burrow characteristics such as diameter and depth (Carotenuto et al., 2020). Many, if not all, subterranean mammals have primary and secondary digging modes (see Stein, 2000), as in pocket-gophers

(Geomyidae, Nevo, 1979; Marcy et al., 2016) and in tuco-tucos (Ctenomyidae; Giannoni et al., 1996; Becerra et al., 2014), which can be used alternatively depending on the characteristics of the soil. Different digging modes are also observed in closely related bathyergid species. Spalacids also comprise very different clades (i.e. Spalacinae, Myospalacinae and Rhizomyinae) for which repeated adaptations to a fossorial life-style are hypothesized (Fournier et al., 2021). As a result, morphological and behavioural variation among rodents does not only reflect phylogenetic relationships. The variation in digging apparatus must rather be viewed as an outcome of complex interactions between phylogenetic history, soil types, and the duration, frequency and nature of surface activities, which have led to convergent behavioural and morphological adaptations (Lessa & Thaeler, 1989; Stein, 2000).

3.2.2 Burrowing Modes and Behaviours

Burrowing is the process of the breaking of soil from the substrate, moving loosened soil below (or along) the body, moving it through the burrow, and finally depositing dirt either above ground into mounds, or backfilling unused burrows. Due to independent colonisations of the subterranean environment, this activity can be realised in different or convergent ways (see Nevo, 1979; Stein, 2000). Scratch digging is widely distributed among mammals. In fossorial rodents, this digging mode is known for most geomyids and ctenomyids, as well as for *Bathyergus* (Bathyergidae), *Myospalax/Eospalax* (Spalacidae), and *Prometheomys* (Cricetidae) among others. Contrastingly, chisel-tooth digging has evolved only in rodents, and is observed in bathyergids (except *Bathyergus*), spalacine and rhizomyine genera (Spalacidae), *Spalacopus* (Octodontidae), *Ellobius* (Cricetidae) and in some species of *Ctenomys* (Ctenomyidae), *Thomomys* (Geomyidae), *Arvicola* (Cricetidae), *Nesokia* and *Bandicota* (Muridae, see Table 3.1). Chisel-tooth digging rodents loosen soil mainly by using their incisors and, as for most scratch diggers, move the soil below the body and kick it vigorously through and out of the burrow. Soil can be loosened by both the incisors and the feet in tuco-tucos (*Ctenomys*) and pushed backward with the feet. After removing soil, some diggers may turn around (180°) and push soil face-first using their head, breast, forefeet, and chin (Airoldi et al., 1976; Stein, 2000).

Only a few studies have thoroughly investigated the digging behaviour and kinematics of fossorial rodents, all of them involving captive specimens (e.g. Jarvis & Sale, 1971; Cuthbert, 1975; Gasc et al., 1985; Lessa, 1987; Laville, 1989; Laville et al., 1989; Gambaryan & Gasc, 1993; Camin et al., 1995; Giannoni et al., 1996; Van Wassenbergh et al., 2017). Apart from data on scratch-diggers, these studies also provide important information regarding the use of incisors during digging and feeding, with different roles being suggested for upper and lower incisors. In *Fukomys micklemi* (Bathyergidae; see Van Wassenbergh et al., 2017), the upper incisors usually remain stationary and play an anchoring role during both digging and gnawing (Fig. 3.2c). In contrast, the lower incisors show upward movements to cut away the soil or to scrape food, and can contribute up to three quarters of the excavating work (Fig. 3.2c). These behaviours were also observed in other bathyergids, such as *Heliophobius*, *Heterocephalus*, and to a lesser extent *Georychus*, as well as in the spalacid *Tachyoryctes* and the cricetid *Arvicola scherman* (Jarvis & Sale, 1971; Cuthbert, 1975; Laville, 1989). In some social mole-rats, such as *Heterocephalus* and *Fukomys*, individual workers can work independently to remove the soil or unite to form so-called digging chains (Jarvis & Sale, 1971; Lacey & Shermann, 1991, RS unpublished observation).

Even if convergence in digging movements has been observed between the fossorial *Arvicola scherman* (Cricetidae) and *Nannospalax ehrenbergi* (Spalacidae; Laville et al., 1989), some differences are evident in the use of incisors and head. The blind mole-rat (*Nannospalax*) scrapes the floor with its incisors, although putatively in combination with its head that is used to push soil up to compact it, which is not the case for *Arvicola* (Gasc et al., 1985; Laville, 1989; Laville et al., 1989; Gambaryan & Gasc, 1993). Head lifting, defined as the "*use of incisors in concert with skull to form a powerful drill and shovel combination that is capable of loosening and removing soil*" (Hildebrand, 1985; Stein, 2000 and references therein) is often considered as the third burrowing type in rodents. However, if the blind mole-rats do use their broad and flat head and nose in effective bulldozing of soil out of burrows, the actual use of the head during digging remains debatable (Zuri et al., 1999; and RS personal observations). Then, pending further evidence, we consider that this digging mode should, rather, be considered as chisel-tooth digging combined with a quite unusual way of soil removal. Some taxa, such as the mole vole *Ellobius* (Cricetidae), often considered as a head lifter, might rather represent typical chisel-tooth diggers that use only their incisors to remove soil (Novikov pers. communication). Head lift digging has, nonetheless, also been reported for zokor *Myospalax* (Spalacidae), and for other mammals, such as the golden mole *Chrysochloris* and the marsupial mole *Notoryctes*, despite these latter species not using incisors for digging (Nevo, 1979; Hildebrand, 1985).

Both modes of digging and their repeated evolution across rodent lineages seem to be strongly related to soil characteristics. The digging mode usually changes depending on the hardness of the soil, with scratch diggers generally being restricted to sandy soils, while chisel-tooth diggers are present in a broader range of soils, as observed in the different species of pocket gophers (Geomyidae) and tuco-tucos (Ctenomyidae; Lessa & Thaeler, 1989; Giannoni et al., 1996; Mora et al., 2003; Marcy et al., 2016; Echeverría et al., 2017). This is particularly well illustrated in the Western Cape region of South Africa where three species of African mole-rats (Bathyergidae) occur sympatrically, but with a microallopatric distribution (Reichman & Jarvis, 1989). The largest species, the Cape dune mole-rat *Bathyergus suillus*, inhabits sandy dune habitats and is a scratch digger, whereas the Cape mole-rat *Georychus capensis* and the common mole-rat *Cryptomys hottentotus* prefer more consolidated soils and are both typical chisel-tooth diggers (Cuthbert, 1975). In these highly specialized rodents, these diverse digging behaviours are strongly associated with morphological adaptations reflecting not only the nature of the soil, but also the modes of digging and removal of soil, and are suggested to drive the repeated evolution of morphological and functional fossorial patterns.

3.3 A Highly Specialized Skull with Massive Masticatory Muscles

3.3.1 Cranial and Mandibular Convergences

Many studies have pinpointed morphological convergences in the skulls of fossorial rodents (e.g. Bekele, 1983; Lessa & Thaeler, 1989; Stein, 2000; Samuels & van Valkenburgh, 2009; Gomes Rodrigues et al., 2016; McIntosh & Cox, 2016a, b; Fournier et al., 2021). Alongside anterior projection of the incisor tips, corresponding to procumbency, all fossorial rodents are characterized by short, flat, but also deep and broad skulls, in association with enlarged zygomatic arches and temporal areas (see Figs. 3.1 and 3.3 for more details). Such cranial similarities

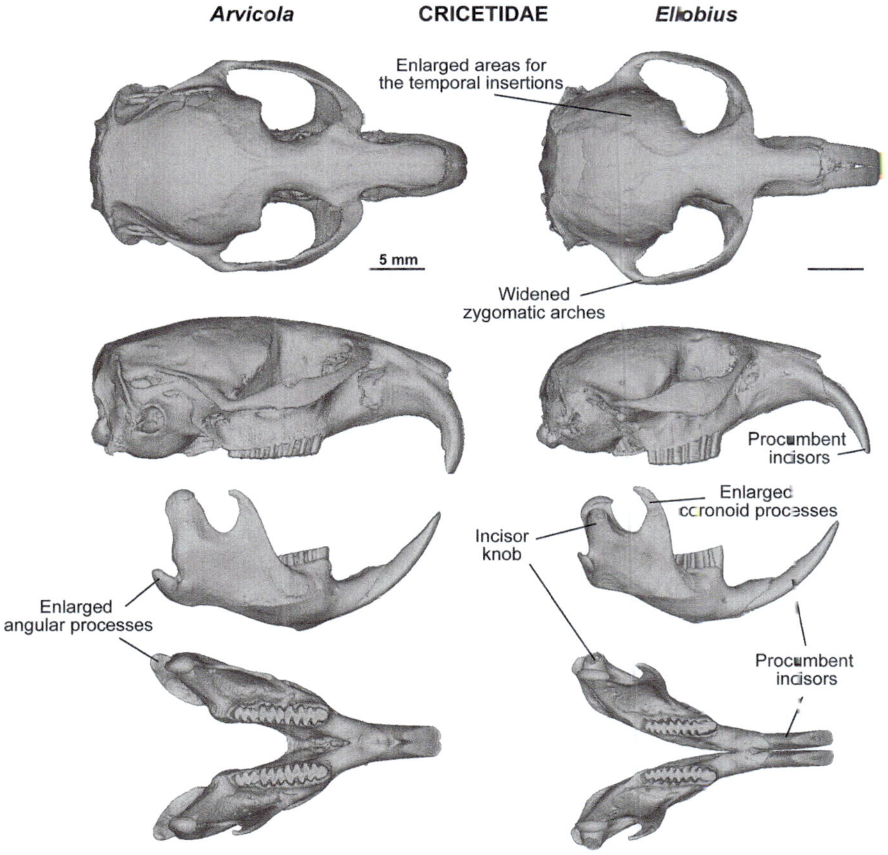

Fig. 3.3 Comparison of skull morphology between a non-fossorial species (*Arvicola amphibius*, BMNH.1937.3.22.48) and a highly specialized chisel-tooth digger (*Ellobius talpinus*, BMNH.34.2.11.30; with the right zygomatic arch partly reconstructed) of the family Cricetidae

are linked to the development of massive and powerful masticatory muscles necessary to enhance incisor biting while digging. These two anatomical and functional aspects are discussed further below. These massive muscles require large areas of attachment on the zygomatic and temporal regions of the skull, as well as on the mandible. All these specializations are the result of successive modifications leading to repeated adaptations to fossorial lifestyles, especially for chisel-tooth digging. Several authors suggested that these adaptive changes toward fossoriality were generally preceded by behavioural changes, since some non-fossorial cricetid and octodontid species (e.g. *Pitymys*, *Octodon*, and *Tympanoctomys* frequently defined as being semi-fossorial), are able to dig complex burrows without significant morphological adaptations for this purpose (Casinos et al., 1983; Lessa et al., 2008). According to the fossil record, such behavioural changes (i.e. digging without morphological adaptations) could be related to the opening of the environment from 40 Ma onward in different areas across the globe (e.g. North American Great Plains, Mongolian Plateau), the need to find new resources, and also the requirement to find alternative shelters for avoiding being preyed upon (Nevo, 1999; Jardine et al., 2012; but see Rodríguez-Serrano et al., 2008 for alternate hypotheses). Morphological specializations are observed in the oldest rodent fossorial lineages, the Cylindrodontidae and Tsaganomyidae (Bryant & MacKenna, 1995), as well as in the Castoridae, Geomyidae, and Aplodontoidea in the last 30 Ma. In contrast, other extant families did not evolve any specific fossorial characteristics prior to 20 Ma (Cook et al., 2000; Hopkins, 2005; Samuels & van Valkenburgh, 2009; Jardine et al., 2012; He et al., 2020; Fournier et al., 2021).

Because of the strong influence of chisel-tooth digging on the morphology of the entire masticatory apparatus, morphological convergences are also evident between species belonging to different genera, which are not exclusively fossorial. The best examples are found among bandicoot rats (Muridae), *Nesokia* and *Bandicota*, species of which show a wide array of behavioural habits, from aquatic to fossorial life (Agrawal, 1967; Kryštufek et al., 2016). Both genera include fossorial species that display similar cranial and mandibular specializations for chisel-tooth digging (Fig. 3.4a; see Kryštufek et al., 2016). Their crania are broad with enlarged zygomatic arches, as well as a wide and short rostrum, when compared to non-fossorial species (Fig. 3.4a, PC2). Their mandibles show the strongest morphological changes, with short and laterally-oriented angular processes, enlarged coronoid processes, as well as prominent alveolar processes at the root of the incisor, forming a knob at the level of the angular process (see Fig. 3.4b, PC1). In general, the mandible shows the strongest ecological imprint, because of its simple morphology consisting of a single bone, the dentary (on each side). Similar morphotypes are also observed in Cricetidae (Durão et al., 2019; Fig. 3.3) and in the Ctenohystrica (Gomes Rodrigues et al., 2016; Fig. 3.4b).

All these morphological characteristics can be found at the intrageneric levels and also at the intraspecific level, although less pronounced, depending on the nature of the soil. When the soil is harder, rodents more frequently use their incisors for digging, which implies similar skull modifications (e.g. short skull with enlarged zygomatic arches, procumbent incisors), as observed in different species of

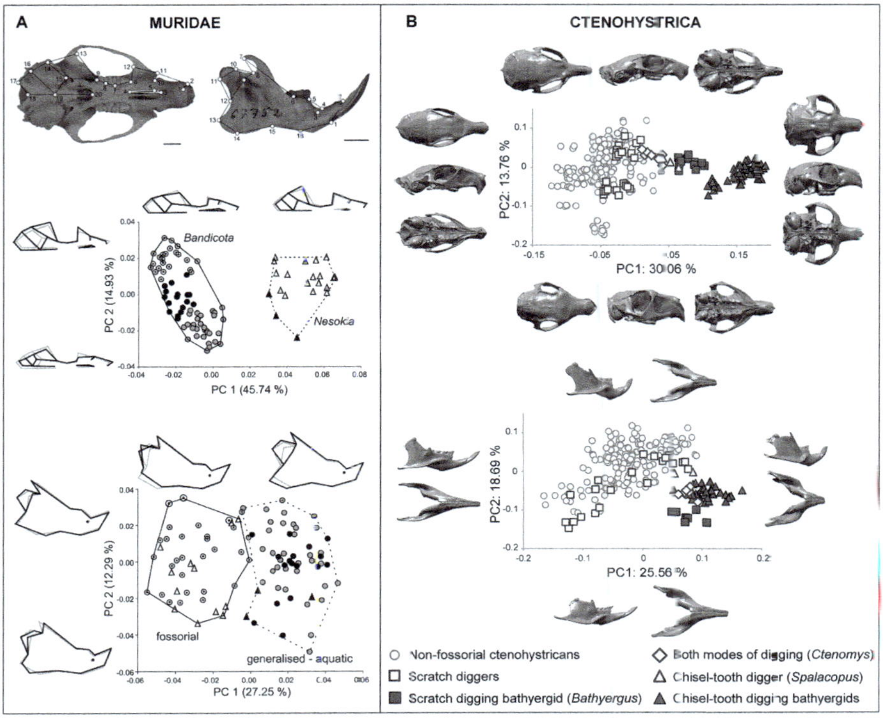

Fig. 3.4 (**a**) Quantification and comparison of cranial and mandibular morphology between convergent fossorial bandicoot rat species (*Bandicota bengalensis* and *Nesokia indica*; open symbols) and non-fossorial species (*Bandicota indica*, *Bandicota savilei*, and *Nesokia bunni*; filled symbols) using geometric morphometric methods (modified from Kryštufek et al., 2016; scale bars: 5 mm). (**b**) Quantification and comparison of cranial and mandibular morphology between fossorial scratch digging, chisel-tooth digging and non-fossorial ctenohystrican rodents using geometric morphometric methods (modified from Gomes Rodrigues et al., 2016)

Tachyoryctes (Beolchini & Corti, 2004), in *Heliophobius argenteocinereus* (Barčiová et al., 2009) and in *Thomomys bottae* (Lessa & Thaeler, 1989). In contrast, soil hardness does not seem to have strongly influenced the skull morphology of *Spalacopus* (Bacigalupe et al., 2002) and *Ctenomys* (Echeverría et al., 2017). At a higher taxonomic level, most chisel-tooth digging species present the above-mentioned specializations in comparison to scratch digging and non-fossorial species. These convergences have been highlighted in a number of geometric morphometric studies dealing either with rodents as a whole (Samuels & van Valkenburgh, 2009; McIntosh & Cox, 2016b) or focusing on specific taxa [e.g., ctenomyids and octodontids in Becerra et al. (2014) and by Gomes Rodrigues et al. (2016) also focusing on bathyergids; geomyids in Marcy et al. (2016); spalacids in Fournier et al. (2021)].

The use of the head for digging, or to push and pack soil, also implies strong but additional cranial adaptations compared to other chisel-tooth diggers. In fossorial taxa that use their head as a shovel (whether for digging or removing soil), the upper incisors are not necessarily highly procumbent. They differ from other chisel-tooth diggers by exhibiting short crania with broader frontal and nasal bones, and have a very broad posterior face due the presence of an enlarged and anteriorly-tilted occipital plate for the insertion of massive neck muscles. These characteristics are evident in *Spalax/Nannospalax* and *Myospalax/Eospalax* (Spalacidae; Fournier et al., 2021), in some extinct taxa (Mylagaulidae, Castoridae; Hopkins, 2005; Samuels & van Valkenburgh, 2009; Fig. 3.1), and in a few other mammals, such as golden moles (Afrosoricida; Hildebrand, 1985). Such a high degree of morphological specialization, repeatedly resulting from selection during the evolution of rodents, is important from an evolutionary viewpoint. It allows us to confidently infer fossorial adaptations in extinct species (e.g. Hopkins, 2005; Mein & Pickford, 2008; Flynn, 2009; Samuels & van Valkenburgh, 2009) and notably to infer a precocious adaptation to chisel-tooth digging in the evolutionary history of rodents, as suggested for the Tsaganomyidae and Cylindrodontidae (Bryant & MacKenna, 1995).

3.3.2 Prominent Adductor Muscles

Although rodents present a wide array of skull morphologies that have been traditionally arranged into four main muscular morphotypes (Wood, 1965), the constraints of fossorial life, especially chisel-tooth digging, are strong enough to have influenced the morphological evolution of the masticatory apparatus. These functional constraints have led to extensive modifications of both the skull structure and anatomy of masticatory muscles, mainly involving the size of the adductor muscles rather than their structural organization (e.g. Lessa & Stein, 1992; Cox et al., 2020). Muscle size is the main factor influencing bite force (Becerra et al., 2014). In rodents, the masseter muscles, especially the superficial and deep layers, and, to a lesser extent, the temporal and pterygoid muscles constitute the most dominant part of the masticatory musculature (Fig. 3.5). In a detailed comparative study, Morlok (1983) described the muscular anatomy in the main fossorial rodent families (Spalacidae, Bathyergidae, Geomyidae, Cricetidae, and Ctenomyidae). However, although the large size of the masseter and temporalis muscles was mentioned, convergent adaptations between families were not discussed with regard to differences in digging modes. These adductor muscles were shown to be very prominent in subterranean rodents, such as bathyergids (Van Daele et al., 2009; Cox et al., 2020), and also in ctenomyids, when compared to less or non-fossorial sister taxa such as the Chinchillidae and Octodontidae (Fig. 3.5; see Becerra et al., 2014). Differences in the size of the musculature are also observed at the intraspecific level in *Thomomys bottae,* depending on the hardness of soils occupied (Lessa & Thaeler, 1989). Specimens found in rocky, clay soils showed larger adductor muscles than those found in sandy soils.

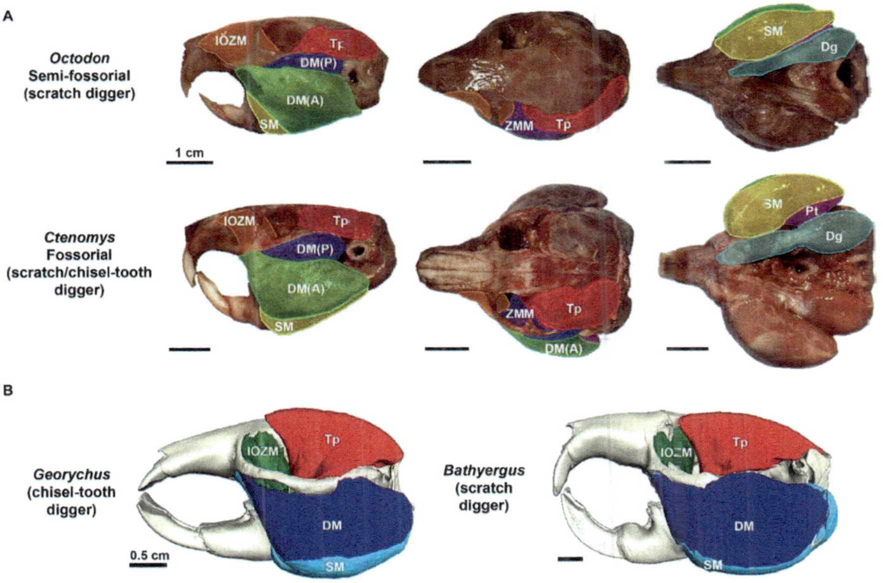

Fig. 3.5 (**a**) Comparison of the anatomy of the masticatory muscles after dissection of a semi-fossorial and a fossorial ctenohystrican rodent in lateral, dorsal, and ventral views (modified from Becerra et al., 2014). (**b**) Anatomy of masticatory muscles in the Bathyergidae (lateral views). Virtual dissection after iodine-based contrast-enhanced microCT imaging (modified from Cox et al., 2020). Abbreviations—*SM* superficial masseter, *DM* deep masseter—(*A*) anterior and (*P*) posterior parts, *ZMM* zygomaticomandibular masseter, *IOZM* infraorbital part of the zygomaticomandibular masseter, *Tp* temporal, *Pt* pterygoid, *Dg* digastric muscle

Among the adductor muscles, the temporal muscle generally consists of several layers; it is highly involved in the jaw-closing motion and, depending on its size, high output force can be produced at the level of the incisors during both feeding and digging (Hiiemae, 1971; Van Daele et al., 2009; McIntosh & Cox, 2016a). This muscle is generally more voluminous in chisel-tooth digging rodents than in scratch diggers, with an attachment closer to the sagittal plane on the cranium, and an insertion on the enlarged coronoid process of the mandible, as shown in some caviomorphs (*Ctenomys vs Octodon*, Becerra et al., 2014, Fig. 3.5a). However, the relative size of the temporal muscle is reduced in ctenomyids (15% of the total mass of masticatory muscles), whereas it reaches approximately 30% in most chisel-tooth digging rodents, such as bathyergids, spalacids, and cricetids (e.g. Morlok, 1983; Cox et al., 2020). Non chisel-tooth diggers, such as *Aplodontia*, *Bathyergus*, and *Castor*, also exhibit a temporal muscle of relatively large size (Druzinsky, 2010; Cox & Baverstock, 2016; Cox et al., 2020) in relation to morphological characteristics inherited from their putative chisel-tooth digging ancestors or to other mechanically-demanding activities (Hopkins, 2005; Samuels & van Valkenburgh, 2009; Gomes Rodrigues et al., 2016). While rarely used for digging, the incisors of non-chisel-tooth diggers are still used to gnaw hard food items, such as geophytes, tubers or tree bark, which can explain why the inherited morpho-functional properties of their masticatory apparatus are conserved. The head-lift digging spalacids, *Spalax/Nannospalax,* show the greatest relative size of the temporal muscle (45%) among the described musculature of fossorial rodents (Morlok, 1983), although this is not associated with procumbent upper incisors but rather with highly procumbent lower incisors. This temporal characteristic, combined with powerful neck muscles (splenius and rhomboideus; Nevo, 1999), might be involved in improving the loosening and removing of soil (Laville et al., 1989), as proposed for extinct aplodontoids and castorids, and, to a lesser extent, for other spalacids (*Myospalax*) and, putatively, for cricetids (e.g. *Ellobius*; see Krapp, 1965).

In highly specialized fossorial rodents, the superficial and deep masseters are massive muscles with large areas of origin on the zygomatic arches and rostrum. They have been shown to contribute extensively to the generation of high bite forces at the level of both the incisors and cheek teeth (Becerra et al., 2014; Cox & Faulkes, 2014). Nonetheless, the superficial masseter generally has an insertion area that is more limited in the posterior part of the mandible in different chisel-tooth digging species, due to its reduced angular process (see Figs. 3.3, 3.4 and 3.5b). This is not the case for tuco-tucos (Ctenomyidae), probably because the incisors are not necessarily the main digging tools in the various species (Becerra et al., 2014; Echeverría et al., 2017). This muscle is considered to be the main protractor of the mandible (Hiiemae, 1971), but its role during the power stroke of the jaw remains to be ascertained. It has been assumed that its limited amount of insertion on the posterior part of the mandible of chisel-tooth diggers might favour a wide gape (McIntosh & Cox, 2016a), so that the temporal muscle can produce a higher output force during incisor biting at the expense of the biomechanical advantage of the superficial masseter (Gomes Rodrigues et al., 2016). The infraorbital part of the zygomaticomandibular masseter is peculiar to rodents and is only present in two "masticatory morphotypes", characterized by their enlarged infra-orbital foramina (Wood, 1965). Interestingly, this muscle tends be less voluminous and less anteriorly expanded in chisel-tooth digging rodents, such as some ctenomyids (Becerra et al., 2014), and more importantly in bathyergids, in which the foramina are highly reduced and the muscle is mostly confined to the orbital region (Morlok, 1983; Gomes Rodrigues et al., 2016; Cox et al., 2020; Fig. 3.5b). This reorganisation is likely related to the reduction of the snout and favours a wider gape, which optimizes incisor output force (McIntosh & Cox, 2016a, b; Cox et al., 2020). More studies on the masticatory musculature of rodents are needed to better understand the anatomical and functional characteristics of each of these muscular layers, as well as their precise contribution during gnawing and incisor-biting and for different digging stages.

3.4 The Incisors: A Powerful Tool for Digging

3.4.1 Highly Specialized Incisors

If the masticatory muscles and skull constitute the machinery producing the energy necessary for gnawing or digging, incisors represent the main tool for the accomplishment of these tasks. Incisors coupled with cheek teeth constitute one of the main functional components of the masticatory apparatus of rodents. All subterranean rodents have convergently evolved high-crowned cheek teeth to cope with high wear resulting from the unintentional ingestion of abrasive particles (i.e. grit and dust) during digging and feeding (Stein, 2000; Gomes Rodrigues, 2015; Gomes Rodrigues & Šumbera, 2015). The most extreme case is observed in the silvery mole-rat, *Heliophobius* (Bathyergidae), which presents a continuous and horizontal replacement of its molars, working like a conveyor belt, assumed to be related to its important digging activity (Gomes Rodrigues et al., 2011). Incisors are ever-growing in rodents and are covered only on their labial surface with enamel for resisting bending stresses. In many rodent species, this enamel layer is enriched with iron oxides, which reinforces the enamel and is responsible for the orange colour of the incisors (mainly the upper incisors; see Fig. 3.2a). This characteristic is, however, not specific to fossorial species as it is observed in many non-fossorial taxa; it can also be absent in some fossorial ones (e.g. Bathyergidae; Gomes Rodrigues, 2015; Fig. 3.2b). This oxide enrichment might be biomechanically linked with procumbency since the upper incisors are usually less coloured when highly procumbent. This observation has been made for the Spalacidae, Geomyidae, Ctenomyidae and Cricetidae (see Stein (2000) for more details).

The deficiency in oxide enrichment of incisors of some chisel-tooth digging rodents can be paralleled by a reinforcement of their enamel microstructure. Previous studies of enamel microstructure have shown that the enamel layer is generally thicker in fossorial rodents, especially the outer enamel layer of the upper incisors. This is nicely exemplified in chisel-tooth digging species such as spalacids, cricetids, geomyids, and the extinct ctenomyids (~35–55% and ~40–60% of the total enamel thickness of the lower and upper incisors respectively; e.g. Flynn et al., 1987; Kalthoff, 2000; Vieytes et al., 2007). Such a difference between fossorial and non-fossorial rodents was also observed between species of *Arvicola*, with the fossorial *Arvicola scherman* showing thicker enamel than the semi-aquatic *A. amphibius and A. sapidus* (Marcolini et al., 2011). This adaptation compensates for high dental wear due to the intense use of incisors during digging. In addition, compared to non-fossorial taxa, fossorial species, such as blind mole rats (Spalacidae), naked mole-rats (Bathyergidae) and pocket-gophers (Geomyidae; Manaro, 1959; Hildebrand, 1985; Zuri et al., 1999; Berkovitz & Faulkes, 2001), exhibit higher growth rates of the incisors, especially the lower ones that are more importantly used during digging. This rate is higher in chisel-tooth diggers than in scratch diggers, as observed for pocket-gophers (scratch digging *Geomys*: 0.35 mm/

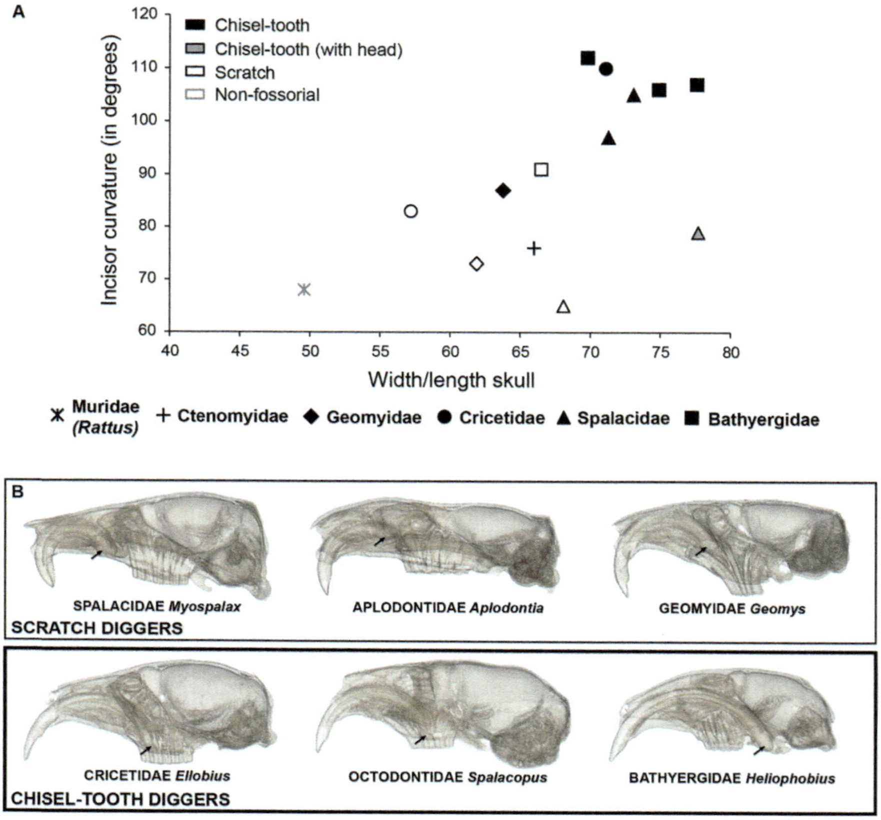

Fig. 3.6 (**a**) Biplot of the width/length ratio of the cranium and incisor curvature in different fossorial rodents (data compiled from Morlok 1983). (**b**) Comparison of incisor root insertion location and procumbency between scratch digging and chisel-tooth digging rodents using translucent X-ray microtomographic reconstruction of the cranium (*Myospalax* sp. BMNH.9.1.1.206; *Aplodontia rufa*, MNHN.ZM.MO1981-683; *Geomys pinetis floridanus*, MNHN.ZM2007-233; *Ellobius talpinus*, BMNH.34.2.11.30; *Spalacopus cyanus*, BMNH.98.1.8.5; *Heliophobius argenteocinereus*, BMNH.68.93)

day *versus* chisel-tooth digging *Thomomys*: 0.5 mm/day; Miller, 1958; Manaro, 1959).

The external protrusion of the upper incisors constitutes one of the main characteristics of fossorial rodents, especially chisel-tooth diggers (see Fig. 3.2). Incisor procumbency is generally greater than 90° relative to the rostral plane in the latter taxa (e.g. *Ellobius*, *Tsaganomys*, *Heliophobius* and *Ctenomys* in Figs. 3.1 and 3.3; Marcy et al., 2016), which defines proodonty (Thomas, 1919; Landry Jr., 1957). The corresponding values of the angle of incisor curvature (or protrusion here) in chisel-tooth digging taxa are always greater than those for scratch diggers and non-fossorial species (Fig. 3.6; Morlok, 1983; Van der Merwe & Botha, 1998; McIntosh & Cox, 2016b). As for the proportion of the skull and the associated mass of main adductor muscles, procumbency tends to increase in relation to soil hardness and to the degree of adaptation to chisel-tooth digging, as observed in several families (Fig. 3.6a) and in different species of *Thomomys* (Geomyidae; Marcy et al., 2016). Incisor procumbency also involves the lower incisors, but to a lesser degree (Van der Merwe & Botha, 1998). This is notably observable in sciurognathous rodents, which are characterized by the absence of lateralization of the angular process of the mandible. In the sciurognathous spalacids, cricetids, murids and geomyids (Figs. 3.1, 3.3 and 3.4a), the incisor root inserts posteriorly and protrudes laterally from the mandible to form a knob close to the articular condyle. The role of the upper incisors is mainly for anchorage to the substrate, whereas the lower incisors are mainly used as a shovel (Jarvis & Sale, 1971; Laville et al., 1989; Van Wassenbergh et al., 2017). Incisor procumbency associated with wide gaping during digging enables rodents to obtain a more effective angle of attack, with the incisor tip always being in contact with the soil (Mora et al., 2003; McIntosh & Cox, 2016b). The external protrusion of the tips of the upper incisors, in addition to the development of protecting folds of the lips, also helps to prevent soil from entering the mouth and nostrils (Agrawal, 1967). More generally, procumbent incisors allow for an increase of mechanical efficiency during digging (Landry Jr., 1957; McIntosh & Cox, 2016b). The lower incisors also serve as manipulators of soil and food, which is also facilitated by the independent movements of the hemi-mandibles, thanks to the presence of an unfused symphysis in most fossorial rodents. These alternate movements are particularly pronounced in bathyergids and spalacids (Landry Jr., 1957; Gomes Rodrigues et al., 2016).

3.4.2 Procumbency and Mechanical Efficiency of the Incisors

Procumbent incisors result in a more efficient bite, but also impose greater biomechanical constraints resulting from important stresses and pressures. Increasing procumbency means increasing the radius of curvature of the incisor, which is generally associated with an enlargement of the incisor, (e.g.; Landry Jr., 1957; McIntosh & Cox, 2016b). An allometric relationship between rostral size and incisor procumbency was emphasized by Lessa and Patton (1989). They noted that incisor procumbency increases with the enlargement of the rostrum, involving an increase of incisor length in many pocket-gophers (Geomyidae), in which the incisor root is inserted in front of the cheek teeth. They also mentioned the occurrence of this allometric pattern related to incisor procumbency in root-rats *Tachyoryctes* (Spalacidae), and it was then generalized to most rodents by Lessa (2000). Incisor procumbency was also suggested to be independent of rostral size as in ctenomyids (Echeverría et al., 2017), but more specifically in bathyergids, in which the incisor root is inserted behind the molars in chisel-tooth digging genera (e.g. *Heliophobius* in Fig. 3.6b; Landry Jr., 1957; McIntosh & Cox, 2016b). Such a departure from allometric "constraints" is also observed in more specialised chisel-tooth digging cricetids, octodontids and spalacids (e.g. *Ellobius*, *Spalacopus*, *Cannomys*) that all display a posterior insertion of the upper incisors at the level of the molars (see Fig. 3.6b). This pattern notably favours incisor lengthening in small-sized subterranean rodents and it contributes to the dissipation of forces during biting (Becerra et al., 2012; McIntosh & Cox, 2016b). It is worth noting that both allometric and non-allometric trends related to procumbency can be observed in the same family and can also depend on the nature of the soil (see Marcy et al. (2016) on species of *Thomomys* for more details).

These structural adjustments may also contribute to the optimization of digging motions and the increase of bite force. Increasing body size is associated with greater muscular strength due to enlarged muscles, a longer rostrum and thus a more proodont incisor, and a resulting higher bite force. This might explain why harder soils are generally inhabited by larger species in a given family, as noticed for instance in the Geomyidae (Marcy et al., 2016), even if this relationship can be more complex when taking into account the burrow architecture (Carotenuto et al., 2020). If this configuration implies greater in-lever arms (i.e. distance from the condyle or fulcrum to the point of muscle attachment), the out-lever arms (distance from the condyle to the incisor tip) are also greater due to procumbent incisors and an enlarged rostrum, which reduces the force applied at the incisors (Bekele, 1983). In rodent species showing a posterior insertion of the incisors and a shortened rostrum, the associated reduction of the out-lever arm would produce a greater bite force owing to a higher mechanical advantage of the temporal muscle. This is typically the case for most chisel-tooth digging ctenohystricans (Gomes Rodrigues et al., 2016), especially bathyergids (McIntosh & Cox, 2016a, b). Different biomechanical configurations may improve bite force, depending on digging mode: from *Thomomys* showing both allometric and non-allometric architectural possibilities (Marcy et al., 2016) to *Ctenomys* showing a combination of an increase of both the adductor muscle size and procumbency without significant shortening or lengthening of the rostrum (Becerra et al., 2014; Echeverría et al., 2017).

3.4.3 Absolute Incisor Bite Force

To sum up, an increase of bite force in rodents is favoured by procumbent incisors combined with the great enlargement of the masticatory muscles and a wide and deep skull, permitting the combination of this force with the wide gape needed for digging (McIntosh & Cox, 2016a, b). Absolute bite force was demonstrated to be correlated to body mass in rodents (e.g. Freeman & Lemen, 2008; Van Daele et al., 2009, 2019; Becerra et al., 2014). Fossorial rodents, especially chisel-tooth diggers, were assumed to have higher bite forces than non-fossorial taxa. For instance, estimated bite force is higher for the fossorial species of *Arvicola* compared to its non-fossorial relatives (Durão et al., 2019). However, this hypothesis was never tested on a large dataset. In order to test whether fossoriality has a convergent impact on bite force in rodents we compiled literature data on absolute bite force (Freeman & Lemen, 2008; Van Daele et al., 2009; Williams et al., 2009; Becerra et al., 2011,

2013, 2014; Kerr et al., 2017; Ginot et al., 2018; Hite et al., 2019). We then gathered data for 456 individuals belonging to 25 species, for which maximal *in vivo* bite forces were available (Table 3.2). In addition, we collected data for two species of fossorial rodent, *Spalacopus cyanus* ($N = 7$) and *Spalax galili* ($N = 36$). In brief, we used an isometric Kistler force transducer connected to a charge amplifier [for details of the experimental setup, see Herrel et al. (1999)]. Animals were taken from their cages, restrained, and allowed to bite the transducer five times. The highest bite force was then retained as an estimate of maximal bite performance.

We found that all fossorial rodents differ significantly from non-fossorial species (see Table 3.2) and have a higher bite force for their body mass according to the ANCOVA ($F_{1,38} = 10.23$, $P = 0.003$; see Fig. 3.7), as previously observed for a few species (e.g. Freeman & Lemen, 2008; Van Daele et al., 2009; Hite et al., 2019). Extensive comparison between chisel-tooth digging and scratch digging species could not be realized since only one scratch digger (*Geomys*) was considered in the analysis and did not significantly differ from the other fossorial species. Among octodontids, the degu (*Octodon*) is a non-specialized scratch digger, which does show a lower bite force than the chisel-tooth digging cururo (*Spalacopus*; Table 3.2). Improvement of the use of incisors and their resistance to bending stresses enable subterranean species not only to generate the high bite force used for digging, but also that needed for the consumption of a wide range of hard geophytes (Van Daele et al., 2009; McIntosh & Cox, 2016b; Vassallo et al., 2021). The high bite force value observed for the non-fossorial squirrel *Sciurus niger* (Table 3.2) was probably similarly related to the gnawing of hard items, such as nutshells (Freeman & Lemen, 2008), and to the large temporal muscles generally observed in sciuromorphous rodents (Ball & Roth, 1995; Cox et al., 2020). By measuring greater resistance to stresses in the skull of chisel-tooth digging caviomorphs and bathyergid species, several studies using finite element analyses (McIntosh & Cox, 2016c; Buezas et al., 2019; Vassallo et al., 2021) have demonstrated that the morpho-functional characteristics of their masticatory apparatus not only allow them to produce high bite forces, but also to be able to sustain them over long periods of time without structural failure (Van Daele et al., 2009; Vassallo et al., 2021).

3.5 Conclusion

From a morpho-functional point of view we demonstrate that fossorial life imposes strong constraints on the rodent skull, especially when the masticatory apparatus is involved in digging through the deployment of powerful incisor biting. This finding alone is sufficient for explaining why the evolution of the masticatory apparatus in fossorial rodents constitutes one of the most striking cases of morphological and functional convergence in mammals. Whereas fossorial and subterranean behaviours might be induced by the opening of landscapes and the search for new shelters and food resources, the main drivers of the evolution of chisel-tooth digging might be related to both the consumption of hard geophytes and living in harder soils. The

Table 3.2 Data on incisor bite force gathered for both fossorial and non-fossorial rodents

Species	*N*	Family	Life style	Method	Body mass (g)	Bite force (N)	References
Apodemus sylvaticus	103	Muridae	Non-fossorial	Piezo	20.4	9.1	Ginot et al. (2018)
Apodemus sylvaticus	11	Muridae	Non-fossorial	Piezo	25.9	9.6	Kerr et al. (2017)
Chinchilla lanigera	10	Chinchillidae	Non-fossorial	Strain gauge	570.0	23.5	Becerra et al. (2014)
Ctenomys australis	10	Ctenomyidae	Fossorial	Strain gauge	360.0	68.7	Becerra et al. (2014)
Ctenomys talarum	15	Ctenomyidae	Fossorial	Strain gauge	146.6	31.7	Becerra et al. (2011)
Ctenomys tuconax	3	Ctenomyidae	Fossorial	Strain gauge	520.0	74.9	Becerra et al. (2013)
Dipodomys ordii	11	Heteromyidae	Non-fossorial	Flexiforce	63.0	14.0	Freeman and Lemen (2008)
Fukomys mechowii	73	Bathyergidae	Fossorial	Piezo	105.6	32.3	Modifed from Van Daele et al. (2009)
Fukomys micklemi	11	Bathyergidae	Fossorial	Piezo	75.9	21.8	Modifed from Van Daele et al. (2009)
Fukomys whytei	8	Bathyergidae	Fossorial	Piezo	82.6	23.0	Modifed from Van Daele et al. (2009)
Geomys bursarius	5	Geomyidae	Fossorial	Flexiforce	153.0	50.6	Freeman and Lemen (2008)
Heterocephalus glaber	10	Bathyergidae *s.l.*	Fossorial	Flexiforce	56.1	21.1	Hite et al. (2019)
Microtus ochrogaster	10	Cricetidae	Non-fossorial	Flexiforce	34.0	12.9	Freeman and Lemen (2008)
Mus spretus	35	Muridae	Non-fossorial	Piezo	13.7	8.3	Ginot et al. (2018)
Neotoma floridana	15	Cricetidae	Non-fossorial	Flexiforce	321.0	30.3	Freeman and Lemen (2008)
Octodon degus	10	Octodontidae	Non-fossorial	Strain gauge	240.0	21.9	Becerra et al. (2014)
Onychomys leucogaster	2	Cricetidae	Non-fossorial	Flexiforce	34.0	11.5	Freeman and Lemen (2008)
Onychomys leucogaster	10	Cricetidae	Non-fossorial	Strain gauge	49.9	13.5	Williams et al. (2009)
Perognathus flavescens	1	Heteromyidae	Non-fossorial	Flexiforce	6.5	4.6	Freeman and Lemen (2008)
Peromyscus leucopus	10	Cricetidae	Non-fossorial	Flexiforce	23.0	10.0	Freeman and Lemen (2008)

(continued)

Table 3.2 (continued)

Species	*N*	Family	Life style	Method	Body mass (g)	Bite force (N)	References
Peromyscus maniculatus	10	Cricetidae	Non-fossorial	Strain gauge	21.2	8.5	Williams et al. (2009)
Peromyscus maniculatus	4	Cricetidae	Non-fossorial	Flexiforce	21.0	8.8	Freeman and Lemen (2008)
Reithrodontomys megalotis	3	Cricetidae	Non-fossorial	Flexiforce	11.5	7.7	Freeman and Lemen (2008)
Sciurus niger	22	Sciuridae	Non-fossorial	Flexiforce	588.0	73.0	Freeman and Lemen (2008)
Sigmodon hispidus	6	Cricetidae	Non-fossorial	Flexiforce	105.0	19.9	Freeman and Lemen (2008)
Spalacopus cyanus	7	Octodontidae	Fossorial	Piezo	96.0	25.7	This study
Spalax galili	36	Spalacidae	Fossorial	Piezo	161.4	43.1	This study
Spermophilus tridecemlineatus	4	Sciuridae	Non-fossorial	Flexiforce	144.0	21.1	Freeman and Lemen (2008)
Zapus hudsonius	1	Zapodidae	Non-fossorial	Flexiforce	24.5	7.6	Freeman and Lemen (2008)

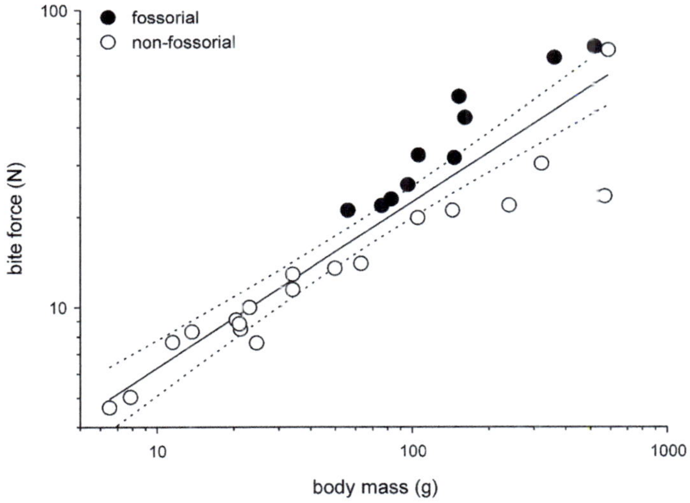

Fig. 3.7 Biplot showing the regression of the body mass against bite force measured in fossorial and non-fossorial rodents

constraints imposed by life underground are such that the whole masticatory apparatus is impacted and such adaptations are evident across different continents. The subsequent specializations are the result of many possible adjustments of the different components of the masticatory apparatus (e.g. skull and adductor muscle architecture, incisor shape and insertion), which in the end lead to similar overall morphologies and functions.

Although these fossorial patterns are now fairly well understood and recognized in different rodent families, the way in which the different muscular morphotypes accommodate such constraints to produce efficient biting and digging motions remains to be more thoroughly assessed. A study of importance investigating intraspecific variation in different fossorial species in relation to the physical properties of the ingested food and soil could be particularly revealing in this respect (e.g. Lessa & Thaeler, 1989; Bacigalupe et al., 2002; Beolchini & Corti, 2004; Barčiová et al., 2009). Ontogenetic aspects should also be more effectively integrated to better capture allometric effects (e.g. Cubo et al., 2006; Durão et al., 2019; Vassallo et al., 2021). Palaeontological and macroevolutionary studies will also play a key role in characterizing the main events leading to fossorial specializations and for defining the main extrinsic drivers (e.g. environment, climate, competition) of their convergent adaptations (see Nevo, 1979; Cook et al., 2000).

Acknowledgement We warmly thank the curators V. Nicolas-Colin, C. Denys (MNHN, Paris), and R. Portela-Miguez (NHM, London) for providing us access to the osteological collections of rodents. We acknowledge staff from the UAR2700 2AD (MNHN, Paris) for providing access to their X-ray microtomography facilities (AST-RX platform). We also acknowledge P.-H. Fabre and Q. Martinez (ISEM, Montpellier) for sending us scans of some rodents from the NHM, and S. Ginot (ISEM, Montpellier) for providing bite force data. We are grateful to J. Okrouhlík and M. Lövy (University of South Bohemia, Ceske Budejovice) for their help in collecting bite force data. We acknowledge the editors V. Bels and A. Russell for having invited us to contribute to this book. We also thank the reviewer, P. Cox, for his helpful comments on the manuscript. The project was supported by the Labex BCDiv (Laboratoire d'Excellence Biological and Cultural Diversities, http://labex-bcdiv.mnhn.fr/) to H.G.R. and by the Czech Science Foundation project GAČR [n. 20-10222S] to R.Š.

References

Agrawal, V. C. (1967). Skull adaptations in fossorial rodents. *Mammalia, 31*, 300–312.

Airoldi, J.-P., Altrocchi, R., & Meylan, A. (1976). Le comportement fouisseur du Campagnol terrestre, *Arvicola terrestris scherman* Shaw. *Revue Suisse de Zoologie, 83*, 282–286.

Bacigalupe, L. D., Iriarte-Díaz, J., & Bozinovic, F. (2002). Functional morphology and geographic variation in the digging apparatus of cururos (Octodontidae: *Spalacopus cyanus*). *Journal of Mammalogy, 83*, 145–152.ù.

Ball, S. S., & Roth, V. L. (1995). Jaw muscles of new world squirrels. *Journal of Morphology, 224*, 265–291.

Barčiová, L., Šumbera, R., & Burda, H. (2009). Variation in the digging apparatus of the subterranean silvery mole-rat, *Heliophobius argenteocinereus* (Rodentia, Bathyergidae): The role of ecology and geography. *Biological Journal of the Linnean Society, 97*, 822–831.

Becerra, F., Echeverría, A. I., Casinos, A., & Vassallo, A. I. (2014). Another one bites the dust: Bite force and ecology in three caviomorph rodents (Rodentia, Hystricognathi). *Journal of Experimental Zoology, 321*, 220–232.

Becerra, F., Echeverría, A. I., Vassallo, A. I., & Casinos, A. (2011). Bite force and jaw biomechanics in the subterranean rodent Talas tuco-tuco (*Ctenomys talarum*) (Caviomorpha: Octodontoidea). *Canadian Journal of Zoology, 89*, 334–342.

Becerra, F., Vassallo, A. I., Echeverría, A. I., & Casinos, A. (2012). Scaling and adaptations of incisors and cheek teeth in caviomorph rodents (Rodentia, Hystricognathi). *Journal of Morphology, 273*, 1150–1162.

Becerra, F., Casinos, A., & Vassallo, A. I. (2013). Biting performance and skull biomechanics of a chisel tooth digging rodent (Ctenomys tuconax; Caviomorpha; Octodontoidea). *The Journal of Experimental Zoology, 319*, 74–85.

Begall, S., Burda, H., & Schleich, C. E. (2007). *Subterranean rodents: News from underground*. Springer.

Begall, S., & Gallardo, M. H. (2000). *Spalacopus cyanus* (Rodentia: Octodontidae): An extremist in tunnel constructing and food storing among subterranean mammals. *Journal of Zoology, 251*, 53–60.

Bekele, A. (1983). The comparative functional morphology of some head muscles of the rodents *Tachyoryctes splendens* and *Rattus rattus*. II. Cervical muscles. *Mammalia, 47*, 549–572.

Beolchini, F., & Corti, M. (2004). The taxonomy of the genus *Tachyoryctes*: A geometric morphometric approach. *Italian Journal of Zoology, 71*, 35–43.

Berkovitz, B. K. B., & Faulkes, C. G. (2001). Eruption rates of the mandibular incisors of naked mole-rats (*Heterocephalus glaber*). *Journal of Zoology, 255*, 461–465.

Brett, R. A. (1991). The ecology of naked mole-rat colonies: Burrowing, food and limiting factors. In P. W. Shermann, J. U. M. Jarvis, & R. D. Alexander (Eds.), *Biology of the naked mole-rat* (pp. 137–184). Princeton University Press.

Bryant, J. D., & MacKenna, M. C. (1995). Cranial anatomy and phylogenetic position of *Tsaganomys altaicus* (Mammalia: Rodentia) from the Hsanda Gol Formation (Oligocene), Mongolia. *American Museum Novitates, 3156*, 1–42.

Buezas, G. N., Becerra, F., Echeverría, A. I., Cisilino, A., & Vassallo, A. I. (2019). Mandible strength and geometry in relation to bite force: a study in three caviomorph rodents. *Journal of Anatomy, 234*, 564–575.

Burda, H., Šumbera, R., & Begall, S. (2007). Microclimate in burrows of subterranean rodents – revisited. In S. Begall, H. Burda, & C. E. Schleich (Eds.), *Subterranean rodents: News from underground* (pp. 21–33). Springer.

Camin, S., Madoery, L., & Roig, V. (1995). The burrowing behavior of *Ctenomys mendocinus* (Rodentia). *Mammalia, 59*, 9–18.

Carotenuto, A. R., Guarracino, F., Šumbera, R., & Fraldi, M. (2020). Burrowing below ground: Interaction between soil mechanics and evolution of subterranean mammals. *Journal of the Royal Society Interface, 17*, 20190521.

Casinos, A., Gasc, J.-P., Renous, S., & Bou, J. (1983). Les modalités de fouissage de *Pitymys duodecimcostatus* (Mammalia, Arvicolidae). *Mammalia, 47*, 27–36.

Cook, J. A., Lessa, E. P., & Hadly, E. A. (2000). Paleontology, phylogenetic patterns, and macroevolutionary processes in subterranean rodents. In E. A. Lacey, J. L. Patton, & G. N. Cameron (Eds.), *Life underground: The biology of subterranean rodents* (pp. 332–369). University of Chicago Press.

Cox, P. G., & Baverstock, H. (2016). Masticatory muscle anatomy and feeding efficiency of the American Beaver, *Castor canadensis* (Rodentia, Castoridae). *Journal of Mammalian Evolution, 23*, 191–200.

Cox, P. G., & Faulkes, C. G. (2014). Digital dissection of the masticatory muscles of the naked mole-rat, *Heterocephalus glaber* (Mammalia, Rodentia). *PeerJ, 2*, e448.

Cox, P. G., Faulkes, C. G., & Bennett, N. C. (2020). Masticatory musculature of the African mole-rats (Rodentia: Bathyergidae). *PeerJ, 8*, e8847.

Cubo, J., Ventura, J., & Casinos, A. (2006). A heterochronic interpretation of the origin of digging adaptations in the northern water vole, *Arvicola terrestris* (Rodentia: Arvicolidae). *Biological Journal of the Linnean Society, 87*, 381–391.

Cuthbert, K. (1975). *Burrowing and the associated modifications in the mole-rats Bathyergus suillus and Georychus capensis - A comparative study* (University of Cape Town). Zoology honours project.

Czaplewski, N. J. (2012). A *Mylagaulus* (Mammalia, Rodentia) with Nasal Horns from the Miocene (Clarendonian) of Western Oklahoma. *Journal of Vertebrate Paleontology, 32*, 139–150.

d'Elia, G., Fabre, P.-H., & Lessa, E. P. (2019). Rodent systematics in an age of discovery: Recent advances and prospects. *Journal of Mammalogy, 100*, 852–871.

Druzinsky, R. E. (2010). Functional anatomy of incisal biting in *Aplodontia rufa* and Sciuromorph rodents – Part 1: Masticatory muscles, skul shape and digging. *Cells, Tissues, Organs, 191*, 510–522.

Durão, A. F., Ventura, J., & Muñoz-Muñoz, F. (2019). Comparative post-weaning ontogeny of the mandible in fossorial and semi-aquatic water voles. *Mammalian Biology, 97*, 95–103.

Echeverría, A. I., Becerra, F., Buezas, G. N., & Vassallo, A. I. (2017). Bite it forward ... bite it better? Incisor procumbency and mechanical advantage in the chisel-tooth and scratch-digger genus *Ctenomys* (Caviomorpha, Rodentia). *Zoology, 125*, 53–68.

Ellerman, J. R. (1956). The subterranean mammals of the world. *Transactions of the Royal Society of South Africa, 35*, 11–20.

Fang, X., et al. (2014). Genome-wide adaptive complexes to underground stresses in blind mole rats *Spalax*. *Nature Communications, 5*, 3966.

Flynn, L. J. (2009). Chapter 4: The antiquity of *Rhizomys* and independent acquisition of fossorial traits in subterranean muroids. In *Systematic Mammalogy: Contributions in Honor of Guy G. Musser, (Bulletin of the American Museum of Natural History)* (pp. 128–156).

Flynn, L. J., Nevo, E., & Heth, G. (1987). Incisor enamel microstructure in blind mole rats: Adaptive and phylogenetic significance. *Journal of Mammalogy, 68*, 500–507.

Fournier, M., Hautier, L., & Gomes Rodrigues, H. (2021). Evolution towards fossoriality and morphological convergence in the skull of Spalacidae and Bathyergidae (Rodentia). *Journal of Mammalian Evolution, 28*, 979–993.

Freeman, P. W., & Lemen, C. A. (2008). A simple morphological predictor of bite force in rodents. *Journal of Zoology, 275*, 418–422.

Gambaryan, P. P., & Gasc, J.-P. (1993). Adaptive properties of the musculoskeletal system in the mole-rat *Myospalax myospalax* (Mammalia, Rodentia), cinefluorographical, anatomical, and biomechanical analyses of burrowing. *Zoologische Jahrbücher, Abteilung für Anatomie und Ontogenie der Tiere, 123*, 363–401.

Gasc, J.-P., Renous, S., Casinos, A., Laville, E., & Bou, J. (1985). Comparison of diverse digging patterns in some small mammals. In H. R. Duncker & G. Fleischer (Eds.), *Fortschritte der Zoologie, Band 30* (pp. 35–38). Gustav Fischer Verlag.

Giannoni, S. M., Borghi, C. E., & Roig, V. G. (1996). The burrowing behavior of *Ctenomys eremophilus* (Ctenomyidae, Rodentia) in relation with substrata hardness. *Mastozoología Neotropical, 3*, 5–12.

Ginot, S., Le Noëne, C., & Cassaing, J. (2018). Comparative bite force in two syntopic murids (Rodentia) suggests lack of competition for food resources. *Canadian Journal of Zoology, 96*, 633–638.

Gomes Rodrigues, H. (2015). The great disparity of dental structures and dynamics in rodents: new insights into their ecological diversity. In P. G. Cox & L. Hautier (Eds.), *Evolution of the rodents: Advances in phylogeny, functional morphology and development* (pp. 424–447). Cambridge University Press.

Gomes Rodrigues, H., Marangoni, P., Šumbera, R., Tafforeau, P., Wendelen, W., & Viriot, L. (2011). Continuous dental replacement in a hyper-chisel tooth digging rodent. *Proceedings of the National Academy of Science United States of America, 108*, 17355–17359.

Gomes Rodrigues, H., & Šumbera, R. (2015). Dental peculiarities in the silvery mole-rat: An original model for studying the evolutionary and biological origins of continuous dental generation in mammals. *PeerJ, 3*, e1233.

Gomes Rodrigues, H., Šumbera, R., & Hautier, L. (2016). Life in burrows channelled the morphological evolution of the skulls in rodents: The case of African mole-rats (Bathyergidae, Rodentia). *Journal of Mammalian Evolution, 23*, 175–189.

He, Y., Hu, S., Ge, D., Yang, Q., Connor, T., & Zhou, C. (2020). Evolutionary history of Spalacidae inferred from fossil occurrences and molecular phylogeny. *Mammal Review, 50*, 11–24.

Herrel, A., Spithoven, L., Van Damme, R., & De Vree, F. (1999). Sexual dimorphism of head size in *Gallotia galloti*: Testing the niche divergence hypothesis by functional analysis. *Functional Ecology, 13*, 289–297.

Heth, G. (1989). Burrow patterns of the mole rat *Spalax ehrenbergi* in two soil types (terra-rossa and rendzina) in Mount Carmel, Israel. *Journal of Zoology, 217*, 39–56.

Hiiemae, K. (1971). The structure and function of the jaw muscles in the rat (*Rattus norvegicus* L.) III. The mechanics of the muscles. *Zoological Journal of the Linnean Society, 50*, 111–132.

Hildebrand, M. (1985). *Chapter 6. Digging of quadrupeds* (pp. 89–109). Harvard University Press.

Hite, N. J., Germain, C., Cain, B. W., Sheldon, M., Perala, S. S. N., & Sarko, D. K. (2019). The better to eat you with: Bite force in the naked mole-rat (*Heterocephalus glaber*) is stronger than predicted based on body size. *Frontiers in Integrative Neuroscience, 13*, 70.

Hopkins, S. S. B. (2005). The evolution of fossoriality and the adaptive role of horns in the Mylagaulidae (Mammalia: Rodentia). *Proceedings of the Royal Society B: Biological Sciences, 272*, 1705–1713.

Jardine, P. E., Janis, C. M., Sahney, S., & Benton, M. J. (2012). Grit not grass: Concordant patterns of early origin of hypsodonty in Great Plains ungulates and Glires. *Palaeogeography, Palaeoclimatology, Palaeoecology, 1–10*, 365–366.

Jarvis, J. U. M., Bennett, N. C., & Spinks, A. C. (1998). Food availability and foraging by wild colonies of Damaraland mole-rats (*Cryptomys damarensis*): Implications for sociality. *Oecologia, 113*, 290–298.

Jarvis, J. U. M., & Sale, J. B. (1971). Burrowing and burrow patterns of East African mole-rats *Tachyoryctes*, *Heliophobius* and *Heterocephalus*. *Journal of Zoology, 163*, 451–479.

Kalthoff, D. (2000). Die Schmelzmikrostruktur in den Incisiven der hamsterartigen Nagetiere und anderer Myomorpha (Rodentia, Mammalia). *Palaeontographica Abteilung A, 259*, 1–193.

Kerr, E., Cornette, R., Gomes Rodrigues, H., Renaud, S., Chevret, P., Tresset, A., & Herrel, A. (2017). Can functional traits help explain the coexistence of two species of *Apodemus*? *Biological Journal of the Linnean Society, 122*, 883–896.

Kim, E. B., Fang, X., Fushan, A. A., Huang, Z., Lobanov, A. V., Han, L., Marino, S. M., et al. (2011). Genome sequencing reveals insights into physiology and longevity of the naked mole rat. *Nature, 479*, 223–227.

Kirby, A. M., Fairman, G. D., & Pamenter, M. E. (2018). Atypical behavioural, metabolic and thermoregulatory responses to hypoxia in the naked mole rat (*Heterocephalus glaber*). *Journal of Zoology, 305*, 106–115.

Korbin, J. P., Bancroft, A., Dunnum, J., & Cook, J. (2020). *X-ray CT Scans - Set 1* (United States).

Krapp, F. (1965). Beobachtungen an Kaumuskulatur und Schädel von *Spalax leucodon* (Nordmann, 1840) (Rodentia, Mammalia). *Revue Suisse de Zoologie, 72*, 504–609.

Kryštufek, B., Janžekovič, F., Hutterer, R., & Klenovšek, T. (2016). Morphological evolution of the skull in closely related bandicoot rats: A comparative study using geometric morphometrics. *Hystrix, the Italian Journal of Mammalogy, 27*, 163–169.

Lacey, E. A., Patton, J. L., & Cameron, G. N. (2000). *Life underground: The biology of subterranean rodents*. University of Chicago Press.

Lacey, E. A., & Shermann, P. W. (1991). Social organization of naked mole-rat colonies: Evidence for divisions of labor. In P. W. Shermann, J. U. M. Jarvis, & R. D. Alexander (Eds.), *Biology of the naked mole-rat* (pp. 275–336). Princeton University Press.

Landry, S. O., Jr. (1957). Factors affecting the procumbency of rodent upper incisors. *Journal of Mammalogy, 38*, 223–234.

Laville, E. (1989). Etude cinématique du fouissage chez *Arvicola terrestris scherman* (Rodentia, Arvicolidae). *Mammalia, 53*, 177–190.

Laville, E., Casinos, A., Gasc, J.-P., Renous, S., & Bou, J. (1989). The mechanism of digging in *Arvicola terrestris* and *Spalax ehrenbergi*: Functional and evolutional studies. *Anatomischer Anzeiger, 169*, 131–144.
Lessa, E. P. (1987). *Functional morphology and allometry of the digging apparatus in pocket gophers of the genera Thomomys, Geomys, and Cratogeomys (Rodentia: Geomyidae)*. New Mexico State University.
Lessa, E. P. (2000). The evolution of subterranean rodents: A synthesis. In E. A. Lacey, J. L. Patton, & G. N. Cameron (Eds.), *Life underground: The biology of subterranean rodents* (pp. 389–420). University of Chicago Press.
Lessa, E. P., & Patton, J. L. (1989). Structural constraints, recurrent shapes, and allometry in pocket gophers (genus *Thomomys*). *Biological Journal of the Linnean Society, 36*, 349–363.
Lessa, E. P., & Stein, B. R. (1992). Morphological constraints in the digging apparatus of pocket gophers (Mammalia: Geomyidae). *Biological Journal of the Linnean Society, 47*, 439–453.
Lessa, E. P., & Thaeler, C. S. (1989). A reassessment of morphological specializations for digging in pocket gophers. *Journal of Mammalogy, 70*, 689–700.
Lessa, E. P., Vassallo, A. I., Verzi, D. H., & Mora, M. S. (2008). Evolution of morphological adaptations for digging in living and extinct ctenomyid and octodontid rodents. *Biological Journal of the Linnean Society, 95*, 267–283.
Luo, Z.-X., & Wible, J. R. (2005). A new Late Jurassic digging mammal and early mammalian diversification. *Science, 308*, 103–107.
Manaro, A. J. (1959). Extrusive incisor growth in the rodent genera *Geomys*, *Peromyscus*, and *Sigmodon*. *Quarterly Journal of the Florida Academy of Sciences, 22*, 25–31.
Manov, I., Hirsh, M., Iancu, T. C., Malik, A., Sotnichenko, N., Band, M., Avivi, A., & Shams, I. (2013). Pronounced cancer resistance in a subterranean rodent, the blind mole-rat, *Spalax*: In vivo and in vitroevidence. *BMC Biology, 11*, 91.
Marcolini, F., Piras, P., Kotsakis, T., Claude, J., Michaux, J., Ventura, J., & Cubo, J. (2011). Phylogenetic signal and functional significance of incisor enamel microstructure in *Arvicola* (Rodentia, Arvicolinae). *Comptes Rendus Palevol, 10*, 479–487.
Marcy, A. E., Hadly, E. A., Sherratt, E., Garland, K., & Weisbecker, V. (2016). Getting a head in hard soils: Convergent skull evolution and divergent allometric patterns explain shape variation in a highly diverse genus of pocket gophers (*Thomomys*). *BMC Evolutionary Biology, 16*, 207.
McIntosh, A. F., & Cox, P. G. (2016a). Functional implications of craniomandibular morphology in African mole-rats (Rodentia: Bathyergidae). *Biological Journal of the Linnean Society, 117*, 447–462.
McIntosh, A. F., & Cox, P. G. (2016b). The impact of digging on craniodental morphology and integration. *Journal of Evolutionary Biology, 29*, 2383–2394.
McIntosh, A. F., & Cox, P. G. (2016c). The impact of gape on the performance of the skull in chisel-tooth digging and scratch digging mole-rats (Rodentia: Bathyergidae). *Royal Society Open Science, 3*, 160568.
Mein, P., & Pickford, M. (2008). Early Miocene Rodentia from the northern Sperrgebiet, Namibia. *Memoir of the Geological Survey of Namibia, 20*, 235–290.
Miller, R. S. (1958). Rate of incisor growth in the mountain pocket gopher. *Journal of Mammalogy, 39*, 380–385.
Mora, M., Olivares, A. I., & Vassallo, A. I. (2003). Size, shape and structural versatility of the skull of the subterranean rodent *Ctenomys* (Rodentia, Caviomorpha): Functional and morphological analysis. *Biological Journal of the Linnean Society, 78*, 85–96.
Morlok, W. F. (1983). Vergleichend- und funktionell-anatomische Untersuchungen an Kopf, Hals und Vorderextremität subterraner Nagetiere (Mammalia, Rodentia). *Forschungsinstitut Senckenberg, 64*, 1–237.
Nevo, E. (1979). Adaptive convergence and divergence of subterranean mammals. *Annual Review of Ecology, Evolution, and Systematics, 10*, 269–308.
Nevo, E. (1999). *Mosaic evolution of subterranean mammals: Regression, progression and global convergence*. Oxford University Press.

Partha, R., Chauhan, B. K., Ferreira, Z., Robinson, J. D., Lathrop, K., Nischal, K. K., Chikina, M., & Clark, N. L. (2017). Subterranean mammals show convergent regression in ocular genes and enhancers, along with adaptation to tunneling. *eLife, 6*, e25884.

Reichman, O. J., & Jarvis, J. U. M. (1989). The influence of three sympatric species of fossorial mole-rats (Bathyergidae) on vegetation. *Journal of Mammalogy, 70*, 763–771.

Rodríguez-Serrano, E., Palma, R. E., & Hernández, C. E. (2008). The evolution of ecomorphological traits within the Abrothrichini (Rodentia: Sigmodontinae): A bayesian phylogenetics approach. *Molecular Phylogenetics and Evolution, 48*, 473–480.

Samuels, J. X., & van Valkenburgh, B. (2009). Craniodental adaptations for digging in extinct burrowing beavers. *Journal of Vertebrate Paleontology, 29*, 254–268.

Šklíba, J., Mazoch, V., Patzenhauerová, H., Hrouzková, E., Lövy, M., Kott, O., & Šumbera, R. (2012). A maze-lover's dream: Burrow architecture, natural history and habitat characteristics of Ansell's mole-rat (*Fukomys anselli*). *Mammalian Biology, 77*, 420–427.

Šklíba, J., Šumbera, R., Chitaukali, W. N., & Burda, H. (2009). Home-range dynamics in a solitary subterranean rodent. *Ethology, 115*, 217–226.

Stein, B. R. (2000). Morphology of subterranean rodents. In E. A. Lacey, J. L. Patton, & G. N. Cameron (Eds.), *Life underground: The biology of subterranean rodents* (pp. 19–61). University of Chicago Press.

Šumbera, R. (2019). Thermal biology of a strictly subterranean mammalian family, the African mole-rats (Bathyergidae, Rodentia) - A review. *Journal of Thermal Biology, 79*, 166–189.

Šumbera, R., Burda, H., Chitaukali, W. N., & Kubová, J. (2003). Silvery mole-rats (*Heliophobius argenteocinereus*, Bathyergidae) change their burrow architecture seasonally. *Naturwissenschaften, 90*, 370–373.

Šumbera, R., Mazoch, V., Patzenhauerová, H., Lövy, M., Šklíba, J., Bryja, J., & Burda, H. (2012) Burrow architecture, family composition and habitat characteristics of the largest social African mole-rat: the giant mole-rat constructs really giant burrow systems. *Acta Theriologica, 57*, 121–130.

Thomas, O. (1919). The method of taking the incisive index in rodents. *Annals and Magazine of Natural History, 4*, 289–290.

Van Daele, P. A. A. G., Desmet, N., Šumbera, R., & Adriaens, D. (2019). Work behaviour and biting performance in the cooperative breeding Micklem's mole-rat *Fukomys micklemi* (Bathyergidae, Rodentia). *Mammalian Biology, 95*, 69–76.

Van Daele, P. A. A. G., Herrel, A., & Adriaens, D. (2009). Biting performance in teeth-digging African mole-rats (*Fukomys*, Bathyergidae, Rodentia). *Physiological and Biochemical Zoology, 82*, 40–50.

Van der Merwe, M., & Botha, A. J. (1998). Incisors as digging tools in molerats (Bathyergidae). *South African Journal of Zoology, 33*, 230–235.

Van Wassenbergh, S., Heindryckx, S., & Adriaens, D. (2017). Kinematics of chisel-tooth digging by African mole-rats. *Journal of Experimental Biology, 220*, 4479.

Vassallo, A. I., Becerra, F., Echeverría, A. I., Díaz, A. O., Longo, M. V., Cohen, M., & Buezas, G. N. (2021). Analysis of the form-function relationship: Digging behavior as a case study. *Journal of Mammalian Evolution, 28*, 59–74.

Vieytes, E. C., Morgan, C. C., & Verzi, D. H. (2007). Adaptive diversity of incisor enamel microstructure in South American burrowing rodents (family Ctenomyidae, Caviomorpha). *Journal of Anatomy, 211*, 296–302.

Wang, B. (2001). On Tsaganomyidae (Rodentia, Mammalia) of Asia. *American Museum Novitates, 3317*, 1–50.

Williams, S., Peiffer, E., & Ford, S. (2009). Gape and bite force in the rodents *Onychomys leucogaster* and *Peromyscus maniculatus*: Does jaw-muscle anatomy predict performance? *Journal of Morphology, 270*, 1338–1347.

Wood, A. E. (1965). Grades and clades among rodents. *Evolution, 19*, 115–130.

Zelová, J., Šumbera, R., Okrouhlík, J., & Burda, H. (2010). Cost of digging is determined by intrinsic factors rather than by substrate quality in two subterranean rodent species. *Physiology & Behavior, 99*, 54–58.

Zuri, I., Kaffe, I., Dayan, D., & Terkel, J. (1999). Incisor adaptation to fossorial life in the blind mole-rax *Spalax ehrenbergi*. *Journal of Mammalogy, 80*, 734–741.

Chapter 4
Testing for Convergent Evolution in Baleen Whale Cochleae

Travis Park, Eric G. Ekdale, Rachel A. Racicot, and Felix G. Marx

Abstract Mysticetes (baleen whales) include the largest animals on Earth and are renowned for their songs and long-distance communication. Even so, the scope and origins of their hearing abilities remain poorly understood. Recent work on their sister clade, the toothed whales (odontocetes), has revealed notably convergent trends in the evolution of their inner ear. Here, we test whether the same applies to baleen whales via SURFACE, a phylogenetic method that fits Ornstein-Uhlenbeck models with stepwise Akaike Information Criterion to identify instances of convergent evolution. We identify a single convergent regime, including minke (*Balaenoptera acutorostrata*) and Bryde's (*Balaenoptera edeni*) whales, which, however, is not statistically significant. We discuss potential reasons for the overall absence of convergence and suggest improvements for future work.

Keywords Convergence · Mysticeti · Inner ear · Ecomorphology · Phylogenetic comparative methods

T. Park (✉)
Science Group, Natural History Museum, London, UK
e-mail: t.park@nhm.ac.uk

E. G. Ekdale
Department of Biology, San Diego State University, San Diego, CA, USA

Department of Paleontology, San Diego Natural History Museum, San Diego, CA, USA

R. A. Racicot
Abteilung Messelforschung und Mammalogie, Senckenberg Forschungsinstitut und Naturkundemuseum, Frankfurt am Main, Germany

Department of Biological Sciences, Vanderbilt University, Nashville, TN, USA

F. G. Marx
Museum of New Zealand Te Papa Tongarewa, Wellington, New Zealand

Department of Geology, University of Otago, Dunedin, New Zealand

V. L. Bels, A. P. Russell (eds.), *Convergent Evolution*, Fascinating Life Sciences,
https://doi.org/10.1007/978-3-031-11441-0_4

4.1 Introduction

Mysticetes are among the largest animals ever and play crucial roles in marine ecosystems as nutrient distributors and mass predators (Croll et al., 2006; Nicol et al., 2010; Roman & McCarthy, 2010). They owe their success to a key innovation—baleen—which they use to filter vast volumes of small prey from seawater (Pivorunas, 1979). They also have highly disparate sensory capabilities, including low-frequency and even infrasonic (<20 Hz) hearing that may predate the appearance of baleen and gigantism (Cummings & Thompson, 1971; Ketten, 2000; Ekdale & Racicot, 2015; Ketten et al., 2016; Park et al., 2017a, b; Ritsche et al., 2018).

The low-frequency hearing capabilities of mysticetes contrast with the high-pitched vocalisations of toothed whales (Ketten & Wartzok, 1990; Geisler et al., 2014; Park et al., 2016, 2019; Churchill et al., 2016; Mourlam & Orliac, 2017). Unlike the latter, mysticetes do not echolocate (Park et al., 2017a). Instead, their low-frequency sounds propagate over long distances and enable them to communicate over hundreds or even thousands of kilometres (Nummela & Yamato, 2018).

Owing to the difficulty of studying such large animals, research on mysticete hearing is still in its infancy. Indirect methods like vocalisation data (Watkins & Wartzok, 1985; Cummings & Thompson, 1971) and the playback technique (Clark & Clark, 1980; Tyack, 1983; Parks, 2003) have provided crucial clues, but are logistically complex. Anatomical studies of auditory structures are an economical alternative and, for example, have revealed the presence of a specialised acoustic funnel in rorquals (Yamato & Pyenson, 2015), bone conduction (Cranford & Krysl, 2015), and acoustic fats that may be homologous to those of toothed whales (Yamato et al., 2012, 2014).

The inner ear—in particular, the cochlea—is the region of the mammalian auditory pathway where sounds are converted into nerve signals. Cochlear anatomy reflects phylogeny, hearing abilities and habitat, and as such provides clues to the ecology of extinct or rare species (Ketten & Wartzok, 1990; Ketten, 2000; Ekdale, 2013; Gutstein et al., 2014; Ekdale & Racicot, 2015; Ekdale, 2016; Park et al., 2016, 2017a; Racicot et al., 2016; Mourlam & Orliac, 2017; Costeur et al., 2018; Ritsche et al., 2018; Racicot et al., 2018; Galatius et al., 2019; Viglino et al., 2021). Thus, for example, the cochleae of extant odontocetes seem to have evolved convergently, likely constrained by the acoustic environment of the deep ocean (Park et al., 2019). Here, we study a broad sample of living and extinct species to test whether mysticete inner ears also show a pattern of convergent evolution.

4.2 Methods

Institutional Abbreviations ChM, Charleston Museum, Charleston, USA; CMM, Calvert Marine Museum, Solomons, USA; HSU, Natural History Museum, Humboldt State University, Arcata, USA; IRSNB, Institut Royal des Sciences Naturelles

Table 4.1 Specimens used in this study, their age and pixel size. μm = micrometers

Taxon	Specimen number	Age	Pixel size (μm)
Aetiocetidae indet.	NMV P229119	Oligocene	47.94
Aglaocetus patulus	USNM V23690	Miocene	24.41
Antwerpibalaena liberatlas	IRSNB M2325	Pliocene	40.92
Balaena mysticetus	LACM 97312	Extant	29.30
Balaenoptera acutorostrata	SDNHM 23642	Extant	29.30
Balaenoptera edeni	NMV P171502	Extant	307.00
Balaenoptera musculus	USNM 269540	Extant	53.62
Balaenoptera physalus	NHMUK 1998.30	Extant	40.05
Balaenopteridae indet.	SDNHM 83695	Pliocene	24.41
Caperea marginata	NMV C28531	Extant	236.00
Cephalotropis coronatus	CMM-V-3636	Miocene	97.66
Cophocetus oregonensis	UO F36450	Miocene	24.41
Eomysticetus whitmorei	ChM PV4253	Oligocene	24.41
Eschrichtiidae indet.	SDNHM 65021	Pliocene	24.41
Eschrichtius robustus	SDNHM24316	Extant	29.30
Eubalaena australis	NHMUK 1873.3.3.1	Extant	21.05
Eubalaena glacialis	LACM 54763	Extant	32.23
Halicetus ignotus	USNM V23636	Miocene	24.41
Herpetocetus morrowi	SDSNH 63690	Pliocene	19.53
Herpetocetus transatlanticus	IRSNB V00372	Pliocene	58.77
Janjucetus sp.	NMV P48867	Oligocene	50.46
Mammalodon colliveri	NMV P199986	Oligocene	51.36
Mammalodontidae indet.	NMVP173220	Oligocene	45.05
Megaptera novaeangliae	HSU VM2776	Extant	29.79
Metopocetus durinasus	USNM V8518	Miocene	24.41
Micromysticetus rothauseni	ChM PV7225	Oligocene	24.41
Norrisanima miocaena	USNM V10300	Miocene	24.41
Parietobalaena securis	SDSNH 61095	Miocene	19.53
Parietobalaena palmeri	USNM 517872	Miocene	28.81
Peripolocetus vexillifer	SDSNH 53999	Miocene	24.41
Toothed mysticete	ChM PV5720	Oligocene	24.41
Zygorhiza kochii	USNM 214433	Eocene	51.35

de Belgique, Brussels, Belgium; LACM, Natural History Museum Los Angeles County, Los Angeles, USA; SDNHM/SDSNH, San Diego Museum Natural History Museum, San Diego, USA; NHMUK, Natural History Museum, London, UK; NMV, Museums Victoria, Melbourne, Australia; UO; Museum of Natural and Cultural History, University of Oregon, Eugene, USA; USNM, National Museum of Natural History, Smithsonian Institution, Washington DC, USA.

Data Collection We compiled microCT scans of the inner ear of 31 living and extinct mysticete specimens (Table 4.1), as well as one stem cetacean. We used only one specimen per species, considering that intraspecific variation is likely

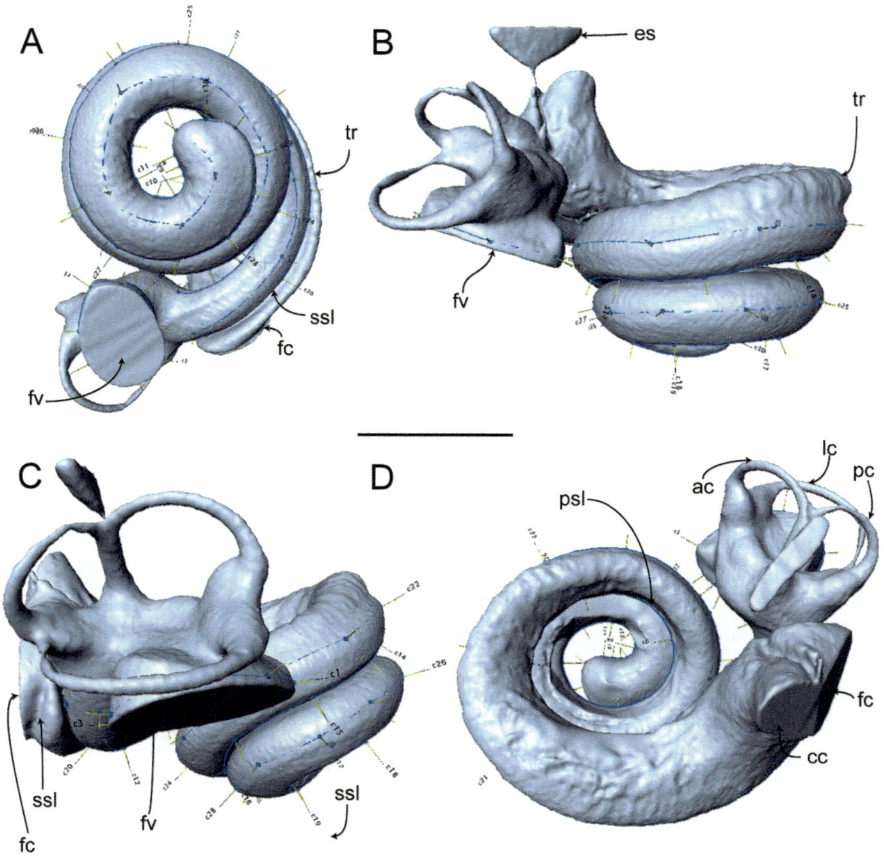

Fig. 4.1 Bony labyrinth of the fin whale, *Balaenoptera physalus* (NHMUK1998.30.2), in: (**a**) vestibular; (**b**) anterior; (**c**) lateral; and (**d**) dorsal views, showing placement of semi-landmarks. Model made from microCT scan data. Scale bar equals 10 mm. *ac* anterior canal, *cc* canaliculus cochleae, *es* endolymphatic sac, *fc* fenestra cochleae, *fv* fenestra vestibuli, *lc* lateral canal, *pc* posterior canal, *psl* primary spiral lamina, *ssl* secondary spiral lamina, *tr* tympanal recess

unproblematic in this context (Martins et al., 2020). Data were collected by imaging museum specimens (Table 4.1) or taken from the literature (Ekdale, 2016; Park et al., 2017a, b).

3D models were constructed via the segmentation and thresholding editors in Avizo 9.2 (Visualisation Sciences Group, 2016), or VGStudio Max 2.2 (Volume Graphics, 2012), cleaned using Geomagic Wrap (3D Systems, 2017), and captured via 30 sliding semi-landmark curves (280 landmarks) in IDAV Landmark (Wiley, 2005) (Fig. 4.1). The position of the curves follows Costeur et al. (2018), except in that none were placed on the dorsal surface of the cochlea owing to tight coiling in some specimens. Landmarking was carried out by a single investigator (TP) to avoid multi-user bias.

Terms of cochlear orientation refer to the spiral itself, rather than the whole body of the animal. We used a pruned version of the time-calibrated metatree of Lloyd and Slater (2021) as a phylogenetic framework.

Geometric Morphometrics Geometric morphometrics were performed in R 4.0.3 (R Core Team, 2020) using the package GEOMORPH (Adams et al., 2018). We used Generalised Procrustes Analysis (GPA) to remove the effects of position, scale and orientation. Next, semi-landmarks were 'slid' along their tangent vectors between adjacent semi-landmarks until their positions minimised the shape difference between specimens as judged by the Procrustes distance criterion (Bookstein et al., 1999; Gunz et al., 2005; Adams et al., 2013). Finally, we summarised the Procrustes coordinates via principal component analysis (PCA) using the 'gm.prcomp' function.

Phylogenetic Signal To determine whether close relatives tend to have similarly shaped cochleae, we estimated the phylogenetic signal in our principal component (PC) scores via the K_{mult} statistic. This method is designed to deal with high-dimensional multivariate data (e.g. landmark configurations) by exploiting the statistical equivalency between covariance-based and distance-based approaches for Euclidean data (Adams, 2014). We calculated this statistic using all PC scores.

Convergent Regimes We used the R package SURFACE (Ingram & Mahler, 2013) to identify putatively convergent regimes in the evolution of cochlear shape without having to define relevant groups *a priori*. SURFACE uses an Ornstein-Uhlenbeck (OU) process—a random walk whereby trait values are pulled back towards a long-term mean—to identify groups that share a similar adaptive peak. Previous studies cautioned against the application of this method to high-dimensional data (Ingram & Mahler, 2013; Zelditch et al., 2017), and we therefore only used it to analyse our first two principal components.

The analysis is split into a forward phase locating regime shifts on a tree, and a backward phase identifying whether shifts are convergent. The forward phase initially assumes a single adaptive regime, and then adds regime shifts one at a time to the origin of each branch. The performance of each model is assessed via the sample size corrected Akaike Information Criterion (AICc) and regime shifts continue to be added until there is no change in AICc (i.e. $\Delta AICc = 0$). In the backward phase, the final model from the forward phase is simplified via pairwise collapses of regimes into one convergent regime and re-assessed using the AICc. Model simplification continues until there is again no change in AICc.

Degree and Significance of Convergence We quantified the degree and significance of the convergent regimes identified in the final SURFACE model using Stayton's (2015) C-metrics, as implemented in the R package CONVEVOL. These distance-based measures define convergence as two lineages growing more similar than their ancestors were (Stayton, 2008, 2015). C1 ranges from 0 to 1, and is the proportion of the maximum phenotypic distance between two lineages (assessed via ancestral state estimation under Brownian Motion) that has been lost through convergence. C2 is similar to C1, but is not scaled and therefore quantifies absolute evolutionary change. C3 and C4 standardise C2 by dividing it by the total amount of phenotypic evolution within the focal clade and the whole phylogeny, respectively. We calculated the C-metrics based on the first 30 PCs, which together represent 95% of the total variation in cochlear shape. Significance was assessed via 1000 Brownian Motion simulations, using the phylogeny and a variance-covariance matrix derived from the PC scores as the rate of evolution. C1–C4 were calculated for each simulated dataset, with the p-value being the proportion of simulations exceeding the observed values for each metric.

4.3 Results

Geometric Morphometrics The PCA reveals rorquals to have the widest range of cochlear shape variation (Fig. 4.2). PC1 accounts for 47.75% of the total variance and correlates with the overall spiral shape of the cochlea (elliptical vs circular). By contrast, PC2 (11.30%) mostly reflects tightness of coiling and the height of the spiral.

Phylogenetic Signal The PCs contain no statistically significant phylogenetic signal (K_{mult} = 0.0652, p-value = 0.105).

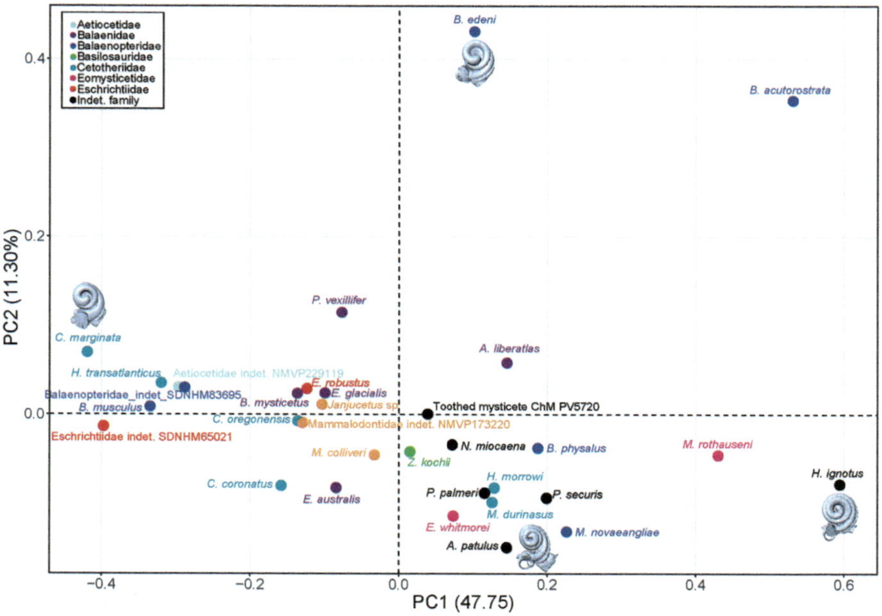

Fig. 4.2 Morphospace of cochlear shape in 31 species of baleen whale, based on a PCA of 280 landmarks. Cochlear models illustrate species at the extremes of the morphospace

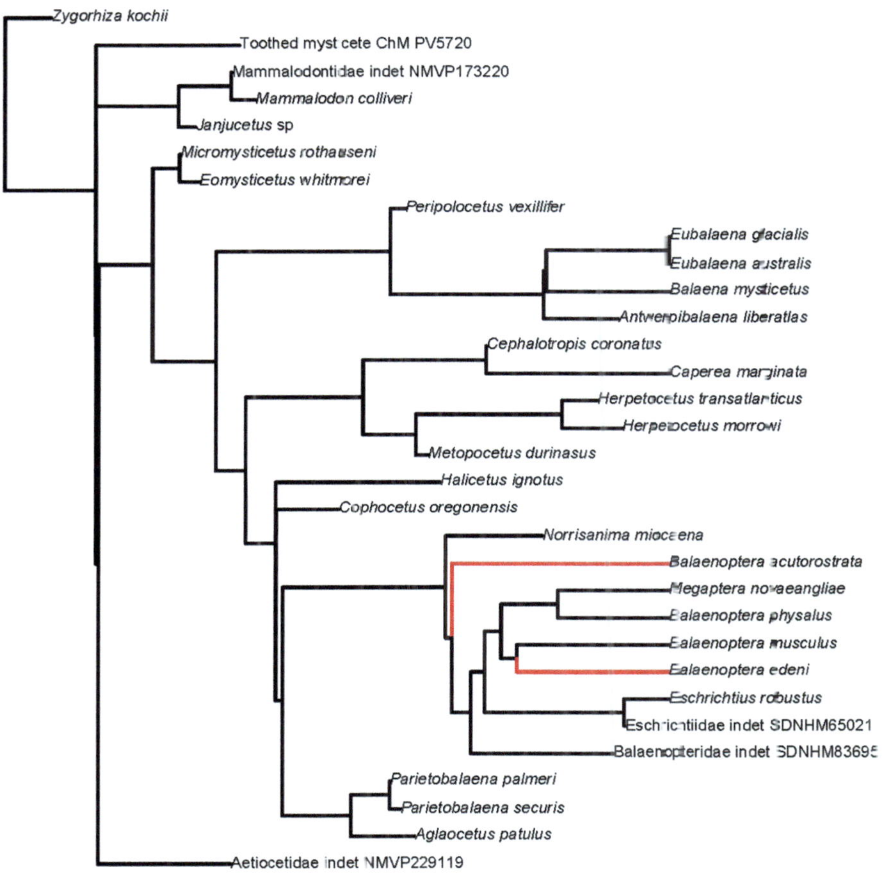

Fig. 4.3 Results of the SURFACE analysis of cochlear shape in 31 species of baleen whale, based on the first two principal components [tree from Lloyd and Slater (2021)]. The single convergent regime is highlighted in red

Convergent Regimes Our final multipeak (OU model has an AICc score of −365.57 and includes two distinct evolutionary regimes, one of which suggests convergence of minke whales, *Balaenoptera acutorostrata*, and Bryde's whales, *B. edeni* (Fig. 4.3; Table 4.2).

Degree and Significance of Convergence None of the C-metrics associated with our single convergent regime are significant (Table 4.3).

Table 4.2 Results of the SURFACE analysis of cochlear shape in 31 species of baleen whale, based on the first two principal components. *α* rate of adaptation to optimum, *t1/2* expected time to evolve halfway to an optimum, *σ2* rate of stochastic evolution, *θ* optimum trait value for each regime

Model outputs	**Value**	
AICc	−65.089	
Phenotypic regimes	2	
Phenotypic regime shifts	3	
Convergent phenotypic regimes	1	
Convergent phenotypic regime shifts	2	
Convergence fraction	0.667	
Parameters	**PC1**	**PC2**
A	0.362	875.472
$t_{1/2}$	1.916	0.001
σ^2	0.039	6.656
θ_a	−0.026	−0.026
θ_b	0.329	0.393

Table 4.3 C1–C4 convergence measures and p-values for each convergent regime

Variable	C1	C2	C3	C4
C-value	0	0	0	0
p-value	0.999	0.999	0.999	0.999

4.4 Discussion

4.4.1 Mysticete Inner Ear Morphospace

Almost half of the total variation in mysticete cochlear shape is explained by PC1, with the remainder spread out among many PCs. A similar pattern occurs in odontocetes (Costeur et al., 2018; Park et al., 2019) and prompts caution when interpreting relationships between PC scores and cochlear shape. Like Ekdale (2016), we did not find a statistically significant phylogenetic signal, suggesting that closely related species do not necessarily have similar cochleae. This is reflected in the absence of clear taxonomic groupings and broad regions of morphospace overlap (Fig. 4.2).

4.4.2 Convergence of Minke and Bryde's Whales

The single (statistically non-significant) regime identified in our analysis includes two extant rorquals, namely, minke and Bryde's whales. Both show rapid convergence towards loose cochlear coiling, as suggested by the high rate of adaptation to an optimum (α) and short expected time to evolve halfway to an optimum (t1/2) associated with PC2 (Table 4.1). The degree of coiling affects how acoustic energy is focused along the outer wall of the cochlear canal and, thus, low-frequency sensitivity (Manoussaki et al., 2008).

Given its functional relevance, tightness of coiling could plausibly be sensitive to various selection pressures and thus change relatively rapidly. It remains unclear, however, how minke and Bryde's whales benefit from the reduced low-frequency sensitivity that results from looser coiling, with their convergent cochlear shapes lacking obvious ecological or acoustic correlates. Thus, minke whales are the smallest of the rorquals (7–9 m long) and range from the tropics to the edge of the ice, whereas Bryde's whales reach lengths of up to 16.5 m and tend to prefer warmer waters (Jefferson et al., 2015). Both species are shallow divers (Christiansen et al., 2015; Kato & Perrin, 2018) and—like other rorquals—lunge feeders (Pivorunas, 1979).

In terms of vocalisations and acoustic abilities, Bryde's whales produce powerful low-frequency tonal and swept calls resembling those of other rorquals, whereas minke whales make a greater variety of sounds, including grunts, thumps and 'boings' (Rankin & Barlow, 2005; Perrin et al., 2018). Based on the equations of Manoussaki et al. (2008), their estimated low-frequency hearing limits are relatively similar at 38.8 and 22.9 Hz, respectively; however, other species also fall within this range (Park et al., 2017a). Apart from being loosely coiled, the cochleae of the two species are not particularly similar, with that of the minke whale being more circular (Fig. 4.2).

Together, these observations perhaps explain why the minke/Bryde's whale convergence is not statistically significant. Nevertheless, future research should further explore this topic by including close relatives of both species, such as the Antarctic minke whales, sei whales, and Omura's whales.

4.4.3 Why Are Baleen Whale Cochleae Not Convergent?

Unlike toothed whales, mysticetes show no obvious phylogenetic, ecological or convergent patterns in their cochlear anatomy. We suggest three partially related reasons for why this may be.

First, toothed whales differ from mysticetes in their ubiquitous use of echolocation (Park et al., 2017a). The functional requirements (i.e. effective frequencies) of this behaviour likely constrain odontocete cochlear anatomy and thus facilitate convergence. By contrast, mysticetes primarily use sound for communication, which may cover a much broader range of frequencies.

Secondly, relevant selection pressures may arise from the physical properties of water itself (Park et al., 2019). The speed of sound in water depends on temperature, salinity, depth and time from source. Sound velocity decreases with temperature to a minimum at around 1000 m depth, but then increases again with the rising pressure and eventually exceeds surface speeds at around 2500 m depth (Urick, 1983). Interestingly, it is in this extreme acoustic environment that odontocetes with convergent cochlear shapes hunt for prey (Park et al., 2019). By contrast, shallow divers tend to show little evidence of convergence.

Finally, changing acoustic environments may hinder convergent evolution of cochlear shape. Most extant mysticetes undergo seasonal migrations between colder feeding and warmer breeding grounds (Geijer et al., 2016). As a result, they are regularly exposed to a variety of acoustic environments, which may prevent adaptation to a specific setting and, thus, convergence.

4.4.4 Challenges and Future Work

Our study is limited by the relative paucity of data on baleen whale hearing. For example, there are still no directly measured mysticete audiograms, with previous papers instead relying on simulations (Tubelli et al., 2012; Cranford & Krysl, 2015). The absence of a phylogenetic signal suggests that variation in mysticete cochlear anatomy is driven by function, or relaxed selection pressures, but the scarcity of observations on different acoustic regimes currently prevents us from testing this. Future studies should also bear in mind potential hybridisation among balaenopterids, which could plausibly confound anatomical patterns (Árnason et al., 2018). Finally, our study would benefit from a broader evolutionary context that includes stem cetaceans once a sufficiently large sample of archaeocete cochleae becomes available (Ekdale & Racicot, 2015; Churchill et al., 2016; Mourlam & Orliac, 2017; Park et al., 2017a, b).

The somewhat extreme position of minke and Bryde's whales in our morphospace (Fig. 4.2) may reflect errors during data processing or analysis, even though all landmarks were rechecked and the scripts used here have been employed in previous studies without obvious problems (Park et al., 2019). Additional specimens of both species would help to confirm our findings.

As with any study, there are also some caveats. The rates of adaptation and diffusion in the OU models of the SURFACE analysis are assumed to be independent, which in reality is not the case (Zelditch et al., 2017). Restricting our study to the first two PCs is a pragmatic choice, but likely affects our results. Intraspecific variation is another potential concern, but probably unproblematic in light of previous studies on odontocetes and terrestrial mammals (Ekdale, 2010; Martins et al., 2020). Finally, alternative landmarking schemes may produce different results, albeit likely not too different from our reasonably comprehensive approach.

4.5 Conclusions

The hearing apparatus of echolocating odontocetes is convergent at both the molecular and morphological level (Li et al., 2010; Liu et al., 2010; Parker et al., 2013; Liu et al., 2014; Park et al., 2019). Here, we tested whether a similar pattern is evident in the cochlear anatomy of baleen whales. We found little evidence of convergent evolution, which may be explained by the absence of constraints associated with echolocation; the absence of selective pressures arising from deep diving; migratory behaviour; and our still limited understanding of mysticete hearing in general. Overall, we still have much to learn about the largest creatures in our oceans.

References

3D Systems. (2017). *Geomagic Wrap*.

Adams, D. C. (2014). A generalized K statistic for estimating phylogenetic signal from shape and other high-dimensional multivariate data. *Systematic Biology, 63*, 685–697.

Adams, D. C., Rohlf, F. J., & Slice, D. E. (2013). A field comes of age: geometric morphometrics in the 21st century. *Hystrix, the Italian Journal of Mammalogy, 24*, 7–14.

Adams, D. C., Collyer, M. L., & Kaliontzopoulou, A. (2018). *Geomorph: Software for geometric morphometric analyses.*

Árnason, Ú., Lammers, F., Kumar, V., Nilsson, M. A., & Janke, A. (2018). Whole-genome sequencing of the blue whale and other rorquals finds signatures for introgressive gene flow. *Science Advances, 4*, eaap9873.

Bookstein, F., Schäfer, K., Prossinger, H., Seidler, H., Fieder, M., Stringer, C., Weber, G. W., Arsuaga, J. L., Slice, D. E., Rohlf, F. J., Recheis, W., Mariam, A. J., & Marcus, L. F. (1999). Comparing frontal cranial profiles in archaic and modern Homo by morphometric analysis. *The Anatomical Record, 257*, 217–224.

Christiansen, F., Lynas, N. M., Lusseau, D., & Tscherter, U. (2015). Structure and dynamics of minke whale surfacing patterns in the gulf of St. Lawrence, Canada. *PLoS One, 10*(5), e0126396.

Churchill, M., Martinez-Caceres, M., de Muizon, C., Mnieckowski, J., & Geisler, J. H. (2016). The origin of high-frequency hearing in whales. *Current Biology, 26*, 2144–2149.

Clark, C. W., & Clark, J. M. (1980). Sound playback experiments with Southern right whales (Eubalaena australis). *Science, 207*, 663–665.

Costeur, L., Grohé, C., Aguirre-Fernández, G., Ekdale, E., Schulz, G., Müller, B., & Mennecart, B. (2018). The bony labyrinth of toothed whales reflects both phylogeny and habitat preferences. *Scientific Reports, 8*, 7841.

Cranford, T. W., & Krysl, P. (2015). Fin whale sound reception mechanisms: Skull vibration enables low-frequency hearing. *PLoS One, 10*, e0116222.

Croll, D. A., Kudela, R., & Tershy, B. R. (2006). Ecosystem impact of the decline of large whales in the North Pacific. In J. A. Estes, D. P. DeMaster, D. F. Doak, T. M. Williams, & R. L. Brownell (Eds.), *Whales, whaling and ocean ecosystems* (pp. 202–214). University of California Press.

Cummings, W. C., & Thompson, P. O. (1971). Underwater sounds from the blue whale, *Balaenoptera musculus. The Journal of the Acoustical Society of America, 50*(4B), 1193–1198.

Ekdale, E. G. (2010). Ontogenetic variation in the bony labyrinth of *Monodelphis domestica* (Mammalia: Marsupialia) following ossification of the inner ear cavities. *The Anatomical Record, 293*, 1896–1912.

Ekdale, E. G. (2013). Comparative anatomy of the bony labyrinth (inner ear) of placental mammals. *PLoS One, 8*, e66624.

Ekdale, E. G. (2016). Morphological diversity among the inner ears of extinct and extant baleen whales (Cetacea: Mysticeti). *Journal of Morphology, 277*, 1599–1615.

Ekdale, E. G., & Racicot, R. A. (2015). Anatomical evidence for low frequency sensitivity in an archaeocete whale: Comparison of the inner ear of Zygorhiza kochii with that of crown Mysticeti. *Journal of Anatomy, 226*, 22–39.

Galatius, A., Olsen, M. T., Steeman, M. E., Racicot, R. A., Bradshaw, C. D., Kyhn, L. A., & Miller, L. A. (2019). Raising your voice: Evolution of narrow-band high-frequency signals in toothed whales (Odontoceti). *Biological Journal of the Linnean Society, 126*, 213–224.

Geijer, C. K., Notarbartolo di Sciara, G., & Panigada, S. (2016). Mysticete migration revisited: Are Mediterranean fin whales an anomaly? *Mammal Review, 46*, 284–296.

Geisler, J. H., Colbert, M. W., & Carew, J. L. (2014). A new fossil species supports an early origin for toothed whale echolocation. *Nature, 508*, 383–386.

Gunz, P., Mitteroecker, P., & Bookstein, F. L. (2005). Semilandmarks in three dimensions. In D. E. Slice (Ed.), *Modern Morphometrics in physical anthropology* (pp. 73–98). Plenum Publishers.

Gutstein, C. S., Figueroa-Bravo, C. P., Pyenson, N. D., Yury-Yañez, R. E., Cozzuol, M. A., & Canals, M. (2014). High frequency echolocation, ear morphology, and the marine–freshwater transition: A comparative study of extant and extinct toothed whales. *Palaeogeography Palaeoclimatology Palaeoecology, 400*, 62–74.

Ingram, T., & Mahler, D. L. (2013). SURFACE: Detecting convergent evolution from comparative data by fitting Ornstein-Uhlenbeck models with stepwise Akaike information criterion. *Methods in Ecology and Evolution, 4*, 416–425.

Jefferson, T. A., Webber, M. A., & Pitman, R. L. (2015). *Marine mammals of the world: A comprehensive guide to their identification*. Academic.

Kato, H., & Perrin, W. F. (2018). Bryde's whale. In B. Würsig, J. G. M. Thewissen, & K. M. Kovacs (Eds.), *Encyclopedia of marine mammals* (3rd ed., pp. 143–145). Academic.

Ketten, D. R. (2000). Cetacean ears. In W. W. L. Au, A. Popper, & R. R. Fay (Eds.), *Hearing by whales and dolphins* (pp. 43–108). Springer.

Ketten, D. R., & Wartzok, D. (1990). Three dimensional reconstructions of the dolphin ear. In J. Thomas & R. Kastelem (Eds.), *Sensory abilities of Cetaceans* (pp. 81–105). Plenum Press.

Ketten, D. R., Arruda, J., Cramer, S., & Yamato, M. (2016). Great ears: Low-frequency sensitivity correlates in land and marine leviathans. In A. N. Popper & A. Hawkins (Eds.), *The effects of noise on aquatic life II* (pp. 529–538). Springer.

Li, Y., Liu, Z., Shi, P., & Zhang, J. (2010). The hearing gene Prestin unites echolocating bats and whales. *Current Biology, 20*, 55–56.

Liu, Y., Cotton, J. A., Shen, B., Han, X., Rossiter, S. J., & Zhang, S. (2010). Convergent sequence evolution between echolocating bats and dolphins. *Current Biology, 20*, 53–54.

Liu, Z., Qi, F. Y., Zhou, X., Ren, H. Q., & Shi, P. (2014). Parallel sites implicate functional convergence of the hearing gene prestin among echolocating mammals. *Molecular Biology and Evolution, 31*, 2415–2424.

Lloyd, G. T., & Slater, G. J. (2021). A total-group phylogenetic metatree for cetacea and the importance of fossil data in diversification analyses. *Systematic Biology, 70*, 922–939.

Manoussaki, D., Chadwick, R. S., Ketten, D. R., Arruda, J., Dimitriadis, E. K., & O'Malley, J. T. (2008). The influence of cochlear shape on low-frequency hearing. *Proceedings of the National Academy of Sciences of the United States of America, 105*, 6162–6166.

Martins, M. C. I., Park, T., Racicot, R., & Cooper, N. (2020). Intraspecific variation in the cochleae of harbour porpoises (*Phocoena phocoena*) and its implications for comparative studies across odontocetes. *PeerJ, 8*, e8916.

Mourlam, M. J., & Orliac, M. J. (2017). Infrasonic and ultrasonic hearing evolved after the emergence of modern whales. *Current Biology, 27*, 1776–81 e9.

Nicol, S., Bowie, A., Jarman, S., Lannuzel, D., Meiners, K. M., & Van Der Merwe, P. (2010). Southern Ocean iron fertilization by baleen whales and Antarctic krill. *Fish and Fisheries, 11*, 203–209.

Nummela, S., & Yamato, M. (2018). Hearing. In B. Würsig, J. G. M. Thewissen, & K. M. Kovacs (Eds.), *Encyclopedia of marine mammals* (3rd ed., pp. 462–470). Academic.

Park, T., Fitzgerald, E. M. G., & Evans, A. R. (2016). Ultrasonic hearing and echolocation in the earliest toothed whales. *Biology Letters, 12*, 20160060.

Park, T., Evans, A. R., Gallagher, S. J., & Fitzgerald, E. M. G. (2017a). Low-frequency hearing preceded the evolution of giant body size and filter feeding in baleen whales. *Proceedings of the Royal Society B, 284*, 20162528.

Park, T., Marx, F. G., Fitzgerald, E. M., & Evans, A. R. (2017b). The cochlea of the enigmatic pygmy right whale *Caperea marginata* informs mysticete phylogeny. *Journal of Morphology, 278*(6), 801–809.
Park, T., Mennecart, B., Costeur, L., Grohé, C., & Cooper, N. (2019). Convergent evolution in toothed whale cochleae. *BMC Evolutionary Biology, 19*(1), 1–11.
Parker, J., Tsagkogeorga, G., Cotton, J. A., Liu, Y., Provero, P., Stupka, E., & Rossiter, S. J. (2013). Genome-wide signatures of convergent evolution in echolocating mammals. *Nature, 502*, 228–231.
Parks, S. E. (2003). Response of North Atlantic right whales (*Eubalaena glacialis*) to playback of calls recorded from surface active groups in both the North and South Atlantic. *Marine Mammal Science, 19*, 563–580.
Perrin, W. F., Mallette, S. D., & Brownell, R. L., Jr. (2018). Minke whales. In B. Würsig, J. G. M. Thewissen, & K. M. Kovacs (Eds.), *Encyclopedia of marine mammals* (3rd ed., pp. 608–613). Academic.
Pivorunas, A. (1979). The feeding mechanisms of baleen whales. *American Scientist, 67*, 432–440.
R Core Team. (2020). *R: A language and environment for statistical computing*. R Foundation for Statistical Computing.
Racicot, R. A., Gearty, W., Kohno, N., & Flynn, J. J. (2016). Comparative anatomy of the bony labyrinth of extant and extinct porpoises (Cetacea: Phocoenidae). *Biological Journal of the Linnean Society, 119*, 831–846.
Racicot, R. A., Darroch, S. A., & Kohno, N. (2018). Neuroanatomy and inner ear labyrinths of the narwhal, *Monodon monoceros*, and beluga, *Delphinapterus leucas* (Cetacea: Monodontidae). *Journal of Anatomy, 233*(4), 421–439.
Rankin, S., & Barlow, J. (2005). Source of the North Pacific "boing" sound attributed to minke whales. *The Journal of the Acoustical Society of America, 118*, 3346–3351.
Ritsche, I. S., Fahlke, J. M., Wieder, F., Hilger, A., Manke, I., & Hampe, O. (2018). Relationships of cochlear coiling shape and hearing frequencies in cetaceans, and the occurrence of infrasonic hearing in Miocene Mysticeti. *Foss Recycling, 21*(1), 33–45.
Roman, J., & McCarthy, J. J. (2010). The whale pump: marine mammals enhance primary productivity in a coastal basin. *PLoS One, 5*, e13255.
Stayton, C. T. (2008). Is convergence surprising? An examination of the frequency of convergence in simulated datasets. *Journal of Theoretical Biology, 252*, 1–14.
Stayton, C. T. (2015). The definition, recognition, and interpretation of convergent evolution, and two new measures for quantifying and assessing the significance of convergence. *Evology, 69*, 2140–2153.
Tubelli, A., Zosuls, A., Ketten, D., & Mountain, D. C. (2012). Prediction of a mysticete audiogram via finite element analysis of the middle ear. In A. N. Popper & A. Hawkins (Eds.), *The effects of noise on aquatic life* (pp. 57–59). Springer.
Tyack, P. (1983). Differential response of humpback whales, *Megaptera novaeangliae*, to playback of song or social sounds. *Behavioral Ecology and Sociobiology, 13*, 49–55.
Urick, R. J. (1983). *Principles of underwater sound for engineers*. Tate McGraw-Hill Education.
Viglino, M., Gaetán, M., Buono, M. R., Fordyce, R. E., & Park, T. (2021). Hearing from the ocean and into the river: the evolution of the inner ear of Platanistoidea (Cetacea: Odontoceti). *Paleobiology*, 1–21.
Visualization Sciences Group—a FEI Company. (2016). *Avizo: 3D analysis software for scientific and industrial data*, standard edition.
Volume Graphics, Part of Hexagon. (2012). *VGStudio Max 2.2*.
Watkins, W. A., & Wartzok, D. (1985). Sensory biophysics of marine mammals. *Marine Mammal Science, 1*, 219–260.
Wiley. (2005). *Landmark Editor 3.0*. Institute for Data Analysis and Visualization.

Yamato, M., & Pyenson, N. D. (2015). Early development and orientation of the acoustic funnel provides insight into the evolution of sound reception pathways in cetaceans. *PLoS One, 10*(3), e0118582.

Yamato, M., Ketten, D. R., Arruda, J., Cramer, S., & Moore, K. (2012). The auditory anatomy of the Minke whale (*Balaenoptera acutorostrata*): A potential fatty sound reception pathway in a baleen whale. *The Anatomical Record, 295*, 991–998.

Yamato, M., Koopman, H., Niemeyer, M., & Ketten, D. (2014). Characterization of lipids in adipose depots associated with minke and fin whale ears: Comparison with "acoustic fats" of toothed whales. *Marine Mammal Science, 30*, 1549–1563.

Zelditch, M. L., Ye, J., Mitchell, J. S., & Swiderski, D. L. (2017). Rare ecomorphological convergence on a complex adaptive landscape: body size and diet mediate evolution of jaw shape in squirrels (Sciuridae). *Evology, 71*, 633–649.

Chapter 5
The Sacro-Iliac Joint of the Felidae and Canidae and Their Large Ungulate Prey: An Example of Divergence and Convergence

Jean-Pierre Pallandre, Franck Lavenne, Eric Pellé, Katia Ortiz, and Vincent L. Bels

Abstract The aim of this chapter is to discuss the evolution of the shape of the sacroiliac joint in two carnivoran lineages (Felidae and Canidae) and their large prey (Ungulata) in the context of divergent and convergent evolution. The significant difference in the angle between the iliac wings of the pelvic girdle in the transverse plane (the interiliac angle) between the Ungulata (>100°) and both carnivoran lineages (30–40°) suggests a divergence in form that relates to the evolution of their feeding behavior over at least 75 Myrs. In the Canidae, the interiliac angle of around 40° and the inner C-shape of the iliac auricular surface congruent with the sacral auricular surface are not influenced either by locomotor nor predatory behavior. Hunting on small or large prey has had no impact on the sacroiliac joint of canids, even though solitary hunting of small prey switches to pack hunting of big prey. A hunting strategy based upon the harassment of large prey individuals could explain why the locking properties of the sacroiliac joint, determined by the interiliac angle, and the inner shape of the articular surface have not been influenced by prey selection. These joint properties are similar to those of felids that select prey with body-mass lower than their own. We suggest that the similarities recorded in canids and these felids result from convergent evolution due to prey selection even though their hunting strategies are different. In contrast, the interiliac angle is significantly

J.-P. Pallandre (✉) · V. L. Bels
Institute of Systematics, Evolution, Biodiversity, ISYEB – UMR 7205 – CNRS/MNHN/EPHE/UA, National Museum of Natural History, Sorbonne University, Paris, France

F. Lavenne
CNRS, INSB, Centre d'Etude et de Recherche Multimodale Et Pluridisciplinaire en Imagerie du Vivant, Bron, France

E. Pellé
Sorbonne Université, Muséum National d'Histoire Naturelle, Direction des Collections, Plateforme de Préparation Ostéologique, Paris, France

K. Ortiz
Réserve Zoologique de la Haute Touche, Muséum national d'Histoire Naturelle, Obterre, France

V. L. Bels, A. P. Russell (eds.), *Convergent Evolution*, Fascinating Life Sciences,
https://doi.org/10.1007/978-3-031-11441-0_5

smaller, and the locking properties of the joint are increased through a strong congruency of the W-shaped inner surface and the outer ridge in solitary big cats that are able to exploit prey with body mass greater than their own, These traits, resulting in a stiff sacroiliac joint, especially during recoil, are probably explained by attributes of the feeding behavior that require a sustained bite during the killing of prey. In lions, the interiliac angle is similar to that of canids, suggesting a relaxation of functional constraints relating to feeding behavior in a species in which individuals organize into social groups for pack-hunting of large prey. This chapter considers the role of divergent and convergent functional evolution of feeding strategies on the morphological traits of the sacroiliac joint that permit us to discuss the "form-function" relationship of this key articulation of the pelvic girdle in the Carnivora.

Keywords Auricular surface · Biting · Congruency · Hunting strategy · Interiliac angle · Killing · Postcranial system · Prey/Predator mass ratio · Retreat movement · Topography

5.1 Introduction

Eutherian mammals play a key role in almost all terrestrial trophic webs at various levels in all the World's eco-regions (Anderson et al., 2016). These tetrapods underwent an explosive radiation approximately 120 My ago after probably originating at the K-Pg transition, although a reliable time-calibrated phylogeny has not yet been generated (Song et al., 2012; Springer et al., 2019). With regard to evolutionary divergence and convergence, the feeding behavior of mammals has benefited from an extremely large number of studies that link the morphological and developmental properties of the trophic (cranial) designs with the behavioral strategies used to catch, manipulate and transport liquid and solid food (Marshall & Pyenson, 2019; Williams, 2019). Over the last 20 years, an increasing number of studies have focused on the morphological and functional links between postcranial and cranial elements in relation to the success of food capture and manipulation in mammals (Cuff et al., 2016a, b; Montuelle & Kane, 2019; Whishaw & Karl, 2019; Pallandre et al., 2020). Such studies, however, deal mainly with the functional integration of these elements in response to proximate ecological factors and habitats (i.e., aquatic *vs* terrestrial habitats, prey size and mass, prey behavior).

About 70–75 My ago in Laurasiaeutherian feeding behavior diverged between the Carnivora (predators), and their perissodactylan and terrestrial cetartiocactylan prey (Kim et al., 2016), along with divergence of their gut microbiomes (Wible et al., 2007; Nishida & Ochman, 2018). This divergence was associated with highly different foraging and feeding behaviors and imposed different behavioral constraints on these predarory and prey mammals. In the terrestrial Cetartiodactyla and Perissodactyla, herbivory became the sole feeding mode (Venter et al., 2019). Compared with carnivorans specialized for hunting, such as the Felidae and Canidae, the postures and movements associated with feeding are stereotyped for grazing and browsing. Indeed, all large ungulates use a typical quadrupedal posture

Fig. 5.1 Typical postures of grazing ungulates. (**a**) *Equus zebra*; (**b**) *Bos gaurus*; (**c**) *Cervus elaphus*; (**d**) *Diceros bicornis*

for grazing and browsing on the abundantly available plant material in all kinds of landscapes and habitats (Fig. 5.1). This posture also permits them to optimize their vigilance to react to activities of both congeners and carnivores (Kröschel et al., 2017). In contrast, terrestrial carnivorans, such as felids and canids, show great diversity in their foraging and feeding abilities, from taxa that are generalists to those that are highly specialized and hypercarnivorous, and these often exhibit an overlap of their habitat and diet in the majority of the World's eco-regions (Figs. 5.2 and 5.3). To be able to chase, kill and feed on living and mobile prey, particularly large prey, these carnivores show complex behaviors involving postures and movements of the cranial and postcranial skeletons (MacDonald, 1983; MacNulty et al., 2007; Montuelle & Kane, 2019; Pallandre et al., 2020).

It is evident that the entire post-cranial skeletal and muscular morphology involved in predatory activities in carnivorans has resulted in a functional trade-off because of the highly different demands imposed by locomotion, foraging, feeding and social interactions. A large number of studies examining form and biomechanics have revealed the functional relationship between the morphological traits of the forelimb and its role in food capture and handling (Iwaniuk et al., 1999; Meachen-Samuels & Van Valkenburgh, 2009; Janis & Figueirido, 2014; Martín-Serra et al., 2016). Such investigations reveal, from the perspective of convergence, the functional trade-off of forelimb morphology that is related to locomotion in various habitats and to successful predation for the gaining of nutrients and energy (Day & Jayne, 2007; Wroe et al., 2008; Meloro et al., 2013; Samuels et al., 2013; Michaud

Fig. 5.2 Variable feeding postures of felids and canids. (**a**) *Panthera tigris*; (**b**) *Panthera leo*; (**c**) *Vulpes vulpes* (Photo credit: Nathalie Gallois); (**d**) *Felis margarita*

Fig. 5.3 Feeding postures of carnivorans that have an omnivorous (**a**) and derived herbivorous (**b**) diet, showing the role of the forelimbs. (**a**) *Nasua* sp.; (**b**) *Ailurus fulgens*

et al., 2020). More recently the properties of the vertebral column have been related to various behavioral abilities, including locomotion in the Felidae and Canidae, these two groups including the top-predators in several of the World's eco-regions (Randau et al., 2016; Randau & Goswami, 2017, 2018). Contrasting with our understanding of the role of forelimbs in predatory behavior, which has been

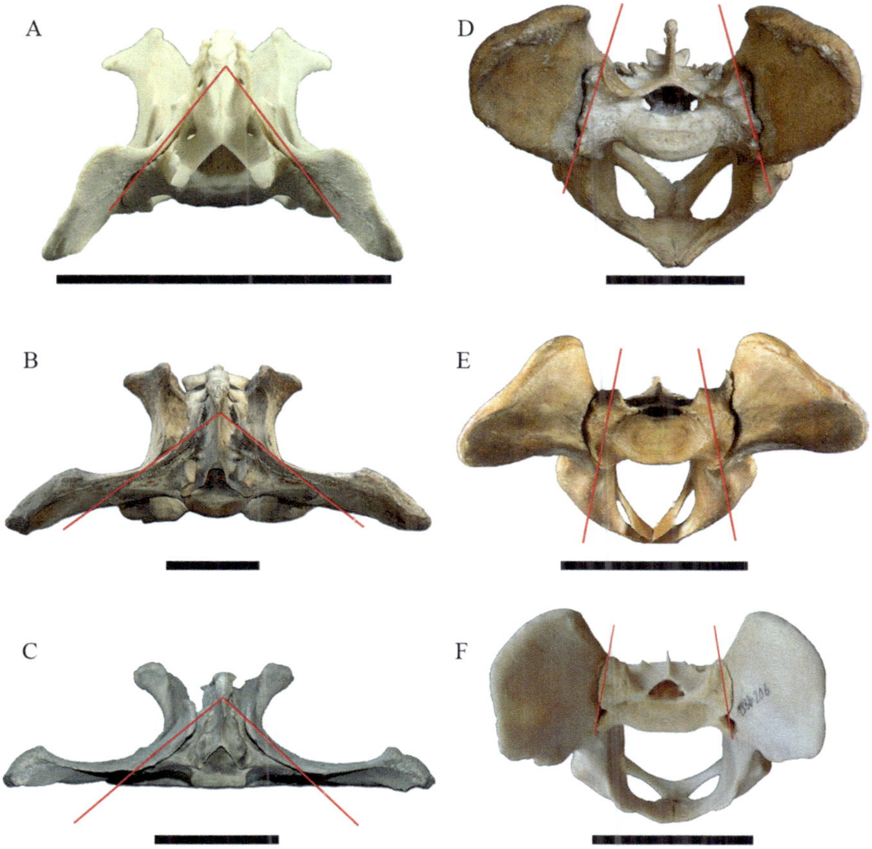

Fig. 5.4 Examples of the pelvic girdle of representative mammals in anterior view, showing the variation in the tightening of iliac wings about the sacrum (interiliac angle). (**a**) *Capreolus capreolus*; (**b**) *Bison bison*; (**c**) *Bubalus bubalis*; (**d**) *Panthera leo*; (**e**) *Ursus arctos* (**f**) *Canis lupus*. Red lines represent the interiliac angle in the transverse plane. Bold black line scale bar = 10 cm. Specimens from the collection "Mammifères et Oiseaux" of the Museum National d'Histoire Naturelle (Paris, France)

extensively studied for members of the Carnivora, the role of hind limbs and pelvic girdle remains rather poorly explored, even though the postures and movements of predators play a key role in the successful capture and killing of prey, as well as in subsequent feeding activities (e.g., food handling; food transportation for its hiding or to permit feeding in an optimal location; plucking pieces of food).

The sacroiliac joint (SIJ) is the key region of the pelvic girdle involved in force transmission from the ground to the vertebral axis. Two main morphological traits determine the functional properties of the SIJ: (1) the interiliac angle (IIA) between both sacroiliac junctions in the transverse plane (Fig. 5.4; Pallandre et al., 2021), and (2) the topography of the auricular surfaces of the iliac and sacral bones, determining

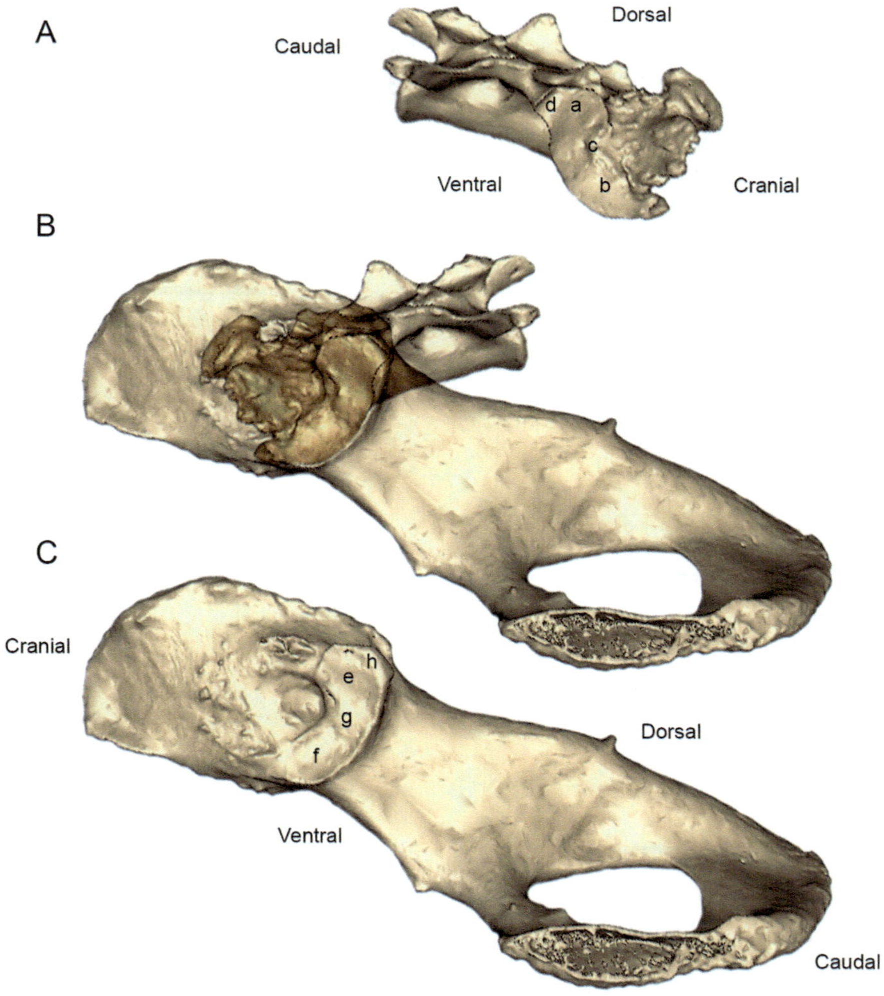

Fig. 5.5 CT-scan reconstruction of the right hemi-pelvis of *Panthera onca*. (**A**) Right lateral view of the sacrum; a, dorsal leg of the auricular surface; b, ventral leg of the auricular surface; c, central depression; d; dorso-caudal notch. (**B**) Ilium and sacrum in articulation. (**C**) Medial view of the right ilium; e, dorsal leg of the auricular surface; f, ventral leg of the auricular surface; g, central eminence; h, dorso-caudal ridge

their congruency (Fig. 5.5), the inner-line and the outer-line of the auricular surface describing the topography of this region (Pallandre et al., 2020).

In this chapter, we consider the evolution of the morphological properties of SIJ in the context of evolutionary divergence and convergence of the feeding behavior between large herbivorous (prey) and carnivorous mammals (predators) from the perspective of the evolutionary relationship between "form and function" (Fig. 5.6).

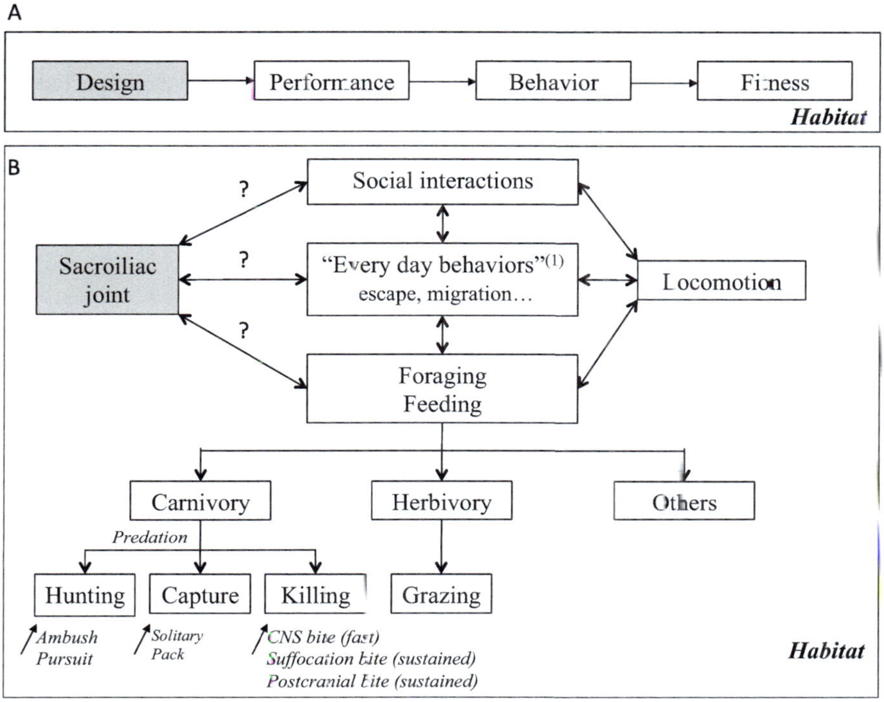

Fig. 5.6 (**a**) Modification of Arnold's paradigm (Arnold, 1983) showing the functional relationships between design (e.g., sacroiliac joint) and fitness under environmental constraints. (**b**) Schematic diagram showing the complex interactions between behavioral and environmental constraints on the traits of the sacroiliac joint

From among the Carnivora we selected two families belonging to two separate lineages, the Felidae (Feliformia) and the Canidae (Caniformia) that diverged within the crown Carnivora between 47 and 53 Mya (Van Valkenburgh et al., 2014; Hassanin et al., 2021). We selected these families because among them their members are able to hunt a variety of prey, including large herbivores. Indeed each family includes species that exploit a wide range of habitats and exhibit great diversity in their ecological traits (e.g., habitat, diet) and social behaviors. Furthermore, these predators can either hunt solitarily or in packs. Within the hypercarnivorous Felidae, the lion (*Panthera leo*) is the only species that pack-hunts with strong interindividual cooperative actions (Pulliam & Caraco, 1984; Gittleman, 1989; Scheel & Packer, 1991; Stander, 1992; Schaller, 2009). Occasional cooperation may occur in solitary species (e.g., cheetah) that may hunt in groups when subjected to high ecological stressors (e.g., availability of prey, climatic constraints; Gittleman, 1989; Caro, 1994a; Radloff & Du Toit, 2004). In contrast, several species of the Canidae of varying body size (e.g., dhole, wolf, lycaon) hunt, and kill their prey within organized social groups (MacDonald, 1983, 2009; Gittleman, 1989; Creel & Creel, 1995).

Firstly our consideration explores the question of SIJ divergence between the Ungulata (large prey) and representative Carnivora (predator) because their very early divergence of feeding behavior during their evolution may have impacted the morphology of the SIJ. Secondly, the impact of feeding behavior on both carnivoran families that we examine, the Felidae and Canidae, is discussed in the context of evolutionary convergence. This discussion is based on additional morphological data to those presented in the previous work of Pallandre et al. (2020, 2021).

5.2 SIJ and Locomotion

In mammals, the forward thrust produced by the hind limb is brought into alignment with the direction of travel and transferred to the vertebral column, whether this direction of travel is in the horizontal or vertical plane (i.e., terrestrial locomotion or climbing) regardless their body form or way of life (Taylor, 1989; Kardong, 2015; Beisiegel and Ades 2002; de Oliveira Calleia et al., 2009; Schwab et al., 2019). The SIJ provides the junction between the hind limb and the vertebral column (Derry, 1911; Barone, 1986; Abitbol, 1987; Pallandre et al., 2020). In the stance phase of locomotion, in which forward thrust is generated, the iliac wings, while moving cranially, get closer to the cranioventrally positioned sacral wings to finally push the sacrum along the SIJ (Fig. 5.4). During this phase, the space between iliac and sacral auricular surfaces is likely shortened and the interosseous ligament probably compressed. In contrast, during the retreat movement, hind limbs produce backward movement of the pelvis. The iliac wings spread away from the sacral wings and the interosseous ligament is stretched (Barone, 1986; Dalin & Jeffcott, 1986a; Pallandre et al., 2020). The IIA value might promote one movement relative to the other (e.g., propulsion *vs* retreat). With increasing speed, the long axis of the pelvic girdle becomes more fully aligned with the long axis of the sacrum (Romer, 1950; Taylor, 1989; Kardong, 2015). A few studies have considered the evolution of the angle between the long axis of the sacrum and the long axis of the ilium in the context of locomotion (Kardong, 2015), and quantitative investigations of this angle have been conducted for the domestic horse (Dalin & Jeffcott, 1986a, b; Erichsen et al., 2002). In this ungulate the IIA is about 120° (Dalin & Jeffcott, 1986b). In contrast, this angle varies between 30° and 50° in hypercarnivorous felids, this range of angles being associated with locomotion and prey selection (Pallandre et al., 2021). To our knowledge this angle has not been investigated for other carnivoran species. Smith and Savage (1956) pointed out that the shape of the iliac bones can be related to differing modes of locomotion in mammals but Lewton et al. (2020) found little evidence of locomotor adaptation in the pelvis of carnivorans, and Pallandre et al. (2020) demonstrated that the shape of SIJ (its inner topography and outer line), is not impacted by the locomotor modes exhibited by felids.

5.3 SIJ and Feeding Behavior

5.3.1 Ungulata (Perissodactyla and Terrestrial Cetartiodactyla)

As illustrated by qualitative descriptions of grazing mammals (Venter et al., 2019), the Perissodactyla (odd-toed ungulates) and terrestrial Cetartiodactyla (even-toed ungulates) use erect postures when acquiring food, regardless of their morphological features (i.e., size, weight, and shape). Nevertheless, various ungulates exploit tree resources by using occasional bipedal posture (e.g., *Nanger* sp.) with or without the help of the forelimbs, and ruminants are known to undergo rumination in the classical resting posture (i.e., sternal recumbency; Pucora et al., 2019).

The antipredator and social interactions of herbivorous mammals also imply varying constraints (Caro, 1994b; 2005; Carter et al., 2019) on the morphological traits of the vertebral system (Van der Linden & Dumont, 2019). It is reasonable to assume that these interactions influence the properties of the sacroiliac joint. During social fights among herbivores (e.g., deer and ovids) each buck primarly stands on all four limbs and is pushed by the other fighter (Fig. 5.7). Biomechanically, the pushing thrust results mainly from the force generated by the hind limbs, to such an extent that sometimes the forelegs leave the ground. In this context, the pushing force generated by the dominant animal is the main cause of the retreat of the weaker one barely exhibiting any pulling actions. This thrust can induce a recession followed by an about-face before fleeing *per se*, confirming that the backward movements exhibited by the Ungulata are permitted via a typical sequence of locomotor recoil (as illustrated in Fig. 5.8). The animal briefly performs only a few recoil steps, then turns and initiates a rapid about-face movement to avoid a potential aggressor.

Fig. 5.7 Example of two *Cervus nippon pseudaxis* bucks fighting, showing the forward thrust during the brawl

Fig. 5.8 Retreat sequence of a *Cervus elaphus* specimen coping with a menace (here a human approach). (**a**) Standing position. (**b**) Backward weight transfer. (**c**) Backward stepping. (**d**) Pivot preparing to flee

5.3.2 Carnivora

The predatory behavior of carnivores is highly complex, involving successive phases: search, approach, watch, attack, and capture (MacNulty et al., 2007; Stanton et al., 2015) regardless of prey type. In this chapter we use the terminology employed by Slater and Friscia (2019): (i) cat-like hypercanivory for the Felidae and mesocarnivory (50–70% of the diet consisting of flesh or meat of another animal; Van Valkenburgh, 2007) for the Canidae. Among the Felidae, cat-like hypercarnivory occurs in species preying on small and large prey (Michaud et al., 2020). Hunting strategies depend on prey size and its ability to avoid (e.g., camouflage, shelter), escape, and fight predators (Gittleman, 1989; MacNulty et al., 2007; Mukherjee & Heithaus, 2013).

Among terrestrial mammals, carnivores of various size and mass employ a large number of movements and postures in hunting, catching and killing prey of various shapes, sizes and mass (Schwab et al., 2019). For example, during the approach phase felids exhibit typical strategies such as crouching that permit observation of the intended prey and preparation for attack, this being considered to be specific for all felids. Capture, including biting and prey control, involves not only the trophic system but also the entire postcranial system (MacNulty et al., 2007; Stanton et al., 2015; Montuelle & Kane, 2019) The forelimbs, due to the use of the claws, play a role in (1) prey manipulation and (2) stabilization of the prey-predator pair during take down in large felids (Gonyea & Ashworth, 1975; Gonyea, 1978; Meachen-Samuels & Van Valkenburgh, 2009, 2010; Meachen-Samuels, 2010; Stanton et al., 2015; Cuff et al., 2016a; Viranta et al., 2016; Michaud et al., 2020. The pelvic girdle has also been suggested to play a key role in subduing prey by enhancing the role of the skull during the application of the lethal bite by large felids (Pallandre et al., 2020, 2021).

The social interactions of canids and felids are highly variable, involving various postures and behaviors during food exploitation (e.g., inter-individual competition; Fig. 5.9).

Fig. 5.9 Competition for food in the Canidae (*Canis lupus*). The animals exert forces on the substratum through the hindlimbs to help food partitioning. Photo credit: Nathalie Gallois

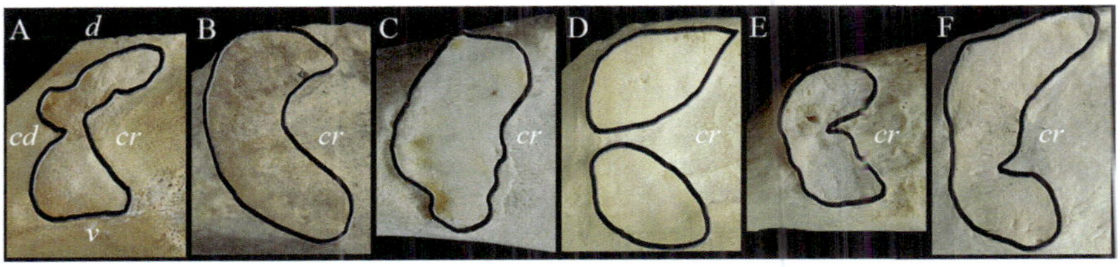

Fig. 5.10 Examples of the iliac auricular surface outline shape found in canids investigated in this study: (**a**) Type 1 (auricle-shaped); (**b**) Type 2 (crescent-shaped); (**c**) Type 3 (spatula-shaped); (**d**) Type 4 (bifoliate-shaped); (**e**) Type 5 (E-shaped); (**f**) Type 6 (Inverted Phrygian cap-shaped). *cr* cranial, *cd* caudal, *d* dorsal, *v* ventral. Canid species are represented as follows: (**a**) and (**d**), *Vulpes vulpes* (solitary hunter); (**b**) *Vulpes bengalensis* (solitary hunter); (**c**) and (**e**) *Speothos venaticus* (pack hunter); (**f**) *Canis lupus* (pack hunter). (**a**), (**b**), (**d**), (**f**), right auricular surface; (**c**), (**e**), left auricular surface. Right auricular surfaces were mirrored for presentation. Note that different types can be recorded in one species (see text for explanations). Photo credit: Layla Adil

5.4 Case Study of the Canidae

5.4.1 General Morphology of the SIJ

Figure 5.10 illustrates the variability of the iliac auricular surface contour in various species of small and large wild dogs that use solitary and pack hunting strategies (Table 5.1). The outline separates the iliac auricular surface into a dorsal and a ventral limb, as described for some other mammal species (Dalin & Jeffcott, 1986a, b; Barone, 1986; Pallandre et al., 2020; Jesse et al., 2017). The central eminence that Pallandre et al. (2020) described for felids, lying at the junction between the two legs at the cranial edge of the joint, is also present in canids.

Table 5.1 Morphological and behavioral characteristics of canid species used in comparisons of their interiliac angle

Species	Number of specimens	Locomotor class	Body mass (kg)[a]	Hunting strategy[b]
Canis aureus	5	Cursorial[c]	8.1	Pack
Canis lupus	8	Cursorial[d]	43.8	Pack
Cuon alpinus javanicus	1	Cursorial[d]	5.5	Pack
Speothos venaticus	6	Terrestrial[c]	6.5	Pack
Vulpes bengalensis	1	Terrestrial[e]	2.5	Solitary
Vulpes lagopus	3	Terrestrial[c]	3.7	Solitary
Vulpes vulpes	7	Cursorial[d]	5.8	Solitary

Body mass and hunting strategies were obtained from the literature: [a]Figueirido et al. (2011), [b]MacDonald (1983), [c]Samuels et al. (2013), [d]Martín-Serra et al. (2014), [e]Gompper and Vanak (2006). Male and female specimens were not treated separately in this study

With a concave border facing cranially, the auricular surface shows complete or incomplete division of the legs in some specimens. As in the Felidae, six types of outline shape can be visually defined: type 1 ("auricle-shape"), type 2 ("crescent-shape"), type 3 ("spatula-shape"), type 4 ("bifoliate-shape"), type 5 ("B-shape"), and type 6 ("Inverted Phrygian cap-shape"). This last shape is inverted compared to the one observed in the Felidae (i.e., Prygian cap-shape; Pallandre et al., 2020). Variation in shape between the right and left iliac surfaces may occur in the same specimen (see variation in the Felidae). The surfaces of the dorsal and ventral legs are generally concave with irregular wave-like striations of various orientations marking the auricular surface, but the area of the central eminence, separating the joint legs is slightly convex. The dorso-caudal ridge, forming a convex articular surface with the sacrum, described for the species of the *Panthera* lineage (Pallandre et al., 2020), was not found in the members of the Canidae investigated in this study.

5.4.2 SIJ Topography

Following the methods of Pallandre et al. (2020) for describing and analyzing the topography of the iliac auricular surface, we calculated the difference in level of landmarks describing the inner-line shape of the surface (Fig. 5.11). In the members of the Canidae we studied, this shape is not impacted either by body mass, locomotion, or by hunting strategy (Table 5.1). Figure 5.11 illustrates the inner-line shape of four canids we investigated. All species show a C-shaped inner-line similar to the shape observed in small cats, there being no significant impact of hunting strategy on any of the landmarks describing the inner-line of iliac auricular surface (Pallandre et al., 2020).

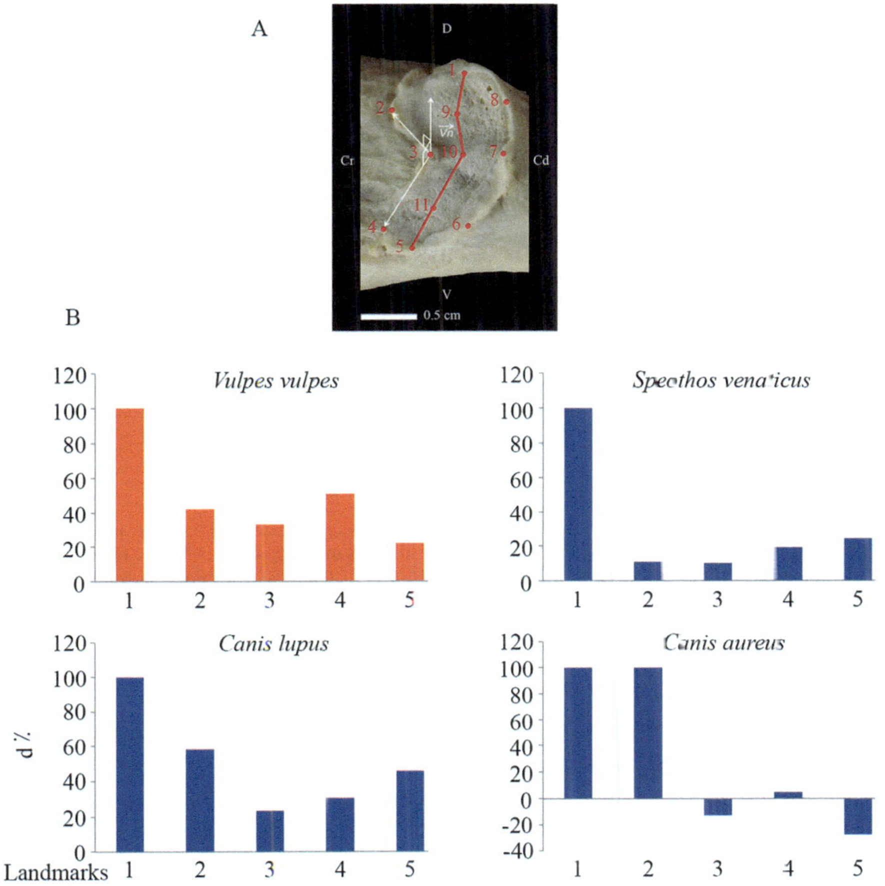

Fig. 5.11 Difference in level (d%) of landmarks 1 (reference landmark = 100%), 9, 10, 11 and 5 that define the inner-line of the iliac surface in canids. (**a**) Landmarks on the iliac auricular surface of a specimen of *Speothos venaticus* after Pallandre et al. (2020). $\overrightarrow{Vn}$ is a vector orthonormal to the plane in which landmarks 2, 3 and 4 lie and is used to calculate the difference in level between other landmarks and this plane (see Appendix 1 for calculation). Red solid line connects landmarks 1, 9, 10, 11 and 5 defines the inner-line. (**b**) Examples of joint congruency in four species with highly different body masses. The difference in level values (d%) of landmarks towards the plane passing through points 2, 3 and 4 relative to the level of landmark 1 (reference point) is plotted for the landmarks defining the inner-line. Orange, solitary hunting species; Blue, pack-hunting species

5.4.3 *Interiliac Angle*

Figure 5.4f illustrates the tightening of the iliac wings around the sacrum at the sacro-iliac joint level (IIA) in *Canis lupus* (see Pallandre et al., 2021 for the analytical method for establishing IIA). The interiliac angle (Table 5.2) is not correlated with body size (R^2 = 0.014; ddl = 1, 27; F = 0.40; p = 0.54).

Table 5.2 Descriptive statistics for the interiliac angle (°) in the Canidae and in all studied Felidae, including *Panthera leo*. The median for *P. leo* was calculated separately because this species is the only pack-hunting felid

Variables	N	Median	Q_1	Q_3	Range
Felidae	59	33.65	30.30	38.65	18.80–49.10
Canidae	31	40.10	34.24	45.94	21.94–54.51
Panthera leo	13	37.90	35.07	46.60	29.50–40.50

N number of specimens, *Q1* first quartile, *Q3* third quartile (See Appendix 2 for felid data)

Table 5.3 Mann-Whitney U-tests used for comparing the interiliac angle within the Canidae and between the Canidae and Felidae

Comparisons	n1	n2	U	p
Cursorial—Terrestrial canids	21	10	166	0.368
Canidae <5.5 kg—Canidae ≥5.5 kg	22	9	87	0.617
Pack—Solitary canids	20	11	124	0.577
Canidae—Felidae	31	59	1292	<0.05
Panthera leo—Canidae	13	31	166	0.368

n1, number of cases for the first term of the comparison; n2, number of cases for the second term of the comparison. As all the recorded data did not follow the assumptions of normal distribution and homoscedasticity, we used non-parametric statistical tests (Sokal & Rohlf, 1981). The level of significance for all of the tests (Bonferroni correction) $p = 0.006$ (Everitt & Dunn, 2001) (See Appendix 2 for felid data)

Table 5.3 shows that this angle is neither significantly influenced by the locomotor mode of dogs, nor by their hunting strategy. Regardless of their hunting strategy, however, the interiliac angle in the canids investigated is significantly greater than that of the felids studied by Pallandre et al. (2021). This significant difference between canids and felids disappears, however, when dogs are compared to *Panthera leo*, the only felid species that practices pack hunting

5.5 Case Study of the Felidae

5.5.1 Topography

The auricular surface of the ilium and the IIA of felids have previously been extensively investigated (Pallandre et al., 2020, 2021). Here, we provide additional topographic properties of the iliac and sacral surfaces in representative felid species to explore the effect of their predatory behavior on the congruency of both surfaces (Figs. 5.5 and 5.12). These data are needed to enable discussion of the evolutionary relationship between the morphological properties of the SIJ and predatory behavior in the Carnivora.

The sacral auricular surface covers the entire surface of the sacral wing, which corresponds to the first sacral vertebra. In physiological position the contour of the

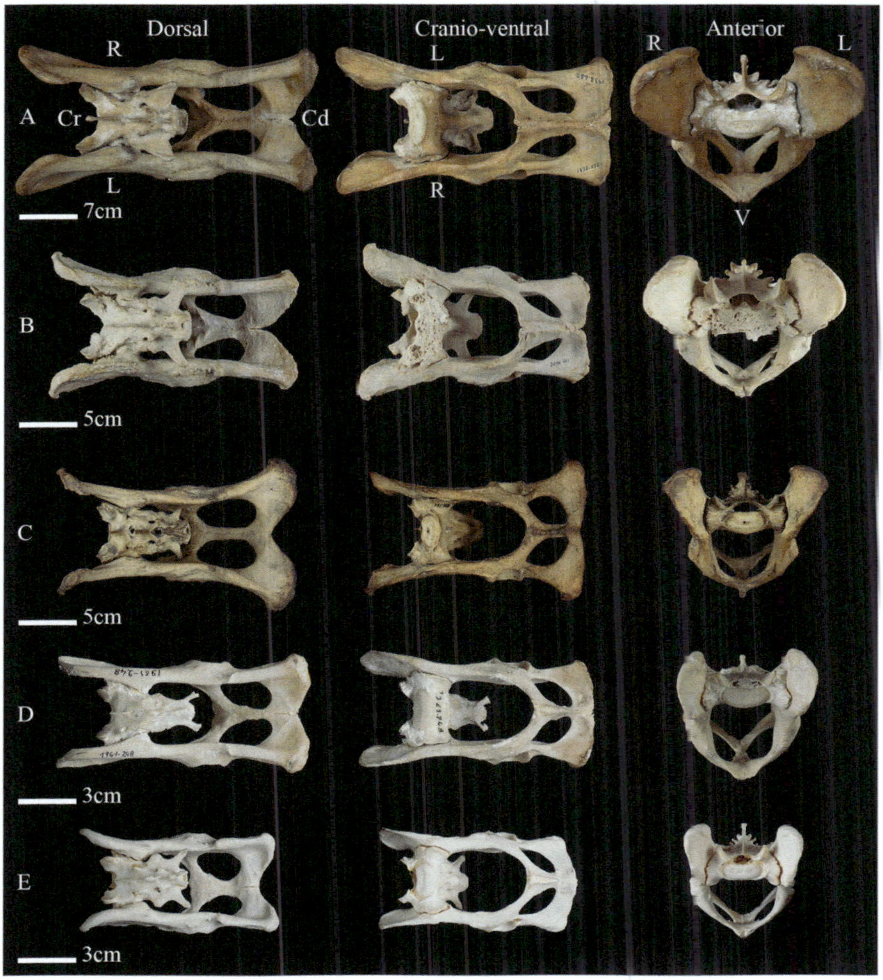

Fig. 5.12 Examples of the pelvic girdle of felids showing the interlocking between the sacrum and the coxal bones in dorsal, ventral and anterior views. (**a**) *Panthera leo*, (**b**) *Panthera onca*, (**c**) *Acinonyx jubatus*, (**d**) *Lynx canadensis*, (**e**) *Felis silvestris*, *Cr* cranial, *Cd* caudal, *L* left, *R* right, *V* ventral

sacral auricular surface perfectly follows that of the contour of the iliac auricular surface. The intra-specific and intra-individual variation in contour shape (Fig. 5.13) recorded for all studied species is, therefore, similar for the sacral and iliac surfaces. The auricular surface of the sacrum mirrors the iliac surface by presenting an inverted topography: the undulations of the two surfaces systematically interdigitate, the depressions of one corresponding to the prominences of the other, as shown for *Panthera onca* in which the outer ridge is clearly more marked than in other *Panthera* species (Figs. 5.5, 5.14 and 5.15).

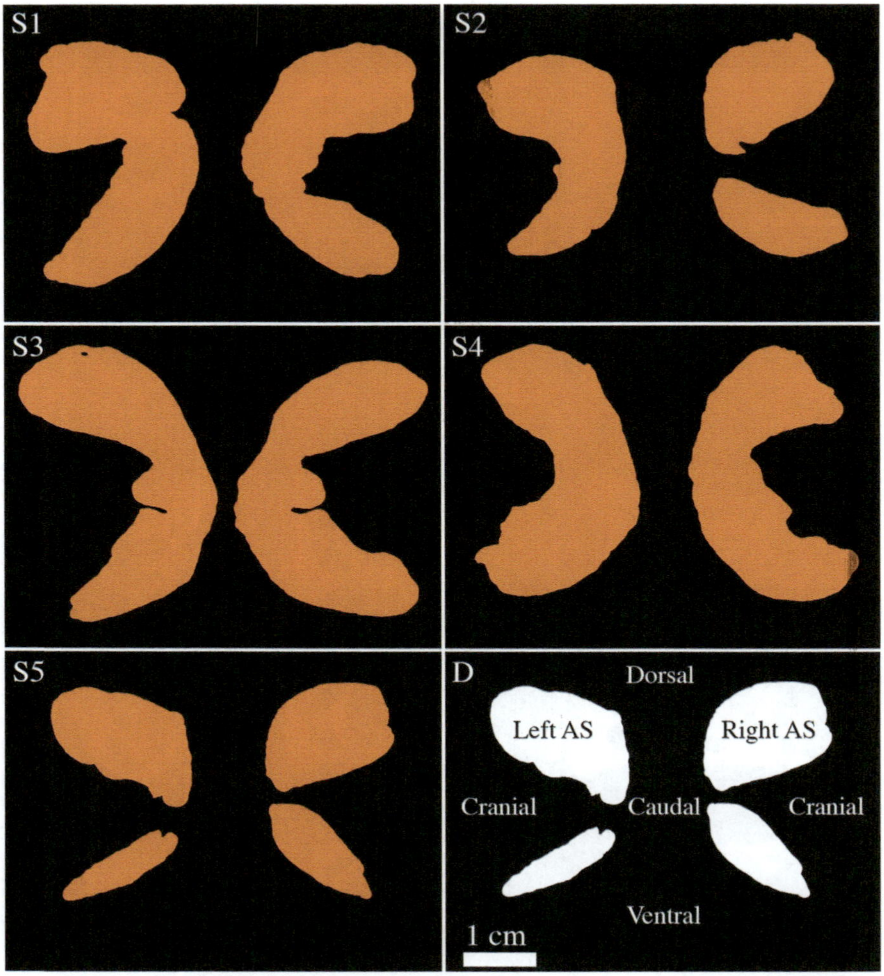

Fig. 5.13 Example of intra-specific and intra-individual variation in five specimens of *Panthera tigris*, showing the outline in ventro-dorsal view of the iliac auricular surface. S1–S5, specimens. *D* directions, *Left AS* left iliac auricular surface, *Right AS* right iliac auricular surface

In accord with previous results (Pallandre et al., 2020; Fig. 5.16), two forms were observed (Fig. 5.17): (1) a regular C-shape for *Felis silvestris, Leptailurus serval, Lynx canadensis, Lynx rufus, Neofelis nebulosa* and *Panthera onca*, and (2) a rugged W-shape *for Acinonyx jubatus, Panthera leo, Panthera pardus, Panthera tigris, Panthera uncia* and *Puma concolor*. Table 5.4 presents the results of the Fisher's exact tests obtained for the distribution of C- and W-shaped articulations in felids. The distribution of the C and W forms of the inner-line connecting landmarks 1, 9, 10, 11 and 5 is only significantly different for the types of bite (i.e., suffocation, crushing of the central nervous system, sustained postcranial bite) and the size of the preferred prey (i.e., "small", "mixed" and "large").

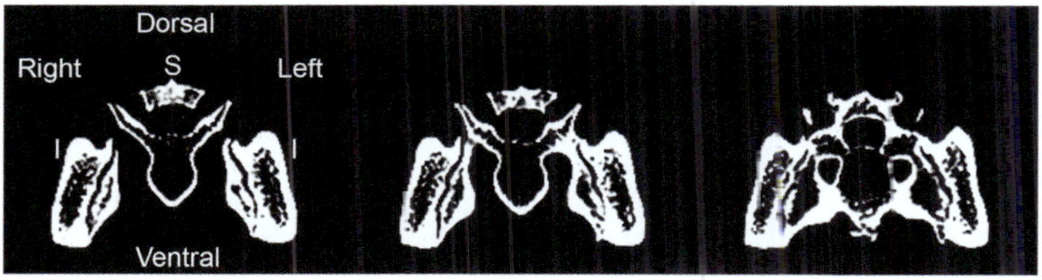

Fig. 5.14 Serial CT-scan cross sections of the sacroiliac joint of *Panthera pardus* showing the congruence between the joint surfaces from cranial (left) to caudal (right). *I* iliac wing, *S* sacrum

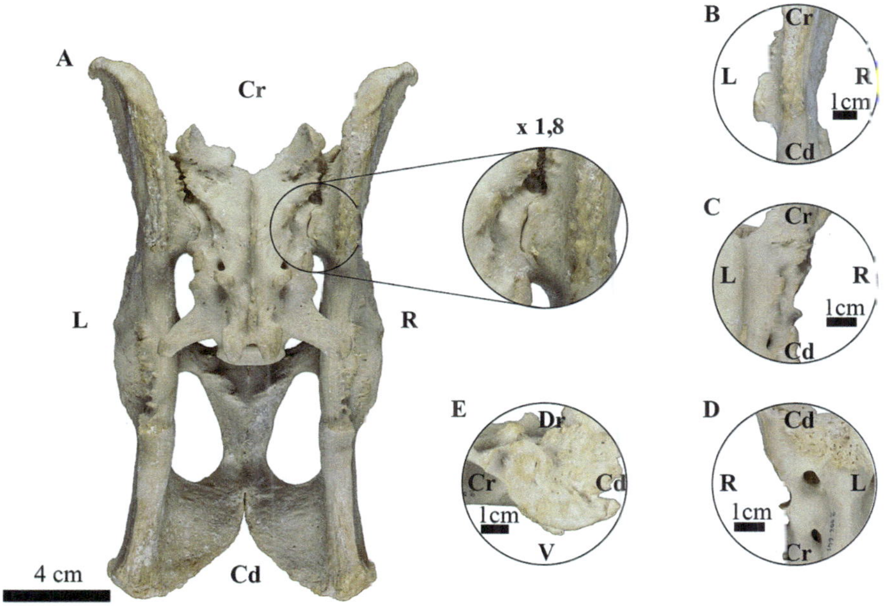

Fig. 5.15 Sacroiliac congruence in a jaguar (*Panthera onca*) showing the dorso-caudal region of the joint. (**a**) Dorsal view of the articulated pelvis. (**b**) Dorsal view of the dorso-caudal crest of the right ilium. (**c**) Dorsal view of the right dorso-caudal notch of the sacrum. (**d**) Ventral view of the right dorso-caudal notch of the sacrum. (**e**) Lateral view of the right articular surface of the sacrum. *Cd* caudal, *Cr* cranial, *Dr* dorsal, *L* left, *R* right, *V* ventral

5.6 Discussion

Intersecting environmental and historical (phylogenetic) factors have acted on the phenotypic traits of the ungulate and carnivoran sacroiliac joint (SIJ). In this chapter we discuss the effect of predator-prey interactions on the functional properties of the SIJ. First, we consider the SIJ in ungulates and then, we compare some

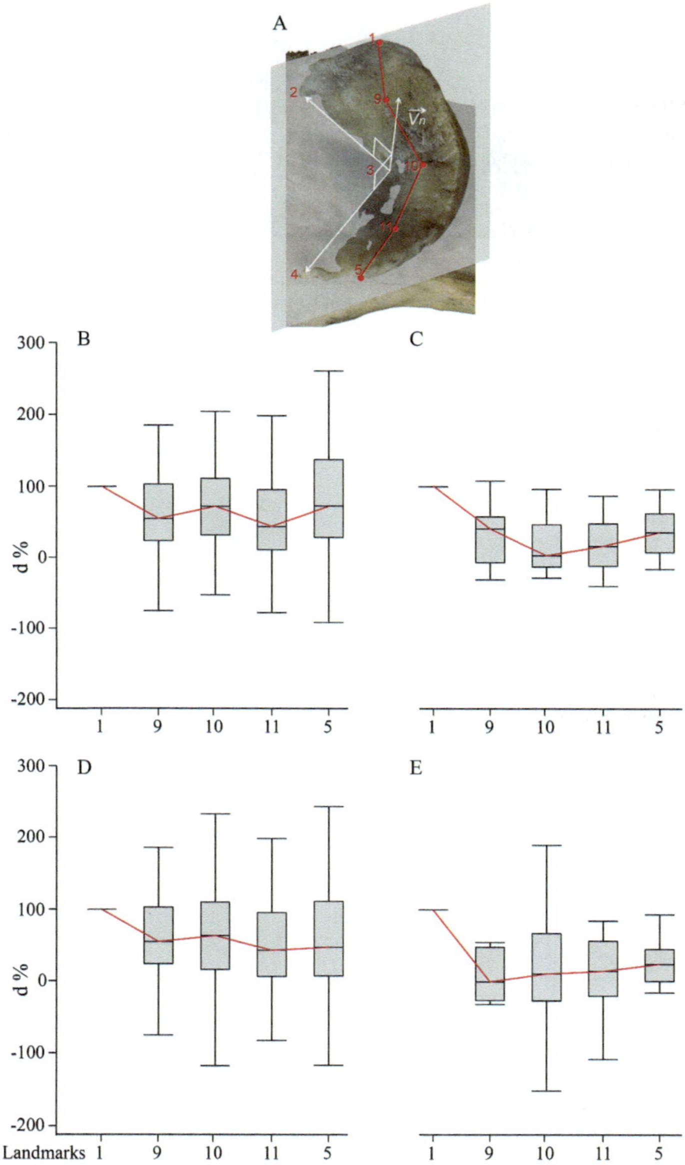

Fig. 5.16 Effect of body mass and prey selection on the topography of the iliac auricular surface in the felids included in this study. (**a**) Measurement of the topography of the right iliac auricular surface of a specimen of *Panthera onca*. $\overrightarrow{Vn}$ is a vector orthonormal to the plane that includes landmarks 2, 3 and 4 and is used to calculate the difference in level (d%) between these and other landmarks and this plane (i.e., topography of the articular surface; Pallandre et al., 2020; Appendix 1). Red solid line connects points 1 (reference landmark = 100%), 9, 10, 11 and 5 defines the inner-line. (**b–d**) Box plots representing the morphological level of selected landmarks along the inner-line of the iliac auricular surface. Each red line connects the medians of the dorsal and ventral landmarks. (**b**) Species with body mass > 5.5 kg; (**c**) Species with body mass < 14.5 kg; (**d**) Species with MPM/PBM* ≥1; (**e**) species with MPM/PBM <1. (*) MPM/PBM represents the ratio between

morphological traits in felids and canids, two eutherian families that hunt a variety of prey, including the largest prey (ungulates) pursued by members of the Carnivora. Then, we discuss these properties in relation to the the hunting and prey-killing strategies employed by felids and canids. To contextualize the effect of these interactions we identified two morphological traits that impact joint mobility: (i) the topography of iliac auricular surface (i.e., its inner shape and outer border), and (ii) the interiliac angle. This discussion cannot be accomplished without considering the potential effects of both locomotion and habitat.

Until recently the SIJ in these tetrapods has been described as a planiform synovial joint, morphologically similar to the amphiarthrotic joints characterized by Winslow (1732), and Bichat (1855). This joint has also been described as a semi-mobile amphi-diarthrodial joint, the structure of which is primarily related to locomotor behavior (Barone, 1986; Kardong, 2015) through the transmission of forces from the ground to the vertebral axis. The diversity of the morphological traits of the SIJ exhibited by some mammals with conspicuously different behaviors (Figs. 5.1–5.17; Tables 5.2, 5.3, and 5.4) suggests that various functional constraints that are related to behavioral traits beyond those of locomotion, such as feeding, playing, and fighting, impact the morphological properties of this joint.

A few studies have described the angle between the long axis of the sacrum and the ilium, mostly in relation to the evolution of locomotion (Smith & Savage, 1956; Kardong, 2015). Quantitative investigations of this angle have largely been pursued with regard to domestic horses (Dalin & Jeffcott, 1986a; Erichsen et al., 2002), in which the angle between both sacroiliac joints in the transverse plane (i.e., the interiliac angle) attains about 120° (Dalin & Jeffcott, 1986a). Qualitatively, this angle is rather similar for all of the large herbivorous mammals, as shown in Fig. 5.4. In contrast, this angle is smaller in various terrestrial and aquatic members of the Carnivora. For example, it attains 25° in *Phoca* sp. (Smith & Savage, 1956) and about 30–40° in felids (Pallandre et al., 2020; Pallandre et al., 2021, this study).

Generally locomotion, for movement within the habitat, migration, approach, escape, and pursuit of prey, has been considered to impose the major constraint on the morphological traits of the pelvic girdle (e.g., SIJ and IIA) of mammals (Fig. 5.18). It is worth noting that ungulates of various sizes constitute abundant prey resources for all large felids (e.g., lions, leopards, and cheetahs) and large canids (e.g., wild dogs) throughout the year (Skogland, 1991) in all of the World's eco-regions (Olson et al., 2001). Qualitative data show that the interiliac angle is large in big ungulates, regardless of their phylogeny and habitat (Fig. 5.4). This joint probably primarily plays a key role in the transmission of propulsive forces from the iliac wings to the sacrum in an anterior direction during all types of posture and locomotor behavior. Indeed, ungulates adopt similar postures for feeding, monitoring the behavior of predators, and escaping from predators and conspecifics

Fig. 5.16 (continued) the maximum average prey mass (MPM) and the average predator body mass (PBM) following Sicuro and Oliveira (2011) and Pallandre et al. (2020). Modified from Pallandre et al. (2020)

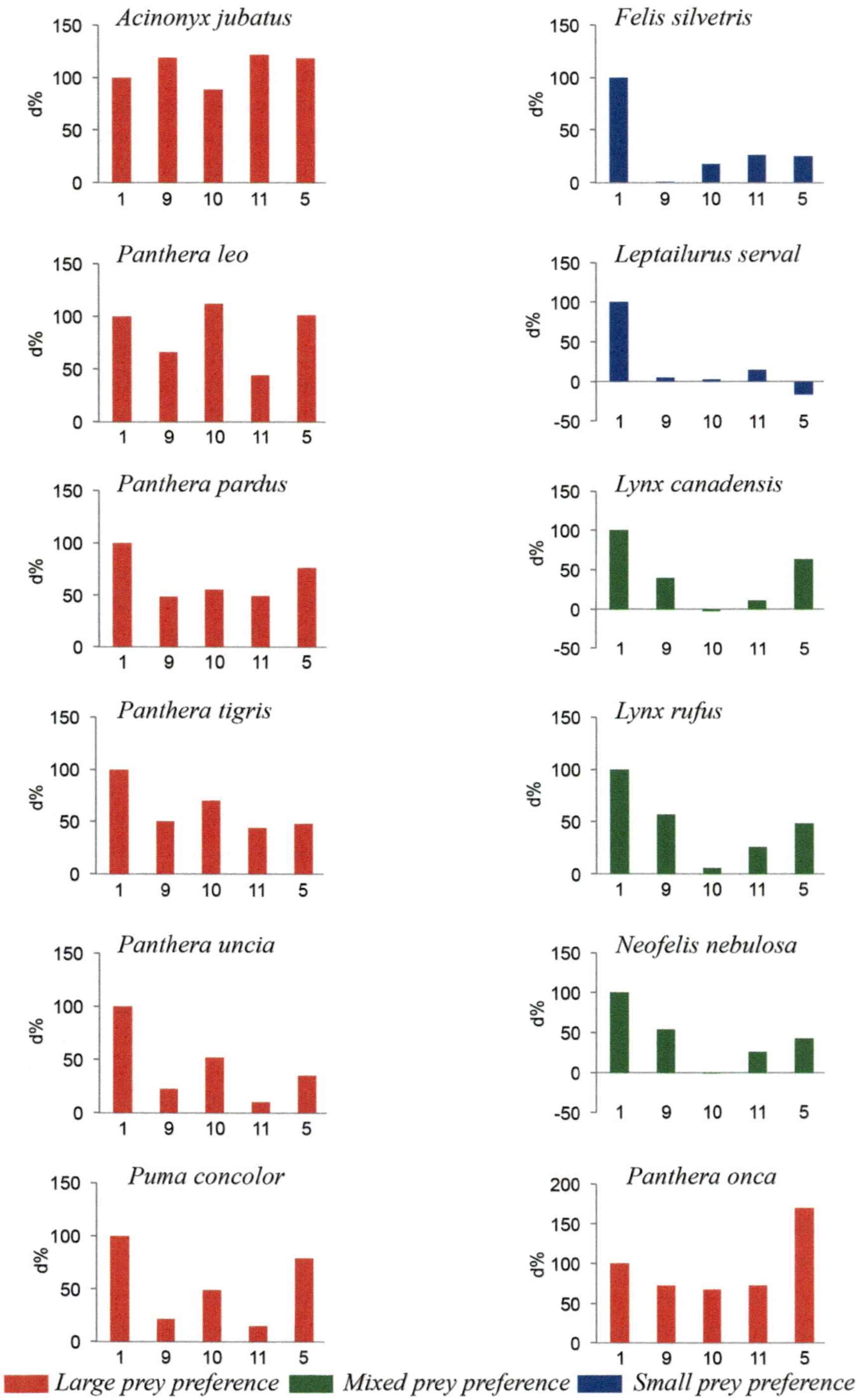

Fig. 5.17 Relative difference in level (d%) of landmarks 1 (reference landmark = 100%), 9, 10, 11, and 5 defining the inner-line of the iliac surface in the felid species investigated (see Figs. 5.19

Table 5.4 Fisher's exact test for approximating the distribution of the C- and W-shapes of the SIJ in the Felidae. The data were calculated as a percentage of the total number of forms measured for each tested factor (significant level: $p < 0.05$)

Factors	p-value
Lineages	0.17
Body mass (<5.5 kg *vs* ≥ 5.5 kg)	0.06
Locomotion	1.00
Habitat	0.24
Solitary *vs* pack	1.00
Ambush *vs* pursuit	0.55
Prey size	0.015
MPM/PBM (1–2–3)	0.55
MPM/PBM (1–2 vs 3)	0.45
Bite type	0.015

Locomotor classes: terrestrial, scansorial, arboreal, cursorial (Pallandre et al., 2020). Habitat: savannah, forest, mountain (Sunquist & Sunquist, 2017; Pallandre et al., 2020). Prey size: large, small, mixed (Meachen-Samuels & Van Valkenburgh, 2009; Pallandre et al., 2021). The bite type classes: suffocation, bite at the spine, bite at the skull (Schaller & Vasconcelos, 1978; Kitchener et al., 2010; Pallandre et al., 2020). *MPM* maximum average prey mass, *PBM* average predator body mass (Sicuro & Oliveira, 2011; Pallandre et al., 2020). MPM/PBM ratio classes 1, MPM/PBM ≥ 1.9; 2, 1 ≤ MPM/PBM ≤ 1.7; 3, MPM/PBM ≤ 0.9 (Pallandre et al., 2020). We tested the ratio MPM/PBM as follows: (1) In the first comparison, each class is compared separately with regard to factor levels, and (2) in the second comparison, classes 1 and 2 are combined and compared with class 3 (See Appendix 3 for felid data)

(Fig. 5.1). Calves take up their posture on four limbs when searching for maternal teats and for suckling, regardless their morphology, the shape and position of the female teats, and maternal behavior (Nowak et al., 2000; Karenina & Giljov, 2018), although occassionaly they support themselves on their wrists for suckling. Postural constraints on the hindlimbs, however, remain similar throughout the life of these tetrapods. To respond to aggression imposed by predation and social interactions they use two different strategies: (1) they face the aggressor and present their weapons (e.g., horns, antlers, and hooves) while advancing towards the aggressor (Fig. 5.7; Caro, 2005; e.g., buffalo that can kill lions; Makacha & Schaller, 1969), or (2) turn away and kick, and/or flee (Fig. 5.8). All large herbivorous mammals ultimately turn away and use forward locomotion to retreat when facing a threat such as that from a predator (Fig. 5.8). Also, during flight a crucial means for a prey animal to avoid a predator is through their manœuvering capacity (Wilson et al.,

Fig. 5.17 (continued) and 5.20 for explanations). Left column, species with a W-shaped inner-line; right column, species with a C-shaped inner-line. The median difference in the level of the landmarks that describe the inner-line shape follows the method of Pallandre et al. (2020) for the topographic study of the iliac auricular surface (Appendix 1) and was calculated within a confidence interval of 95% for each studied species

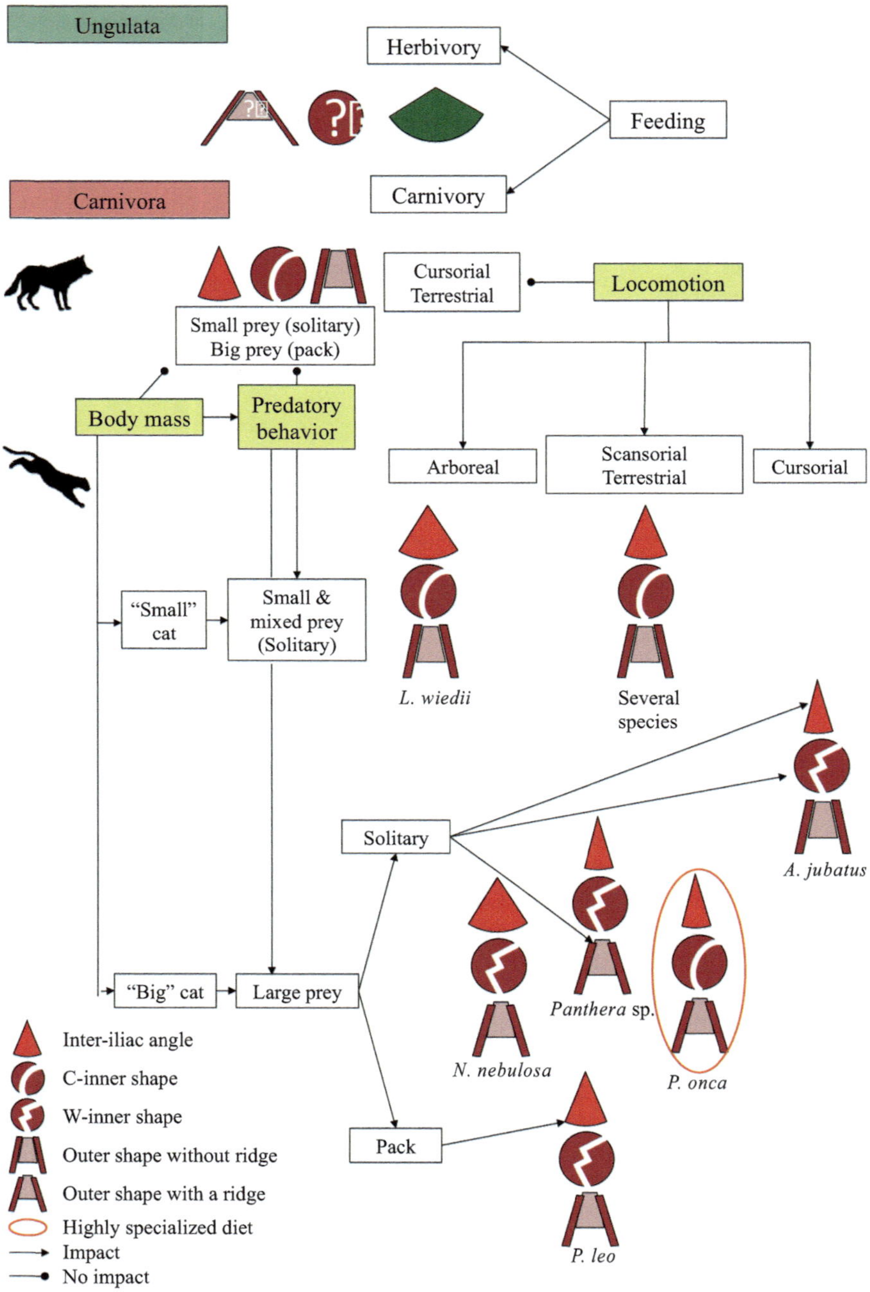

Fig. 5.18 Schematic summary of the hypothesized relationships between the properties of the SIJ and the functional constraints imposed by locomotion and feeding behavior on terrestrial eutherians. The divergence of feeding behavior between the Ungulata and Carnivora has impacted the evolution of the IIA. Among Carnivora, the Felidae and Canidae separated 70 Mya and show an IIA varying from around 30–50°. The SIJ in extant representative species of both families shows either a

2018). Big ungulates (e.g., zebra and impala), although slower than their predators (e.g., lions an cheethas), are able to move unpredictably to evade the predator while remaining only a step or two ahead (Wilson et al., 2018). These movements play a key role in survival of these mammals and have probably imposed functional constraints on the shape of the IIA. We suggest that the large IIA observed in ungulates may well promote their manoeuverability by enhancing their capacity for sudden directional change. The morphology of the articular surface and the IIA remain to be investigated to confirm that locomotor postures and abilities to feed and survey for predator activities represent major selective forces acting on the SIJ in all herbivorous species throughout their life. This hypothesis remains to be tested.

The shape of SIJ is particularly impacted by the performance of predatory behavior in the Carnivora, as demonstrated by felids and canids (Figs. 5.11, 5.15, 5.17, and 5.18; Tables 5.2, 5.3, and 5.4). Based upon previous data for the Felidae (Pallandre et al., 2020), we hypothesized that predator-prey interactions and not locomotion have been the major selective factors operating on the SIJ of carnivores subsequent to the divergence, some 120 Mya, between ungulates (prey) and the Carnivora (predators). The impact of social interactions (e.g., play, sexual interactions, struggles for food and partners) could also probably have played functional roles relating to the morphological traits of the SIJ, but such factors remain to be investigated and discussed (Fig. 5.6).

In all of the felids and canids investigated in this study the IIA is much smaller (Fig. 5.4) than that so far described for ungulates, suggesting that the articulation can be more easily locked in the former, the sacrum being squeezed between the iliac wings under the action of the muscular system (Pallandre et al., 2020, 2021; this chapter). Previously detailed studies of the SIJ have shown that the topography of the inner- and outer-lines of the auricular iliac articular surface can variably promote stiffness of the SIJ (Pallandre et al., 2020, 2021). Because the auricular surface of the sacrum interdigitates with that of the ilium, morphological traits such as a W-shaped inner-line, an outer ridge, and the central eminence described for the iliac auricular surface do not favor translation at the articular surface. Such stiffness provides firm support for force transmission for efficient resistance to the struggles of prey and forces imposed during killing. These traits largely decrease the potential movements between the ilium and sacrum, resulting in a complex functional trade-off between diverse behavioral activities, primarily associated with predatory behavior and locomotion, as shown in Fig. 5.18.

Considered as a cursorial representative, the cheetah, with a small interiliac angle (median = 27.5°) differs significantly from arboreal felids that exhibit the widest angle (median = 49.1°). Terrestrial and scansorial felids cluster in a group with an intermediate angle (Pallandre et al., 2021). It is interesting to note that this segregation between cursorial and terrestrial felids also separates the species with the fastest

Fig. 5.18 (continued) C- or a W-shaped inner-line, and there is a dorso-caudal ridge on the outer line in all *Panthera* species except *P. leo*, *P. onca*, and *Neofelis nebulosa* and several other species: *Felis silvestris*, *Leptailurus serval*, *Lynx canadensis* and *Lynx rufus*

recorded speed among terrestrial animals (i.e., *A. jubatus*: around 110 km/h; Taylor et al., 1974; Garland & Janis, 1993; Hudson et al., 2012) from all other terrestrial species that attain sprint speeds up to 69 km/h (e.g., 59 km/h for the lion and 56 km/h for the tiger; Garland & Janis, 1993). Furthermore, the IIA of cursorial canids, as exhibited by the wolf and the jackal, respectively reaching a maximum speed of 64 km/h and of 56 km/h (Garland & Janis, 1993) is not different from that of other studied canids. Based on these data, the correlation between speed and the sacroiliac angle remains to be investigated to uncover the effect of this aspect of locomotor performance on IIA (Fig. 5.6).

For the Canidae the results presented in this paper reveal no significant correlation between the IIA and body size, regardless of locomotor mode and predatory behavior. All studied canids share some common morphological traits (e.g., long body, often two cusps on the carnassial teeth; MacDonald, 2009), although skull size and shape are strongly linked to diet (Meloro et al., 2015). Regardless of their body mass, hunting strategies are very diverse among the Canidae, ranging from mesocarnivorous to hypo-carnivorous solitary species that take small prey (e.g., arthopods, eggs, birds) to hyper-carnivorous species able to catch large prey through pack hunting (Fig. 5.19; Gittleman, 1985; Sheldon, 2013). In these carnivores pack-

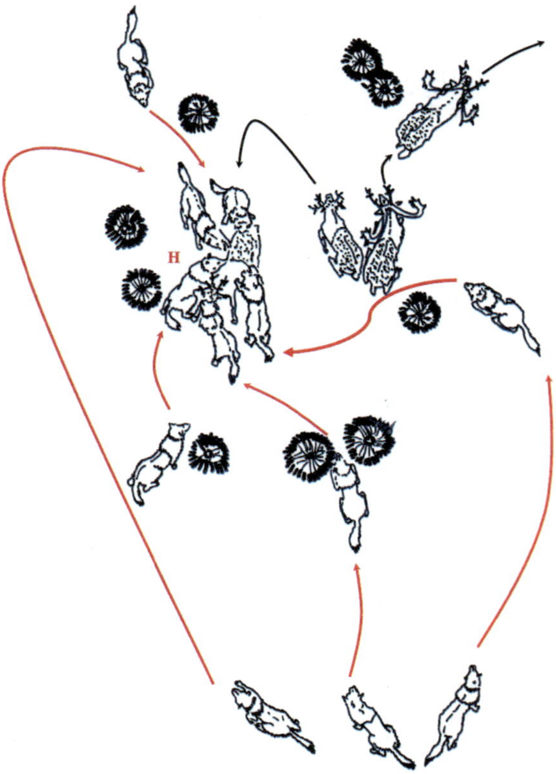

Fig. 5.19 Example of pack-hunting strategy in the Canidae (Wolves) showing the cooperation (red arrows) between individuals harassing (H) the prey (reindeer). The prey is either killed by biting (including a suffocation bite) and/or by being disemboweled alive. Modified from Chinery and Lambiotte (1983)

hunting *vs* solitary hunting does not impact the topography of the SIJ, which is C-shaped in all of the species studied. Futhermore, the outer shape of the joint never exhibits any specialized structures that foster increased congruency, in contrast to that of big felids (Fig. 5.18).

In contrast, body weight and predatory behavior crucially impact the properties of the SIJ in the Felidae. Pallandre et al. (2020) demonstrated that the 5.5 kg predator body mass threshold has an impact on these properties in solitary felids hunting large prey, implying the need to optimize the congruency of the joint for predation. With increasing body mass, when switching from small and mixed prey to the selection of large prey, the inner shape of the iliac and sacral auricular surfaces shifts from C- to W-shaped. This last trait increases the interlocking properties of the SIJ and probably bestows benefit in association with the lethal sustained bite performed by large felids when struggling with prey (Pallandre et al., 2021). The only exception is *P. onca* that mainly preys upon capybara and caimans employing a crushing bite to the back of the skull (Rodríguez-Alba et al., 2019; Flores, 2018; Eisenberg, 2014; Aranda & Sánchez-Cordero, 1996; Ávila–Nájera et al. 2018; Azevedo & Verdade, 2012; Da Silveira et al., 2010). It does not depend exclusively on large prey for its survival but takes both medium-sized prey (e.g., pecarries) and large-size prey (Ávila-Nájera et al. 2018; Aranda & Sánchez-Cordero, 1996). The lack of stiffness of the joint due to the C-shaped inner-line of the auricular surface may be compensated for by (1) the most pronounced ridge observed among all studied species (Figs. 5.5, 5.14 and 5.15), and (2) biting performance (Christiansen & Adolfssen, 2005; Sicuro & Oliveira, 2011; Wroe et al., 2005; Hartstone-Rose et al., 2012). It would also be interesting to investigate the time a jaguar needs to kill its prey by a neural injury compared to the time needed to kill a big prey item by suffocation. According to Pallandre et al. (2020, 2021), the SIJ lock could play a key role during the application of the sustained bite by stabilizing the prey-predator couple during the struggle. However, the outer ridge of the SIJ is present in every *Panthera* species, thereby optimizing the congruency between the ilium and the sacrum by preventing ventral slippage of the sacrum during recoil (Figs. 5.5, 5.14 and 5.15; Pallandre et al., 2020). With the selection of large prey the IIA decreases, and felids selecting prey much bigger than themselves show an IIA of around 10° less than that of other felids (Pallandre et al., 2021). The exception is the cooperatively-hunting lion which has an IIA similar to that of canids (Fig. 5.20, Tables 5.2 and 5.3). We suggest that pack hunting has imposed a major selective force that has driven the evolution of the morphological traits of the SIJ traits in species practicing this strategy (Figs. 5.18, 5.19, and 5.20, Tables 5.2 and 5.3).

Among canids, several species hunt in packs. For example, wolves, lycaons (*Lycaon pictus*) and dholes (*Cuon alpinus*) hunt exclusively large prey, which they pursue in groups consisting of a widely variable number of individuals (Creel & Creel, 1995; Van Valkenburgh, 1996; Radloff & Du Toit, 2004). Environmental factors can greatly influence the number of individuals involved. Wolves participate in prolonged pursuit, harassing the selected large prey individual (Muro et al., 2011). *Lycaon pictus* individuals are also involved in long collaborative pursuits of a single prey individual on the grassy plains of East Africa. In contrast, in mixed woodland

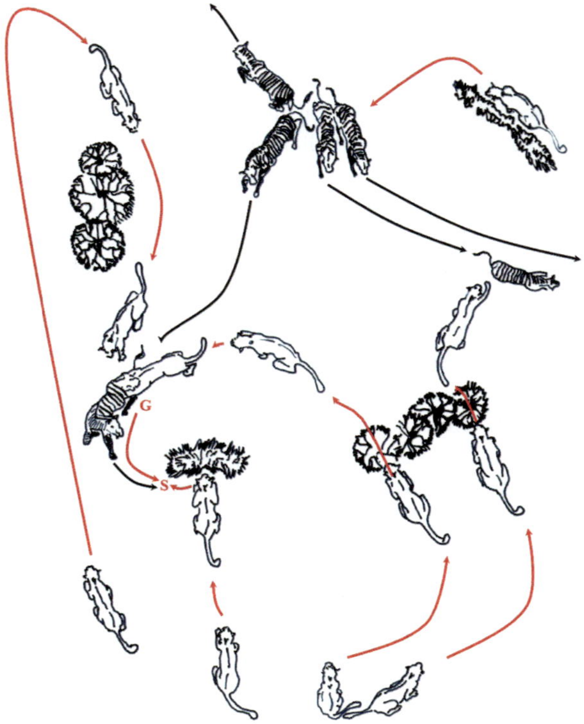

Fig. 5.20 Example of the deployment (red arrows) of pack-hunting strategy by *Panthera leo*. The prey is startled (S) by an individual of the hunting group lying in ambush along the trajectory of the prey (black arrows). Grabbing (G) involves the forelimbs and the jaw. A lethal bite at the throat or muzzle smothers the prey. Modified from Chinery and Lambiotte (1983)

savannah these wild dogs show simultaneous, opportunistic, and short-distance pursuit of whatever prey they can find (Hubel et al., 2016). The switch to cooperative hunting brings benefits in the reduction in hunting distance, increase in kill rate and the capture of larger prey (Eaton, 1969; Taylor et al., 1971; Liebenberg, 2006; Carbone et al., 2007; Bailey et al., 2013; Hubel et al., 2016). To wear big prey down dogs are known to exhaust them through harassment (Kleiman, 1972; Kleiman & Eisenberg, 1973; Bailey et al., 2013; Mukherjee & Heithaus, 2013; Hayward et al., 2014; Dar & Khan, 2016; Muro et al., 2011). They do not use a sustained suffocating lethal bite like big felids do. Instead, the fight is shared among individuals and the subduing of the prey individual is via a prolonged struggle compared to the fast killing practiced by felids (Kitchener et al., 2010). Large canids also often disembowel the prey and start to consume it before it is dead (Kleiman, 1972; Kleiman & Eisenberg, 1973; Bailey et al., 2013; Mukherjee & Heithaus, 2013; Hayward et al., 2014; Dar & Khan, 2016). We hypothesize that in large canids hunting strategies do not impact the interiliac joint because the success of cooperatively hunting big prey is mainly related to pursuit and harassement strategies. In contrast the morphology of the SIJ joint permits the complex reverse movements that occur during individual-to-individual struggles (Fig. 5.9).

For lions, their hunting strategy that is based on social cooperation (Fig. 5.20; Chizzola et al., 2018; Scheel & Packer, 1991) results in a decrease of the functional constraints otherwise associated with killing large prey (e.g., buffalo) by felids. By hunting as a group the greatest limitation placed upon each individual is lessened (MacNulty et al., 2007; Mukherjee & Heithaus, 2013; Scheel, 1993; Scheel & Packer, 1991). This decrease in functional demand related to the predatory behavior of lions could help to explain the wider IIA in this species, even though the W-shaped articulation and the dorso-caudal ridge are maintained (Fig. 5.18). We suggest that the IIA in lions reflects a relaxation of the functional demands related to the killing of prey, thereby driving convergence of the IIA between lions and canids under the influence of the social mode of hunting and killing of large prey. This is supported (1) by the emergence of *P. leo* very late within felid phylogeny (about 3 Mya, Bagatharia et al., 2013), and (2) the hypothesis that the most recent common ancestor of *Panthera* was a solitary hunter able to subdue large prey, had a postcranial morphology similar to that of members of the genus that are solitary hunters, and had a IIA similar to that of the IIA of solitary-hunting members of the genus. This hypothesis remains to be tested using paleontological data gathered from the ancestors of *Panthera*.

In conclusion, we suggest that the feeding performance (i.e., movements and postures) of ungulates and their predators diverged about 120 Mya and that this segregation resulted in a key functional dichotomy that helps to explain the morphological divergence of the joint between the ilium and the sacrum (SIJ) in these two lineages. This divergence of the structure of the SIJ, in the context of prey-predator interactions, is responsible for the difference in the interiliac angle. The common ancestor of the Carnivora probably had an acute interiliac angle, ranging between 30° and 50°, and an SIJ that was required, biomechanically, to lock, permitting force transmission from this region to the skull. Simultaneously, the topography and the outer border of the articulation became more complex to decrease the propensity for translation between iliac and sacral auricular surfaces. From a common ancestor approximately 75 Mya, the Canidae and Felidae probably developed their ability to kill large prey. Of course, ontogenetic change could have occured during the evolutionary trajectory of both of these families. Our study does not permit us to discuss this effect, which is recognized as one of the major factors related to evolutionary convergence. Along their evolutionary pathways both families exploited large ungulates but developed highly different hunting-killing strategies. Hypercarnivorous felids have remained solitary hunters whereas mesocarnivorous and hypercarnivorous canids pursue their large prey in groups. The development of these highly different hunting and killing strategies exposed the morphological properties of the joint between the ilium and the sacrum to different selective demands. Associated with this, the personality of the individuals and their effect on gene flow in the populations could had have played an important role because the abilities of each individual probably played a key role in its survival and its potential reproductive success. For the first time, we demonstrate inter- and intra-individual variability of the surface of the SIJ articuation and, *de facto,* its congruence. This is particularly true for solitary species. But the effect of such variability on the evolutionary trajectories of felids (e.g., *Panthera*) remains to be investigated. To be a solitary hunter of large prey requires particular morphological traits permitting strong interlocking properties of the SIJ that are involved in force transmission from the hind limb to the skull along the vertebral column during the subduing and killing of prey. This is indeed the case for large felids that remain strongly attached to their prey during the struggle, regardless of the habitat and locomotor abilities of the species. In contrast, all canids that chase and kill large prey use a completely different strategy: harassment. This divergence helps to explain the significant difference in their interiliac angle. Along the felid evolutionary pathway lions developed group living and a regular pack hunting strategy within *Panthera*. This strategy decreased the functional constraints imposed upon the SIJ articulation, particularly on the interiliac angle, for each individual. We suggest that cooperative hunting in the Canidae and *P. leo* has resulted in convergence of the morphological properties of the SIJ. This hypothesis remains to be tested in a comparative analysis of SIJ morphology encompassing all lineages of the Carnivora.

Acknowledgements The authors thank the following for financial support of this study (2014–2018): (i) ATMs—Muséum national d'Histoire naturelle 2013–2015 "*Formes possibles, Formes réalisées...*" (Dir: Pr Vincent Bels & Pr Pierre-Henri Gouyon), and (ii) UMR 7205 (Dir: Dr. Philippe Grancolas). We are grateful to Pr Luc Zimmer who allowed us to use the "*Centre d'Etude et de Recherche multimodale en imagerie du Vivant*" (CERMEP, CNRS – INSB) for all of the CT-scans employed in this study, and to Franck Lamberton for his help and coordination with the platform. This study was assisted by Marie-Ange Pierre† who worked as a technician in the FORCE team (UMR7205) during data collection. Many thanks to Roland Simon and Patrick Roux for allowing us to photograph mammals during feeding at the "*Réserve Zoologique de la Haute-Touche*" (MNHN, Obterre, France). Warm thanks to Anthony Russell for reviewing this chapter. The specimen samples are held in the "Mammifères et Oiseaux" collection at the Museum national d'Histoire naturelle in Paris (MNHN). A list of specimens is available from jeanpierre.pallandre@wanadoo.fr

Authors Contribution JPP conducted the study and wrote the paper. FL generated the CT-scans. EP provided the support and help needed to work with the collections of the Museum. KO organized the mammals feeding protocols. VB participated the study and wrote the paper.

Appendix 1: Geometric Calculation to Determine the Topography of the Articular Surface (Figs. 5.12, 5.16 and 5.17)

For the representative canid and felid species, for each iliac auricular surface, we define the plane including landmark 2 ($Pt2(x_2, y_2, z_2)$), landmark 3 ($Pt3(x_3, y_3, z_3)$) and landmark 4 ($Pt4(x_4, y_4, z_4)$).

We consider Pt3 to be the origin of the new coordinate system.

Vectors $\overrightarrow{Pt3Pt2} = (Pt2 - Pt3)$ and $\overrightarrow{Pt3Pt4} = (Pt4 - Pt3)$ are calculated as follows:

$$\overrightarrow{Pt3Pt2}(x_{32}, y_{32}, z_{32})$$

where $x_{32} = x_2 - x_3$
$y_{32} = y_2 - y_3$
$z_{32} = z_2 - z_3$

$$\overrightarrow{Pt3Pt4}(x_{34}, y_{34}, z_{34})$$

where $x_{34} = x_4 - x_3$
$y_{34} = y_4 - y_3$
$z_{34} = z_4 - z_3$

We calculated the coordinates (X, Y, Z) of a vector $\vec{V}$ normal to vectors $\overrightarrow{Pt3Pt2}$ and $\overrightarrow{Pt3Pt4}$:

$$X = y_{32}z_{34} - z_{32}y_{34}$$

$$Y = z_{32}x_{34} - x_{32}z_{34}$$

$$Z = x_{32}y_{34} - y_{32}x_{34}$$

In order to find the coordinates (X_n, Y_n, Z_n) of this normalized vector $\overrightarrow{Vn}$ we first have to calculate its length (L):

$$L = \sqrt{X^2 + Y^2 + Z^2}$$

The coordinates of $\overrightarrow{Vn}$ orthonormal to the plan are calculated:

$$X_n = X : L$$

$$Y_n = Y : L$$

$$Z_n = Y : L$$

The distance (d) of each point (n, m, p) from the plane that includes Pt2, Pt3 and Pt4 is given by the equation:

$$d = (n - x_3)\, X_n + (m - y_3)\, Y_n + (p - z_3)\, Z_n$$

To compare the distance of each point to the plane within various sized auricular surfaces, the relative distance of each point from the plane was given in percentage ($d_{\%}$) of the distance of landmark 1 (d_1) from the plane. Landmark 1 is selected because it is the most dorsal point of each articulation regardless of their size and shape. For each landmark, $d_{\%}$ is given by:

$$d_{\%} = (d : d_1)100$$

$d_{\%}$ measures the difference in level of each landmark relative to the plane that includes landmarks 2, 3 and 4. According to our calculation $d_{\%} = 0$ for landmarks 2, 3 and 4 and $d_{\%} = 100$ for landmark 1.

Appendix 2: Data Set Used for the Study of the Interiliac Angle (Fig. 5.4; Tables 5.2 and 5.3)

Species	Number of specimens	Locomotor class[a]	Hunting strategy[b]	Body mass (kg)[c]
Acinonyx jubatus	7	Cursorial	Solitary	53.5
Felis silvestris	2	Scansorial	Solitary	5.5
Leopardus wiedii	1	Arboreal	Solitary	3.3
Leptailurus serval	1	Terrestrial	Solitary	13.4
Lynx canadensis	2	Terrestrial	Solitary	11.2
Lynx rufus	2	Scansorial	Solitary	11.2
Neofelis nebulosa	1	Arboreal	Solitary	19.5
Panthera leo	13	Terrestrial	Pack	185.0
Panthera onca	5	Scansorial	Solitary	105.7
Panthera pardus	13	Scansorial	Solitary	59.0
Panthera tigris	10	Terrestrial	Solitary	185.5
Panthera uncia	2	Scansorial	Solitary	50.0

[a]Samuels et al. (2013)
[b]Sunquist and Sunquist (2017)
[c]Sicuro and Oliveira (2011)

Appendix 3: Data set for the Felidae Used for the SIJ Topographic Study (Figs. 5.5, 5.13, 5.14, 5.15, 5.16, and 5.17; Table 5.4)

Species	Number of specimens	Body mass (kg)[(1)]	Locomotor class[(2)]	Habitat[(3)]	MPM/ PBM [(1)] classes	Foraging strategy[(3)] [(4)]	Hunting strategy[(3)]	Bite[(5)(6)(7)]
Acinonyx jubatus	7	53.5	Cursorial	Savannah	2 (1.0)	Pursuit	Solitary	Suffocation
Felis silvestris	6	5.5	Scansorial	Forest	3 (0.7)	Ambush	Solitary	Spine
Leptailurus serval	2	13.4	Terrestrial	Savannah	3 (0.4)	Pursuit	Solitary	Spine
Lynx canadensis	2	11.2	Terrestrial	Forest	2 (1.2)	Ambush	Solitary	Spine

(continued)

Species	Number of specimens	Body mass (kg)[1]	Locomotor class[2]	Habitat[3]	MPM/PBM [1] classes	Foraging strategy[3] [4]	Hunting strategy[3]	Bite[5][6][7]
Lynx rufus	3	11.2	Scansorial	Forest	1 (2.4)	Ambush	Solitary	Spine
Neofelis nebulosa	2	19.5	Arboreal	Forest	1 (2.7)	Ambush	Solitary	Suffocation
Panthera leo	14	185.0	Terrestrial	Savannah	1 (2.3)	Pursuit	Pack	Suffocation
Panthera onca	5	105.7	Scansorial	Forest	1 (2.0)	Ambush	Solitary	Back of skull
Panthera pardus	12	59.0	Scansorial	Savannah	1 (2.0)	Ambush	Solitary	Suffocation
Panthera tigris	11	185.5	Terrestrial	Forest	1 (2.7)	Ambush	Solitary	Suffocation
Panthera uncia	2	50.0	Scansorial	Mountain	1 (1.9)	Pursuit	Solitary	Suffocation
Puma concolor	2	67.5	Scansorial	Forest	2 (1.7)	Ambush	Solitary	Suffocation

[a]Sicuro and Oliveira (2011)
[b]Samuels et al. (2013)
[c]Sunquist and Sunquist (2017)
[d]MacNulty et al. (2007)
[e]Kitchener et al. (2010)
[f]Schaller and Vasconcelos (1978)
[g]Palmeira et al. (2008)

References

Abitbol, M. M. (1987). Evolution of the sacrum in hominoids. *American Journal of Physical Anthropology, 74*(1), 65–81.

Anderson, T. M., White, S., Davis, B., Erhardt, R., Palmer, M., Swanson, A., Kosmala, M., & Packer, C. (2016). The spatial distribution of African savannah herbivores: Species associations and habitat occupancy in a landscape context. *Philosophical Transactions of the Royal Society B: Biological Sciences, 371*(1703), 20150314.

Aranda, M., & Sánchez-Cordero, V. (1996). Prey spectra of jaguar (*Panthera onca*) and puma (*Puma concolor*) in tropical forests of Mexico. *Studies on Neotropical Fauna and Environment, 31*(2), 65–67.

Arnold, S. J. (1983). Morphology, performance and fitness. *American Zoologist, 23*(2), 347–361.

Ávila-Nájera, D. M., Palomares, F., Chávez, C., Tigar, B., & Mendoza, G. D. (2018). Jaguar (*Panthera onca*) and puma (*Puma concolor*) diets in Quintana Roo, Mexico. *Animal Biodiversity and Conservation, 41*(2), 257–266.

Azevedo, F. C. C., & Verdade, L. M. (2012). Predator–prey interactions: jaguar predation on caiman in a floodplain forest. *Journal of Zoology, 286*(3), 200–207.

Bagatharia, S. B., Joshi, M. N., Pandya, R. V., Pandit, A. S., Patel, R. P., Desai, S. M., Sharma, A., Panchal, O., Jasmani, F. P., & Saxena, A. K. (2013). Complete mitogenome of asiatic lion resolves phylogenetic status within *Panthera*. *BMC Genomics, 14*(1), 1–9.

Bailey, I., Myatt, J. P., & Wilson, A. M. (2013). Group hunting within the Carnivora: physiological, cognitive and environmental influences on strategy and cooperation. *Behavioral Ecology and Sociobiology, 67*(1), 1–17.

Barone, R. (1986). *Anatomie comparée des Mammifères domestiques, tome 2, Arthrologie et myologie* (p. 1986). Vigot Freres.

Beisiegel, B. d. M., & Ades, C. (2002). The behavior of the bush dog (*Speothos venaticus* Lund, 1842) in the field: A review. *Revista de Etologia, 4*(1), 17–23.

Bichat, X. (1855). *Traité d'anatomie descriptive* Delahays.

Carbone, C., Teacher, A., & Rowcliffe, J. M. (2007). The costs of carnivory. *PLoS Biology, 5*(2), e22.

Caro, T. (1994a). *Cheetahs of the Serengeti Plains: Group living in an asocial species*. University of Chicago Press.

Caro, T. M. (1994b). Ungulate antipredator behaviour: Preliminary and comparative data from African bovids. *Behaviour, 128*(3–4), 189–228.

Caro, T. (2005). *Antipredator defenses in birds and mammals*. University of Chicago Press.

Carter, R. N., Romanow, C. A., Pellis, S. M., & Lingle, S. (2019). Play for prey: Do deer fawns play to develop species-typical antipredator tactics or to prepare for the unexpected? *Animal Behaviour, 156*, 31–40.

Chinery, M., & Lambiotte, A.-F. (1983). *Les prédateurs et leurs proies*. Delachaux et Niestlé.

Chizzola, M., Belton, L., Ganswindt, A., Greco, I., Hall, G., Swanepoel, L., & Dalerum, F. (2018). Landscape level effects of lion presence (*Panthera leo*) on two contrasting prey species. *Frontiers in Ecology and Evolution, 6*, 191.

Christiansen, P., & Adolfssen, J. S. (2005). Bite forces, canine strength and skull allometry in carnivores (Mammalia, Carnivora). *Journal of Zoology, 266*(2), 133–151.

Creel, S., & Creel, N. M. (1995). Communal hunting and pack size in African wild dogs, *Lycaon pictus*. *Animal Behaviour, 50*(5), 1325–1339.

Cuff, A. R., Sparkes, E. L., Randau, M., Pierce, S. E., Kitchener, A. C., Goswami, A., & Hutchinson, J. R. (2016a). The scaling of postcranial muscles in cats (Felidae) I: Forelimb, cervical, and thoracic muscles. *Journal of Anatomy, 229*(1), 128–141.

Cuff, A. R., Sparkes, E. L., Randau, M., Pierce, S. E., Kitchener, A. C., Goswami, A., & Hutchinson, J. R. (2016b). The scaling of postcranial muscles in cats (Felidae) II: Hindlimb and lumbosacral muscles. *Journal of Anatomy, 229*(1), 142–152.

Da Silveira, R., Ramalho, E. E., Thorbjarnarson, J. B., & Magnusson, W. E. (2010). Depredation by jaguars on caimans and importance of reptiles in the diet of jaguar. *Journal of Herpetology, 44*(3), 418–424.

Dalin, G., & Jeffcott, L. B. (1986a). Sacroiliac joint of the horse 1. Gross morphology. *Anatomia, Histologia, Embryologia, 15*(1), 80–94.

Dalin, G., & Jeffcott, L. B. (1986b). Sacroiliac joint of the horse 2. Morphometric features. *Anatomia, Histologia, Embryologia, 15*(2), 97–107.

Dar, S. A., & Khan, J. A. (2016). Food habits of dhole Cuon alpinus in tropical forests of southern India. *Current Science, 111*(10), 1701–1705.

Day, L. M., & Jayne, B. C. (2007). Interspecific scaling of the morphology and posture of the limbs during the locomotion of cats (Felidae). *Journal of Experimental Biology, 210*(4), 642–654.

de Oliveira Calleia, F., Rohe, F., & Gordo, M. (2009). Hunting strategy of the margay (*Leopardus wiedii*) to attract the wild pied tamarin (*Saguinus bicolor*). *Neotropical Primates, 16*(1), 32–34.

Derry, D. E. (1911). Note on accessory articular facets between the sacrum and ilium, and their significance. *Journal of Anatomy and Physiology, 45*(Pt 3), 202.

Eaton, R. L. (1969). Cooperative hunting by cheetahs and jackals and a theory of domestication of the dog. *Mammalia, 33*(1), 87–92.

Eisenberg, C. (2014). Jaguar (*Panthera onca*). In: *The carnivore way*. Springer, 217–240.

Erichsen, C., Berger, M., & Eksell, P. (2002). The scintigraphic anatomy of the equine sacroiliac joint. *Veterinary Radiology and Ultrasound, 43*(3), 287–292.

Everitt, B. S., & Dunn, G. (2001). *Applied multivariate data analysis*. Wiley Online Library.

Figueirido, B., Pérez-Claros, J. A., Hunt, R. M., Jr., & Palmqvist, P. (2011). Body mass estimation in amphicyonid carnivoran mammals: A multiple regression approach from the skull and skeleton. *Acta Palaeontologica Polonica, 56*(2), 225–246.

Flores, J. S. P. (2018). Predation of an adult female Morelet's crocodile (*Crocodylus moreletii*) by a jaguar (*Panthera onca*) in the Calakmul region, Mexico. *Herpetology Notes, 11*, 613–616.

Garland, T., & Janis, C. M. (1993). Does metatarsal/femur ratio predict maximal running speed in cursorial mammals? *Journal of Zoology, 229*(1), 133–151.

Gittleman, J. L. (1985). Carnivore body size: ecological and taxonomic correlates. *Oecologia, 67*(4), 540–554.

Gittleman, J. L. (1989). *Carnivore group living: Comparative trends. In: Carnivore behavior, ecology, and evolution* (pp. 183–207). Springer.

Gompper, M. E., & Vanak, A. T. (2006). Vulpes bengalensis. *Mammalian Species, 2006*, 1–5.

Gonyea, W. J. (1978). Functional implications of felid forelimb anatomy. *Cells, Tissues, Organs, 102*(2), 111–121.

Gonyea, W., & Ashworth, R. (1975). The form and function of retractile claws in the Felidae and other representative carnivorans. *Journal of Morphology, 145*(2), 229–238.

Hartstone-Rose, A., Perry, J. M., & Morrow, C. J. (2012). Bite force estimation and the fiber architecture of felid masticatory muscles. *The Anatomical Record: Advances in Integrative Anatomy and Evolutionary Biology, 295*(8), 1336–1351.

Hassanin, A., Veron, G., Ropiquet, A., Jansen van Vuuren, B., Lécu, A., Goodman, S. M., Haider, J., & Nguyen, T. T. (2021). Evolutionary history of Carnivora (Mammalia, Laurasiatheria) inferred from mitochondrial genomes. *PLoS One, 16*(2), e0240770.

Hayward, M. W., Lyngdoh, S., & Habib, B. (2014). Diet and prey preferences of dholes (*Cuon alpinus*): Dietary competition within Asia's apex predator guild. *Journal of Zoology, 294*(4), 255–266.

Hubel, T. Y., Myatt, J. P., Jordan, N. R., Dewhirst, O. P., McNutt, J. W., & Wilson, A. M. (2016). Additive opportunistic capture explains group hunting benefits in African wild dogs. *Nature Communications, 7*, 11033.

Hudson, P. E., Corr, S. A., & Wilson, A. M. (2012). High speed galloping in the cheetah (*Acinonyx jubatus*) and the racing greyhound (*Canis familiaris*): Spatio-temporal and kinetic characteristics. *Journal of Experimental Biology, 215*(14), 2425–2434.

Iwaniuk, A. N., Pellis, S. M., & Whishaw, I. Q. (1999). The relationship between forelimb morphology and behaviour in North American carnivores (Carnivora). *Canadian Journal of Zoology, 77*(7), 1064–1074.

Janis, C. M., & Figueirido, B. (2014). Forelimb anatomy and the discrimination of the predatory behavior of carnivorous mammals: the thylacine as a case study. *Journal of Morphology, 275*(12), 1321–1338.

Jesse, M. K., Kleck, C., Williams, A., Petersen, B., Glueck, D., Lind, K., & Patel, V. (2017). 3D Morphometric analysis of normal sacroiliac joints: A new classification of surface shape variation and the potential implications in pain syndromes. *Pain Physician, 20*.

Kardong, K. V. (2015). *Vertebrates: Comparative anatomy, function, evolution*. McGraw-Hill.

Karenina, K., & Giljov, A. (2018). Mother and offspring lateralized social behavior across mammalian species. *Progress in Brain Research, 238*, 115–141.

Kim, S., Cho, Y. S., Kim, H.-M., Chung, O., Kim, H., Jho, S., Seomun, H., Kim, J., Bang, W. Y., & Kim, C. (2016). Comparison of carnivore, omnivore, and herbivore mammalian genomes with a new leopard assembly. *Genome Biology, 17*(1), 1–12.

Kitchener, A. C., Van Valkenburgh, B., Yamaguchi, N., Macdonald, D. W., & Loveridge, A. J. (2010). Felid form and function. *Biology and Conservation of Wild Felids, 2010*, 83–106.

Kleiman, D. G. (1972). Social behavior of the maned wolf (*Chrysocyon brachyurus*) and bush dog (Speothos venaticus): a study in contrast. *Journal of Mammalogy, 53*(4), 791–806.

Kleiman, D. G., & Eisenberg, J. F. (1973). Comparisons of canid and felid social systems from an evolutionary perspective. *Animal Behaviour, 21*(4), 637–659.

Kröschel, M., Reineking, B., Werwie, F., Wildi, F., & Storch, I. (2017). Remote monitoring of vigilance behavior in large herbivores using acceleration data. *Animal Biotelemetry, 5*(1), 10.

Lewton, K. L., Brankovic, R., Byrd, W. A., Cruz, D., Morales, J., & Shin, S. (2020). The effects of phylogeny, body size, and locomotor behavior on the three-dimensional shape of the pelvis in extant carnivorans. *PeerJ, 8*, e8574.

Liebenberg, L. (2006). Persistence hunting by modern hunter-gatherers. *Current Anthropology, 47*(6), 1017–1026.
MacDonald, D. W. (1983). The ecology of carnivore social behaviour. *Nature, 301*(5899), 379.
MacDonald, D. (2009). *The encyclopedia of mammals*. Oxford University Prss.
MacNulty, D. R., Mech, L. D., & Smith, D. W. (2007). A proposed ethogram of large-carnivore predatory behavior, exemplified by the wolf. *Journal of Mammalogy, 88*(3), 595–605.
Makacha, S., & Schaller, G. B. (1969). Observations on lions in the lake Manyara National Park, Tanzania. *African Journal of Ecology, 7*(1), 99–103.
Marshall, C.D., & Pyenson, N.D. (2019). Feeding in aquatic mammals: An evolutionary and functional approach. In: *Feeding in vertebrates* (pp. 743–785). Springer.
Martín-Serra, A., Figueirido, B., & Palmqvist, P. (2014). A three-dimensional analysis of the morphological evolution and locomotor behaviour of the carnivoran hind limb. *BMC Evolutionary Biology, 14*(1), 129.
Martín-Serra, A., Figueirido, B., & Palmqvist, P. (2016). In the pursuit of the predatory behavior of Borophagines (Mammalia, Carnivora, Canidae): Inferences from forelimb morphology. *Journal of Mammalian Evolution, 23*(3), 237–249.
Meachen-Samuels, J. (2010). Comparative scaling of humeral cross-sections of felids and canids using radiographic images. *Journal of Mammalian Evolution, 17*(3), 193–209.
Meachen-Samuels, J., & Van Valkenburgh, B. (2009). Forelimb indicators of prey-size preference in the Felidae. *Journal of Morphology, 270*(6), 729–744.
Meachen-Samuels, J. A., & Van Valkenburgh, B. (2010). Radiographs reveal exceptional forelimb strength in the sabertooth cat, *Smilodon fatalis*. *PLoS one, 5*(7), e11412.
Meloro, C., Elton, S., Louys, J., Bishop, L. C., & Ditchfield, P. (2013). Cats in the forest: Predicting habitat adaptations from humerus morphometry in extant and fossil Felidae (Carnivora) Morphometric habitat-adaptation. *Paleobiology, 39*(3), 323–344.
Meloro, C., Clauss, M., & Raia, P. (2015). Ecomorphology of Carnivora challenges convergent evolution. *Organisms Diversity and Evolution, 15*(4), 711–720.
Michaud, M., Veron, G., & Fabre, A.-C. (2020). Phenotypic integration in feliform carnivores: Covariation patterns and disparity in hypercarnivores versus generalists. *Evolution, 2020*.
Montuelle, S. J., & Kane, E. A. (2019). Food capture in vertebrates: A complex integrative performance of the cranial and postcranial systems. In V. Bels & I. Wishaw (Eds.), *Feeding in vertebrates* (pp. 71–137). Springer.
Mukherjee, S., & Heithaus, M. R. (2013). Dangerous prey and daring predators: a review. *Biological Reviews, 88*(3), 550–563.
Muro, C., Escobedo, R., Spector, L., & Coppinger, R. P. (2011). Wolf-pack (*Canis lupus*) hunting strategies emerge from simple rules in computational simulations. *Behavioural Processes, 88*(3), 192–197.
Nishida, A. H., & Ochman, H. (2018). Rates of gut microbiome divergence in mammals. *Molecular Ecology, 27*(8), 1884–1897.
Nowak, R., Porter, R. H., Lévy, F., Orgeur, P., & Schaal, B. (2000). Role of mother-young interactions in the survival of offspring in domestic mammals. *Reviews of Reproduction, 5*(3), 153–163.
Olson, D. M., Dinerstein, E., Wikramanayake, E. D., Burgess, N. D., Powell, G. V., Underwood, E. C., D'amico, J. A., Itoua, I., Strand, H. E., & Morrison, J. C. (2001). Terrestrial Ecoregions of the World: A New Map of Life on EarthA new global map of terrestrial ecoregions provides an innovative tool for conserving biodiversity. *Bioscience, 51*(11), 933–938.
Pallandre, J.-P., Cornette, R., Placide, M.-A., Pelle, E., Lavenne, F., Abad, V., Ribaud, M., & Bels, V. L. (2020). Iliac auricular surface morphofunctional study in Felidae. *Zoology, 138*, 125714.
Pallandre, J.-P., Lavenne, F., Pellé, E., Breton, G., Ribaud, M., & Bels, V. (2021). Variation in the sacroiliac joint in Felidae. *PeerJ, 9*, e11116.
Palmeira, F. B., Crawshaw, P. G., Jr., Haddad, C. M., Ferraz, K. M. P., & Verdade, L. M. (2008). Cattle depredation by puma (Puma concolor) and jaguar (Panthera onca) in central-western Brazil. *Biological conservation, 141*(1), 118–125.

Pucora, E., Schiffmann, C., & Clauss, M. (2019). Resting postures in terrestrial mammalian herbivores. *Journal of Mammalogy, 100*(2), 552–563.
Pulliam, H. R., & Caraco, T. (1984). Living in groups: Is there an optimal group size. *Behavioural ecology: An evolutionary approach, 2*, 122–147.
Radloff, F. G. T., & Du Toit, J. T. (2004). Large predators and their prey in a southern African savanna: A predator's size determines its prey size range. *Journal of Animal Ecology, 73*(3), 410–423.
Randau, M., & Goswami, A. (2017). Morphological modularity in the vertebral column of Felidae (Mammalia, Carnivora). *BMC Evolutionary Biology, 17*(1), 133.
Randau, M., & Goswami, A. (2018). Shape covariation (or the lack thereof) between vertebrae and other skeletal traits in felids: the whole is not always greater than the sum of parts. *Evolutionary Biology, 45*(2), 196–210.
Randau, M., Goswami, A., Hutchinson, J. R., Cuff, A. R., & Pierce, S. E. (2016). Cryptic complexity in felid vertebral evolution: Shape differentiation and allometry of the axial skeleton. *Zoological Journal of the Linnean Society, 178*(1), 183–202.
Rodríguez-Alba, J. J., Linares-Matás, G., & Yravedra, J. (2019). First assessments of the taphonomic behaviour of jaguar (*Panthera onca*). *Quaternary International, 517*, 88–95.
Romer, A. S. (1950). *The vertebrate body*. WB Saunders Company.
Samuels, J. X., Meachen, J. A., & Sakai, S. A. (2013). Postcranial morphology and the locomotor habits of living and extinct carnivorans. *Journal of Morphology, 274*(2), 121–146
Schaller, G. B. (2009). *The Serengeti lion: a study of predator-prey relations*. University of Chicago Press.
Schaller, G. B., & Vasconcelos, J. M. C. (1978). Jaguar predation on capybara. *Z. Säugetierk, 43*, 296–301.
Scheel, D. (1993). Profitability, encounter rates, and prey choice of African lions. *Behavioral Ecology, 4*(1), 90–97.
Scheel, D., & Packer, C. (1991). Group hunting behaviour of lions: A search for cooperation. *Animal Behaviour, 41*(4), 697–709.
Schwab, J. A., Kriwet, J., Weber, G. W., & Pfaff, C. (2019). Carnivoran hunting style and phylogeny reflected in bony labyrinth morphometry. *Scientific Reports, 9*(1), 1–9.
Sheldon, J. W. (2013). *Wild dogs: The natural history of the nondomestic Canidae*. Elsevier.
Sicuro, F. L., & Oliveira, L. F. B. (2011). Skull morphology and functionality of extant Felidae (Mammalia: Carnivora): A phylogenetic and evolutionary perspective. *Zoological Journal of the Linnean Society, 161*(2), 414–462.
Skogland, T. (1991). What are the effects of predators on large ungulate populations? *Oikos, 1991*, 401–411.
Slater, G. J., & Friscia, A. R. (2019). Hierarchy in adaptive radiation: A case study using the Carnivora (Mammalia). *Evolution, 73*(3), 524–539.
Smith, J. M., & Savage, R. J. (1956). Some locomotory adaptations in mammals. *Zoological Journal of the Linnean Society, 42*(283), 603–622.
Sokal, R. R., & Rohlf, F. J. (1981). *Biometry. The principles and practice of statistics in biological research*, 1981.
Song, S., Liu, L., Edwards, S. V., & Wu, S. (2012). Resolving conflict in eutherian mammal phylogeny using phylogenomics and the multispecies coalescent model. *Proceedings of the National Academy of Sciences, 109*(37), 14942–14947.
Springer, M. S., Foley, N. M., Brady, P. L., Gatesy, J., & Murphy, W. J. (2019). Evolutionary models for the diversification of placental mammals across the KPg boundary. *Frontiers in Genetics, 10*, 1241.
Stander, P. E. (1992). Cooperative hunting in lions: The role of the individual. *Behavioral Ecology and Sociobiology, 29*(6), 445–454.
Stanton, L. A., Sullivan, M. S., & Fazio, J. M. (2015). A standardized ethogram for the Felidae: A tool for behavioral researchers. *Applied Animal Behaviour Science, 173*, 3–16.
Sunquist, M., & Sunquist, F. (2017). *Wild cats of the world*. University of Chicago Press.

Taylor, M. E. (1989). Locomotor adaptations by carnivores. In: *Carnivore behavior, ecology, and evolution* (382–409). Springer.

Taylor, C. R., Schmidt-Nielsen, K., Dmi'el, R., & Fedak, M. (1971). Effect of hyperthermia on heat balance during running in the African hunting dog. *American Journal of Physiology-Legacy Content, 220*(3), 823–827.

Taylor, C. R., Shkolnik, A., Dmi'el, R., Baharav, D., & Borut, A. (1974). Running in cheetahs, gazelles, and goats: Energy cost and limb configuration. *American Journal of Physiology-Legacy Content, 227*(4), 848–850.

Van Valkenburgh, B. (1996). Feeding behavior in free-ranging, large African carnivores. *Journal of Mammalogy, 77*(1), 240–254.

Van Valkenburgh, B. (2007). Déjà vu: The evolution of feeding morphologies in the Carnivora. *Integrative and Comparative Biology, 47*(1), 147–163.

Van Valkenburgh, B., Pang, B., Bird, D., Curtis, A., Yee, K., Wysocki, C., & Craven, B. A. (2014). Respiratory and olfactory turbinals in feliform and caniform carnivorans: The influence of snout length. *The Anatomical Record, 297*(11), 2065–2079.

Vander Linden, A., & Dumont, E. R. (2019). Intraspecific male combat behaviour predicts morphology of cervical vertebrae in ruminant mammals. *Proceedings of the Royal Society B, 286*(1915), 20192199.

Venter, J. A., Vermeulen, M. M., & Brooke, C. F. (2019). Feeding ecology of large browsing and grazing herbivores. In: *The ecology of browsing and grazing II* (127–153). Springer.

Viranta, S., Lommi, H., Holmala, K., & Laakkonen, J. (2016). Musculoskeletal anatomy of the Eurasian lynx, *Lynx lynx* (Carnivora: Felidae) forelimb: Adaptations to capture large prey? *Journal of Morphology, 277*(6), 753–765.

Whishaw, I. Q., & Karl, J. M. (2019). The evolution of the hand as a tool in feeding behavior: The multiple motor channel theory of hand use. In V. Bels & I. Wishaw (Eds.), *Feeding in vertebrates* (pp. 159–186). Springer.

Wible, J. R., Rougier, G. W., Novacek, M. J., & Asher, R. J. (2007). Cretaceous eutherians and Laurasian origin for placental mammals near the K/T boundary. *Nature, 447*(7147), 1003–1006.

Williams, S. H. (2019). Feeding in mammals: Comparative, experimental, and evolutionary insights on form and function. In V. L. Bels & I. Wishaw (Eds.), *Feeding in vertebrates* (pp. 695–742). Springer.

Wilson, A. M., Hubel, T. Y., Wilshin, S. D., Lowe, J. C., Lorenc, M., Dewhirst, O. P., Bartlam-Brooks, H. L., Diack, R., Bennitt, E., & Golabek, K. A. (2018). Biomechanics of predator–prey arms race in lion, zebra, cheetah and impala. *Nature, 554*(7691), 183.

Winslow, J. B. (1732). *Exposition anatomique de la structure du corps humain* chez. Guillaume Desprez et Jean Desessartz.

Wroe, S., McHenry, C., & Thomason, J. (2005). Bite club: Comparative bite force in big biting mammals and the prediction of predatory behaviour in fossil taxa. *Proceedings of the Royal Society of London B: Biological Sciences, 272*(1563), 619–625.

Wroe, S., Lowry, M. B., & Anton, M. (2008). How to build a mammalian super-predator. *Zoology, 111*(3), 196–203.

Chapter 6
Aquatic Feeding in Lissamphibia

Egon Heiss and Patrick Lemell

Abstract Modern amphibians are referred to as Lissamphibia and comprise the three extant groups: Anura (frogs and toads), Caudata (salamanders and newts) and Gymnophiona (caecilians). From a phylogenetic point of view, lissamphibians are considered the sister taxon of extant amniotes (sauropsids and mammals). Lissamphibians have a long evolutionary history, reaching back into the Late Paleozoic and most probably originated within a temnospondyl clade. One of the most conspicuous features of lissamphibians is their aquatic larval stage. Many lissamphibians have, however, secondarily reduced the free-living larval stage and are direct developers. Direct development is a secondary feature and might be seen as an adaptation to terrestrial life. Given that the aquatic larval stage is the ancestral condition for lissamphibians, adaptations to aquatic feeding might also be seen to be the ancestral condition, at least for lissamphibian larvae. After metamorphosis, some lissamphibians become terrestrial, others adopt a semiterrestrial/semiaquatic life-style, while others remain fully aquatic. Accordingly, although in many lissamphibian cases the secondary nature of aquatic adaptations might be obvious, a strict distinction between secondary and primary adaptations is less clear in others. Examples of secondarily aquatic lissamphibians are aquatic frogs and toads, as well as some desmognathid salamanders that have definitely reinvaded aquatic trophic habitats during their evolutionary history. In contrast, some salamandrid and ambystomatid salamanders continuously switch between aquatic and terrestrial lifestyles after metamorphosis and it is not obvious whether their (semi)aquatic lifestyle is retained from their larval condition (i.e. primary) or has evolved de novo. In fact, many adaptations to aquatic feeding in lissamphibians might represent a combination of both primary and secondary features, defying a strict dichotomy. In this chapter we summarize aquatic feeding strategies in all three extant groups of

E. Heiss (✉)
Institute of Zoology and Evolutionary Research, Friedrich-Schiller-University Jena, Jena, Germany
e-mail: heissegon@hotmail.com

P. Lemell
Department of Evolutionary Biology, Unit for Integrative Zoology, University of Vienna, Vienna, Austria

V. L. Bels, A. P. Russell (eds.), *Convergent Evolution*, Fascinating Life Sciences,
https://doi.org/10.1007/978-3-031-11441-0_6

lissamphibians and highlight homologous and convergent features where appropriate.

Keywords Ampullary organs · Bidirectional suction flow · Branchial filtering · Electroreception · Jaw prehension · Lingual prehension · Mechanoreception · Paedomorphosis · Suction feeding · Unidirectional suction flow

6.1 Caudata (Salamanders)

6.1.1 Introduction

The Caudata, also refered to as salamanders, includes approximately 600 species and is considered to be the sister group of anurans (Pyron & Wiens, 2011). Many species of salamander remain aquatic throughout their life and only occasionally leave the water. Others are semiaquatic, semiterrestrial, or entirely terrestrial. While semiaquatic and semiterrestrial species can shift permanently or seasonally between water and land, entirely terrestrial salamanders seek aquatic habitats only during their reproductive period, in those taxa that have retained the aquatic larval stage. A prominent trait that has evolved several times independently in salamanders is the retention of larval features into adulthood, referred to as "paedomorphosis" (Gould, 1977; Denoël & Joly, 2000; Wiens et al., 2005). Paedomorphic salamanders do not—or only partly—undergo metamorphosis and retain larval characters as adults, such as the larval skull, hyobranchial musculoskeletal system, external gills, gill slits and fin folds (Gould, 1977; Lauder & Shaffer, 1986; Reilly, 1986; Trueb, 1993; Ivanović et al., 2014). In contrast, individuals that undergo metamorphosis are referred to as "metamorphs". While for some salamander species paedomorphosis is an obligate ontogenetic process, others show facultative paedomorphosis, wherein paedomorphic and metamorphic adults can coexist in the same population (Denoël et al., 2005; Wiens et al., 2005; Heiss, 2017). Paedomorphic and metamorphic individuals show fundamentally different morphologies of their feeding apparatus, which in turn impacts some aspects of their feeding ecology. All salamanders are opportunistic predators and feed on small to large prey, including conspecifics in some cases.

6.1.2 Food Detection

Salamanders rely on vision, olfaction, electro- and mechanoreception to detect and localize prey (Przyrembel et al., 1995). Although terrestrial salamanders are restricted to vision and olfaction for prey detection, aquatic salamanders (larvae, paedomorphs and aquatic metamorphs) additionally use electro- and mechanoreception.

Regardless of their developmental stage (larvae, paedomorphs, metamorphs), salamanders have good eyesight and are capable of trichromic colour vision (Przyrembel et al., 1995). Interestingly, aquatic salamanders are more sensitive to light of longer wavelengths (Himstedt, 1973a, b) than terrestrial salamanders: an adaptation that enhances vision under water. Salamanders do not use vision only in daylight and experiments have shown that salamanders can localize prey stimuli at low illumination levels such as, for example, those measured during an average moonless night (Himstedt, 1982). Even if vision works well in water and in relative darkness, visually oriented salamanders only react to moving prey and immobile prey are largely ignored unless detected by olfactory cues (Himstedt, 1982; Roth, 1987; Schülert & Dicke, 2002). Olfaction is indeed an important sense for prey localization for all salamanders, but this sense interferes with vision, in that the visual system inhibits the olfactory guidance of feeding behaviour (Roth, 1987). Accordingly, it is not surprising that olfaction is especially prominently developed in blind cave salamanders, such as *Proteus*. In fact, as a result of convergent sensory adaptations to life in light-less habitats, *Proteus* responds readily to immobile, dead prey in complete darkness (Roth, 1987; Uiblein et al., 1992), just like other cave dwelling vertebrates (Uiblein et al., 1992; Parzefall, 1993).

The lateral line system is present in virtually all aquatic salamanders regardless of their developmental stage. The two main sensory organs of the lateral line system are neuromasts and ampullary organs. While neuromasts can be found in the head region, lateral body wall and tail, the ampullary organs are restricted to the head. Neuromasts are mainly mechanosensory organs, specialized for detecting minute water currents and pressure changes, as for example those generated by moving prey (Münz et al., 1984; Park et al., 2008). Ampullary organs by contrast are electrosensitive organs and can detect electrical fields generated by muscle contractions of any living animal (Himstedt, 1982). Additionally, neuromasts are capable of subtle electrosensation and ampullary organs can also detect mechanical stimuli to some degree (Münz et al., 1984). The lateral line system is an ancestral feature found in Cyclostomata (hagfish and lampreys), Chondrichthyes (sharks and rays), Actinopterygii (ray-finned fishes) and Dipnoi (lungfishes). Never the less, because of remarkable structural differences of ampullary organs between vertebrate groups, it has been suggested that electroreception evolved several times independently amongst vertebrates (Bullock et al., 1993).

6.1.3 Food Capture

Once detected, localized and approached, most aquatic salamanders use suction feeding to catch prey (Deban & Wake, 2000; Heiss et al., 2013b; Lauder & Reilly, 1994; Lauder & Shaffer, 1985, 1986; Reilly & Lauder, 1988; Shaffer & Lauder, 1985) regardless of their developmental stage (Fig. 6.1). However, two very different flow patterns are used by different developmental stages. Specifically, salamanders with larval morphology (i.e. larvae and paedomorphs) use unidirectional suction flow in which the engulfed water from a suction strike is released through

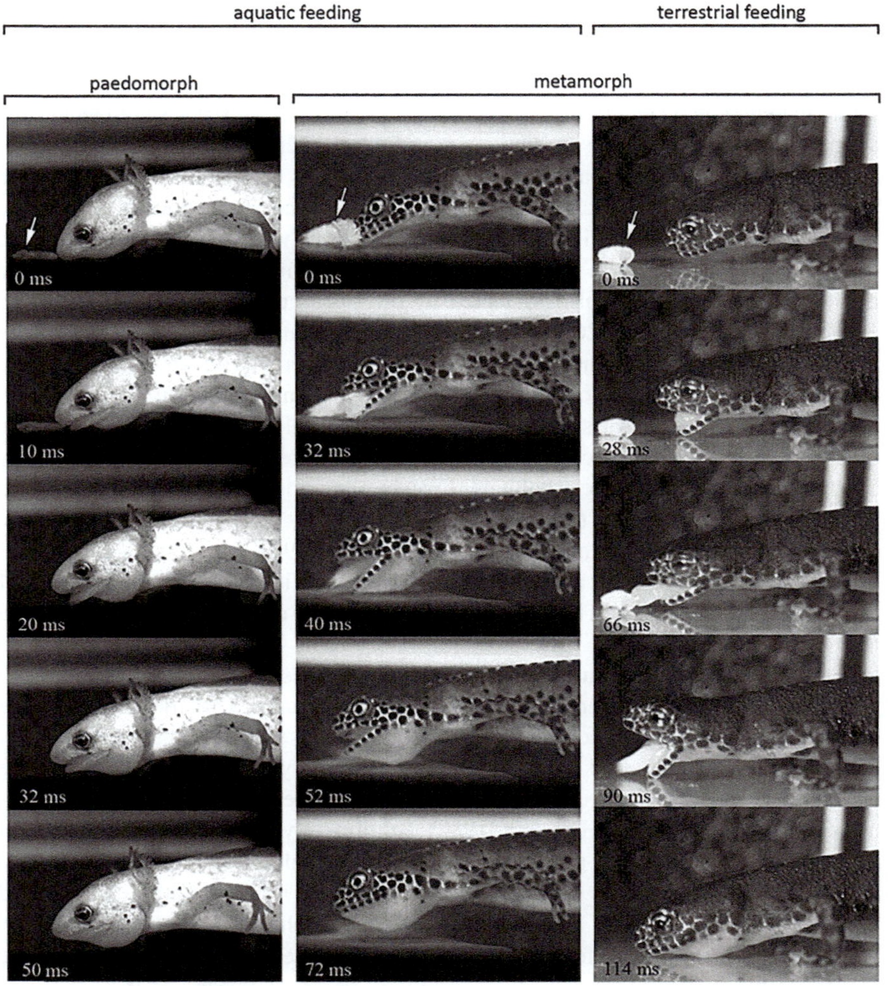

Fig. 6.1 Frames from high-speed-recordings showing three different prey capture mechanisms in the Alpine newt *Ichthyosaura alpestris*. Left and middle column: aquatic prey capture in a paedomorphic individual using unidirectional suction flow (left) and a metamorphic individual using bidirectional suction flow (middle). Right: terrestrial prey capture in a metamorphic individual (the same individual as shown in the middle column). Note the similar movement patterns in the two aquatic feeding events despite diverging flow regimes and cranial morphologies between paedomorphs and metamorphs (see also Fig. 6.2). The arrows in the first frames of each sequence indicate the position of the prey item (maggot for the metamorphs, red blood worm for the paedomorph). *ms* milliseconds. Modified from Heiss and Grell (2019) and Heiss et al. (2013a)

the open gill slits (Lauder & Shaffer, 1986). By contrast, salamanders with postmetamorphic morphology (i.e. "metamorphs") lack gill slits and the engulfed water is temporarily stored in the oropharyngeal cavity and released through the slightly opened mouth while prey is retained (Lauder & Shaffer, 1986; Miller &

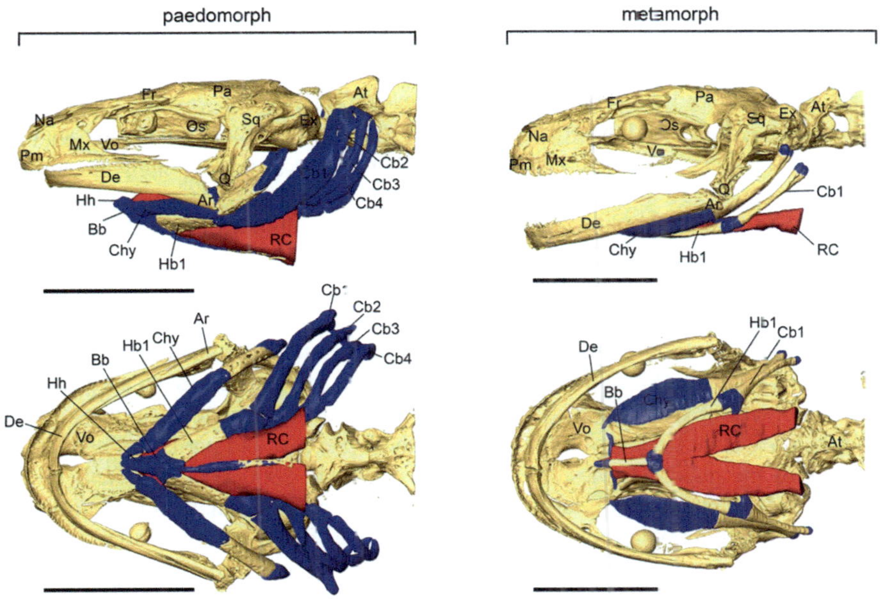

Fig. 6.2 Digital 3D reconstructions of cephalic structures, including the skull, first cervical vertebra and hyobranchial skeleton of a paedomorph and a metamorph individual of the Alpine newt, *Ichthyosaura alpestris*, based on µCT scans. Top row: lateral views, bottom row: ventral views. Note the conspicuous morphological differences that involve the entire head skeleton. The red ventral structures are the rectus cervicis muscles, one of the main muscles involved in suction generation in salamanders; blue structures represent the cartilaginous skeleton of the hyobranchium. *Ar* articular, *At* atlas (cervical vertebra), *Bb* basibranchial, *Cb1–4* ceratobranchial 1–4, *Chy* ceratohyal, *De* dentary, *Ex* exoccipital, *Fr* frontal, *Hb1–2* hyobranchial 1–2, *Hh* hypohyal (=radial), *Mx* maxillary, *Na* nasal, *Os* orbitosphenoid, *Pa* parietal, *Pm* premaxillary, *Q* quadrate, *RC* rectus cervicis muscles, *Sq* squamosal, *Vo* vomerine. Scale bars: 5 mm. Modified from Heiss and Grell (2019)

Larsen Jr, 1989; Heiss et al., 2013b). Interestingly, despite the different flow patterns and different morphology of the feeding apparatus (Fig. 6.2), motor patterns and kinematics of the suction strikes show only minor differences between salamanders with larval and postmetamorphic morphology (Lauder & Shaffer, 1988; Shaffer & Lauder, 1988; Heiss & Grell, 2019) (Fig. 6.1). Although behaviourally similar, unidirectional flow regimes might be regarded as being ancestral as they are found in virtually all lissamphibian larvae as well as non-tetrapod vertebrates to some degree; any adaptations to bidirectional flow regimes might be viewed as secondary adaptations.

Regardless of whether uni- or bidirectional flow patterns are used, the inward flow is generally induced by extremely fast oropharyngeal volume expansion that in turn relies on a combination of fast jaw opening and hyobranchial depression (Lauder & Shaffer, 1985; Reilly & Lauder, 1992). This generalized pattern can vary to some degree. For example, broad-headed salamanders, such as the giant salamanders (Cryptobranchidae), induce rapid accelerations of prey-laden water into the mouth cavity by fast jaw opening (Heiss et al., 2013b). Extensive volumetric expansion of the pharyngeal region, which follows maximum gape, is partly passive due to the momentum of the inflowing water. Thus, the main driver of suction flow in giant salamanders is not hyobranchial depression, as it is the case for most other salamanders, but fast jaw opening. Similar (and convergently evolved) suction strikes based on rapid separation of broad jaw plates, are also found in aquatic frogs (Dean, 2003) and might also have been used in some early tetrapod clades that, similarly to giant salamanders, had characteristically broad heads (Heiss et al., 2013b; Smithson et al., 2017).

A secondary anatomical adaptation that tunes suction efficacy in salamanders is the occlusion of the corners of the mouth that helps to concentrate the flow at the centre of the mouth, where suction feeders generally position their prey (Skorczewski et al., 2012). In many suction-feeding salamanders, such an occlusion of the corners of the mouth is achieved with the aid of skin folds, referred to as "labial lobes" that stretch from the upper to the lower jaw and partly seal the lateral aspects of the gape (Matthes, 1934; Deban & Wake, 2000; Van Wassenbergh & Heiss, 2016). Interestingly, labial lobes of metamorphosed salamanders are structurally similar but not homologous to the labial lobes found in larval and paedomorphic salamanders. Accordingly, they must have evolved *de novo* in the postmetamorphic stages (Matthes, 1934; O'Reilly et al., 2002). In postmetamorphic newts that seasonally shift between aquatic and terrestrial habitats, these labial lobes grow as animals invade the aquatic habitat and are resorbed as the newts leave the water again (Matthes, 1934; Van Wassenbergh & Heiss, 2016). Experiments and computational fluid dynamic simulations have shown that labial lobes can increase water flow in front of the mouth of suction-feeding newts by 30% (Van Wassenbergh & Heiss, 2016), which clearly shows an increase in suction-feeding efficacy by sealing the corners of the mouth. Lateral sealing of the corners of the mouth is not exclusive to salamanders but is also common among suction-feeding fishes. Specifically, many bony fishes are able to temporarily close the corners of their mouth by swinging the maxillary bones anteriorly, which are attached to lateral labial membranes that seal the gape (Van Dobben, 1937; Muller & Osse, 1984; Gosline, 1987). This anterior swinging of the maxillary bones is mechanically coupled to jaw opening. Although labial lobes (salamanders) and anteriorly-swinging maxillary bones (fishes) serve to seal the corners of the mouth, they have evolved independently to increase suction feeding efficacy.

Suction feeding is certainly the most common prey capture strategy used by aquatic salamanders, but some are not capable of suction feeding and instead use different strategies. For example, no postmetamorphic plethodontid salamander has yet been reported to employ suction feeding, although some postmetamorphic plethodontids do feed in aquatic conditions. Deban and Marks (2002) showed that some plethodontids use a forward lunge and grasp prey with the jaws (i.e. jaw prehension) without generating any meaningful suction flow. In other cases postmetamorphic plethodontids may even use lingual prehension for aquatic prey capture. Both techniques, jaw prehension and lingual prehension, are mechanisms otherwise almost exclusively found in terrestrially feeding salamanders, but plethodontids have also found a way to employ and adapt these mechanisms for underwater prey capture.

6.1.4 Intraoral Transport, Processing and Swallowing

Once in the oral cavity, salamanders use actively induced water flows or cyclic tongue movements to transport prey items posteriorly towards the oesophagus (Reilly & Lauder, 1991; Gillis & Lauder, 1994; Reilly, 1996). The movement patterns used to generate intraoral transport flow are similar to suction feeding strikes, although slower in most cases. Along with water flow or cyclic tongue movements, a further mechanism of intraoral food transport in salamanders involves the retraction of the eye bulbs into the oral cavity through two openings in the skeletal part of the roof of the mouth (Deban & Wake, 2000). The retracting bulbs push the food posteriorly towards the oesophagus. Additionally, only anurans (see below) make (and potentially some temnospondylians made) use of their retracting eyeballs for intraoral food transport (Regal, 1966; Regal & Gans, 1976; Duellman & Trueb, 1994; Levine et al., 2004; Witzmann & Werneburg, 2017). Accordingly, intraoral transport and swallowing using eye bulb retraction seems to have evolved several times independently. In addition to pure transport, salamanders are also able to process (i.e. mechanically disintegrate) food before swallowing. A simple form of intraoral processing is the use of a series of forceful bites to crush or kill a prey item (Fortuny et al., 2015). The plethodontid salamander *Desmognathus*, for example, employs cyclic head bobbing movements, with the prey held between the jaws to deliver a series of powerful bites to the prey. Specifically, head depression stretches the massive, pulley-like atlantomandibular ligaments (running between the cervical vertebra and the lower jaw) to transmit force generated by head flexion to assist the jaw adductor muscles with gape closure and the amplification of bite force (Dalrymple et al., 1985; Schwenk & Wake, 1993; Deban & Richardson, 2017).

A different chewing behaviour is used by the paedomorphic salamander *Siren intermedia*. As shown by using biplanar high-speed X-ray analysis, *Siren* employs propalinal (back and forth) movements of the lower jaw to rasp prey items against the palatal dentition (Schwarz et al., 2020a). The palatal denticle field represents the functional upper jaw. Interestingly, a similar processing behaviour has also been observed in paedomorphic salamandrids (Schwarz et al., 2020b). Metamorphosed semiaquatic salamandrids, by contrast, use cyclic head bobbing movements in combination with cyclic looping motions of the tongue to move the prey and rasp it across and against the palatal dentition (Heiss et al., 2019; Schwarz et al., 2020c). Although prey capture mechanisms differ fundamentally between terrestrial and aquatic feeding events in salamandrids (Fig. 6.1), the processing mechanisms do not (Schwarz et al., 2020c). Similar processing mechanisms to those encountered in metamorph salamandrids (the tongue rasping food items against rough palatal structures) are also found in such phylogenetically distant groups as the echidna *Tachyglossus aculeatus* (Doran & Baggett, 1972; Griffiths, 1978; Schwenk &

Rubega, 2005), sirenians (Werth, 2000) and possibly some extinct early amphibian and amniote clades (Matsumoto & Evans, 2017). Accordingly, tongue-to-palate rasping might reflect convergent processing solutions when chewing using the dentition of the lower jaw is not possible due to morphological constraints (e.g. unsuitable or reduced jaw dentition or limited mobility of the jaw joint).

Salamanders, like frogs and possibly caecilians, swallow by using a combination of tongue movements, pharyngeal constriction and eye retraction (Regal, 1966; Duellman & Trueb, 1994; Deban & Wake, 2000; Levine et al., 2004). Specifically, cyclic tongue movements, in concert with contractions of vertically-oriented throat muscles (e.g. intermandibularis, interhyoideus, gularis, quadratopectoralis) can be used to push food objects into the oesophagus (Deban & Wake, 2000). Eyebulb retraction, as described above, helps during the swallowing process by pushing the food backwards.

6.2 Anura (Frogs and Toads)

6.2.1 Introduction

The order Anura includes more than 7500 species, which makes up 88% of all extant lissamphibians (https://amphibiansoftheworld.amnh.org/index.php/Amphibia/Anura visited on November 11, 2022). Anurans inhabit almost every type of environment except for very cold regions, those that are extremely dry, and oceanic islands (Pough et al., 2004). Anuran lifestyles include purely aquatic, semiaquatic, terrestrial, arboreal, and fossorial. Tadpoles feed primarily on decaying vegetation, algae and plankton in ponds and streams (Grzimek, 2003). By contrast, adult anurans are carnivorous, typically feeding on small invertebrates such as insects, worms and snails. Large frogs with wide gapes tend to eat larger prey, which may include other frogs, lizards, small snakes, birds, or small mammals. Anurans are unique among lissamphibians and all other vertebrates in having a broad head, large mouth, large eyes, short body, and usually no tail. The hind limbs are typically elongated and modified for jumping.

6.2.2 Food Detection

As on land, the two major foraging methods in water are sit-and-wait (ambush) predation and active foraging. Terrestrial frogs depend heavily on vision to locate prey, but larval stages (i.e. tadpoles) and the truly aquatic pipids use their lateral line system as their main mechanosensory receptors to locate prey in water–just as do many other aquatic amphibians (see the sections on salamanders and gymnophiones, this chapter). Frogs, in general, have an architecturally complex nasal cavity consisting of three chambers and a large vomeronasal organ (Vitt & Caldwell,

2009). Some species can locate prey on the basis of olfactory cues alone (Shinn & Dole, 1978; Dole et al., 1981). Except for the pipids, *Occidozyga laevis*, and *Lepidobatrachus laevis*, the lateral line system is restricted to tadpoles and is completely lost during metamorphosis. The aforementioned species retain their lateral line system as adults (Fritzsch et al., 1987). The lateral line system exhibits three types of receptors: mechanosensory neuromasts, pit organs (Northcutt, 1992) and electroreceptive ampullary organs. Superficial neuromasts, embedded in the epidermis, are found in both amphibians and fish, whereas canal neuromasts, protected in grooves, are exclusive to fish (Northcutt, 1992; Quinzio & Fabrezi, 2014). All anuran larvae have two orbital (supra- and infraorbital) and three mandibular (oral, angular, and preopercular) lateral line nerves, as well as trunk and tail nerves (Quinzio & Fabrezi, 2014).

6.2.3 *Food Capture*

6.2.3.1 Aquatic Feeding in Tadpoles

The literature commonly states that tadpoles are herbivorous, because many species scrape algae and other plant materials. Food items usually are small and are sucked into the mouth using water flow created by rhythmic pumping movements of the hyobranchial apparatus. In the buccal cavity, branchial filtering epithelia and mucus secretions trap food items and water is released through the gill slits or a spiracle. This feeding mode allows exploitation of a wide variety of food resources: from small, freely floating organisms to periphyton, carrion, or detritus (Altig et al., 2007). There are several studies documenting the relations between buccal, skeletal, and muscular characteristics and feeding habits (for review see: Vera Candioti, 2006). Macrophagy and carnivory are rare and appear mostly in combination with suction feeding (e.g. *Lepidobatrachus*, Fabrezi & Lobo, 2009; *Hymenochirus*, Deban & Olson, 2002; *Occidozyga*, Haas et al., 2014; *Hyla nana*, Vera Candioti et al., 2004). Tadpoles of other species simply bite off flesh with their large beaks. In biting tadpoles the jaws and hyoid musculature are well developed (e.g. *Ceratophrys*, Vera Candioti, 2005; *Hoplobatrachus*, Grosjean et al., 2004; *Spea*, Bragg, 1964). Through her morphometric studies, Vera Candioti (2006) showed that the variable development of the branchial apparatus of tadpoles allowed categorization into at least four distinct clusters of feeding ecomorphologies: macrophagous, megalophagous, generalized, and microphagous (Fig. 6.3). Species placed in the macrophagous group (*Ceratophrys*, *Hyla*) are characterized mainly by the reduced size of the branchial basket and large ceratohyals and hypobranchial plates. In the megalophagous *Lepidobatrachus* the hyobranchial skeleton has large lateral extensions for widening the entire buccal cavity during suction. Generalized tadpoles possess a large branchial basket, and microphagous tadpoles have an even more expanded filtering area that is increased by transverse connections between the ceratobranchials. *Occidozyga baluensis,* placed within the microphagous category

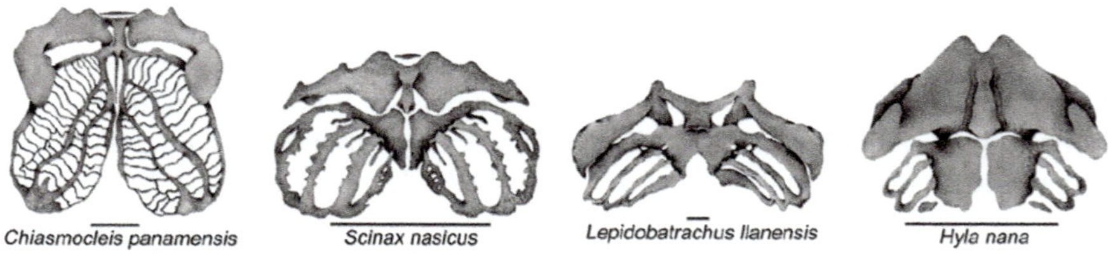

Fig. 6.3 From left to right, the hyobranchial skeleton of microphagous, generalized, macrophagous, and megalophagous tadpoles. Scale bar is 1 mm. Modified from Vera Candioti (2006)

(Haas et al., 2014), has an additional characteristic feature in the form of an anteriorly-directed horseshoe-shaped mouth lacking keratodonts, a morphology also described for *Lepidobatrachus* (Haas, 2003). Some obligately carnivorous tadpoles, however, such as *Hymenochirus boettgeri* (Deban & Olson, 2002), *Ceratophrys cranwelli*, *Hyla nana* (Vera Candioti, 2006), *Lepidobatrachus* (Fabrezi & Lobo, 2009), and *Occidozyga baluensis* (Haas et al., 2014) are known to use suction feeding mechanisms. The suction feeders with the most highly derived morphologies are the tadpoles of *Hymenochirus*, and presumably *Pseudhymenochirus*. During prey capture, their tubular mouth is protruded, this coinciding with an explosive expansion of the buccal volume. Mouth protrusion is completed within 2 ms and the prey is engulfed within 4 ms; prey items are drawn into the mouth at 0.6 m/s (in comparison, adult pipids engulf prey within about 30 ms – measurements have been roughly estimated from Figures in Carreno & Nishikawa, 2010). The feeding mechanism is convergent with that of teleosts (Sokol, 1969; Deban & Olson, 2002), which also employ suction feeding using a combination of rapid mouth protrusion, hyobranchial depression and cranial elevation. *Hymenochirus* and teleosts also share a hydrodynamically advantageous round mouth opening (Skorczewski et al., 2012).

6.2.3.2 Aquatic Feeding Modes of Adult Frogs

An aquatic lifestyle and feeding behaviour have evolved independently in several anuran families. The majority of aquatic-feeding anurans use "terrestrial" methods, such as jaw, tongue or forelimb prehension, for aquatic prey capture (Dean, 2003). Nevertheless, a few species, mainly from the family Pipidae, have developed a suction feeding mechanism. This feeding method is dependent on the physical constraints of the aquatic medium and is the most common means of aquatic feeding in vertebrates for capturing prey and further manipulating or transporting it to the oesophagus (Wainwright et al., 2015).

Overall, pipid anurans are the most specialized and most thoroughly investigated with regard to aquatic prey capture (Sokol, 1969; O'Reilly et al., 2002; Dean, 2003; Carreno & Nishikawa, 2010; Cundall et al., 2017; Fernandez et al., 2017). Their adaptations for aquatic feeding are the retention of the lateral line organs on the head

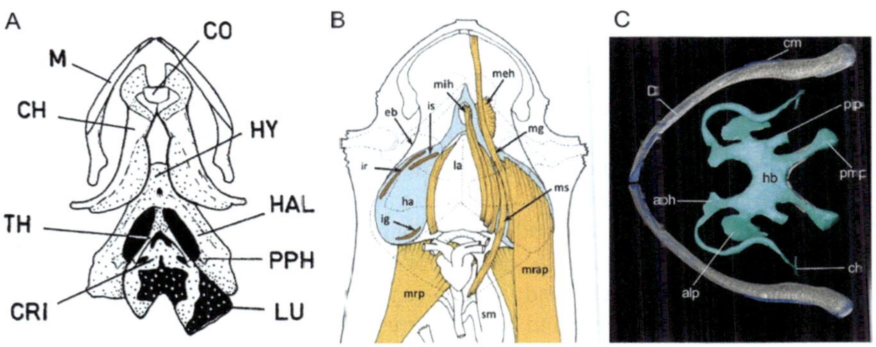

Fig. 6.4 Comparison of the hyolaryngeal apparatus of two pipid frogs (**a**, **b**) with that of a generalized aquatic anuran (**c**). A) *Hymenochirus boettgeri* from Sokol (1969). *CH* ceratohyal, *CO* copula, *CRI* cricoid, *HAL* hyoid ala, *HY* corpus hyoidei, *LU* lung, *M* mandible, *PPH* passage of the anterior head of the petrohyoideus muscle, *TH* thyrohyal. (**b**) from Cundall et al. (2017). Ventral view of the resting positions of the hyoid alae, major viscera (heart, stomach) and limbs and girdles of *Pipa pipa*. *eb* edge of buccopharyngeal cavity, *ha* hyoid alae, *ig* insertion of geniohyoid muscle, *ir* insertion region of rectus abdominis, *is* insertion region of sternohyoid muscle, *la* larynx, *meh* external hyoglossus muscle, *mg* geniohyoid muscle, *mih* internal hyoglossus muscle, *mrap* rectus abdominis, *mrp* pharyngeal retractor muscle, *ms* sternohyoid muscle, *sm* stomach. Hyoid apparatus in blue, muscles in yellow. (**c**) ventral view of the hyoid apparatus with osseous (grey) and cartilaginous (blue) structures of *Calyptocephalella gayi*. *alp* anteriolateral process of the hyoid, *aph* anterior process of hyoid, *ch* ceratohyale, *cm* cartilago meckeli, *D* Dentale, *hb* hyoid body, *plp* posteriolateral process of the hyoid, *pmp* posteriomedial process of hyoid

and body sides, their suction feeding mode and the complete absence of a tongue. Sokol (1969) was the first to describe the morphology and function of the African pipid *Hymenochirus boettgeri*. Its skull is flattened and elongate and it has a large hyoid skeleton with ossified ceratohyals that are separated from the rest of the hyolaryngeal structures. The hyolarynx is not notably different from that of *Xenopus* and *Pipa*, having a corpus hyoidei with large lateral hyoid alae (Fig. 6.4). The complete hyoid skeleton is retracted and lowered by large portions of the sternohyoid musculature. All these features point to a good capacity for suction feeding, but Sokol (1969) described the feeding behaviour as being best adapted for the capture of small, weakly-swimming prey. When feeding, the animals lunge forward, empowered by action of the hind legs (with simultaneous retraction of the fore legs), prior to mouth opening (Dean, 2003). Initial movement of prey items into the frog's mouth is induced by hyoid depression, but jaw opening alone also seems to be sufficient for drawing food into the mouth.

All other non-pipid suction feeding frogs such as *Telmatobius rubigo* (Barrionuevo, 2016), possibly *Lankanectes corrugatus* (Pethiyagoda et al., 2014), *Lepidobatrachus laevis* and *L. llanensis* (Fabrezi & Lobo, 2009) and *Calyptocephalella gayi* (Wiesinger, 2017; Kunisch et al., 2021) have, at least, a broadened hyoid apparatus with heavier musculature when compared to more generalized frogs. On the other hand, enlarged hyoids are also encountered in the Discoglossidae and Bombinatoridae (see Trewavas, 1933), but their aquatic feeding habits have not yet been studied in detail. Nevertheless, the functional aspects of the hyoid are difficult to assess, since, as well as its role in suction feeding it plays one or more roles in various other functions, such as breathing, vocalization, and closing the nares (Regal & Gans, 1976; Emerson, 1977; Trueb, 1993).

Other important features in suction feeding vertebrates for stabilizing the inflow of water are a small gape with labial folds, a flat and smooth palate, and a small tongue. Pipids have developed the most derived solution to this end and lack a functional tongue. The other aquatic anurans that have been investigated show, at the least, reduction in tongue size and weaker tongue musculature compared to related non-aquatic species. *Telmatobius rubigo,* for example, has a small, flat tongue with thin hyoglossal muscles and only superficial genioglossal musculature. *T. oxycephalus,* on the other hand, a more terrestrial species, has a larger and thicker tongue with bulky hyoglossal muscles and genioglossal muscles that enter the tongue. Tongue reduction with poorly developed intrinsic musculature is similar to that found in other secondarily aquatic vertebrates, such as salamanders (Stinson & Deban, 2017) and turtles (Winokur, 1988; Beisser et al., 1995; Lemell et al., 2002). Regarding a flat, smooth palate, pipid species all have an extremely streamlined skull shape (Carreno & Nishikawa, 2010; Fernandez et al., 2017) and the palate (even if not yet described) also looks flat. Kunisch et al. (2021) described the palate of the aquatic species *Calyptocephallela gayi* as being flat and streamlined, in contrast to the more vaulted form of the terrestrial *Leptodactylus pentadactylus*. The same condition is also found in turtles (e.g. *Chelus fimbriatus*, Lemell et al., 2002). Natchev et al. (2009) compared the feeding kinematics and anatomy of two closely related turtle species with different habitat preferences, the predominantly aquatic *Cuora amboinensis* and the predominantly terrestrial *C. flavomarginata*. They found a flat palate in the predominantly aquatic turtle compared to the vaulted palate in the predominantly terrestrial turtle. Additionally, the tongue of *C. amboinensis* is smaller, and lingual papillae in aquatic turtles, if present at all, are usually broader and shorter than in terrestrial species (Beisser & Weisgram, 2001). Finally, labial lobes as, for example, those found in aquatic salamanders, are absent from anurans. Nevertheless, frogs are able to bend their mandibles, a phenomenon that was hypothesized to be involved in the occlusion of the lateral regions of the gape (Deban et al., 2001). In anterior view, mandibular bending occludes the lateral areas of the open mouth and thus creates a more circular gape. Using pressure recordings, the most recent work (Carreno & Nishikawa, 2010) shows, that all four investigated pipid species exhibit a drop in buccopharyngeal pressure when feeding. These pressure drops were not as high as those in other high-performance suction feeding vertebrates but were similar to those in aquatic salamanders. They are related to the feeding habits of pipids when hunting for generally more immobile prey. *Pseudhymenochirus merlini* was found to feed similarly to *Hymenochirus*, whereas the feeding behaviour of *Pipa pipa* and *Xenopus laevis* differs significantly. *P. pipa* initiates the gape cycle at a greater distance from prey than do all the others; nevertheless, *Xenopus* exhibits about twice the speed of the pressure drop during suction feeding when compared to *Pipa* (60 kPa/s vs 32 kPa/s; for comparison: *Hymenochirus*: 16 kPa/s, *Pseudhymenochirus*: 19 kPa/s).

Suction feeding has been shown to be highly efficient for the capture of elusive aquatic prey, but a large volume of water must be engulfed and handled along with

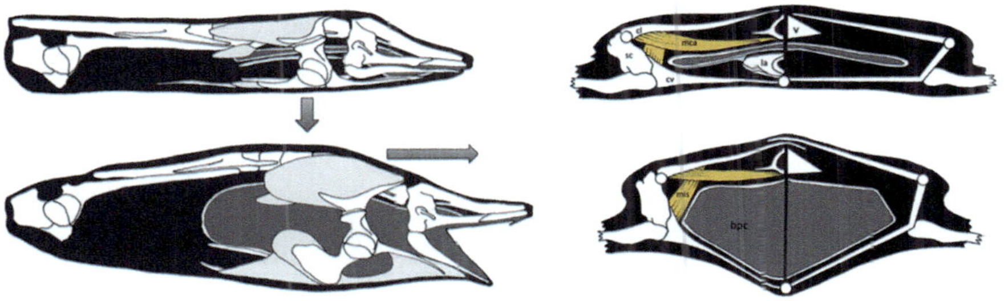

Fig. 6.5 Lateral view and transverse section showing the skeletal relationships between resting and peak expansion during food capture by *Pipa pipa*. Contraction of the cleithral adductor (mca) pulls the scapular-cleithral joint medially toward the vertebrae (v), causing the suprascapulae to overlap (or increase their extent of overlap) in the midline, while the clavicles (cv) and coracoids are pushed ventrally, flexing the epicoracoids around their midline collagenous joint. The larynx (la), hyoid alae, and buccopharyngeal lining are pulled caudally during buccopharyngeal expansion. *bpc* buccopharyngeal cavity, *cl* cleithrum, *mis* interscapular muscle, *sc* scapula. Modified from Cundall et al. (2017)

the prey. Unlike salamanders with larval morphologies, or fish, where engulfed water passes through the gill slits in a unidirectional flow system, metamorphosed salamanders, frogs and other secondarily aquatic vertebrates must reverse water flow during the feeding cycle, with the engulfed water being reversed and released through the mouth opening (Lauder & Shaffer, 1986). Pipids (Fig. 6.5) have solved the problem of temporarily storing large volumes of water by having an extremely expandable "post-glottal" pharynx, described by Sokol (1969) for *H. boettgeri*. This indicates a potential water storage system that enables the prolongation of the incoming water flow. A convergent solution was described for the aquatic specialized turtle *Chelus fimbriatus* (Lemell et al., 2002), which has a highly distensible oesophagus in which engulfed water is temporarily stored.

Fernandez et al. (2017) studied the suction feeding mechanism of *Pipa pipa* when catching live fish and found that it ingests a large volume of water by depressing the ventral elements of the pectoral girdle (clavicles, coracoids; Fig. 6.5). As for the expandable "post-glottal" pharynx of *Hymenochirus*, *Pipa* is able to increase the volume of the buccopharynx and a large part of the anterior trunk region. The viscera (hyoid and larynx, heart, lungs, liver, oesophagus, and stomach) are arranged so as to allow them to shift caudally by about one third of the length of the trunk. During prey capture *P. pipa* generates initial suction by mouth opening, but the oral cavity is quite small and prey does not exhibit much movement towards and into the mouth until the frog begins to expand the anterior region of its trunk. Consistent with suction feeding in other vertebrate groups, the success rate of prey capture is dependent on size of the prey and the distance from which suction begins. Fernandez et al. (2017) also found that during one fourth of capture events the left and right sides of the lower jaw are asynchronously depressed, depending on the position of the prey. This has previously been demonstrated only for an aquatic salamander (*Cryptobranchus alleganiensis*; Cundall et al., 1987; Lorenz-Elwood & Cundall,

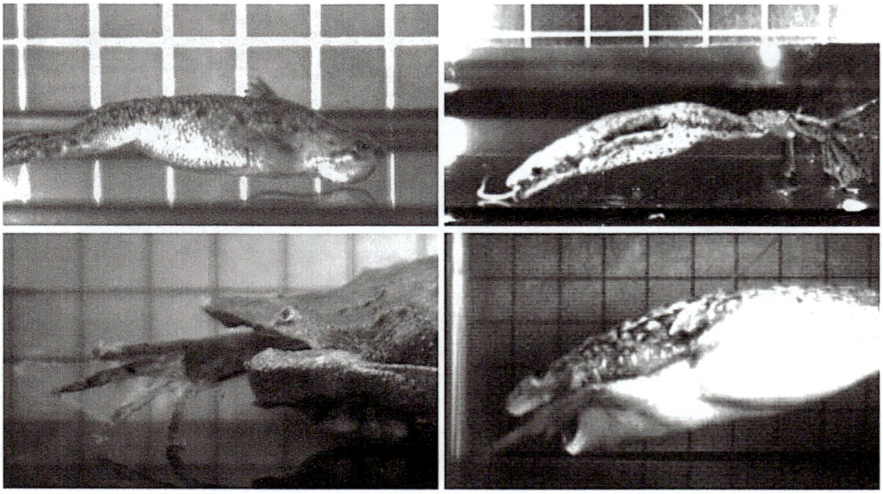

Fig. 6.6 Still frames during prey capture by *Hymenochirus boettgeri*, *Pseudhymenochirus merlini* (top from left to right), *Pipa pipa* and *Xenopus laevis* (bottom from left to right). Note the posterior adduction of the forelimbs in *H. boettgeri* and *P. merlini*, whereas the other species use forearm movements for prey capture. We are grateful to Carrie Carreño-Zingaro for providing these pictures

1994), and Carreno & Nishikawa (2010) noted that asynchronous mandibular depression is also used in *Pipa* during prey manipulation phases. Furthermore, asynchronous mandible movements are clearly evident in the feeding sequences of *Xenopus* presented by Anzeraey et al. (2017). An additional feature shared by *Pipa* and *Xenopus* is the use of the forelimbs during feeding (Carreno & Nishikawa, 2010; Fernandez et al., 2017). *Pipa* uses its forelegs mostly in combination with suction to corral elusive prey, the fingers usually touching the prey and restricting its possibility of escaping. *Xenopus* preferentially uses its fingers to scoop prey into the mouth, in contrast to the other two studied pipids, *Hymenochirus* and *Pseudhymenochirus*, which do not use their forelegs during feeding (Fig. 6.6). Gray et al. (1997) identified five behaviour patterns related to the movement of the forelimbs for capturing and transporting prey items in water and on land: (1) scooping: the back of the hand pushes prey items into the mouth, (2) wiping: the palm of the forelimb moves prey items that protrude laterally from the mouth towards the midline, (3) prey stretching: the hand holds one end of the prey item in a stationary position while the jaws pull the other end upwards, (4) grasping: the fingers wrap around prey items while the forelimbs move it towards the mouth and (5) hand grasping (with wrist rotating): the palms facing the mouth. Anzeraey et al. (2017) investigated *Xenopus laevis* with regard to its grasping and manipulation capabilities, but scooping was never observed, in contrast to what was stated by Avila and Frye (1977, 1978). It must be mentioned, however, that Avila and Frye (1977, 1978) did not distinguish between backhand and forehand usage of the forelegs as Gray et al. (1997) did. Two other feeding patterns were added by Anzeraey et al. (2017): (6) pushing:

movement of hand toward the mouth without touching the prey and (7) shredding: use of the hind limbs to tear large prey items apart. Furthermore, they identified which fingers are involved during grasping. Wiping was the most common behaviour, being employed in 41.3% of prey capture events, followed by pushing (34.8%) and suction (18.3%). When using the fingers, the longest digits (II–III) were the most frequently involved (37% of prey capture bouts). *Xenopus laevis* displays many different prey manipulation strategies, characterizing its overall complex prey capture behaviour. The morphology of the forelimb of *Xenopus* is different from that of other taxa, but they still exhibit prehension skills similar to those of primates that lack an opposable thumb (Anzeraey et al., 2017).

Apart from studies of the pipids, there are just two investigations about aquatic feeding frogs within the Neobatrachia: *Lankanectes corrugatus* (Pethiyagoda et al., 2014) and *Telmatobius rubigo* (Barrionuevo, 2016). Data on a third species, *Calyptocephalella gayi*, are not yet published. *Lankanectes corrugatus* does not practice suction feeding; after detecting prey via tactile stimuli the frogs dived or sank towards it with the mouth open, performing a scooping behaviour. A depression of the hyoid is discernible in the available frames, which probably results passively from the effects of the inflowing volume of water. *Telmatobius rubigo*, on the other hand, never uses its forelimbs for prey capture but employs suction feeding in contrast to the more terrestrial *Telmatobius oxycephalus*. A further variation shown by the aquatic species is a wide gape, opening up to 90°, but no obvious mandibular bending. Barrionuevo (2016) compared the hyoid apparatus and tongue morphology of these species and revealed a greater level of ossification of the hyoid plate in the aquatic *T. rubigo*, and broader insertion areas of the sternohyoid muscle, which serves as a hyoid retractor. The tongue of the predominantly terrestrial species was broader and thicker and showed some surface structures. The intrinsic tongue musculature is also more massive in the predominantly terrestrial species. *Calyptocephalella gayi* is a purely aquatic frog, but is known to also feed on land (Nishikawa, 1999). During prey capture, *C. gayi* opens its mouth widely and some initial suction is evident as a result of the mouth opening; and there is always a scooping behaviour to corral the prey, without touching it (Wiesinger, 2017). The hyoid apparatus also shows similar features to those found in *T. rubigo*.

6.2.4 Intraoral Transport, Processing and Swallowing

Later anuran feeding stages, such as intraoral transport, processing and swallowing, are scarcely reported in the literature. In most cases the tongue serves as the main manipulative organ for shifting food within the buccal cavity. Selective movements of the tongue pad, perhaps against the palate (sometimes furnished with vomerine teeth), shift food objects (Regal & Gans, 1976). Furthermore, the large retracting eye bulbs push the food posteriorly towards the oesophagus, as described above. Intraoral transport of food particles in water is possible (involving inertial feeding on suspended objects) utilizing the pumping effects of the buccal musculature and the abrasive action of pharyngeal and more anterior teeth, when present (e.g. Regal, 1966). These considerations functionally explain the tonguelessness of pipids, given that the condition in the Aglossa is most probably a secondary feature.

6.3 Gymnophiona

6.3.1 Introduction

The Gymnophiona or "caecilians" represent the third branch of extant lissamphibians and are characterized by an elongated, snake-like body, and, convergently with snakes, a complete reduction of their limbs (Duellman & Trueb, 1994). There has been some debate about the phylogeny and evolutionary history of gymnophiones. Some morphological features may indicate that they arose from lepospondylians (Carroll & Currie, 1975; Anderson et al., 2008; Huttenlocker et al., 2013) and would, therefore, be more closely related to amniotes than to lissamphibians. More recent molecular phylogenetic studies, along with employment of molecular clock analyses, point to lissamphibian monophyly and place gymnophiones as the sister group of the branch containing salamanders and anurans (Roelants et al., 2007; San Mauro, 2010; Pyron & Wiens, 2011).

Most caecilians are fossorial and are found throughout the humid tropics, except for New Guinea and Australia (Nussbaum & Wilkinson, 1989). However, some caecilians have entered aquatic habitats and either only occasionally leave the water or are permanently aquatic (Moodie, 1978; Nussbaum & Wilkinson, 1989). Those species with an aquatic lifestyle have an obligate aquatic, gill bearing larval stage, while most terrestrial forms are direct developers or are viviparous (Parker, 1956; Wake, 1977a; Wilkinson & Nussbaum, 1998). All caecilians are carnivores and their diet includes a wide range of invertebrates, such as earthworms, beetle larvae, termites and shrimps, as well as small vertebrates (O'Reilly, 2000).

6.3.2 Food Detection

Caecilians localize and detect prey by using visual, olfactory, electro- and mechanoreception (O'Reilly, 2000). Amongst these four sensory systems, vision seems to be the least effective in caecilians and probably plays a minor role (if any role at all) in prey detection (O'Reilly, 2000). In fact, their eyes are generally small and hidden under the skin or even under dermal bones (Wake, 1985; Himstedt, 1995). In contrast to the visual system, olfaction is well-developed in caecilians. Both aquatic and terrestrial caecilians use buccal pumping to flush, respectively, air or water over the nasal epithelia. Mechanical cues can be detected by several sensory systems, such as the tentacle organ and the lateral line system. The tentacle organ is unique to caecilians: it is positioned anterior to the eye and contains tactile sensory cells (Fox,

1985). The lateral line system is developed in the head and trunk regions of larval and aquatic adult caecilians, whereas terrestrial stages completely lack the lateral line system (Fritzsch & Wake, 1986). Where present, the lateral line system consists of two sensory organs: neuromasts and ampullary organs. Neuromasts are responsible for the detection of vibrations, water currents and pressure changes, whereas ampullary organs are mainly used for detecting electrical fields (Himstedt & Fritzsch, 1990).

6.3.3 Food Capture

Caecilians might be the least investigated vertebrate group, and this is especially so for their aquatic feeding biology which remains only superficially studied. A terrestrial fossorial lifestyle and, accordingly, terrestrial feeding can be considered to be ancestral for extant postmetamorphic caecilians (Wake, 1977a, b, 1992; Wilkinson & Nussbaum, 1999; Deban et al., 2001; O'Reilly et al., 2002). Some branches, however, became secondarily aquatic or semiaquatic and feed in water (O'Reilly, 2000). Although many caecilians are direct developers, the aquatic larva is considered to be ancestral and is still found in many branches that have aquatic, semiaquatic or terrestrial adults. Larval caecilians are agile animals, can catch elusive prey and are able to use suction feeding (O'Reilly, 2000; Kleinteich, 2010). The suction strikes of caecilian larvae are rapid and the movement patterns of the feeding apparatus are very similar to those found in suction feeding salamanders (O'Reilly, 2000). In detail, suction strikes in caecilian larvae involve rapid gape opening followed immediately by hyobranchial depression. Given the striking morphological similarities of the hyobranchial system of salamander and caecilian larvae (Kleinteich & Haas, 2011), which appears tuned to suction feeding performance, a suction feeding larva is most parsimoniously considered to be the ancestral state in the most recent common ancestor of the two groups (O'Reilly et al., 2002). By contrast, adult aquatic caecilians have not retained the ability for powerful suction strikes, but rather use their ancestral terrestrial feeding behaviour, with only minor adjustments, for feeding in aquatic conditions (O'Reilly et al., 2002) (Fig. 6.7). Although inertial suction strikes have not been observed in any adult caecilians, some species might use secondarily evolved compensatory suction to avoid bow waves (that could push floating prey away) anterior to the head when lunging towards the prey (O'Reilly, 2000) (Fig. 6.7).

Fig. 6.7 Frames from a high-speed video recording showing *Typhlonectes natans* catching a prey item (maggot) under water. In this feeding event, the animal approached the prey from a vertical position. The entire prey-capture event lasted approximately 60 ms. Courtesy of Daniel Schwarz and Katja Söhnel

6.3.4 Intraoral Transport, Processing and Swallowing

After prey capture, gymnophiones can employ different mechanisms for intraoral transport, depending on the type of prey and the medium in which feeding occurs (water vs. air) (O'Reilly, 2000). Larvae and aquatic postmetamorphic caecilians use water flow to transport prey intraorally. Additionally, aquatic postmetamorphic caecilians are able to use inertial transport, by which food items are shifted posteriorly by using the substrate as the counterpoint of the forwardly-moving, gaping mouth, or by using cyclic movements of their tongue to transport food intraorally (O'Reilly, 2000). Postmetamorphic caecilians possess a unique jaw closing system that allows them to generate a substantial bite force despite their restricted head width and small external jaw adductor muscles (Measey & Herrel, 2006; Kleinteich et al., 2008). Specifically, the head width is restricted due to the functional constraints imposed by their ancestral fossorial lifestyle, and their external jaw adductor muscles are greatly reduced (Bemis et al., 1983; Nussbaum, 1983; O'Reilly, 2000). However, their derived jaw closing-system suggests that caecilians can generate high bite forces (Measey & Herrel, 2006). Accordingly, prey items might be mechanically reduced, killed, or crushed by using a series of strong bites, similar to the mechanisms employed by other lissamphibians (Schwenk & Wake, 1993; Fortuny et al., 2015; Lukanov et al., 2016). Most caecilians, however, use forceful longitudinal spinning instead to subdue struggling prey or to tear off chunks of flesh from larger prey (Bemis et al., 1983; O'Reilly, 2000). Large arthropods may also be rubbed against the substrate in order to kill them (O'Reilly, 2000). Abrasion of prey by holding it between the jaws and rubbing it against the substrate is a feature also known for salamanders and lizards (Natchev et al., 2015; Lukanov et al., 2016) and has evolved several times independently. There is no reliable literature devoted to swallowing in caecilians. However, making deductions from their anatomy and comparisons with their sister-group, it is most likely that coordinated contractions of the pharyngeal constrictor muscles, and possibly cyclic tongue movements, push prey backwards into the oesophagus. In contrast to salamanders and frogs, eye bulb retraction is not used for intraoral transport and swallowing in caecilians.

6.4 Concluding Remarks

Amphibians are excellent model organisms for studying convergent features of the feeding system. On the one hand, aquatic larval stages, as well as paedomorphs, display unidirectional flow patterns, whereas metamorphs with an aquatic lifestyle are bidirectional feeders. Although behaviourally similar, unidirectional flow regimes might be regarded as being ancestral as they are found in virtually all lissamphibian larvae as well as non-tetrapod vertebrates. Many adaptations to bidirectional flow regimes might be viewed as secondary adaptations. For example, aquatic metamorphs can be found within anurans as well as desmognathid salamanders, which have definitely reinvaded aquatic habitats. On the other hand, some salamandrid and ambystomatid salamanders switch between habitats after metamorphosis, so it remains unclear as to whether their lifestyle is retained from their larval condition or not. Due to the physical constraints of water, the kinematics (such as timing of gape and hyobranchial depression) of aquatic prey capture show many similarities in vertebrates in general. Typical convergent solutions that can be seen within secondarily aquatic lissamphibians are (1) a flattened skull shape with a large hyoid skeleton that is partly ossified in association with the generation of high suction forces, (2) the occlusion of the corners of the mouth, which is achieved with labial lobes of salamanders and pipid frogs, where in both they seal the lateral aspects of the gape, (3) a small tongue that does not compromise the inward flow of water during suction, which is most fully expressed in the aglossan pipids, and (4) a water storage system that enables the prolongation of the incoming water flow in bidirectional feeders. All of these major features have evolved independently within the three extant lissamphibian lineages in association with increasing suction feeding efficacy.

Acknowledgements The authors thank Martin Lehmann, Daniel Schwarz and Katja Söhnel for providing video recordings of aquatic feeding in *Typhlonectes natans* and Carrie Carreño-Zingaro for providing pictures of feeding sequences of pipid frogs.

References

Altig, R., Whiles, M., & Taylor, C. (2007). What do tadpoles really eat? Assessing the trophic status of an understudied and imperiled group of consumers in freshwater habitats. *Freshwater Biology, 52*, 386–395.

Anderson, J. S., Reisz, R. R., Scott, D., Fröbisch, N. B., & Sumida, S. S. (2008). A stem batrachian from the Early Permian of Texas and the origin of frogs and salamanders. *Nature, 453*, 515–518.

Anzeraey, A., Aumont, M., Decamps, T., Herrel, A., & Pouydebat, E. (2017). The effect of food properties on grasping and manipulation in the aquatic frog *Xenopus laevis*. *The Journal of Experimental Biology, 220*, 4486–4491.

Avila, V. L., & Frye, P. G. (1977). Feeding behavior in the African clawed frog (*Xenopus laevis* Daudin). *Herpetologica, 33*, 152–161.

Avila, V. L., & Frye, P. G. (1978). Feeding behavior of the African clawed frog (*Xenopus laevis* Daudin): (Amphibia, Anura, Pipidae): effect of the prey type. *Journal of Herpetology, 12*, 391–396.

Barrionuevo, J. S. (2016). Independent evolution of suction feeding in Neobatrachia: Feeding mechanisms of two species of *Telmatobius* (Anura: Telmatobiidae). *The Anatomical Record, 299*, 181–196.

Beisser, C. J., & Weisgram, J. (2001). Dorsal tongue morphology and lingual glands in Chelonians. *Journal of Morphology, 248*, 205.

Beisser, C. J., Weisgram, J., & Splechtna, H. (1995). Dorsal lingual epithelium of *Platemys pallidipectoris* (Pleurodira, Chelidae). *Journal of Morphology, 226*, 267–276.

Bemis, W. E., Schwenk, K., & Wake, M. (1983). Morphology and function of the feeding apparatus in *Dermophis mexicanus* (Amphibia: Gymnophiona). *Zoological Journal of the Linnean Society London, 77*, 75–96.

Bragg, A. N. (1964). Further study of predation and cannibalism in spadefoot tadpoles. *Herpetologica, 20*, 17–24.

Bullock, T. H., Bodznick, D. A., & Northcutt, R. G. (1993). The phylogenetic distribution of electroreception: evidence for convergent evolution of a primitive vertebrate sense modality. In *How do brains work?* (581–602). Springer.

Carreno, C. A., & Nishikawa, K. C. (2010). Aquatic feeding in pipid frogs: The use of suction for prey capture. *The Journal of Experimental Biology, 213*, 2001–2008.

Carroll, R. L., & Currie, P. J. (1975). Microsaurs as possible apodan ancestors. *Zoological Journal of the Linnean SocietyLondon, 57*, 229–247.

Cundall, D., Lorenz-Elwood, J., & Groves, J. D. (1987). Asymmetric suction feeding in primitive salamanders. *Experientia, 43*, 1229–1231.

Cundall, D., Fernandez, E., & Irish, F. (2017). The suction mechanism of the pipid frog, Pipa pipa (Linnaeus, 1758). *Journal of Morphology, 278*(9), 1229–1240. https://doi.org/10.1002/jmor.20707

Dalrymple, G. H., Juterbock, J. E., & La Valley, A. L. (1985). Function of the atlanto-mandibular ligaments of desmognathine salamanders. *Copeia*, 254–257.

Dean, M. (2003). Suction feeding in the pipid frog, *Hymenochirus boettgeri:* Kinematic and behavioral considerations. *Copeia, 2003*(4), 879–886.

Deban, S. M., & Marks, S. B. (2002). Metamorphosis and evolution of feeding behaviour in salamanders of the family Plethodontidae. *Zoological Journal of the Linnean Society London, 134*, 375–400.

Deban, S. M., & Olson, W. M. (2002). Suction feeding by a tiny predatory tadpole. *Nature, 420*, 41–42.

Deban, S. M., & Richardson, J. C. (2017). A peculiar mechanism of bite-force enhancement in lungless salamanders revealed by a new geometric method for modeling muscle moments. *The Journal of Experimental Biology, 220*, 3588–3597.

Deban, S., & Wake, D. (2000). Aquatic feeding in salamanders. In K. Schwenk (Ed.), *Feeding: Form, function and evolution in tetrapod vertebrates* (pp. 65–94). Academic.

Deban, S. M., O'Reilly, J. C., & Nishikawa, K. C. (2001). The evolution of the motor control of feeding in amphibians. *American Zoologist, 41*, 1280–1298.

Denoël, M., & Joly, P. (2000). Neoteny and progenesis as two heterochronic processes involved in paedomorphosis in *Triturus alpestris* (Amphibia: Caudata). *Proceedings of the Royal Society of London Series B, 267*, 1481–1485.

Denoël, M., Joly, P., & Whiteman, H. H. (2005). Evolutionary ecology of facultative paedomorphosis in newts and salamanders. *Biological Reviews, 80*, 663–671.

Dole, W., Rose, B. B., & Tachiki, K. H. (1981). Western toads (*Bufo boreas*) learn odor of prey insects. *Herpetologica, 37*, 63–68.

Doran, G., & Baggett, H. (1972). The specialized lingual papillae of *Tachyglossus aculeatus* I. Gross and light microscopic features. *The Anatomical Record, 172*, 157–165.

Duellman, W. E., & Trueb, L. (1994). *Biology of amphibians*. JHU Press.

Emerson, S. B. (1977). Movement of the hyoid in frogs during feeding. *The American Journal of Anatomy, 149*, 115–120.
Fabrezi, M., & Lobo, F. (2009). Hyoid skeleton, its related muscles, and morphological novelties in the frog *Lepidobatrachus* (Anura, Ceratophyridae). *The Anatomical Record, 292*, 1700–1712.
Fernandez, E., Irish, F., & Cundall, D. (2017). How a frog, *Pipa pipa*, succeeds or fails in catching fish. *Copeia, 105*(1), 108–119.
Fortuny, J., Marcé-Nogué, J., Heiss, E., Sanchez, M., Gil, L., & Galobart, À. (2015). 3D Bite modeling and feeding mechanics of the largest living amphibian, the Chinese Giant Salamander *Andrias davidianus* (Amphibia: Urodela). *PLoS One, 10*, e0121885.
Fox, H. (1985). The tentacles of *Ichthyophis* (Amphibia: Caecilia) with special reference to the skin. *Journal of Zoology, 205*, 223–234.
Fritzsch, B., & Wake, M. H. (1986). The distribution of ampullary organs in Gymnophiona. *Journal of Herpetology, 20*, 90–93.
Fritzsch, B., Drewes, R. C., & Ruibal, R. (1987). The retention of the lateral line nucleus in adult anurans. *Copeia, 1987*, 127–135.
Gillis, G., & Lauder, G. (1994). Aquatic prey transport and the comparative kinematics of *Ambystoma tigrinum* feeding behaviors. *The Journal of Experimental Biology, 187*, 159–179.
Gosline, W. A. (1987). Jaw structures and movements in higher teleostean fishes. *Japanese Journal of Ichthyology, 34*, 21–32.
Gould, S. J. (1977). *Ontogeny and phylogeny*. Harvard University Press.
Gray, L. A., O'Reilly, J. C., & Nishikawa, K. C. (1997). Evolution of forelimb movement patterns for prey manipulation in anurans. *Journal of Experimental Zoology, 277*, 417–424.
Griffiths, M. (1978). *The biology of the monotremes*. Academic.
Grosjean, S., Vences, M., & Dubois, A. (2004). Evolutionary significance of oral morphology in the carnivorous tadpoles of tiger frogs, genus *Hoplobatrachus* (Ranidae). *Biological Journal of the Linnean Society, 81*, 171–181.
Grzimek. (2003). Grzimek's animal life encyclopedia. In M. Hutchins, W. E. Duellman, & N. Schlager (Eds.), *Amphibians* (Vol. 6, 2nd ed.). Gale Group.
Haas, A. (2003). Phylogeny of frogs as inferred from primarily larval characters (Amphibia: Anura). *Cladistics, 19*, 23–89.
Haas, A., Pohlmeyer, J., McLeod, D. S., Kleinteich, T., Hertwig, S. T., Das, I., & Buchholz, D. R. (2014). Extreme tadpoles II: the highly derived larval anatomy of *Occidozyga baluensis* (Boulenger, 1896), an obligate carnivorous tadpole. *Zoomorphology, 133*, 321–342.
Heiss, E. (2017). The Alpine "axolotl": A remarkable example of phenotypic plasticity in *Ichthyosaura alpestris* (Amphibia: Salamandridae). *Salamandra, 53*, 137–141.
Heiss, E., & Grell, J. (2019). Same but different: aquatic prey capture in paedomorphic and metamorphic Alpine newts. *Zoological Letters, 5*, 24.
Heiss, E., Aerts, P., & Van Wassenbergh, S. (2013a). Masters of change: Seasonal plasticity in the prey-capture behavior of the Alpine newt *Ichthyosaura alpestris* (Salamandridae). *The Journal of Experimental Biology, 216*, 4426–4434.
Heiss, E., Natchev, N., Gumpenberger, M., Weissenbacher, A., & Van Wassenbergh, S. (2013b). Biomechanics and hydrodynamics of prey capture in the Chinese giant salamander reveal a high-performance jaw-powered suction feeding mechanism. *The Journal of the Royal Society Interface, 10*, 20121028.
Heiss, E., Schwarz, D., & Konow, N. (2019). Chewing or not? Intraoral food processing in a salamandrid newt. *The Journal of Experimental Biology, 2019*, 222.
Himstedt, W. (1973a). Die spektrale Empfindlichkeit von Urodelen in Abhängigkeit von Metamorphose, Jahreszeit und Lebensraum. *Zoologische Jahrbücher Abteilung für allgemeine Zoologie, 77*(2), 246–274.
Himstedt, W. (1973b). Die spektrale Empfindlichkeit von *Triturus alpestris* (Amphibia, Urodela) während des Wasser- und Landlebens. *Pflügers Archiv, 341*, 7–14.
Himstedt, W. (1982). Prey selection in salamanders. In D. J. Ingle, M. A. Goodale, & R. J. M. Mansield (Eds.), *Analysis of visual behavior*. MIT Press.

Himstedt, W. (1995). Structure and function of the eyes in the caecilian *Ichthyophis kohtaoensis* (Amphibia, Gymnophiona). *Zoology, 99*, 81–94.

Himstedt, W., & Fritzsch, B. (1990). Behavioural evidence for electroreception in larvae of the caecilian *Ichthyophis kohtaoensis* (Amphibia, Gymnophiona). *Zoologische Jahrbücher Abteilung für allgemeine Zoologie, 94*, 486–492.

Huttenlocker, A. K., Pardo, J. D., Small, B. J., & Anderson, J. S. (2013). Cranial morphology of recumbirostrans (Lepospondyli) from the Permian of Kansas and Nebraska, and early morphological evolution inferred by micro-computed tomography. *Journal of Vertebrate Paleontology, 33*, 540–552.

Ivanović, A., Cvijanović, M., Denoël, M., Slijepčević, M., & Kalezić, M. L. (2014). Facultative paedomorphosis and the pattern of intra- and interspecific variation in cranial skeleton: lessons from European newts (*Ichthyosaura alpestris* and *Lissotriton vulgaris*). *Zoomorphology, 133*, 99–109.

Kleinteich, T. (2010). Ontogenetic differences in the feeding biomechanics of oviparous and viviparous caecilians (Lissamphibia: Gymnophiona). *Zoology, 113*, 283–294.

Kleinteich, T., & Haas, A. (2011). The hyal and ventral branchial muscles in caecilian and salamander larvae: Homologies and evolution. *Journal of Morphology, 272*, 598–613.

Kleinteich, T., Haas, A., & Summers, A. P. (2008). Caecilian jaw-closing mechanics: Integrating two muscle systems. *The Journal of the Royal Society Interface, 5*, 1491–1504.

Kunisch, S., Schwaha, T., Blüml, V., Beisser, C. J., Handschuh, S., & Lemell, P. (2021). Digital dissection of the head of the frogs *Calyptocephalella gayi* and *Leptodactylus pentadactylus* with emphasis on the feeding apparatus. *Journal of Anatomy, 00*, 1–14. https://doi.org/10.1111/joa.13426

Lauder, G., & Reilly, S. (1994). Amphibian feeding behavior: Comparative biomechanics and evolution. In V. Bels, M. Chardon, & P. Vandewalle (Eds.), *Biomechanics of feeding in vertebrates* (pp. 163–195). Springer.

Lauder, G. V., & Shaffer, H. B. (1985). Functional morphology of the feeding mechanism in aquatic ambystomatid salamanders. *Journal of Morphology, 185*, 297–326.

Lauder, G. V., & Shaffer, H. B. (1986). Functional design of the feeding mechanism in lower vertebrates: unidirectional and bidirectional flow systems in the tiger salamander. *Zoological Journal of the Linnean Society London, 88*, 277–290.

Lauder, G. V., & Shaffer, H. B. (1988). Ontogeny of functional design in tiger salamanders (*Ambystoma tigrinum*): Are motor patterns conserved during major morphological transformations? *Journal of Morphology, 197*, 249–268.

Lemell, P., Lemell, C., Snelderwaard, P., Gumpenberger, M., Wocheslánder, R., & Weisgram, J. (2002). Feeding patterns of *Chelus fimbriatus* (Pleurodira: Chelidae). *The Journal of Experimental Biology, 205*, 1495–1506.

Levine, R. P., Monroy, J. A., & Brainerd, E. L. (2004). Contribution of eye retraction to swallowing performance in the northern leopard frog, *Rana pipiens*. *The Journal of Experimental Biology, 207*, 1361–1368.

Lorenz-Elwood, J. R., & Cundall, D. (1994). Morphology and feeding behavior of the feeding apparatus in *Cryptobranchus alleganiensis* (Amphibia: Caudata). *Journal of Morphology, 220*, 47–70.

Lukanov, S., Tzankov, N., Handschuh, S., Heiss, E., Naumov, B., & Natchev, N. (2016). On the amphibious food uptake and prey manipulation behavior in the Balkan-Anatolian crested newt (*Triturus ivanbureschi*, Arntzen and Wielstra, 2013). *Zoology, 119*, 224–231.

Matsumoto, R., & Evans, S. E. (2017). The palatal dentition of tetrapods and its functional significance. *Journal of Anatomy, 230*, 47–65.

Matthes, E. (1934). Bau und Funktion der Lippensäume wasserlebender Urodelen. *Zeitschrift für Morphologie und Ökologie der Tiere, 28*, 155–169.

Measey, G. J., & Herrel, A. (2006). Rotational feeding in caecilians: Putting a spin on the evolution of cranial design. *Biology Letters, 2*, 485–487.

Miller, B. T., & Larsen, J. H., Jr. (1989). Feeding performance in aquatic postmetamorphic newts (Urodela: Salamandridae): Are bidirectional flow systems necessarily inefficient? *Canadian Journal of Zoology, 67*, 2414–2421.

Moodie, G. (1978). Observations on the life history of the caecilian *Typhlonectes compressicaudus* (Dumeril and Bibron) in the Amazon basin. *Canadian Journal of Zoology, 56*, 1005–1008.

Muller, M., & Osse, J. (1984). Hydrodynamics of suction feeding in fish. *Transactions of the Zoological Society of Londo, 37*, 51–135.

Münz, H., Claas, B., & Fritzsch, B. (1984). Electroreceptive and mechanoreceptive units in the lateral line of the axolotl *Ambystoma mexicanum*. *Journal of Comparative Physiology. A, 154*, 33–44.

Natchev, N., Heiss, E., Lemell, P., Stratev, D., & Weisgram, J. (2009). Analysis of prey capture and food transport kinematics in two Asian box turtles, *Cuora amboinensis* and *Cuora flavomarginata* (Chelonia, Geoemydidae), with emphasis on terrestrial feeding patterns. *Zoology, 112*, 113–127.

Natchev, N., Tzankov, N., Vergilov, V., Kummer, S., & Handschuh, S. (2015). Functional morphology of a highly specialised pivot joint in the cranio-cervical complex of the minute lizard *Ablepharus kitaibelii* in relation to feeding ecology and behaviour. *Contributions to Zoology, 84*, 13–S3.

Nishikawa, K. C. (1999). Neuromuscular control of prey capture in frogs. *Philosophical Transactions of the Royal Society of London Series B, 354*, 941–954.

Northcutt, R. G. (1992). Distribution and innervation of lateral line organs in the axolotl. *The Journal of Comparative Neurology, 325*, 95–123.

Nussbaum, R. A. (1983). The evolution of a unique dual jaw-closing mechanism in caecilians (Amphibia: Gymnophiona) and its bearing on caecilian ancestry. *Journal of Zoology, 199*, 545–554.

Nussbaum, R. A., & Wilkinson, M. (1989). On the classification and phylogeny of caecilians (Amphibia: Gymnophiona), a critical review. *Herpetological Monographs, 3*, 1–42.

O'Reilly, J. (2000). Feeding in caecilians. In K. Schwenk (Ed.), *Feeding: Form, function and evolution in tetrapod vertebrates* (pp. 149–166). Academic.

O'Reilly, J. C., Deban, S. M., & Nishikawa, K. C. (2002). Derived life history characteristics constrain the evolution of aquatic feeding behavior in adult amphibians. In P. Aerts, K. D'Août, A. Herrel, & R. Van Damme (Eds.), *Topics in functional and ecological vertebrate morphology* (pp. 153–190). Shaker Publishing.

Park, D., Lee, J.-H., Ra, N.-Y., & Eom, J. (2008). Male salamanders *Hynobius leechii* respond to water vibrations via the mechanosensory lateral line system. *Journal of Herpetology, 42*, 615–625.

Parker, H. (1956). Viviparous caecilians and amphibian phylogeny. *Nature, 178*, 250–252.

Parzefall, J. (1993). Behavioural ecology of cave-dwelling fishes. In T. J. Pitscher (Ed.), *Behaviour of teleost fishes* (pp. 573–606). Chapman & Hall.

Pethiyagoda, R., Manamendra-Arachchi, K., & Sudasinghe, H. (2014). Underwater and terrestrial feeding in the Sri Lankan wart-frog, *Lankanectes corrugatus*. *The Ceylon Journal of Science (Biological Sciences), 43*(2), 79–82.

Pough, F. H., Andrews, R. M., Cadle, J. E., Crump, M. L., Savitzky, A. H., & Wells, K. D. (2004). *Herpetology* (3rd ed.). Prentice Hall.

Przyrembel, C., Keller, B., & Neumeyer, C. (1995). Trichromatic color vision in the salamander (*Salamandra salamandra*). *Journal of Comparative Physiology. A, 176*, 575–586.

Pyron, R. A., & Wiens, J. J. (2011). A large-scale phylogeny of Amphibia including over 2800 species, and a revised classification of extant frogs, salamanders, and caecilians. *Molecular Phylogenetics and Evolution, 61*, 543–583.

Quinzio, S., & Fabrezi, M. (2014). The lateral line system in anuran tadpoles: Neuromast morphology, arrangement, and innervation. *The Anatomical Record, 297*, 1508–1522.

Regal, P. J. (1966). Feeding specializations and the classification of terrestrial salamanders. *Evolution*, 392–407.

Regal, P. J., & Gans, C. (1976). Functional aspects of the evolution of frog tongues. *Evolution, 30*, 718–734.

Reilly, S. M. (1986). Ontogeny of cranial ossification in the eastern newt, *Notophthalmus viridescens* (Caudata: Salamandridae), and its relationship to metamorphosis and neoteny. *Journal of Morphology, 188*, 315–326.

Reilly, S. M. (1996). The metamorphosis of feeding kinematics in *Salamandra salamandra* and the evolution of terrestrial feeding behavior. *The Journal of Experimental Biology, 199*, 1219–1227.

Reilly, S. M., & Lauder, G. V. (1988). Ontogeny of aquatic feeding performance in the eastern newt, *Notophthalmus viridescens* (Salamandridae). *Copeia*, 87–91.

Reilly, S. M., & Lauder, G. V. (1991). Prey transport in the tiger salamander: quantitative electromyography and muscle function in tetrapods. *The Journal of Experimental Zoology, 260*, 1–17.

Reilly, S. M., & Lauder, G. V. (1992). Morphology, behavior, and evolution: Comparative kinematics of aquatic feeding in salamanders. *Brain, Behavior and Evolution, 40*, 182–196.

Roelants, K., Gower, D. J., Wilkinson, M., Loader, S. P., Biju, S., Guillaume, K., Moriau, L., & Bossuyt, F. (2007). Global patterns of diversification in the history of modern amphibians. *Proceedings of the National Academy of Sciences of the United States of America, 104*, 887–892.

Roth, G. (1987). *Visual behavior in salamanders*. Springer.

San Mauro, D. (2010). A multilocus timescale for the origin of extant amphibians. *Molecular Phylogenetics and Evolution, 56*, 554–561.

Schülert, N., & Dicke, U. (2002). The effect of stimulus features on the visual orienting behaviour of the salamander *Plethodon jordani*. *The Journal of Experimental Biology, 205*, 241–251.

Schwarz, D., Konow, N., Roba, Y. T., & Heiss, E. (2020a). A salamander that chews using complex, three-dimensional mandible movements. *The Journal of Experimental Biology, 223*, jeb220749.

Schwarz, D., Konow, N., Porro, L. B., & Heiss, E. (2020b). Ontogenetic plasticity in cranial morphology is associated with a change in the food processing behavior in Alpine newts. *Frontiers in Zoology, 17*, 1–16.

Schwarz, D., Gorb, S. N., Kovalev, A., Konow, N., & Heiss, E. (2020c). Flexibility of intraoral food processing in the salamandrid newt *Triturus carnifex*: Effects of environment and prey type. *The Journal of Experimental Biology, 223*, jeb232868.

Schwenk, K., & Rubega, M. (2005). Diversity of vertebrate feeding systems. In J. M. Starck & T. Wang (Eds.), *Physiological and ecological adaptations to feeding in vertebrates* (pp. 1–41). Science Publishers.

Schwenk, K., & Wake, D. B. (1993). Prey processing in *Leurognathus marmoratus* and the evolution of form and function in desmognathine salamanders (Plethodontidae). *Biological Journal of the Linnean Society, 49*, 141–162.

Shaffer, H. B., & Lauder, G. V. (1985). Patterns of variation in aquatic ambystomatid salamanders: Kinematics of the feeding mechanism. *Evolution, 39*(1), 83–92.

Shaffer, H. B., & Lauder, G. V. (1988). The ontogeny of functional design: Metamorphosis of feeding behaviour in the tiger salamander (*Ambystoma tigrinum*). *Journal of Zoology, 216*, 437–454.

Shinn, E. A., & Dole, J. W. (1978). Evidence for a role for olfactory cues in the feeding response of leopard frogs, *Rana pipiens*. *Herpetologica, 34*, 167–172.

Skorczewski, T., Cheer, A., & Wainwright, P. C. (2012). The benefits of planar circular mouths on suction feeding performance. *The Journal of the Royal Society Interface, 9*, 1767–1773.

Smithson, T. R., Browne, M. A., Davies, S. J., Marshall, J. E., Millward, D., Walsh, S. A., & Clack, J. A. (2017). A new Mississippian tetrapod from Fife, Scotland, and its environmental context. *Papers in Palaeontology, 3*, 547–557.

Sokol, O. M. (1969). Feeding in the pipid frog *Hymenochirus boettgeri* (Tornier). *Herpetologica, 25*(1), 9–24.

Stinson, C. M., & Deban, S. M. (2017). Functional trade-offs in the aquatic feeding performance of salamanders. *Zoology, 125*, 69–78.

Trewavas, E. (1933). The hyoid and larynx of the anura. *Philosophical Transactions of the Royal Society of London Series B, 222*, 433–493.

Trueb, L. (1993). Patterns of cranial diversity among the lissamphibia. In J. Hanken & B. K. Hall (Eds.), *The skull – Volume 2: Patterns of structural and systematic diversity* (pp. 255–343). University of Chicago Press.

Uiblein, F., Durand, J., Juberthie, C., & Parzefall, J. (1992). Predation in caves: The effects of prey immobility and darkness on the foraging behaviour of two salamanders, *Euproctus asper* and *Proteus anguinus*. *Behavioural Processes, 28*, 33–40.

Van Dobben, W. H. (1937). Über den Kiefermechanismus der Knochenfische. *Archives Néerlandaises de Zoologie, 2*, 1–72.

Van Wassenbergh, S., & Heiss, E. (2016). Phenotypic flexibility of gape anatomy fine-tunes the aquatic prey-capture system of newts. *Scientific Reports, 6*, 29277.

Vera Candioti, M. F. (2005). Morphology and feeding in tadpoles of *Ceratophrys cranwelli* (Anura: Leptodactylidae). *Acta Zoologica, 86*, 1–11.

Vera Candioti, M. F. (2006). Ecomorphological guilds in anuran larvae: an application of geometric morphometric methods. *Herpetological Journal, 16*, 149–162.

Vera Candioti, M. F., Lavilla, E. O., & Echeverria, D. D. (2004). Feeding mechanisms in two treefrogs, *Hyla nana* and *Scinax nasicus* (Anura: Hylidae). *Journal of Morphology, 261*, 206–224.

Vitt, L. J., & Caldwell, J. P. (2009). *Herpetology. An introductory biology of amphibians and reptiles* (3rd ed.). Academic.

Wainwright, P. C., McGee, M. D., Longo, S. J., & Hernandez, P. L. (2015). Origins, innovations, and diversification of suction feeding in vertebrates. *Integrative and Comparative Biology, 55*(1), 134–145.

Wake, M. H. (1977a). The reproductive biology of caecilians: an evolutionary perspective. In E. Taylor & S. Guttman (Eds.), *The reproductive biology of amphibians* (pp. 73–101). Springer.

Wake, M. H. (1977b). Fetal maintenance and its evolutionary significance in the Amphibia: Gymnophiona. *Journal of Herpetology, 11*, 379–386.

Wake, M. H. (1985). The comparative morphology and evolution of the eyes of caecilians (Amphibia, Gymnophiona). *Zoomorphology, 105*, 277–295.

Wake, M. H. (1992). Reproduction in caecilians. In W. C. Hamlett (Ed.), *Reproductive biology of South American vertebrates* (pp. 112–120). Springer.

Werth, A. (2000). Feeding in marine mammals. In K. Schwenk (Ed.), *Feeding: Form, function and evolution in tetrapod vertebrates* (pp. 475–514). Academic.

Wiens, J. J., Bonett, R. M., & Chippindale, P. T. (2005). Ontogeny discombobulates phylogeny: Paedomorphosis and higher-level salamander relationships. *Systematic Biology, 54*, 91–110.

Wiesinger, V. (2017). Feeding mechanisms of *Calyptocephalella gayi*. Diploma thesis at University of Vienna.

Wilkinson, M., & Nussbaum, R. A. (1998). Caecilian viviparity and amniote origins. *Journal of Natural History, 32*, 1403–1409.

Wilkinson, M., & Nussbaum, R. A. (1999). Evolutionary relationships of the lungless caecilian *Atretochoana eiselti* (Amphibia: Gymnophiona: Typhlonectidae). *Zoological Journal of the Linnean Society of London, 126*, 191–223.

Winokur, R. M. (1988). The buccopharyngeal mucosa of the turtles (Testudines). *Journal of Morphology, 196*, 33–52.

Witzmann, F., & Werneburg, I. (2017). The palatal interpterygoid vacuities of temnospondyls and the implications for the associated eye-and jaw musculature. *The Anatomical Record, 300*, 1240–1269.

Chapter 7
Convergence of Aquatic Feeding Modes in the Sauropsida (Crocodiles, Birds, Lizards, Snakes and, Turtles)

Egon Heiss, Paul M. Gignac, Laura B. Porro, and Patrick Lemell

Abstract The Sauropsida includes the extant crocodiles, birds, turtles, lizards and snakes. With roughly 30,000 described species, it is not only the largest phylogenetic group within Amniota, but the largest extant group within all tetrapods. Like many other tetrapod branches, sauropsids have evolved many adaptations to aquatic lifestyles: from species that occasionally feed in aquatic habitats to fully aquatic forms that only rise to the water's surface for breathing. As amniotes, sauropsids can safely be considered primarily terrestrial vertebrates and any adaptations to aquatic life and feeding can be regarded as secondary features. Sauropsids show a very broad spectrum of convergently-evolved adaptations for aquatic feeding, from crocodylian apex predators to high-performance suspension feeding birds, suction feeding in turtles and alga-scraping in marine iguanas. Adaptations for aquatic feeding in sauropsids have evolved multiple times independently, both between and within groups. For example, suction feeding has evolved independently in turtles and birds; extremely fast forward strikes by straightening of the curved postcranial vertebral column in birds and snakes; and suspension feeding in mallards, flamingoes and sea-birds. In the following sections, we summarize the diverse adaptations to aquatic feeding in crocodylians, birds, lepidosaurs and turtles and highlight convergence and homologies where appropriate.

E. Heiss (✉)
Institute of Zoology and Evolutionary Research, Friedrich-Schiller-University Jena, Jena, Germany
e-mail: heissegon@hotmail.com

P. M. Gignac
Department of Cellular and Molecular Medicine, University of Arizona, Tucson, AZ, USA

L. B. Porro
Department of Cell and Developmental Biology, Centre for Integrative Anatomy, University College London, London, UK

P. Lemell
Department of Evolutionary Biology, Unit for Integrative Zoology, University of Vienna Vienna, Austria

V. L. Bels, A. P. Russell (eds.), *Convergent Evolution*, Fascinating Life Sciences,
https://doi.org/10.1007/978-3-031-11441-0_7

Keywords Aquatic feeding · Bidirectional flow system · Biomechanical modelling · Bite force · Finite element analysis · Inertial feeding · Jaw prehension · Ram feeding · Suction feeding · Suspension feeding

7.1 Crocodylia

7.1.1 Introduction

Today, crocodylians (Archosauria: Crocodylia) are medium to large-bodied, semi-aquatic reptiles with snap-trap jaws and impressive bite-force capacities, inhabiting tropical zones around the world (Gignac et al., 2019). Long considered to be "living fossils", it is now understood that crocodylians, along with their crocodyliform precursors, previously exhibited a wide range of body plans and skull shapes corresponding to diverse locomotor, dietary, and habitat specializations, including adaptations to terrestrial and marine niches (Stubbs et al., 2013; Mannion et al., 2015). Numerous studies have demonstrated that crocodyliforms have been shaped extensively by convergent evolution (von Huene, 1933; Buffetaut, 1982; Brochu, 2001; Wroe & Milne, 2007; Jones, 2008; Pierce et al., 2009; Wilberg et al., 2019). These morphological shifts have helped this group invade new adaptive zones (Erwin, 1992; Wainwright & Price, 2016) and dominate predatory niches in and around the water for the past 200 million years (Wilberg et al., 2019).

Fish-eating, or piscivory, is a common behavior of crocodylians, owing to the abundance of such prey in the world's rivers, lakes and oceans. One species in particular, the Indian gharial (*Gavialis gangeticus*), is routinely singled out as the most piscivorous of extant crocodylians (Pooley, 1989; Whitaker & Basu, 1982; Thorbjarnarson, 1990; Trutnau & Sommerlad, 2006; but also see Forsyth, 1910; Shortt, 1921; Biswas, 1970), capable of securing agile aquatic prey with its elongated jaws and > 100 needle-like teeth (Taylor, 1987; Singh, 2015; Ballell et al., 2019). Several other extant crocodylians, including the semi-piscivorous Malay ("false") gharial (*Tomistoma schlegelii*), the African slender-snouted crocodile (*Mecistops cataphractus*), the Orinoco crocodile (*Crocodylus intermedius*), and the Australian freshwater crocodile (*C. johnstoni*) (Pooley & Gans, 1976; Pooley, 1989; Webb & Manolis, 1989), while not strictly piscivorous, do consume an abundance of fish (Brochu, 2001; Erickson et al., 2012). Notably, all share a series of independently derived features, including slender snouts (see Fig. 7.1), that appear to permit the capture of highly elusive aquatic prey, especially fish (Brochu, 2001; Erickson et al., 2012), by enabling wide head sweeps encompassing large strike zones (Erickson et al., 2012; Ballell et al., 2019) and rapid jaw closure under water (McHenry et al., 2006).

Such rostrodental traits are shared by other secondarily aquatic fish-eating specialists, including river dolphins (e.g., *Inia geoffrensis*; Walmsley et al., 2013; McCurry et al., 2017a), suggesting that piscivorous crocodylian taxa are excellent modern analogs for addressing the convergent evolution of fish-eating and the re-invasion of freshwater and marine niches by now extinct tetrapods, especially

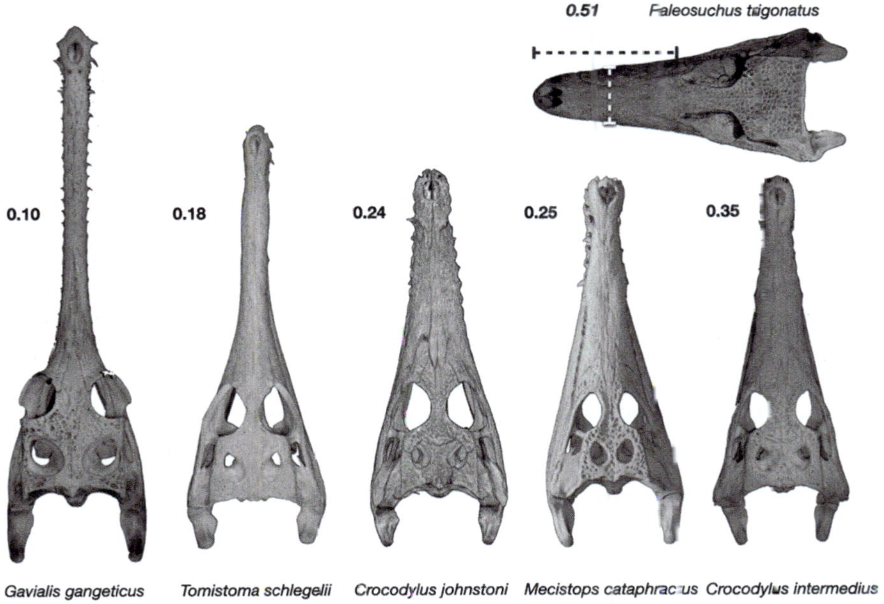

Fig. 7.1 Images of adult skulls in dorsal view exemplifying snout proportions in *Paleosuchus trigonatus* (Field Museum of Natural History specimen no. [FMNH] 68,379; upper right), which has a snout aspect ratio—mid-rostral width [white dashed line] divided by snout length [black dashed line]—of 0.51, near the interspecific mean of 0.53 for the clade Crocodylia (Erickson et al., 2012). The five semi- and highly-piscivorous species with narrow snouts are ordered from left to right by ascending value of snout aspect ratio: *Gavialis gangeticus* (FMNH 82861) and *Tomistoma schlegelii* (University of California Museum of Paleontology specimen no. 31702) are considered longirostrine (Brochu, 2001), whereas *Crocodylus johnstoni* (Texas Memorial Museum specimen no. [TMM] M-6807), *Mecistops cataphractus* (TMM M-3529), and *C. intermedius* (FMNH 75658) are considered to be slender-snouted. All skull images are scaled to the same head width for easier comparison, following O'Brien et al. (2019). Photographs taken by P.M.G.; TMM M-6807 and M-3529 are dorsal view, 3D projections based on computed tomography scans completed by the University of Texas at Austin

reptiles (Massare, 1987; Hua & De Buffrenil, 1996; McHenry et al., 2006; Pierce & Benton, 2006; Pierce et al., 2009; Young et al., 2010; Salas-Gismondi, 2016; McCurry et al., 2017a, b; Ballell et al., 2019). In fact, crocodyliforms from the Jurassic and Cretaceous exhibit multiple independent forays into freshwater and marine environments, as evidenced by characteristic rostral, dental, and body-shape features found in fossils preserved within aquatic and marine depositional environments (Young et al., 2010; Bronzati et al., 2015; Ballell et al., 2019; Wilberg et al., 2019). Additionally, the appearance of conspicuously long-snouted (longirostrine) morphologies is linked to increases in body size in aquatic taxa (Godoy, 2020). Paleobiogeographic reconstructions indicate that members of the Thalattosuchia, Pholidosuridae, and Dyrosauridae share comparable, or more extreme, adaptations to aquatic and marine lifestyles than those observed in extant crocodylians (Wilberg et al., 2019). The most highly derived examples occurred among thalattosuchians, a group of fully pelagic, marine crocodyliforms featuring tail flukes, manual and pedal flippers, and hydrodynamic body shapes (Young et al., 2010; Foffa et al., 2018; Ballell et al., 2019; Schwab et al., 2020). Across many of these groups, both extant and extinct, shifts between freshwater and marine environments appear to have been common, indicating that such specializations were essential for land-to-water transitions (Wilberg et al., 2019).

Studying the aquatic feeding mechanics and cranial sensory systems of crocodylians has led to major advances in our understanding of how adaptation to aquatic niches can shape reptile bauplans (Massare, 1987; Hua & De Buffrenil, 1996; Pierce et al., 2009; Young et al., 2010; Schwab et al., 2020). Below, we detail these specializations as well as discuss how crocodylians detect, capture, and consume aquatic prey, including how this understanding can help to further advance the study of extreme convergence in the crocodyliform fossil record.

7.1.2 Food Detection

Semi-aquatic taxa commonly possess visual acuity in subaerial and subaquatic environments (Howland & Sivak, 1984; Underwood, 1970). Extant crocodylians, however, have visual systems equipped for functional focus and accommodation in subaerial environments only. Like many birds, they possess nasotemporally elongate foveae, five types of photoreceptors (one rod, one double-cone, and three single cones), and tapeta lucida (Nagloo et al., 2016; Soares & Bierman, 2018). These traits are advantageous for focusing along the length of shorelines, seeing in trichromatic color vision, and maintaining low-light vision by increasing the light available to stimulate photoreceptors. As a result, visual acuity is generally good in air (Nagloo et al., 2016); however, this is not the case underwater (Fleishman et al., 1988). Subaquatic vision is highly reduced in all species studied to date, including *G. gangeticus* (Fleishman et al., 1988). Although crocodylians reduce pupil size when they dive to adjust focal depth, they do not achieve the wide accommodation range necessary for subaquatic visual acuity (Fleishman et al., 1988). They likewise lack the flattened corneas, typical of water fowl (Sivak, 1980), needed to reduce refraction underwater. Thus, even highly piscivorous taxa are severely farsighted (hyperopia) when hunting aquatic prey (Fleishman et al., 1988).

However, crocodylians are highly capable subaquatic hunters, even in the absence of subaquatic visual acuity (Neill, 1971; Schaller & Crawshaw Jr., 1982). Instead of relying on vision, crocodylians detect prey in the water using integumentary sensory organs (a.k.a., dome pressure receptors; Brazaitis, 1987; Soares, 2002; Leitch & Catania, 2012). These neurosensory structures evolved independently from the lateral line systems of fishes and amphibians (e.g., Coombs et al., 2012) but share broadly similar roles for mechanically interpreting the direction of pressure waves in the water column (Soares, 2002). Remarkably, these integumentary sensory organs enable greater high-resolution mechanosensitivity than is achievable by the primate fingertip (Leitch & Catania, 2012), thus providing the dense sensory input necessary for precise orienting behaviors, even in the absence of light (Singh, 1976; Soares, 2002). Alligators and caimans restrict integumentary sensory organs to the head, whereas other crocodylians harbor them on scales across most of the body surface (Leitch & Catania, 2012). Presumably, this broader sensory field enables more accurate directional assessment for the source of water displacement caused by potential prey, conspecifics, and abiotic factors (also see Jackson et al., 1996; Jackson & Brooks, 2007).

7.1.3 Food Capture

Crocodylians are jaw-prehension feeders, and most species are mesorostrine, having medium-length snouts (McCurry et al., 2017b). Species such as *Crocodylus rhombifer* and *Paleosuchus trigonatus* (see Fig. 7.1) exemplify the mean crocodylian snout aspect ratio (snout width divided by snout length; Erickson et al., 2012) of approximately 0.53 with most other taxa falling within one standard deviation of this value. These species have snouts that are relatively longer than those of other reptiles but are of average length for crocodylians. A few forms have slender snouts that are not exceptionally long but are notably narrow, represented today by *C. crocodylus*, *C. intermedius,* and *Mecistops cataphractus* (Brochu, 2001; McHenry et al., 2006; Erickson et al., 2012). Two other species, *Gavialis gangeticus* and *Tomistoma schlegelii*, are characterized by exceptionally narrow and elongate jaws (i.e., a longirostrine morphology), giving their pre-orbital skulls a tubular appearance (Brochu, 2001). All of these species are comparably slender-snouted (Fig. 7.1), but *C. crocodylus*, *C. intermedius,* and *M. cataphractus* have snout aspect ratios of 0.35–0.24, whereas *G. gangeticus* and *T. schlegelii* have ratios of 0.10 and 0.18, respectively (Erickson et al., 2012).

When capturing and consuming fish, generalist semi-aquatic crocodylians, such as *Alligator mississippiensis* and *Crocodylus niloticus* employ lateral swipes of the head and rapid jaw closure (Taylor, 1987). This approach appears to minimize pressure drag induced by broad rostra by moving the snout mediolaterally, along the axis of its lowest profile (Liem et al., 2001). This behavior displaces less water than would be the case by attempting to raise or lower the snout within the water column, thereby incurring lower resistance to motion (McHenry et al., 2006; Pierce et al., 2008; McCurry et al., 2017b). Slender-snouted forms appear to be adapted for minimizing drag along the mediolateral as well as the dorsoventral axis, enabling opportunities to elevate and depress the jaw while also moving the rostrum from side-to-side without eliciting substantial additional pressure drag. *Gavialis gangeticus*, for example, routinely makes rapid, 180° sweeps with its head underwater and is highly successful at lateral strikes that are ≤90° (Thorbjarnarson, 1990). In addition, longirostrine taxa are proposed to achieve the most rapid closure at the distal end of the jaws (Alexander, 1983; Taylor, 1987; McHenry et al., 2006), further improving their chances of prey capture success.

Jaw elongation, however, may come with trade-offs. McHenry et al. (2006) calculated drag moments for crocodylian jaws based on standardized skull dimensions and found that longirostrine ecomorphs incurred substantially greater drag moments than shorter, narrow-snouted forms. This is because the more rapidly a rigid beam rotates through water, the more pressure drag it incurs, and longer skulls distribute more material further from the center of rotation (e.g., cranial cervical joint, quadrate articular joint; McHenry et al., 2006). Lever mechanics dictate that the distal end of a longer snout necessarily moves more rapidly than that of a shorter snout, all else being equal (Cochran, 1982). Thus, pressure drag increases greatly (i.e., as a quadratic function; McHenry et al., 2006) along with rostral elongation. This drag disadvantage is presumably balanced somewhat by the narrower snout aspect ratios of longirostrine ecomorphs (McHenry et al., 2006; Erickson et al., 2012; Fig. 7.1), causing the mechanical benefits of narrowing and elongation to be common contributors to convergent skull evolution among highly aquatic and pelagic crocodyliforms (McHenry et al., 2006; Pierce & Benton, 2006; Pierce et al., 2009; Young et al., 2010; Walmsley et al., 2013; McCurry et al., 2017a, b; Ballell et al., 2019). These findings help elucidate why the highly plastic rostra of crocodyliforms (Iordansky, 1973; Langston, 1973; Brochu, 2001) converge in predominantly aquatic environments. Indeed, the evolution of superficially similar longirostrine phenotypes even appears attainable via multiple potential developmental pathways (Morris et al., 2019).

In addition to slender and sometimes elongate jaws, piscivorous crocodyliforms have tended to evolve relatively long retroarticular processes (e.g., dyrosaurs, thalattosuchians, gavialoids; Gignac & O'Brien, 2016). As an in-lever for the two largest jaw elevator muscles (*Musculus pterygoideus dorsalis* and *ventralis*; Holliday & Witmer, 2007), this feature might be taken to be an adaptation for higher maximum bite forces (Gignac & Erickson, 2016). However, the retroarticular process also serves as the in-lever for the *M. depressor mandibulae*, the sole jaw-opening muscle (Holliday & Witmer, 2007). As a result, convergently elongate retroarticular processes among aquatic and marine crocodyliforms may serve to enhance the force of subaquatic jaw opening, which is otherwise resisted by drag forces and the viscosity of water (Gignac et al., 2019). This hypothesis requires further evaluation but may represent an important potential functional integration within the cranio-dental complex of highly-piscivorous crocodyliforms (Gignac & O'Brien, 2016; Gignac et al., 2019).

Elongation of the jaws also impacts the mechanical response of the skull to loads imposed during prey capture, both by elevator muscles acting to close the jaws as well as reaction forces at bite points and jaw joints. Bone-strain experiments and biomechanical modeling of mesorostrine species (particularly for the model taxon *Alligator mississippiensis*) have demonstrated that during biting the snout is subjected to upward bending and twisting (Metzger et al., 2005). Interestingly, the upper jaw does not appear to be optimized for resisting the high feeding forces imposed by the jaw muscles and bite forces, suggesting competing functional demands (e.g., streamlining for stealth and lateral snapping movements in water) also impact skull shape. The lower jaws experience primarily dorsoventral bending and twisting about their long axes (Porro et al., 2013) with mediolateral bending constrained by the presence of enlarged pterygoid flanges (Porro et al., 2011).

Experimental and modeling results from *Alligator mississippiensis* form a baseline against which skull mechanical behavior during biting in other, rarer taxa can be compared. For example, even though geometric morphometric analyses show that the longirostrine taxa *Gavialis gangeticus* and *Tomistoma schlegelii* occupy disparate areas of morphospace—implying major skull-shape differences despite the fact that both are superficially "long snouted" (Brochu, 2001; Pierce et al., 2008)—two-dimensional finite element analysis (FEA) indicates that the upper jaws of both experience substantially higher stresses under feeding loads compared to shorter-snouted crocodylians (Pierce et al., 2008). Results from comparative three-dimensional FEA of the skulls of multiple crocodylian taxa also support these findings. McHenry et al. (2006) observed that the narrow snout of adult *Crocodylus johnstoni* experienced higher strains during biting than those of five blunt-snouted (brevirostrine) taxa. Similar patterns were exhibited by the lower jaws of crocodylians across a range of snout aspect ratios: the mandibles of blunt-snouted forms exhibited lower strains when subjected to simulated biting, shaking, and torsional loads compared to the mandibles of narrow- and long-snouted species (Walmsley et al., 2013).

Biomechanical modeling also reveals broad similarities in skull mechanical behavior across other clades exhibiting longirostrine morphologies (and piscivorous diets either documented or inferred based on convergent feeding functional morphology and habitat). For example, comparable strain differences were reported for short and long-snouted taxa of crocodylians and odontocetes during biting, shaking, and twisting, particularly at anterior bite points. Regardless of phylogenetic affinity, brevirostrine taxa experience lower strains than longirostrine ones, suggesting analogous form-function relationships even among unrelated clades (McCurry et al., 2017a). Convergence of skull shape and performance with those of longirostrine crocodylians has also been reported for Triassic phytosaurs, Jurassic thalattosuchians, and Cretaceous dinosaurs (Rayfield et al., 2007; Stubbs et al., 2013; Lemanis et al., 2019). This phylogenetic menagerie has also allowed researchers to probe how subtle shape differences and inferred loading behaviors between longirostrine species can result in variations in mechanical performance. For instance, initial simulation studies on the elongate snout of the spinosaurid dinosaur *Baryonyx* indicated that it performed more similarly to that of *Gavialis gangeticus* than *Alligator mississippiensis* in terms of its mechanical response to feeding loads (Rayfield et al., 2007). Further analyses, however, suggest that the response of the snout of the spinosaurid *Spinosaurus* seems to resemble that of *G. gangeticus* more so, whereas the mechanical performance of *Baryonyx* diverged from both that of the gharial and the closely related *Spinosaurus* (Cuff & Rayfield, 2013). Similarly, the long-snouted thalattosuchian *Pelagosaurus typus*, while sharing a skull morphology and general stress distribution patterns with *G. gangeticus,* exhibited lower mechanical resistance to simulated feeding forces, suggesting this fossil taxon may have specialized further for the consumption of compliant prey (Ballell et al., 2019).

As with their jaws, the teeth of slender-snouted crocodylians are also elongate. Tall, narrow, and conical teeth are thought to allow these ecomorphs to more readily seize elusive prey by spearing them during underwater jaw closure, as compared to their blunter-toothed, brevirostrine counterparts (Pooley, 1989; Grenard, 1991; Grigg et al., 2001). Although the posterior teeth are the most robust in the jaw, even in *Gavialis gangeticus*, they are considerably less molarifom than those of other crocodylians, which seize prey items with the anterior teeth and crush them posteriorly (Erickson et al., 2012). Because aquatic prey is typically swallowed whole, employing crushing bites via the posterior tooth row is rarely observed (Taylor, 1987). As a result, slender-snouted species are capable of using almost their entire jawline to spear prey (e.g., Thorbjarnarson, 1990), making them highly effective underwater hunters.

It was long suspected that semi- and highly-piscivorous taxa are bite-force limited and that evolution of their narrow and elongate jaws favored a trade-off from high-force to high-velocity biting (see e.g., McHenry et al., 2006). The latter, it is reasoned, provides advantages for rapid jaw closure to enable the securing of elusive prey that is capable of swimming off in any direction (Gignac et al., 2019). In 2012, Erickson and colleagues tested whether bite forces are correlated with dietary preferences. Of the five recognized slender-snouted crocodylians alive today (*Crocodylus johnstoni, C. intermedius, Gavialis gangeticus, Mecistops cataphracus*, and *Tomistoma schlegelii*), only *G. gangeticus* was an outlier for the clade (i.e., a quantitatively low-force biter; Erickson et al., 2012). Adults of the other four species reliably generate bite forces indistinguishable from those of all other remaining adult crocodylians, relative to their body size (Erickson et al., 2012, 2014). Notably, these four slender-snouted species maintain the pennate muscle-fiber arrangement and lateralized jaw insertion of the largest jaw elevator muscle (e.g., *Musculus pterygoideus ventralis;* Endo et al., 2002) common to all other crocodylians (Iordansky, 1964; Endo et al., 2002; Holliday & Witmer, 2007). By contrast, the lineage of *G. gangeticus* evolved fusiform-fibered jaw muscles with medially shifted insertion points (e.g., *M. pterygoideus ventralis;* Endo et al., 2002), which are advantageous for rapid—at the expense of forceful—jaw closure (Porro et al., 2011). In addition, the jaw musculature of *G. gangeticus* features a remarkably well-developed *M. pseudotemporalis superficialis* when compared to other slender-snouted crocodylians (Endo et al., 2002). Hypertrophy of this muscle suggests that it plays a greater role in jaw elevation in *G. gangeticus,* potentially compensating, to some extent, for the evolutionary shift away from the high-force generating, plesiomorphic arrangement of crocodylian jaw-closing musculature (Holliday & Witmer, 2009).

Exceptional bite-force capabilities would seem to be a peculiar feature of fish-eating taxa, not only because of their apparently delicate jaws and teeth but also because it allows them to generate exceptionally high tooth contact pressures (= maximum bite force/tooth contact area) when they engage with prey. Coupling high bite-force capacities with narrow tooth cross sections causes these pressures routinely to be more than ten-fold greater than those required to indent cortical bone, which is among the stiffest of vertebrate tissues (Carter & Beaupré, 2001; Turner et al., 2001; Erickson et al., 2012). That needle-toothed fish-eaters are capable of generating tooth pressures far in excess of those necessary to puncture their aquatic and marine prey (Erickson et al., 2012) suggests that semi- and highly-piscivorous crocodylians are mechanically over-capable for their feeding niches (Erickson et al., 2012, 2014; Gignac et al., 2019). Theoretically, this places their jaws and teeth closer to rupture when feeding on large prey, as demonstrated by beam theory and finite element analyses (Busbey, 1995; McHenry et al., 2006; Porro et al., 2011, 2013; Ballell et al., 2019). Why this should be the case likely has more to do with the phylogenetic inertia of evolving from large-bodied, generalist ancestors than it does with the functional value of driving delicate teeth through the compliant bodies of aquatic prey (Erickson et al., 2012). These forms seem to avoid damaging their svelte rostrodental features by behaviorally electing to consume relatively smaller and more compliant food resources (Erickson et al., 2012, 2014). Thus, although their slender and elongate rostrodental morphologies appear to have been shaped by their environments, convergence did not necessarily alter the full suite of their feeding capabilities (Erickson et al., 2012, 2014; Gignac & O'Brien, 2016).

What constitutes convergence of the crocodyliform jaw apparatus upon principally aquatic feeding modes? The relationships outlined above support phylogenetically broad, deep-time sampling that suggest rostra and dentitions have been plesiomorphically decoupled from the post-orbital region of the skull of crocodylians (Gignac & O'Brien, 2016) and their evolutionary precursors (Felice et al., 2019; Gignac et al., 2019; O'Brien et al., 2019). If generally applicable, then a common evolutionary sequence—pieced together from extant trait combinations—emerges: crocodyliforms may first converge upon needle-toothed and slender or long-snouted prey-capture morphologies without similarly extreme alteration of their musculoskeletal apparatus (e.g., *Crocodylus johnstoni, Mecistops cataphractus, Tomistoma schlegelii*; Iordansky, 1964; Endo et al., 2002); this may be followed by retroarticular process elongation and jaw elevator muscle evolution in the most specialized lineages (e.g., Gavialoidea; Brochu, 2001; Endo et al., 2002) as the trade-off of plesiomorphic, high bite-force capacities gives way to apomorphic, rapid jaw-closing velocities. *Gavialis gangeticus* is unlike all other crocodylians in that it possesses the most extreme of these character states (Iijima, 2017). A future focus on addressing the development, function, and evolution of gavialoid feeding biomechanics will go a long way to further clarifying convergence in the aquatic feeding of crocodyliforms, foremost, and secondarily aquatic tetrapods more generally.

7.1.4 *Intraoral Transport, Processing and Swallowing*

After a prey item is seized in the anterior part of the jaws, it is shifted posteriorly in the oral cavity using inertial feeding. Gravity alone, or the use of rapid, jerking head movements in coordination with jaw elevation and depression (Cleuren & De Vree, 1992), accelerates the prey item, harnessing its inertia to shift it backwards in the oral cavity (Taylor, 1987). The tongue is affixed to the floor of the oral cavity and does not participate substantially in intraoral prey transport (Busbey, 1989; Cleuren & De Vree, 1992). While feeding in water, crocodylians elevate the head above the water line to employ inertial feeding (Abercromby, 1913; Johnson, 1973; Taylor, 1987). A prey item may first be shifted to the posterior region of the dentition for the application of additional, more forceful bites, but ultimately it is moved towards the gular valve and esophagus for swallowing (Gans, 1969; Pooley & Gans, 1976).

Crocodylians are capable of capturing submerged prey without swallowing water owing to a soft palatine flap that forms a seal against the tongue (Fleming & Fontenot, 2015). This gular valve excludes water from entering the esophagus and trachea. It is assumed that for most crocodylians the valve also prevents swallowing of prey while submerged because even semi- and highly piscivorous species are seen to elevate their heads above the water line to achieve deglutition (Taylor, 1987). However, subaquatic swallowing is routinely and directly observed in captive individuals (e.g., *Crocodylus johnstoni*, *C. porosus*, *Tomistoma schelegelli*; St. Augustine Alligator Farm Zoological Park, 2020). Deglutition occurs by using cyclical muscular contractions of the gular region (Cleuren & De Vree, 1992). Prey items are pushed past the gular valve and into the esophagus using active protraction and retraction of the hyoid apparatus in small, repetitive orbits (Busbey, 1989; Cleuren & De Vree, 1992). This behavior is often assisted by forward thrusts of the head (especially underwater), gravity, or jaw depression and elevation to exaggerate the range of hyoid motion (Taylor, 1987; Cleuren & De Vree, 1992). Fish are typically, but not always, maneuvered so that they enter the esophagus head first (Thorbjarnarson, 1990; Sharma et al., 2013).

7.2 Birds

7.2.1 Introduction

Birds are characterized by the presence of feathers, the modification of the forelimbs to wings, toothless beaked jaws, hard-shelled eggs, a high metabolic rate along with a specialized, high-performance respiratory system, and a four-chambered heart. Birds comprise the most diverse terrestrial vertebrate group, with more than 18,000 described species (Barrowclough et al., 2016). Based on fossil, morphological, physiological and molecular biological evidence, birds are considered extant theropod dinosaurs–and accordingly, represent the extant sister group of crocodylians (Janke & Arnason, 1997; Prum, 2002). Birds are primarily terrestrial amniotes and most have retained the ability to fly, but some branches have adapted to aquatic lifestyles to different degrees and have evolved, on multiple independent occasions, strategies for exploiting aquatic food sources (Schwenk & Rubega, 2005; Rico-Guevara et al., 2019), including mechanisms as diverse as suspension feeding (filter feeding), surface skimming, scything, pursuit fishing, spearing of prey and, in at least one case, suction feeding.

7.2.2 Food Detection

Birds primarily use vision to detect food (Goldsmith, 1990). In fact, birds might be the most visually dependent group amongst vertebrates, reflected by the relatively large size of their eyes compared to those of other vertebrates of similar mass (Zeigler & Bischof, 1993). Most birds are tetrachromatic, with green, red, blue and ultraviolet (UV) sensitive cone photoreceptors in the retina (Wilkie et al., 1998). The photoreceptors of birds bear colored oil droplets which narrow spectral sensitivity and reduce the overlap in sensitivity between cone types, which in turn is hypothesized to improve color discrimination (Govardovskii, 1983; Vorobyev, 2003; Olsson et al., 2015). Bird eyes have adapted to a wide spectrum of functional demands (Zeigler & Bischof, 1993). Many aquatic birds, for instance, have evolved very flexible lenses that allow accommodation of their eyes to air and water (Gill, 1995). Next to vision, tactile cues play a central role for food detection in many aquatic feeding birds, and the number and distribution of mechanoreceptors on the beak and tongue are correlated with the respective feeding behavior (Gottschaldt, 1985). For example, mechanoreceptors are concentrated on the beak tip in shorebirds that engage in probing (von Bolze, 1968; Pettigrew & Frost, 1985). By contrast mechanoreceptors are concentrated on the tip and the lateral ridges of the beak, as well as on the fleshy tongue in mallards that use suspension feeding (Berkhoudt, 1979). A correlation of the distribution of mechanoreceptors and feeding mode allows fast feedback-responses to fine-tune the feeding behavior to a given situation. The role of olfaction has, for a long time, been underestimated in birds, but empirical studies have shown that the sense of smell plays an important role in food detection in many avian groups (Roper, 1999). Olfactory information is processed in the olfactory bulbs of the central nervous system and anatomical investigations have revealed that the olfactory bulbs are significantly enlarged in birds using olfactory cues to detect prey compared to non-olfactory-oriented birds (Bang & Cobb, 1968), a trait that has evolved several times independently in different bird branches. For example, some vultures, such as the turkey vulture, have large olfactory bulbs and are known to localize food by smell (Smith & Paselk, 1986). Similarly, it was shown experimentally that odoriferous baits attract various sea-bird species. For instance, cod-liver oil slicks deployed on the water surface induced specific search behaviors in storm-petrels and other sea-birds (Lequette et al., 1989; Verheyden & Jouventin, 1994). Acoustic location of prey has been shown to be employed by (mostly nocturnal) hawks and owls (Rice, 1982). Although the auditory systems also seem to be well-developed in aquatic feeding birds and acoustic communication is important to them, prey detection by acoustic cues seems to play a minor role. A sonar system for prey detection was once hypothesized for penguins (Poulter, 1969), but no morphological or physiological evidence has been found to support this idea (Wever et al., 1969).

7.2.3 Food Capture

Convergently with other groups of aquatic and marine vertebrates, such as cetaceans, tadpoles, actinopterygians, chondrichthyans and cyclostomes, birds have evolved elaborate mechanisms for filtering small food particles from the water by suspension feeding. Suspension feeding birds, such as mallards, flamingoes and some sea-birds, are equipped with rows of fine keratinized lamellae along the margins of their bill, which can strain small particles from the water (Jenkin, 1957; Zweers & Wouterlood, 1973; Kooloos et al., 1989; Sanderson & Wassersug, 1990). To induce water flow into the mouth, mallards and flamingoes oscillate their piston-like tongue in an anterior-posterior direction as the mouth opens and closes with each tongue cycle. Ingested water is released through the lateral lamellae of the beak where food particles are retained (Fig. 7.2). Keratinized spines located on the posterior margin of the tongue finally dislodge the entrapped food particles from the lamellae and draw them toward the esophagus as the tongue is retracted (Jenkin, 1957; Zweers et al., 1977; Kooloos et al., 1989; Sanderson & Wassersug, 1990). Similarly, some prions (*Pachyptila desolata*, *P. vittata*, *P. salvini*) are likely to use cyclic movements of their large muscular tongue to induce water flows across their filter apparatus (Klages & Cooper, 1992). Additionally, prions, along with the short-tailed shearwater (*Puffinus tenuirostris*), have also been reported to use a method whereby they

Fig. 7.2 Foraging flamingoes (Phoenicopteridae). Flamingoes typically forage by partly submerging their beak which is equipped with rows of fine keratinized lamellae along the upper bill. Oscillations of their piston-like tongue induce water flow across the fine lamellae whereupon food particles are entrapped. Photo by David Hensley on Unsplash

swim with their head and gaping beak patially submerged, allowing water to continuously flow in through the anterior region of the gape and leave at the posterior corners of the beak with food objects being retained at the palatal papillae in prions, or within the filter apparatus consisting of lateral lingual papillae that overlap with the lateral palatal papillae in the short-tailed shearwater (Prince, 1980; Morgan & Ritz, 1982; Klages & Cooper, 1992). The suspension feeding mode in prions shows striking convergence of both form and function with the mode used by right whales, *Eubalena* sp., shedding light on the origin of their trivial name: "whale-birds" (Sanderson & Wassersug, 1990).

Suspension feeding does not rely on visual detection of prey, but rather on mechanosensation (and possibly gustation, see Berkhoudt, 1985), where sensitive beaks and tongues allow fast feedback responses during feeding bouts (Berkhoudt, 1979). Similarly, scything and skimming rely on tactile cues but in contrast to suspension feeding, these feeding modes are used to catch larger prey (Becker et al., 2002; Swennen & Yu, 2005). Scything has evolved independently in a few branches of wading birds, including spoonbills (*Platalea* sp.) and avocets (*Recurvirostra* sp.), that typically feed in shallow waters (Becker et al., 2002). Although spoonbills and avocets have very different beak morphologies, their scything mechanism shows striking similarities. Scything birds submerge their slightly opened elongated beaks and sweep their heads from side to side while wading through the water (Becker et al., 2002). As the slightly opened beak contacts potential prey objects, mostly fish and crustaceans, they are quickly captured by the closing beak and swallowed (Swennen & Yu, 2005).

Skimming is a unique feeding behavior where the skimmer flies straight and close over the water surface with the mouth slightly open and the lower beak partially submerged. Food objects that are contacted by the lower beak are seized as a result of a fast reflex beak closure (Tomkins, 1951; Martin et al., 2007). Skimming is found in the scissorbills (*Rynchops* sp.) that are characterized by a special morphological adaptation for skimming, a substantially-elongated lower beak, which might largely prevent feeding methods other than skimming (Tomkins, 1951; Black & Harris, 1983).

In contrast to suspension feeding, scything and skimming, vision becomes essential for birds that target and strike at individual prey items. Herons (Ardeidae), for example, are mostly ambush predators and target their prey (usually fish, crustaceans or amphibians) while standing still in shallow water, or standing on platforms close to water (Kushlan, 1976). After estimating the position of the prey accurately, they strike by a sudden, rapid straightening of the long neck that thrusts the head forwards and downwards towards the prey (Katzir & Intrator, 1987; Lotem et al., 1991). The head can reach mean velocities of over 270 cm/s and, immediately before the bill contacts the prey, the beak is slightly opened to seize it (Katzir & Intrator, 1987). While smaller prey items are usually grasped by the closing beak, larger prey items are often stabbed and speared by the sharply pointed beak tips (Forbes, 1982).

Similarly (and convergently evolved) to herons, snakebirds (*Anhinga* sp.) possess long necks, elongated sharply pointed beaks and also capture aquatic prey by rapidly

Fig. 7.3 Snakebird (*Anhinga* sp.) with speared prey. Snakebirds are skilled divers and are equipped with a long neck and elongated, sharply pointed beak. After stalking prey, it is targeted and captured by rapidly straightening the curved neck to finally spear it. Photo by R. Mac Wheeler on Unsplash

straightening their neck to grasp–or more commonly–spear it (Owre, 1967) (Fig. 7.3). Just like herons, snakebirds strike at prey with a slightly open beak. However, in contrast to herons and egrets, snakebirds do not strike from an emergent position relative to their aquatic prey, but are instead skilled foot-propelled divers that quietly ambush their prey (usually fish) underwater to catch it using a sudden strike (Owre, 1967).

Probably the most spectacular prey capture mode amongst aquatic feeding birds is the plunge-dive. In short, a plunge-dive is a hunting strategy whereby birds dive head-first from the air into water (Duffy et al., 1986; Carl, 1987; Chang et al., 2016), and adaptations to it have convergently evolved in sulids (with gannets and boobies), terns (Sterninae), kingfishers (Alcedinidae), pelicans (Pelecanidae) and cormorants (Phalacrocoracidae). The northern gannet (*Morus bassanus*), for example, can dive from heights of up to 45 m, attaining speeds at the impact with water of more than 20 m/s. Such a dive would most probably be lethal to humans, but plunge-divers exhibit kinematic and morphological adaptations, such as a sharp, arrow-like body posture with a long and slender beak that, together, minimize drag and keep impact forces relatively low (Chang et al., 2016; Crandell et al., 2019). In fact, the sum of adaptations in plunge-divers results in only very low decelerations when hitting the water surface (Ropert-Coudert et al., 2004). Once submerged, the birds immediately grasp the targeted prey with their beak or use their momentum to travel underwater to attain a desired depth of up to 10 m (Ropert-Coudert et al., 2004). If the initial plunge-dive does not immediately result in successful prey capture, the birds can actively pursue the targeted prey item for a short time by using their feet or wings for propulsion (Ropert-Coudert et al., 2004). After a prey item is captured, the birds rise to the water surface, mostly passively due to their high buoyancy.

Other aquatic birds actively hunt prey underwater without plunge-diving. In fact, pursuit feeding is one of the most common feeding modes and has evolved many times independently in different branches of birds (Shealer, 2002). In pursuit hunts, prey (usually fish or crustaceans) is typically detected visually, stalked and grasped by the beak after a short chase. Pursuit hunters are fast and skilled divers that, depending on the species, use paddling by webbed feet, wing beats, or a combination thereof to propel themselves forwards (Townsend, 1909; Owre, 1967; Raikow et al., 1988). Many pursuit hunters show morphological adaptations to efficiently seize slippery prey, such as a terminal hook on the upper beak (Owre, 1967; Anderson et al., 1974; Shealer, 2002) or keratinous papillae and spines on beaks, palate and tongue (Kobayashi et al., 1998; Matsumoto & Evans, 2017). However, fast approaches to a prey might induce a positive pressure gradient in front of the birds' head that could push floating prey away (i.e. a "bow wave") or at least alert the prey organism to the approach of a predator (Taylor, 1987). How pursuit hunting birds circumvent hydrodynamic effects imposed by fast accelerations towards aquatic prey has not yet been studied in detail, but the generally streamlined body posture, in combination with a long and slender beak, might keep negative hydrodynamic effects low (Parfitt & Vincent, 2005; Crandell et al., 2019). Other aquatic vertebrates often use suction feeding to overcome negative hydrodynamic effects (see the section on turtles in this chapter). As of yet, only one bird species has been reported to use suction feeding: the little auk, *Alle alle,* a 150 g diving seabird of the North Atlantic (Enstipp et al., 2018), which actively chases small prey (copepods) underwater and engulfs them in the final stage by actively induced suction flows. Their suction feeding mechanism has not yet been studied in detail, but video recordings show that gular depression (sub-lingual pouch extension) shortly follows beak opening and induces prey ingestion (Enstipp et al., 2018). This movement pattern is very similar to the general pattern observed in other suction feeding vertebrates, including chondrichthyans, actinopterygians, dipnoans, lissamphibians, turtles and mammals: jaw opening is followed by gular (hyobranchial) depression (Lauder, 1980; Bemis & Lauder, 1986; Deban & Wake, 2000; Lemell et al., 2002; Wilga & Sanford, 2008; Kane & Marshall, 2009). Accordingly, the little auk nicely shows once more that suction feeding has evolved multiple times independently by convergence upon particular motion patterns.

Although suction feeding seems to be an important adaptation for aquatic predators, the masters of underwater pursuit hunting in birds, the penguins (Spheniscidae), are unlikely to use suction feeding (Charrassin et al., 2001; Takahashi et al., 2004). Penguins are capable of long and deep dives during which they catch prey such as shrimp and fish that are seized by the beak. Not much is known of the biomechanics of penguin feeding, but experiments with transponders have shown that penguins not only catch prey during their dives but also swallow prey under water (Charrassin et al., 2001; Takahashi et al., 2004).

7.2.4 Food Transport, Processing and Swallowing

After successful food capture, most aquatic-feeding birds must raise their head above the water level and use terrestrial transport mechanisms, such as fast dorsally directed head rotation while loosening the grip upon the prey item to throw it from the beak tip to the back of the oral cavity for swallowing (Owre, 1967; Forbes, 1982; Swennen & Yu, 2005) (i.e. inertial transport, see also the parts of this chapter dealing with lizards and crocodylians). Extensive intraoral processing is rare in birds, given that most of the physical processing action is performed in the gizzard (Van Gils et al., 2003; Fritz et al., 2011). Still, some aquatic feeding birds do direct a series of bites to the prey held in the beak. For example, herons have been reported to "chew" fishes so as to break their spines, scales and other protective mechanical adaptations (Forbes, 1982). Similarly, the high frequency beak movements of mallards might, next to their role in suspension feeding, be used to mechanically reduce food items before swallowing. However, all these processing functions occur with the head raised out of the water. Only a few birds are known to intraorally transport and swallow food underwater. The little auk catches large numbers of copepods individually, but how aquatic transport and swallowing is accomplished remains unknown. Penguins are exceptional as they might, next to auks, be the only birds that can intraorally transport and swallow underwater. Key to allowing intraoral transport in penguins is the interplay between the tongue and the palate (Matsumoto & Evans, 2017). Both palate and tongue are studded with large, sharply pointed keratinous papillae pointing rearwards and once prey is seized by the beak, cyclic pro- and retraction of the tongue moves any food object posteriorly (Kobayashi et al., 1998; Matsumoto & Evans, 2017).

7.3 Lepidosauria

7.3.1 Introduction

The taxon Lepidosauria contains over 9900 species (Uetz, 2010) and includes two orders, the Squamata and Rhynchocephalia. The Squamata comprises lizards, snakes and amphisbaenids, while the Rhynchocephalia is represented solely by the extant genus *Sphenodon*. Although body size, shape, and lifestyle varies significantly within lepidosaurs, they all possess overlapping keratinous scales. Some lepidosaurs are ferocious predators that chase prey larger than themselves, while others are ambush predators, insectivores, scavengers, omnivores or herbivores (Schwenk, 2000a). Some lepidosaurs are fast runners and skilled climbers, others live a fossorial lifestyle. Limbs can be well-developed, but have been lost independently in many groups, such as in snakes, amphisbaenids, anguids and pygopodids. The ancestral lifestyle of lepidosaurs is certainly terrestrial, but some groups live close to water, are semiaquatic or have evolved fully aquatic lifestyles.

7.3.2 Food Detection

Lepidosaurs have a full arsenal of sensory systems for detecting food at their disposal (Schwenk, 2000a) but vision might be the major sensory system for detecting food sources in most cases. Diurnal lepidosaurs are assumed to be capable of color vision. In fact, some studies show preferences of some lizards for food items of a certain color (Benes, 1969; McGovern et al., 1984).

Next to vision, most lepidosaurs rely heavily on chemosensory cues to detect food (Schwenk, 2000a). Chemosensation is achieved via three main systems: (1) gustation, (2) olfaction, and (3) the vomeronasal system. Gustation is mostly used for the discrimination of food once items have been seized (Berkhoudt, 1985; Schwenk, 1985). Gustatory cues are transmitted by taste buds which are located in the oropharyngeal cavity, including on the tongue of many species. By contrast, olfaction is used to detect more volatile chemicals; that is, to detect food from a distance (Kratzing, 1975; Bull et al., 1999). Olfaction is mediated by the olfactory epithelia that cover the nasal cavities. The vomeronasal system consists of the paired vomeronasal organs that lie dorsal to the anterior portion of the palate. Each vomeronasal organ houses a cavity that opens into the oral cavity and is lined with a chemosensory epithelium (Parsons, 1970), and it is this epithelium that is stimulated by environmental chemicals gathered by the tongue during a behavior known as tongue flicking (Halpern & Kubie, 1980; Schwenk, 1995, 2000a). Tongue flicking is best known for snakes and some lizards, but it is employed by virtually all lepidosaurs (Schwenk, 1995, 2000a). Infrared organs are known from snakes and are used for the perception of electromagnetic waves with a length of 8000–12,000 nm (Grace et al., 1999), which corresponds with the wavelength radiated from the surface of endothermic animals, such as mammals and birds (Goris, 2011). However, such infrared "vision" is so far known only for terrestrially-feeding snakes.

Aquatic snakes possess specialized mechanoreceptors (scale sensilla) that detect water motion and are likely used for prey detection (Van Der Kooij & Povel, 1996; Westhoff et al., 2005; Catania et al., 2010; Crowe-Riddell et al., 2016). Scale sensilla of aquatic snakes and the lateral line system of fishes and aquatic lissamphibians might be regarded as convergently evolved mechanosensitive systems. Although lepidosaurs show an impressive range of sensory systems available for food detection, mechanoreception, vision, and chemoreception are likely the most important sensory systems for detecting and localizing food under aquatic conditions (Drummond, 1983; Kutsuma et al., 2018).

7.3.3 Food Capture

Aquatic feeding is exhibited by many squamate clades. Snakes, in particular, are known to have secondarily evolved aquatic or semiaquatic lifestyles and aquatic food uptake (Cundall & Greene, 2000; Moon et al., 2019), but several lizards have convergently acquired semiaquatic lifestyles and are capable of aquatic feeding (Carpenter, 1966; Mayes et al., 2005; Mesquita et al., 2006; Langner, 2017).

Aquatic feeding has evolved multiple times, and independently in most major snake groups. It is found among boas, pythons, elapids, viperids and colubrids (Young, 1991; Cundall & Greene, 2000; Bilcke et al., 2006). Accordingly, snakes show a broad spectrum of convergent morphological, behavioral and physiological adaptations for catching prey under water. Several levels of aquatic commitment are known among snakes: from snakes that occasionally strike aquatic prey from land, to semiaquatic snakes that regularly enter aquatic habitats in search of prey, to permanently aquatic snakes that forage exclusively under water (Drummond, 1983). Regardless, whether a snake is semi- or fully aquatic, it has to overcome the same functional challenge: because of their morphological constraints, snakes, in general, are not capable of suction feeding. Such morphological constraints include a reduced hyobranchial skeleton and associated musculature (McDowell, 1972; Alfaro, 2002; Herrel et al., 2008) that is used in other aquatic vertebrates for rapid oropharyngeal volume expansion in suction feeding. As observed in other tetrapods that forage on elusive aquatic prey, at least some amount of suction feeding has been hypothesized to be advantageous for avoiding the bow wave generated in front of the accelerating head that would push prey away or alert prey of the approaching predator (Taylor, 1987). Nonetheless, the evolutionary success of aquatic feeding in snakes implies that they are capable of efficient aquatic feeding in the absence of being able to suction feed. So how do aquatic snakes strike prey and how do they cope with the physical constraints imposed by a medium that is about 850 times as dense and 50 times as viscous as air?

The ability of some snakes to execute very fast strikes underwater suggests that bow waves do not impose a universal constraint on aquatic feeding behavior (Alfaro, 2002). In other words, snakes have not evolved strategies to completely avoid bow waves during prey strikes, but they have evolved strategies to limit negative hydrodynamic effects to small enough values to permit the successful catching of prey (Herrel et al., 2008; Van Wassenbergh et al., 2009; Segall et al., 2019). Two main strategies are used by aquatic-feeding snakes to catch elusive prey: lateral and frontal strikes (Figs. 7.4 and 7.5). Lateral strikes are achieved by swinging the head to the side. Sideways movements can be continued until the head is directed 180° from its original orientation (Drummond, 1983). Depending on the snake species and the foraging situation, lateral strikes can be (1) slow and repetitive or (2) sudden and fast (Young, 1991; Alfaro, 2003). Repetitive sideways movements of the head with the gape open, referred to as "lateral head sweeping", are mostly used by snakes that feed in water with high prey densities and use tactile cues to search for prey (Drummond, 1983; Alfaro, 2002, 2003). In contrast, fast lateral strikes are used if the snake directs its attack to a specific prey item and can achieve peak velocities comparable to those of high-performance terrestrially-striking snakes (Smith et al., 2002; Alfaro, 2003; Catania, 2009). As the lateral strike is performed with the mouth gaping, the hydrodynamic disadvantage (e.g., pressure drag and bow wave) is reduced (Fig. 7.5c) compared to the situation with the mouth closed (Young,

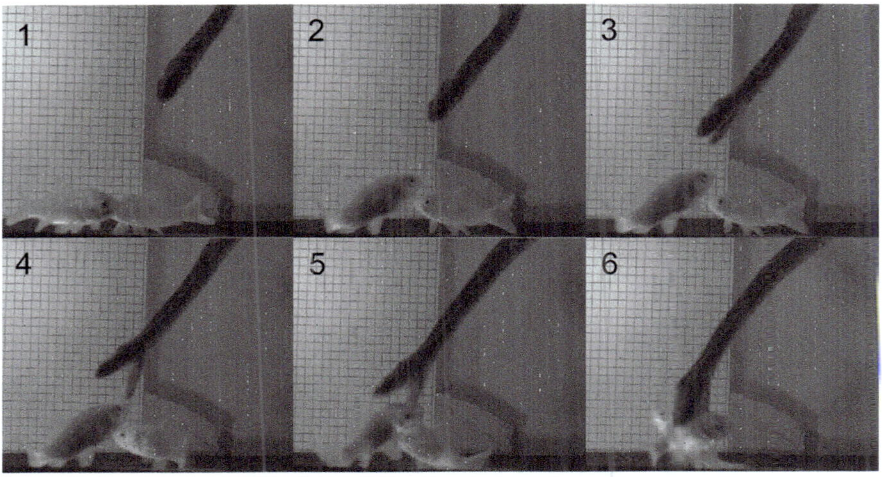

Fig. 7.4 Frame shots from a high-speed recording showing the semiaquatic snake *Natrix tessellata* striking frontally at a goldfish. The strike lasts approximately 100 ms. Courtesy of Sam Van Wassenbergh and Jonathan Brecko

1991; Braun & Cundall, 1995). Additionally, as surface area exposed to the fluid consists only of the lateral head area in lateral strikes, this capture mode is also feasible for snakes with relatively large and wide heads (Young, 1991; Vincent et al., 2009). Compared to frontal strikes, lateral strikes are usually not particularly accurate, probably because of the lack of visual overlap between the left and right eye, which makes estimates of prey distance difficult (Herrel et al., 2008). However, a particularly elaborate and effective lateral strike mechanism has been reported for the tentacled snake, *Erpeton tentaculatum*, which is a typical ambush predator and exploits the typical escape response (C-start maneuver) of fish to catch them efficiently. To do so, the tentacled snake feints with its trunk to elicit a C-start-response in a nearby fish, which startles if towards the snakes' approaching jaws, or a position the snake anticipates and strikes toward (Smith et al., 2002; Catania, 2009, 2010).

Frontal strikes in aquatic snakes are based on a fast forward acceleration of the head with the jaws open (Drummond, 1983; Van Wassenbergh et al., 2009). The forward acceleration of the head results from fast straightening of the curved trunk and neck, and prey is captured by the closing jaws (Drummond, 1983; Alfaro, 2003) (Fig. 7.4). Previously, it had been hypothesized that underwater strikes with open jaws may be hindered by drag and may generate bow waves that displace prey, making this method of capture more challenging (Young, 1991; Vincent et al., 2005; Moon et al., 2019). However, *in silico* (Van Wassenbergh et al., 2009) (Fig. 7.5b, c) and experimental (Segall et al., 2019) studies have shown that the hydrodynamic disadvantages are limited and that the head shapes of forward-striking snakes minimize the hydrodynamic constraints. Van Wassenbergh et al. (2009) showed that hydrodynamic drawbacks are minimized when snakes strike at large prey, as the

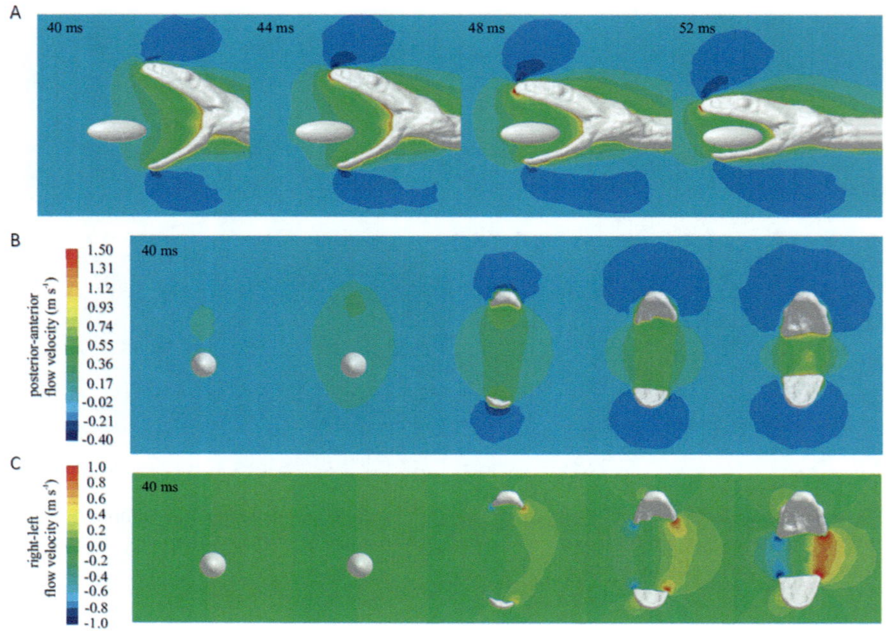

Fig. 7.5 Computational fluid dynamic (CFD) simulation of a frontal (**a**, **b**) and a lateral aquatic strike of *Natrix tessellata* along the posterior to anterior axis of the midsagittal plane (**a**), and for a series of frontal view planes (4 mm interval) at one time instant, showing anterior–posterior flow velocity (**b**) and right–left flow velocity (**c**). In this CFD simulation, the snake model translated with a forward velocity of 1 m/s and started to close its mouth at time = 40 ms. The prey contacted the lower jaw at time = 53 ms. The velocity scale is the same for (**a**) and (**b**). Modified from Van Wassenbergh et al. (2009)

inertia of large prey reduces the effect of the bow wave. Furthermore, precise aiming prevents the prey from deviating to a path that eludes the corners of the mouth. Indeed, most aquatic snakes forage for relatively large prey organisms (e.g., fish or amphibians) and frontal strikers have excellent underwater vision that allows precise aiming (Schaeffel & de Queiroz, 1990; Alfaro, 2002). Segall et al. (2019) showed that hydrodynamic drawbacks of frontal striking snakes can be further minimized by morphological adaptations, such as having narrow (Van Wassenbergh et al., 2009) or short (Segall et al., 2020) heads, if such modifications decrease the area exposed to the fluid. In fact, aquatic snakes have evolved, independently and on multiple occasions, narrower anterior regions of the head (Segall et al., 2016) and a shorter head (Segall et al., 2020).

Other strategies employed by aquatic snakes to overcome hydrodynamic effects imposed by the inability to perform suction feeding comprise trapping burrowing fishes in their burrows or crevices (Voris & Voris, 1983; Young, 1991) or very slow feeding modes, as for example found in the turtle-headed snake that feeds on fish eggs (Shine et al., 2004).

Fig. 7.6 Frame shots of a *Shinisaurus crocodilurus* individual that raises its head above the water line to intraorally transport and swallow the fish that was previously caught under water. Courtesy of Marvin Mulder

In contrast to snakes, aquatic feeding in lizards remains only superficially studied. The best-known example of aquatic feeding lizards is probably the marine iguana, *Amblyrhynchus cristatus*, from the Galapagos Islands. Marine iguanas spend most of their time on land but undertake prolonged dives to scrape algae from submerged rocks (Carpenter, 1966; Wikelski & Trillmich, 1994). The mechanics of its scraping behavior have not been studied in detail but based on the high inertia of algae tightly fixed to the substrate, it might be assumed that scraping algae from submerged rocks is not fundamentally different from grazing on land. Varanids, in contrast, are largely carnivorous and a few species are able to exploit aquatic food sources, including elusive prey such as fish, shrimps, or crabs (Mayes et al., 2005; Kulabtong & Mahaprom, 2014). These varanids locate aquatic prey by chemical cues and actively chase them (Mayes et al., 2005). The aquatic ingestion mode of varanids is unknown, but due to adaptations of the hyobranchial musculoskeletal system, it probably does not involve suction feeding (Smith, 1986). Instead, varanids might use laterally- or frontally-directed strikes similar to those of aquatic snakes or crocodylians (see the respective sections of this chapter), or a modified mechanism of suction feeding based on fast mouth opening, analogous to the method used by the Chinese giant salamander (Heiss et al., 2013). The earless monitor, *Lanthanotus borneensis*, is largely terrestrial, but it is known to be a skilled swimmer that regularly visits creeks to prey on fish and crustaceans (Harrisson, 1961; Harrisson & Haile, 1961; Langner, 2017). Its feeding mechanism has, to date, not been studied in any detail, but *Lanthanotus* might use similar strategies to the closely related varanids. Other aquatic foraging lizards include the teiids *Crocodilurus amazonicus* and *Dracaena guianensis* (Mesquita et al., 2006), the scincids *Tropidophorus hainanus* and *Sphenomorphus cryptotis* and the shinisaurid *Shinisaurus crocodilurus* (Fig. 7.6) (Ziegler et al., 2008), but their aquatic feeding mechanisms have not yet been studied. However, lizards might have evolved multiple ways of feeding underwater. For example, it has been shown that sometimes only small alterations of a terrestrial behavioral repertoire are necessary for exploiting aquatic food sources, as exemplified by Hawaiian *Anolis* lizards that have learned to use fast forward lunges to catch guppies that swim to the water surface of artificial fish tanks (Hawaii Hobbyist, 2019). In theory, the behavior exemplified by the Hawaiian *Anolis* lizards might be just a few functional steps away from more elaborate aquatic prey capture strategies.

7.3.4 Food Transport, Processing and Swallowing

Following prey capture and subjugation, snakes usually use a mechanism referred to as the "pterygoid walk", where alternate pro- and retraction of left and right jaws, plus the respective pterygoid bone, pull the prey item posteriorly (Kardong, 1977; Moon, 2000). The pterygoid walk can be used equally well in terrestrial and aquatic conditions. This snake-specific intraoral transport mechanism is more efficient in animals with relatively longer quadrate bones as the width and height of the posterior part of the head impacts the length of the lever arm involved in the pterygoid walk: the wider the head, the more efficient are intraoral transport and swallowing (Young, 1991; Vincent et al., 2009). Accordingly, aquatic snakes seem to be subject to a functional trade-off between prey capture and intraoral transport: narrow or short heads might reduce hydrodynamic drawbacks during a frontal strike, but render intraoral transport and swallowing slow (Vincent et al., 2009). On the other hand, broad or elongated heads, with longer quadrate bones, allow rapid intraoral transport and swallowing, but make frontal strikes hydrodynamically more challenging. Aquatic snakes might have solved this trade-off in two ways: lateral strikers can possess a wide head without suffering hydrodynamic drawbacks (Vincent et al., 2009) and frontally striking snakes have a streamlined, narrow anterior, but a wider posterior region of the head and/or longer quadrate bones (Segall et al., 2016; Rhoda et al., 2020). Anyway, we might still be far from fully understanding the form-function relationships and functional trade-offs characteristic of aquatic feeding snakes. Specifically, while many studies have focused on semiaquatic snakes, such as natricines (e.g. Bilcke et al., 2006; Van Wassenbergh et al., 2009; Vincent et al., 2009), others have taken a wider phylogenetic approach and have included semi-aquatic and fully aquatic species (e.g. Segall et al., 2016, 2019, 2020). The functional demands imposed on the semiaquatic species might differ fundamentally from those acting on the fully aquatic ones. In fact, the fully aquatic species tend to have short heads and the more semiaquatic species have long and narrow heads, suggesting that, in addition to the hydrodynamic constraints, semi-terrestrial habits induce additional constraints (M. Segall, pers. comm., June 2021). Phylogenetic mapping of morphological and functional solutions for overcoming the functional trade-off between aquatic strike and intraoral transport show that they have evolved multiple times independently (Segall et al., 2016).

Aquatic-feeding lizards probably use slightly modified terrestrial intraoral transport mechanisms with a submerged or emergent head, and move food posteriorly towards the esophagus by employing cyclic tongue loops (Schwenk, 2000a). Similar terrestrial feeding styles have been shown to be used by some semiaquatic turtles for intraoral transport (Natchev et al., 2010). Alternatively, lizards might raise their heads above the water line to make use of inertial transport (Smith, 1982, 1986) (Fig. 7.6) by quick dorsal or lateral head movements while temporarily releasing the grip on the prey, for example, literally throwing the prey through the mouth to the esophagus–analogous to mechanisms employed by crocodylians (Cleuren & De Vree, 1992, 2000). Hypotheses on aquatic transport in lizards remain speculative at this point, signaling the urgent need for empirical studies.

7.4 Testudines

7.4.1 Introduction

Turtles are one of the oldest known reptile orders, appearing about 240 million years ago, shortly after the Permian–Triassic extinction event. The phylogenetic position of turtles is still not fully resolved because morphological, developmental and genetic studies have been unable to reach a consensus on the relationships of the group within sauropsids (e.g., Anquetin, 2011; Bever et al., 2015). *Eunotosaurus africanus* is generally accepted as a stem turtle (Bever et al., 2015), followed by *Pappochelys rosinae* (Schoch & Sues, 2015) and *Odontochelys semitestatcea* (Li et al., 2008). In *Pappochelys* the plastron is not yet evident, but robust gastralia are present, indicating the origin of the plastron through fusion of the ventral ribs (Gilbert et al., 2001). *Odontochelys* represents the next step in shell evolution and has a fully developed plastron, with the dorsal shell consisting of neural plates and expanded ribs. This kind of evolutionary step (broadening of dorsal ribs) is also recognizable in the embryonic development of extant turtles (e.g., Sheil & Greenbaum, 2005; Scheyer et al., 2013).

Also unresolved is the question of what was the original habitat of the stem turtles. *Eunotosaurus* probably lived in terrestrial habitats (Bever et al., 2015), *Pappochelys* and *Odontochelys* have been described as being semiaquatic, living along lake shores with frequent visits to water (Schoch & Sues, 2015). However, Joyce (2015) argued that *Odontochelys* was likely a fully terrestrial stem turtle, and, at most, an inhabitant of swampy freshwater environments. The oldest known completely-shelled turtles (i.e. the Upper Triassic *Proterochersis* and *Proganochelys*), were likely semiaquatic (Gaffney, 1990); however, Joyce and Gauthier (2004) and Scheyer and Sander (2007) argued for a terrestrial habitat preference for *Proterochersis* and *Proganochelys* because of their forelimbs bearing short hands (Joyce & Gauthier, 2004) and because of similarities in shell bone histology to that of extant terrestrial turtles (Scheyer & Sander, 2007). In sum, turtles most likely had a terrestrial origin, thus representing an important taxon for the study of feeding in secondarily aquatic vertebrates.

Turtles are among the most morphologically specialized vertebrates. They have evolved an unusual body plan, with most of their body encased in a protective box of bone and keratin. Collectively, within the two testudinian suborders, Pleurodira

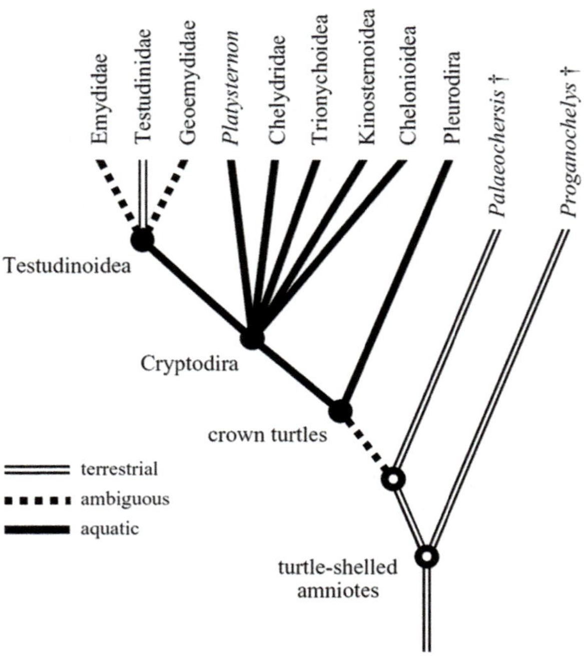

Fig. 7.7 Cladogram modified from Joyce and Gauthier (2004) illustrating the assumed habitat preferences of the major clades of crown turtles and their hypothetical ancestors. Joyce and Gauthier (2004) advocated terrestriality as the ancestral lifestyle for the entire clade, with convergent acquisition of aquatic habits in sauropterygians and crown turtles, and a subsequent reversal to terrestrial habits in the Testudinidae

(side-necked turtles) and Cryptodira (hidden-necked turtles), there are 14 extant families and around 470 species (Rhodin et al., 2017). These show adaptations to different lifestyles and are found from marine to freshwater to terrestrial habitats, as well as from temperate to tropical regions. All pleurodirans and many cryptodirans are fully aquatic, and only the cryptodiran superfamily Testudinoidea (Emydidae, Geoemydidae, and Testudinidae) has successfully reconquered terrestrial habitats (Fig. 7.7). Dietary preferences range from completely carnivorous to completely herbivorous, but most extant species are omnivorous. The two major foraging methods are sit-and-wait foraging and active foraging; in water typical aquatic feeding modes range from ram feeding to suction feeding.

7.4.2 *Food Detection*

Among turtles, visual, chemical, and tactile cues may be involved in food detection, with visual and olfactory senses being predominantly used in water. Turtles have photoreceptors containing colored oil droplets that appear to play a role in contrast enhancement and in protection from glare. The turtle eye is especially sensitive to red light (Granda & Dvorak, 1977). Similarly to tadpoles, aquatic turtles have retinal visual pigments (porphyropsins), which leads to a red shift and a considerable improvement of sensitivity in water (Reuter & Peichl, 2008). Marine turtles are emmetropic (normal sighted) in water and myopic (nearsighted) in air. Freshwater turtles show the opposite trend, being emmetropic in air but their eyes have a sufficiently developed accommodative range to be able to fully compensate for the lack of refraction at the cornea in water (Kröger & Katzir, 2008). Accordingly, the turtle eye is well suited for food detection under aquatic conditions.

Chemical cues can be detected via olfaction, vomeronasal chemoreception and gustation. Underwater nasal chemoreception is in general used for exploratory behavior, food location and discrimination, as well as reproductive behavior (Schwenk, 2008). The nasal cavity in turtles typically consists of a dorsal chamber containing the olfactory epithelium and a ventral "intermediate region" in which the vomeronasal epithelium lies (Tucker, 1971). Sea turtles (Cheloniidae and Dermochelyidae) are known to rely heavily on chemical cues to detect food sources and, in contrast to most marine animals, they surface to breathe and thus potentially have access to olfactory cues in both air and in water (Endres et al., 2009). Gustation is mediated by taste buds that are located within the oropharyngeal cavity. Taste buds are developed to a variable degree in the various turtle branches. For example, semiaquatic or terrestrial species are well equipped with taste buds (Heiss et al., 2008, 2011; Lintner et al., 2012), whereas highly aquatic species lack them on their poorly developed tongue (Lemell et al., 2002; Beisser et al., 2004) and sea turtles appear to lack taste buds completely (Iwasaki et al., 1996a, b). When present, taste buds can be distributed randomly throughout the oropharyngeal cavity (Heiss et al., 2011) or may show patterns of regional concentration that are correlated with the respective mode of food prehension. For instance, the semiaquatic turtle *Cuora amboinensis* shows aggregations of taste buds on the praechoanal palate and grasps food by its jaws (Heiss et al., 2008) Accordingly, the first contact with the food item occurs at the tip of the beak (i.e. the praechoanal region) that has high taste bud densities, allowing rapid feedback response and the avoidance of unpalatable items.

Next to vision and chemosensation, some aquatic turtles use mechanosensitive elements, such as mechanosensitive skin flaps and barbels on the anteroventral areas of the neck and head, to detect prey. Such mechanosensitive elements are sensitive to water motions and are, for instance, found in *Chelus fimbriatus*, a typical aquatic ambush predator (Wise et al., 1989) The skin flaps on the ventral region of the neck and the barbels (ventral to the mandibles) are innervated by peripheral nerve fibers and have been reported to be sensitive to small disturbances (Hartline, 1967).

7.4.3 Food Capture

When feeding in water, suction feeding is the predominant mode employed. Although most fish and larval salamanders use a suction feeding mechanism with uni-directional flow (water flows in through the mouth opening and out through the gill openings), reptiles and other secondarily aquatic vertebrates rely on a bidirectional flow system, wherein water flows in through the mouth but is then expelled through the nostrils or slightly opened jaws (Lauder & Prendergast, 1992). Such a bidirectional system implies a decrease in feeding performance due to lower negative pressure generated within the buccal cavity (Lauder & Shaffer, 1986). This impairment must be overcome by morphological specializations in secondarily aquatic lineages that lack gill openings. Such specializations in turtles comprise a flat and streamlined skull, as found in trionychids and chelids, which is very effective for fast forward movement of the head during capture as a bow wave can be minimized. Another trait of aquatic feeders is the enlarged supraoccipital bone which supports the jaw adductor musculature, along with an enlargement of the upper temporal area. Such a configuration can be used for generating the high bite forces (Herrel et al., 2002) typical of durophagous or carnivorous biting specialists. Additionally, such traits are advantageous for closing the mouth against water resistance during rapid forward movement of the head (e.g., in *Chelus fimbriatus*, Lemell et al., 2002). Further, morphological features typical of vertebrates related to feeding in aquatic habitats (Bramble & Wake, 1985) include: a large, rigid and well-ossified hyoid skeleton with massive hyoid musculature for rapid depression of the hyoid apparatus during the oropharyngeal expansion phase of the suction strike; a flat and smooth palate; a small tongue; and a short gape with labial folds for stabilizing the water flow inside the mouth cavity. All of these features are exhibited by freshwater turtles. However, marine turtles do not feature such morphologies, which is related to their feeding preferences. The skulls of marine turtles are more suited for forceful biting while feeding on relatively slow moving and sometimes armored prey (Jones et al., 2012). The tongue usually does not play a role during food manipulation in purely aquatic feeders. In such turtles the tongue tends to be reduced in size and in surface structure so as not to obstruct suction kinematics. *Chelus fimbriatus* is a good example of this configuration (Lemell et al., 2010) because its tongue is reduced to a tiny evagination anterior to the larynx and lacks any dorsal morphological differentiation. In more generalist aquatic to semiaquatic species, the tongue plays no role during prey capture but is used in some cases for further manipulation phases, similarly to more terrestrial species (Fig. 7.8).

In semi-aquatic species, skull design more closely resembles that of terrestrial species, being relatively tall and narrow. But some features indicate adaptation to an aquatic medium; for example, many of these taxa are moderate suction feeders, using suction to compensate for the bow wave forming in front of the head as it is accelerated forwards towards the prey (Natchev et al., 2009). Aquatically-adapted features include a flat palate, as in the predominantly aquatic *Cuora amboinensis,* in contrast to a vaulted one, as in the predominantly terrestrial *Cuora galbinifrons* (Natchev et al., 2009, 2010). The tongue of amphibious species is simple and poorly endowed with glands, and bears low to moderately high papillae that lack intrinsic muscles. These papillae are longer in more terrestrial species. Such traits are also obvious in salamanders, in which the surface topography of the tongue changes seasonally with the shift from an aquatic to a terrestrial lifestyle (Heiss et al., 2017).

When turtles feed under water, the kinematics (such as the timing of gaping and hyobranchial depression) show many similarities with other suction feeding vertebrates. A typical feeding sequence requires multiple gape cycles from ingestion to swallowing, so there is always some variation in kinematic patterns during prey

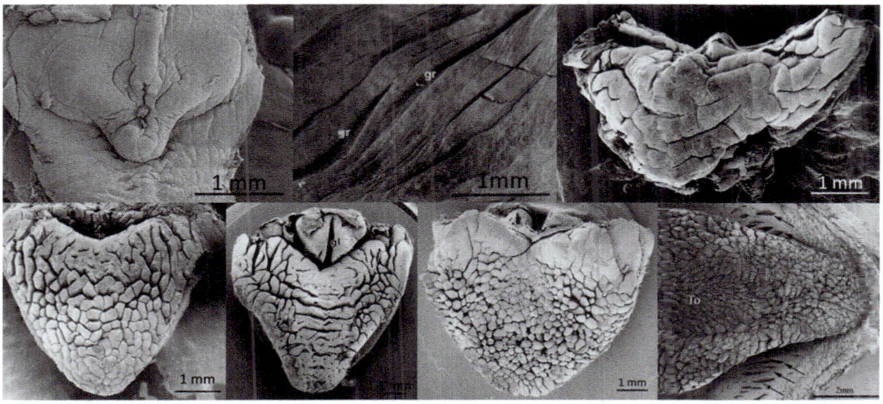

Fig. 7.8 Scanning electron microscopic images of the tongue surface of strictly aquatic (top) to semiaquatic to terrestrial (bottom) turtles. From left to right: top row *Chelus fimbriatus* (Lemell et al., 2010), *Acanthochelys pallidipectoris* (Beisser et al., 1995), *Pelusios castaneus* (Lemell et al., 2000); bottom row *Cuora amboinensis* (provided by CJ Beisser), *Rhinoclemmys pulcherrima* (Beisser et al., 2004), *Cuora galbinifrons* (Natchev et al., 2010), *Manouria emys emys* (Heiss et al., 2011). Compare the increase of size and number of dorsal papillary structures from highly aquatic to terrestrial

capture and manipulation phases. Although these movements are probably driven by a central pattern generator, sensory feedback is important for any adjustments associated with the respective feeding phase and the type and properties of the food (hard, soft) (Schwenk, 2000b). An absolutely necessary feature for successful ingestion is the anteroposterior sequence in peak excursions of head elements, which are mouth opening, hyoid depression and esophageal expansion. In aquatic-feeding turtles the hyoid apparatus has become enlarged to increase the volume of the mouth cavity during the expansion phase, as occurs with the suspensorium of teleost fish; the tongue, in contrast, has become reduced in size to minimize turbulence during prey uptake. Esophageal expansion appears to be convergent with the expansion of the opercular cavity of teleost fish, and serves to maintain the unidirectional flow posteriorly until the jaws have closed. This esophageal bulging commences via lateral dilatation of the second ceratobranchials, with further expansion achieved passively by the force of the incoming water. Pipid frogs show a similar modification for temporarily storing water sucked in during prey capture. The post-glottal pharynx of *Hymenochirus boettgeri* (Sokol, 1969) and the highly extensible buccopharyngeal region of *Pipa pipa* (Fernandez et al., 2017) serve as a temporary reservoir for a large volume of water.

The most specialized prey capture strategy in aquatic turtles may be that of the luring mechanism employed by *Macrochelys temminckii* (Drummond & Gordon, 1979; Spindel et al., 1987). *Macrochelys* has developed an unusual sit and wait feeding strategy: while remaining motionless they present a red, wiggling lure situated at the center of their widely opened jaws. This lure is a highly mobile, vermiform appendage of the tongue. If curious potential prey animals (usually fish or other turtles) approach closely enough they are caught by a sudden and violent strike.

7.4.4 Intraoral Transport, Processing and Swallowing

Further prey processing stages of aquatic turtles, such as intraoral transportation and swallowing, have not yet been sufficiently investigated. In aquatic feeding specialists, the prey is either reduced or sucked further backwards, or it is transported from the esophagus directly to the stomach by waves of contraction. The kinematic pattern of manipulation and intraoral transport phases is essentially identical to that typical of the capture phase, albeit much slower. Inertial feeding (*sensu* Gans, 1969) is typically used for intraoral transport (Lemell & Weisgram, 1997; Van Damme & Aerts, 1997). During transport, the jaws release the object and the head shifts anteriorly while the inertia of the object restricts its propensity to move. Hyobranchial depression supports these head movements by holding the prey in place or sucking it further backwards. During these manipulation cycles, reduced material, such as cracked shells of mollusks, can be expelled. The jaws are held slightly open to allow flushing of the residual water, with or without any content. Lemell et al. (2002) described two transport modes for *Chelus*, based upon the analysis of X-ray videos. They distinguished between two suction mechanisms: in the first the complete hyobranchial system undergoes several slight movements to carry the prey further backwards toward the esophagus. The second transport mode is characterized by a single, massive hyobranchial depression with much slower velocity, resulting in the prey item being sucked further in, as far as the posterior end of the pharynx, where it is held in place by the horns of the second branchial arch while the water is expelled. The tongue usually does not play a role during manipulation in purely aquatic feeders, but in semiaquatic species it does. Entirely lingually-based aquatic transport, with patterns of jaw, head and hyolingual movements resembling those of terrestrial prey positioning/transport, likely driven by the same motor-program, is executed in the manner described for the predominantly terrestrial *C. galbinifrons* (Bels et al., 2008; Natchev et al., 2010). Bels et al. (1998) compared food ingestion of the estuarine *Malaclemys terrapin* with that of the truly marine *Dermochelys coriacea*. Whereas *Malaclemys* is able to modulate at least the manipulation phases according to the particular prey type by using the tongue for further transport, *Dermochelys* uses actively generated water flow along with tongue movements. Rhythmic movements of the hyolingual apparatus carry food to the posterior end of the pharynx where it is swallowed by the action of the pharyngeal constrictors and moved to the stomach by peristaltic contractions of the esophageal musculature.

7.5 Conclusions

With numerous independent invasions of aquatic environments and trophic niches, sauropsids exemplify successful alternatives to a life on land. Retaining predominantly jaw-prehension feeding modes, aquatic lizards, snakes, crocodilians, and birds tend to use grasping as the means of food acquisition. On one hand, evolution of specializations that enable subaquatic head acceleration (especially in piscivorous reptiles) such as expansive necks and slender jaws seem to dominate convergent predatory phenotypes. On the other hand, many aquatic bird species have modified their approach to jaw prehension by harvesting small objects suspended in the water column via filter feeding. To induce a flow of water for filtration, these sauropsids evolved piston-like tongues that oscillate as the jaws open and close, enabling beak lamellae to capture food particles as water is expelled. Turtles deserve special attention as a stand-out reptile group because of their use of suction feeding, which is achieved by expansion of oropharyngeal volume—a feature shared among other sauropsids only by the little Auk.

The feeding strategies and functional morphologies explored in this chapter, however, are certainly not unique to sauropsids. For example, highly elaborate mechanisms for generating suction flows by rapid oropharyngeal volume expansion have evolved independently in such diverse groups as cartilaginous and bony fishes, salamanders, caecilians, anurans, cetaceans, and pinnipeds. Similarly, suspension feeding using complex food-particle trapping systems can be found in agnathans, cartilaginous and bony fishes, anuran tadpoles, and cetaceans. What seems so remarkable is not necessarily that sauropsids have converged on these tried-and-true feeding systems, but that such phylogenetically diverse groups have evolved similar approaches to food capture so regularly. In this regard, sauroposids are yet another attestation of the role that aquatic environments have played in shaping the body systems of even distantly related vertebrates.

Acknowledgements E.H. thanks R. Mac Wheeler, D. Hensley, M. Mulder, S. Van Wassenbergh and J. Brecko for providing the material used in Figs. 7.2, 7.3, 7.4, 7.5 and 7.6. P.M.G. and L.B.P. thank A. Resetar, P. Makovicky, and the Field Museum of Natural History for collections access. They also thank H. O'Brien, G. Erickson, K. Vliet, G. Webb, C. Ross, C. Holliday, K. Sellers, E. Rayfield, C. Brochu, A. Turner, J. Brueggen, E. Wilberg, A. Watanabe, S. Steppan and A. Lappin for more than a decade of thoughtful feedback about the evolution of crocodyliform feeding biomechanics. Support was provided to P.M.G. by the Department of Anatomy and Cell Biology at the Oklahoma State University Center for Health Sciences and the National Science Foundation (DEB 1754659).

References

Abercromby, A. F. (1913). How a crocodile feeds. *Spolia Zeylanica, 9*, 147–148.
Alexander, R. M. N. (1983). *Animal mechanics* (2nd ed.). Blackwell Scientific Publications.

Alfaro, M. E. (2002). Forward attack modes of aquatic feeding garter snakes. *Functional Ecology, 16*, 204–215.

Alfaro, M. E. (2003). Sweeping and striking: A kinematic study of the trunk during prey capture in three thamnophiine snakes. *The Journal of Experimental Biology, 206*, 2381–2392.

Anderson, B. W., Reeder, M. G., & Timken, R. L. (1974). Notes on the feeding behavior of the Common Merganser (*Mergus merganser*). *Condor, 76*, 472–476.

Anquetin, J. (2011). Reassessment of the phylogenetic interrelationships of basal turtles (Testudinata). *Journal of Systematic Palaeontology, 10*(1), 3–45.

Ballell, A., Moon, B. C., Porro, L. B., Benton, M. J., & Rayfield, E. J. (2019). Convergence and functional evolution of longirostry in crocodylomorphs. *Palaeontology, 62*(6), 867–887.

Bang, B. G., & Cobb, S. (1968). The size of the olfactory bulb in 108 species of birds. *Auk, 85*, 55–61.

Barrowclough, G. F., Cracraft, J., Klicka, J., & Zink, R. M. (2016). How many kinds of birds are there and why does it matter? *PLoS One, 11*, e0166307.

Becker, M., Rubega, M. A., & Oring, L. (2002). Development of feeding mechanics in growing birds: scything behavior in juvenile American avocets (*Recurvirostra americana*). *Bird Behavior, 15*, 1–10.

Beisser, C. J., Weisgram, J., & Splechtna, H. (1995). Dorsal lingual epithelium of *Platemys pallidipectoris* (Pleurodira, Chelidae). *Journal of Morphology, 226*, 267–276.

Beisser, C. J., Lemell, P., & Weisgram, J. (2004). The dorsal lingual epithelium of *Rhinoclemmys pulcherrima incisa* (Chelonia, Cryptodira). *The Anatomical Record, 277A*, 227–235.

Bels, V. L., Davenport, J., & Renous, S. (1998). Food ingestion in the estuarine turtle *Malaclemys terrapin*: Comparison with the marine leatherback turtle *Dermochelys coriacea*. *Journal of the Marine Biological Association of the UK, 78*, 953–972.

Bels, V. L., Baussart, S., Davenport, J., Shorten, M., O'Riordan, R. M., Renous, S., & Devneport, J. (2008). Functional evolution of feeding behaviour in turtles. In J. Wyneken, M. H. Godfrey, & V. Bels (Eds.), *Biology of turtles* (pp. 189–212). CRC Press Taylor & Francis Group.

Bemis, W., & Lauder, G. V. (1986). Morphology and function of the feeding apparatus of the lungfish, *Lepidosiren paradoxa* (Dipnoi). *Journal of Morphology, 187*, 81–108.

Benes, E. S. (1969). Behavioral evidence for color discrimination by the whiptail lizard, Cnemidophorus tigris. *Copeia, 1969*(4), 707–722.

Berkhoudt, H. (1979). The morphology and distribution of cutaneous mechanoreceptors (Herbst and Grandry corpuscles) in bill and tongue of the mallard (*Anas platyrhynchos* L.). *Netherlands Journal of Zoology, 30*, 1–34.

Berkhoudt, H. (1985). Structure and function of avian taste receptors. In A. S. King & J. McLelland (Eds.), *Form and function in birds* (pp. 463–496). Academic.

Bever, G. S., Lyson, T. R., Field, D. J., & Bhullar, B. S. (2015). Evolutionary origin of the turtle skull. *Nature, 525*, 239–242.

Bilcke, J., Herrel, A., & Van Damme, R. (2006). Correlated evolution of aquatic prey-capture strategies in European and American natricine snakes. *Biological Journal of the Linnean Society, 88*, 73–83.

Biswas, S. (1970). A preliminary survey of gharial in the Koshi River. *Indian Forester, 96*(9), 705–710.

Black, B. B., & Harris, L. D. (1983). Feeding habitat of Black Skimmers wintering on the Florida Gulf coast. *Wilson Bulletin, 95*(3), 404–415.

Bramble, D. M., & Wake, D. B. (1985). Feeding mechanisms of lower tetrapods. In M. Hildebrand, D. M. Bramble, K. F. Liem, & D. B. Wake (Eds.), *Functional vertebrate morphology* (Vol. 13, pp. 230–261). Harvard University Press.

Braun, T., & Cundall, D. (1995). Hydrodynamics of fishing in snakes (*Nerodia*). *American Zoologist, 35*, 105A.

Brazaitis, P. (1987). Identification of crocodilian skins and products. In G. J. Webb, S. C. Manolis, & P. J. Whitehead (Eds.), *Wildlife Management: Crocodiles and Alligators* (pp. 373–386). Surrey Beatty and Sons Pty Ltd.

Brochu, C. A. (2001). Crocodylian snouts in space and time: Phylogenetic approaches toward adaptive radiation. *American Zoologist, 41*(3), 564–585.

Bronzati, M., Montefeltro, F. C., & Langer, M. C. (2015). Diversification events and the effects of mass extinctions on Crocodyliformes evolutionary history. *Royal Society Open Science, 2*(5), 140385.

Buffetaut, E. (1982). Radiation évolutive, paléoécologie et biogéographie des crocodiliens mésosuchiens. *Mémoires de la Société Géologique de France, 142*, 1–88.

Bull, C. M., Griffin, C. L., & Johnston, G. R. (1999). Olfactory discrimination in scat-piling lizards. *Behavioral Ecology, 10*, 136–140.

Busbey, A. B. I. I. I. (1989). Form and function of the jaw musculature of *Alligator mississippiensis*. *Journal of Morphology, 202*, 99–127.

Busbey, A. B. (1995). The structural consequences of skull flattening in crocodilians. In J. J. Thomas (Ed.), *Functional morphology in vertebrate paleontology* (pp. 173–192). Cambridge University Press.

Carl, R. A. (1987). Age-class variation in foraging techniques by Brown Pelicans. *Condor, 89*(3), 525–533.

Carpenter, C. C. (1966). The marine iguana of the Galapagos Islands: Its behavior and ecology. *Proceedings of the California Academy of Sciences, 34*, 329–376.

Carter, D. R., & Beaupré, G. S. (2001). *Skeletal function and form: Mechanobiology of skeletal development, aging, and regeneration*. Cambridge University Press.

Catania, K. C. (2009). Tentacled snakes turn C-starts to their advantage and predict future prey behavior. *Proceedings of the National Academy of Sciences of the United States of America, 106*, 11183–11187.

Catania, K. C. (2010). Born knowing: Tentacled snakes innately predict future prey behavior. *PLoS One, 5*, e10953.

Catania, K. C., Leitch, D., & Gauthier, D. (2010). Function of the appendages in tentacled snakes (*Erpeton tentaculatus*). *The Journal of Experimental Biology, 213*, 359–367.

Chang, B., Croson, M., Straker, L., Gart, S., Dove, C., Gerwin, J., & Jung, S. (2016). How seabirds plunge-dive without injuries. *Proceedings of the National Academy of Sciences of the United States of America, 113*, 12006–12011.

Charrassin, J.-B., Kato, A., Handrich, Y., Sato, K., Naito, Y., Ancel, A., Bost, C.-A., Gauthier-Clerc, M., Ropert-Coudert, Y., & Le Maho, Y. (2001). Feeding behaviour of free–ranging penguins determined by oesophageal temperature. *Proceedings of the Royal Society London B, 268*, 151–157.

Cleuren, J., & De Vree, F. (1992). Kinematics of the jaw and hyolingual apparatus during feeding in *Caiman crocodilus*. *Journal of Morphology, 212*, 141–154.

Cleuren, J., & De Vree, F. (2000). Feeding in crocodilians. In K. Schwenk (Ed.), *Feeding: Form, function, and evolution in tetrapod vertebrates* (pp. 337–358). Academic.

Cochran, G. V. B. (1982). *A primer of orthopaedic biomechanics*. Churchill Livingstone.

Coombs, S., Görner, P., & Münz, H. (2012). *The mechanosensory lateral line: Neurobiology and evolution*. Springer.

Crandell, K., Howe, R., & Falkingham, P. (2019). Repeated evolution of drag reduction at the air–water interface in diving kingfishers. *Journal of the Royal Society Interface, 16*, 20190125.

Crowe-Riddell, J. M., Snelling, E. P., Watson, A. P., Suh, A. K., Partridge, J. C., & Sanders, K. L. (2016). The evolution of scale sensilla in the transition from land to sea in elapid snakes. *Open Biology, 6*, 160054.

Cuff, A. R., & Rayfield, E. J. (2013). Feeding mechanics in spinosaurid theropods and extant crocodilians. *PLoS One, 8*(5), e65295.

Cundall, D., & Greene, H. W. (2000). Feeding in snakes. In K. Schwenk (Ed.), *Feeding: Form, function, and evolution in tetrapod vertebrates* (pp. 293–333). Academic.

Deban, S., & Wake, D. (2000). Aquatic feeding in salamanders. In K. Schwenk (Ed.), *Feeding: Form, function and evolution in tetrapod vertebrates* (pp. 65–94). Academic.

Drummond, H. (1983). Aquatic foraging in garter snakes: A comparison of specialists and generalists. *Behaviour, 86*, 1–30.
Drummond, H., & Gordon, E. R. (1979). Luring in the neonate alligator snapping turtle (*Macroclemys temmincki*): Description and experimental analysis. *Zeitschrift für Tierpsychologie, 50*, 136–152.
Duffy, D. C., Wilson, R. P., Wilson, M. P., & Velasquez, C. R. (1986). Plunge-diving by olivaceous cormorants in Chile. *Wilson Bulletin, 98*, 607–608.
Endo, H., Aoki, R., Taru, H., Kimura, J., Sasaki, M., Yamamoto, M., Arishima, K., & Hayashi, Y. (2002). Comparative functional morphology of the masticatory apparatus in the long-snouted crocodiles. *Anatomia, Histologia, Embryologia, 31*(4), 206–213.
Endres, C. S., Putman, N. F., & Lohmann, K. J. (2009). Perception of airborne odors by loggerhead sea turtles. *The Journal of Experimental Biology, 212*, 3823–3827.
Enstipp, M. R., Descamps, S., Fort, J., & Grémillet, D. (2018). Almost like a whale–first evidence of suction feeding in a seabird. *The Journal of Experimental Biology, 221*, jeb182170.
Erickson, G. M., Gignac, P. M., Steppan, S., Lappin, A. K., Vliet, K. A., Brueggen, J. D., Inouye, B. D., Kledzik, D., & Webb, G. J. W. (2012). Insights into the ecology and evolutionary success of crocodilians revealed through bite force and tooth pressure experimentation. *PLoS One, 7*, e31781.
Erickson, G. M., Gignac, P. M., Lappin, A. K., Vliet, K. A., Brueggen, J. D., & Webb, G. J. W. (2014). A comparative analysis of ontogenetic bite-force scaling among Crocodylia. *The Journal of Zoology London, 292*(1), 48–55.
Erwin, D. H. (1992). A preliminary classification of evolutionary radiations. *Historical Biology, 6*(2), 133–147.
Felice, R. N., Watanabe, A., Cuff, A. R., Noirault, E., Pol, D., Witmer, L. M., Norell, M. A., O'Connor, P. M., & Goswami, A. (2019). Evolutionary integration and modularity in the archosaur cranium. *Integrative and Comparative Biology, 59*(2), 371–382.
Fernandez, E., Irish, F., & Cundall, D. (2017). How a frog, *Pipa pipa*, succeeds or fails in catching fish. *Copeia, 105*(1), 108–119.
Fleishman, L. J., Howland, H. C., Howland, M. J., Rand, A. S., & Davenport, M. L. (1988). Crocodiles don't focus underwater. *The Journal of Comparative Physiology A: Neuroethology, Sensory, Neural, and Behavioral Physiology, 163*(4), 441–443.
Fleming, G. J., & Fontenot, D. K. (2015). Crocodilians (crocodiles, alligators, caiman, gharial). In R. E. Miller & M. E. Fowler (Eds.), *Fowler's zoo and wild animal medicine* (Vol. 8, pp. 38–49). Elsevier.
Foffa, D., Young, M. T., Stubbs, T. L., Dexter, K. G., & Brusatte, S. L. (2018). The long-term ecology and evolution of marine reptiles in a Jurassic seaway. *Nature Ecology and Evolution, 2*(10), 1548–1555.
Forbes, L. (1982). Prey manipulation in the great blue heron. *The Murrelet, 63*, 89–91.
Forsyth, H. W. (1910). The food of crocodiles. *The Journal of the Bombay Natural History Society, 20*, 1160.
Fritz, J., Hummel, J., Kienzle, E., Wings, O., Streich, W. J., & Clauss, M. (2011). Gizzard vs. teeth, it's a tie: Food-processing efficiency in herbivorous birds and mammals and implications for dinosaur feeding strategies. *Paleobiology, 37*, 577–586.
Gaffney, E. S. (1990). The comparative osteology of the triassic turtle *Proganochelys*. *Bulletin of the American Museum of Natural History, 194*, 1–268.
Gans, C. (1969). Comments on inertial feeding. *Copeia, 1969*(4), 855–857.
Gignac, P. M., & Erickson, G. M. (2016). Ontogenetic bite-force modeling of *Alligator mississippiensis*: Implications for dietary transitions in a large-bodied vertebrate and the evolution of crocodylian feeding. *Journal of Zoology, 299*(4), 229–238.
Gignac, P. M., & O'Brien, H. D. (2016). Suchian feeding success at the interface of ontogeny and macroevolution. *Integrative and Comparative Biology, 56*(3), 449–458.
Gignac, P. M., O'Brien, H. D., Turner, A. H., & Erickson, G. M. (2019). Feeding in crocodylians and their relatives: Functional insights from ontogeny and evolution. In V. Bels & I. Q. Wishaw

(Eds.), *Feeding in vertebrates: Evolution, morphology, behavior, biomechanics* (pp. 575–610). Springer.

Gilbert, S. F., Loredo, G. A., Brukman, A., & Burke, A. C. (2001). Morphogenesis of the turtle shell: The development of a novel structure in tetrapod evolution. *Evolution and Development, 3*, 47–58.

Gill, F. B. (1995). *Ornithology*. Macmillan.

Godoy, P. L. (2020). Crocodylomorph cranial shape evolution and its relationship with body size and ecology. *Journal of Evolutionary Biology, 33*(1), 4–21.

Goldsmith, T. H. (1990). Optimization, constraint, and history in the evolution of eyes. *The Quarterly Review of Biology, 65*, 281–322.

Goris, R. C. (2011). Infrared organs of snakes: An integral part of vision. *Journal of Herpetology, 45*, 2–14.

Gottschaldt, K. (1985). Structure and function of avian somatosensory receptors. In A. S. King & J. McLelland (Eds.), *Form and function in birds* (Vol. 3, pp. 375–461). Academic.

Govardovskii, V. I. (1983). On the role of oil drops in colour vision. *Vision Research, 23*, 1739–1740.

Grace, M. S., Church, D. R., Kelly, C. T., Lynn, W. F., & Cooper, T. M. (1999). The Python pit organ: imaging and immunocytochemical analysis of an extremely sensitive natural infrared detector. *Biosensors & Bioelectronics, 14*, 53–59.

Granda, A. M., & Dvorak, C. A. (1977). Vision in turtles. The visual system in vertebrates. In F. Crescitelli (Ed.), *Handbook of sensory physiology* (Vol. 7/5). Springer.

Grenard, S. (1991). *Handbook of alligators and crocodiles*. Krieger Publishing.

Grigg, G. C., Seebacher, F., & Franklin, C. E. (2001). *Crocodilian biology and evolution*. Surrey Beatty.

Halpern, M., & Kubie, J. L. (1980). Chemical access to the vomeronasal organs of garter snakes. *Physiology & Behavior, 24*, 367–371.

Harrisson, B. (1961). *Lanthanotus borneensis*-habits and observations. *The Sarawak Museum Journal, 10*, 286–292.

Harrisson, T., & Haile, N. (1961). A rare earless monitor lizard from Borneo. *Nature, 190* 1213–1213.

Hartline, P. (1967). The unbelievable fringed turtle. *International Tortoise and Turtle Society Journal, 1*, 24–29.

Hawaii Hobbyist. (2019, September 14). *Lizard eats guppy for lunch* [video file]. YouTube. Accessed December 2, 2020, from https://www.youtube.com/watch?v=pWq3r20jjIM

Heiss, E., Plenk, H. J., & Weisgram, J. (2008). Microanatomy of the palatal mucosa of the semiaquatic Malayan box turtle, *Cuora amboinensis*, and functional implications. *The Anatomical Record, 291*, 876–885.

Heiss, E., Natchev, N., Schwaha, T., Salaberger, D., Lemell, P., Beisser, C., & Weisgram, J. (2011). Oropharyngeal morphology in the basal tortoise *Manouria emys* emys with comments on form and function of the testudinid tongue. *Journal of Morphology, 272*, 1217–1229.

Heiss, E., Natchev, N., Gumpenberger, M., Weissenbacher, A., & Van Wassenbergh, S. (2013). Biomechanics and hydrodynamics of prey capture in the Chinese giant salamander reveal a high-performance jaw-powered suction feeding mechanism. *Journal of the Royal Society Interface, 10*(82), 20121028.

Heiss, E., Handschuh, S., Aerts, P., & Van Wassenbergh, S. (2017). A tongue for all seasons: Extreme phenotypic flexibility in salamandrid newts. *Scientific Reports, 7*, 1006.

Herrel, A., O'Reilly, J. C., & Richmond, A. M. (2002). Evolution of bite performance in turtles. *Journal of Evolutionary Biology, 15*, 1083–1094.

Herrel, A., Vincent, S., Alfaro, M., Van Wassenbergh, S., Vanhooydonck, B., & Irschick, D. (2008). Morphological convergence as a consequence of extreme functional demands: Examples from the feeding system of natricine snakes. *Journal of Evolutionary Biology, 21*, 1438–1448.

Holliday, C. M., & Witmer, L. M. (2007). Archosaur adductor chamber evolution: integration of musculoskeletal and topological criteria in jaw muscle homology. *Journal of Morphology, 268*(6), 457–484.

Holliday, C. M., & Witmer, L. M. (2009). The epipterygoid of crocodyliforms and its significance for the evolution of the orbitotemporal region of eusuchians. *The Journal of Vertebrate Paleontology, 29*(3), 715–733.

Howland, H. C., & Sivak, J. (1984). Penguin vision in air and water. *Vision Research, 24*(12), 1905–1909.

Hua, S., & De Buffrenil, V. (1996). Bone histology as a clue in the interpretation of functional adaptations in the Thalattosuchia (Reptilia, Crocodylia). *The Journal of Vertebrate Paleontology, 16*(4), 703–717.

Iijima, M. (2017). Assessment of trophic ecomorphology in non-alligatoroid crocodylians and its adaptive and taxonomic implications. *Journal of Anatomy, 231*(2), 192–211.

Iordansky, N. N. (1964). The jaw muscles of the crocodiles and some relating structures of the crocodilian skull. *Anatomischer Anzeiger, 115*, 256–280.

Iordansky, N. N. (1973). The skull of the Crocodilia. In C. Gans & T. S. Parson (Eds.), *The biology of the Reptilia* (Vol. 4, pp. 210–262). Academic.

Iwasaki, S. I., Asami, T., & Wanichanon, C. (1996a). Fine structure of the dorsal lingual epithelium of the juvenile hawksbill turtle, *Eretmochelys imbricata bissa*. *The Anatomical Record, 244*, 437–443.

Iwasaki, S. I., Wanichanon, C., & Asami, T. (1996b). Histological and ultrastructural study of the lingual epithelium of the juvenile Pacific ridley turtle, *Lepidochelys olivacea* (Chelonia, Cheloniidae). *Annals of Anatomy, 178*, 243–250.

Jackson, K., & Brooks, D. R. (2007). Do crocodiles co-opt their sense of "touch" to "taste"? A possible new type of vertebrate sensory organ. *Amphibia-Reptilia, 28*(2), 277–285.

Jackson, K., Butler, D. G., & Youson, J. H. (1996). Morphology and ultrastructure of possible integumentary sense organs in the estuarine crocodile *Crocodylus porosus*. *Journal of Morphology, 229*(3), 315–324.

Janke, A., & Arnason, U. (1997). The complete mitochondrial genome of *Alligator mississippiensis* and the separation between recent archosauria (birds and crocodiles). *Molecular Biology and Evolution, 14*, 1266–1272.

Jenkin, P. M. (1957). The filter-feeding and food of flamingoes (Phoenicopteri). *Philosophical Transactions of the Royal Society London B Biological Sciences, 240*, 401–493.

Johnson, C. R. (1973). Behaviour of the Australian crocodiles, *Crocodylus johnstoni* and *C. porosus*. *Zoological Journal of the Linnean Society, 52*(4), 315–336.

Jones, M. E. H. (2008). Skull shape and feeding strategy in Sphenodon and other Rhynchocephalia (Diapsida: Lepidosauria). *Journal of Morphology, 269*(8), 945–966.

Jones, M. E. H., Werneburg, I., Curtis, N., Penrose, R., O'Higgins, P., Fagan, M. J., & Evans, S. E. (2012). The head and neck anatomy of sea turtles (Cryptodira: Chelonioidea) and skull shape in testudines. *PLoS One, 7*(11), e47852.

Joyce, W. G. (2015). The origin of turtles: a paleontological perspective. *The Journal of Experimental Zoology, 324*(3), 181–193.

Joyce, W. G., & Gauthier, J. A. (2004). Palaeoecology of Triassic stem turtles sheds new light on turtle origins. *Proceedings of the Royal Society of London B, 271*, 1–5.

Kane, E., & Marshall, C. (2009). Comparative feeding kinematics and performance of odontocetes: Belugas, Pacific white-sided dolphins and long-finned pilot whales. *The Journal of Experimental Biology, 212*, 3939–3950.

Kardong, K. V. (1977). Kinesis of the jaw apparatus during swallowing in the cottonmouth snake, *Agkistrodon piscivorus*. *Copeia, 1977*, 338–348.

Katzir, G., & Intrator, N. (1987). Striking of underwater prey by a reef heron, *Egretta gularis schistacea*. *Journal of Comparative Physiology A, 160*, 517–523.

Klages, N. T., & Cooper, J. (1992). Bill morphology and diet of a filter-feeding seabird: The broad-billed prion *Pachyptila vittata* at South Atlantic Gough Island. *Journal of Zoology, 227*, 385–396.

Kobayashi, K., Kumakura, M., Yoshimura, K., Inatomi, M., & Asami, T. (1998). Fine structure of the tongue and lingual papillae of the penguin. *Archives of Histology and Cytology, 61*, 37–46.

Kooloos, J. G. M., Kraaijeveld, A. R., Langenbach, G. E. J., & Zweers, G. A. (1989). Comparative mechanics of filter feeding in *Anas platyrhynchos*, *Anas clypeata* and *Aythya fuligula* (Aves, Anseriformes). *Zoomorphology, 108*, 269–290.

Kratzing, J. E. (1975). The fine structure of the olfactory and vomeronasal organs of a lizard (*Tiliqua scincoides scincoides*). *Cell and Tissue Research, 156*, 239–252.

Kröger, R. H. H., & Katzir, G. (2008). Comparative anatomy and physiology of vision in aquatic tetrapods. In J. G. M. Thewissen (Ed.), *Sensory evolution on the threshold – Adaptations in secondarily aquatic vertebrates*. University of California Press.

Kulabtong, S., & Mahaprom, R. (2014). Observation on food items of Asian water monitor, *Varanus salvator* (Laurenti, 1768) (Squamata Varanidae), in urban eco-system, Central Thailand. *Biodiversity Journal, 6*, 695–698.

Kushlan, J. A. (1976). Feeding behavior of North American herons. *Auk, 93*, 86–94.

Kutsuma, R., Sasai, T., & Kishida, T. (2018). How snakes find prey underwater: Sea snakes use visual and chemical cues for foraging. *Zoological Science, 35*, 483–486.

Langner, C. (2017). Of an enigmatic lizard: First records of habitat use, behavior, and food items of *Lanthanotus borneensis* Steindachner, 1878 in its natural habitat. *Russian Journal of Herpetology, 24*, 1–10.

Langston, W. J. (1973). The crocodilian skull in historical perspective. In C. Gans & T. S. Parson (Eds.), *The biology of the Reptilia* (Vol. 4, pp. 263–284). Academic.

Lauder, G. V. (1980). Evolution of the feeding mechanism in primitive actinopterygian fishes: A functional anatomical analysis of *Polypterus*, *Lepisosteus*, and *Amia*. *Journal of Morphology, 163*, 283–317.

Lauder, G. V., & Prendergast, T. (1992). Kinematics of aquatic prey capture in the snapping turtle *Chelydra serpentina*. *The Journal of Experimental Biology, 164*, 55–78.

Lauder, G. V., & Shaffer, H. B. (1986). Functional design of the feeding mechanism in lower vertebrates: Unidirectional and bidirectional flow systems in the tiger salamander. *Zoological Journal of the Linnean Society, 88*, 277–290.

Leitch, D. B., & Catania, K. C. (2012). Structure, innervation and response properties of integumentary sensory organs in crocodilians. *Journal of Experimental Biology, 215*(23), 4217–4230.

Lemanis, R., Jones, A. S., Butler, R. J., Anderson, P. S., & Rayfield, E. J. (2019). Comparative biomechanical analysis demonstrates functional convergence between slender-snouted crocodilians and phytosaurs. *PeerJ Preprints, 7*, e27476v1.

Lemell, P., & Weisgram, J. (1997). Feeding patterns of *Pelusios castaneus*. *Netherlands Journal of Zoology, 47*(4), 429–441.

Lemell, P., Beisser, C. J., & Weisgram, J. (2000). Morphology and function of the feeding apparatus of *Pelusios castaneus* (Chelonia; Pleurodira). *Journal of Morphology, 244*, 127–135.

Lemell, P., Lemell, C., Snelderwaard, P., Gumpenberger, M., Wocheslander, R., & Weisgram, J. (2002). Feeding patterns of *Chelus fimbriatus* (Pleurodira: Chelidae). *The Journal of Experimental Biology, 205*, 1495–1506.

Lemell, P., Beisser, C. J., Gumpenberger, M., Snelderwaard, P., Gemel, R., & Weisgram, J. (2010). The feeding apparatus of *Chelus fimbriatus* (Pleurodira; Chelidae) – Adaptation perfected? *Amphibians and Reptiles, 31*, 97–107.

Lequette, B., Verheyden, C., & Jouventin, P. (1989). Olfaction in subantarctic seabirds: Its phylogenetic and ecological significance. *Condor, 91*, 732–735.

Li, C., Wu, X. C., Rieppel, O., Wang, L., & Zhao, L. (2008). An ancestral turtle from the Late Triassic of southwestern China. *Nature, 456*, 497–501.

Liem, K. R., Bemis, W. E., Walker, W. F., Jr., & Grande, L. (2001). *Functional anatomy of the vertebrates: An evolutionary perspective*. Thomson and Brooks/Cole.

Lintner, M., Weissenbacher, A., & Heiss, E. (2012). The oropharyngeal morphology in the semiaquatic giant Asian pond turtle, *Heosemys grandis*, and its evolutionary implications. *PLoS One, 7*(9), e46344.

Lotem, A., Schechtman, E., & Katzir, G. (1991). Capture of submerged prey by little egrets, *Egretta garzetta garzetta*: strike depth, strike angle and the problem of light refraction. *Animal Behaviour, 42*, 341–346.

Mannion, P. D., Benson, R. J., Carrano, M. T., Tennant, J. P., Judd, J., & Butler, R. J. (2015). Climate constrains the evolutionary history and biodiversity of crocodylians. *Nature Communications, 6*(1), 8438.

Martin, G. R., Mcneil, R., & Rojas, L. M. (2007). Vision and the foraging technique of skimmers (Rynchopidae). *Ibis, 149*, 750–757.

Massare, J. A. (1987). Tooth morphology and prey preference of Mesozoic marine reptiles. *The Journal of Vertebrate Paleontology, 7*(2), 121–137.

Matsumoto, R., & Evans, S. E. (2017). The palatal dentition of tetrapods and its functional significance. *Journal of Anatomy, 230*, 47–65.

Mayes, P., Thompson, G., & Withers, P. (2005). Diet and foraging behaviour of the semi-aquatic *Varanus mertensi* (Reptilia: Varanidae). *Wildlife Research, 32*, 67–74.

McCurry, M. R., Evans, A. R., Fitzgerald, E. M. G., Adams, J. W., Clausen, P. D., & McHenry, C. R. (2017a). The remarkable convergence of skull shape in crocodilians and toothed whales. *Proceedings of the Royal Society B, 284*(1850), 20162348.

McCurry, M. R., Walmsley, C. W., Fitzgerald, E. M. G., & McHenry, C. R. (2017b). The biomechanical consequences of longirostry in crocodilians and odontocetes. *Journal of Biomechanics, 56*, 61–70.

McDowell, S. B. (1972). The evolution of the tongue of snakes, and its bearing on snake origins. In T. Dobzhansky, M. K. Hecht, & W. C. Steere (Eds.), *Evolutionary biology* (pp. 191–273). Springer.

McGovern, G. M., Mitchell, J. C., & Knisley, C. B. (1984). Field experiments on prey selection by the whiptail lizard, *Cnemidophorus inornatus*, in Arizona. *Journal of Herpetology, 18*, 347–349.

McHenry, C. R., Clausen, P. D., Daniel, W. J. T., Meers, M. B., & Pendharkar, A. (2006). Biomechanics of the rostrum in crocodilians: A comparative analysis using finite-element modeling. *The Anatomical Record Part A, 288A*, 827–849.

Mesquita, D. O., Colli, G. R., Costa, G. C., França, F. G., Garda, A. A., & Péres, A. K., Jr. (2006). At the water's edge: Ecology of semiaquatic teiids in Brazilian Amazon. *Journal of Herpetology, 40*, 221–229.

Metzger, K. A., Daniel, W. J. T., & Ross, C. F. (2005). Comparison of beam theory and finite-element analysis with in vivo bone strain data from the alligator cranium. *The Anatomical Record, 283A*(2), 331–348.

Moon, B. R. (2000). The mechanics of swallowing and the muscular control of diverse behaviours in gopher snakes. *The Journal of Experimental Biology, 203*, 2589–2601.

Moon, B. R., Penning, D. A., Segall, M., & Herrel, A. (2019). Feeding in snakes: Form, function, and evolution of the feeding system. In V. Bels & I. Q. Whishaw (Eds.), *Feeding in vertebrates* (pp. 527–574). Springer.

Morgan, W. L., & Ritz, D. A. (1982). Comparison of the feeding apparatus in the muttonbird, *Puffinus tenuirostris* (Temminck) and the fairy prion, *Pachyptila turtur* (Kuhl) in relation to the capture of the krill, *Nyctiphanes australis* Sars. *Journal of Experimental Marine Biology and Ecology, 59*, 61–75.

Morris, Z. S., Vliet, K. A., Abzhanov, A., & Pierce, S. E. (2019). Heterochronic shifts and conserved embryonic shape underlie crocodylian craniofacial disparity and convergence. *Proceedings of the Royal Society B, 286*(1897), 20182389.

Nagloo, N., Collin, S. P., Hemmi, J. M., & Hart, N. S. (2016). Spatial resolving power and spectral sensitivity of the saltwater crocodile, *Crocodylus porosus*, and the freshwater crocodile, Crocodylus johnstoni. *Journal of Experimental Biology, 219*(9), 1394–1404.

Natchev, N., Heiss, E., Lemell, P., Stratev, D., & Weisgram, J. (2009). Analysis of prey capture and food transport kinematics in two Asian box turtles, *Cuora amboinensis* and *Cuora flavomarginata* (Chelonia, Geoemydidae), with emphasis on terrestrial feeding patterns. *Zoology, 112*(2), 113–127.

Natchev, N., Lemell, P., Heiss, E., Beisser, C. J., & Weisgram, J. (2010). Aquatic feeding in a terrestrial turtle: a functional-morphological study of the feeding apparatus in the Indochinese box turtle *Cuora galbinifrons* (Testudines, Geoemydidae). *Zoomorphology, 129*, 111–119.

Neill, W. T. (1971). *The Last of the Ruling Reptiles: Alligators*. Columbia University Press, Ithaca.

O'Brien, H. D., Lynch, L. M., Vliet, K. A., Brueggen, J., Erickson, G. M., & Gignac, P. M. (2019). Crocodylian head width allometry and phylogenetic prediction of body size in extinct crocodyliforms. *Integrative Organismal Biology, 1*(1), obz006.

Olsson, P., Lind, O., & Kelber, A. (2015). Bird colour vision: Behavioural thresholds reveal receptor noise. *The Journal of Experimental Biology, 218*, 184–193.

Owre, O. T. (1967). Adaptations for locomotion and feeding in the Anhinga and the Double-crested Cormorant. *Ornithological Monographs, 1967*, 1–138.

Parfitt, A. R., & Vincent, J. F. V. (2005). Drag reduction in a swimming Humboldt Penguin, *Spheniscus humboldtii*, when the boundary layer is turbulent. *Journal of Bionic Engineering, 2*(2), 57–62.

Parsons, T. S. (1970). The nose and Jacobson's organ. In C. Gans & T. S. Parsons (Eds.), *Biology of the reptilia* (Vol. 2, pp. 99–191). Academic Press.

Pettigrew, J., & Frost, B. (1985). A tactile fovea in the Scolopacidae? *Brain, Behavior and Evolution, 26*, 185–195.

Pierce, S. E., & Benton, M. J. (2006). *Pelagosaurus typus* Bronn, 1841 (Mesoeucrocodylia: Thalattosuchia) from the Upper Lias (Toarcian, Lower Jurassic) of Somerset, England. *The Journal of Vertebrate Paleontology, 26*(3), 621–635.

Pierce, S. E., Angielczyk, K. D., & Rayfield, E. J. (2008). Patterns of morphospace occupation and mechanical performance in extant crocodilian skulls: A combined geometric morphometric and finite element modeling approach. *Journal of Morphology, 269*(7), 840–864. https://doi.org/10.1002/jmor.10627

Pierce, S. E., Angielczyk, K. D., & Rayfield, E. J. (2009). Shape and mechanics in thalattosuchian (Crocodylomorpha) skulls: Implications for feeding behaviour and niche partitioning. *Journal of Anatomy, 215*(5), 555–576.

Pooley, A. C. (1989). Food and feeding habits. In C. A. Ross (Ed.), *Crocodiles and alligators* (pp. 76–91). Facts on File.

Pooley, A. C., & Gans, C. (1976). The Nile crocodile. *Scientific American, 234*(4), 114–125.

Porro, L. B., Holliday, C. M., Anapol, F., Ontiveros, L. C., Ontiveros, L. T., & Ross, C. F. (2011). Free body analysis, beam mechanics, and finite element modeling of the mandible of *Alligator mississippiensis*. *Journal of Morphology, 272*(8), 910–937.

Porro, L. B., Metzger, K. A., Iriarte-Diaz, J., & Ross, C. F. (2013). In vivo bone strain and finite element modeling of the mandible of *Alligator mississippiensis*. *Journal of Anatomy, 223*(3), 195–227.

Poulter, T. C. (1969). Sonar of penguins and fur seals. *Proceedings of the California Academy of Sciences, 3*, 363–380.

Prince, P. A. (1980). The food and feeding ecology of blue petrel (*Halobaena caerulea*) and dove prion (*Pachyptila desolata*). *Journal of Zoology, 190*, 59–76.

Prum, R. O. (2002). Why ornithologists should care about the theropod origin of birds. *Auk, 119*, 1–17.

Raikow, R. J., Bicanovsky, L., & Bledsoe, A. H. (1988). Forelimb joint mobility and the evolution of wing-propelled diving in birds. *Auk, 105*, 446–451.

Rayfield, E. J., Milner, A. C., Xuan, V. B., & Young, P. G. (2007). Functional morphology of spinosaur 'crocodile-mimic' dinosaurs. *The Journal of Vertebrate Paleontology, 27*(4), 892–901.

Reuter, T., & Peichl, L. (2008). Structure and function of the retina in aquatic tetrapods. In J. G. M. Thewissen (Ed.), *Sensory evolution on the threshold – Adaptations in secondarily aquatic vertebrates*. University of California Press.
Rhoda, D., Polly, P. D., Raxworthy, C., & Segall, M. (2020). Morphological integration and modularity in the hyperkinetik feeding system of aquatic-foraging snakes. *Evolution, 75*(1), 56–72.
Rhodin, A. G. J., Iverson, J. B., Bour, R., Fritz, U., Georges, A., Shaffer, H. B., & van Dijk, P. P. (2017). Turtles of the World: Annotated checklist and atlas of taxonomy, synonymy, distribution, and conservation status (8th Ed.). In Rhodin, A. G. J., Iverson, J. B., van Dijk, P. P., Saumure, R. A., Buhlmann, K. A., Pritchard, P. C. H., & Mittermeier, R. A. (Eds.). *Conservation biology of freshwater turtles and tortoises: A compilation project of the IUCN/SSC tortoise and freshwater turtle specialist group* (Vol. 7, pp. 1–292). Chelonian Research Foundation and Turtle Conservancy.
Rice, W. R. (1982). Acoustical location of prey by the marsh hawk: Adaptation to concealed prey. *The Auk, 99*, 403–413.
Rico-Guevara, A., Sustaita, D., Gussekloo, S., Olsen, A., Bright, J., Corbin, C., & Dudley, R. (2019). Feeding in birds: Thriving in terrestrial, aquatic, and aerial niches. In V. Bels & I. Q. Whishaw (Eds.), *Feeding in vertebrates* (pp. 643–693). Springer.
Roper, T. J. (1999). Olfaction in birds. *Advances in the Study of Behavior, 28*, 247–332.
Ropert-Coudert, Y., Grémillet, D., Ryan, P., Kato, A., Naito, Y., & Le Maho, Y. (2004). Between air and water: the plunge dive of the Cape Gannet *Morus capensis*. *Ibis, 146*, 281–290.
Salas-Gismondi, R. (2016) The evolution of orbital shape in gavialoid crocodylians is associated with habits and visual foraging strategies. *76th Annual Meeting for the Society of Vertebrate Paleontology Abstracts*, 32.
Sanderson, S. L., & Wassersug, R. (1990). Suspension-feeding vertebrates. *Scientific American, 262*, 96–102.
Schaeffel, F., & de Queiroz, A. (1990). Alternative mechanisms of enhanced underwater vision in the garter snakes *Thamnophis melanogaster* and *T. couchii*. *Copeia, 1990*, 50–58.
Schaller, G. B., & Crawshaw, P. G., Jr. (1982). Fishing behavior of paraguayan caiman (*Caiman crocodilus*). *Copeia, 1982*(1), 66–72.
Scheyer, T. M., & Sander, P. M. (2007). Shell bone histology indicates terrestrial palaeoecology of basal turtles. *Proceedings of the Royal Society of London Series B, 274*, 1885–1893.
Scheyer, T. M., Werneburg, I., Mitgutsch, C., Delfino, M., & Sanchez-Villagra, R. (2013). Three ways to tackle the turtle: Integrating fossils, comparative embryology, and microanatomy. In D. B. Brinkman, P. A. Holroyd, & J. D. Garder (Eds.), *Morphology and evolution of turtles* (pp. 63–70). Springer.
Schoch, R. R., & Sues, H. D. (2015). A Middle Triassic stem-turtle and the evolution of the turtle body plan. *Nature, 523*, 584–587.
Schwab, J. A., Young, M. T., Neenan, J. M., Walsh, S. A., Witmer, L. M., Herrera, Y., Allain, R., Brochu, C. A., Choiniere, J. N., Clark, J. M., Dollman, K. N., Etches, S., Fritsch, G., Gignac, P. M., Ruebenstahl, A., Sachs, S., Turner, A. H., Vignaud, P., Wilberg, E. W., Xu, X., Zanno, L. E., & Brusatte, S. L. (2020). Inner ear sensory system changes as extinct crocodylomorphs transitioned from land to water. *Proceedings of the National Academy of Sciences of the United States of America, 117*(19), 10422–10428.
Schwenk, K. (1985). Occurrence, distribution and functional significance of taste buds in lizards. *Copeia, 1985*, 91–101.
Schwenk, K. (1995). Of tongues and noses: Chemoreception in lizards and snakes. *Trends in Ecology & Evolution, 10*, 7–12.
Schwenk, K. (2000a). Feeding in lepidosaurs. In K. Schwenk (Ed.), *Feeding: Form, function and evolution in tetrapod vertebrates* (pp. 175–291). Academic.
Schwenk, K. (2000b). An introduction to tetrapod feeding. In K. Schwenk (Ed.), *Feeding: Form, function and evolution in tetrapod vertebrates* (pp. 21–61). Academic.

Schwenk, K. (2008). Comparative anatomy and physiology of chemical senses in nonavian aquatic reptiles. In J. G. M. Thewissen (Ed.), *Sensory evolution on the threshold – Adaptations in secondarily aquatic vertebrates* (pp. 65–81). University of California Press.

Schwenk, K., & Rubega, M. (2005). Diversity of vertebrate feeding systems. In J. M. Starck & T. Wang (Eds.), *Physiological and ecological adaptations to feeding in vertebrates* (pp. 1–41). Science Publishers.

Segall, M., Cornette, R., Fabre, A.-C., Godoy-Diana, R., & Herrel, A. (2016). Does aquatic foraging impact head shape evolution in snakes? *Proceedings of the Royal Society of London. Series B, 283*, 20161645.

Segall, M., Herrel, A., & Godoy-Diana, R. (2019). Hydrodynamics of frontal striking in aquatic snakes: Drag, added mass, and the possible consequences for prey capture success. *Bioinspiration & Biomimetics, 14*, 1–9.

Segall, M., Cornette, R., Godoy-Diana, R., & Herrel, A. (2020). Exploring the functional meaning of head shape disparity in aquatic snakes. *Ecology and Evolution, 10*(14), 6993–6700.

Sharma, B. K., Kulshreshtha, S., & Rahmani, A. R. (2013). *Faunal heritage of Rajasthan, India: Conservation and management of vertebrates* (Vol. 2). Springer.

Shealer, D. A. (2002). Foraging behavior and food of seabirds. In E. A. Schreiber & J. Burger (Eds.), *Biology of marine birds* (pp. 137–177). CRC Press.

Sheil, C. A., & Greenbaum, E. (2005). Reconsideration of skeletal development of *Chelydra serpentina* (Reptilia: Testudinata: Chelydridae): Evidence for intraspecific variation. *Journal of Zoology, 265*, 235–267.

Shine, R., Bonnet, X., Elphick, M., & Barrott, E. (2004). A novel foraging mode in snakes: Browsing by the sea snake *Emydocephalus annulatus* (Serpentes, Hydrophiidae) *Functional Ecology, 18*, 16–24.

Shortt, W. H. O. (1921). A few hints on crocodile shooting. *Journal, Bombay Natural History Society, 29*, 77.

Singh, L. A. K. (1976). Rearing gharial in captivity. *Hamadryad, 1*(2), 5–6

Singh, A. (2015). Save the gharials. *Science India, 18*, 33–35.

Sivak, J. G. (1980). Avian mechanisms for vision in air and water. *Trends in Neurosciences, 3*(12), 314–317. https://doi.org/10.1016/S0166-2236(80)80182-2

Smith, K. K. (1982). An electromyographic study of the function of the jaw adducting muscles in *Varanus exanthematicus* (Varanidae). *Journal of Morphology, 173*, 137–158.

Smith, K. K. (1986). Morphology and function of the tongue and hyoid apparatus in *Varanus* (Varanidae, Lacertilia). *Journal of Morphology, 187*, 261–287.

Smith, S. A., & Paselk, R. A. (1986). Olfactory sensitivity of the turkey vulture (*Cathartes aura*) to three carrion-associated odorants. *Auk, 103*, 586–592.

Smith, T. L., Povel, G. D. E., & Kardong, K. V. (2002). Predatory strike of the tentacled snake (*Erpeton tentaculatum*). *Journal of Zoology, 256*, 233–242.

Soares, D. (2002). An ancient sensory organ in crocodilians. *Nature, 417*(6886), 241–242.

Soares, D., & Bierman, H. (2018). Crocodilian sensory systems. In J. Vonk & T. Shackelford (Eds.), *Encyclopedia of animal cognition and behavior* (pp. 1–6). Springer International Publ.

Sokol, O. M. (1969). Feeding in the pipid frog *Hymenochirus boettgeri* (Tornier). *Herpetologica, 25*(1), 9–24.

Spindel, E. L., Dobie, J. L., & Buxton, D. F. (1987). Functional mechanism and histological composition of the lingual appendage in the Alligator Snapping Turtle, *Macroclemys temminckii* (Troost) (Testudines: Chelydridae). *Journal of Morphology, 194*, 287–301.

St. Augustine Alligator Farm Zoological Park. (2020). *Crocodilians Swallowing Underwater* [Online]. Accessed December 21, 2020, from https://www.alligatorfarm.com/blog-item/crocodilians-swallowing-underwater/

Stubbs, T. L., Pierce, S. E., Rayfield, E. J., & Anderson, P. S. L. (2013). Morphological and biomechanical disparity of crocodile-line archosaurs following the end-Triassic extinction. *Proceedings of the Royal Society B, 280*(1770), 20131940.

Swennen, C. K., & Yu, Y.-T. (2005). Food and feeding behavior of the black-faced spoonbill. *Waterbirds, 28*, 19–27.

Takahashi, A., Dunn, M., Trathan, P., Croxall, J., Wilson, R. P., Sato, K., & Naito, Y. (2004). Krill-feeding behaviour in a chinstrap penguin compared to fish-eating in Magellanic penguins: a pilot study. *Marine Ornithology, 32*, 47–54.

Taylor, M. A. (1987). How tetrapods feed in water: A functional analysis by paradigm. *Zoological Journal of the Linnean Society London, 91*(2), 171–195.

Thorbjarnarson, J. B. (1990). Notes on the feeding behavior of the gharial (*Gavialis gangeticus*) under semi-natural conditions. *Journal of Herpetology, 24*(1), 99–100.

Tomkins, I. T. (1951). Method of feeding of the black skimmer, *Rynchops nigra*. *Auk, 68*, 236–239.

Townsend, C. W. (1909). The use of the wings and feet by diving birds. *Auk, 26*, 234–248.

Trutnau, L., & Sommerlad, R. (2006). *Crocodilians: Their natural history and captive husbandry*. Edition Chimaira.

Tucker, D. (1971). Nonolfactory responses from the nasal cavity: Jacobson's organ and the trigeminal system. In L. M. Beidler (Ed.), *Handbook of sensory physiology, Vol. IV*, Chemical senses, Part 1, Olfaction (pp. 151–181). Springer.

Turner, C. H., Wang, T., & Burr, D. B. (2001). Shear strength and fatigue properties of human cortical bone determined from pure shear tests. *Calcified Tissue International, 69*(6), 373–378.

Uetz, P. (2010). The original descriptions of reptiles. *Zootaxa, 2334*, 59–68.

Underwood, G. (1970). The eye. In C. Gans (Ed.), *Biology of the Reptilia* (Vol. 2, pp. 1–79). Academic.

Van Damme, J., & Aerts, P. (1997). Kinematics and functional morphology of aquatic feeding in Australian snake-necked turtles (Pleurodira; Chelodina). *Journal of Morphology, 233*, 113–125.

Van Der Kooij, J., & Povel, D. (1996). Scale sensillae of the file snake (Serpentes: Acrochordidae) and some other aquatic and burrowing snakes. *Netherlands Journal of Zoology, 47*, 443–456.

Van Gils, J. A., Piersma, T., Dekinga, A., & Dietz, M. W. (2003). Cost-benefit analysis of mollusc-eating in a shorebird II. Optimizing gizzard size in the face of seasonal demands. *The Journal of Experimental Biology, 206*, 3369–3380.

Van Wassenbergh, S., Brecko, J., Aerts, P., Stouten, I., Vanheusden, G., Camps, A., Van Damme, R., & Herrel, A. (2009). Hydrodynamic constraints on prey-capture performance in forward-striking snakes. *Journal of the Royal Society Interface, 7*, 773–785.

Verheyden, C., & Jouventin, P. (1994). Olfactory behavior of foraging procellariiforms. *Auk, 111*, 285–291.

Vincent, S. E., Herrel, A., & Irschick, D. J. (2005). Comparisons of aquatic versus terrestrial predatory strikes in the pitviper, Agkistrodon piscivorus. *Journal of Experimental Zoology Part A, 303*, 476–488.

Vincent, S. E., Brandley, M., Herrel, A., & Alfaro, M. (2009). Convergence in trophic morphology and feeding performance among piscivorous natricine snakes. *Journal of Evolutionary Biology, 22*, 1203–1211.

von Bolze, G. (1968). Anordnung und Bau der herbstschen Körperchen in Limicolenschnabeln im Zusammenhang mit Nahrungsfindung. *Zoologischer Anzeiger, 181*, 313–355.

von Huene, F. (1933). Ein Versuch zur Stammesgeschichte der Krokodile. *Centralblatt für Mineralogie, Geologie und Paläontologie, Abteilung B, 11*, 577–585.

Voris, H. K., & Voris, H. H. (1983). Feeding strategies in marine snakes: An analysis of evolutionary, morphological, behavioral and ecological relationships. *American Zoologist, 23*, 411–425.

Vorobyev, M. (2003). Coloured oil droplets enhance colour discrimination. *Proceedings of the Royal Society of London Series B, 270*, 1255–1261.

Wainwright, P. C., & Price, S. A. (2016). The impact of organismal innovation on functional and ecological diversification. *Integrative and Comparative Biology, 56*(3), 479–488.

Walmsley, C. W., Smits, P. D., Quayle, M. R., McCurry, M. R., Richards, H. S., Oldfield, C. C., Wroe, S., Clausen, P. D., & McHenry, C. R. (2013). Why the long face? The mechanics of mandibular symphysis proportions in crocodiles. *PLoS One, 8*, e53873.

Webb, G., & Manolis, C. (1989). *Crocodiles of Australia*. Reed.

Westhoff, G., Fry, B. G., & Bleckmann, H. (2005). Sea snakes (*Lapemis curtus*) are sensitive to low-amplitude water motions. *Zoology, 108*, 195–200.

Wever, E. G., Herman, P. N., Simmons, J. A., & Hertzler, D. R. (1969). Hearing in the blackfooted penguin, *Spheniscus demersus*, as represented by the cochlear potentials. *Proceedings of the National Academy of Sciences of the United States of America, 63*, 676–680.

Whitaker, R., & Basu, D. (1982). The gharial (*Gavialis gangeticus*): A review. *Journal Bombay Natural History Society, 79*(3), 531–548.

Wikelski, M., & Trillmich, F. (1994). Foraging strategies of the Galapagos marine iguana (*Amblyrhynchus cristatus*): Adapting behavioral rules to ontogenetic size change. *Behaviour, 128*, 255–279.

Wilberg, E. W., Turner, A. H., & Brochu, C. A. (2019). Evolutionary structure and timing of major habitat shifts in Crocodylomorpha. *Scientific Reports, 9*(1), 1–10.

Wilga, C. D., & Sanford, C. P. (2008). Suction generation in white-spotted bamboo sharks *Chiloscyllium plagiosum*. *The Journal of Experimental Biology, 211*, 3128–3138.

Wilkie, E. S., Vissers, P. M., Das, D., DeGrip, J. W., Bowmaker, K. J., & Hunt, M. D. (1998). The molecular basis for UV vision in birds: Spectral characteristics, cDNA sequence and retinal localization of the UV-sensitive visual pigment of the budgerigar (*Melopsittacus undulatus*). *The Biochemical Journal, 330*, 541–547.

Wise, S. C., Formanovic, D. R., & Brodie, E. D. (1989). Matamata turtles ambush but do not herd prey. *Journal of Herpetology, 23*, 297–299.

Wroe, S., & Milne, N. (2007). Convergence and remarkably consistent constraint in the evolution of carnivore skull shape. *Evolution, 61*(5), 1251–1260.

Young, B. (1991). The influences of the aquatic medium on the prey capture system of snakes. *Journal of Natural History, 25*, 519–531.

Young, M. T., Brusatte, S. L., Ruta, M., & de Andrade, M. B. (2010). The evolution of Metriorhynchoidea (Mesoeucrocodylia, Thalattosuchia): an integrated approach using geometric morphometrics, analysis of disparity, and biomechanics. *Zoological Journal of the Linnean Society, 158*(4), 801–859.

Zeigler, H. P., & Bischof, H.-J. (1993). *Vision, brain, and behavior in birds*. MIT Press.

Ziegler, T., Quyet, L. K., Thanh, V. N., Hendrix, R., & Böhme, W. (2008). A comparative study of crocodile lizards (*Shinisaurus crocodilurus* AHL, 1930) from Vietnam and China. *The Raffles Bulletin of Zoology, 56*, 181–187.

Zweers, G., & Wouterlood, F. (1973). Functional anatomy of the feeding apparatus of the mallard (*Anas platyrhynchos* L.). In *Proceedings of the Third European Anatomical Congress*. Manchester (pp. 88–89).

Zweers, G., Gerritsen, A. C., & van Kranenburg-Voogd, P. (1977). Mechanics of feeding of the mallard (*Anas playrhyncos* L., Aves, Anseriformes). *Contributions to Vertebrate Evolution, 3*, 1–109.

Chapter 8
Convergent Evolution of Secondarily Aquatic Feeding in Mammals

Alexander J. Werth and Christopher D. Marshall

Abstract Several mammalian lineages, most notably cetaceans, sirenians, and pinnipeds, have independently reverted to the marine environment of their long-ago, pre-mammalian ancestors. Other mammals have also adapted to coastal, estuarine, or freshwater habitats. These include various members of the Carnivora and Rodentia, along with some other living and extinct mammals. Because water is dense, heavy, viscous, and incompressible, feeding in water poses challenges, especially for animals whose ancestors evolved in terrestrial settings. Many secondarily aquatic mammals separately adopted similar functional and structural solutions to acquire, ingest, and process food, particularly suction feeding, filter feeding, raptorial ("seizing") grasping of prey, or adaptations to remove prey from benthic sediments. This led to striking examples of convergence with other mammals or with other aquatic animals, including sharks, bony fishes, marine reptiles, and birds. Most instances of convergence involve close similarities in jaws, dentition, and musculature, overall shape of the head and mouth, methods for separating food from water, and neural and behavioral adaptations to locate and capture prey. Following discussion of basic principles underlying aquatic mammalian feeding, we outline numerous examples of convergence in extant and extinct taxa.

Keywords Baleen · Biting · Dentition · Filter feeding · Hydraulic jetting · Jaws · Mastication · Rostrum · Suction feeding · Tongue

---

A. J. Werth (✉)
Department of Biology, Hampden-Sydney College, Hampden-Sydney, VA, USA
e-mail: awerth@hsc.edu

C. D. Marshall
Department of Marine Biology, Texas A&M University, Galveston, TX, USA

Department of Ecology and Conservation Biology, Texas A&M University, Galveston, TX, USA
e-mail: marshalc@tamug.edu

V. L. Bels, A. P. Russell (eds.), *Convergent Evolution*, Fascinating Life Sciences,
https://doi.org/10.1007/978-3-031-11441-0_8

8.1 Introduction: General Evolution of and Strategies for Aquatic Feeding

Aquatic mammals are a taxonomically and ecologically diverse group with over 130 living members, including marine mammals such as cetaceans (whales, dolphins, and porpoises) and sirenians ("sea cows": manatees and dugongs), as well as many taxa within the Order Carnivora, including obligate marine mammals such as pinnipeds (seals, sea lions, and walruses). The first cetaceans and sirenians evolved about 50 Ma in the Eocene, whereas early pinnipeds arose in the late Oligocene around 25 Ma (Berta et al., 2015; Fordyce, 2018; Marshall & Pyenson, 2019; Werth, 2020). Other aquatic fissiped carnivoran mammals include polar bears, which are typically considered to be marine mammals based on their exclusively marine trophic ecology and habitat, sea otters and lesser known marine otters (and recently extinct sea minks), and several non-marine but freshwater aquatic species, including about ten species of river otters. In addition there are numerous species of tenrecs, shrews, moles, and rodents ranging from beavers, capybaras, and muskrats to small water voles. Other mammals including duck-billed platypuses, water opossums, hippopotamuses and pygmy hippos, and even water bats, are often included within lists of aquatic mammals. The extent to which these are obligately or facultatively aquatic species, as in the differing cases of whales versus river otters, varies widely. Numerous "semiaquatic" mammals divide their time between land and water. Some, like platypuses, feed in water and take shelter on land; others, like hippos, do the reverse.

However, all aquatic mammals are distinguished by their terrestrial origins. Their secondary reversion to aquatic habitats of long-ago (pre-tetrapod) ancestors is striking and often involves remarkable transformation from the typical quadrupedal body form of most mammals. As a medium for feeding, locomotion, and other vital activities, water is wholly distinct from air, with fundamental differences in density, viscosity, oxygen levels, and conduction of heat, light, sound, and electricity (Nybakken & Bertness, 2005). In addition to basic physical and chemical differences, aquatic habitats also differ from terrestrial/aerial habitats in crucial ways involving ecosystem dynamics. For example, terrestrial ecosystems are dominated by a background of large, long-lived plants supporting a community of smaller, shorter-lived fauna. However, autotrophs are typically small and short-lived in aquatic ecosystems, whereas animals are larger and, if not more abundant, more conspicuous and dominant. This is why one speaks of oak forests or tallgrass prairies, but in seas of coral reefs, mussel beds, or oyster shoals. Terrestrial plants have rigid bodies made of complex, long-chain polysaccharides, particularly cellulose, whereas aquatic ecosystems are dominated by animals (usually with soft bodies) made predominantly of proteins. Marine food chains tend to be longer than those in terrestrial ecosystems, with an average of five links (steps) between autotrophs and an apex predator, compared to three on land (Nybakken & Bertness, 2005); this difference arises from increased energy transfer efficiency between marine trophic levels. All of these distinctions have consequences for aquatic

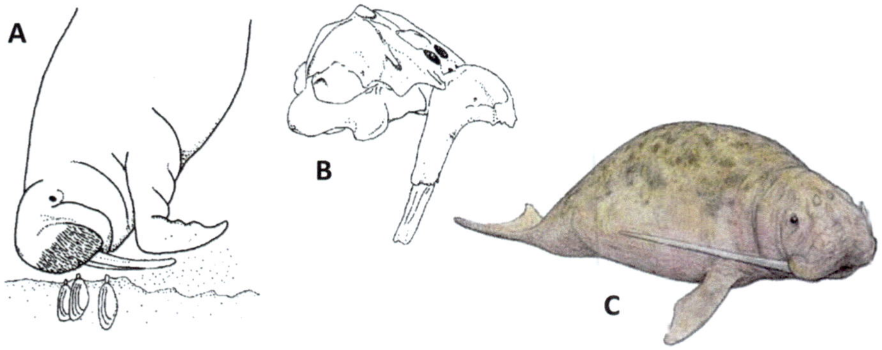

Fig. 8.1 Striking convergence of marine mammals that suction feed on benthic bivalves: the living walrus *Odobenus* **a** and skull **b**, and reconstruction **c** of Pliocene toothed whale *Odobenocetops* (Credit: **a**, **b**: **a**. Werth, 2000b used by permission; **c**: Wikimedia Commons, published under CC BY 2.0, https://creativecommons.org/licenses/by/2.0/, no changes)

feeding, including the possibility of bulk filter feeding and the greater likelihood of "trophic blurring," in which organisms feed at multiple trophic levels rather than at a single level, which is more common in terrestrial ecosystems.

Another major consideration of aquatic ecosystems is that food resources tend to be extremely patchy in both time and space: here in one moment or place but gone the next. These patches usually involve rapidly appearing (and rapidly disappearing) swarms of phytoplankton and zooplankton, followed by increasingly larger fishes and other consumers, often culminating in mammals and birds. This patchiness holds true for benthic communities but especially for open water columns of pelagic and neritic zones. For these reasons, marine mammals, like all aquatic animals, have evolved means of locating and ingesting food quickly and efficiently.

Given that the evolutionary transformation from fully terrestrial to fully or semi-aquatic mammals unfolded only over the past 40–50 million years (Kelley & Pyenson, 2015), and that it involved independent reversion from land-based back to water-based lifestyles in several distinct lineages, it should not be surprising that many instances of convergent evolution can be found within diverse aquatic mammals. Some examples involve convergence *within* aquatic mammals, such as the remarkable similarities between walruses, *Odobenus*, and the walrus-like Pliocene fossil odontocete *Odobenocetops* (Fig. 8.1; De Muizon, 1993), as will be described later. Other examples involve convergence with non-mammalian aquatic animals, such as between platypuses, *Ornithorhynchus anatinus* (with the generic name meaning "bird snout" and specific name meaning "duck-like"), and, of course, ducks.

Perhaps the most classic, iconic example of convergent evolution involves the torpedo-like, fusiform body of dolphins and similar aquatic apex predators including sharks and ichthyosaurs, all of which have generally similar tails, dorsal fins for stability, and tapered snouts bearing rows of sharp teeth (Fig. 8.2). Given that marine

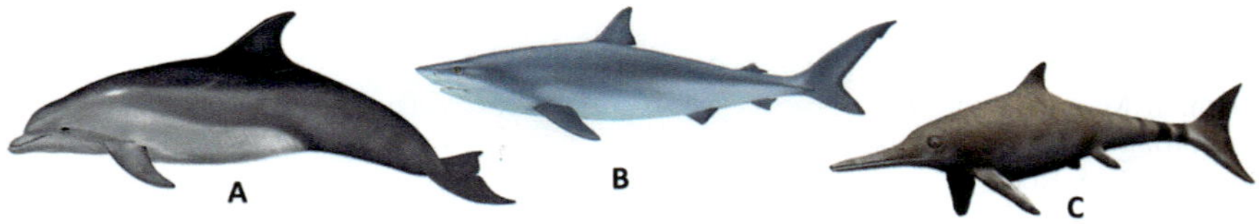

Fig. 8.2 Convergent bodies of apex marine predators: a dolphin (**a**), shark (**b**), and ichthyosaur (**c**). (Credit **a**: CETASea; **b**: Timothy Knepp USFWS; **c**: Nobu Tamura/http://spinops.blogspot.com, all Creative Commons use, published under CC BY 2.0, https://creativecommons.org/licenses/by/2.0/, no changes)

and other aquatic mammals relatively recently re-adapted their terrestrial body plans to compete for ecological niches that in many cases were already filled by existing species in aquatic habitats (from shallow rivers and estuaries to the open ocean and deep seas), it would be startling indeed if convergent evolution did not continually recur.

The existence, and frequency, of convergent evolution has long been recognized. Darwin himself commented on this phenomenon in his landmark *Origin of Species* (1859). However, more recently there has been debate about the extent to which convergent evolution is a chance rarity or whether it is an inevitable and regular occurrence. In *Wonderful Life*, Gould (1989) claimed that evolution's path depends on random contingencies, such that if one were to "roll back and replay the tape," the outcome would likely be entirely different. In concluding that there is little probability that hominins or other intelligent bipeds would evolve, Gould (1989) argued that humans, and indeed all species, are the end result of many fortuitous accidents. In contrast, Conway Morris (Conway Morris, 2003), Losos (2018), and others have argued that the continued prevalence of convergence in so many diverse taxa points to an inexorable predictability of evolutionary outcomes, with the same results reappearing again and again. Contra Gould, these biologists hold that convergence is not only not unusual, but also that it is inescapable.

No doubt the extent of convergent evolution's prominence depends on the eye of the observer, but there are particular constraints that tilt the playing field toward similar outcomes. Each organism's phenotype (including its morphology, physiology, behavior, and so on) results from an interplay of intrinsic and extrinsic forces. Intrinsic (internal) drivers of, and constraints on, phenotype are the product of genetic and epigenetic instructions that together generate and regulate gene expression: inherited genes that specify the construction and operation of proteins, which in turn lead to the construction and operation of other biological materials and structures. Extrinsic (external) or environmental factors similarly shape and select the aspects of expressed phenotype. But although intrinsic and extrinsic factors combine to create new possibilities, they also sharply constrain potential phenotypes by limiting possibilities. Both determinants of form create new possibilities, but both factors likewise limit possibilities. This is why, as Kardong (2019) explains, no animals move with wheels or metal wings, even if both solutions would be strongly selected for. Environmental, developmental, and evolutionary constraints make convergence more likely.

Functional morphologists recognize the ways that intrinsic and extrinsic factors are interrelated, and how they interact to generate convergent outcomes. Phenotypic disparity arises not merely from gradual changes in structural genes, but often from mutations in regulatory genes that govern expression of entire gene "families." In ontogenetic terms, these gene controls commonly tweak developmental pathways that influence precise timing and placement of structures—for example, a bone, a muscle, or a tendon linking bone to muscle—to create diverse phenotypic forms. In recent decades we have learned that the total number of genes in any organism, even in large, complex vertebrates, is far lower than once imagined, and that mutation of a single gene, promoter, or transcription factor generally leads to multiple changes in gene expression. In sum, the number of potential outcomes is not exactly unlimited, and we might expect to see the same sorts of phenotypic changes (e.g., longer or shorter bones, or more or less abundant clusters of nerves or sensory receptors) that seem perennially to lead to recurring evolutionary outcomes and thus patterns of convergence.

At the same time, the extrinsic (environmental) factors that drive evolution seem also to work to some extent in lockstep with these arrays of polymorphism, again leading to outcomes that are, if not predictable, at least somewhat less variable than might be imagined. As Jacques Monod (1971) famously explained, a mix of "chance and necessity" underlie evolutionary change. Features (whether anatomical, physiological, or behavioral) that impinge on basic activities—for example, the ability to locate and acquire food, or to detect and avoid predators—will prove adaptive in predictable ways. In the end, when there are only so many ways to obtain calories effectively and economically in an aquatic environment, we should not be surprised to see the same functional solutions, such as pincer-like jaws or sieve-like filters, arising again and again via convergent evolution.

Selection pressures for feeding adaptations are among the strongest in biology (Schwenk, 2000) because feeding not only relates strongly to fitness but is one of its strongest determinants. Furthermore, aquatic habitats pose strong selection pressures because of the severe ways that water's physicochemical properties constrain vital activities, such that feeding may be linked to convergence in locomotion (Boessenecker et al., 2020). Investigations of aquatic feeding are of particular interest because many early innovations are established and modified by aquatic forms of life.

The mechanics of food acquisition in aquatic mammals is best viewed as a collection of multimodal behaviors incorporating one or more of the following main feeding modes: raptorial biting and durophagous biting, suction feeding, and filter feeding. A number of key studies have presented and reviewed basic feeding modes of aquatic mammals (Pivorunas, 1979; Sanderson & Wassersug, 1990, 1993; Marshall et al., 1998a, b, 2003, 2008, 2014, 2015; Werth, 2000a, b, 2001, 2004a, b, 2006a, b, 2007, 2013; Bloodworth & Marshall, 2005, 2007; Kane & Marshall, 2009; Johnston & Berta, 2011; Marshall & Goldbogen, 2015; Timm-Davis et al. 2015, 2017; Kienle & Berta, 2016; Werth & Potvin, 2016; Werth et al., 2016b, 2018a, b, c,

2019a, 2020a; Goldbogen et al., 2017; Potvin & Werth, 2017; Werth & Ito, 2017; Marshall, 2018; Kosma et al., 2019; Marshall & Pyenson, 2019; Potvin et al., 2020; Werth & Sformo, 2020). Several monographs also include extended sections on aquatic mammalian feeding (Kellogg, 1928; Howell, 1930; Nemoto, 1959; Slijper, 1962; Gaskin, 1982; King, 1983; Riedman, 1990). Major contributions on marine mammal evolution also discuss the importance of feeding (Repenning, 1976; Fordyce, 1980; Mchedlidze, 1984; Fordyce & Barnes, 1994; Thewissen, 1998; Reynolds & Rommel, 1999; Thewissen & Williams, 2002; Reidenberg, 2007; Thewissen et al., 2009; Fitzgerald, 2010; Uhen, 2010; Berta et al., 2015; Marx & Fordyce, 2015; Marx et al., 2016b).

Potential means of obtaining food in an aquatic habitat with a mammalian body plan are sharply limited (Liem, 1990; Heiss et al., 2018). The same small number of possible feeding methods apply to all consumers, whether herbivorous, carnivorous, or omnivorous. Although some aquatic mammals are highly specialized feeders, many are opportunistic generalists. The small number of possible aquatic feeding methods and specialized niches derive both from the limited types of prey available (for example, few opportunities to graze on macrophytes) and from physical limits (e.g., buoyancy) imposed by the water these prey organisms inhabit. Thus some kinds of foraging, notably suction feeding and bulk filter feeding, are highly effective in water and common in a wide range of aquatic animals, yet virtually nonexistent for consumers living in terrestrial settings.

A key difference between ingestion, oral transport, and processing of food in air versus water is that gravity plays a major role in air yet it is negligible in water, where most food items are of neutral buoyancy given that biological tissues have density similar to water. Further, water is not only 830 times denser than air (Nybakken & Bertness, 2005) but also incompressible, unlike air. Together, these factors enable bulk collection of prey items suspended in water, which can greatly aid aquatic animals. On the other hand, these same factors also hinder aquatic feeding, as prey is likely to be pushed ahead and out of the reach of the mouth by a compressive bow wave (an anteriorly directed pressure surge) generated by forward locomotion, or simply from the motion of the head or jaws.

One way to circumvent these constraints is to draw water into the mouth via suction, by generating a subambient pressure differential. Although water is heavy and thus energetically costly to move, this suction solution is common among aquatic organisms (Wainwright et al., 2015). Another solution is to filter food from water. This can be achieved either by moving a net-like filter through water, or conversely by moving water through a filter. Many invertebrates use a filter external to the body, often on long appendages, but vertebrates have generally evolved internal filters, often related to water flow through a perforated pharynx and/or along gills for gas exchange. Indeed, filter feeding was the original feeding mode of the earliest ancestral chordates (Kardong, 2019), and water flow through pharyngeal perforations was exploited to exapt gills for respiratory gas exchange, enabling vertebrates to achieve higher activity levels and metabolic rates, along with larger bodies with more complex tissues and organ systems, than their aquatic invertebrate ancestors.

Although some aquatic animals raptorially seize and bite prey, this is often combined with ram ingestion (where the predator acquires food by overtaking it via rapid locomotion) and by striking with protrusible body parts such as a flexible neck or jaws. Still, many general reviews of aquatic mammal feeding (Werth, 2000b; Marshall & Goldbogen, 2015; Hocking et al., 2014; Hocking et al., 2017a, b; Kienle et al., 2017; Marshall & Pyenson, 2019) rightly point out that aquatic animals frequently utilize a combination of mechanisms to ingest, transport, and process (and expel excess water from) prey. For example, gray whales, *Eschrichtius robustus*, are unique among baleen whales in using intraorally generated suction to draw prey into the mouth, but then like other mysticetes they trap their prey and purge unwanted water via filter feeding. Gray whales are clearly *both* filter feeders as well as suction feeders. Likewise many aquatic mammals that seize prey via dentition then use suction to transport the food item to the rear of the oral cavity prior to swallowing. Ram feeding is used by phylogenetically and functionally diverse aquatic predators (from giant lunge-feeding whales to small seals and dolphins) to approach and capture prey. In short, aquatic mammals are often resourceful opportunists, but even when specialized their foraging often combines multiple means of collecting and processing prey items, complicating simplistic schemes of feeding classification.

Hocking et al. (2017a) argued that aquatic mammalian feeding strategies fall along a behavioral continuum that may reflect evolutionary history, with terrestrial feeding preceding semi-aquatic feeding, followed by increasingly specialized (in terms of form, function, and behavioral ecology) raptorial, suction, and filter feeding, but others (notably Kienle et al., 2017) dispute the likelihood of evolution following such a linear sequence. Crucially, Hocking et al. (2017b) distinguish foraging strategies (such as raptorial feeding) where water can be seen as an encumbrance from more specialized strategies (suction and filtration) where water is an essential tool needed to acquire food.

Interestingly, suction feeding and filter feeding are basal, primitive feeding modes for many aquatic animals, whereas raptorial (seizing or grasping) feeding methods are more derived (Werth, 2000b). However, the opposite is the case for aquatic mammals: raptorial feeding is simpler, more common, and more basal, whereas suction and filter feeding, although highly effective for aquatic prey acquisition and ingestion, are more highly derived for mammals. This holds true for an obvious, basic reason: whereas aquatic habitats were the original home for many animals, they are for mammals, as for all tetrapod vertebrates (except the very first tetrapods, such as *Tiktaalik*), a secondary home. That is, all mammals evolved from fully terrestrial ancestors, although some lineages "chose," for various reasons (such as food or shelter from predators) to live and/or feed in water, as their pre-mammalian ancestors did.

Another factor distinguishing aquatic feeding from terrestrial feeding is the aquatic medium's complex three-dimensional nature. Not only can prey escape in all directions, but it is also difficult for predators to conceal themselves for ambush. However, interfaces at the upper and lower boundaries of the water column (i.e., between the water's surface and overlying air, and the underlying benthos) serve as barriers against which predators can trap and corral prey. Additionally, predators can catch prey at or near the surface and bring food into the air to process and swallow it more easily, without interference from or the consumption of water. Some cetaceans, including bottlenose dolphins and killer whales, momentarily propel themselves onto shore to capture fish and pinnipeds, respectively.

In short, because of the extrinsic (environmental) constraints imposed by the challenges of feeding in water, it is not at all surprising—and perhaps it is inevitable—that multiple phylogenetic lineages have independently arrived at the same, or nearly the same, feeding methods and mechanisms. Thus, convergence is not unexpected in the case of aquatic mammal feeding. It is to be anticipated. This is why it is easy to find books (like Dougal Dixon's *After Man*, 1981) and television programs (like the Discovery Channel's *The Future Is Wild*; Dixon & Adams, 2003) in which biologists build on functional principles, as well as evolutionary history, to speculate about potential future species, including many taxa that resemble—purely by convergent evolution—today's marine mammals. Dixon (1981) described rats evolving into walrus-like creatures and penguins evolving to fill niches of dolphins or even giant filter-feeding whales. Such speculation is based on sound science: no one should have been surprised when the walrus-like whale *Odobenocetops* (Fig. 8.1) was described (De Muizon, 1993). Undoubtedly, many exciting new examples of convergent evolution involving aquatic mammal feeding will continue to be described as new findings and fossils emerge. Therefore, this chapter, although thorough, cannot be considered exhaustive, but it presents, along with general conclusions regarding convergence, an array of well-known and lesser known examples.

Kelley and Motani (2015) concluded that morphological convergence in aquatic tetrapods is driven by trophic convergence, which offers both opportunities and constraints. They found similarities in skulls due to similar diets. However, Antarctic krill support a tremendous diversity of aquatic tetrapods ranging from blue whales and crabeater seals to penguins, albatrosses, and other sea birds (as well as many fish and squid species), which exhibit varied form as well as foraging methods. Following a different approach, Foote et al. (2015) described numerous convergent gene mutations between cetaceans, pinnipeds, and sirenians. Often these involve genes related to moving and diving in water—for example, with positive selection for genes related to cardiac muscle development, blood coagulation, and bone growth. Genes related to the expression of keratin might relate to integumentary changes in marine mammals, but they might also be associated with feeding if, for example, they contributed to the origin of baleen, a unique oral tissue used by mysticetes for filtering prey from water. Once again we see that the complex interplay of intrinsic and extrinsic factors, working against each other, but also in concert, establishes a limited number of solutions to the problems posed by feeding in water.

Although homoplasy is common in nature, not all instances reflect convergence. One must understand phylogenetic relationships and ancestral conditions to be able to determine whether similarity arises from parallel evolution of closely related taxa or truly convergent evolution via analogous derivation of common functional solutions via shared phenotypic adaptations. As with homology, the extent to which similarity involves convergence may be a matter of degree. Instances of convergent evolution involving aquatic mammal feeding can involve broad categories or specific details. Despite the shortcomings inherent in relying on simple categories, for organizational sake we focus in this chapter on instances of convergent evolution among major feeding strategies employed by aquatic mammals. We include in our account examples of convergence in both extant and extinct aquatic mammals.

8.2 Raptorial (Seizing) Biting

Biting-based foraging includes many different strategies used by diverse predators to capture and ingest varied prey (Adam & Berta, 2002; Thewissen et al., 2011; Churchill & Clementz, 2015). Such strategies are often lumped together as raptorial (i.e., seizing) feeding due to strikes of the head, jaws, or forelimbs. We use the term "biting" to encompass a wide array (Hocking et al., 2017a) of behaviors, including pierce seizure or jaw snapping, grip and tear feeding (including head shaking and grappling with forelimbs), and crushing (as exemplified by the flat molars of sea otters which are used to break hard prey). Following a few general remarks about aquatic feeding modes and methods that involve biting, we distinguish raptorial biting (in this section) from biting on vegetation (in a following section).

Dentition figures prominently in prey ingestion and processing with all biting-based foraging (Werth et al., 2019b), unlike other feeding modes (particularly suction and filter feeding). The general loss of heterodont (differentiated) teeth and their replacement with simple homodont cones throughout many cetacean lineages (Armfield et al., 2013) clearly demonstrates convergent evolution. Dental ridges evolved convergently (McCurry et al., 2020), likely for strengthening teeth or improving grip. Biting can involve processing food into small pieces for easier swallowing (e.g., slicing with dentition, shaking or twisting the head or body, holding and tearing via forelimbs, or using tools). Alternatively, prey that are seized or grasped between tooth rows can be swallowed whole, often with little or no processing. Jaws can be snapped shut anteriorly or swung laterally, as in many side-swimming river dolphins. Food can be manipulated and transported by gravity, inertia, or suction, or by movements of the forelimbs or of the tongue, lips, and whiskers.

Clearly, the biting category of aquatic feeding encompasses a broad range of form, function, and behavior, all of which are widely distributed through a diversity of aquatic mammals, including small rodents and insectivorans that are only barely semiaquatic to large, highly specialized pinnipeds and cetaceans that primarily capture/ingest, transport, and process prey via teeth (Fig. 8.3). For such animals, water presents challenges for prey acquisition and manipulation. In these aquatic mammals, morphology and behavior are often more closely allied with terrestrial feeding strategies than with suction or filter feeding strategies. Feeding morphology and behavior of these animals are more closely allied with those of terrestrial

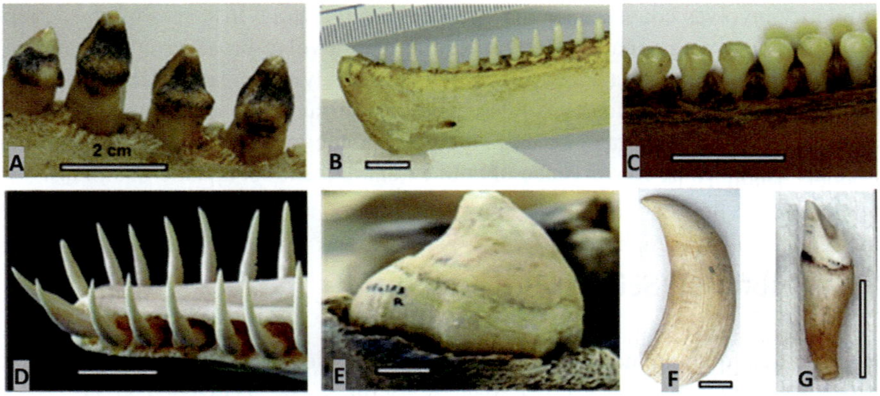

Fig. 8.3 Dental diversity of living odontocetes (all scale bars = 2 cm): (**a**) wrinkled teeth with shelf-like cusp of Amazon river dolphin, *Inia geoffrensis*; (**b**) smooth conical pegs of dusky dolphin, *Lagenorhynchus obscurus*; (**c**) spade-shaped teeth of harbor porpoise, *Phocoena phocoena*; (**d**) sharp, narrow teeth of pygmy sperm whale, *Kogia breviceps*; (**e**) flat, broad, mostly unerupted and cementum-covered tooth of Blainville's beaked whale, *Mesoplodon densirostris*; (**f**) large, wrinkled, enamel-free cone of sperm whale, *Physeter macrocephalus*; (**g**) highly worn peg of beluga, *Delphinapterus leucas* (Credit for all images: Werth et al., 2019b used by permission)

Fig. 8.4 Some archaeocete whales: crocodile-like *Ambulocetus* (**a**) and *Remingtonocetus* (**b**), and later shark-like *Dorudon* (**c**) and *Dorudon* skull (**d**) (Credit **a–b**: Nobu Tamura/http://spinops.blogspot.com; D: Doug Boyer/Wikimedia Commons, all Creative Commons use, published under CC BY 2.0, https://creativecommons.org/licenses/by/2.0/, no changes)

mammalian ancestors than with the form and function of more distant, pre-mammalian ancestors.

In this regard it should be noted that biting-based foraging is the basal condition, and thus the default feeding mode, for all aquatic and semi-aquatic mammals.

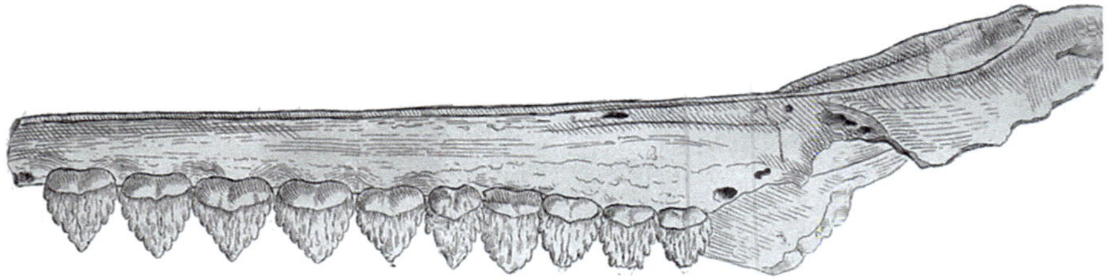

Fig. 8.5 Shark-like teeth and jaws of the extinct toothed whale *Squalodon* (Credit: J.P S. Grateloup 1840/Wikimedia Commons, Creative Commons use, published under CC BY 2.0, https://creativecommons.org/licenses/by/2.0/, no changes)

Therefore there are many instances of parallelism among disparate taxa but really no examples of genuine convergence with other aquatic mammals. There are, however, numerous instances of these mammals converging with other aquatic vertebrates, including both fishes and non-mammalian tetrapods, such as crocodiles and other reptiles.

The oldest known cetaceans are the archaeocetes of the early and middle Eocene, appearing about 49 Ma: the Pakicetidae, Ambulocetidae, Remingtonocetidae, and Protocetidae (Uhen, 2010). These amphibious early whales (Fig. 8.4) gave rise to much larger and more recent archaeocetes, the fully marine Basilosauridae and Kekenodontidae. Fossils, as well as geochemical and isotopic analysis of fossil remains, reveal that archaic odontocetes converged on crocodilians in form, function, and lifestyle, and especially feeding habits. Like crocodilians, the early archaeocetes had flattened, long-snouted jaws bearing numerous large, sharp teeth. The dorsally-facing eyes, like the nostrils at the tip of the snout, protruded above the water's surface. Jaw and neck muscles were large and bones were heavy and dense (pachyostotic), presumably to counteract buoyancy forces. Although the short, squat legs and stout tail were strong, it is likely that members of early archaeocete families were ambush predators that waited and stalked prey near the shallow shores of lakes and rivers, again like crocodilians.

In contrast, the larger basilosaurid archaeocetes (Fig. 8.4c, d) were much better swimmers and retained only vestigial hindlimbs. Their jaws and dentition at first suggest more shark-like feeding on fish. However, the combination of a long, slim, flexible body and large head with long, pointed jaws bearing many sharp, triangular teeth strongly suggests that basilosaurids closely converged on the body plan of mosasaurs. The similar size and proportions of the head, trunk, and tail, and the size of the orbits and dentition, suggest that, like mosasaurs, these archaeocetes were fearsome predators on many types of aquatic prey (Uhen, 2004).

Some extinct sperm whales (e.g., *Livyatan* and several related genera) were giant macroraptorial predators (Lambert et al., 2010, 2017) and, as such, converged in body form and feeding habits on living killer whales, *Orcinus orca*, particularly in their skull and dentition. They also likely resembled, in terms of feeding habits, huge extinct sharks such as *Carcharocles (=Otodus) megalodon*.

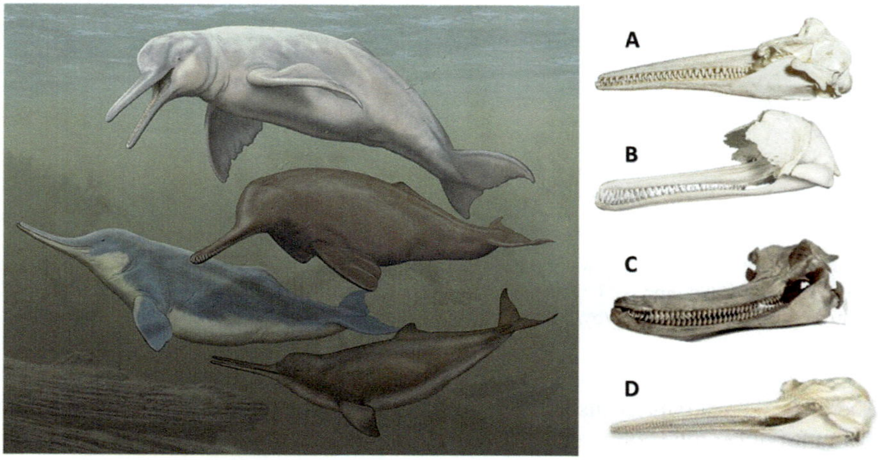

Fig. 8.6 Convergent river dolphins from around the world, top to bottom with associated skulls (**a**–**d**): Amazon river dolphin (**a**), *Inia geoffrensis*, Ganges river dolphin (**b**) *Platanista gangetica*, Chinese white flag dolphin or baiji (**c**), the now-extinct *Lipotes vexillifer*, and La Plata River dolphin or franciscana of South America (**d**) *Pontoporia blainvillei* (Painting credit: Jiaming Liu, Creative Commons use, all Creative Commons use, published under CC BY 2.0, https://creativecommons.org/licenses/by/2.0/, no changes)

The extant squalodontid odontocetes are named for their obviously shark-like teeth and similarly strong jaws, which suggests an evident instance of convergence (Fig. 8.5). Known from several genera from the Eocene through Pliocene, squalodonts were large, widely dispersed predators, with fossils found in marine deposits on all continents. As members of an early odontocete radiation, they may have been closely related to the modern South Asian river dolphins of the genus *Platanista*. Comparably named squalodelphinids were also noted for their somewhat shark-like teeth, but the Squalodelphinidae is more likely closely related to the ancestry of Platanistidae.

"River dolphins" provide one of the best examples of convergent evolution among extant odontocetes (Fig. 8.6). Once classified in a single family (Platanistidae), they are now recognized as being distantly related members of multiple families (Hamilton et al., 2001), including the Platanistidae (genus *Platanista*), Lipotidae (the very recently extinct *Lipotes*), Iniidae (*Inia*), Pontoporiidae (*Pontoporia*), as well as riverine and coastal genera of the Delphinidae (*Sotalia*, *Orcaella*). Although the evolutionary history of modern and extinct river dolphins remains murky, it is obvious that the four genera of non-delphinid river dolphins have closely converged on a basic body plan focused on snapping, pincer-like polydont jaws bearing the greatest number of teeth (as many as 300) of all cetaceans, a condition far exceeding the normal eutherian maximum of 42 adult teeth. The body is generally small and lithe, with large flippers, small (even rudimentary) eyes, and a markedly flexible neck. Both upper and lower jaws can be many times longer than the braincase, and the conical teeth are often

Fig. 8.7 Billfish-like bony sword rostrum of *Eurhinodelphis* (Credit: skull **a**: R. Kellogg 1925; reconstruction **b**: Prehistopedia.fandom.com, both Creative Commons use, published under CC BY 2.0, https://creativecommons.org/licenses/by/2.0/, no changes)

Fig. 8.8 Narwhal with spiraled tusk and shark-like pygmy sperm whale (Credit **a**: Piotr Siedlecki; **b**: NOAA.gov, both Creative Commons use, published under CC BY 2.0, https://creativecommons org/licenses/by/2.0/, no changes)

narrow and sharply pointed for catching and holding fish and other prey items. However, the posterior teeth of the Amazon river dolphin or boto, *Inia*, have a flattened molariform cusp or shelf (the only instance of differentiated dentition in modern Cetacea) that appears to have converged on the typical molar cheek teeth of terrestrial mammals, and which presumably functions similarly in prey processing given that *Inia*'s diet often involves hard or spiny prey (Layne, 1959).

Extinct odontocetes of the family Eurhinodelphinidae (including multiple genera known from the Eocene through Pliocene) are named for their long "nose," a remarkably lengthy upper jaw which greatly protruded beyond the lower jaw, and which, while resembling a tusk, bore no teeth for much of its length (Fig. 8.7). This bony protrusion may have been used to stir up or excavate benthic sediments to uncover and flush out prey, as was presumed to be the case for the extinct porpoise *Semirostrum ceruttii* (named for its "half beak," although unlike eurhinodelphinids it was the lower jaw that protruded further in *Semirostrum*; Racicot et al. 2014). However, it is equally likely that eurhinodelphinids converged on the body form of billfish, such as marlins, sailfish, and swordfish, and used their very long, partially edentulous upper jaw as billfish do, to swing at schools of prey fish, either slicing or hitting and stunning prey before ingesting them.

The long, spiraled tusk of narwhals, *Monodon monoceros* (Fig. 8.8), typically an upper left canine (Best, 1981b), inspires much speculation. Nweeia et al. (2009) suggested the tusk is a sensory organ based on its highly innervated network of nerve endings, but all mammalian teeth are sensitive to multiple stimuli (tactile, thermal, and otherwise). The tusk is likely used for display or possibly male-male interaction as a secondary sexual feature, but another possible convergent function that has been posited is its use in a similar fashion to the sword-like upper jaw of billfish, as in eurhinodelphinids (Fig. 8.7).

Living pygmy and dwarf sperm whales, *Kogia breviceps* (Fig. 8.8) and *K. simus*, are often compared to sharks in general form and feeding morphology, largely due to their thin, underslung (ventrally positioned) lower jaws which bear numerous needle-like teeth, unlike those of other extant odontocetes. Stomach contents indicate a diet of cephalopods, fish, and crustaceans, which may be ingested via suction (Bloodworth & Marshall, 2005, 2007), as also occurs in some sharks. The presence of a lightly colored "false gill" posterior to the eyes in both *Kogia* species completes the remarkable shark-like convergence.

One of the canonical examples of convergence involves dolphins, sharks, and ichthyosaurs, all of which share a fusiform (torpedo-shaped) body with similar dorsal, pectoral, and caudal fins, albeit disposed vertically in the latter two and horizontally in the former (Fig. 8.2). This convergence goes deeper than mere correspondence in locomotion, given further likeness in jaws and teeth related to shared foraging ecology as apex predators of the sea.

The long rostra of pelagic and river dolphins have also been likened to those of living and extinct crocodilians, particularly narrow-jawed gharials, with resemblance in the form and function of jaws and dentition, and perhaps with similarly large pterygoid musculature (Massare, 1987; Taylor, 1987; Kelley & Motani, 2015; McCurry et al., 2017a, b, c). It appears that this phenotype, with long-snouted, snapping jaws, has evolved multiple times in diverse lineages as a means of capturing small prey, such as fish, as well for grasping and tearing larger prey in water or on shore. Even the rostra and dentition of semi-aquatic murid rodents (Rowe et al., 2014) demonstrate convergence with this long-snouted form.

8.3 Suction Feeding

Whereas raptorial feeders take advantage of the buoyancy and density of water by using suction to transport grasped prey toward the posterior region of the oral cavity for easier swallowing, other aquatic animals, including numerous pinnipeds and cetaceans, have eliminated the grasp-and-transport step and instead use suction directly to capture and ingest prey. This has been closely observed or documented experimentally for several species of whales, dolphins, porpoises, seals, and for walruses (Fay, 1982; Kastelein et al., 1994; Werth, 2000a; Bloodworth & Marshall, 2005; Marshall et al., 2008, 2014, 2015; Kane & Marshall, 2009; Johnston & Berta, 2011; Hocking et al., 2013, 2014). In yet other species there is strong evidence of suction feeding as deduced from morphology or stomach contents, specifically relating to the type, size, and condition of prey, this often being whole and lacking bite marks, or the presence of benthic debris (Werth, 2000b).

Because suction feeding represents a sharp deviation from the dentition-based mammalian basal condition of biting, it is easy to demonstrate examples of

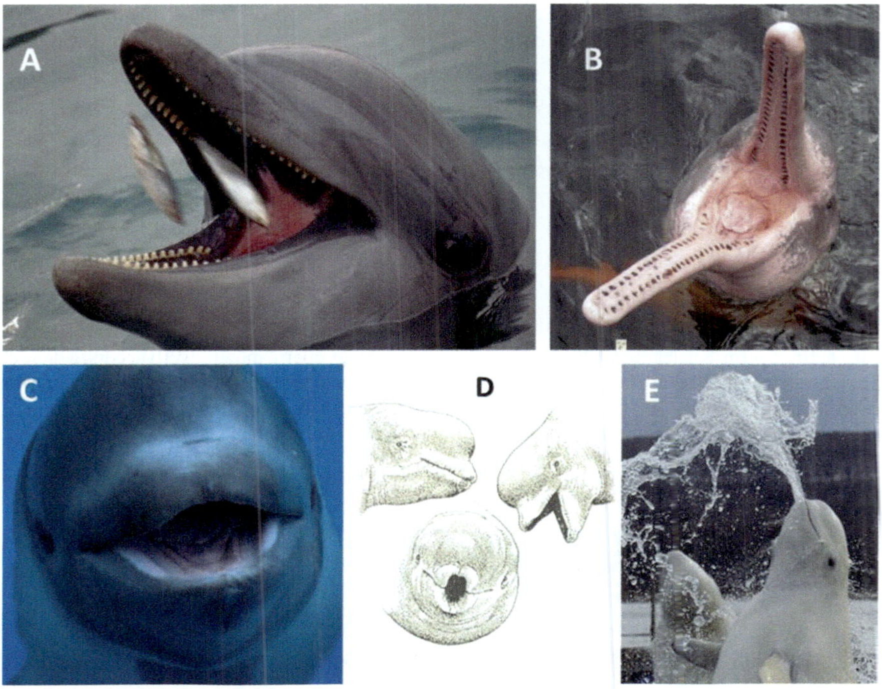

Fig. 8.9 Diversity of the jaws and mouth of odontocetes: notch-like gape of the bottlenose dolphin *Tursiops truncatus* (**a**) and wider gape and longer jaws of the Amazon river dolphin (**b**). Pursing of lips to form a round oral opening can be seen in beluga calves (**c**) and adults (**d**), and this species can also engulf and forcefully squirt out engulfed water (**e**). (Credit A: pxfuel; B: Jorge Andrade; C: NOAA.gov; D: Werth 1992 used by permission; E: Youri Smityuk, all Creative Commons use, published under CC BY 2.0, https://creativecommons.org/licenses/by/2.0/, no changes)

independent (convergent) derivation of suction in distinct lineages. Marked reduction, or even near-total loss, of dentition is a prominent characteristic of suction feeders, as are a suite of alterations of the skull. Mainly these involve shorter, broader rostra with wider jaws (Werth, 2006a, b). In place of a long triangular "notched" gape, the oral opening instead becomes a more rounded orifice, often with soft tissues of the cheek and lips being modified to create a circular mouth (Fig. 8.9). Such an oral opening improves the effectiveness of water intake from a point directly anterior to the head, while at the same time limiting lateral ingestion of water, thus decreasing the overall volume of water necessary to be sucked in to ingest prey items.

As in suction-feeding fishes, suction is generated in aquatic mammals by expansion of the oral and pharyngeal spaces, mainly by rapid depression and retraction of a piston-like tongue by robust musculature of a pronounced hyoid apparatus (Reidenberg & Laitman, 1994). Unlike bony and cartilaginous fishes, however, the engulfed water cannot be purged via unidirectional flow directly through gill slits;

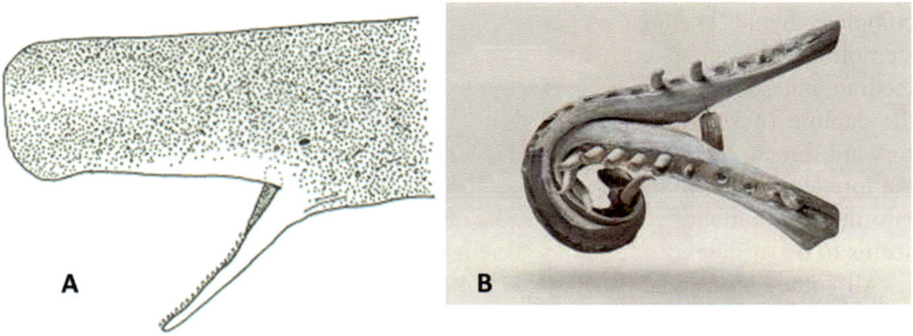

Fig. 8.10 Sperm whales have sharp lower teeth and a wide open "mouth" yet successfully suck prey through a circular oropharyngeal opening (**a**), even if jaws develop with a congenital curve (**b**). (Credit whale: Werth, 2004a used by permission, B, jaws: London Natural History Museum, Creative Commons use, published under CC BY 2.0, https://creativecommons.org/licenses/by/2.0/, no changes)

Fig. 8.11 In Blainville's beaked whale only two teeth erupt (and only in males), and they project outside the oral cavity and are commonly covered with barnacles (Credit: Robin Baird/Cascadia Research, Creative Commons use, published under CC BY 2.0, https://creativecommons.org/licenses/by/2.0/, no changes)

instead, engulfed water must be expelled back through the anterior oral opening by which it entered, necessitating a bidirectional water flow pathway. Nonetheless, suction feeding is the presumed default feeding method for many aquatic tetrapods, particularly amphibians, which, like mammals, lack patent pharyngeal openings as adults, and thus employ bidirectional water flow, making salamanders a better convergent analogue of the numerous mammalian suction feeding whales, dolphins, and seals.

The teeth of many suction-feeding marine mammals may become a secondary sexual feature, as in beaked whales (Ziphiidae), in which teeth often erupt solely in

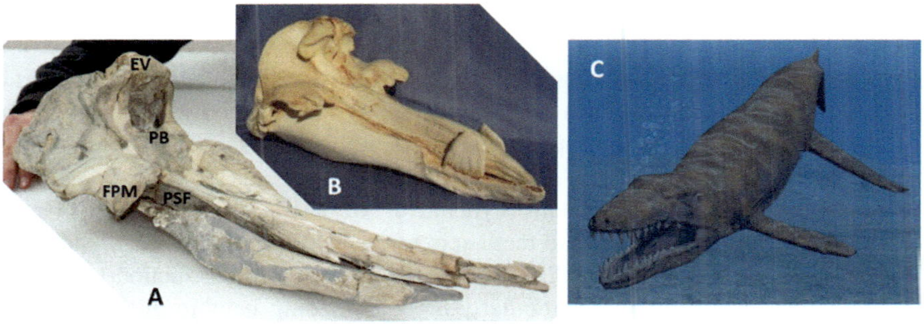

Fig. 8.12 Convergent skulls of fossil *Australodelphis* (**a**) and suction feeding beaked whale (**b**) The extinct marine reptile *Dakosaurus* (**c**) resembled a killer whale but may have been a suction feeder. (Credit A, B: R.E. Fordyce used by permission, C: CCY Nobu Tamura/http://spinops.blogspot.com, Creative Commons use, published under CC BY 2.0, https://creativecommons.org/licenses/by/2.0/, no changes)

adult males (Fig. 8.3), and which appear to be used for display or male-male competition, as evidenced by scratches, scars, and parallel rake marks on the integument, mainly of males. Teeth of suction feeders may lack complex structural prisms of the enamel (which normally resist shearing forces of mastication), or the enamel may be very thin or wholly lacking. Teeth of suction feeders are often greatly reduced in number, or are even functionally absent altogether, as in narwhals. Teeth may be present only in the lower jaw, as in sperm whales (*Physeter macrocephalus*, Fig. 8.10) and Risso's dolphins (*Grampus griseus*). In some aquatic mammal suction feeders, teeth may erupt outside the oral cavity and be covered with barnacles or other epizoic organisms; these conditions occur in several beaked whale species (Fig. 8.11). The presence of healthy animals with congenitally deformed or broken and poorly healed jaws, especially in sperm whales, all attest to the limited involvement of jaws and teeth in suction feeding (Werth, 2004a).

Suction feeding arose independently as a convergent suite of functional attributes in various lineages of extant odontocetes, including sperm whales (Physeteridae), beaked whales (Ziphiidae), narwhals and belugas (Monodontidae), and to varying extents in several porpoises (Phocoenidae) and dolphins (Delphinidae). Teuthophagy (predation on cephalopods, especially those found in deep waters) is a common theme among odontocete suction feeders (Heyning & Mead, 1996).

Not only have many living toothed whales converged on the suction feeding morphotype and ecotype, but also some extinct forms likewise exhibit the same adaptations. One of the best examples is that of *Australodelphis mirus*, an entirely toothless Miocene delphinid discovered in Antarctica (Fordyce et al., 2002; Fig. 8.12). The skull suture patterns, basicranial sinuses, and ear bones of *Australodelphis* all reveal a close phylogenetic relationship with living long-snouted oceanic dolphins. However, the overall form of the skull of *Australodelphis*, with its narrow, triangular shape and wholly edentulous upper and lower jaws, is strikingly reminiscent of beaked whales, particularly of the genus *Mesoplodon*. Further, the elevated vertex (labeled EV in Fig. 8.12), flanged premaxilla and maxilla (FPM), prenarial basin (PB), and pterygoid sinus fossae (PSF) are all much like those of a ziphiid skull. These features, together with the absence of teeth, strongly support a remarkable case of convergence of *Australodelphis* with beaked whales in both its specific cranial osteology and in its generally teuthophagous, suction-feeding ecotype (Heyning & Mead, 1996).

An extinct crocodyliform metriorhynchid of the late Jurassic and early Cretaceous, *Dakosaurus maximus*, had teeth somewhat like that of a killer whale but a study (Young et al., 2012; Fig. 8.12) concluded it was a likely a suction feeder. Of course, some extant reptiles, particularly ambush-feeding benthic turtles (notably snapping turtles, Chelydridae), also ingest prey via suction, as is the case for certain fossil turtles that are known to have been suction feeders.

Many pinnipeds are highly specialized and well adapted for suction feeding. These include, in particular, walruses (*Odobenus rosmarus*) and bearded seals (*Erignathus barbatus*), as outlined in the next section.

8.4 Durophagous Biting, Herbivory, and Benthic Foraging

Sea otters, *Enhydra lutris*, use a variant of biting that emphasizes durophagy (Timm-Davis et al., 2017): feeding on hard (often shelled) items such as bivalve mollusks, crabs, and sea urchins (Fig. 8.13). Durophagy has been proposed for various fossil mammals, including *Gomphotaria pugnax*, a late Miocene dusignathine walrus (Barnes & Raschke, 1991) whose worn and broken teeth suggest convergence on the sea otter feeding mechanism. An early Miocene ursid carnivoran, *Kolponomos*, also had broad, heavy, flattened molars and a downturned snout that together suggest

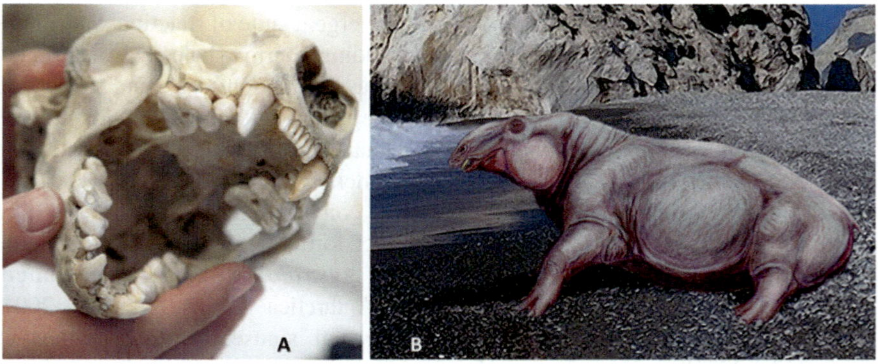

Fig. 8.13 Dentition of sea otter, *Enhydra lutris* (**a**, credit Burke Museum used by permission); extinct desmostylian (**b**, CCY Dmitry Bogdanov dmitrchel@mail.ru/Wikimedia Commons, Creative Commons use, published under CC BY 2.0, https://creativecommons.org/licenses/by/2.0/, no changes)

Fig. 8.14 (**a**, **b**) Benthic feeding in dugong, *Dugong dugon* (Credit: M. Marsa Alam/Wikimedia Commons), and (**c**) West Indian manatee, *Trichechus manatus* (Credit: Phil pix, all Creative Commons use, published under CC BY 2.0, https://creativecommons.org/licenses/by/2.0/, no changes)

foraging on coastal mollusks in the style of sea otters (Tseng et al., 2016). There are intriguing hints that the jaws of *Kolponomos* also converged on those of sabretooth cats such as *Smilodon* (Tseng et al., 2016).

The taxonomic affiliation and feeding method of desmostylians (Fig. 8.13), the only order of aquatic mammals that is now entirely extinct, are unclear, but features of the jaws, dentition, and skeleton suggest that desmostylians probably fed at least in part on kelp, sea grasses, or other aquatic vegetation. In this way desmostylians likely would have converged at least weakly on the herbivorous sirenians (manatees and dugongs).

As members of Tethytheria, sirenians themselves (Fig. 8.14) are related to proboscideans (elephants), and have evolved convergently or in parallel with these and other large terrestrial herbivores, not only in their large body form with tough hides and little hair, but in many aspects of their feeding (Best, 1981a). Some sirenians (dugongs) have tusks, which, although considerably shorter than those of living and extinct proboscideans, are likewise used in social displays and interactions; however, only among extinct dugongs is there evidence for the use of tusks to dig up or manipulate roots, rhizomes, and other vegetation. Sirenians and elephants both wear through many large cheek teeth as they masticate tough, abrasive grasses and other plant matter. Manatees have no incisors, canines, or premolars, but at any time have 8–10 molars which gradually move anteriorly via conveyor belt-like horizontal tooth replacement (Domning & Hayek, 1984), whereas elephants and dugongs have molar progression in which new teeth emerge to replace cheek teeth worn away by heavy use (Lanyon & Sanson, 2006). Although various aquatic mammals, such as pinnipeds, use teeth for oral processing of food, sirenians are the only aquatic mammals to retain true mastication, with precise dental occlusion and power strokes (Marsh et al., 1999).

The extinct Steller's sea cow, *Hydrodamalis gigas* (Fig. 8.15) was a gigantic (9 m, 8–10 ton) dugong found in the Bering Sea. It was discovered in 1741 and had become extinct (due to hunting) by 1768. It had adaptations to the cold waters of the North Pacific, including thick blubber, and fed on kelp. This whale-like species converged on terrestrial herbivores, including elephants and some sauropod dinosaurs, in its giant body size and foraging mechanics. Unlike elephants and extant

Fig. 8.15 Reconstruction of the extinct giant Steller's sea cow, *Hydrodamalis gigas*, showing 1 meter scale bar (Credit: Biodiversity Heritage Library, published under CC BY 2.0, https://creativecommons.org/licenses/by/2.0/, cropped, originally published in Extinct monsters, Chapman & Hall, 1896, London)

sirenians, but like some dinosaurs, it fed, without teeth, on large quantities of vegetation. Instead, *Hydrodamalis* had rough, rasp-like keratinous plates or masticatory pads on its jaw and palate, with which it grasped and crushed vegetation. It also had a dense array of bristles on its lips, as do other sirenians (Fig. 8.16).

Sirenians are also involved in another broad and well supported case of convergent evolution in secondarily aquatic mammals that involves benthic foraging, which is used by many marine mammals. Similar traits have evolved in disparate taxa, such as sirenians and some pinnipeds, for effective and efficient benthic foraging (Fig. 8.16). Using walruses, bearded seals and sirenians as a model system,

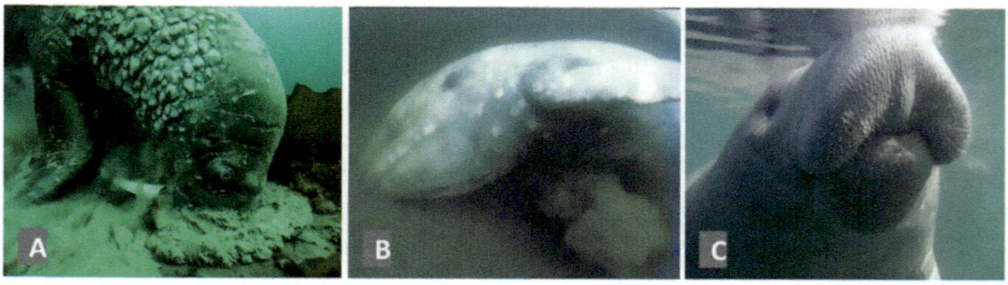

Fig. 8.16 Mud plumes from benthic feeding by walrus (**a**) and gray whale (**b**); bristled muzzle of manatee (**c**). (Credit A: Paul Nicklen/National Geographic.com, B: Wikimedia Commons, C: U.S. Fish and Wildlife Service, Creative Commons use, published under CC BY 2.0, https://creativecommons.org/licenses/by/2.0/, no changes)

we can begin to compare and contrast functional suites of attributes of their feeding morphology, performance and ecology.

Benthic and epibenthic foraging necessitates the contact of the feeding apparatus with the seafloor or freshwater benthos. Walruses (*Odobenus rosmarus*) and, to a lesser degree, bearded seals (*Erignathus barbatus*) specialize on foraging upon infaunal bivalves. Similarly, sirenians, particularly dugongs (*Dugong dugon*) and West Indian manatees (*Trichechus manatus*), spend considerable time on the benthos consuming seagrasses and other submerged aquatic vegetation (brackish and freshwater) among other food items. All are primarily or obligatorily benthic foragers and feeding occurs in relatively shallow waters. All perturbate the substrate from which they feed (Fig. 8.16), and in the case of walruses and dugongs, the excavation of infaunal bivalves and rhizomes results in significant bioturbation and signatures of benthic feeding records (furrows, pits, pit furrows, trails, etc.) that can be attributed to species with particular feeding biomechanics (Anderson & Birtles, 1978; Nerini & Oliver, 1983; Oliver et al., 1985). "No other biological or physical process can account for the record of excavations and discarded shells [from walrus feeding]" (Oliver et al., 1983). This benthic disturbance can have positive impacts upon the benthic community and increase biodiversity of the benthic community upon which aquatic mammals feed.

Walruses, bearded seals and sirenians converge on a suite of morphological features and benthic behaviors (Fay, 1982; Kastelein & Mosterd, 1989; Kastelein et al., 1994; Marshall et al., 1998a, 2000, 2003, 2006). This includes well-developed facial muscles (Domning, 1977; Fay, 1982, Kastelein et al., 1991, 1993; Marshall et al., 1998a; Marshall et al., 2008), elaborate vibrissae (Dosch, 1915; Bryden et al., 1978; Kamiya & Yamasaki, 1981; Fay, 1982; Reep et al., 1998; Marshall et al., 2006), and complex innervation for fine motor control of vibrissae as well as sensitive vibrotactile sensation from mystacial vibrissae (Kastelein & Van Gaalen, 1988; Marshall et al., 1998b; Bachteler & Dehnhardt, 1999; Reep et al., 2001; Marshall et al., 2007; Bauer et al., 2012). Other convergent morphological features within this specialized niche include increased width of the rostrum, a broad cranium, relatively large body size compared to the size of the head, and relatively small eyes (Kastelein et al., 1993; Murie, 1872; Fay, 1982).

Walruses, bearded seals and sirenians use their vibrissae as exploratory, and sometimes manipulative, tools during feeding. The mystacial vibrissae are greatly modified, typically by an increase in number, or by changes in length (longer in pinnipeds but shorter sirenians; Reep et al., 1998; Fay, 1982; Marshall et al., 2006). Sirenians depart from pinnipeds in using their mystacial whiskers in conjunction with modified facial muscles to form a muscular-vibrissal complex to physically grasp vegetation and substrate for excavation (Marshall et al., 1998a, 2000, 2003).

The rostrum and muzzle are broadest in walruses and dugongs (Fay, 1982; Kastelein et al., 1991; Marshall et al., 2003), which are benthic specialists (Marshall & Pyenson, 2019) but it is still comparatively broad in manatees (Reep et al., 1998; Marshall et al., 2003) and bearded seals (Marshall et al., 2006, 2008; Marshall et al., 2008). Although the anatomy of their facial muscles varies, the broad and expansive nature of the rostrum is due to hypertrophy of the orofacial musculature, which forms a muscular hydrostat (*sensu* Kier & Smith, 1985; walruses: Fay, 1982; Kastelein et al., 1991; sirenians: Domning, 1977, 1978; Marshall et al., 1998b; bearded seals: Marshall et al., 2008). Muscular hydrostats are characterized as three-dimensional arrays of muscles of constant volume, with a capability of highly controlled, detailed, and varied movements (Kier & Smith, 1985). The feeding and associated oral anatomy and behaviors of the vibrissal-muscular complex of walruses, bearded seals and sirenians fall within the definitions of a muscular hydrostat. This muscular hydrostat is used both for movement of vibrissae and also to purse the anterior and lateral regions of the lips. In pinnipeds the orofacial muscles are crucial for specialization for suction feeding (Marshall et al., 2015) whereas in sirenians it is crucial for motor and sensory use of vibrissae/bristles to manipulate (oripulate *sensu* Reep et al., 2001) vegetation and to excavate rhizomes within the benthic substrate (Marshall et al., 1998a, 2000, 2003).

As alluded to above, the precise method of food acquisition differs between pinnipeds and sirenians. Both walruses and bearded seals are suction feeding specialists. In many ways, bearded seals are the phocid eco-type equivalent of walruses. Whereas walruses specialize on infaunal bivalves, bearded seals are generalist foragers that will consume infaunal bivalves, marine worms, and also epibenthic invertebrates and fish (e.g., Vibe, 1950; Kosygin, 1971; Lowry et al., 1980; Finley & Evans, 1983; Hjelset et al., 1999; Dehn et al. 2006). Both walruses and bearded seals have developed a technique for excavating prey that incorporates suction and hydraulic jetting (Fay, 1982; Kastelein & Mosterd, 1989; Marshall et al., 2008).

Suction feeding oral morphologies are similar to those used for benthic feeding in general. In addition to a broad and flat muzzle, the palate is smooth and arched (more so in walruses than bearded seals), presumably to increase the intraoral volume. The tongue is large and piston-like and the robust extrinsic lingual musculature (styloglossus, hyoglossus, and genioglossus) retracts the tongue to produce rapidly-generated and powerful, negative intraoral pressures (Fay, 1982; Gordon, 1984; Kastelein et al., 1994; Marshall et al., 2008). In pinnipeds the anterior portion of the lips are "pursed to form a narrow pipette-like opening and the lateral lips are pursed to occlude lateral gape, which maintains subambient pressures generated intraorally" (Marshall et al., 2008). Subambient intraoral pressure generation by an adult walrus was recorded to be as great as 87.9 kPa in air and 118.8 kPa underwater (Kastelein & Mosterd, 1989; Kastelein et al., 1994). The maximum subambient pressure measured for bearded seals was 91.9·kPa (Marshall et al., 2008), which is comparable to that of walruses (91.2·kPa by Fay, 1982; 51–118·kPa by Kastelein et al., 1994). Feeding performance data for pinnipeds and cetaceans shows that even among non-suction specialists, suction is the preferred feeding mode of most marine mammals and many taxa converge upon these morphological traits to generate and maintain subambient intraoral pressures. While pinniped suction feeding specialists are able to generate up to an atmosphere or more of subambient pressure, generalist pinniped foragers can also produce substantial subambient pressures during feeding events (Marshall & Pyenson, 2019).

Feeding methods used by walruses, bearded seals, sirenians, and other benthic foragers (e.g., gray whales) influence the structure of the benthic community upon which they feed. The excavation technique of walruses "tills" the seafloor sediment and is a source of substantial bioturbation for the benthic community. As noted above, such bioturbation by groups of foraging marine mammals alters the structure of the benthic community and can increase benthic invertebrate diversity and productivity (Oliver et al., 1983, 1985; Shull, 2009; Van Blaricom, 1982).

This comparison among walruses, bearded seals and sirenians for acquiring prey can be expanded upon. Other marine mammals that converge upon these morphological and behavioral traits include other suction specialists for which data are available, such as beluga whales (*Delphinapterus leucas*), likely finless porpoises (*Neophocaena phocaenoides*), and some unlikely benthic foragers such as gray whales. A similar suite of convergent feeding morphologies and behaviors can probably be found in other habitats, such as mesopelagic feeding (Northern fur seals, *Callorhinus ursinus*; Marshall et al., 2015), feeding under the ice (Weddell seals, *Leptonychotes weddellii*) and feeding in more general open water (epipelagic and mesopelagic) habitats.

For example, beaked whales (Ziphiidae) and elephant seals (*Mirounga* spp.) have converged on similar features of anatomy (Bianucci et al., 2016), physiology, and behavior related to prolonged (>60 min) and very deep (>2000 m) dives for schooling fish and squid. Such features include not only convergent respiratory and related neurological adaptations but also general aspects of overall body form related to descending from and ascending to the surface, and similar features of the lips and tongue for creating a circular oral orifice for more efficient ingestion of prey and expulsion of ingested water. Note that despite these convergent similarities, there are also major disparities between these taxa, such as the fact that deep-diving odontocetes locate prey via echolocation, whereas deep-diving pinnipeds utilize tactile vibrissae and large eyes for enhanced low-light vision.

As for the Cetacea, presumed epipelagic and benthic suction feeding forms can be found in extinct pinnipeds (Kienle & Berta, 2016). Nonetheless, one of the most extreme and compelling examples of convergent evolution in aquatic mammals involves living walruses (*Odobenus rosmarus*) and the extinct odontocete *Odobenocetops*, whose very name translates to "walrus-like whale" (or, more literally, "cetacean that seems to walk on its teeth," as the generic name of walruses, *Odobenus*, refers to 'tooth-walking," because the tusks were once thought to aid walruses in hauling out onto ice floes). Apart from the basicranium, which is clearly cetacean, the entire skull of *Odobenocetops* looks much like that of a walrus. Unlike other odontocetes, *Odobenocetops* possessed a very short, broadly rounded rostrum bearing two large tusk-like teeth. *Odobenocetops* is closely related to monodontid narwhals, but in cases where one *Odobenocetops* tusk is markedly larger than the other, this tusk erupts on the right side of the skull (unlike the tusk-like left tooth of narwhals) and projects posteriorly. As in walruses, the external nares are located near the tip of the rostrum, the eyes are directed dorsally (so that the animal could look forward with its head down), and the arched, vaulted palate is toothless. Prominent muscle scars indicate a strong, mobile upper lip, and its cervical vertebrae suggest a flexible neck. Together, these features reveal an extraordinary convergence with the benthic suction feeding habits of walruses, with the asymmetrical tusks of *Odobenocetops* used either to stir up bivalve mollusks on the seafloor or to guide the head, in sled-like fashion, so that the mouth could be positioned close to, but above, the seafloor for sucking thin-shelled mollusk bodies directly from their then-discarded shells, as occurs in walruses. Not only is the extreme specialization of this odontocete species "unprecedented" (De Muizon, 1993; de Muizon & Domning, 2002), but the degree of morphological convergence with *Odobenus* is truly astonishing.

8.5 Filter Feeding

For air-breathing mammals, one of the best (most effective) answers to the challenges of reverting to aquatic feeding is to separate prey items from water by filtration. This is a common, and in some ways simple, solution adopted by many aquatic organisms including countless diverse invertebrates, although invertebrates generally rely on movement of water-borne prey around branched, antenna-like appendages or tentacles, called lophophores. Such lophophores have sufficient surface area (and often stickiness) to collect suspended food particles, either in the form of tiny whole organisms or simply organic detritus. However, mammals, like other filter feeders of large body size and high metabolic rate, rely instead on internal filtration: they use various mechanisms to bring water-borne prey into the body, where it is separated from the water and the filtered water then purged (Sanderson & Wassersug, 1990, 1993). Tunicates, cephalochordates and fishes filter using a perforated pharynx which, in fishes, houses the gas exchange organs (gills), enabling both a steady source of incurrent water and a ready pathway for unidirectional flow. Like some other filter feeders, mammals use an intraoral filter derived from specialized dentition or, in the case of mysticete whales, a novel keratinous oral tissue, baleen, that wholly replaces teeth but functions somewhat analogously in collecting and holding food while allowing water to be expelled (Pivorunas, 1979; Werth, 2000b; Goldbogen et al., 2006, 2017; Werth & Ito, 2017).

There are several clear advantages to filter feeding (Fordyce, 1980; Werth, 2000b). Although the energetic costs of moving large volumes of water can be prohibitive, these costs are often linked to an animal's forward locomotion, with elastic tissues in and around the throat (Shadwick et al., 2013) recovering some of the energy expended for locomotion. Although filtration can target individual items, it is generally used to obtain large quantities of small prey (relative to the feeder's body size) via bulk or batch feeding. This enables whales and other filter feeders to feed near the base of a trophic pyramid, consuming large amounts of biomass and energy and thus attaining and sustaining very large body size (Pyenson & Vermeij, 2016). Large body size is important because food in aquatic ecosystems is often patchy, so a predator can (1) rapidly engulf a large volume of prey before competitors can get to it, (2) undertake travel over long distances between prey patches, and

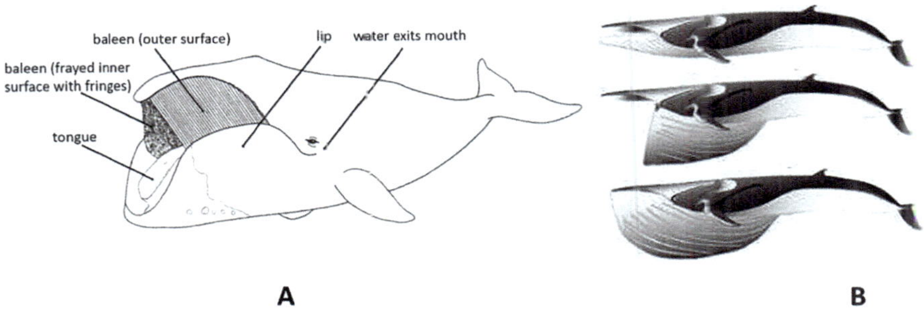

Fig. 8.17 Morphology of continuous skim feeding in right whale, (**a**) versus intermittent lunge feeding in rorqual fin whale, (**b**) (Credit A: A. Werth used by permission, B: J. Goldbogen used by permission)

(3) live off stored energy if food is unavailable for long periods of time, such as winter months when primary productivity is low or near absent.

Not surprisingly, then, aquatic filter feeders of all kinds are the largest animals in the sea, both now and in prehistoric seas (Bianucci et al., 2019). Whale sharks (*Rhincodon*), basking sharks (*Cetorhinus*), and manta rays (*Mobula*) are all gigantic living chondrichthyans. Many abundant filter feeding bony fishes can be found around the world today (particularly the clupeoid sardine, herrings, and anchovies), but during the Mesozoic Era, large pachycormid bony fish such as *Leedsichthys* may have reached lengths of 9–10 m. Many diverse forms of marine reptiles existed during the Mesozoic, yet filtering forms are conspicuously absent from the fossil record, perhaps due to the success of large filtering fish. Even the early Paleozoic anomalocaridid invertebrate *Aegirocassis* grew to large size (a then-immense 2 m) by filtering plankton.

The most basic kind of filtering is simple sieving, in which the filter's pore size is sufficiently small to trap and hold desired prey (Werth, 2012, 2013; Werth et al., 2016a, b). However, studies (Werth & Potvin, 2016; Potvin & Werth, 2017) indicate that balaenid mysticetes, the bowhead and right whales, use tangential rather than "dead-end" or throughput filtration, with water and prey flowing along rather than perpendicularly through the filter; this both prevents clogging of the filter and causes the filtrate (the captured items) to move down the filter and accumulate in a location, presumably near the oropharyngeal opening, where it can be more easily swallowed. This sort of filtration is possible in balaenids because, unlike other whales, they use continuous rather than intermittent filtration, as do gray whales and groove-throated rorquals, including humpback, fin, and blue whales (Goldbogen et al., 2006, 2017; Simon et al., 2012; Potvin et al., 2020). The intermittent filterers engulf and filter prey from a single discrete mouthful of water (Goldbogen et al., 2006; Simon et al., 2012). Not only are there multiple means by which mysticete filters can catch prey, but different whales use multiple means to bring water-borne prey into the mouth (Fig. 8.17). Rorquals are lunge feeders: they quickly expand the mouth with a ram mechanism from rapid forward fluking, and the loose, saccular tongue inverts and

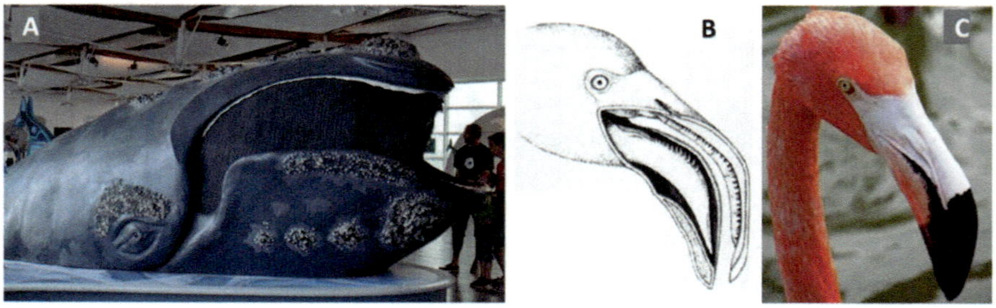

Fig. 8.18 Comparison of arched lamellar filter of right whale (**a**) versus flamingo (**b**, **c**) (Credit: A Oceanarium France, B P. Jenkin Wikimedia Commons, C H. Zell Wikimedia Commons, all Creative Commons use, published under CC BY 2.0, https://creativecommons.org/licenses/by/2.0/, no changes)

occupies an expanded ventral pouch. Accordion-like external throat pleats also expand, so that a normally slim whale can momentarily, before the engulfed water is expelled, look like a bloated tadpole or frog with a bulging mouth (Shadwick et al., 2013). Sei whales are rorqual species that specialize on small zooplankton, and can skim feed like balaenids (which also feed on tiny plankton) in addition to lunging (Werth et al., 2018b).

In contrast, gray whales (*Eschrichtius robustus*) normally feed on benthic plankton. They use orally-generated suction from rapid mouth expansion (caused by rapid tongue depression and retraction) to draw in a mouthful of water (Ray & Schevill, 1974). Even though gray whales have a filter of stiff, rough baleen plates and worn hair-like fringes, this filtering mechanism involves ingestion of much benthic sediment, which abrades and wears down gray whale baleen (especially on the right side of the mouth, which the whales often turn downward as they suck in muck). In addition, this benthic suction ingestion also leaves large pits that scar the coastal seafloor where gray whales feed; this act of bioturbation (Nerini & Oliver, 1983; Oliver et al., 1983, 1985; Shull, 2009) could in some ways degrade benthic habitats but may enhance benthic biodiversity and productivity.

In this manner of filter feeding, gray whales have converged on the filtering mechanisms of aquatic birds, including ducks and flamingos (Fig. 8.18), which also use an oral sieve of bristle-like lamellae to separate zooplankton and organic debris from mouthfuls of ingested benthic sediment (Jenkin, 1957; Zweers et al., 1995). Most ducks, geese, and swans filter feed by "dabbling" with a strainer of 50–70 lamellae in each side of their bill, but filtering specialists like the northern shoveler, *Anas clypeata*, have as many as 400 lamellae, making them more like balaenid whales. On the other hand, ducks like mergansers (*Mergus* sp.), which feed mainly on fish or other macroscopic prey, lack oral lamellae entirely (Sanderson & Wassersug, 1990, 1993).

Field et al. (2011) concluded that balaenopterid (rorqual) whales have converged in several aspects of their feeding ecology and morphology with pelicans, namely in the bending curves and cross-sectional profile of the mandibles, which rotate to

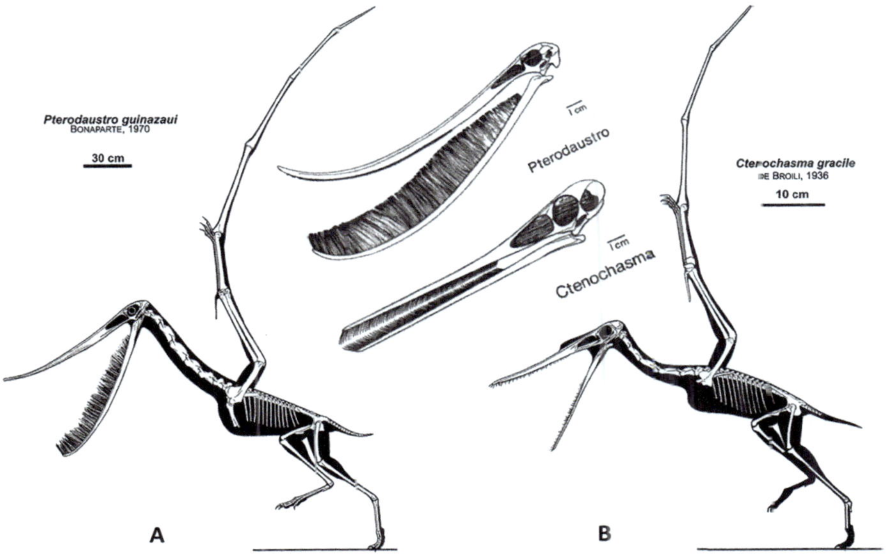

Fig. 8.19 Two pterosaurs: *Pterodaustro* (**a**) was a filter feeder; *Ctenochasma* (**b**) may have been one (Credit both: J.A. Headden, http://qilong.deviant.com/art, Creative Commons use, published under CC BY 2.0, https://creativecommons.org/licenses/by/2.0/, no changes)

accommodate a stretchy oral floor used to engulf capacious volumes of water during capture of prey.

Continuous filtration by bowhead and right whales (Werth, 2004b; Simon et al., 2009; Werth & Potvin, 2016; Potvin & Werth, 2017; Van der Hoop et al., 2019; Werth & Sformo, 2020), which stands in contrast to intermittent filtration by other mysticetes, relies on a steady-state, unidirectional flow of water, which is often observed during skimming on the surface but which also occurs (as confirmed by biotagging) at all levels of the water column, including near the bottom where copepods may gather before vertical migration upward at night. Unlike other mysticetes, balaenids have a space between the paired left and right baleen racks (the subrostral gap) through which water and prey enter. The water then splits, forming a Y-like pathway, and flows along the fringed medial face of each baleen rack (Werth & Potvin, 2016; Potvin & Werth, 2017). Filtered water is expelled from an oval-shaped opening on each side of the mouth, behind the semicircular lip at the angle of the mouth (Werth, 2004b, 2013). This continuous, unidirectional flow is highly specialized for cetaceans but has converged on the basic filter feeding mode seen in filter-feeding bony fish and sharks. In fact, the presumed crossflow filtration of balaenids appears to have converged on similar flow pathways and tangential filtration mechanisms that has been described for various sharks, rays, and bony fishes (Cheer et al., 2012; Paig-Tran & Summers, 2014; Sanderson et al., 2016; Divi et al., 2018). Olson and Feduccia (1980) and Sanderson and Wassersug (1993)

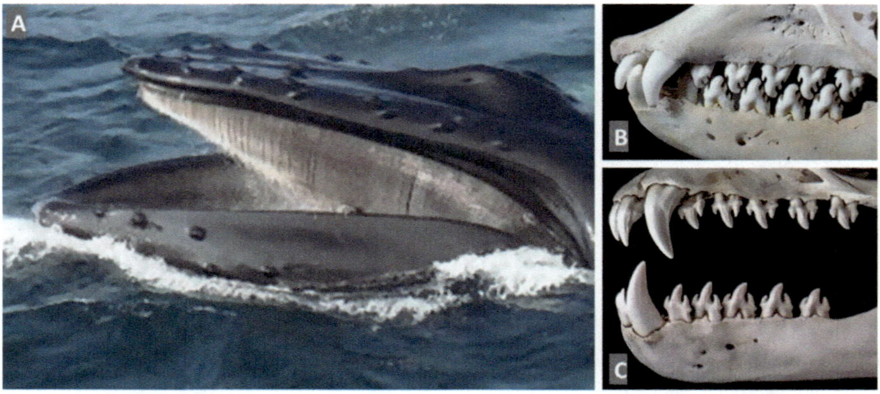

Fig. 8.20 Comb-like oral baleen filter of humpback whale (**a**) (*Megaptera novaeangliae*) compared to dental filters of crabeater seal (**b**) (*Lobodon carcinophaga*) and leopard seal (**c**) (*Hydrurga leptonyx*). (Credit: A, Whale & Dolphin Conservation Society, Creative Commons use, published under CC BY 2.0, https://creativecommons.org/licenses/by/2.0/, no changes; B, C: Museum Victoria, Melbourne used by permission)

pointed out the remarkably convergent curvature of the jaws of flamingos and balaenids, although the jaws are moved side-to-side during flamingo filtration and dorsoventrally in all whale feeding events.

Although filter-feeding aquatic reptiles appear to have been nonexistent—at least among swimming forms—one notable exception involved flying reptiles of the Mesozoic that, based on morphology, clearly show specialization for aquatic filtration (Fig. 8.19). *Pterodaustro*, a large pterosaur, had lower jaws bearing paired series of baleen-like combs of approximately 100 very thin needles. These structures were not initially thought to be teeth, but close examination revealed that they were indeed very thin teeth that formed a very fine filter, presumably for catching marine plankton which the pterosaurs could skim from the sea surface. Although *Pterodaustro*'s combs projected upward from its lower jaws instead of hanging downward from its upper jaws as in all whales, the fineness of the flying reptile's filter indicates a close convergence with that of balaenids, although the inverted nature of the filter suggests convergence with flamingos. Another pterosaur, *Ctenochasma*, may have been a filter feeder, as it possessed numerous thin, fine, and likely flexible maxillary and mandibular teeth along its long, narrow upper and lower jaws.

Crabeater seals, *Lobodon carcinophaga*, which should more appropriately be called krilleater seals due to their sole focus on filter feeding on krill (especially large Antarctic krill, *Euphausia superba*), have elaborate cheek teeth bearing cusps with large gaps through which excess water is expelled (Fig. 8.20). These phocid pinnipeds ingest krill via suction, then purge water via tongue elevation; although they can filter individual prey items, crabeater seals normally ingest and filter many krill at once (Ross et al., 1976). Leopard seals, *Hydrurga leptonyx*, have a reputation for feeding predominantly on penguins, but they too have highly lobed teeth, reminiscent of, yet distinct from, those of *Lobodon* (Fig. 8.20); when krill are abundant in Antarctic waters, *Hydrurga* feeds mainly via filtering (Hocking et al., 2013). Australian sea lions, *Neophoca cinerea*, have small, simple cheek teeth that might also be used for filter feeding on krill or other small prey. Crabeater and leopard seals are closely related but likely independently evolved their filter feeding morphology and ecology, making this a possible example of convergent evolution. Some biologists presume that mysticete filtration evolved from a stage in which ancestral mysticetes filtered through their teeth, but this is far from certain. Studies of tooth sharpness and dental wear patterns (Hocking et al., 2017c) indicate that baleen whale ancestors did not go through a dental-based filtering stage, suggesting that pinniped dental filtration is not, as often alleged, a true instance of convergence with Mysticeti (Deméré et al., 2008; Gatesy et al., 2013; McGowen et al., 2014; Berta et al., 2016; Marx et al., 2016a).

8.6 Conclusions

Aquatic mammals secondarily reverted to marine or freshwater habitats from terrestrial ancestry, and therefore have modified multiple aspects of form, function, and behavior to survive (Fordyce, 2018; Werth, 2020). Aquatic habitats impose strong selection pressures and constitute a strikingly different environment from the one that terrestrial mammals inhabit, because water is dense, heavy, and incompressible, yet it is also buoyant and can support large bodies and large aggregations of suspended matter or prey organisms. Aquatic mammals have succeeded in adapting to the challenges posed by the need to obtain food in water by adopting similar solutions to those exhibited by other (non-mammalian) aquatic organisms, including fishes, amphibians, turtles, waterfowl, and invertebrates. Because aquatic mammals have repeatedly evolved the same foraging methods and feeding modes—notably raptorial seizing, benthic feeding, suction feeding, and filter feeding—it is not surprising that they continually express the same suites of morphological traits seen within diverse groups of aquatic mammals as well as in other types of aquatic organisms. However, the precise manners and structures with which feeding methods of aquatic mammals have evolved to ingest, transport, process, and swallow prey are, while strikingly reminiscent of those found in other taxa, often novel and unique, such as the keratinous oral pads of Steller's sea cows and the baleen tissue of mysticete whales.

References

Adam, P. J., & Berta, A. (2002). Evolution of prey capture strategies and diet in the Pinnipedimorpha (Mammalia, Carnivora). *Oryctos, 4*, 83–107.

Anderson, P. K., & Birtles, A. (1978). Behaviour and ecology of the dugong, *Dugong dugon* (Sirenia): Observations in Shoalwater and Cleveland Bays, Queensland. *Australian Wildlife Research, 5*, 1–23.
Armfield, B. A., Zheng, Z., Bajpai, S., Vinyard, C. J., & Thewissen, J. G. M. (2013). Development and evolution of the unique cetacean dentition. *PeerJ, 1*, e24.
Bachteler, D., & Dehnhardt, G. (1999). Active touch performance in the Antillean manatee: Evidence for a functional differentiation of the facial tactile hairs. *Zoology, 102*, 61–69.
Barnes, L. G., & Raschke, R. E. (1991). *Gomphotaria pugnax*, a new genus and species of Late Miocene Dusignathine Otariid Pinniped (Mammalia: Carnivora) from California. *Natural History Museum of Los Angeles County Contributions in Science, 426*, 1–16.
Bauer, G. B., Gaspard, J. C., III, Colbert, D. E., Leach, J. B., Stamper, S. A., Mann, D. A., & Reep, R. L. (2012). Tactile discrimination of textures by Florida Manatees (*Trichechus manatus latirostris*). *Marine Mammal Science, 28*, 456–471.
Berta, A., Sumich, J. L., & Kovacs, K. (2015). *Marine mammals: Evolutionary biology 3e*. Elsevier.
Berta, A., Lanzetti, A., Ekdale, E. G., & Deméré, T. A. (2016). From teeth to baleen and raptorial to bulk filter feeding in mysticete cetaceans: The role of paleontological, genetic, and geochemical data in feeding evolution and ecology. *Integrative and Comparative Biology, 56*, 1271–1284.
Best, R. C. (1981a). Foods and feeding habits of wild and captive Sirenia. *Mammal Review, 11*, 3–29.
Best, R. C. (1981b). The tusk of the narwhal (*Monodon monoceros* L.): Interpretation of its function (Mammalia: Cetacea). *Canadian Journal of Zoology, 59*, 2386–2393.
Bianucci, G., Di Celma, C., Urbina, M., & Lambert, O. (2016). New beaked whales from the late Miocene of Peru and evidence for convergent evolution in stem and crown Ziphiidae (Cetacea, Odontoceti). *PeerJ, 4*, e2479.
Bianucci, G., Marx, F. G., Collareta, A., DiStefano, A., Landini, W., Morigi, C., & Varola, A. (2019). Rise of the titans: Baleen whales became giants earlier than thought. *Biology Letters, 15*, e0175.
Bloodworth, B., & Marshall, C. D. (2005). Feeding kinematics of *Kogia* and *Tursiops* (Odontoceti: Cetacea): characterization of suction and ram feeding. *The Journal of Experimental Biology, 208*, 3721–3730.
Bloodworth, B., & Marshall, C. D. (2007). A functional comparison of the hyolingual complex in pygmy and dwarf sperm whales (*Kogia breviceps* & *K. sima*), and bottlenose dolphins (*Tursiops truncatus*). *Journal of Anatomy, 211*, 78–91.
Boessenecker, R. W., Churchill, M., Buchholtz, E. A., Beatty, B. L., & Geisler, J. H. (2020). Convergent evolution of swimming adaptations in modern whales revealed by a large microphagous dolphin from the Oligocene of South Carolina. *Current Biology, 30*, 1–7.
Bryden, M. M., Marsh, H., & MacDonald, B. W. (1978). The skin and hair of the dugong, *Dugong dugon*. *Journal of Anatomy, 126*, 637–638.
Cheer, A., Cheung, S., Hung, T.-C., Piedrahita, R. H., & Sanderson, S. L. (2012). Computational fluid dynamics of fish gill rakers during crossflow filtration. *Bulletin of Mathematical Biology, 74*, 981–1000.
Churchill, M., & Clementz, M. T. (2015). Functional implications of variation in tooth spacing and crown size in Pinnipedimorpha (Mammalia: Carnivora). *The Anatomical Record, 298*, 878–902.
Conway Morris, S. (2003). *Life's solution: Inevitable humans in a lonely universe*. Cambridge University Press.
Darwin, C. R. (1859). *On the origin of species by means of natural selection*. John Murray.
De Muizon, C. (1993). Walrus-like feeding adaptation in a new cetacean from the Pliocene of Peru. *Nature, 365*, 745–748.
De Muizon, C., & Domning, D. P. (2002). The anatomy of *Odobenocetops* (Delphinoidea, Mammalia), the walrus-like dolphin from the Pliocene of Peru and its palaeobiological implications. *Zoological Journal of the Linnean Society, 134*, 423–452.

Dehn, L.-A., Sheffield, G., Follman, E. H., Duffy, L., Thomas, D. L., & O'Hara, T. M. (2006). Feeding ecology of phocid seals and some walrus in the Alaskan and Canadian Arctic as determined by stomach contents and stable isotope analysis. *Polar Biology, 30*(2), 167–181

Deméré, T. A., McGowen, M. R., Berta, A., & Gatesy, J. (2008). Morphological and molecular evidence for a stepwise evolutionary transition from teeth to baleen in mysticete whales. *Systematic Biology, 57*, 15–37.

Divi, R. V., Strother, J. A., & Paig-Tran, E. W. M. (2018). Manta rays feed using ricochet separation, a novel nonclogging filtration mechanism. *Science Advances, 4*, e9533.

Dixon, D. (1981). *After man: A zoology of the future*. St. Martin's.

Dixon, D., & Adams, J. (2003). *The future is wild: A natural history of the future*. Firefly Books.

Domning, D. P. (1977). Observations on the myology of *Dugong dugon* (Müller). *Smithsonian Contributions to Zoology, 226*, 1–57.

Domning, D. P. (1978). The myology of the Amazonian manatee, *Trichechus inunguis* (Natterer) (Mammalia: Sirenia). *Acta Amaz, 8*(Suppl. 1), 1–81.

Domning, D. P., & Hayek, L. A. C. (1984). Horizontal tooth replacement in the Amazonian manatee (*Trichechus inunguis*). *Mammalia, 48*, 105–127.

Dosch, F. (1915). Bau und Entwicklung des Integuments der Sirenen (Structure and development of the integument of Sirenia, translated by D.A. Sinclair). Tech. Trans. No. 1626. National Research Council of Canada, Ottawa, 1973. *Jenaische Zeitschrift, 53*, 805–854.

Fay, F. (1982). Ecology and biology of the Pacific walrus, *Odobenus rosmarus divergens* Illiger. *North American Fauna* Number 74. Washington, DC, U.S. Fish and Wildlife Service, 279.

Field, D. J., Lin, S. C., Ben-Zvi, M., Goldbogen, J. A., & Shadwick, R. E. (2011). Convergent evolution driven by similar feeding mechanics in balaenopterid whales and pelicans. *The Anatomical Record, 294*(8), 1273–1282.

Finley, K. J., & Evans, C. R. (1983). Summer diet of the bearded seal (*Erignathus barbatus*) in the Canadian High Arctic. *Arctic, 36*, 82–89.

Fitzgerald, E. M. G. (2010). The morphology and systematics of *Mammalodon colliveri* (Cetacea: Mysticeti), a toothed mysticete from the Oligocene of Australia. *Zoological Journal of the Linnean Society, 158*, 367–476.

Foote, A. D., Liu, Y., Thomas, G. W. C., Vinar, T., Alfoldi, J., Deng, J., et al. (2015). Convergent evolution of the genomes of marine mammals. *Nature Genetics, 47*, 272–275.

Fordyce, R. E. (1980). Whale evolution and Oligocene southern ocean environments. *Palaeogeography, Palaeoclimatology, Palaeoecology, 31*, 319–336.

Fordyce, R. E. (2018). Cetacean evolution. In B. Würsig, H. G. M. Thewissen, & K. Kovacs (Eds.), *Encyclopedia of marine mammals 3e* (pp. 180–185). Elsevier.

Fordyce, R. E., & Barnes, L. G. (1994). The evolutionary history of whales and dolphins. *Annual Review of Earth and Planetary Sciences, 22*, 419–455.

Fordyce, R. E., Quilty, P. G., & Daniels, J. (2002). *Australodelphis mirus*, a bizarre new toothless ziphiid-like fossil dolphin (Cetacea: Delphinidae) from the Pliocene of Vestfold Hills, East Antarctica. *Antarctic Science, 14*(1), 37–54.

Gaskin, D. E. (1982). *The ecology of whales and dolphins*. Heinemann.

Gatesy, J., Geiser, J. H., Chang, J., Buell, C., Berta, A., Meredith, R. W., Spring, M. S., & McGowen, M. R. (2013). A phylogenetic blueprint for a modern whale. *Molecular Phylogenetics and Evolution, 66*, 479–506.

Goldbogen, J. A., Calambokidis, J., Shadwick, R. E., Oleson, E. M., McDonald, M. A., & Hildebrand, J. A. (2006). Kinematics of diving and lunge-feeding in fin whales. *The Journal of Experimental Biology, 209*, 1231–1244.

Goldbogen, J. A., Cade, D., Calambokidis, J. A., Friedlaender, A. S., Potvin, J., Segre, P. S., & Werth, A. J. (2017). How baleen whales feed: the biomechanics of engulfment and filtration. *Annual Review of Marine Science, 9*, 367–386.

Gordon, K. R. (1984). Models of tongue movement in the walrus (*Odobenus rosmarus*). *Journal of Morphology, 182*(2), 179–196.

Gould, S. J. (1989). *Wonderful life: The burgess shale and the nature of history*. W.W. Norton.

Hamilton, H., Caballero, S., Collins, A. G., & Brownell, R. L. (2001). Evolution of river dolphins. *Proceedings of the Royal Society B, 268*, 549–556.
Heiss, E., Aerts, P., & Wassenburgh, S. V. (2018). Aquatic-terrestrial transitions of feeding systems in vertebrates: a mechanical perspective. *The Journal of Experimental Biology, 221*, e154427.
Heyning, J. E., & Mead, J. G. (1996). Suction feeding in beaked whales: morphological and observational evidence. *Contributions in Science, Museum of Natural History, Los Angeles County, 464*, 1–12.
Hjelset, A. M., Anderson, M., Gjertz, I., Lydersen, C., & Gulliksen, B. (1999). Feeding habits of bearded seals (*Erignathus barbatus*) from the Svalbard area, Norway. *Polar Biology, 21*, 186–193.
Hocking, D. P., Evans, A. R., & Fitzgerald, E. M. G. (2013). Leopard seals (*Hydrurga leptonyx*) use suction and filter feeding when hunting small prey underwater. *Polar Biology, 36*, 211–222.
Hocking, D. P., Salverson, M., Fitzgerald, E. M. G., & Evans, A. R. (2014). Australian fur seals (*Arctocephalus pusillus doriferus*) use raptorial biting and suction feeding when targeting prey in different foraging scenarios. *PLoS One, 9*, e112521.
Hocking, D. P., Marx, F. G., Park, T., Fitzgerald, E. M. G., & Evans, A. R. (2017a). A behavioural framework for the evolution of feeding in predatory aquatic mammals. *Proceedings of the Royal Society B, 284*, e2016750.
Hocking, D. P., Marx, F. G., Park, T., Fitzgerald, E. M. G., & Evans, A. R. (2017b). Reply to comment by Kienle et al. 2017. *Proceedings of the Royal Society B, 284*, e20171836.
Hocking, D., Marx, F. G., Fitzgerald, E. M. G., & Evans, A. R. (2017c). Ancient whales did not filter feed with their teeth. *Biology Letters, 13*(8), e20170348.
Hocking, D. P., Marx, F. G., Fitzgerald, E. M. G., & Evans, A. R. (2017). Ancient whales did not filter feed with their teeth. *Biology Letters, 13*(8), 20170348. https://doi.org/10.1098/rsbl.2017.0348
Howell, A. B. (1930). *Aquatic mammals: Their adaptations to life in the water*. Charles C. Thomas.
Jenkin, P. M. (1957). The filter-feeding and food of flamingoes (Phoenocopteri). *Philosophical Transactions of the Royal Society B, 240*, 401–493.
Johnston, C., & Berta, A. (2011). Comparative anatomy and evolutionary history of suction feeding in cetaceans. *Marine Mammal Science, 27*, 493–513.
Kamiya, T., & Yamasaki, F. (1981). A morphological note on the sinus hair of the dugong. In H. Marsh (Ed.), *The Dugong* (pp. 111–113). Department of Zoology, James Cook University of North Queensland.
Kane, E. A., & Marshall, C. D. (2009). Comparative feeding kinematics and performance of odontocete belugas, Pacific white-sided dolphins, and long-finned pilot whales. *The Journal of Experimental Biology, 212*, 3939–3950.
Kardong, K. V. (2019). *Vertebrates: Comparative anatomy, function, evolution 8e*. McGraw-Hill.
Kastelein, R. A., & Mosterd, P. (1989). The excavation technique for molluscs of Pacific walruses (*Odobenus rosmarus divergens*) under controlled conditions. *Aquatic Mammals, 15*, 3–5.
Kastelein, R. A., & Van Gaalen, M. A. (1988). The sensitivity of the vibrissae of a Pacific walrus (*Odobenus rosmarus divergens*) Part 1. *Aquatic Mammals, 14*, 23–133.
Kastelein, R. A., Gerrits, N. M., & Dubbeldam, J. L. (1991). The anatomy of the walrus head (*Odobenus rosmarus*): part 2. Description of the muscles and of their role in feeding and haul-out behavior. *Aquatic Mammals, 17*(3), 156–180.
Kastelein, R. A., Zweypfenning, R. C. V. J., Spekreijse, H., Dubbeldam, J. L., & Born, E. W. (1993). The anatomy of the walrus head (*Odobenus rosmarus*). Part 3: The eyes and their function in walrus ecology. *Aquatic Mammals, 19*, 61–92.
Kastelein, R. A., Muller, M., & Terlouw, A. (1994). Oral suction of a Pacific walrus (*Odobenus rosmarus divergens*) in air and under water. *Zeitschrift fur Saugetierkunde, 59*, 105–115.
Kelley, N. P., & Motani, R. (2015). Trophic convergence drives morphological convergence in marine tetrapods. *Biology Letters, 11*, 20140709.
Kelley, N. P., & Pyenson, N. D. (2015). Evolutionary innovation and ecology in marine tetrapods from the Triassic to the Anthropocene. *Science, 348*(6232), aaa3716.

Kellogg, R. (1928). The history of whales: Their adaptation to life in the water. *The Quarterly Review of Biology, 3*, 29–76.
Kienle, S. S., & Berta, A. (2016). The better to eat you with: The comparative feeding morphology of phocid seals (Pinnipedia, Phocidae). *Journal of Anatomy, 228*, 396–413.
Kienle, S. S., Law, C. J., Costa, D. P., Berta, A., & Mehta, R. S. (2017). Revisiting the behavioral framework of feeding in predatory aquatic mammals. *Proceedings of the Royal Society B, 284*, e2017035.
Kier, W. M., & Smith, K. K. (1985). Tongues, tentacles and trunks: the biomechanics of movement in muscular-hydrostats. *Zoological Journal of the Linnean Society, 83* 307–324.
King, J. E. (1983). *Seals of the world 3e*. Cornell University Press.
Kosma, M. M., Werth, A. J., Szabo, A J., & Straley, J. M. (2019). Pectoral herding: an innovative tactic for humpback whale foraging. *Royal Society Open Science, 6*(10), e191104.
Kosygin, G. M. (1971). Feeding of the bearded seal *Erignathus barbatus nauticus* (Pallas) in the Bering Sea during the spring-summer period. *Izvestiya TINRO, 75*, 144–151.
Lambert, O., Bianucci, G., Post, K., de Muizon, C., Salas-Gismondi, R., Urbina, M., & Reumer, J. (2010). The giant bite of a new raptorial sperm whale from the Miocene epoch of Peru. *Nature, 466*, 105–108.
Lambert, O., Bianucci, G., & de Muizon, C. (2017). Macroraptorial sperm whales (Cetacea Odontoceti, Physeteroidea) from the Miocene of Peru. *Zoological Journal of the Linnean Society London, 179*, 404–474.
Lanyon, J. M., & Sanson, G. D. (2006). Degenerate dentition of the dugong (Dugong dugon), or why a grazer does not need teeth: morphology, occlusion and wear of mouthparts. *Journal of Zoology, 268*(2), 133–152.
Layne, J. N. (1959). Feeding adaptations and behavior of a freshwater dolphin, Inia geoffrensis. *The Anatomical Record, 134*, 598.
Liem, K. F. (1990). Aquatic versus terrestrial feeding modes: possible impacts on the trophic ecology of vertebrates. *American Zoologist, 30*, 209–221.
Losos, J. B. (2018). *Improbable destinies: Fate, chance, and the future of evolution*. Penguin Random House.
Lowry, L. F., Frost, K. J., & Burns, J. J. (1980). Feeding of bearded seals in the Bering and Chukchi seas and trophic interaction with Pacific walruses. *Arctic, 33*, 330–342.
Marsh, H., Beck, C. A., & Vargo, T. (1999). Comparison of the capabilities of dugongs and West Indian manatees to masticate seagrasses. *Marine Mammal Science, 15*(1), 250–255.
Marshall, C. D. (2018). Feeding morphology. In B. Würsig, H. G. M. Thewissen, & K. Kovacs (Eds.), *Encyclopedia of marine mammals 3e* (pp. 349–354). Elsevier.
Marshall, C. D., & Goldbogen, J. A. (2015). Marine mammal feeding mechanisms. In M. A. Castellini & J. A. Mellish (Eds.), *Marine mammal physiology: Requisites for ocean living* (pp. 95–117). CRC Press.
Marshall, C. D., & Pyenson, N. D. (2019). Feeding in aquatic mammals: an evolutionary and functional approach. In V. Bels & I. Q. Whishaw (Eds.), *Feeding in vertebrates: Anatomy, biomechanics, evolution* (pp. 743–785). Springer.
Marshall, C. D., Huth, G. D., Edmonds, V. M., Halin, D. L., & Reep, R. L. (1998a). Prehensile use of perioral bristles during feeding and associated behaviors of the Florida manatee (*Trichechus manatus latirostris*). *Marine Mammal Science, 14*, 274–289.
Marshall, C. D., Clark, L. A., & Reep, R. L. (1998b). The muscular hydrostat of the Florida manatee (*Trichechus manatus latirostris*) and its role in the use of perioral bristles. *Marine Mammal Science, 14*, 290–303.
Marshall, C. D., Huth, G. D., Edmonds, V. M., Halin, D. L., & Reep, R. L. (2000). Food-handling ability and feeding-cycle length of manatees feeding on several species of aquatic plants. *Journal of Mammalogy, 81*, 649–658.
Marshall, C. D., Maeda, H., Iwata, M., Furuta, M., Asano, A., Rosas, F., & Reep, R. L. (2003). Orofacial morphology and feeding behaviour of the dugong, Amazonian West African and

Antillean manatees (Mammalia: Sirenia): functional morphology of the muscular-vibrissal complex. *Journal of Zoology (London), 259*, 1–16.

Marshall, C. D., Amin, H., Kovacs, K., & Lydersen, C. (2006). Microstructure and innervation of the vibrissal follicle-sinus complex in the bearded seal, *Erignathus barbatus* (Pinnipedia: Phocidae). *The Anatomical Record, 288a*, 13–25.

Marshall, C. D., Vaughn, S. D., Sarko, D., & Reep, R. L. (2007). Topographical organization of the facial motor nucleus in Florida manatees (*Trichechus manatus latirostris*). *Brain, Behavior and Evolution, 70*, 164–173.

Marshall, C. D., Kovacs, K., & Lydersen, C. (2008). Feeding kinematics, suction, and hydraulic jetting capabilities in bearded seals (*Erignathus barbatus*). *The Journal of Experimental Biology, 211*, 699–708.

Marshall, C. D., Wieskotten, S., Hanke, W., Hanke, F. D., Marsh, A., Kot, B., & Dehnhardt, G. (2014). Feeding kinematics, suction, and hydraulic jetting performance of harbor seals (*Phoca vitulina*). *PLoS One, 9*, e86710.

Marshall, C. D., Rosen, D., & Trites, A. W. (2015). Feeding kinematics and performance of basal otariid pinnipeds, Steller sea lions (*Eumetopias jubatus*), and northern fur seals (*Callorhinus ursinus*): Implications for the evolution of feeding modes. *The Journal of Experimental Biology, 218*, 3229–3204.

Marx, F. G., & Fordyce, R. E. (2015). Baleen boom and bust: A synthesis of mysticete phylogeny, diversity, and disparity. *Royal Society Open Science, 2*, e140434.

Marx, F. G., Hocking, D. P., Park, T., Ziegler, T., Evans, A. R., & Fitzgerald, E. M. G. (2016a). Suction feeding preceded filtering in baleen whale evolution. *Memoirs of Museum Victoria, 75*, 71–82.

Marx, F., Lambert, O., & Uhen, M. D. (2016b). *Cetacean paleobiology*. Wiley-Blackwell.

Massare, J. A. (1987). Tooth morphology and prey preference of Mesozoic marine reptiles. *The Journal of Vertebrate Paleontology, 7*, 121–137.

McCurry, M. R., Evans, A. R., Fitzgerald, E. M. G., Adams, J. W., Clausen, P. D., & McHenry, C. R. (2017a). The remarkable convergence of skull shape in crocodilians and toothed whales. *Proceedings of the Royal Society B, 284*, 20162348.

McCurry, M. R., Fitzgerald, E. M. G., Evans, A. R., Adams, J. W., & McHenry, C. R. (2017b). Skull shape reflects prey size niche in toothed whales. *Biological Journal of the Linnean Society, 121*(4), 936–946.

McCurry, M. R., Walmsley, C. W., Fitzgerald, E. M. G., & McHenry, C. R. (2017c). The biomechanical consequences of longirostry in crocodilians and odontocetes. *Journal of Biomechanics, 55*, 61–70.

McCurry, M. R., Evans, A. R., Fitzgerald, E. M. G., McHenry, C. R., Bevitt, J., & Pyenson, N. D. (2020). The repeated evolution of dental apicobasal ridges in aquatic-feeding mammals and reptiles. *Biological Journal of the Linnean Society, 127*, 245–259.

McGowen, M. R., Gatesy, J., & Wildman, D. E. (2014). Molecular evolution tracks macroevolutionary transitions in Cetacea. *Trends in Ecology & Evolution, 29*, 336–346.

Mchedlidze, G. A. (1984). *General features of the paleobiological evolution of Cetacea*. Oxonian.

Monod, J. (1971). *Chance and necessity: Essay on the natural philosophy of modern biology*. Knopf.

Murie, J. (1872). On the walrus. *Transactions of the Zoological Society of London, 7*, 411–464.

Nemoto, T. (1959). Food of baleen whales with reference to whale movements. *The Scientific Reports of the Whales Research institute, 14*, 149–290.

Nerini, M. K., & Oliver, J. S. (1983). Gray whales and the structure of the Bering Sea benthos. *Oecologia, 59*, 224–225.

Nweeia, M. T., Eichmiller, F. C., Nutarak, C., Eidelman, N., Giuseppetti, A. A., Quinn, J., Mead, J. G., K'issuk, K., Hauschka, P. V., & Tyler, E. M. (2009). Considerations of anatomy, morphology, evolution, and function for narwhal dentition. In M. A. Lang & S. E. Miller (Eds.), *Smithsonian at the poles: Contributions to international polar year science* (pp. 223–240). Smithsonian.

Nybakken, J. W., & Bertness, M. D. (2005). *Marine biology: An ecological approach 6e*. Pearson Benjamin Cummings.
Oliver, J. S., Slattery, P. N., O'Connor, E. F., & Lowry, L. F. (1983). Walrus, *Odobenus rosmarus*, feeding in the Bering Sea: A benthic perspective. *Fishery Bulletin U.S., 81*, 501–512.
Oliver, J. S., Kvitek, R. G., & Slattery, P. N. (1985). Walrus feeding disturbance: Scavenging habits and recolonisation of the Bering Sea benthos. *The Journal of Experimental Biology, 91*, 233–246.
Olson, S. L., & Feduccia, A. (1980). Relationships and evolution of flamingos (Aves: Phoenicopteridae). *Smithsonian Contributions to Zoology, 316*, 1–73.
Paig-Tran, E. W. M., & Summers, A. P. (2014). Comparison of the structure and composition of the branchial filters in suspension feeding elasmobranchs. *The Anatomical Record, 297*, 701–715.
Pivorunas, A. (1979). The feeding mechanisms of baleen whales. *American Scientist, 67*, 432–440.
Potvin, J., & Werth, A. J. (2017). Oral cavity hydrodynamics and drag production in balaenid whale suspension feeding. *PLoS One, 12*, e175220.
Potvin, J., Cade, D. E., Werth, A. J., Shadwick, R. E., & Goldbogen, J. A. (2020). A perfectly inelastic collision: bulk prey engulfment by baleen whales and dynamical implications for the world's largest cetaceans. *American Journal of Physics, e1771.*
Pyenson, N. D., & Vermeij, G. J. (2016). The rise of ocean giants: Maximum body size in Cenozoic marine mammals as an indicator for productivity in the Pacific and Atlantic Oceans. *Biology Letters, 12*, 20160186.
Racicot, R. A., Deméré, T. A., Beatty, B. L., & Boessenecker, R. W. (2014). Unique feeding morphology in a new prognathous extinct porpoise from the Pliocene of California. *Current Biology, 24*(7), 774–779.
Ray, G. C., & Schevill, W. E. (1974). Feeding of a captive gray whale, *Eschrichtius robustus*. *Marine Fisheries Review, 36*, 31–38.
Reep, R. L., Marshall, C. D., Stoll, M. L., & Whitaker, D. M. (1998). Distribution and innervation of facial bristles and hairs in the Florida *manatee (Trichechus manatus latirostris). Marine Mammal Science, 14*, 257–273.
Reep, R. L., Stoll, M. L., Marshall, C. D., Homer, B. L., & Samuelson, D. A. (2001). Microanatomy of facial vibrissae in the Florida manatee: The basis for specialized sensory function and oripulation. *Brain, Behavior and Evolution, 58*, 1–14.
Reidenberg, J. (2007). Anatomical adaptations of aquatic mammals. *The Anatomical Record, 290*, 507–513.
Reidenberg, J. S., & Laitman, J. T. (1994). Anatomy of the hyoid apparatus in Odontoceti (toothed whales): Specializations of their skeleton and musculature compared with those of terrestrial mammals. *The Anatomical Record, 240*, 598–624.
Repenning, C. A. (1976). Adaptive evolution of sea lions and walruses. *Systematic Biology, 25*, 375–390.
Reynolds, J. E., & Rommel, S. A. (1999). *Biology of marine mammals*. Smithsonian.
Riedman, M. (1990). *The pinnipeds: Seals, sea lions, and walruses*. University of California Press.
Ross, G. J. B., Ryan, F., Saayman, G. S., & Skinner, J. (1976). Observations on two captive crabeater seals at the Port Elizabeth Oceanarium. *International Zoo Yearbook, 16*, 160–164.
Rowe, K. C., Achmadi, A. S., & Esselstyn, J. A. (2014). Convergent evolution of aquatic foraging in a new genus and species (Rodentia: Muridae) from Sulawesi Island, Indonesia. *Zootaxa, 3815*, 541–564.
Sanderson, S. L., & Wassersug, R. (1990). Suspension feeding vertebrates. *Scientific American, 262*, 96–101.
Sanderson, S. L., & Wassersug, R. (1993). Convergent and alternative designs for vertebrate suspension feeding. In J. Hanken & B. K. Hall (Eds.), *The skull, Volume 3: Functional and evolutionary mechanisms* (pp. 37–112). University of Chicago Press.
Sanderson, S. L., Roberts, E., Lineburg, J., & Brooks, H. (2016). Fish mouths as engineering structures for vortical cross-step filtration. *Nature Communications, 7*, e11092.

Schwenk, K. (Ed.). (2000). *Feeding: Form, function and evolution in tetrapod vertebrates*. Academic.

Shadwick, R. E., Goldbogen, J. A., Potvin, J., Pyenson, N. D., & Vogl, A. W. (2013). Novel muscle and connective tissue design enables high extensibility and controls engulfment volume in lunge feeding rorqual whales. *The Journal of Experimental Biology, 216*, 2691–2701.

Shull, D. H. (2009). Bioturbation. *Encyclopedia of Ocean Sciences, 220*(4), 395–400.

Simon, M., Johnson, M., Tyack, P., & Madsen, P. T. (2009). Behaviour and kinematics of continuous ram filtration in bowhead whales (*Balaena mysticetus*). *Proceedings of the Royal Society B, 276*, 3819–3828.

Simon, M., Johnson, M., & Madsen, P. T. (2012). Keeping momentum with a mouthful of water: Behavior and kinematics of humpback whale lunge feeding. *The Journal of Experimental Biology, 215*, 3786–3798.

Slijper, E. J. (1962). *Whales*. Basic Books.

Taylor, M. A. (1987). How tetrapods feed in water: A functional analysis by paradigm. *Zoological Journal of the Linnean Society, 91*, 171–195.

Thewissen, J. G. M. (1998). *The emergence of whales: Evolutionary patterns in the origin of Cetacea*. Plenum.

Thewissen, J. G. M., & Williams, E. N. (2002). The early radiation of Cetacea (Mammalia): Evolutionary pattern and developmental correlations. *Annual Review of Ecology and Systematics, 33*, 73–90.

Thewissen, J. G. M., Cooper, L. N., George, J. C., & Bajpai, S. (2009). From land to water: The origin of whales, dolphins, and porpoises. *Evolution: Education and Outreach, 2*, 272–288.

Thewissen, J. G. M., Sensor, J., Clementz, M., & Bajpai, S. (2011). Evolution of dental wear and diet during the origin of whales. *Paleobiology, 37*, 655–669.

Timm-Davis, L. L., DeWitt, T. J., & Marshall, C. D. (2015). Divergent skull morphology supports two tropic specializations in otters (Lutrinae). *PLoS One, 10*, e0143236.

Timm-Davis, L. L., Davis, R., & Marshall, C. D. (2017). Durophagous biting in sea otters (*Enhydra lutris*) differs kinematically from raptorial biting of other marine mammals. *The Journal of Experimental Biology, 220*, 4703–4710.

Tseng, Z. J., Grohé, C., & Flynn, J. J. (2016). A unique feeding strategy of the extinct marine mammal *Kolponomos*: convergence on sabretooths and sea otters. *Proceedings of the Royal Society of London, 283*, N20160044.

Uhen, M. D. (2004). Form, function, and anatomy of *Dorudon atrox* (Mammalia, Cetacea): An archaeocete from the middle to late Eocene of Egypt. *Papers on Paleontology University of Michigan, 34*, 1–222.

Uhen, M. D. (2010). The origin(s) of whales. *Annual Review of Earth and Planetary Sciences, 38*, 189–219.

Van Blaricom, G. R. (1982). Experimental analysis of structural regulation in a marine sand community exposed to oceanic swell. *Ecological Monographs, 52*, 283–305.

Van der Hoop, J. M., Nousek-McGregor, A. E., Nowacek, D. P., Parks, S. E., Tyack, P., & Madsen, P. T. (2019). Foraging rates of ram-filtering North Atlantic right whales. *Functional Ecology, 33*(7), 1290–1306.

Vibe, C. (1950). The marine mammals and the marine fauna in the Thule district (Northwest Greenland) with observations on ice conditions in 1939-41. *Meddelelser Om Gronland, 150*, 54–64.

Wainwright, P. C., McGee, M. D., Longo, S. J., & Hernandez, L. P. (2015). Origins, innovations, and diversification of suction feeding in vertebrates. *Integrative and Comparative Biology, 55*(1), 134–145.

Werth, A. J. (1992). *Anatomy and evolution of odontocete suction feeding*. PhD thesis. Harvard University.

Werth, A. J. (2000a). A kinematic study of suction feeding and associated behavior in the long-finned pilot whale, *Globicephala melas* (Traill). *Marine Mammal Science, 16*, 299–314.

Werth, A. J. (2000b). Marine mammals. In Schwenk, K. (ed.). *Feeding: Form, function and evolution in tetrapod vertebrates* (pp. 475–514). Academic.

Werth, A. J. (2001). How do mysticetes remove prey trapped in baleen? *Bulletin of the Museum of Comparative Zoology at Harvard College, 156*, 189–203.

Werth, A. J. (2004a). Functional morphology of the sperm whale (*Physeter macrocephalus*) tongue, with reference to suction feeding. *Aquatic Mammals, 30*, 405–418.

Werth, A. J. (2004b). Models of hydrodynamic flow in the bowhead whale filter feeding apparatus. *The Journal of Experimental Biology, 207*, 3569–3580.

Werth, A. J. (2006a). Mandibular and dental variation and the evolution of suction feeding in Odontoceti. *Journal of Mammals, 87*, 579–588.

Werth, A. J. (2006b). Odontocete suction feeding: Experimental analysis of water flow and head shape. *Journal of Morphology, 267*, 1415–1428.

Werth, A. J. (2007). Adaptations of the cetacean hyolingual apparatus for aquatic feeding and thermoregulation. *The Anatomical Record, 290*, 546–568.

Werth, A. J. (2012) Hydrodynamic and sensory factors governing response of copepods to simulated predation by baleen whales. *International Journal of Ecology*, e1155.

Werth, A. J. (2013). Flow-dependent porosity of baleen from the bowhead whale (*Balaena mysticetus*). *The Journal of Experimental Biology, 216*, 1152–1159.

Werth, A. J. (2020). Cetaceans as exemplars of evolution and evolutionary ecology. *Oceans, 1*(2), 56–76.

Werth, A. J., & Ito, H. (2017). Sling, scoop, squirter: Anatomical features facilitating prey transport concentration, and swallowing in rorqual whales (Mammalia: Mysticeti). *The Anatomical Record, 3001*, 2070–2086.

Werth, A. J., & Potvin, J. (2016). Baleen hydrodynamics and morphology of cross-flow filtration in balaenid whale suspension feeding. *PLoS One, 11*, e0150106.

Werth, A. J., & Sformo, T. L. (2020). Anatomy and function of feeding. In J. C. George & J. G. M. Thewissen (Eds.), *The Bowhead Whale, Balaena mysticetus: Biology and human interactions* (pp. 213–223). Academic Press.

Werth, A. J., Harriss, R. W., Rosario, M. V., George, J. C., & Sformo, T. L. (2016a). Hydration affects the physical and mechanical properties of baleen tissue. *Royal Society Open Science, 3*(10), e160591.

Werth, A. J., Straley, J., & Shadwick, R. E. (2016b). Baleen wear reveals intraoral water flow patterns of mysticete filter feeding. *Journal of Morphology, 277*, 453–471.

Werth, A. J., Lillie, M. A., Piscitelli, M., Vogl, A. W., & Shadwick, R. E. (2018a). Slick, stretchy fascia underlies sliding tongue of rorquals. *The Anatomical Record, 302*, 735–744.

Werth, A. J., Potvin, J., Shadwick, R. E., Jensen, M. M., Cade, D. E., & Goldbogen, J. A. (2018b). Filtration area scaling and evolution in mysticetes: Trophic niche partitioning and the curious cases of the sei and pygmy right whales. *Biological Journal of the Linnean Society, 125*, 264–279.

Werth, A. J., Rita Espada, D., Moore, M. J., & Sformo, T. L. (2018c). How do baleen whales stow their filter: a comparative biological analysis. *The Journal of Experimental Biology, 221*, 189233.

Werth, A. J., Kosma, M. A., Chenoweth, E. M., & Straley, J. M. (2019a). New views of humpback whale flow dynamics and morphology during prey engulfment. *Marine Mammal Science, 35*(4), 1556–1578.

Werth, A. J., Loch, C., & Fordyce, R. E. (2019b). Enamel microstructure in Cetacea: A case study in evolutionary loss of complexity. *Journal of Mammalian Evolution, 27*(4), 789–807.

Werth, A. J., Ito, H., & Ueda, K. (2020a). Multiaxial movements of the minke whale temporomandibular joint. *Journal of Morphology, 281*, 402–412.
Werth, A. J., Rita Espada, D., Lysiak, N. S., George, J. C., & Sformo, T. L. (2020b). Baleen turnover and gut transit in mysticete whales and its environmental implications. *Polar Biology, 43*(6), 707–723.
Young, M. T., Brusatte, S. L., de Andrade, M. B., Desojo, J. B., Beatty, B. L., Steel, L., et al. (2012). The cranial osteology and feeding ecology of the metriorhynchid crocodylomorph genera *Dakosaurus* and *Plesiosuchus* from the Late Jurassic of Europe. *PLoS One, 7*(9), e44985.
Zweers, G., de Jong, F., Berkhoudt, H., & Vanden Berge, J. C. (1995). Filter feeding in flamingos (*Phoenicopterus ruber*). *Condor, 97*(2), 297–324.

Chapter 9
Solutions to a Sticky Problem: Convergence of the Adhesive Systems of Geckos and Anoles (Reptilia: Squamata)

Anthony P. Russell and Austin M. Garner

Abstract The modes and mechanisms of organismal attachment are numerous and diverse. Terrestrial vertebrates, however, achieve robust and releasable attachment to both abiotic and biotic substrata in three chief ways: hook-like anchors, such as claws, permit temporary attachment to surfaces via mechanical interlocking and/or frictional interactions with surface asperities; attachment organs releasing glandular secretions (e.g., the toe pads of hylid frogs, suction cups of disc-winged bats) achieve attachment via wet adhesion and/or suction; subdigital pads of some lineages of lizards possess filamentous outgrowths that induce friction and/or adhesion via molecular interactions. Lizards are the largest organisms to employ fibrillar-based attachment, but only the adhesive subdigital pads of geckos and anoles are sufficiently adhesively competent to support forces in excess of their body mass. The adhesive systems of geckos and anoles have long been considered convergent, but beyond general statements to this effect, convergence has not been rigorously assessed. Here we review what is known of the adhesive apparatus of both gekkotan and anoline lizards within the context of two hierarchically stratified domains: (1) adhesive attachment and the structure of setae and setal fields, and (2) the higher-order anatomical specializations that control the operation of the setae. We employ this information to identify the physical and organismic drivers of convergence of fibrillar adhesive systems, thereby enabling us to assess the particular, rather than superficially general, extent of convergence of the adhesive system of geckos and anoles.

Our synopsis of gekkotan and anoline setae, setal fields, and their adhesive systems reveals numerous physical and organismic constraints, perceived as the drivers of convergent evolution, that have led to similar morphological and functional outcomes. We posit that the setae and setal fields of geckos and anoles are

---

A. P. Russell (✉)
Department of Biological Sciences, University of Calgary, Calgary, AB, Canada
e-mail: arussell@ucalgary.ca

A. M. Garner
Department of Biology, Syracuse University, Syracuse, NY
e-mail: agarner@syr.edu

V. L. Bels, A. P. Russell (eds.), *Convergent Evolution*, Fascinating Life Sciences,
https://doi.org/10.1007/978-3-031-11441-0_9

convergent structures that enhance effective attachment to diverse substrata. Setae exhibit deep homology, arising from the convergently evolved spinulate Oberhäutchen of the epidermis. Following the initial exaptation of spinules as van der Waals adhesion-promoting setae, those of geckos and anoles followed somewhat different evolutionary pathways as the setae became organized into integrated setal fields. These pathways are reflective of differences in how the biomechanical control of the setal fields, during their application and release from the substratum, is achieved. Although anoles seemingly exhibit only a single evolutionary origin of the adhesive system, that of geckos has arisen on multiple independent occasions, with a broad range of expression of anatomical configurations that characterize the functional system. A broad survey of such configurations among geckos reveals that some are morphologically (and probably behaviorally) more similar to those of anoles than are others. Our assessment of the extent of convergence of the adhesive apparatuses of geckos and anoles identifies gekkotan taxa with an adhesive apparatus that most closely resembles that of anoles and explores what is minimally necessary to promote reversible attachment via molecular interactions. Our findings should contribute not only to ongoing investigations of the functional morphology of these adhesive systems but also should be informative to those who design biomimetic fibrillar adhesives intended to operate similarly to their natural counterparts.

Keywords Adhesion · Biomechanical control · Friction · Lamellae · Material properties · Molecular bonding · Scansors · Setae · Toe pads · van der Waals forces

9.1 Attachment Systems of Vertebrates and the Filament-Based Adhesion of Geckos and Anoles

The ways in which organisms attach themselves to components of their environment (both biotic and abiotic) are numerous and diverse (Nachtigall, 1974), and such attachment may be permanent or temporary and releasable. If the latter, attachment may be long-term, momentary, or anywhere in between (even for a given individual under differing circumstances). Terrestrial vertebrates employ three main ways of achieving releasable attachment, and most often, but not exclusively, it is their digits (Tornier, 1899; Nachtigall, 1974; Rosenberg & Rose, 1999) that carry the attachment devices (*sensu* Vogel, 1988). Hook-like anchors, such as claws, are widespread among amniote vertebrates (Maddin et al., 2009) and rely upon frictional interlocking interactions (Garner et al., 2017). Attachment organs employing glandular secretions, such as the "suckers" of hylid frogs (Green, 1981; Barnes et al., 2006), disc-winged bats (Wimsatt & Villa-R, 1970; Riskin & Fenton, 2001), mouse lemurs, tarsiers (Nachtigall, 1974), and feathertail gliders (Rosenberg & Rose, 1999), rely chiefly upon a combination of suction and wet adhesion. Molecular bonding via filamentous outgrowths of the integument relies upon attraction between surfaces at nano-scale separation distances and occurs in few lineages of lizards (Maderson, 1970). The latter mode of attachment is accomplished in the absence of secretions and independently of, although perhaps cooperatively with, claws (Naylor & Higham, 2019), if present. Similar fibrillar adhesives are widely distributed amongst insect and arachnid arthropods (Home, 1816; Gorb & Beutel, 2001; Federle, 2006). In some cases this occurs in combination with glandular secretions that enhance attachment (Gorb & Beutel, 2001; Federle, 2006; Gorb, 2008) whereas in others, as for lizards, attachment is achieved via dry adhesion that relies solely upon the properties of the extremely fine tips of their filaments (Federle, 2006). Dry filamentous adhesives permit exploitation of surfaces not available to taxa without them (Russell et al., 2015: Higham et al., 2017b; Pinto et al., 2018).

Organism-surface interactions using any one of the above-mentioned methods are not necessarily mutually exclusive (Riskin & Fenton, 2001; Barnes et al., 2006; Langowski et al., 2018), but the dominant attachment mode is reflected in the overall morphology of the system.

The largest organisms to employ attachment-enhancing integumentary filaments occur in certain lineages of squamate reptiles (Arzt et al., 2003; Labonte et al., 2016). Squamate scales are characteristically covered with keratinized epidermis that provides both mechanical protection and an effective barrier against water loss (Maderson et al., 1978). The β-keratin of the outer scale surfaces is stiffer and less flexible than the α-keratin that invests the hinge region between scales (Sawyer et al., 2000; Autumn et al., 2006a). The attachment filaments of squamates are derived from β-keratin (Alibardi, 2009) that is geometrically modified such that their effective elastic modulus is lessened (Autumn et al., 2006a), promoting contact with the substratum and enhancing friction, or in some cases, inducing adhesion.

The mode and magnitude of attachment in squamates is lineage dependent. Some clades of scincid lizard effect relatively weak attachment using elaborations of the outer epidermal generation (Irschick et al., 1996). The digits and tail tip of some chameleons bear friction-enhancing (but not demonstrably adhesive) seta-like filaments (Lange, 1931; Khannoon et al., 2014; Spinner et al., 2014). Filament-dependent adhesion is most evidently expressed, both morphologically and functionally, among gekkotan and anoline (dactyloid) lizards that carry their epidermal outgrowths on modified scales (Fig. 9.1)—the scansors of geckos and lamellae of anoles (Russell & Eslinger, 2017). The adhesively competent digits of geckos and anoles (Fig. 9.1) are generally regarded as being convergent (Hagey et al., 2017), but the functional and structural extent of such convergence has not previously been detailed.

9.2 Review of the Gekkotan and Anoline Adhesive Systems

To explore the extent of convergence of the adhesive system exhibited by geckos and anoles (Fig. 9.1), we review what is known of the system in each clade in the context of two hierarchically-stratified domains: (1) adhesive attachment and the structure of setae and setal fields, and (2) the higher-order anatomical specializations that control their operation. In so-doing, we attempt to identify what is necessary and

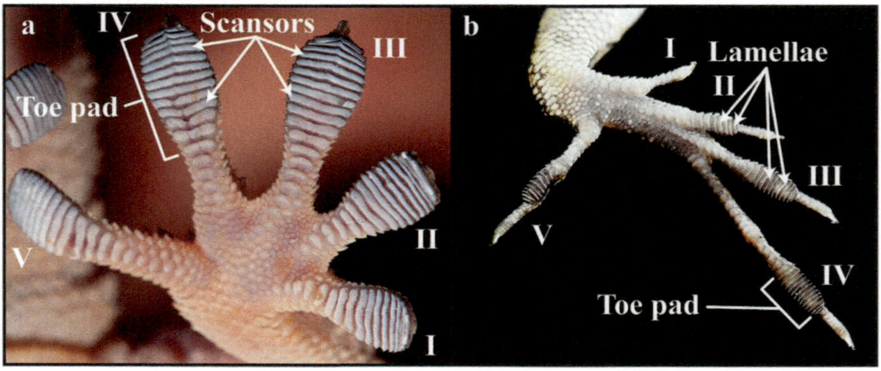

Fig. 9.1 (**a**) Ventral view of the right pes of *Tarentola mauritanica* showing the symmetrical disposition of the digits with the toe pads arranged in a fan-like fashion and lying essentially directly adjacent to each other around a broad arc. All five digits bear a toe pad. The toe pads in this taxon extend beneath the intermediate and distal phalanges and there is no free distal portion of the digits beyond the toe pads. (**b**) Ventral view of the left pes of *Anolis sagrei* showing the staggered disposition of the toe pads, the absence of a toe pad on digit I, the free distal portions of the digits extending distal to the toe pads, and the tightly clustered bases of the first four digits, with the fifth being markedly deviant from them. Note that in both taxa the scansors/lamellae become shorter from proximal to distal along the digits. The toe pads in both taxa are elliptical, with the scansors/lamellae being widest at about the midpoint of the toe pad

sufficient for reversible, locomotor-integrated adhesion in squamates and use this to assess the extent of convergence between geckos and anoles.

9.2.1 Gekkotan Setae and Setal Fields

9.2.1.1 Form and Variability of Gekkotan Subdigital Epidermal Outgrowths

Epidermal outgrowths of the subdigital pad surface of gekkotans vary considerably in form (Delannoy, 2005; Russell et al., 2007; Koppetsch et al., 2020). Delannoy (2005) described a morphotypic series of the epidermal outgrowths occurring on the subdigital scales of *Gekko gecko* (Fig. 9.2a): (1) true setae, outgrowths between 47.4 and 130 μm in length that are cleft into several sequential distal subdivisions, resulting in multiple branches that carry expanded, triangular, plate-like tips called spatulae, (2) seta-like outgrowths, branched fibrils with round, hook-like termini that range in length between 13.3 and 36.4 μm, (3) branched prongs with curved, rounded ends that range between 7.6 and 28.2 μm in length, and (4) hooked spines that range between 1.2 and 4.3 μm in length. The latter three may be involved in the enhancement of friction and are generally found on proximally situated lamellae, but the true setae, located on the scansors, are responsible for gekkotan adhesive attachment (Autumn et al., 2000).

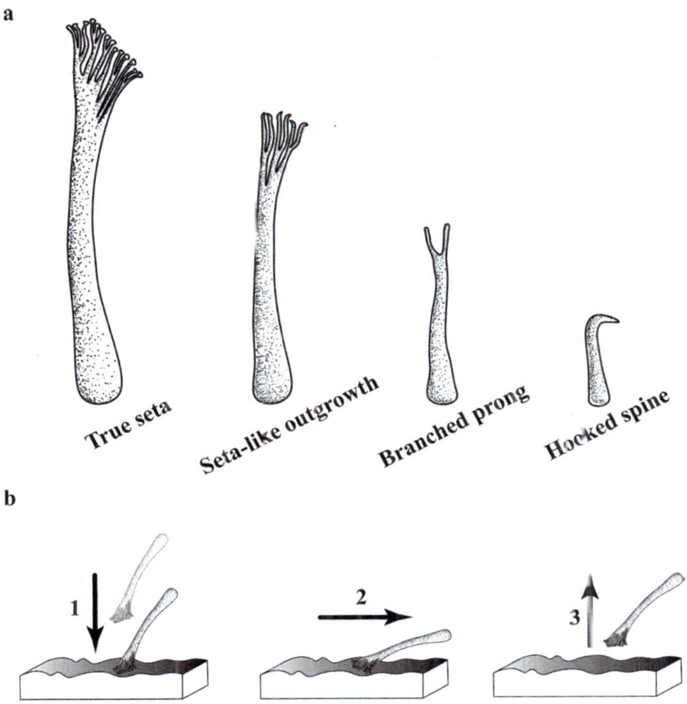

Fig. 9.2 (**a**) The clinal series of epidermal outgrowths occurring on the subdigital scales of *Gekko gecko* [as documented by Delannoy (2005)]. True setae are multiply branched outgrowths between 47.4 and 130 μm in length that carry expanded, triangular, plate-like tips called spatulae. These are present on true scansors and are responsible for the majority of gecko attachment forces. Seta-like outgrowths are, like true setae, multiply branched filaments, but terminate in multiple rounded, hook-like tips in lieu of spatulae and are 13.3–36.4 μm in length. Branched prongs are outgrowths with curved, rounded tips and are 7.6–28.2 μm in length. Hooked spines range in length from 1.2 to 4.3 μm. Seta-like outgrowths, branched prongs, and hooked spines may enhance frictional interactions with the substratum but are not thought to contribute to the majority of gecko attachment capacity. (**b**) The attachment and detachment mechanics of isolated gekkotan setae as revealed by Autumn et al. (2000). Gekkotan setae approach a substratum at an incident angle greater than 30°. A normal (perpendicular) preload is applied to the setae, pressing its spatulae onto the surface of the substratum (1). A shear (parallel) load is then applied, reorienting the triangular-shaped faces of the spatulae to make intimate contact with the surface, promoting the induction of van der Waals intermolecular forces. The shear load also results in the depression of the setal shaft angle to below 30° (2). Setal shafts are raised beyond 30°, at which point the spatulae detach from the surface of the substratum (3)

The term "seta" has historically been reserved for the elaborations of the squamate spinulate Oberhäutchen that carry expanded, spatulate tips capable of inducing adhesion (e.g., Ruibal & Ernst, 1965; Ruibal, 1968; Williams & Peterson, 1982; Peterson, 1983a, b). A recent study (Koppetsch et al., 2020), however, introduced an alternative definition of "setae" that encompassed all filaments constituting the spinulate outer epidermal generation. This stands in opposition to the extensive literature describing epidermal micro-ornamentation of squamates (e.g., Lange, 1931; Ruibal, 1968; Maderson, 1970; Stewart and Daniel, 1972; Peterson, 1983a, b) and makes no distinction between outgrowths that vary considerably in form and function. It essentially simply describes epidermal outgrowths as being unbranched or branched, and terminology for this is already recognized by more conventional nomenclature (e.g., branched prongs, see Fig. 9.2). We continue with the more conventional and broadly understood definition of "setae" that restricts its usage to filaments carrying expanded, spatulate tips capable of inducing adhesion (Ruibal & Ernst, 1965; Ruibal, 1968; Williams & Peterson, 1982; Peterson, 1983a, b; Autumn et al., 2002).

9.2.1.2 Adhesion Mechanics and Properties of Individual Gekkotan Setae

To generate reliable and reversible adhesion by means of van der Waals intermolecular forces, setal tips must be able to make and break intimate contact with the substratum (Autumn et al., 2002). Setae, in their default, unengaged state, rest at angles greater than 30° relative to the plane of the epidermal surface of the scansor, with their spatulae randomly arranged (Fig. 9.2b) (Autumn et al., 2000; Autumn & Hansen, 2006). As the setae are brought into contact with the substratum's surface, a normal (perpendicular) load presses the setal tips onto the surface (Fig. 9.2b). Subsequent application of a shear (parallel) load reorients the randomly arranged spatulae so that their expanded, triangular-shaped faces are placed into intimate contact with the surface, inducing van der Waals intermolecular forces (Fig. 9.2b) (Autumn et al., 2000, 2002; Autumn & Peattie, 2002). The normal and shear forces depress the setae to angles below 30°, which theoretically increases the adhesive force potential (Kendall, 1975; Autumn et al., 2000). Setal detachment occurs when the shear load is relaxed and the setal shaft angle is increased to at least 30°, at which point spatulae detach from the surface likely resulting from stress concentrations at the trailing edge of the spatulae (Fig. 9.2b) (Autumn et al., 2000; Autumn & Peattie, 2002).

The hierarchical structure of gekkotan setae (their subdivision into branches and the carriage of multiple spatulae) has been hypothesized to improve adhesion capacity (Peattie & Full, 2007; Murphy et al., 2009; Garner et al., 2021). Contact splitting (the subdivision of an adhesive area into many smaller adhesive contacts) was initially advanced to describe the benefit that accrues from breaking down an adhesive organ into many fibrils rather than it consisting of a single contact (e.g., one soft pad). The Johnson-Kendall-Roberts (JKR) theory of elastic contact predicts that the overall adhesive pull-off force is increased relative to a single contact when contact splitting occurs (Arzt et al., 2003). The adhesive pads of gekkotans are broken up into smaller, denser setae, with these, in turn, being subdivided into smaller, denser spatulae. In light of this, some authors (Peattie & Full, 2007) have hypothesized that geckos reap the benefits of contact splitting at two levels (via subdivision of the pad and the setae). Based on the theory of contact splitting, the subdivision of setae into multiple, smaller spatulae should increase the adhesive force potential compared to unbranched fibrils of similar dimensions (Garner et al., 2021). This prediction has been validated for synthetic fibrillar systems (e.g., Murphy et al., 2009). Structural hierarchy of setae may also reduce the potential for the formation of cracks, allowing the spatulae to detach uniformly (flaw tolerance) (Yao & Gao, 2006), and may allow for robust adhesion because for a branched fibril adhesive failure of a single tip is unlikely to impact adhesion of the entire fibril (Arzt et al., 2003; Peressadko & Gorb, 2004).

The form of gekkotan setae thus endows them with many properties that enhance their function in unpredictable and challenging environments. They are self-cleaning—when incapacitated by particulate matter (e.g., dust, sand) they can be passively cleaned by contacting a surface free of additional contaminants, whereby the energy imbalance between the substratum and the spatulae on one hand and the particle and substratum on the other, results in the shedding of the particle to the substratum (Hansen & Autumn, 2005). The efficacy of self-cleaning is actively enhanced by the disengagement of the setae from the substratum during subdigital pad peeling (mechanisms discussed later), whereupon the release of sufficient stored elastic energy propels particulate matter from the setal array (Hu et al., 2012). Structural hierarchy has also been suggested to improve adhesion to unpredictably rough and undulant substrata by increasing conformation to surface asperities (Persson, 2003; Bhushan et al., 2006; Kim & Bhushan, 2007).

9.2.1.3 Configuration of Gekkotan Setal Fields

Gekkotan setae are clustered in groups of four (tetrads) that form clearly distinguishable rows (proximodistally) and ranks (mediolaterally) on the scansors (Fig. 9.3a, b) (Ruibal & Ernst, 1965; Delannoy, 2005). The tetrads are uniformly spaced apart (Fig. 9.3a) with the branched distal tips (Fig. 9.3b) filling the spaces evident at their bases (Fig. 9.3b). The presence of tetrads and the organization of the setal arrays, however, varies depending upon the scansor on which they are carried (Delannoy, 2005). Many studies of the mechanics of individual setae describe setal dimensions as if they are invariant, but for any species there is seemingly no typical seta. Instead, setal characteristics (and form) vary predictably along the proximodistal axis of gekkotan subdigital pads (Fig. 9.4) (Delannoy, 2005; Russell et al., 2007; Johnson & Russell, 2009; Russell & Johnson, 2014). In general, the length of gekkotan setae increases proximodistally both within and between scansors (Fig. 9.4). Within scansors, setal basal diameter decreases proximodistally (Fig. 9.4). Such variation in setal length and basal diameter effectively results in setae with greater flexibility distally because of their higher aspect ratio. Scansor length (Fig. 9.1a) and the number of setal rows, however, decrease proximodistally (Fig. 9.4). Overall, setal density is greatest on the proximalmost scansors of the subdigital pad, but within the confines of individual scansors density increases proximodistally (except for the proximalmost scansors) (Fig. 9.4). This patterning of setal morphometrics is consistent across several genera of gekkotans and suggests that such variability is integral

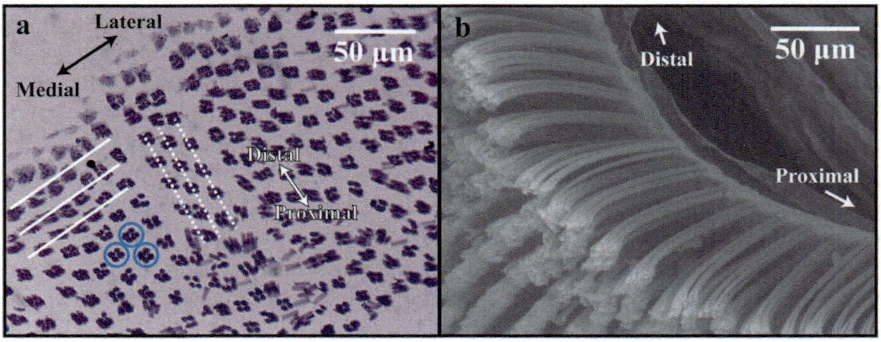

Fig. 9.3 Arrangement of the setae on the scansors of *Gekko gecko*. (**a**) Frontal section through the setal stalks close to their bases on a distal scansor of digit III, right manus. Section cut at 10 μm and stained with haematoxylin. The setal stalks are arranged in tetrads (outlined by blue circles) and are arrayed in rows proximodistally (indicated by dashed white lines connecting tetrads) and ranks mediolaterally (indicated by solid white lines connecting tetrads). The spacing between tetrads is approximately equal proximodistally and mediolaterally. (**b**) Scanning electron micrograph of a longitudinal section through a distal scansor of digit IV, left pes. This section cuts a transect along a single row of setae and shows the increasing length of the setae from proximal to distal along the scansor. The equidistant spacing between tetrads of fibrils is evident at their bases and the branching of the setae into a plethora of spatulate tips is evident towards their free ends. Extending into the plane of the picture are distinct ranks of setae that are aligned mediolaterally across the width of the scansor

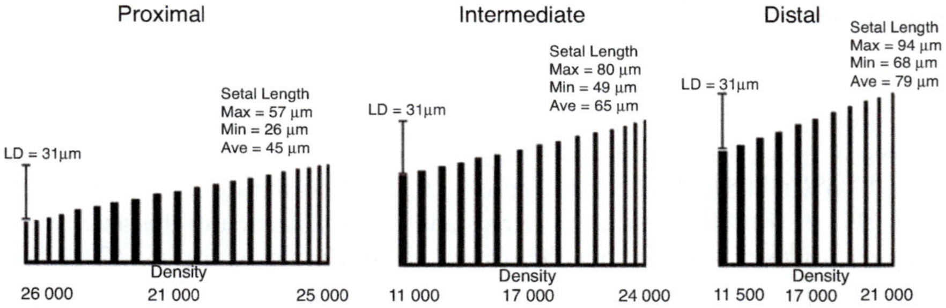

Fig. 9.4 Proximodistal variation in setal morphometrics of *Rhoptropus* geckos [data from Johnson and Russell (2009)]. Setal length increases proximodistally within and between scansors, whereas setal base diameter decreases proximodistally within scansors. Scansor length decreases proximodistally. Setal density is greatest on the proximal scansors of the subdigital pad. Within the confines of individual scansors, setal density increases proximodistally (except for the most proximal scansors). The discrepancy between the maximum and minimum length of setae on any given scansor remains constant (LD=31 μm)

to the effective operation of the integrated setal fields (Johnson & Russell, 2009). For effective attachment and detachment of the setal arrays, interference of individual setae should be minimized and the variability in setal length (Fig. 9.4) may allow for this (Johnson & Russell, 2009).

9.2.1.4 Attachment and Properties of Gekkotan Setal Arrays

Arrangement of setae into setal fields leads to a number of emergent properties beyond those of individual setae. The hydrophobic nature of the phospholipid coating of gekkotan setae (Alibardi et al., 2011; Hsu et al., 2012), in addition to the surface roughness of the closely packed hierarchical fibers (Fig. 9.3b), results in superhydrophobicity (strong water-repellence) of the setal field (Fig. 9.5a) (Pesika et al., 2009). This is critical for adhesion of the subdigital pads to wet, hydrophobic substrata, since dry contact is effected through expulsion of water from the contact interface (Stark et al., 2013). This state, however, is metastable; the subdigital pad's surface can transition to a hydrophilic wetting state under some conditions (Fig. 9.5b) which can reduce adhesion by interfering with the intimate contact needed to sustain van der Waals interactions (Stark et al., 2012). Much like the self-cleaning capability of gecko setae and setal arrays, geckos taking steps on clean, dry substrata can rapidly dry their subdigital pads by shedding water to the substratum and thus regaining maximum adhesive capacity and the subdigital pad's innate superhydrophobic state (Stark et al., 2014; Garner et al., 2019b).

In many gekkotans with well-developed subdigital pads, detachment of the setal fields occurs by distoproximal hyperextension of the digits driven by contraction of the interossei dorsales muscles (Russell, 1975). Other taxa with adhesively competent digits, but with less well-elaborated toe pads, employ the ancestral lacertilian pattern of digital roll-off whereby the digits are raised proximodistally onto their tips (Russell et al., 2015; Higham et al., 2017a, b; Russell & Gamble, 2019) and hyperextended in this fashion. Peeling of the subdigital pad, either distoproximally or proximodistally, is thought to raise the setal shaft angles to above 30°, resulting in

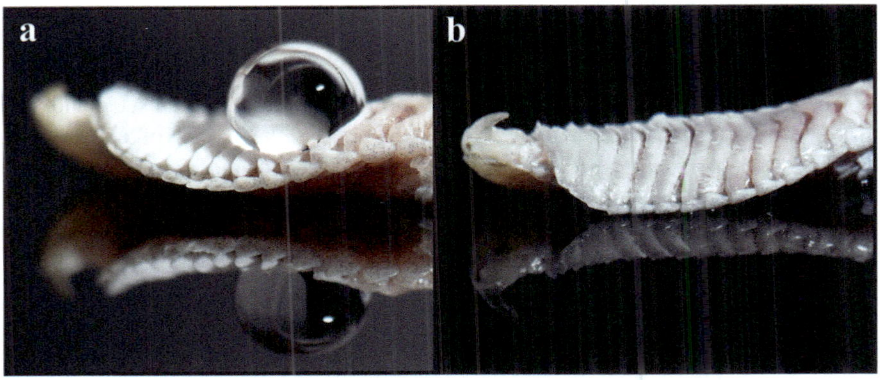

Fig. 9.5 (**a**) The superhydrophobicity (strong water-repellence) of *Gekko gecko* setal arrays. This property allows geckos to adhere underwater or to wet surfaces when the substratum is hydrophobic (Stark et al., 2013). Water contact angles of pristine setal arrays are generally greater than that observed in the image; water droplets often roll-off the setal arrays immediately after placement (this subdigital pad was only partially wetted to permit the water droplet to remain in place). (**b**) The superhydrophobic nature of gekkotan setal arrays is a meta-stable state; setal arrays can transition to a hydrophilic wetting state under certain circumstances (Stark et al., 2012)

the release of the spatulae from the substratum (Autumn et al., 2000; Autumn & Peattie, 2002). Step-like variation in setal length (Figs. 9.3b and 9.4) permits all the setae on a single scansor to reach the critical detachment angle simultaneously, resulting in the instantaneous release of the entire scansor (Johnson & Russell, 2009). Contraction of the digital flexor muscles results in proximodistal unfurling of the subdigital pad onto the substratum's surface, driving the setae and their spatulae into contact (Fig. 9.2b) (Russell, 1975) and contributing the normal and shear displacements critical for spatular attachment (Autumn & Peattie, 2002; Russell, 2002).

Most studies of the attachment capabilities of the gekkotan adhesive system (and that of *Anolis*) have been conducted using relatively smooth substrata (e.g., glass, acrylic) (Niewiarowski et al., 2016). Real-world substrata utilized by free-ranging gekkotans, however, likely vary in surface roughness across multiple length scales (Higham et al., 2019; Niewiarowski et al., 2019). Studies of uniform, elastic materials in contact with rough substrata reveal a dependence of adhesion on the material properties of the adhesive (e.g., elastic modulus); pressure-sensitive adhesives with low elastic modulus are capable of better deforming to accommodate to surface asperities, increasing the apparent contact area and subsequent adhesive force (Fuller & Tabor, 1975). Although gekkotan setae are composed of β-keratin, which has a bulk elastic modulus of about 1–3 GPa (Autumn et al., 2006a), the high aspect ratio of the fibrils (Fig. 9.4) results in a lowering of the effective elastic modulus (Autumn et al., 2006a) to that falling within the range of pressure-sensitive adhesives (e.g., Sylgard 184, polydimethylsiloxane; Khanafer et al., 2008; Bartlett et al., 2012). The low effective elastic modulus of gekkotan setal arrays, and the presence of structural hierarchy, promote the ability to conform to rough substrata (Persson, 2003). Furthermore, the patterning of setal length along the proximodistal axis of gekkotan subdigital pads (Fig. 9.4) provides the potential for allowing adhesion to be maintained on substrata that vary unpredictably in surface roughness (Johnson & Russell, 2009).

The results of studies detailing the form and function of gekkotan setae and setal arrays furnishes information that allows us to assess the extent of convergence exhibited independently by their anoline counterparts. Since these structures depend upon governance mechanisms to be effective, however, we must also consider the structure of the adhesive system that carries them to fully appreciate the level of fidelity of convergence.

9.2.2 The Gekkotan Adhesive System

9.2.2.1 General Characteristics

Although often subsumed under the simplified umbrella term "toe pads" (Fig. 9.1a) (Liu et al., 2015; Hagey et al., 2017; Harrington & Reeder, 2017), there are many variants of the anatomical composition of the adhesive apparatus among geckos

(Russell & Gamble, 2019: Table 1). The functional adaptive complex (Russell, 1976) constituting adhesive toe pads has arisen independently on multiple occasions within the Gekkota (Gamble et al., 2012, 2017; Russell & Delaugerre, 2017; Russell & Gamble, 2019). The attributes of the adhesion mechanics of setae (see above) have thus been integrated with anatomical systems of varying configuration that control their application and release.

The structure of the gekkotan adhesive apparatus has been most extensively studied for the Tokay gecko (*Gekko gecko*) (Russell, 1975). Russell (2002) documented many of the hierarchically integrated mechanical units (*sensu* Gans, 1969) of its digital adhesive apparatus and discussed how these are coordinated to bring about effective substratum contact and release of the setal arrays. Lauff et al. (1993) provided information about cutaneous sensilla on the Tokay's digits, thereby indicating how sensory feedback is inculcated into the operation of the adhesive apparatus.

Russell and Gamble (2019) mapped 34 digital characters and their states onto a well-supported, time-calibrated phylogeny of the Gekkota, and analyzed combinations of features that characterize the hierarchical structure of the adhesive system across all gekkotan families (except for the limbless pygopodids) Complimentarily, Russell et al. (2015) and Higham et al. (2017b) investigated evolutionary transitions from adhesively non-competent to adhesively competent digits in geckos, indicating that such a shift requires only relatively minor alterations of epidermal filament form, scale structure, digit proportions, and phalangeal morphology (Fig. 9.6a–c). These studies established that many of the features characterizing the structurally more complex adhesive pads of geckos, such as *Gekko gecko*, are not necessary for the establishment of adhesive competency. Occurrences of incipient toe pads (Russell, 1976; Gamble et al., 2012; Russell & Gamble, 2019) in various gekkotan taxa corroborate the hypothesis that the transition to adhesive competency has occurred independently many times within the Gekkota and provides the necessary information for establishing the assembly rules (Zweers, 1979; Haefner, 1988) minimally necessary and sufficient for the deployment of reversible adhesion in geckos. Such data also provide us with the context for determining to what degree the independently-evolved adhesive system of *Anolis* has converged upon that of geckos.

9.2.2.2 Selection of an Appropriate Gekkotan Model for Comparison with Anolis

Russell and Gamble (2019: Table 1) detailed modifications of skeletal, compliance-related, tendinous and muscular features associated with the gekkotan adhesive system. These were shown to be numerous for the Tokay gecko (*Gekko gecko*) and its relatives. Contrastingly, Russell and Gamble (2019: Table 1) noted that the incipient adhesive pads of *Gonatodes humeralis* (Fig. 9.6a–c; Russell et al. 2015; Higham et al., 2017b) express few such anatomical modifications, these being mostly associated with the phalanges. Compared to its congeners, *G. humeralis*

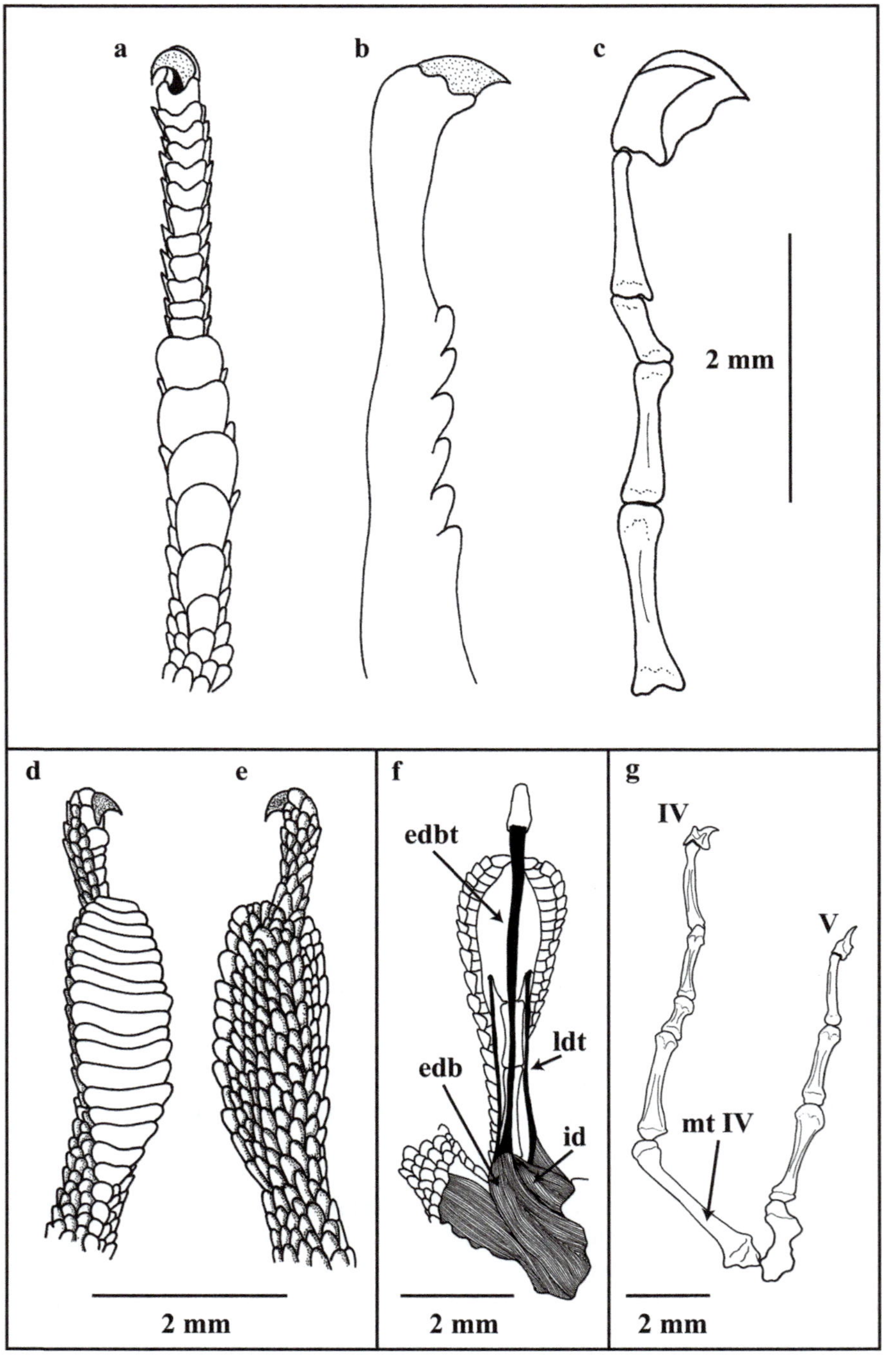

Fig. 9.6 (**a**) Ventral aspect of digit IV, right pes (Natural History Museum, London BMNH 1971.1047); (**b**) lateral profile of digit IV, left pes (BMNH 1971.1049); and (**c**) lateral view of the skeleton of digit IV, right pes (BMNH 1971.1049) of the sphaerodactylid gekkotan *Gonatodes*

exhibits additional subtle changes of digit proportions, phalangeal orientation, patterns of scalation and epidermal micro-ornamentation, with setae being present in the region of the incipient toe pads (Russell et al., 2015). Collectively the modified features of *G. humeralis* enable adhesion-based locomotion on smooth, low-friction vertical surfaces (Higham et al., 2017a, b) while using ancestral (Brinkman, 1980; Rewcastle, 1980, 1983) locomotor digital kinematics. Its adhesively-competent digits are detached by proximal-to-distal digital hyperextension, rather than the distal-to-proximal hyperextension pattern typically associated with gekkotan adhesive release (Autumn, 2006).

Gonatodes humeralis, although adhesively competent, lacks conspicuously obvious subdigital toe pads (Fig. 9.6a, b) and lacks lateral digital tendons (Russell, 1986), features that are clearly evident in *Anolis* (Russell & Gamble, 2019: Table 1). Among basally-padded geckos (Russell & Gamble, 2019: Table 1) the least complex manifestation of visibly recognizable toe pads is encountered in the *Aristelliger* lineage of the Sphaerodactylidae (Fig. 9.6d–g). Beyond the changes documented above for *G. humeralis*, *Aristelliger* possesses an arcuate penultimate phalanx that carries the distal part of the digit beyond and above the toe pad and provides the ungual phalanx and claw with a steep angle of attack relative to the substratum (Fig. 9.6d–g). *Aristelliger* has a subphalangeal cushioning structure in the form of a central vascular sinus and reticular networks that permeate the proximal regions of the scansors (Russell, 1981). It also has prominent lateral digital tendons (Russell, 1986) (Fig. 9.6f) that merge directly with the stratum compactum of the dermis of the scansors. The bellies of the dorsal interossei muscles do not extend beyond the metapodial-phalangeal joint capsules (Fig. 9.6f), and their insertion tendon does not extend along the digit in the fashion seen in *Gekko* and many other pad-bearing geckos (Fig. 9.7). *Aristelliger* thus appears to lack a mechanism for distal-to-proximal digital hyperextension (Russell, 2002). Setal structure in *Aristelliger* (Ruibal & Ernst, 1965) is relatively simple by gecko standards. Compared to the *Luperosaurus-Gekko* lineage (Russell & Gamble, 2019: Table 1), which includes *Gekko gecko* (Fig. 9.7a), anatomical modifications associated with the operation of

Fig. 9.6 (continued) *humeralis*. These panels show the enlarged plates (**a** and **b**) beneath the intermediate region of the digit that bear adhesively-competent setae and the modified phalangeal structure (**c**) related to a flattening of the proximal region of the digit associated with the location of the adhesively-competent enlarged scales. (**d**–**g**) Subdigital pad structure in the sphaerodactylid gekkotan genus *Aristelliger*. (**d**) and (**e**) The ventral and dorsal aspects, respectively, of digit IV, left pes of *Aristelliger lar* showing the expanded subdigital adhesive pad occupying the basal region of the digit and the free distal portion, bearing the terminal claw, extending beyond the toe pad. (**f**) Dorsal view of digit IV left pes of *Aristelliger lar* dissected to reveal the extent of the musculature and tendons associated with the control of the adhesive apparatus. (**g**) Dorsal view of the skeleton of digits IV and V of the right pes of *Aristelliger praesignis* (drawn from an Alizarin-stained preparation) showing the elongate proximal phalanges, the modified intermediate phalanges associated with the toe pads, and the penultimate and ungual phalanges supporting the free distal portion of the digit that extends beyond the toe pad. *edb* extensor digitorum brevis muscle, *edbt* tendon of the extensor digitorum brevis muscle, *id* interossei dorsalis muscle, *ldt* lateral digital tendon, *mt IV and V* fourth and fifth metatarsals, *IV and V* digits IV and V

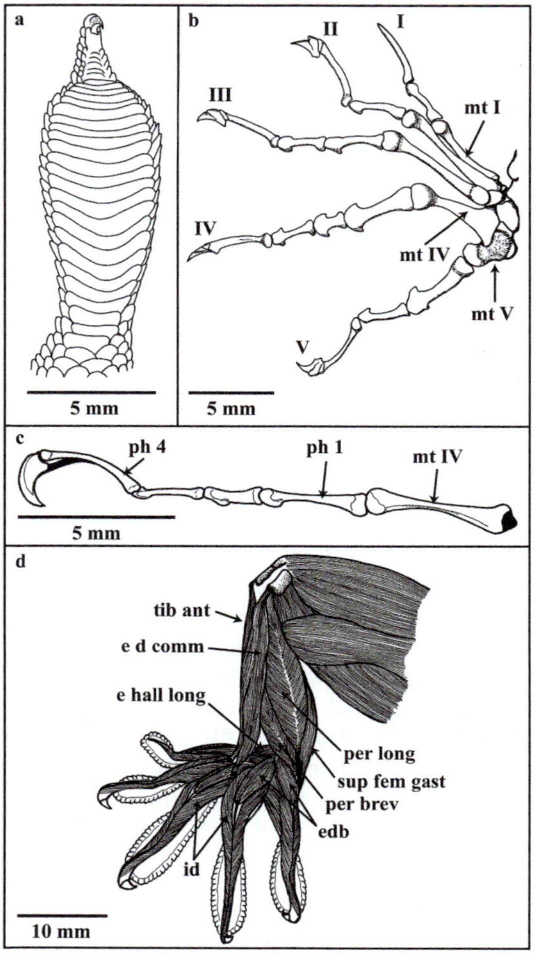

Fig. 9.7 Digit structure in the gekkonid gekkotan genus *Gekko*. (**a**) Ventral aspect of digit IV, right pes of *Gekko smithii* (Natural History Museum, London BMNH 91.8.29.23) showing the extensive toe pad that encroaches far distally along the digit. (**b**) Dorsal aspect of the skeleton of the right pes of *Gekko gecko* (drawn from an Alizarin-stained preparation) showing the modified, short, depressed phalanges associated with the more proximal parts of the toe pads and the elongated penultimate phalanges of digits II–V that are enveloped within the confines of the toe pads, with only the ungual phalanges extending beyond the pad (see panel **d**). Note the unusual phalangeal structure of digit I, in which the ungual phalanx does not carry a claw but is instead elongated to support the toe pad on this digit. (**c**) Medial aspect of the articulated skeleton of digit IV, right pes of *Gekko gecko* (BMNH 1910.4.26.14A) showing the depressed and shortened intermediate phalanges and the strongly arched penultimate phalanx (ph4). (**d**) Superficial musculature of the left crus and pes of *Gekko gecko* showing the incursion of muscle bellies along the length of the digits as far distally as the ungual phalanx. *edb* extensor digitorum brevis muscle, *e d comm* extensor digitorum communis muscle, *e hall long* extensor hallucis longus muscle, *id* interossei dorsalis muscle, *mt I, IV, V* first, fourth and fifth metatarsals, *per brev* peroneus brevis muscle, *per long* peroneus longus muscle, *ph1, ph4* first and fourth phalanges, *sup fem gast* superficial femoral gastrocnemius muscle, *tib ant* tibialis anterior muscle, *I, V* digits I and V

the toe pads are much less extensive in *Aristelliger*. The digital patterns expressed by *Gonatodes humeralis*, *Aristelliger*, and *Gekko*, together with other expressions of anatomical modification documented by Russell and Gamble (2019: Table 1) provide comparative information appropriate for assessing the extent of anatomical convergence between the variously expressed adhesive system of geckos and that of anoles.

9.2.3 *Anoline Setae and Setal Fields*

9.2.3.1 Anoline Setal Form and Variability

Not surprisingly, the epidermal outgrowths of the subdigital pads of *Anolis* also vary considerably in form. Peterson and Williams (1981) described a morphotypic series of epidermal outgrowths of *Anolis* (Fig. 9.8a), which ultimately led to similar observations in gekkotans by Delannoy (2005). Five morphotypes were identified by Peterson and Williams (1981): (1) true setae, outgrowths 10–30 μm in length with expanded, spatulate tips, (2) seta-prong intermediates, outgrowths 5–20 μm in length with flattened tips, (3) prongs, outgrowths 5–20 μm in length terminating in blunt tips with a slight taper, (4) spikes, outgrowths 5–15 μm in length with straight or recurved, pointed tips, and (5) spines, outgrowths up to 5 μm in length with recurved, pointed tips. All lamellae (scales bearing true setae) of *Anolis* with well-developed subdigital pads are thought to display a proximodistal clinal gradation of all 5 morphotypes (from spines to setae) on each lamella. As in gekkotans, it is the true setae and their spatulae that are responsible for the majority of adhesive force capacity in *Anolis*.

9.2.3.2 Anoline Setal Field Configuration

Table 9.1 (reproduced from Garner et al., 2019a) summarizes what was known about the morphometrics of exemplar anoline setae up to 2019. Although these data have been of crucial comparative value for describing and comparing the fibrillar adhesive outgrowths of squamates, they are drawn from single exemplar setae and therefore do not indicate whether the inter- and intraspecific variability of setal morphometrics of gekkotans (Delannoy, 2005; Russell et al., 2007; Johnson & Russell, 2009; Russell & Johnson, 2014) are also evident in anoles. Garner et al. (2021) explored this possibility by examining the morphometrics of the setal arrays of the Cuban knight anole (*Anolis equestris*), a dactyloid of similar size to *Gekko gecko*. That study revealed that the setae of *A. equestris* increase in length and decrease in basal diameter proximodistally along regions (proximal, intermediate, distal) of the subdigital pad (Fig. 9.8b). Within a single lamella, however, setal length is greatest in the intermediate region, whereas setal basal diameter decreases proximodistally (Fig. 9.8b). The length of lamellae decreases proximodistally along regions of the

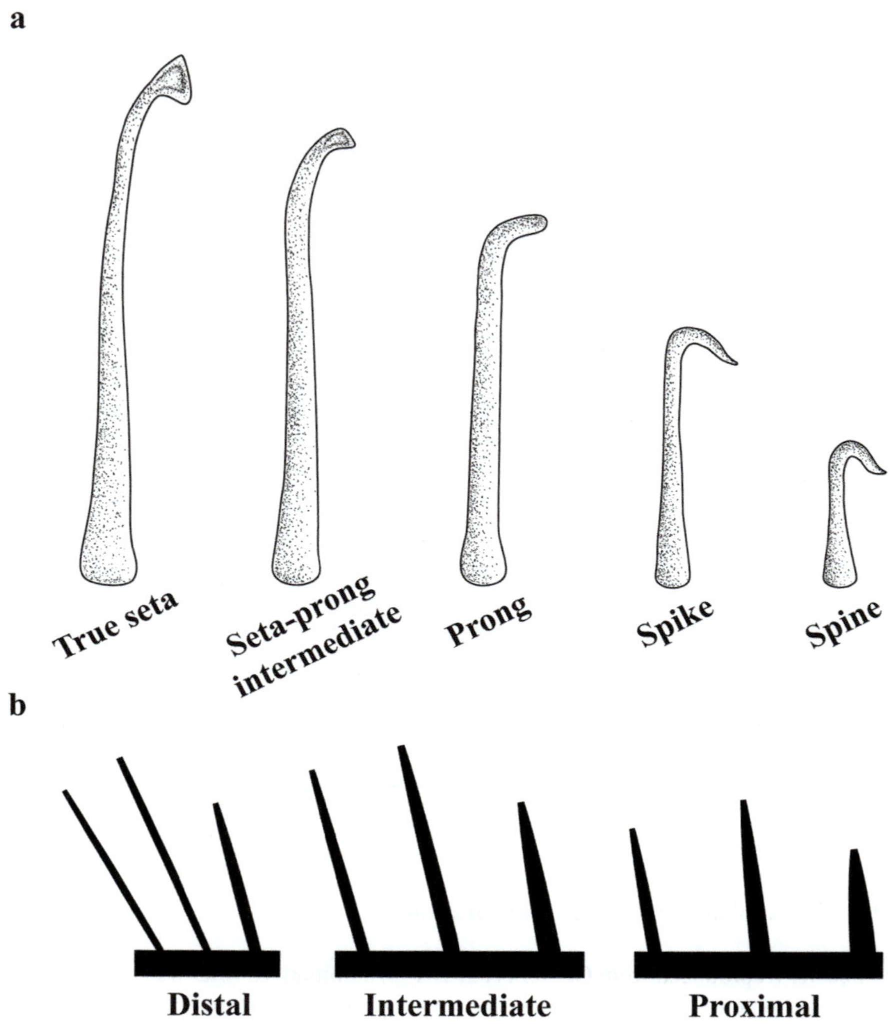

Fig. 9.8 (**a**) Clinal series of the epidermal outgrowths present on the subdigital pad of anoline lizards (as described by Peterson and Williams, 1981). True setae are outgrowths 10–30 μm in length that carry a single, expanded spatulate tip. Seta-prong intermediates are outgrowths 5–20 μm in length with flattened tips. Prongs terminate in a blunt tip with a slight taper and are 5–20 μm in length. Spikes possess straight or recurved, pointed tips and are 5–15 μm in length. Spines are outgrowths up to 5 μm in length with pointed recurved tips. (**b**) Trends of setal morphometrics along subdigital pad regions and lamellar zones of *Anolis equestris* (as reported by Garner et al., 2021). Setal length increases and setal base diameter decreases proximodistally along pad regions. Setal length is maximal in the intermediate zones of lamellae. Setal base diameter decreases proximodistally along lamellar zones. Lamella length decreases proximodistally

Table 9.1 A summary of the known average setal morphometrics of anoline lizards (Garner et al., 2019a)

Species	Density (setae/μm^2)	Length (μm)	Diameter (μm)	Tip width (μm)	Tip area (μm^2)	Sources
Anolis chameleontides	1	18.4–18.6	0.56–0.58	–	0.206–0.377	(1)
Anolis porcus	1	22.3	0.53	–	0.264	(1)
Anolis barbouri	0.5–0.6	5.0–8.4	0.47–0.56	–	0.6–1.472	(1)
Anolis heterodermus	1.2	13.7–14.8	0.41–0.46	–	0.214–0.303	(1)
Anolis cuvieri	0.9–1.4	22.4–27.2	0.51–0.65	–	0.184–0.253	(1)
*Anolis cuvieri***	1	22.4	0.51	0.729	0.229	(2)
Anolis sheplani	1.1–1.2	11.8–12.6	0.39–0.41	–	0.220–0.279	(1)
Anolis occultus	1.4	11	0.49	–	0.237	(1)
Anolis sp. n. near *eulaemus*	1	20.4	0.57	–	0.593	(1)
Anolis valencienni	1.1–1.4	15.3–17.2	0.40–0.47	–	0.171–0.209	(1)
Anolis carolinensis	0.83	21	0.5	0.87	–	(3,4)
Anolis equestris	0.7	30	–	–	–	(3,4)
Anolis lineatopus	0.51	–	–	1	–	(3,4)
Anolis sagrei	1.7	20	–	0.75	–	(3,4)
Anolis homolechis	–	20	–	–	–	(3)
*Anolis annectens***	1.0–2.0	20	0.48	0.73	0.211	(5)

Note that only 15 species are represented by these data and that the complete set of setal characters have only been reported for two of these species (denoted by **). Sources: (1) Peterson (1983b), (2) Williams and Peterson (1982), (3) Ruibal and Ernst (1965), (4) Peattie and Full (2007), and (5) Peterson and Williams (1981). Table reproduced from Garner et al. (2019a). Going out on a limb: how investigation of the anoline adhesive system can enhance our understanding of fibrillary adhesion. *Integr Comp Biol*, 59, 61–69 by permission of the Society for Integrative and Comparative Biology

subdigital pad (Fig. 9.1b). Setal density remains relatively consistent along the entire subdigital pad and along regions (proximal, intermediate, distal) of individual lamellae. Garner et al. (2021) also estimated the effective elastic modulus (E_{eff}) of the setal arrays of *A. equestris* and found that in most regions of the setal fields this falls below 2 MPa, suggesting that anoline setal fields can deform to surfaces to maximize contact area and adhesion

9.2.3.3 Attachment Mechanics and Functional Morphology of Anoline Setae and Setal Arrays: What We Do and Don't Know

Currently the attachment mechanics of anoline setae and setal arrays remain relatively understudied when compared to those of gekkotans (Garner et al., 2019a). Given the similarity of setal form and material properties, however, the setae of *Anolis* and gekkotans presumably require similar loading conditions (normal load followed by shear load), because measurements of whole animal adhesive performance of geckos and anoles indicate that the adhesive systems of each can be engaged in the same manner (Irschick et al., 1996). The means by which setal loading and unloading occurs in anoles, however, differs from that typically attributed to geckos. Russell and Bels (2001) examined the kinematics of *Anolis sagrei* running on an inclined acrylic surface and found that subdigital pad placement and retraction occurs proximodistally (as opposed to the distoproximal pattern of some geckos), with the digits rolling off onto their distal tips, a pattern typical of the digital kinematics of lizards in general (Brinkman, 1980; Rewcastle, 1980, 1983). This pattern of release of the setae is consistent with the notion that the step-like patterning of setal length in setal arrays assists in the effective detachment of the adhesive apparatus. The patterning of setal length and subdigital pad peeling of *Anolis* (Garner et al., 2021) are effectively identical to those taken to be characteristic of geckos (Johnson & Russell, 2009), but are expressed in the opposite direction.

The measurement of the adhesive forces of single gekkotan seta by Autumn et al. (2000) resulted in a deluge of interdisciplinary research focusing on the mechanisms, mechanics, and properties of this system (Niewiarowski et al., 2016). Investigation of the adhesion mechanics of anole setae has not, however, been explored in like fashion and adhesion in these lizards has largely been explored by biologists alone (Garner et al., 2019a). Three approaches have predominated: (1) examination of gross aspects of setae and setal field morphology (Ruibal & Ernst, 1965; Peterson & Williams, 1981; Williams & Peterson, 1982; Peterson, 1983a, b; Peattie & Full, 2007); (2) investigation of adhesion of intact animals to smooth substrata (Irschick et al., 1996; Elstrott & Irschick, 2004; Bloch & Irschick, 2005; Irschick et al., 2005; Garner et al., 2017); and (3) exploration of correlations between morphology and performance of the subdigital adhesive pads as a whole (Macrini et al., 2003; Elstrott & Irschick, 2004; Irschick et al., 2005; Donihue et al., 2018). When introducing *Anolis* as a model system for the study of fibrillar adhesion, Garner et al. (2019a) acknowledged the invaluable contribution that such studies have made, but concomitantly recognised clear gaps in our knowledge of the form and function of anoline setae and setal fields.

Although anoline setae likely operate under similar principles and conditions to those of geckos, differences in their overall form and dimensions may result in differential attachment mechanics. For example, the critical angle of detachment of gekkotan setae is about 30°. Do anoline setae also detach from surfaces at such angles or does the lack of structural hierarchy and presence of greater tip sizes result in different detachment angles? Preliminary data measuring detachment of entire anoline subdigital pads suggest that anoline setae detach at angles of less than 30° (Hagey, 2013), but more conclusive work with isolated single setae is necessary to validate this. Furthermore, the properties of anoline setae and setal arrays have not been investigated. For example, do they exhibit the same wetting and self-cleaning properties as gekkotan setal arrays? Anoline setae, like those of geckos, are clad in phospholipids (Alibardi et al., 2011), and the roughness introduced by the surface profile of the batteries of setae should enable them to be superhydrophobic. Additionally, there is no reason to suspect that anoline setal arrays are not self-cleaning, but self-cleaning performance could differ between the two groups as a result of differences in tip density and area.

There has been considerable attention paid to the investigation of gekkotan adhesive system in ecologically relevant conditions (Russell, 2002; Autumn et al., 2014; Niewiarowski et al., 2016, 2017) and such studies are increasing. This, however, is not the case for anoles, even though they are model organisms for evolutionary ecological studies (Losos, 2011). How the differences in setal form between geckos and anoles might relate to differential adhesive performance in ecologically relevant circumstances is an area ripe for future research.

9.2.4 The Anoline Adhesive System

The anatomical modifications associated with the expression of the adhesive system in geckos are modestly represented by *Anolis*, being confined to skeletal (phalangeal) and tendinous features (Russell & Gamble, 2019: Table 1). The toe pads of *Anolis* are visibly evident (Figs. 9.1b and 9.9a) and are located basally on the digits, ventral to the location of the digital inflection (between phalanges 2 and 3 of digit IV—Fig. 9.9a; Russell et al., 2015). The penultimate phalanx is arcuate and extends distally beyond the toe pad. The claw is thus carried some distance beyond the distal extremity of the pad (Figs. 9.1b and 9.9a).

The metapodial-phalangeal joints of *Anolis* are unicondylar (Fig. 9.9b), permitting a greater range of movement than the ginglymous joints of lizards without subdigital pads (including other iguanoids) (Russell & Bauer, 2008). Although the metacarpals are widely divergent in the manus (Fig. 9.9a), a situation widespread among lizards, the first four metatarsals of the pes are subparallel, with the pes (Fig. 9.1b) having the asymmetrical form of lizards in general rather than exhibiting the secondary symmetry evident in geckos (Fig. 9.1a; Russell et al., 1997). In both the manus and pes the structure of the phalanges and the joints between them are highly modified. This is particularly so for the phalanges associated directly with the toe pads (for example, phalanges 2 and 3 of digit IV of the manus and pes—Fig. 9.9a, b) when compared to those for lizards ancestrally lacking toe pads (Russell & Bauer, 2008). Their cross-sectional profile is depressed and widened, with a ventral excavation that conducts the tendon of the flexor digitorum longus muscle (which inserts on the ungual phalanx). In the manus digits III and IV are subequal in length (Fig. 9.1b) as a result of metacarpal III being relatively elongated and phalanx

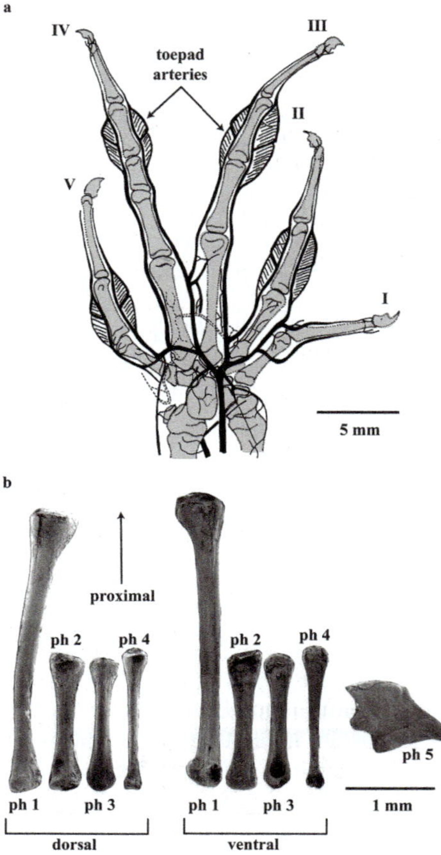

Fig. 9.9 (**a**) Dorsal view of the left manus of *Anolis garmani* showing the carpals, metacarpals and phalanges in grey and the arterial vessels in black. Note the pattern of arterial vessels that outline the contours of the toe pads on digits II–V. Drawn from a cleared and Alizarin-stained Microfil-injected specimen. (**b**) Phalanges of digit IV, right pes of *Anolis garmani* in dorsal (left) and ventral (right) aspects (except for the ungual phalanx, which is shown in lateral view). Note the greatly elongated first phalanx and the relatively short, somewhat depressed third phalanx that supports the majority of the toe pad (see panel a for a depiction of the location of the phalanges in relation to the toe pads). *ph1–ph5* phalanges 1–5, *I–V* digits I–V

2 of digit IV being relatively short. In the pes, in contrast, phalanx I of digit IV is greatly elongated (Fig. 9.9b), resulting in the relative distal displacement of the toe pad on that digit (Fig. 9.1b). Phalanx 2 on pedal digit IV is longer than its corresponding manual phalanx. Thus, the osteology of the manus and pes of *Anolis* both deviate from the ancestral pattern (Russell & Bauer, 2008) but exhibit differences from each other reflective of the symmetrical disposition of the digits in the manus and their asymmetrical carriage in the pes. The accommodation of the toe pads and their potential to impinge upon one another due to their breadth has been accomplished differently in the manus and pes.

Associated with the toe pads of *Anolis* are lateral digital tendons (Fig. 9.10a) similar to those of geckos. These course along the lateral and medial borders of the phalanges of each digit (except the first which lacks a toe pad, due to constraints imposed by phalangeal number—Russell & Bauer, 1990) and branch to serve each lamella (Fig. 9.10a). As in geckos the dense, collagen rich connective tissue of the lateral digital tendons is continuous with the stratum compactum of the dermis of the lamellae (Fig. 9.10b) and the lamellar dermis lacks a stratum laxum. Thus, as in

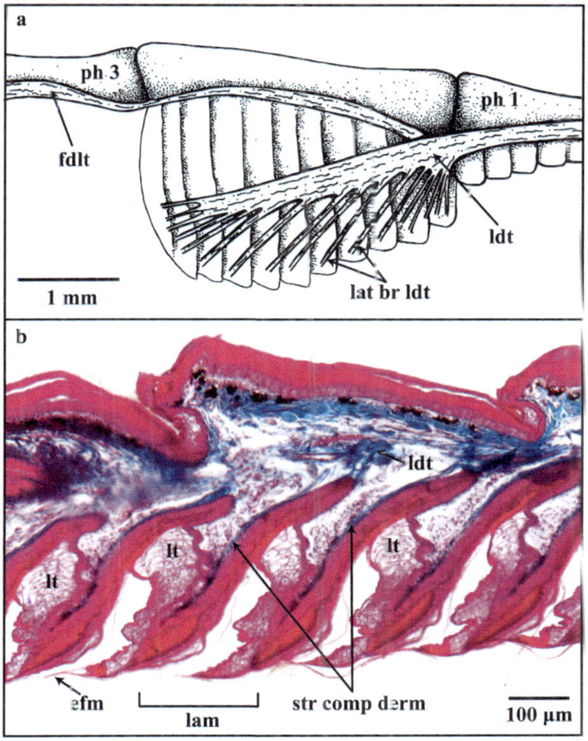

Fig. 9.10 (**a**) Medial aspect of a dissection of digit III, right pes of *Anolis garmani* showing the lateral digital tendons and their association with the lamellae. (**b**) Parasagittal section through the toe pad of digit IV, right pes of *Anolis garmani* showing the relationship of the epidermal lacunar cells to the configuration of the lamellae and their epidermal free margins that carry the setae. Section stained with Masson's trichrome, section thickness 8 μm. *efm* epidermal free margin of lamella, *fdlt* tendon of the flexor digitorum longus muscle, *lam* lamella, *lat br ldt* lateral branch of the lateral digital tendon, *ldt* lateral digital tendon, *lt* lacunar tissue, *ph1, ph3* phalanges 1 and 3, *str comp derm* stratum compactum of the dermis

geckos (Russell, 1986) the lateral digital tendon/lamellar dermis complex furnishes a tensile skeleton that provides connectivity between the setae and skeleton at the metapodial-phalangeal joint capsules. This chain reinforces the junction between the integument and the underlying tissues and permits the tensile load imposed on the setae to be channeled to points of resistance deep within the manus and pes, and also permits active tensile loading to be imposed upon the setae via the muscles acting on the metapodial-phalangeal joint capsules. The digits of *Anolis* exhibit little in the way of muscular modifications (Fig. 9.11) akin to those exhibited by at least some gecko lineages (Russell & Gamble, 2019) (Fig. 9.7d).

The above-mentioned features of the anole adhesive system are those that Russell and Gamble (2019: Table 1) documented in their broad comparative survey of the adhesive system of geckos. Other gecko-like modifications are also present, however. Although there is no vascular-based cushioning structure (Russell 1981) associated with the lamellae (Fig. 9.10b), there is something akin to this. *Anolis* lamellae lack a perfect resting stage of the epidermis (Lillywhite & Maderson, 1968) and the lacunar epidermal cells of the inner faces of the lamellae show precocious development in stage 1 of the shedding cycle and by stage 4 are hypertrophied (Fig. 9.10b). This expansion of lacunar cells as far distal as the base of the epidermal free margin of the lamellae results in a continuous band of such cells that lies dorsal

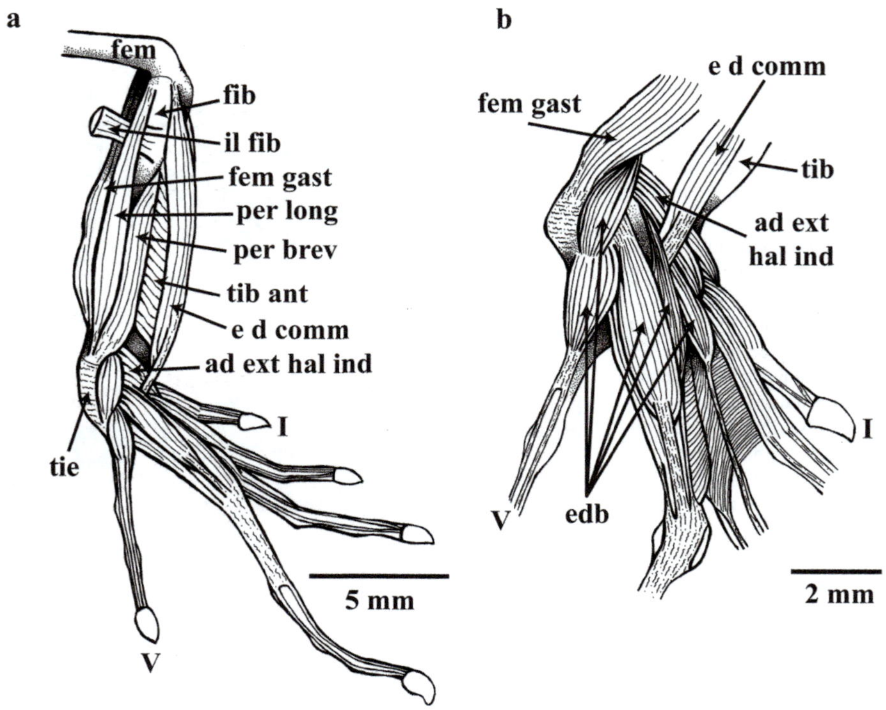

Fig. 9.11 Musculature of the crus and pes of *Anolis garmani*. (**a**) Lateral aspect of the right crus and pes showing the superficial extensor musculature; (**b**) Superficial extensor musculature of the distal end of the right crus and adjacent pes. *ad ext hal ind* adductor et extensor hallucis et indicus muscle, *edb* extensor digitorum brevis muscle, *e d comm* extensor digitorum communis muscle, *fem* femur, *fem gast* femoral gastrocnemius muscle, *fib* fibula, *il fib* iliofibularis muscle, *per brev* peroneus brevis muscle, *per long* peroneus longus muscle, *tib* tibia, *tib ant* tibialis anterior muscle, *tie* tie tendon, *I, V* digits I and V

to the setae and overlies the entire setal field when the toe pad is flattened against the locomotor surface (Russell, 2016), suggesting that it serves to transmit pressure to the underlying setae, thereby assisting with compliance of the setal fields with the substratum (although this has yet to be biomechanically confirmed). The vascularity of the toe pads is not markedly different from that of lizards in general (Fig. 9.9a; Russell, 2016), but includes arteries that branch from the main supply to the digits and serve the borders of the pad and, by branching further, the individual lamellae (Fig. 9.9a). All of these vessels are nutritive and play no role in hydrostatic support (Russell, 1981) of the lamellar system.

9.2.5 Comparison of Clinging Performance in Geckos and Anoles

Only two reports compare the adhesive performance of geckos and anoles. Ruibal and Ernst (1965) observed gecko and anole adhesive locomotion on a vertical raceway and noted no obvious differences in performance. More comprehensive work by Irschick et al. (1996) corroborated this finding and found that static clinging performance of geckos and anoles is not markedly different. These results are surprising given the marked differences in morphology and anatomy of the adhesive apparatuses of the two groups.

As noted above, the structural hierarchy present in the fibrils of the gekkotan adhesive system should result in greater adhesive force production than unbranched fibrils of similar size (Peattie & Full, 2007; Murphy et al., 2009; Garner et al., 2021). Therefore, if geckos and anoles only differed in the presence of structural hierarchy of their setae, gekkotan setae should induce greater adhesive forces. Gekkotan and anoline setae and setal fields, however, differ in setal size, the number of spatulae per seta, and the number of fibrils present per unit area (Garner et al., 2021). Application of the Johnson-Kendall-Roberts (JKR) theory of elastic contact to the assessment of adhesive performance of gekkotan and anoline setal arrays, in conjunction with the comprehensive morphological data obtained about the setal fields of *Gekko gecko* (Delannoy, 2005) and *Anolis* (Garner et al., 2021), reveals that adhesive performance of gekkotan and anoline setal arrays should theoretically be similar when taking the differences in the morphology and configuration of their setae and setal fields into consideration. Anoles possibly compensate for the lack of structural hierarchy of their setae by possessing greater setal density compared to geckos. Structural hierarchy of setae may, however, be important in other aspects of the gekkotan adhesive system beyond those that directly affect adhesion (Persson, 2003; Yao & Gao, 2006; Persson, 2003). The impact of morphological and anatomical disparities on the function of gekkotan and anoline adhesive apparatuses may become more apparent in dynamic and/or more ecologically relevant circumstances.

9.3 Fundamental Factors Affecting the Form and Function of Fibrillar Adhesive Systems

Based on our review above, it is clear that organisms employing dry, fibrillar, molecularly-based adhesive systems are challenged by many environmental and organismic demands. These collectively influence the structure and deployment of the adhesive organs and drive them towards functionally and structurally similar outcomes (Federle, 2006). In accord with this, we compartmentalize our consideration of the drivers of convergence of the gecko and anoline adhesive system into those related to (1) the physical interactions between the filaments and the substratum, and (2) the means by which attachment and detachment of the filaments is controlled by morphological adaptations of the organs that bear them.

9.3.1 Physical Constraints

Physical strictures are imposed upon fibrillar systems that chiefly rely upon van der Waals forces (Autumn et al., 2002) for attachment. To be effective in supporting the animal's body mass during station holding and locomotion (Higham et al., 2017a), the potential contact area of the fibril tips must be appropriately extensive and the separation distances minute enough to induce van der Waals interactions. Exploitable surfaces, for example, must fall within a range of asperity configurations (roughness) to enable sufficient access of the fibrillar arrays (Johnson & Russell, 2009; Russell & Johnson, 2007, 2014; Gillies et al., 2014); sufficient contact area with the surface is a function of the length of the fibrils, dimensions of their tips, and their material properties (Ruibal & Ernst, 1965; Persson, 2003; Federle, 2006). Release of the fibrils from contact with the substratum must be able to be repeatedly and rapidly accomplished without loss of adhesive capacity (which would occur if the filaments were damaged in this process).

The default state of the filaments in the unloaded, unattached state must be non-sticky (Autumn & Hansen, 2006) to avoid them attaching to each other, thereby rendering them ineffective. Furthermore, because minute, loose particles may adhere to the tips of the filaments during their operation, they must be self-cleaning (Hansen & Autumn, 2005; Hu et al., 2012), promoting the shedding of contaminants that may otherwise foul the system and result in loss of adhesive capacity (Russell & Delaugerre, 2017). Other physical environmental parameters, such as temperature, humidity and the presence of surface water, also impact the effectiveness of fibrillar adhesives (Bergmann & Irschick, 2005; Chen & Gao, 2010; Puthoff et al., 2010; Prowse et al., 2011; Stark et al., 2012, 2013, 2016; Garner et al., 2019b; Stark & Mitchell, 2019). At the most fundamental level, therefore, the organism-environment interface is the critical juncture at which selection operates to determine the configuration (Russell et al., 2007, 2015) and material properties (Greiner et al., 2009) of the adhesive system.

9.3.2 Structural Constraints

The organismally-related set of challenges facing molecularly-based fibrillar adhesives concern the mechanics of their operation. The surface area able to be devoted to the adhesive apparatus (Fig. 9.1) must be compatible with the configuration of the locomotor system and must be able to compensate for increases in body mass as growth occurs (Webster et al., 2009). Beyond the simple relationship of available surface area relative to volume (and hence mass), the fibrils must be arranged, arrayed, and controlled such that sufficient contact with the substratum can be made to effect body support under the various demands of static clinging and active locomotion (Autumn et al., 2006b; Russell & Oetelaar, 2016; Higham et al., 2017a; Song et al., 2020).

The small size of the fibrils and the relatively high tensile loadings imposed upon them render them potentially easily damaged and suggest that careful control during their deployment and release is essential (Russell et al., 2019). The fibrils are products of the epidermis and the loading that they experience must be able to be transmitted to deeper layers of the integument and other parts of the locomotor system (Russell, 1986) to avoid integumentary rupture. Thus, reinforcement of the bond between the skin and underlying tissues of tetrapod vertebrates, which generally allows movement between it and the underlying subdermis (Roach, 2003), is essential.

Following the initial evolutionary establishment of adhesive effectiveness of subdigital fibrils, as determined by the physical demands outlined above, selection may further operate on structural aspects of the anatomy that carries and operates them (Russell et al., 2015). This may lead to the elaboration of more complex control and monitoring mechanisms that may add to the structural complexity (and possibly functional versatility) of the adhesive system.

9.4 Assessing the Fidelity of Convergence Between Gekkotan and Anoline Adhesive Systems

The hierarchical dependency outlined above has led to overall general convergence (the evolutionarily independent reoccurrence of form and function—Kuhn et al., 2020) in both the physical and structural domains pertaining to the fibrillar adhesive systems of geckos and anoles. There are, however, because of multiple transitions to adhesively-competent digits within the Gekkota, differences of expression of the structure of the adhesive system in various gecko lineages (Russell & Gamble, 2019). Thus, assessment of the convergence between anoles and geckos must take the variation within the Gekkota into account.

Both geckos and anoles have accomplished filament-based adhesion via the elaboration of epidermal outgrowths present on the surface of their integument. The spinulate Oberhäutchen of these squamates has been exapted (Gould & Vrba, 1982; Russell et al., 2015) for adhesion using intermolecular forces. Selective pressures for robust attachment to vertical and inverted substrata ultimately led to the elaboration of epidermal outgrowths (setae) capable of supporting the lizard's body mass via van der Waals intermolecular forces (Russell et al., 2015; Higham et al., 2017a, b). Several gekkotans possess adhesive structures on the ventral surface of their tail tips (Bauer, 1998), providing evidence that transitions to adhesive competence can occur in any region of the integument that makes regular and close contact with the substratum. The spinulate Oberhäutchen of the digits of some other squamate clades, such as chameleons and some skinks, has also undergone elaboration to support enhanced friction (or modest adhesive capacity, in the case of skinks) (Williams & Peterson, 1982; Irschick et al., 1996; Khannoon et al., 2014). Fundamentally, the adhesive setae of geckos and anoles are convergent structures that arose to support reversible attachment for static clinging and effective locomotion on heterogeneous and complex substrata via molecular interactions. Nevertheless, these two groups of squamates achieve this with structures that differ in form and, in some cases, function (Garner et al., 2021).

After the initial transformation of relatively simple fibrillar outgrowths to those that could effect robust organismal attachment via intermolecular forces, the evolutionary elaboration of setae and setal field configuration differed in geckos and anoles. The setae of geckos became further elaborated into multiply branched structures [even in one of the simplest manifestations, *Gonatodes humeralis* (Russell et al., 2015)] whereas those of anoles remained unbranched. Recent work has also revealed that the dimensions of the setae of *Anolis equestris* differ markedly from those of *Gekko gecko*, being between 3 and 5 times shorter and having bases between 2 and 7 times narrower (Garner et al., 2021). The differences in morphology and dimensions of setae in anoles and geckos are accompanied by differences in the manner in which the setal fields are assembled. The clinal pattern of setal length of *Anolis equestris* is similar to that of geckos but effectively opposite in direction (Garner et al., 2021), indicating that the pattern of setal length is largely driven by differences in the mechanics of subdigital pad peeling in these two clades. The majority of anoline setae are located on the epidermal free margin, a thin, flexible extension of the lamella (Ruibal & Ernst, 1965; Ernst & Ruibal, 1966; Maderson, 1970; Peterson, 1983a, b; Russell & Eslinger, 2017). An epidermal free margin may be present on the scansors of some gekkotans, but it is not as extensive (Alibardi et al., 2007). Functionally, the epidermal free margin of anoline lamellae is thought to promote conformation with the substratum to increase setal contact fraction (Russell & Eslinger, 2017), although this has not been investigated empirically. Thus, although gekkotan and anoline setae seem to operate utilizing similar principles and mechanics, differences in setal form and dimensions relate to the performance and properties of the adhesive apparatus as a whole.

Geckos exhibit multiple origins of adhesively-competent digits within their ranks (Russell & Gamble, 2019) whereas anoles have originated such a system only once (Losos, 2011). The strictures imposed upon the interactions between keratinous dry fibrillar adhesives and surfaces to which they can attach determine the fundamental mechanics of their deployment. Initially the adhesive competence of setae was incorporated into pre-existing locomotor repertoires, as indicated by gekkotan taxa with incipient toe pads, such as *Gonatodes humeralis* (Russell et al., 2015; Higham et al., 2017a, b). These incipient toe pads rely upon proximodistal hyperextension of the digits for detachment of the adhesive filaments, as do those of *Anolis* (Russell & Bels, 2001). Neither *Gonatodes humeralis* nor *Anolis* (Russell & Gamble, 2019: Table 1) exhibit modifications of their digital musculature (Russell, 1975, 2002) associated with driving distoproximal digital hyperextension.

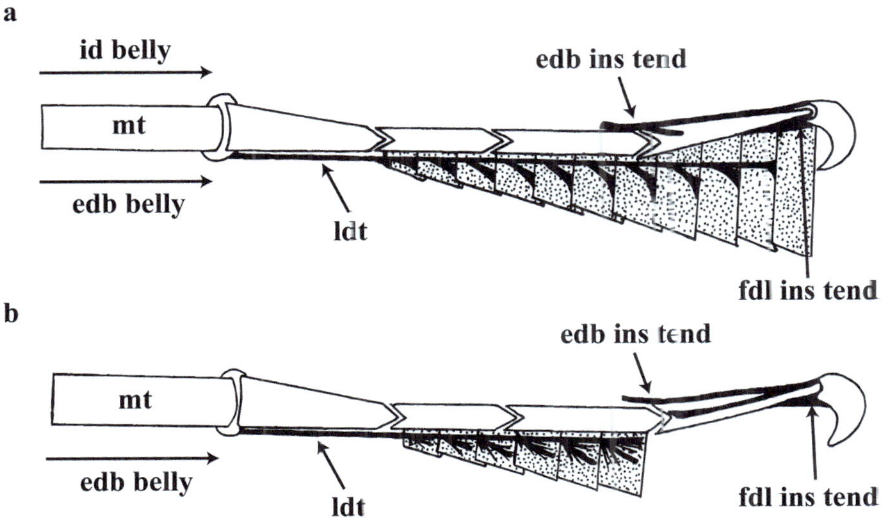

Fig. 9.12 Diagrammatic representation of the arrangement of the major mechanical units of digit IV of the pes of the sphaerodactylid gekkotan *Aristelliger* (**a**) and *Anolis* (**b**). Both diagrams start at the left with the metatarsal (mt). Scansors and lamellae are represented by stippled overlapping plates. The extent of the bellies of the extensor digitorum brevis (edb) and interossei dorsalis (id) muscles are indicated by arrows, and the insertion of the tendons of the extensor digitorum brevis (edb ins tend) and flexor digitorum longus (fdl ins tend) muscles are indicated by arrows pointing to the respective tendons. The lateral digital tendons (ldt) and their branches that meld with the stratum compactum of the scansors and lamellae are shown. Joints modified for hyperextension are indicated by chevron shapes (>>)

The initial critical changes of the subdigital integument, phalanges and overall conformation and proportions of the digits of *Gonatodes humeralis* are quite subtle when compared to those of their congeneric non-adhesive relatives (Gamble et al., 2012; Russell et al., 2015). In contrast, *Anolis* exhibits fully expressed toe pads (Fig. 9.1b) and modifications of the intermediate phalanges (Fig. 9.9) that enhance both the pressing of the toe pads onto the substratum during attachment and their hyperextension during release. It also possesses a lateral digital tendon system that merges with the dermis of the lamellae (Fig. 9.10) and transmits tensile loading from the setae to points of resistance deeper within the autopodium (Russell, 1986). Furthermore, *Anolis* incorporates a compliance-promoting cushioning system (Fig. 9.10) into its toe pads, in the form of hypertrophied lacunar cells of the epidermis (Russell, 2016). This cushioning system is unique to *Anolis* but is analogous and positionally similar to the vascular (Russell, 1981) and adipose tissue (Russell & Bauer, 1988) compliance structures found among geckos. Based on this, the gekkotan configuration structurally most closely resembling that of *Anolis* is that of the sphaerodactylid genus *Aristelliger* (Fig. 9.12). Compared to those of *Gekko gecko*, the setae of *Aristelliger* are relatively short and slender, have a modest branching pattern and terminate in relatively broad spatulate tips (Ruibal & Ernst, 1965). Whether *Aristelliger* employs proximodistal or distoproximal digit hyperextension during toe pad peeling is not known, but it lacks modifications of the digital musculature (Figs. 9.6f, 9.11, and 9.12a) associated with the latter (Fig. 9.7d).

9.5 General Conclusions Relating to Convergence of Gekkotan and Anoline Adhesive Systems

The spinulate outer epidermal generation of geckos and anoles was derived independently and served as the precursor of adhesively interactive, deeply homologous setae in each [originating multiple times in the former (Gamble et al., 2012; Russell & Gamble, 2019), but once only in the latter (Losos, 2011)]. The physical principles governing substratum adhesive attachment and release of setae have resulted in similar (but not identical) configurations in the two clades. The integration of the operational strictures governing seta-substratum interactions into the locomotor repertoires of the taxa that bear them has resulted in regimented patterning of setae within the setal fields and the development of basic mechanisms for controlling these interactions. There is no genus of gecko that shows precise convergence on the anole configuration, but at least one approaches it closely (Fig. 9.12). Physical principles have determined the way in which keratinous filamentous adhesives are able to be successfully deployed and ancestry has determined the particular form that the controlling anatomical superstructure takes (Russell & Gamble, 2019).

Recent years have witnessed a multitude of studies that have investigated the gekkotan adhesive system (Russell et al., 2019) and these provide clear evidence of how productive and enlightening interdisciplinary and transdisciplinary research can be. Such studies, however, have focused on relatively few taxa (most frequently *Gekko gecko*) and investigation and modelling has been largely based on taxa with the most complex configurations of the adhesive system. Although such research has greatly influenced the design and fabrication of biologically inspired fibrillar adhesives, recognition that there are simpler structural configurations exhibited by anoles and certain gekkotan lineages provides background for potential refinement of biomimetic applications based upon more basic structural and operational principles (Garner et al., 2019a). The common elements of convergence between anoles and geckos points to new possibilities for investigating and exploiting fibrillar adhesives. Comparison of the functional and structural aspects of the adhesive apparatus of both clades yields information about what is minimally necessary and sufficient for its effective operation and potentially simplifies approaches that can be taken in the development of biomimetic derivatives.

References

Alibardi, L. (2009). Cell biology of adhesive setae in gecko lizards. *Zoology, 112*(6), 403–424.

Alibardi, L., Toni, M., & Dalla Valle, L. (2007). Expression of beta-keratin mRNAs and proline uptake in epidermal cells of growing scales and pad lamellae of gecko lizards. *Journal of Anatomy, 211*(1), 104–116.

Alibardi, L., Edward, D. P., Patil, L., Bouhenni, R., Dhinojwala, A., & Niewiarowski, P. H. (2011). Histochemical and ultrastructural analyses of adhesive setae of lizards indicate that they contain lipids in addition to keratins. *Journal of Morphology, 272*(6), 758–768.

Arzt, E., Gorb, S., & Spolenak, R. (2003). From micro to nano contacts in biological attachment devices. *Proceedings of the National Academy of Sciences of the United States of America, 100*(19), 10603–10606.

Autumn, K. (2006). Principles and parameters of the gecko adhesive system. In A. M. Smith & J. A. Calow (Eds.), *Biological adhesives* (pp. 225–256). Springer.

Autumn, K., & Hansen, W. (2006). Ultrahydrophobicity indicates a non-adhesive default state in gecko setae. *Journal of Comparative Physiology A, 192*(11), 1205–1212.

Autumn, K., & Peattie, A. M. (2002). Mechanisms of adhesion in geckos. *Integrative and Comparative Biology, 42*(6), 1081–1090.

Autumn, K., Liang, Y. A., Hsieh, S. T., Zesch, W., Chan, W. P., Kenny, T. W., Fearing, R., & Full, R. J. (2000). Adhesive force of a single gecko foot-hair. *Nature, 405*(6787), 681–685.

Autumn, K., Sitti, M., Liang, Y. A., Peattie, A. M., Hansen, W. R., Sponberg, S., Kenny, T. W., Fearing, R., Israelachvili, J. N., & Full, R. J. (2002). Evidence for van der Waals adhesion in gecko setae. *Proceedings of the National Academy of Sciences of the United States of America, 99*(19), 12252–12256.

Autumn, K., Majidi, C., Groff, R. E., Dittmore, A., & Fearing, R. (2006a). Effective elastic modulus of isolated gecko setal arrays. *The Journal of Experimental Biology, 209*(18), 3558–3568.

Autumn, K., Hsieh, S. T., Dudek, D. M., Chen, J., Chitaphan, C., & Full, R. J. (2006b). Dynamics of geckos running vertically. *The Journal of Experimental Biology, 209*, 260–272.

Autumn, K., Niewiarowski, P. H., & Puthoff, J. B. (2014). Gecko adhesion as a model system for integrative biology, interdisciplinary science, and bioinspired engineering. *Annual Review of Ecology, Evolution, and Systematics, 45*(1), 445–470.

Barnes, W. J. P., Oines, C., & Smith, J. M. (2006). Whole animal measurements of shear and adhesive forces in adult tree frogs: insights into underlying mechanisms of adhesion obtained from studying the effects of size and scale. *Journal of Comparative Physiology A, 192*, 1179–1191.

Bartlett, M. D., Croll, A. B., King, D. R., Paret, B. M., Irschick, D. J., & Crosby, A. J. (2012). Looking beyond fibrillar features to scale gecko-like adhesion. *Advanced Materials, 24*(8), 1078–1083.

Bauer, A. M. (1998). Morphology of the adhesive tail tips of carphodactyline geckos (Reptilia: Diplodactylidae). *Journal of Morphology, 235*(1), 41–58.

Bergmann, P. J., & Irschick, D. J. (2005). Effects of temperature on maximum clinging ability in a diurnal gecko: Evidence for a passive clinging mechanism? *The Journal of Experimental Zoology A, 303*(9), 785–791.

Bhushan, B., Peressadko, A. G., & Kim, T.-W. (2006). Adhesion analysis of two-level hierarchical morphology in natural attachment systems for "smart adhesion". *Journal of Adhesion Science and Technology, 20*(13), 1475–1491.

Bloch, N., & Irschick, D. J. (2005). Toe-clipping dramatically reduces clinging performance in a pad-bearing lizard (*Anolis carolinensis*). *Journal of Herpetology, 39*(2), 288–293.

Brinkman, D. B. (1980). Structural correlates of tarsal and metatarsal functioning in *Iguana* (Lacertilia: Iguanidae) and other lizards. *Canadian Journal of Zoology, 58*, 277–289.

Chen, B., & Gao, H. (2010). An alternative explanation of the effect of humidity in gecko adhesion: Stiffness reduction enhances adhesion on a rough surface. *International Journal of Applied Mechanics, 2*, 1–9.

Delannoy, S. M. (2005). *Subdigital setae of the Tokay gecko (Gekko gecko): Variation in form and implications for adhesion*. MSc Thesis. Department of Biological Science, University of Calgary.

Donihue, C. M., Herrel, A., Fabre, A.-C., Kamath, A., Geneva, A. J., Schoener, T. W., Kolbe, J. J., & Losos, J. B. (2018). Hurricane-induced selection on the morphology of an island lizard. *Nature, 560*(7716), 88–91.

Elstrott, J., & Irschick, D. J. (2004). Evolutionary correlations among morphology, habitat use and clinging performance in Caribbean *Anolis* lizards. *Biological Journal of the Linnean Society, 83*(3), 389–398.

Ernst, V., & Ruibal, R. (1966). The structure and development of the digital lamellae of lizards. *Journal of Morphology, 120*(3), 233–265.

Federle, W. (2006). Why are so many adhesive pads hairy? *The Journal of Experimental Biology, 209*, 2611–2621.

Fuller, K. N. G., & Tabor, D. (1975). The effect of surface roughness on the adhesion of elastic solids. *Proceedings of the Royal Society of London. Series A, Mathematical and Physical Sciences, 345*(1642), 327–342.

Gamble, T., Greenbaum, E., Jackman, T. R., Russell, A. P., & Bauer, A. M. (2012). Repeated origin and loss of adhesive toepads in geckos. *PLoS One, 7*, e39429.

Gamble, T., Greenbaum, E., Jackman, T. R., Russell, A. P., & Bauer, A. M. (2017). Repeated evolution of digital adhesion in geckos: A reply to Harrington and Reeder. *Journal of Evolutionary Biology, 30*, 1429–1436.

Gans, C. (1969). Functional components versus mechanical units in descriptive morphology. *Journal of Morphology, 128*, 365–368.

Garner, A. M., Lopez, S. M., & Niewiarowski, P. H. (2017). Brown anole (*Anolis sagrei*) adhesive forces remain unaffected by partial claw clipping. *Acta Herpetologica, 12*, 133–137.

Garner, A. M., Wilson, M. C., Russell, A. P., Dhinojwala, A., & Niewiarowski, P. H. (2019a). Going out on a limb: how investigation of the anoline adhesive system can enhance our understanding of fibrillary adhesion. *Integrative and Comparative Biology, 59*, 61–69.

Garner, A. M., Buo, C., Piechowski, J. M., Pamfilie, A. M., Stefanovic, S. R., Dhinojwala, A., & Niewiarowski, P. H. (2019b). Digital hyperextension has no influence on the active self-drying of gecko adhesive subdigital pads. *Journal of Experimental Zoology Part A, 332*, 118–125.

Garner, A. M., Wilson, M. C., Wright, C., Russell, A. P., Niewiarowski, P. H., & Dhinojwala, A. (2021). The same but different: setal arrays of anoles and geckos indicate alternative approaches to achieving similar adhesive effectiveness. *Journal of Anatomy, 238*, 1143–1155.

Gillies, A. G., Henry, A., Lin, H., Ren, A., Shiuan, K., Fearing, R. S., & Full, R. J. (2014). Gecko toe and lamellar shear adhesion on macroscopic engineered rough surfaces. *The Journal of Experimental Biology, 217*, 283–289.

Gorb, S. N. (2008). Biological attachment devices: Exploring nature's diversity for biomimetics. *Philosophical Transactions of the Royal Society A, 366*(1870), 1557–1574.

Gorb, S. N., & Beutel, R. G. (2001). Evolution of locomotory attachment pads in hexapods. *Naturwissenschaften, 88*, 530–534.

Gould, S. J., & Vrba, E. S. (1982). Exaptation – A missing term in the science of form. *Paleobiology, 8*, 4–15.

Green, D. M. (1981). Adhesion and the toe-pads of tree frogs. *Copeia, 1981*(4), 790–796.

Greiner, C., Spolenak, R., & Arzt, E. (2009). Adhesion design maps for fibrillar adhesives: The effect of shape. *Acta Biomaterialia, 5*(2), 597–606.

Haefner, J. W. (1988). Assembly rules for Greater Antillean *Anolis* lizards. *Oecologia, 74*, 551–565.

Hagey, T. J. (2013). *Mechanics, diversity, and ecology of gecko adhesion*. Doctoral dissertation. University of Idaho.

Hagey, T. J., Uyeda, J. C., Crandell, K. E., Cheney, J. A., Autumn, K., & Harmon, L. J. (2017). Tempo and mode of performance evolution across multiple independent origins of adhesive toe pads in lizards. *Evolution, 71*, 2344–2358.

Hansen, W. R., & Autumn, K. (2005). Evidence for self-cleaning in gecko setae. *Proceedings of the National Academy of Sciences of the United States of America, 102*(2), 385–389.

Harrington, S., & Reeder, T. W. (2017). Rate heterogeneity across Squamata, misleading ancestral state reconstruction and the importance of proper null model specification. *Journal of Evolutionary Biology, 30*, 313–325.

Higham, T. E., Russell, A. P., & Niklas, K. J. (2017a). Leaping lizards landing on leaves: Escape-induced jumps in the rainforest canopy challenge the adhesive limits of geckos. *Journal of the Royal Society Interface, 14*, 20170156.

Higham, T. E., Gamble, T., & Russell, A. P. (2017b). On the origin of frictional adhesion in geckos: small morphological changes lead to a major biomechanical transition in the genus *Gonatodes*. *Biological Journal of the Linnean Society, 120*, 503–517.

Higham, T. E., Russell, A. P., Niewiarowski, P. H., Wright, A., & Speck, T. (2019). The ecomechanics of gecko adhesion: Natural surface topography, evolution, and biomimetics. *Integrative and Comparative Biology, 59*(1), 148–167.

Home, E. (1816). Some account of the feet of those animals whose progressive motion can be carried on in opposition to gravity. *Philosophical Transactions of the Royal Society, 106*, 149–155 + plates VII and VIII.

Hsu, P. Y., Ge, L., Li, X., Stark, A. Y., Wesdemiotis, C., Niewiarowski, P. H., & Dhinojwala, A. (2012). Direct evidence of phospholipids in gecko footprints and spatula-substrate contact interface detected using surface-sensitive spectroscopy. *The Journal of the Royal Society Interface, 9*(69), 657–664.

Hu, S., Lopez, S., Niewiarowski, P. H., & Xia, Z. (2012). Dynamic self-cleaning in gecko setae via digital hyperextension. *The Journal of the Royal Society Interface, 9*(76), 2781–2790.

Irschick, D. J., Austin, C. C., Petren, K., Fisher, R. N., Losos, J. B., & Ellers, O. (1996). A comparative analysis of clinging ability among pad-bearing lizards. *Biological Journal of the Linnean Society, 59*(1), 21–35.

Irschick, D. J., Carlisle, E., Elstrott, J., Ramos, M., Buckley, C., Vanhooydonck, B., Meyers, J. A. Y., & Herrel, A. (2005). A comparison of habitat use, morphology, clinging performance and escape behaviour among two divergent green anole lizard (*Anolis carolinensis*) populations. *Biological Journal of the Linnean Society, 85*(2), 223–234.

Johnson, M. K., & Russell, A. P. (2009). Configuration of the setal fields of *Rhoptropus* (Gekkota: Gekkonidae): Functional, evolutionary, ecological and phylogenetic implications of observed pattern. *Journal of Anatomy, 214*(6), 937–955.

Kendall, K. (1975). Thin-film peeling-the elastic term. *Journal of Physics D: Applied Physics, 8*(13), 1449–1452.

Khanafer, K., Duprey, A., Schlicht, M., & Berguer, R. (2008). Effects of strain rate, mixing ratio, and stress–strain definition on the mechanical behavior of the polydimethylsiloxane (PDMS) material as related to its biological applications. *Biomedical Microdevices, 11*(2), 503.

Khannoon, E. R., Endlein, T., Russell, A. P., & Autumn, K. (2014). Experimental evidence for friction-enhancing integumentary modifications of chameleons and associated functional and evolutionary implications. *Proceedings of the Royal Society of London. Series B, Biological Sciences, 281*(1775), 20132334.

Kim, T. W., & Bhushan, B. (2007). Adhesion analysis of multi-level hierarchical attachment system contacting with a rough surface. *Journal of Adhesion Science and Technology, 21*(1), 1–20.

Koppetsch, T., Böhme, W., Büsse, S., & Gorb, S. N. (2020). Comparative epidermal microstructure anatomy and limb and tail osteology of eyelid geckos (Squamata: Eublepharidae): Implications of ecomorphological adaptations. *Zoologischer Anzeiger*. https://doi.org/10.1016/j.jcz.2020.05.005

Kuhn, A., Skipwith, P., & Overcast, I. (2020). Digest: an emerging model system for understanding ecomorphological convergence. *Evolution, 74*, 696–697.

Labonte, D., Clemente, C. J., Dittrich, A., Kuo, C. Y., Crosby, A. J., Irschick, D. J., & Federle, W. (2016). Extreme positive allometry of animal adhesive pads and the size limits of adhesion-based climbing. *Proceedings of the National Academy of Sciences of the United States of America, 113*(5), 1297–1302.

Lange, B. (1931). Integument der Sauropsiden. In L. Bolk, E. Göppert, E. Kallius, & W. Lubosch (Eds.), *Handbuch der vergleichenden Anatomie der Wirbeltiere* (pp. 375–448). Urban und Schwarzenberg.

Langowski, J. K., Dodou, D., Kamperman, M., & van Leeuwen, J. L. (2018). Tree frog attachment: Mechanisms, challenges, and perspectives. *Frontiers in Zoology, 15*(1), 32.

Lauff, R. F., Russell, A. P., & Bauer, A. M. (1993). Topography of the digital cutaneous sensilla of the Tokay gecko, *Gekko gecko* (Reptilia, Gekkonidae), and their potential role in locomotion. *Canadian Journal of Zoology, 71*, 2462–2472.

Lillywhite, H. B., & Maderson, P. F. A. (1968). Histological changes in the epidermis of the subdigital lamellae of *Anolis carolinensis* during the shedding cycle. *Journal of Morphology, 125*, 379–401.

Liu, Y., Zhou, Q., Wang, Y., Luo, L., Yang, J., Yang, L., Liu, M., Li, Y., Quian, T., Zheng, Y., Li, M., Li, J., Gu, Y., Han, Z., Man, X., Wang, Y., Zhu, C., Yu, B., Yang, Y., Ding, F., Jiang, J., Yang, H., & Gu, X. (2015). *Gekko japonicus* genome reveals evolution of adhesive toe pads and tail regeneration. *Nature Communications, 6*, 10033.

Losos, J. B. (2011). *Lizards in an evolutionary tree: Ecology and adaptive radiation of anoles*. University of California Press.

Macrini, T. E., Irschick, D. J., & Losos, J. B. (2003). Ecomorphological differences in toepad characteristics between mainland and island anoles. *Journal of Herpetology, 37*(1), 52–58.

Maddin, H. C., Eckhart, L., Jaeger, K., Russell, A. P., & Ghannadan, M. (2009). The anatomy and development of the claws of *Xenopus laevis* (Amphibia: Anura) reveal alternate pathways of structural evolution of the integument in tetrapods. *Journal of Anatomy, 214*, 607–619.

Maderson, P. F. A. (1970). Lizard glands and lizard hands: Models for evolutionary study. *Forma et Functio, 3*, 179–204.

Maderson, P. F. A., Zucker, A. H., & Roth, S. I. (1978). Epidermal regeneration and percutaneous water loss following cellophane stripping of reptile epidermis. *The Journal of Experimental Zoology, 204*, 11–32.

Murphy, M. P., Kim, S., & Sitti, M. (2009). Enhanced adhesion by gecko-inspired hierarchical fibrillar adhesives. *ACS Applied Materials and Interfaces, 1*(4), 849–855.

Nachtigall, W. (1974). *Biological mechanisms of attachment: The comparative morphology and bioengineering of organs for linkage, suction and adhesion*. Springer.

Naylor, E. R., & Higham, T. E. (2019). Attachment beyond the adhesive system: The contribution of claws to gecko clinging and locomotion. *Integrative and Comparative Biology, 58*, 168–181.

Niewiarowski, P. H., Stark, A. Y., & Dhinojwala, A. (2016). Sticking to the story: Outstanding challenges in gecko-inspired adhesives. *The Journal of Experimental Biology, 219*(7), 912–919.

Niewiarowski, P. H., Stark, A. Y., & Dhinojwala, A. (2017). A bibliometric analysis of gecko adhesion: A view of its origins and current directions. In L. Heepe, L. Xue, & S. N. Gorb (Eds.), *Bio-inspired structured adhesives* (pp. 1–19). Springer.

Niewiarowski, P. H., Dhinojwala, A., & Garner, A. M. (2019). A physical model approach to gecko adhesion opportunity and constraint: How rough could it be? *Integrative and Comparative Biology, 59*(1), 203–213.

Peattie, A. M., & Full, R. J. (2007). Phylogenetic analysis of the scaling of wet and dry biological fibrillar adhesives. *Proceedings of the National Academy of Sciences of the United States of America, 104*(47), 18595–18600.

Peressadko, A., & Gorb, S. N. (2004). When less is more: Experimental evidence for tenacity enhancement by division of contact area. *The Journal of Adhesion, 80*, 247–261.

Persson, B. N. J. (2003). On the mechanism of adhesion in biological systems. *The Journal of Chemical Physics, 118*(16), 7614–7621.

Pesika, N. S., Zeng, H., Kristiansen, K., Zhao, B., Tian, Y., Autumn, K., & Israelachvili, J. (2009). Gecko adhesion pad: A smart surface? *Journal of Physics. Condensed Matter, 21*(46), 464132.

Peterson, J. A. (1983a). The evolution of the subdigital pad in *Anolis*. I. Comparisons among the anoline genera. In *Advances in herpetology and evolutionary biology. Essays in honor of Ernest E. Williams*. Museum of Comparative Zoology, Harvard University.

Peterson, J. A. (1983b). The evolution of the subdigital pad of *Anolis* 2. Comparisons among the iguanid genera related to the anolines and a view from outside the radiation. *Journal of Herpetology*, 371–397.

Peterson, J. A., & Williams, E. E. (1981). A case history in retrograde evolution: the *onca* lineage in anoline lizards: II. Subdigital fine structure. *Bulletin of the Museum of Comparative Zoology at Harvard College, 149*, 215–268.

Pinto, B., Colli, G., Higham, T., Russell, A. P., Scantlebury, D., Vitt, L., & Gamble, T. (2018). Population genetic structure and species delimitation of a widespread, Neotropical dwarf gecko. *Molecular Phylogenetics and Evolution, 133*, 54–66.

Prowse, M. S., Wilkinson, M., Puthoff, J. B., Mayer, G., & Autumn, K. (2011). Effects of humidity on the mechanical properties of gecko setae. *Acta Biomaterialia, 7*, 733–738.

Puthoff, J. B., Prowse, M. S., Wilkinson, M., & Autumn, K. (2010). Changes in materials properties explain the effects of humidity on gecko adhesion. *The Journal of Experimental Biology, 213*, 3699–3704.

Rewcastle, S. C. (1980). Form and function of the lacertilian knee and mesotarsal joints – A contribution to the analysis of sprawling locomotion. *Journal of Zoology, 191*, 147–170.

Rewcastle, S. C. (1983). Fundamental adaptations in the lacertilian hind limb: A partial analysis of the sprawling limb posture and gait. *Copeia, 1983*, 476–487.

Riskin, D. K., & Fenton, M. B. (2001). Sticking ability in Spix's disc-winged bat, *Thryroptera tricolor* (Microchiroptera: Thryropteridae). *Canadian Journal of Zoology, 79*, 2261–2267.

Roach, M. (2003). *Stiff: The curious lives of human cadavers*. W.W. Norton and Company.

Rosenberg, H. I., & Rose, R. (1999). Volar adhesive pads of the feathertail glider, *Acrobates pygmaeus* (Marsupialia; Acrobatidae). *Canadian Journal of Zoology, 77*, 233–248.

Ruibal, R. (1968). The ultrastructure of the surface of lizard scales. *Copeia, 1968*(4), 698–703.

Ruibal, R., & Ernst, V. (1965). The structure of the digital setae of lizards. *Journal of Morphology, 117*(3), 271–293.

Russell, A. P. (1975). A contribution to the functional analysis of the foot of the Tokay, *Gekko gecko* (Reptilia: Gekkonidae). *Journal of Zoology, 176*(4), 437–476.

Russell, A. P. (1976). Some comments concerning interrelationships amongst gekkonine geckos. In Bellairs, A.d'A., Cox, C.B. (Eds.) *Morphology and biology of reptiles*. Linnean Symposium Series No. 3 (pp. 217–244). Academic.

Russell, A. P. (1981). Descriptive and functional anatomy of the digital vascular system of the Tokay, *Gekko gecko*. *Journal of Morphology, 169*, 293–323.

Russell, A. P. (1986). The morphological basis of weight-bearing in the scansors of the Tokay gecko (*Gekko gecko*). *Canadian Journal of Zoology, 64*, 948–955.

Russell, A. P. (2002). Integrative functional morphology of the gekkotan adhesive system (Reptilia: Gekkota). *Integrative and Comparative Biology, 42*(6), 1154–1163.

Russell, A. P. (2016). The structure of anoline (Reptilia: Dactyloidae: *Anolis*) toe pads in relation to substratum conformity. *Acta Zoologica, 98*, 300–309.

Russell, A. P., & Bauer, A. M. (1988). Paraphalangeal elements of gekkonid lizards: a comparative survey. *Journal of Morphology, 197*, 221–240.

Russell, A. P., & Bauer, A. M. (1990). Digit I in pad-bearing gekkonine geckos: Alternate designs and the potential constraints of phalangeal number. *Memoirs of the Queensland Museum, 29*, 453–472.

Russell, A. P., & Bauer, A. M. (2008). The appendicular locomotor apparatus of *Sphenodon* and normal-limbed squamates. In C. Gans, A. S. Gaunt, & K. Adler (Eds.), *Biology of the Reptilia* (Vol. 1, pp. 1–466). Society for the Study of Amphibians and Reptiles.

Russell, A. P., & Bels, V. (2001). Digital hyperextension in *Anolis sagrei*. *Herpetologica, 57*(1), 58–65.

Russell, A. P., & Johnson, M. K. (2007). Real-world challenges to, and capabilities of, the gekkotan adhesive system: contrasting the rough and the smooth. *Canadian Journal of Zoology, 85*, 1228–1238.

Russell, A. P., & Delaugerre, M.-J. (2017). Left in the dust: Differential effectiveness of the two adhesive pad configurations in geckos (Reptilia: Gekkota). *Journal of Zoology, 301*, 61–68.

Russell, A. P., & Eslinger, A. (2017). A whole lamella perspective on the origin of the epidermal free margin of *Anolis* (Reptilia: Dactyloidae) toe pads. *Journal of Morphology, 278*(3), 360–368.

Russell, A. P., & Gamble, T. (2019). Evolution of the gekkotan adhesive system: Does digit anatomy point to one or more origins? *Integrative and Comparative Biology, 59*, 131–147.

Russell, A. P., & Johnson, M. K. (2014). Between a rock and a soft place: Microtopography of the locomotor substrate and morphology of the setal fields of Namibian day geckos (Gekkota: Gekkonidae: *Rhoptropus*). *Acta Zoologica, 95*, 299–318.

Russell, A. P., & Oetelaar, G. S. (2016). Limb and digit orientation during vertical clinging in Bibron's gecko, *Chondrodactylus bibronii* (A. Smith, 1846) and its bearing on the adhesive capabilities of geckos. *Acta Zoologica, 97*, 345–360.

Russell, A. P., Bauer, A. M., & Laroiya, R. (1997). Morphological correlates of the secondarily symmetrical pes of gekkotan lizards. *Journal of Zoology, 241*, 767–790.

Russell, A. P., Johnson, M. K., & Delannoy, S. M. (2007). Insights from studies of gecko-inspired adhesion and their impact on our understanding of the evolution of the gekkotan adhesive system. *Journal of Adhesion Science and Technology, 21*(12–13), 1119–1143.

Russell, A. P., Baskerville, J., Gamble, T., & Higham, T. E. (2015). The evolution of digit form in *Gonatodes* (Gekkota: Sphaerodactylidae) and its bearing on the transition from frictional to adhesive contact in gekkotans. *Journal of Morphology, 276*, 1311–1332.

Russell, A. P., Stark, A. Y., & Higham, T. E. (2019). The integrative biology of gecko adhesion: historical review, current understanding and grand challenges. *Integrative and Comparative Biology, 59*, 101–116.

Sawyer, R. H., Glenn, T., French, J. O., Mays, B., Shames, R. B., Barnes, G. L., Jr., Rhodes, W., & Ishikawa, Y. (2000). The expression of beta (β) keratins in the epidermal appendages of reptiles and birds. *American Zoologist, 40*, 530–539.

Song, Y., Dai, Z., Wang, Z., & Full, R. J. (2020). Role of multiple, adjustable toes in distributed control shown by sideways wall-running in geckos. *Proceedings of the Royal Society B, 287*(1926), 20200123.

Spinner, M., Westhoff, G., & Gorb, S. N. (2014). Subdigital setae of chameleon feet: friction-enhancing microstructures for a wide range of substrate roughness. *Scientific Reports, 4*, 5481.

Stark, A. Y., & Mitchell, C. T. (2019). Stick or slip: Adhesive performance of geckos and gecko-inspired synthetics in wet environments. *Integrative and Comparative Biology, 59*, 214–226.

Stark, A. Y., Sullivan, T. W., & Niewiarowski, P. H. (2012). The effect of surface water and wetting on gecko adhesion. *The Journal of Experimental Biology, 215*(17), 3080–3086.

Stark, A. Y., Badge, I., Wucinich, N. A., Sullivan, T. W., Niewiarowski, P. H., & Dhinojwala, A. (2013). Surface wettability plays a significant role in gecko adhesion underwater. *Proceedings of the National Academy of Sciences of the United States of America, 110*(16), 6340–6345.

Stark, A. Y., Wucinich, N. A., Paoloni, E. L., Niewiarowski, P. H., & Dhinojwala, A. (2014). Self-drying: A gecko's innate ability to remove water from wet toe pads. *PLoS One, 9*(7), e101885.

Stark, A. Y., Klittich, M. R., Sitti, M., Niewiarowski, P. H., & Dhinojwala, A. (2016). The effect of temperature and humidity on adhesion of gecko-inspired adhesives: implications for the natural system. *Scientific Reports, 6*, 30936.

Stewart, G. R., & Daniel, R. S. (1972). Scales of the lizard Gekko gecko: surface structure examined with the scanning electron microscope. *Copeia, 1972*, 252–257.

Tornier, G. (1899). Ein Eidechsenschwanz mit Saugscheibe. *Biologisches Zentralblatt, 19*, 549–552.

Vogel, S. (1988). *Life's devices: The physical world of animals and plants*. Princeton University Press.
Webster, N. B., Johnson, M. K., & Russell, A. P. (2009). Ontogenetic scaling of scansorial surface area and setal dimensions of *Chondrodactlus bibronii* (Gekkota: Gekkonidae): Testing predictions derived from cross-species comparisons of gekkotans. *Acta Zoologica, 90*, 18–29.
Williams, E. E., & Peterson, J. A. (1982). Convergent and alternative designs in the digital adhesive pads of scincid lizards. *Science, 215*(4539), 1509–1511.
Wimsatt, W. A., & Villa-R, B. (1970). Locomotor adaptations of the disc-winged bat *Thryroptera tricolor*. I. Functional organization of the adhesive discs. *The American Journal of Anatomy, 129*, 89–120.
Yao, H., & Gao, H. (2006). Mechanics of robust and releasable adhesion in biology: Bottom-up designed hierarchical structures of gecko. *Journal of the Mechanics and Physics of Solids, 54*(6), 1120–1146.
Zweers, G. A. (1979). Explanation of structure by optimization and systemization. *Netherlands Journal of Zoology, 29*, 418–440.

Chapter 10
Convergent Evolution of Animal Adhesive Pads

Thies H. Büscher and Stanislav N. Gorb

Abstract Functional systems that evolve as a response to specific environmental challenges often exhibit convergent traits. Organs adapted for attachment to a surface are tuned to a general requirement independent of the phylogenetic position of the organism. The different strategies employed for solving similar problems often represent the same physical principles, and that is why the morphology of attachment structures (and also many other functional systems) is channelled by physical rules. Different animals, therefore, employ similar mechanisms to attach to the broad variety of substrates with different surface conditions. There are two main types of attachment devices that occur on animal legs: hairy and smooth. They differ greatly in their morphology and ultrastructure, but both solve the same problem of proper mechanical adaptation to the variety of natural roughnesses by maximising real contact area with them. Adaptation to specific surface conditions within these groups resulted in several different solutions to the specific ecological surroundings the lineages radiated into. As the conditions are similar in the discrete environments, the adaptations of the attachment systems of different animal groups reveal similar mechanisms. For this reason, on the one hand, similar attachment organs evolved in different lineages of animals, and, on the other hand, different attachment organ modifications occur within the same lineages. In this chapter we summarise the data published in the literature on the structural and functional principles of hairy and smooth attachment pads with a special focus on insects. We describe ultrastructure, surface patterns, the origin of different pads and their evolution, discuss the results of mechanical testing of material properties (elasticity, viscoelasticity, adhesion, friction) and basic physical forces contributing to adhesion, show the influence of different factors, such as substrate roughness and pad stiffness, on contact forces, and review the chemical composition of pad fluids, which are an important component of adhesive function. The structure of this chapter is a kind of fractal. It starts with the omnipresence of the pads in animals. Then we zoom into the phylogeny focusing on insects as the largest animal group on earth, showing convergent

T. H. Büscher (✉) · S. N. Gorb
Department of Functional Morphology and Biomechanics, Zoological Institute, Kiel University, Kiel, Germany
e-mail: tbuescher@zoologie.uni-kiel.de; sgorb@zoologie.uni-kiel.de

V. L. Bels, A. P. Russell (eds.), *Convergent Evolution*, Fascinating Life Sciences,
https://doi.org/10.1007/978-3-031-11441-0_10

evolution of attachment pads therein. In the subsequent step we further zoom in on the phylogeny of one insect group, Phasmatodea, and explore convergent evolution of attachment pads at an even finer scale. Such a hierarchical structure of the chapter helps us to show that convergent evolution occurs at different levels within the animal tree of life. Since convergent events might be potentially interesting for engineers in revealing a kind of optimal solution by nature. Finally, the biomimetic implications of the discussed results are briefly presented.

Keywords Attachment devices · Locomotion · Frictional adhesion · Biomechanics · Substrate adaptation

10.1 The Role of Adhesion for Animals

Adhesion in the broader sense is of major importance for most living animals. Secure attachment to various surfaces is essential for many animals, for example, to maintain access to nutrients, but also to support locomotion on any terrain. The properties of the specific surfaces in their natural environments shape the morphology and function of their attachment devices independent of their phylogenetic position. The characteristics of similar habitats resulted in similar selective pressures upon various animal groups. Attachment devices are omnipresent in animals (Fig. 10.1). The morphological and ultrastructural backgrounds of the ability of animals to attach to, and walk on, vertical surfaces and ceilings have been studied in detail in many animal taxa, including insects (Beutel & Gorb, 2001, 2006; Gorb, 2001; Gorb & Beutel, 2001), arachnids (Homann, 1957; Kesel et al., 2003; Gorb et al., 2006; Niederegger & Gorb, 2006; Wolff & Gorb, 2016), tree frogs (Hanna &

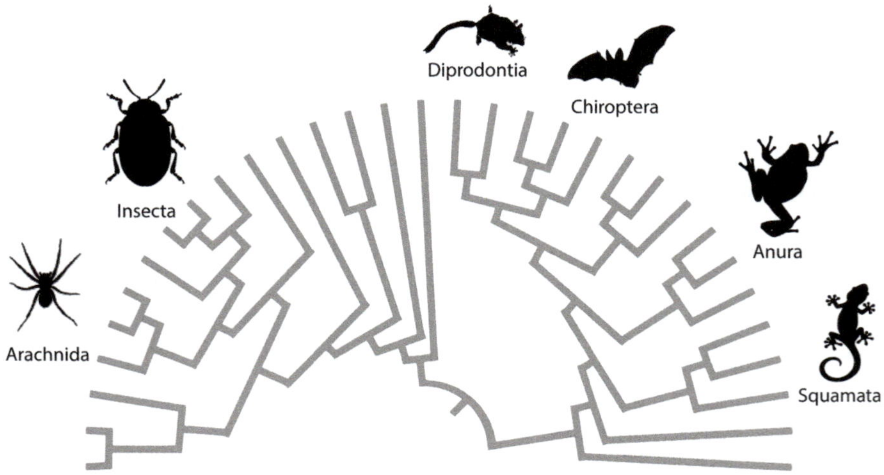

Fig. 10.1 Omnipresence of attachment devices in animals: examples shown in the tree of life. Reproduced from Büscher and Gorb (2021)

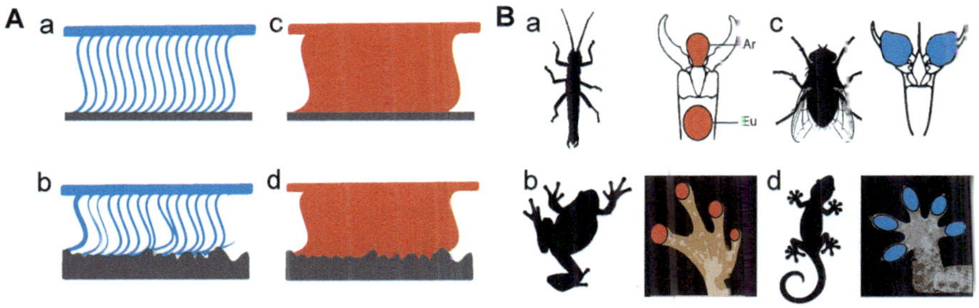

Fig. 10.2 Attachment systems of animals. (**A**) Schematic representation of hairy (a, b) and smooth (c, d) attachment pads. Both systems provide adjustment to the surface profile of smooth (a, c) and corrugated (b, d) substrates. (**B**) Examples of smooth and hairy adhesive systems. (a) Smooth arolium (Ar) and euplantulae (Eu) of Phasmatodea. (b) Hairy pulvilli of Diptera. (c) Smooth toe pads of Anura. (d) Hairy toe pads of Gekkota. Colours correspond to the type of attachment pad (cyan = hairy; orange = smooth). Figure (**A**) and tarsi from (**B**) modified from Beutel and Gorb (2001). Reproduced from Büscher and Gorb (2021)

Barnes, 1990; Ba-Omar et al., 2000), arboreal salamanders (Green & Alberch, 1981), lizards (Hiller, 1968; Autumn et al., 2000; Autumn & Peattie, 2002; Gao et al., 2005; Huber et al., 2005a, b; Autumn, 2006; Rizzo et al., 2006; Koppetsch et al., 2020), echinoderms (Feder, 1955; Flammang, 1996), and mammals (Schliemann, 1970, 1971, 1974, 1975, 1983; Schliemann & Rehn, 1980; Thewissen & Etnier, 1995; Rosenberg & Rose, 1999). These studies show that during the course of biological evolution, animals have convergently developed two distinctly different types of structures to attach themselves to a variety of substrates: hairy (setose) pads and smooth pads. In the following we focus on attachment systems used for terrestrial locomotion based on the publication of Büscher and Gorb (2021). Aquatic adhesive systems are discussed in more detail in a subsequent chapter (see Chap. 16).

Hairy pads are covered with hairs, setae, acanthae and microtrichia (Richards & Richards, 1979), fine cuticular surface outgrowths, which due to their flexibility can maximise the extent of contact with a wide range of microscopically rough substrate profiles (Fig. 10.2) and, due to the low bending stiffness of their terminal plates, can even adapt to substrates with roughness at a sub-nanometer scale (Beutel & Gorb, 2001; Gorb, 2001; Gorb & Beutel, 2001; Gorb et al., 2002). Smooth pads can also maximise their contact areas with a variety of substrates due to their specialised material structure and properties (Gorb, 2008a). Interestingly, in the course of biological evolution, both functional solutions have originated many times independently in different animal groups.

10.2 Adhesion in Insects

Represented by more than one million described species, insects constitute the majority of animals on earth. With their astonishing diversity, they are one of the most remarkable lineages in the 3.5 billion years of life history on this planet (Engel, 2015). Insects are, in terms of diversity, biomass and organismic interactions, indisputably one of the most important groups of animals (Grimaldi & Engel, 2005). The chitinous exoskeleton is often mentioned as the basis of structural diversification and is considered to be a key innovation for the success of insects (Wilson, 1990; Gillot, 2005; Grimaldi & Engel, 2005). The versatility of the cuticular integument provides a broad array of tools for various functional demands (Gunderson & Schiavone, 1989; Gorb, 2001). Specifically, the wings are considered important to facilitate mobility, dispersal and escape from predators (Wilson, 1969; Wootton, 1992; Wagner & Liebherr, 1992; Roff, 1994; Roff & Fairbairn, 2007; Dudley, 2000). Furthermore, the ability to move efficiently in different environments promotes niche diversity and subsequently species diversity in insects (Ritzmann et al., 2004). A key feature for mobility, next to the evolution of wings, is the evolution of a segmented leg in arthropods (Hoyle, 1976). These paired, articulated appendages, in combination with the hardened exoskeleton, served, for both Arthropoda in general and insects in particular, as a tool for becoming ubiquitous in nearly all habitats on earth (Lankester, 1904; Boudreaux, 1979; Weygoldt, 1986). Besides exploiting the advantages of tagmosis, the adaptability of the jointed limb enabled settlement in different habitats and furnished numerous opportunities for adaptive radiation (Gillot, 2005).

The legs of insects are usually used for walking, but show adaptations for locomotion on different terrains (Pearson & Franklin, 1984; Gillot, 2005). Even without morphological specialisations, such as those for jumping or digging, legged motion is very diverse. Some groups specialized towards very specific substrates: water striders for example run on the surface of the water (Brinkhurst, 1959; Darnhofer-Demar, 1969; Hu et al., 2003) and ectoparasitic flies are highly modified for remaining attached to their hosts and moving on them (Petersen et al., 2018). There are numerous other functional modifications of regions of insect legs, including those associated with silk production (e.g. Büsse et al., 2019) or prey capture (Prete & Hamilton, 1999), but one is of major importance for nearly all insects: the attachment system.

During the evolution of insects, the acquisition of flight and its consequences strongly influenced the evolution of attachment devices. Wings enabled dispersal and colonization of different environments, and flying necessitated insects be able to land and attach to several different, often unpredictable substrates (Beutel & Gorb, 2001, 2006, 2008; Friedemann et al., 2014a). Many insects are phytophagous, often being strongly associated with a narrow spectrum of angiosperm plants (Grimaldi & Engel, 2005; Thorpe & Caudle, 1938; Brower, 1958; Dethier, 1959; Ward & Spalding, 1993; Kester & Barbosa, 1994; Friedemann et al., 2015). Additionally, plants are not only used for nutrition; but represent sites for foraging, mating and deposition of offspring (Voigt & Gorb, 2010; Al Bitar et al., 2012, 2014; Hilker & Fatouros, 2015; Voigt et al., 2019). Hence, the attachment system appears to be highly influenced by the strong connection that insects have had with plants throughout their evolution.

The coevolution of angiosperms and insects is suggested to have resulted in an extensive adaptive co-radiation (Ehrlich & Raven, 1964; Wiebes, 1979; Farrell & Mitter, 1994). Besides chemical weapons against insect herbivory, a large variety of surface adaptations have been evolved among plants. As a response, insects developed different strategies to enable them to walk on, and attach to, those surfaces (Beutel & Gorb, 2001). Consequently, a plethora of different attachment devices have been evolved by insects.

Although insects are extremely diverse, the design of their attachment devices can be divided into the two fundamental types: hairy and smooth attachment pads (Beutel & Gorb, 2001; Gorb, 2001), similar to the aforementioned categorization of attachment devices in the animal kingdom as a whole. In insects, the hairy structures consist of deformable adhesive setae, typically originating from the tarsus itself (Fig. 10.2). Hairy attachment pads occur in different groups of insects. They are common in Coleoptera (e.g. Stork, 1980a, b, c, 1983a, b, Ishii, 1987, Kölsch & Betz, 1998, Gorb & Gorb, 2002, Betz, 2002, 2003, 2006a, b, Zurek et al., 2007, 2015, 2017, Voigt et al., 2008, 2017a, Bullock & Federle, 2011a, b, Liu & Liang, 2016, Gnaspini et al., 2017), Dermaptera (e.g. Haas & Gorb, 2004), Megaloptera (Maki, 1936; Theischinger, 1991; Beutel & Gorb, 2001), Strepsiptera (Kinzelbach, 1971; Beutel & Gorb, 2001), Mantophasmatodea (Beutel & Gorb, 2006, 2008), and Diptera (e.g. Bauchhenss & Renner, 1977; Bauchhenss, 1979; Walker et al., 1985; Röder, 1986; Gorb, 1998; Gorb et al., 2001; Niederegger et al., 2002; Niederegger & Gorb, 2003; Friedemann et al., 2014a; Petersen et al., 2018). In stick insects (Phasmatodea) (Büscher et al., 2018a, 2019) and true bugs (Heteroptera) (Gillett & Wigglesworth, 1932; Edwards & Tarkanian, 1970; Schuh, 1976; Weirauch, 2005, 2007; Gorb & Gorb, 2004; Gorb et al., 2004; Voigt et al., 2007, 2008; Friedemann et al., 2014b; Salerno et al., 2017, 2018; Rebora et al., 2018) some species are reported to have hairy attachment structures, although the majority of included taxa have smooth ones (Fig. 10.3). Although very similar in shape and identical in function, these structures evolved convergently in different insect groups.

Smooth attachment systems on the other hand, comprise soft cuticular pads lacking large fibrillar outgrowths. Usually, such attachment pads, similarly to hairy ones, are ventrally located on the tarsus (e.g. euplantulae) or on the pretarsus (e.g. single arolia or paired pulvilli). In some cases, attachment structures are present on the tibia as well (Lees & Hardie, 1988; Dixon et al., 1990; Beutel & Gorb, 2001; Friedemann et al., 2015; Wang & Liang, 2015; Büscher et al., 2019). Smooth attachment pads are found in most groups of insects, for example in Orthoptera (Slifer, 1950; Kendall, 1970; Henning, 1974; Gorb & Scherge, 2000; Jiao et al., 2000; Gorb et al., 2000; Perez-Goodwyn et al., 2006; Grohmann et al., 2015), Siphonaptera (Beutel & Gorb, 2001), Phthiraptera (Beutel & Gorb, 2001; Cruz & Mateo, 2009), Mantodea (Beutel & Gorb, 2001; Wieland, 2013), Hymenoptera (Holway, 1935; Snodgrass, 1956; Brainerd, 1994; Federle et al., 2000, 2001,

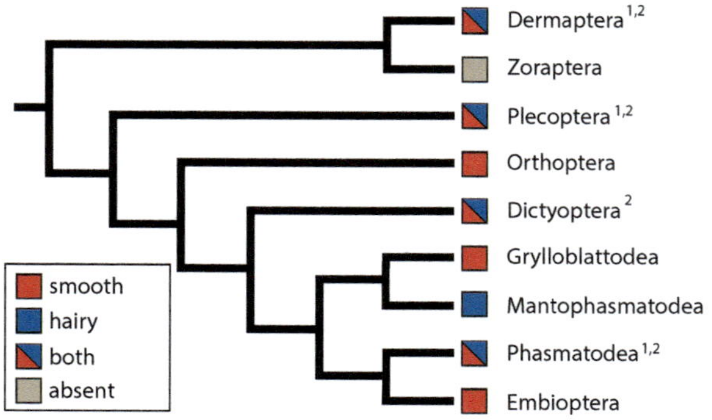

Fig. 10.3 Phylogeny of Polyneoptera (following Wipfler et al., 2019). Coloured squares indicate the type of attachment pads. [1]different pad types on the same tarsus, [2]in different species. Reproduced from Büscher and Gorb (2021)

2002; Beutel & Gorb, 2001; Orivel et al., 2001; Federle, 2002; Frantsevich & Gorb, 2002, 2004; Schulmeister, 2003; Federle & Endlein, 2004; Endlein & Federle, 2008), Embioptera (Ross, 1991, 2000; Büscher et al., 2019), Ephemeroptera, in form of a claw pad, (Beutel & Gorb, 2001), Thysanoptera (Heming, 1971, 1972, 1973; Beutel & Gorb, 2001), Blattodea (Roth & Willis, 1952; Arnold, 1974; Clemente & Federle, 2008), Phasmatodea (Beutel & Gorb, 2006, 2008; Scholz et al., 2008; Bußhardt et al., 2012; Gottardo & Heller, 2012; Gottardo et al., 2015; Büscher & Gorb, 2017, 2019; Büscher et al., 2018a, b, 2019, 2020a), Stenorrhyncha (White & Carver, 1971; Lees & Hardie, 1988; Dixon et al., 1990; Carver et al., 1991; Friedemann et al., 2015), Auchennorrhyncha (Lee et al., 1986; Frantsevich et al., 2008; Clemente et al., 2017; Goetzke et al., 2019), and some Mecoptera (Röder, 1986; Byers, 1991; Beutel & Gorb, 2001).

In many groups, however, the type of attachment system is not necessarily uniform throughout the entire group (e.g. Phasmatodea, Büscher et al., 2019). Unfortunately, broad comparative analyses based on several species per group have not been conducted for most insect lineages. In addition, the same anatomical structure might be hairy or smooth in different representatives of the same group, for example, the pulvilli of flies (Friedemann et al., 2014a) and the plantulae of Hymenoptera (Schulmeister, 2003). Furthermore, structurally similar organs, which have been ambiguously discussed in terms of homology in different orders, for example, the pulvilli of flies and true bugs (Beutel & Gorb, 2001; Friedemann et al., 2014a, b), or the plantulae of Hymenoptera (e.g. Bohart & Menke, 1976) and euplantulae of other insects (e.g. Schedl, 1991), can be either hairy or smooth in different groups. Consequently, the majority of these attachment structures are most probably not homologous, but evolved independently multiple times (Breidbach, 1980; Beutel & Gorb, 2001; Labonte et al., 2014; Schulmeister, 2003). In Acercaria [Psocodea, Hemiptera and Thysanoptera, according to Börner (1919)], the pulvillus evolved independently at least twice (Friedemann et al., 2014b). In some groups, a combination of the two types is found on the tarsus, such as the smooth arolia and hairy soles or hairy euplantulae of Mantophasmatodea (Beutel & Gorb, 2006, 2008), Tipulidae (Diptera) (Rees & Ferris, 1939; Henning, 1973; Beutel & Gorb, 2001;

Friedemann et al., 2014a), Plecoptera (Beutel & Gorb, 2001; Nelson, 2009) and Lepidoptera (Oseto & Helms, 1976; Faucheux, 1985; Nielsen & Common, 1991; Beutel & Gorb, 2001; Al Bitar et al., 2009).

10.3 Polyneoptera, a Striking Example for Convergent Traits

The Polyneoptera is a group of insects comprising around 40.000 extant species and includes the majority of the hemimetabolous insects (Wipfler et al., 2019). Although the number of taxa is much less than that of other groups of insects, such as the Diptera or Coleoptera, many details of their evolution, such as changes in morphology, behaviour or lifestyle remain unresolved. One reason lies in the notable differences in the subgroups of Polyneoptera and their strong ecological differentiation that has impeded a reliable reconstruction of the internal relationships for many years (Bradler, 2009; Beutel et al., 2013; Wipfler et al., 2019). Attachment pad morphology is both an indicator and a result of the complex mesh of adaptations. The presence of hairy and smooth attachment structures within Polyneoptera is accounted for by convergence on different levels (Fig. 10.3). While adhesive structures are absent and probably secondarily lost in Zoraptera (Beutel & Gorb, 2008), several other groups possess hairy structures. Adhesive hairs evolved independently in Dermaptera (Haas & Gorb, 2004), Plecoptera (Nelson, 2009), Phasmatodea (Büscher et al., 2018a, 2019), Dictyoptera (Arnold, 1974) and Mantophasmatodea (Beutel & Gorb, 2006, 2008). This convergence is not limited to the presence of hairy attachment devices in the respective groups, but also reveals different stages of reversal or repetitive origins within several of these. While all mantophasmids possess hairy euplantulae, and only very few species of Dictyoptera, Plecoptera and Phasmatodea possess hairy attachment structures, the hairy structures within Dermaptera alone arose independently multiple times (Haas & Gorb, 2004). Polyneoptera is not the only striking example for convergent presence of the primary types of attachment structures, and additionally functional microstructures on the adhesive devices of many groups have converged also (Grohmann et al., 2015).

Smooth attachment pads are not always absolutely smooth. Sometimes, they can bear surface microstructures with particular functions (Grohmann et al., 2015). In many polyneopteran species, the attachment pads have been described as being smooth, but they are covered with cuticular patterns or protuberances (Grohmann et al., 2015). These outgrowths have been differentiated from setae/acantae because of their low aspect ratio (height-to-width ratio). In contrast to hairy (seta-like) protuberances with high aspect ratios (greater than 10), many species exhibit smaller sized cuticular nubs (aspect ratio usually <5). Nubs and other surface patterns are reported to be adaptations for tuning the contact conformation of smooth attachment pads to specific substrate conditions (Bußhardt et al., 2012; Grohmann et al., 2015; Gottardo et al., 2015; Büscher & Gorb, 2019). Similar attachment microstructures

(AMS) are convergently found in different polyneopteran groups, in species with similar ecology (Grohmann et al., 2015; Büscher & Gorb, 2017; Büscher et al., 2018a, b, 2019). In general, because of the lack of broad comparative studies of many taxa with smooth pads, the distribution of different AMS within Polyneoptera is not well resolved. However, stick and leaf insects (Phasmatodea) recently became identified as a fascinating model group for attachment-related evolutionary questions.

10.4 Adhesive Microstructures in Phasmatodea

Stick and leaf insects are an impressive model group for the exploration of many evolutionary questions, especially convergence. Limited spatial dispersion, and extensive adaptive radiation, have led to a high degree of convergent traits in Phasmatodea, such as those related to visual camouflage (Bradler, 2003, 2009, 2015; Bradler & Buckley, 2018), oviposition techniques (Carlberg, 1983, 1987; Sellick, 1997a, b; Goldberg et al., 2015; Robertson et al., 2018; Büscher et al., 2019), different degrees of wing loss (Bradler et al., 2003; Whiting et al., 2003; Zeng et al., 2019), and ecomorphs (morphological forms with similar ecological occupancy) with specific vertical stratification within the vegetation (Büscher et al., 2018a; Buckley et al., 2010; Bank et al., 2021). Phasmids are predominantly nocturnal insects that are distributed nearly worldwide and exclusively feed on plants (Bradler, 2003, 2009, 2015; Bradler & Buckley, 2018; Robertson et al., 2018; Brock et al., 2020). Their greatest diversity occurs in the tropics. The majority of species inhabit shrubs and trees of most tropical and temperate ecosystems (Günther, 1953; Bedford, 1978). As suggested by their name, many stick and leaf insects are impressively well camouflaged in these environments and visually blend with their surroundings due to their outward appearance (Bradler, 2009, 2015; Bradler & Buckley, 2018; Robertson et al., 2018; Cumming et al., 2020). This visual camouflage evolved prior to the emergence of angiosperms, when gymnosperms represented the majority of plant diversity (Wang et al., 2014; Yang et al., 2020). The oldest fossil occurrence of stick insects dates back to 165 mya (Jurassic) and morphological specializations related to mimicry were already present (Yang et al., 2020). Another fossil stick insect from the Cretaceous, *Cretophasmomima melanogramma* Wang et al. (2014), had already impressively copied the visual appearance of plants. Its tegminal colouration visually mimics characteristics of the gymnosperm *Membranifolia admirabilis* Sun & Zheng, 2001, comprising a common component of the Cretaceous flora of the same formation (Wang et al., 2014). Subsequently phasmids and plants probably co-radiated, when stick insects began to imitate their floral surroundings to avoid predators (Wedmann et al., 2007; Buckley et al., 2009). During the emergence of angiosperms, and their major radiation (Bell et al., 2010; Magallón & Castillo, 2009), stick insects evolved at a similar pace (Buckley et al., 2009, 2010; Bradler et al., 2015; Goldberg et al., 2015; Simon et al., 2019), possibly in response to the burgeoning diversity of plants and their corresponding adaptations (Robertson

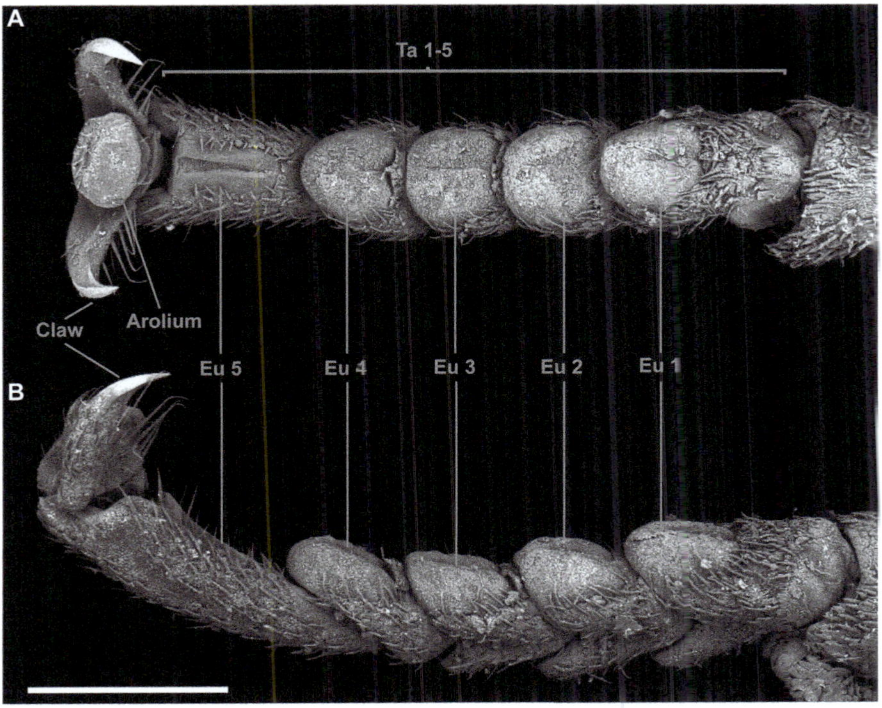

Fig. 10.4 Scanning electron microscopy (SEM) images of a typical phasmatodean tarsus. *Orestes draegeri* Bresseel & Constant, 2018. (**a**) Ventral view. (**b**) Lateral view. Ta1–5, tarsomeres; Eu1–5, euplantulae. Scale bars: 1 mm. Reproduced from Büscher and Gorb (2021)

et al., 2018; Büscher et al., 2020b, c). This not only resulted in a strong host-specific mimicry response for many recent phasmids, but also led to several counter-adaptations against herbivory on the plant side (e.g. Knoll, 1926, 1930; Vogel, 1965; Singer, 2002). While plants evolved defensive strategies to repel the herbivorous stick insects, the latter evolved strategies to overcome these (Peccoud & Simon, 2010; Åhman, 1990; Brennan & Weinbaum, 2001; Brennan et al., 2001; Eigenbrode & Espelie, 1995; Federle, 1999; Rebora et al., 2020).

The ongoing arms race between stick insects and their host plants led to adaptations in the attachment system of phasmids to the features of different plant surfaces (Bußhardt et al., 2012; Büscher & Gorb, 2017, 2019; Büscher et al., 2018a, b, 2019), which, in turn, resulted in the strong association between plants and herbivorous insects. A typical phasmatodean tarsus (Fig. 10.4) consists of five tarsomeres. It is equipped with two claws and an arolium on the pretarsus, as well as euplantulae on the proximal four to five tarsomeres (Beutel & Gorb, 2001, 2006; Vallotto et al., 2016; Büscher & Gorb, 2017, 2019; Büscher et al., 2019). Except for the euplantulae of some members of the Aschiphasmatini (Aschiphasmatinae) that are covered with adhesive setae (Büscher et al., 2019), all attachment pads of Phasmatodea are

Fig. 10.5 Diversity of stick insect ecomorphs and their respective euplantular attachment microstructures. (**a**) *Eurycantha calcarata*, female, smooth. (**b**) *Dajaca monilicornis*, male, hexagonal. (**c**) *Necroscia annulipes*, female, flat pads. (**d**) *Pseudophasma velutinum*, female, small nubs. (**e**) *Leiophasma* sp., couple, maze. (**f**) *Megacrania phelaus*, couple, ridges. (**g**) *Orestes mouhotii*, female, long nubs. (**h**) *Timema* sp., couple, acanthae, image provided by S. Büsse. (**i**) *Dinophasma saginatum*, female, hairy. Scale bars: 1 cm. Modified after Büscher et al. (2018a). Reproduced from Büscher and Gorb (2021)

smooth, but covered with functional surface microstructures. The arolia are rather uniform in their morphology, but their surface microstructure reflects the basal sister-group relationship of *Timema* and Euphasmatodea (all remaining Phasmatodea). The arolia of *Timema* are covered with acanthae (unicellular cuticular outgrowths). According to Richards and Richards, (1979) those of Euphasmatodea are completely smooth (Beutel & Gorb, 2001, 2006).

The euplantulae, that is the proximal tarsal attachment pads, reveal a great diversity of attachment microstructures (AMS), which is suggested to be of functional relevance in terms of adaption to surfaces found in the species-specific environments (Beutel & Gorb, 2001, 2006; Bußhardt et al., 2012; Büscher et al., 2018a, b, 2019). A comparative study of a large number of stick insect species yielded twelve different types of AMS on the euplantulae (Fig. 10.5), including one lineage with adhesive setae (Büscher et al., 2019). Previous studies hypothesized a phylogenetic signal in the characters of the tarsal attachment system. Those were discussed regarding the relationships of Phasmatodea within insects (Beutel & Gorb, 2001, 2006).

Other studies attempted to make use of AMS for the reconstruction of the internal systematics of Phasmatodea (Gottardo, 2011; Gottardo & Heller, 2012; Büscher & Gorb, 2017). These studies revealed distinctive features above the species level in the form of the AMS of the euplantulae; however, these features apparently represent adaptations to similar habitats and do not indicate phylogenetic relationships, as suggested by character mapping based on molecular data (Büscher et al., 2018a).

The distribution of the euplantular AMS suggests a high correlation between the microstructure and the habitat of the species. Ground-dwelling stick insects from different clades reveal convergent nubby microstructures, and many unrelated canopy-dwelling species possess smooth structures, without cuticular patterns on the surface (Büscher et al., 2018a). The various types of AMS within phasmids stand out in comparison to other insects, judging by the diversity of attachment structures reported in the literature (Friedemann et al., 2014a; Haas & Gorb, 2004; Nelson, 2009). The great degree of convergence in the AMS of phasmids probably indicates adaptations to the surfaces encountered in the environment (Büscher et al., 2018a), as hypothesized in the literature (Bußhardt et al., 2012, Büscher & Gorb, 2017, 2019, Büscher et al., 2018a, b, 2019). The disparity of AMS among the phylogenetic lineages, however, does not reveal the clear clustering of species with the same AMS as suggested by previous authors (Beutel & Gorb, 2001; Gottardo, 2011; Gottardo & Heller, 2012; Gottardo et al., 2015; Büscher & Gorb, 2017). The convergent presence of the same microstructures, in contrast, is a result of similar demands for adhesion in the respective habitats, which means that the physical rules of contact mechanics have a very strong influence on the adaptive evolution of attachment structures in general. The reason is that similar AMS provide similar properties to optimize attachment to particular surfaces. Different AMS follow specific functional principles, which are beneficial in specific environments and therefore occur convergently within phasmids faced with similar eco-morphological demands.

10.5 Versatile Adaptability Promotes Convergences

Stick and leaf insects evidently evolved in close association with the evolution of plants (Wang et al., 2014). The extensive diversity of AMS indicates a potential for rapid evolution. The versatile solutions to different attachment problems arose convergently (Büscher et al., 2019). As the degree of convergence in the AMS is extensive and different AMS types coexist in the same groups, the adaptation to the corresponding natural surfaces probably took place in a comparatively short period of time (Büscher et al., 2018b). Using a mathematical model, the potential of self-assembly of the structures observed in the AMS of phasmids have recently been evaluated, based on a reaction-diffusion-model considering a two-morphogen-interaction. The self-formation of different patterns in nature can be explained by Alan Turing's reaction–diffusion model (Turing, 1952). This model has previously been employed to model similar patterns on insects as well as to investigate evolutionary scenarios, for example the patterns of nano-coatings on the corneae of different lineages (Blagodatski et al., 2015). Employing this mathematical model to assess self-formation and transformations of the euplantular AMS of phasmids yielded the prediction of stable patterns of functional AMS on the euplantulae of phasmids. The transitions observed in the simulations were used to evaluate adaptability of the structures, transitions between the structural patterns that could reflect the evolutionary processes, and to re-evaluate the potential ancestral state of stick insect AMS and suggest a rapid response and versatile adaptability of the AMS over a relatively short evolutionary time period (Büscher et al., 2018b). The Turing model indicates a fast response, when animals face changes in surface composition, contributing to flexible adaptability of the functionalized attachment surfaces. Similar changes in the surface geometry of functional microstructures have previously been shown to have arisen within less than 5000 years (Kryuchkov et al., 2017a, b). The ecomorphological specialization, influenced by the ability of the insects to securely attach to the surface of a specific plant, contributes to their specialization to the plant (Kennedy, 1986). Nevertheless, this also plays a role in host fidelity and, potentially, even speciation. Although the acquisition of flight supposedly induced the diversification of attachment structures in insects (e.g. Beutel & Gorb, 2001), it is likely that in phasmids the convergent loss of flight ability (Bradler et al., 2003; Vallotto et al., 2016) enhanced host plant dependence. Adaptation to specific plant surfaces, due to strong coevolution with plants, enhanced diversity even more. This resulted in frequent independent origins of the same AMS in different lineages of phasmids, and aided in achievement of the adhesive properties demanded by their respective environments. Consequently, the convergent presence of the same AMS is primarily a result of the same environmental conditions and the similar physical constraints required for attachment.

10.6 General Functional Principles for Adhesion

Studies of different groups of insects have shown that claws generally contribute to attachment on rough surfaces due to friction and mechanical interlocking (Betz, 2002; Bullock & Federle, 2011a; Zurek et al., 2017; Cartmill, 1979, 1985; Dai et al., 2002; Bußhardt et al., 2014; Pattrick et al., 2018). The performance of claws depends on the radius of the claw tip in relation to the curvature of the surface irregularities (Betz, 2002; Dai et al., 2002; Federle et al., 1997; Song et al., 2016). However, in combination with claws, attachment pads provide adhesion to surfaces with different roughnesses (Song et al., 2016; Büscher & Gorb, 2019). This ability has numerous contact mechanical demands (called "functional principles" below), which evolved under similar boundary conditions in different groups and hence reveal convergent results. Below we discuss the following functional principles: (1) Adaptation to fractal substrate surfaces due to hierarchical organization and a thin surface layer, (2) Surface pattern and contact splitting, (3) Pad material (structure) that is soft in compression but strong in tension, (4) Anisotropy in fibre orientation, and (5) Presence of a fluid in the contact area.

Adaptation to Fractal Substrate Surfaces Due to Hierarchical Organization Hairs with high aspect ratios within hairy systems, and internal fibers/filaments of smooth systems, bend during establishment of contact with the substrate (Fig. 10.6a, b). The pad can therefore work as a damper under high- speed-deformation episodes during jumping or landing. More importantly, in terms of contact mechanics, the deformability functions as a basis for replicating a complex substrate profile during contact formation. Also, the hair- or rod-like organization of the pad architecture allows independent local load distribution over the area of contact between pad and substrate. This aids in an enhancement of the adaptation of the pad to surface

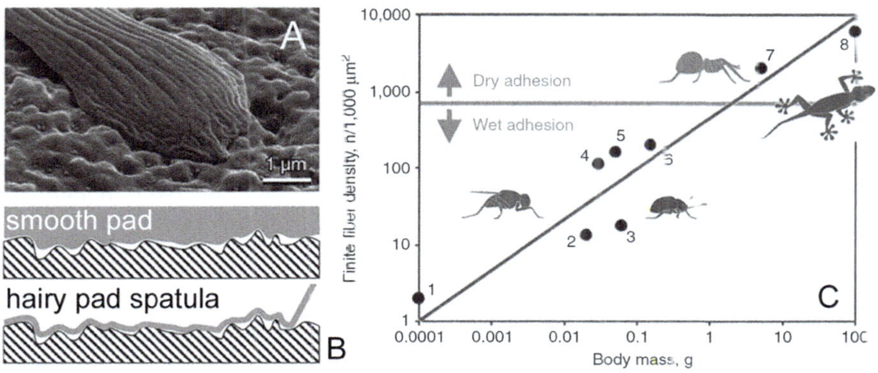

Fig. 10.6 Compliance of adhesive structures with the substrate (**a**, **b**) and contact splitting (**c**). (**a**) Contact of a spatula of the beetle *Gastrophysa viridula* with a micro-rough surface. (**b**) A soft smooth pad requires additional load to form adhesive contact (**b**, upper image), whereas adhesion interaction pulls the elastic thin film of the spatula into complete contact with the rough substrate surface [From Gorb (2011)]. (**c**) Dependence of the contact density of terminal contacts on the body mass in fibrillar pad systems in representatives from diverse animal groups: 1, 2, 4, 5, flies; 3 beetle; 6 bug; 7 spider; 8 gekkonid lizard [Adapted from Scherge and Gorb (2001)]. The systems, situated above the solid horizontal line, preferentially rely on van der Waals forces (dry adhesion), whereas the rest rely mostly on capillary and viscous forces (wet adhesion). Composition reproduced from Büscher and Gorb (2021)

irregularities of non-smooth natural substrata. However, the convergent basic architecture of the pads can be tuned in the evolution of individual animal groups and species in association with different needs and constraints. The structural principle, based on branching rods, in a smooth pad may, for example, additionally contribute to maintaining the shape of the pad.

An important structural feature of both types of attachment pads is the presence of a thin superficial film contributing to the compliance of hairy and smooth attachment devices. In smooth systems the epicuticle covers the fibrous material of the pad, and spatulae forming the tips of the cuticlar outgrowths of hairy systems form the superficial film. These films are responsible for proper contact formation with the substrate due to their low bending stiffness at the minimum of normal load (Persson & Gorb, 2003). The film/spatula is able to conform to the surface profile and to replicate surface irregularities over certain length scales. The range of length scales to which adaptation is possible depends upon the stiffness of the film. Spatulae are able to adapt to even nanoscale roughness (Eimüller et al., 2008; Gorb, 2011). Thick films within the smooth pads of the bush cricket, *Tettigonia viridissima* and the locust, *Locusta migratoria* (Gorb & Scherge, 2000; Perez-Goodwyn et al., 2006) adapt to the microscale roughness: however, the latter species has lesser adaptability to surface roughness because of the much thicker superficial film compared to that of *T. viridissima*. In smooth pads, films terminating fibres which are sometimes of an extreme high aspect ratio, prevent lateral collapse (condensation, conglutination) of fibres (Jagota & Bennison, 2002; Spolenak et al., 2005a, b). They would otherwise agglomerate with each other and not work as separate springs (Schargott et al., 2006). The film also delimits the smooth pad as reservoir filled with fluid and, under certain pressure, maintains the pad as a stable unit (Gorb, 2008a, b). The thicker superficial film of desert species may also minimize water loss (Perez-Goodwyn et al., 2006) and presumably prevents fragile fibrous material from wearing out during walking (Slifer, 1950; Kendall, 1970). However, in thicker/stiffer films, the adhesive properties will concomitantly be reduced because of the reduced ability to make close contact with rough substrata.

Surface Pattern and Contact Splitting The function of hairs/setae in hairy pads has partially been discussed in the previous paragraphs. Based on the studies of different animal groups with hairy pads, an interesting correlation between the geometrical properties of setal tips and animal weight was found (Fig. 10.6c): the heavier the animals, the smaller and more densely packed are the terminal contact elements (Scherge & Gorb, 2001). This scaling effect was explained by introducing the principle of contact splitting, according to which splitting up the contact into finer subcontacts increases adhesion on a flat substrate (Johnson et al., 1971; Arzt et al., 2003). This relationship holds because animals cannot increase the area of attachment devices in proportion to body weight due to the different scaling rules for mass and surface area. Therefore, the increase of attachment strength in hairy systems is achieved by increasing the number of single contact points, that is, by increasing the hair/setal density. However, it was later shown that, for multiple reasons, this trend varies within each single lineage of organisms (Peattie & Full, 2007). The fundamental importance of contact splitting for adhesion to rough substrates has been explained by a small effective elastic modulus of the array of hairs (Persson, 2003) contributing to the formation of larger real contact area under the same applied load compared to unstructured material. Data from broad comparative analyses of different animals suggest that animal lineages that rely on dry adhesion (lizards, spiders, mites) exhibit a much higher density of terminal contact elements (of smaller size) than systems using a wet adhesive mechanism (insects).

Additionally, most smooth pads are not ideally smooth but are rather wrinkled or, in some cases, patterned at micron or submicron levels (Gorb, 2008a; Büscher et al., 2019). The upper sides of surface patterns in contact with a substrate may approach their counterpart surface very closely. In this case, solid–solid interactions occur between pad material and substrate. Under a particular load a fluid is pressed out of the contact into the gaps between outgrowths. The non-ideal smooth surface of the pad, similar to a tyre profile, prevents aquaplaning and enhances solid–solid interactions, which are not only important for adhesion enhancement due to van der Waals forces, but also for enhancement of friction (Varenberg & Gorb, 2009). Fluid trapped in gaps might be additionally used in the next step cycle. Prevention of aquaplaning is especially important for walking on wet surfaces in rain forest or temperate area environments. Also nubby pad microstructures can generate additional frictional grip on rough surfaces (Bußhardt et al., 2012; Büscher & Gorb, 2019).

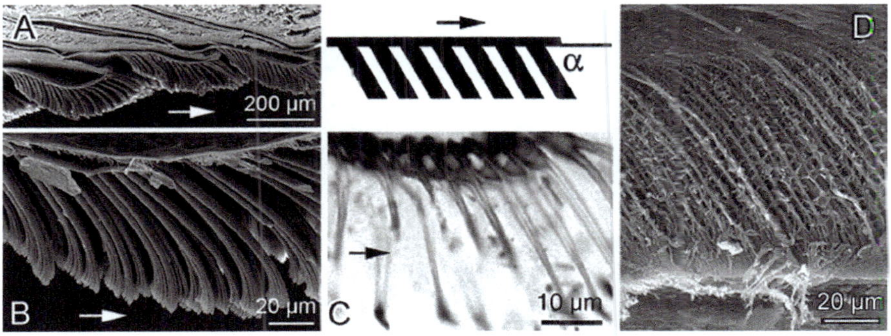

Fig. 10.7 Convergent evolution of an asymmetry of structural features (scheme is given in the inset) of animal attachment pads, leading to functional anisotropy in contact. Longitudinal sections of pads are visualised by SEM (**a**, **b**, **d**) and by light microscopy (**c**). (**a**, **b**). Tokay *Gekko gecko* (hairy pad). (**c**) Fly *Calliphora vicina* (hairy pad). (**d**) Bush cricket *Tettigonia viridissima* (smooth pad). The arrows indicate distal direction in all pads. (**a**–**c**). From Gorb (2011). (**d**) From Gorb and Scherge (2000). Composition reproduced from Büscher and Gorb (2021)

Pad Material (Structure) that Is Soft in Compression But Strong in Tension It is well known that an array of thin fibres is soft in compression, but exceptionally strong in tension (Neville, 1993). This is the key principle of both pad architectures: hairy and smooth (Fig. 10.7). The specific external (hairs/setae) or internal (fibres/foam) structure of attachment pads is not only responsible for their softness in compression, but also for stiffness in tension. The specific orientation of thin hairs or fibres aligned with the direction of tensile forces, acting on the pad in contact, aids in resisting such forces, when the animal is attached to the ceiling or wall, or sliding along the substrate. The relatively high tensile strength of a soft material would not be possible without such a fibre-like reinforcement. Fibrillar organisation of smooth pads represents their main structure-functional similarity to hairy pads.

Anisotropy in Fibre Orientation Since fibres are normally not oriented perpendicularly to the pad surface (Fig. 10.7), but rather at some angle (45–60°) and sloped in the distal direction, they do not buckle but rather bend under load, which makes the pad material even more flexible. Structural anisotropy of the pad material is also responsible for frictional anisotropy (Gorb & Scherge, 2000). Friction is greatest while the pad is sliding in a proximal direction because the fibres of smooth pads or hairs of hairy pads can be more easily recruited in this case. Such a mechanism may secure a stable position of an animal on the ceiling. As the shear forces are applied proximally, towards the body, in such a situation, and because of greater friction in this direction due to intimate contact between the membranes/spatulae, pad sliding can be prevented. Fibre anisotropy may also be additionally involved in the detachment mechanism of the pad (Niederegger & Gorb, 2003).

Presence of Fluid in the Contact Area Fluid is reported to be secreted into the contact area in the smooth pads of cockroaches (Roth & Willis, 1952), orthopterans (Jiao et al., 2000; Vötsch et al., 2002), aphids (Lees & Hardie, 1988; Dixon et al.,

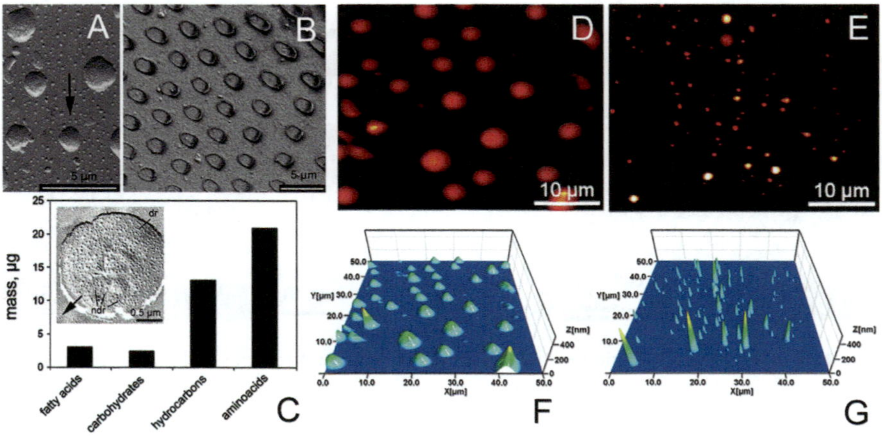

Fig. 10.8 Fluid in animal attachment pads. (**a**) SEM image of a carbon-platinum replica of frozen and coated droplets of the fly *Calliphora vicina* (black arrow indicates direction of coating). Note pattern of nanodrops on the surface of the major droplets [From Gorb (2001)]. (**b**) Menisci formed around single terminal contact elements of the setae of *C. vicina*. The fly's leg was frozen in contact with smooth glass, carefully removed, and the fluid residues were viewed using a cryo-SEM [From Gorb (2011)]. (**c**) Chemical composition (absolute concentration of substance groups) of the pad secretion of the euplantulae of *Locusta migratoria* [From Vötsch et al. (2002)]. (**d**–**g**) Atomic force microscopy (AFM) height images of the footprint droplets of the beetle *Coccinella septempunctata* (**d**, **f**) and the fly *Calliphora vicina* (**e**, **g**). (**d**) and (**e**) share the same colour scale. Brighter pixels correspond to higher z values. (**f**, **g**) are three-dimensional impressions of the images shown in (**d**) and (**e**), respectively. From Peisker and Gorb (2012). Composition reproduced from Büscher and Gorb (2021)

1990), pentatomid bugs (Hasenfuss, 1977, 1978, Ghasi-Bayat & Hasenfuss, 1980a, b) and the hairy pads of reduviid bugs (Edwards & Tarkanian, 1970), flies (Bauchhenss, 1979; Walker et al., 1985), coccinellid (Ishii, 1987; Kosaki & Yamaoka, 1996), and chrysomelid (Eisner & Aneshansley, 2000) beetles. Footprints can be easily observed with the light microscope, especially under phase contrast. The hairy pad secretion has been chemically studied mostly for representatives of the Coleoptera. It contains a non-volatile, lipid-like substance that can be observed in footprints stained with Sudan Black. It has been shown that the pad adhesive secretion of ladybird beetles (Coccinellidae) consists of hydrocarbons and true waxes (Ishii, 1987, Kosaki & Yamaoka, 1996), which correspond well with the composition of the covering of the cuticle. Similar data have been obtained for the chrysomelid beetle *Hemisphaerota cyanea* (Chrysomelidae, Cassidinae) (Attygalle et al., 2000). In smooth insect pads the secretion consists of a water-soluble and a lipid-soluble part (Vötsch et al., 2002). Data obtained from shock-freezing, carbon–platinum coating, and replica preparation show that the secretory droplets contain nano-droplets on their surfaces (Fig. 10.8). These results led authors to suggest that the pad secretion is an emulsion consisting of lipoid nanodroplets dispersed in an aqueous liquid.

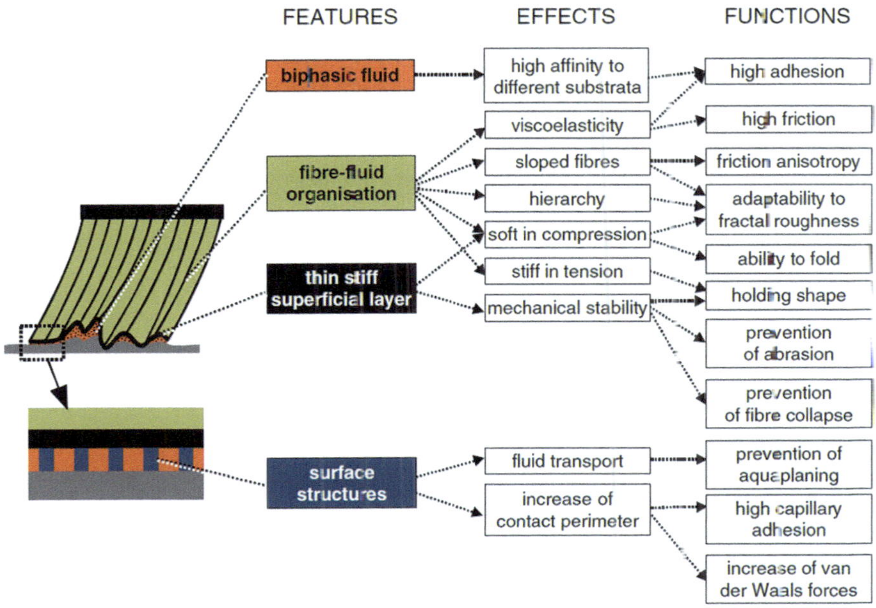

Fig. 10.9 Diagram summarizing convergently evolved structural features of smooth attachment pads responsible for particular functional effects. From Gorb (2008a)

The fluid within the smooth pad contributes to the viscoelastic behaviour of the pad because the fluid is able to flow through the gaps between the rods when the pad deforms (Gorb et al., 2000; Gorb & Scherge, 2000). Fluid, which is pressed out of the smooth pad or from the insect hairy pads into the contact area, may serve several functions (Fig. 10.9). It can enhance the contact initialisation due to capillary forces, which represent long-range interactions. The capillary forces themselves contribute to adhesion (Langer et al., 2004). Fluid can also fill nano-scale gaps on the surface and thus improve contact conformation on non-smooth substrata. Since the fluid consists of two phases it has higher affinity for substrata with various physico-chemical properties (hydrophilic and hydrophobic). In other words, the fluid may be a kind of coupling agent, promoting and strengthening adhesion between otherwise incompatible materials by providing the proximity of contact for intermolecular forces. At relatively high separation the contribution of viscous forces to adhesion and friction will be enhanced due to the presence of a thin fluid layer in the contact area (Scherge & Gorb, 2001).

10.7 Transfer to Biomimetics

Since the structure-function relationships discussed above are based on fundamental physical principles and are mostly related to the geometry of the structure, they must also hold for artificial surfaces with a similar geometry. This, in turn, means that ideas from biology can be potentially used for engineering applications (Fig. 10.10). Since convergent events are indications of a kind of optimal solution, or even a single solution developed in the course of biological evolution, broad comparative

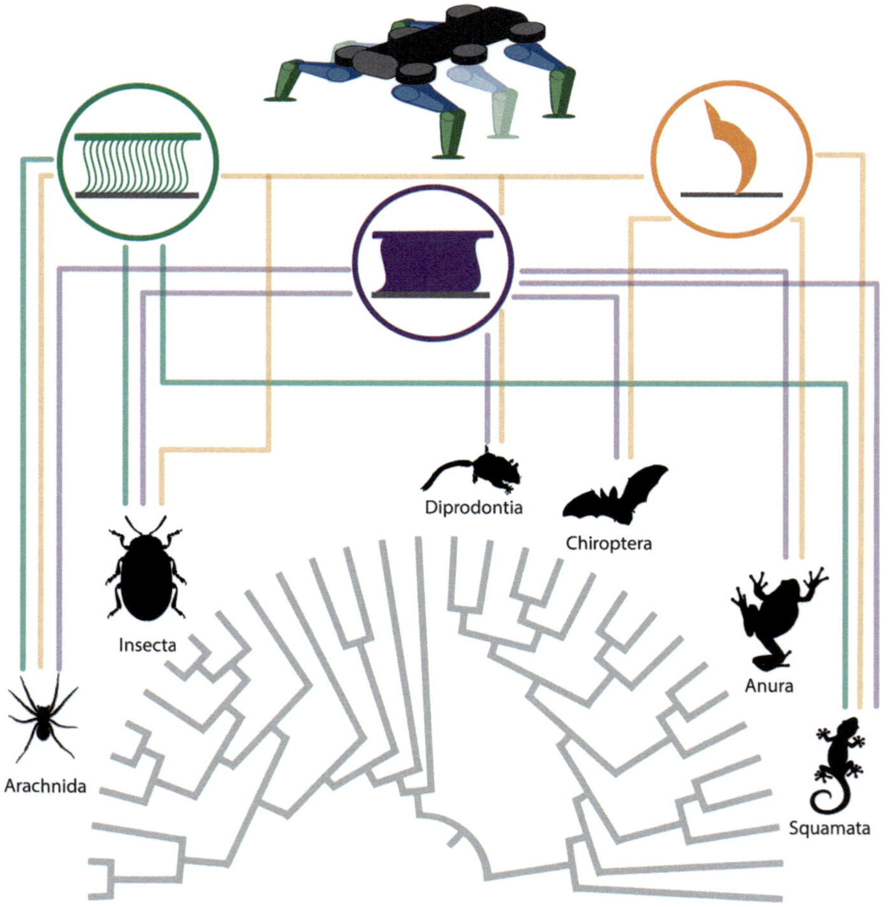

Fig. 10.10 Sources of bioinspiration for attachment systems from the animal tree of life. Shown are examplar groups in which attachment systems occur, as depicted in Fig. 10.1. The functional aspects (1) hairy attachment pads (green), (2) smooth attachment pads (purple) and claws (orange), that can inspire technical applications for a four or six legged robot are linked by colour mapping to the groups, which include species that represent examples of the mechanism and may serve as biological sources of inspiration. Reproduced from Büscher and Gorb (2021)

studies of animal attachment devices can be a great approach for further biomimetic innovations. For example, further research on the material-function-relationship of the attachment pads can be useful for technical applications for artificial attachment systems with either surface-specific use or for providing universal solutions for unpredictable surfaces. The dependence of the adhesive devices on leg movements and body kinematics can be useful for basic research and applications in the field of robotics (Full & Tu, 1991; Dirks et al., 2012; Jayaram & Full, 2016; Di Canio et al., 2016). In return, robotic systems can provide insights into the regulation and temporal resolution of attachment, which can solidify experimental results. Characterization of attachment solutions in nature can, furthermore, support the development of bioinspired gripping devices (Thor et al., 2018, Ignasov et al., 2018, Gorb et al., 2007. Voigt et al., 2012, b).

References

Åhman, I. (1990). Plant-surface characteristics and movements of two *Brassica*-feeding aphids, *Lipaphis erysimi* and *Brevicoryne brassicae*. *Symposia Biologica Hungarica, 39*, 119–125.

Al Bitar, L., Voigt, D., Zebitz, C. P., & Gorb, S. N. (2009). Tarsal morphology and attachment ability of the codling moth *Cydia pomonella* L. (Lepidoptera, Tortricidae) to smooth surfaces *Journal of Insect Physiology, 55*, 1029–1038.

Al Bitar, L., Gorb, S. N., Zebitz, C. P., & Voigt, D. (2012). Egg adhesion of the codling moth *Cydia pomonella* L. (Lepidoptera, Tortricidae) to various substrates: I. Leaf surfaces of different apple cultivars. *Arthropod-Plant Interactions, 6*, 471–488.

Al Bitar, L., Gorb, S. N., Zebitz, C. P., & Voigt, D. (2014). Egg adhesion of the codling moth *Cydia pomonella* L. (Lepidoptera, Tortricidae) to various substrates: II. Fruit surfaces of different apple cultivars. *Arthropod-Plant Interactions, 8*, 57–77.

Arnold, J. (1974). Adaptive features on the tarsi of cockroaches (Insecta: Dictyoptera). *International Journal of Insect Morphology and Embryology, 3*, 317–334.

Arzt, E., Gorb, S., & Spolenak, R. (2003). From micro to nano contacts in biological attachment devices. *Proceedings of National Academy of Sciences of the United Science of America, 100*, 10603.

Attygalle, A. B., Aneshansley, D. J., Meinwald, J., & Eisner, T. (2000). Defence by foot adhesion in a chrysomelid beetle (*Hemisphaerota cyanea*): Characterization of the adhesive oil *Zoology, 103*, 1–6.

Autumn, K. (2006). How gecko toes stick. *American Scientist, 94*, 124–132.

Autumn, K., & Peattie, A. M. (2002). Mechanisms of adhesion in geckos. *Integrative and Comparative Biology, 42*, 1081–1090.

Autumn, K., Liang, Y. A., Hsieh, S. T., Zesch, W., Chan, W. P., Kenny, T. W., Fearing, R., & Full, R. J. (2000). Adhesive force of a single gecko foot-hair. *Nature, 405*, 681–685.

Bank, S., Buckley, T. R., Büscher, T. H., Bresseel, J., Constant, J., De Haan, M., Dittmar, D., Dräger, H., Kahar, R. S., Kang, A., Kneubühler, B., Langdon-Myers, S., & Bradler, S. (2021). Reconstructing the nonadaptive radiation of an ancient lineage of ground-dwelling stick insects (Phasmatodea: Heteropterygidae). *Systematic Entomology, 46*, 487–507.

Ba-Omar, T. A., Downie, J. R., & Barnes, W. J. P. (2000). Development of adhesive toe-pads in the tree-frog (*Phyllomedusa trinitatis*). *Journal of Zoology London, 250*, 267.

Bauchhenss, E. (1979). Die Pulvillen von *Calliphora erythrocephala* Meig. (Diptera, Brachycera) als Adhäsionsorgane. *Zoomorphologie, 93*, 99–123.

Bauchhenss, E., & Renner, M. (1977). Pulvillus of *Calliphora erythrocephala* Meig. (Diptera; Calliphoridae). *International Journal of Insect Morphology and Embryology, 6*, 225–227.
Bedford, G. O. (1978). Biology and ecology of the Phasmatodea. *Annual Review of Entomology, 23*, 125–149.
Bell, C. D., Soltis, D. E., & Soltis, P. S. (2010). The age and diversification of the angiosperms re-revisited. *American Journal of Botany, 97*, 1296–1303.
Betz, O. (2002). Performance and adaptive value of tarsal morphology in rove beetles of the genus *Stenus* (Coleoptera, Staphylinidae). *The Journal of Experimental Biology, 205*, 1097–1113.
Betz, O. (2003). Structure of the tarsi in some *Stenus* species (Coleoptera, Staphylinidae): External morphology, ultrastructure, and tarsal secretion. *Journal of Morphology, 255*, 24–43.
Betz, O. (2006a). Ecomorphology: Integration of form, function, and ecology in the analysis of morphological structures. *Mitteilungen der Deutschen Gesellschaft für Allgemeine und Angewandte Entomologie, 15*, 409–416.
Betz, O. (2006b). Der Anpassungswert morphologischer Strukturen: Integration von Form, Funktion und Ökologie am Beispiel der Kurzflügelkäfer-Gattung *Stenus* (Coleoptera, Staphylinidae). *Entomologie heute, 18*, 3–26.
Beutel, R. G., & Gorb, S. N. (2001). Ultrastructure of attachment specializations of hexapods (Arthropoda): Evolutionary patterns inferred from a revised ordinal phylogeny. *Journal of Zoological Systematics and Evolutionary Research, 39*, 177–207.
Beutel, R. G., & Gorb, S. N. (2006). A revised interpretation of the evolution of attachment structures in Hexapoda with special emphasis on Mantophasmatodea. *Arthropod Systematics and Phylogeny, 64*, 3–25.
Beutel, R. G., & Gorb, S. N. (2008). Evolutionary scenarios for unusual attachment devices of Phasmatodea and Mantophasmatodea (Insecta). *Systematic Entomology, 33*, 501–510.
Beutel, R. G., Wipfler, B., Gottardo, M., & Dallai, R. (2013). Polyneoptera or "Lower Neoptera"-- new light on old and difficult phylogenetic problems. *Atti Accademia nazionale italiana di entomologia, 61*, 113–142.
Blagodatski, A., Sergeev, A., Kryuchkov, M., Lopatina, Y., & Katanaev, V. L. (2015). Diverse set of Turing nanopatterns coat corneae across insect lineages. *Proceedings of the National Academy of Sciences of the United States of America, 112*, 10750–10755.
Bohart, R. M., & Menke, A. S. (1976). *Sphecid wasps of the world*. University of California Press.
Börner, C. (1919). Stammesgeschichte der Hautflügler. *Biologisches Zentralblatt, 39*, 145–186.
Boudreaux, H. B. (1979). *Arthropod phylogeny with special reference to insects*. Wiley.
Bradler, S. (2003). Phasmatodea, Gespenstschrecken. In H. H. Dathe (Ed.), *Kaestner, Lehrbuch der speziellen Zoologie, I, 5* (pp. 251–261). Gustav Fischer/Spektrum Akademischer Verlag.
Bradler, S. (2009). Phylogenie der Stab- und Gespenstschrecken (Insecta: Phasmatodea). *Species, Phylogeny and Evolution, 2*, 3–139.
Bradler, S. (2015). The Phasmatodea tree of life: Surprising facts and open questions in the evolution of stick and leaf insects. *Entomologie heute, 27*, 1–23.
Bradler, S., & Buckley, T. R. (2018). Biodiversity of Phasmatodea. In R. G. Foottit & P. H. Adler (Eds.), *Insect biodiversity: Science and society* (Vol. 2, pp. 281–313). Wiley.
Bradler, S., Whiting, M. F., & Klug, R. (2003). Basal Diversification and the evolution of wings within stick insects (Phasmatodea). *Entomologische Abhandlungen, 61*, 132–133.
Bradler, S., Cliquennois, N., & Buckley, T. R. (2015). Single origin of Mascarene stick insects: Ancient radiation on sunken islands? *BMC Evolutionary Biology, 15*, 196.
Brainerd, E. (1994). Adhesion force of ants on smooth surfaces. *American Zoologist, 34*, 128A.
Breidbach, O. (1980). Die Tarsen von Insekten - ein schönes Beispiel von Konvergenz. *Mikrokosmos, 69*, 200–201.
Brennan, E. B., & Weinbaum, S. A. (2001). Effect of epicuticular wax on adhesion of psyllids to glaucous juvenile and glossy adult leaves of *Eucalyptus globulus* Labillardière. *Australian Journal of Entomology, 40*, 270–277.

Brennan, E. B., Hrusa, G. F., Weinbaum, S. A., & Levison, W., Jr. (2001). Resistance of *Eucalyptus* species to *Glycaspis brimblecombei* (Homoptera: Psyllidae) in the San Francisco Bay area. *Pan-Pacific Entomologist, 77*, 249–253.

Brinkhurst, R. O. (1959). Studies on the functional morphology of *Gerris najas* De Geer (Hem. Het. Gerridae). *Proceedings of the Zoological Society of London, 133*, 531–559.

Brock, P. D., Büscher, T. H., & Baker, E. (2020). Phasmida species file (Version 5.0). In Y. Roskov, G. Ower, T. Orrell, D. Nicolson, N. Bailly, P. M. Kirk, T. Bourgoin, R. E. DeWalt W. Decock, E. van Nieukerken, & L. Penev (Eds.), *Species 2000 & ITIS catalogue of life, Leiden: Species 2000*. Naturalis.

Brower, L. P. (1958). Bird predation and foodplant specificity in closely related procryptic insects *The American Naturalist, 92*, 183–118.

Buckley, T. R., Attanayake, D., & Bradler, S. (2009). Extreme convergence in stick insect evolution: Phylogenetic placement of the Lord Howe Island tree lobster. *Proceedings of the Royal Society B, 276*, 1055–1062.

Buckley, T. R., Attanayake, D., Nylander, J. A. A., & Bradler, S. (2010). The phylogenetic placement and biogeographical origins of the New Zealand stick insects (Phasmatodea). *Systematic Entomology, 35*, 207–225.

Bullock, J. M., & Federle, W. (2011a). The effect of surface roughness on claw and adhesive hair performance in the dock beetle *Gastrophysa viridula*. *Insect Science, 18*, 298–304.

Bullock, J. M., & Federle, W. (2011b). Beetle adhesive hairs differ in stiffness and stickiness: in vivo adhesion measurements on individual setae. *Naturwissenschaften, 98*, 381–387.

Büscher, T. H., & Gorb, S. N. (2017). Subdivision of the neotropical Prisopodinae Brunner von Wattenwyl, 1893 based on features of tarsal attachment pads (Insecta, Phasmatodea). *ZooKeys, 645*, 1–11.

Büscher, T. H., & Gorb, S. N. (2019). Complementary effect of attachment devices in stick insects (Phasmatodea). *The Journal of Experimental Biology, 222*, jeb209833.

Büscher, T. H., & Gorb, S. N. (2021). A review: Physical constraints lead to parallel evolution of micro- and nanostructures of animal adhesive pads. *Beilstein Journal of Nanotechnology, 12*, 725–743.

Büscher, T. H., Buckley, T. R., Grohmann, C., Gorb, S. N., & Bradler, S. (2018a). The evolution of tarsal adhesive microstructures in stick and leaf insects (Phasmatodea). *Frontiers in Ecology and Evolution, 6*, 69.

Büscher, T. H., Kryuchkov, M., Katanaev, V. L., & Gorb, S. N. (2018b). Versatility of Turing patterns potentiates rapid evolution in tarsal attachment microstructures of stick and leaf insects (Phasmatodea). *Journal of the Royal Society Interface, 15*, 20180281.

Büscher, T. H., Grohmann, C., Bradler, S., & Gorb, S. N. (2019). Tarsal attachment pads in Phasmatodea (Hexapoda: Insecta). *Zoologica, 164*, 1–94.

Büscher, T. H., Becker, M., & Gorb, S. N. (2020a). Attachment performance of stick insects (Phasmatodea) on convex substrates. *The Journal of Experimental Biology, 223*(17), jeb226514.

Büscher, T. H., Quigley, E., & Gorb, S. N. (2020b). Adhesion performance in the eggs of the Philippine leaf insect *Phyllium philippinicum* (Phasmatodea: Phylliidae). *Insects, 11*(7), 400.

Büscher, T. H., Lohar, R., Kaul, M. C., & Gorb, S. N. (2020c). Multifunctional adhesives on the eggs of the leaf insect *Phyllium philippinicum* (Phasmatodea: Phylliidae): Solvent influence and biomimetic implications. *Biomimetics, 5*(4), 66.

Büsse, S., Büscher, T. H., Kelly, E. T., Heepe, L., Edgerly, J. S., & Gorb, S. N. (2019). Pressure-induced silk spinning mechanism in webspinners (Insecta: Embioptera). *Soft Matter, 15*, 9742–9750.

Bußhardt, P., Wolf, H., & Gorb, S. N. (2012). Adhesive and frictional properties of tarsal attachment pads in two species of stick insects (Phasmatodea) with smooth and nubby euplantulae. *Zoology, 115*, 135–141.

Bußhardt, P., Kunze, D., & Gorb, S. N. (2014). Interlocking-based attachment during locomotion in the beetle *Pachnoda marginata* (Coleoptera, Scarabaeidae). *Scientific Reports, 4*, 1–8.

Byers, G. W. (1991). Mecoptera (Scorpionflies, hangingflies). In CSIRO (Ed.), *The Insects of Australia* (Vol. 2, pp. 696–716). Cornell University Press.

Carlberg, U. (1983). A review of different types of egglaying in the Phasmida in relation to the shape of the eggs and with a discussion on their taxonomic importance (Insecta). *Biologisches Zentralblatt, 102*, 587–602.

Carlberg, U. (1987). Evolutionary and ecological aspects on ovarian diversity in Phasmida. In B. Baccetti (Ed.), *Evolutionary biology of orthopteroid insects* (pp. 174–176). Ellis Horwood Ltd.

Cartmill, M. (1979). The volar skin of primates: Its frictional characteristics and their functional significance. *American Journal of Physical Anthropology, 50*, 497–509.

Cartmill, M. (1985). Climbing. In *Functional vertebrate morphology* (pp. 73–88). Belknap Press.

Carver, M., Gross, G. F., & Woodward, T. E. (1991). Hemiptera (true bugs, leafhoppers, cicadas, aphids, scale insects etc.). In CSIRO (Ed.), *The insects of Australia* (2nd ed., pp. 429–509). Melbourne University Press.

Clemente, C. J., & Federle, W. (2008). Pushing versus pulling: Division of labour between tarsal attachment pads in cockroaches. *Proceedings of the Royal Society B, 275*, 1329–1336.

Clemente, C. J., Goetzke, H. H., Bullock, J. M., Sutton, G. P., Burrows, M., & Federle, W. (2017). Jumping without slipping: Leafhoppers (Hemiptera: Cicadellidae) possess special tarsal structures for jumping from smooth surfaces. *Journal of the Royal Society Interface, 14*, 20170022.

Cruz, M. S., & Mateo, M. M. (2009). Scanning electron microscopy of legs of two species of sucking lice (Anoplura: Phthiraptera). *Micron, 40*, 401–408.

Cumming, R. T., Le Tirant, S., Teemsma, S. N., Hennemann, F. H., Willemse, L., & Büscher, T. H. (2020). Lost lovers linked at long last: elusive female *Nanophyllium* mystery solved after a century of being placed in a different genus (Phasmatodea, Phylliidae). *ZooKeys, 969*, 43.

Dai, Z., Gorb, S. N., & Schwarz, U. (2002). Roughness-dependent friction force of the tarsal claw system in the beetle *Pachnoda marginata* (Coleoptera, Scarabeidae). *The Journal of Experimental Biology, 205*, 2479–2488.

Darnhofer-Demar, B. (1969). Zur Fortbewegung des Wasserläufers *Gerris lacustris* L, auf der Wasseroberfläche. *Zoologischer Anzeiger Supplement, 32*, 430–439.

Dethier, V. G. (1959). Food-plant distribution and density and larval dispersal as factors affecting insect populations. *Canadian Entomologist, 91*, 581–596.

Di Canio, G., Stoyanov, S., Larsen, J. C., Hallam, J., Kovalev, A. E., Kleinteich, T., Gorb, S. N., & Manoonpong, P. (2016). A robot leg with compliant tarsus and its neural control for efficient and adaptive locomotion on complex terrains. *Artificial Life and Robotics, 21*, 274–281.

Dirks, J. H., Li, M., Kabla, A., & Federle, W. (2012). In vivo dynamics of the internal fibrous structure in smooth adhesive pads of insects. *Acta Biomaterialia, 8*, 2730–2736.

Dixon, A., Croghan, P., & Gowing, R. (1990). The mechanism by which aphids adhere to smooth surfaces. *The Journal of Experimental Biology, 152*, 243–253.

Dudley, R. (2000). *The biomechanics of insect flight*. Princeton University of Press.

Edwards, J. S., & Tarkanian, M. (1970). The adhesive pads of Heteroptera: A re-examination. *Proceedings of the Royal Entomological Society of London, Series A, 45*, 1–5.

Ehrlich, P. R., & Raven, P. H. (1964). Butterflies and plants: A study in coevolution. *Evolution, 18*, 586–608.

Eigenbrode, S., & Espelie, K. (1995). Effects of plant epicuticular lipids on insect herbivores. *Annual Review of Entomology, 40*, 171–194.

Eimüller, T., Guttmann, P., & Gorb, S. N. (2008). Terminal contact elements of insect attachment devices studied by transmission X-ray microscopy. *The Journal of Experimental Biology, 211*, 1958–1963.

Eisner, T., & Aneshansley, D. J. (2000). Defense by foot adhesion in a beetle (*Hemisphaerota cyanea*). *Proceedings of the National Academy of Sciences of the United States of America, 97*, 6568–6573.

Endlein, T., & Federle, W. (2008). Walking on smooth or rough ground: Passive control of pretarsal attachment in ants. *Journal of Comparative Physiology. A, 194*, 49–60.

Engel, M. S. (2015). Insect evolution. *Current Biology, 25*, R868–R872.

Farrell, B. D., & Mitter, C. (1994). Adaptive radiation in insects and plants: Time and opportunity. *American Zoologist, 34*, 57–69.

Faucheux, M. J. (1985). Structure of the tarso-pretarsal chordotonal organ in the imago of *Tineola bisselliella* (Lepidoptera, Tineidae). *International Journal of Insect Morphology and Embryology, 14*, 147–154.

Feder, H. M. (1955). On the method used by the starfish *Pisaster ochraceus* in opening three types of bivalve molluscs. *Ecology, 36*(4), 764.

Federle, W. (1999). Significance of epicuticular wax barriers in antplants of the genus *Macaranga* (Euphorbiaceae). *Zoology, 102*, 27.

Federle, W. (2002). Comment on: "Relationships between pretarsus morphology and arboreal life in ponerine ants of the genus *Pachycondyla* (Formicidae: Ponerinae)". *Annals of the Entomological Society of America, 95*, 141–142.

Federle, W., & Endlein, T. (2004). Locomotion and adhesion: Dynamic control of adhesive surface contact in ants. *Arthropod Structure & Development, 33*, 67–75.

Federle, W., Maschwitz, U., Fiala, B., Riederer, M., & Hölldobler, B. (1997). Slippery ant-plants and skilful climbers: Selection and protection of specific ant partners by epicuticular wax blooms in *Macaranga* (Euphorbiaceae). *Oecologia, 112*, 217–224.

Federle, W., Rohrseitz, K., & Hölldobler, B. (2000). Attachment forces of ants measured with a centrifuge: Better "wax-runners" have a poorer attachment to a smooth surface. *The Journal of Experimental Biology, 203*, 505–512

Federle, W., Brainerd, E., McMahon, T., & Hölldobler, B. (2001). Biomechanics of the movable pretarsal adhesive organ in ants and bees. *Proceedings. National Academy of Sciences. United States of America, 98*, 6215–6220.

Federle, W., Riehle, M., Curtis, A., & Full, R. (2002). An integrative study of insect adhesion: Mechanics and wet adhesion of pretarsal pads in ants. *Integrative and Comparative Biology, 42*, 1100–1106.

Flammang, P. (1996). Adhesion in echinoderms. *Echinoderm Studies, 5*(1), 1–14.

Frantsevich, L., & Gorb, S. N. (2002). Arcus as a tensegrity structure in the arolium of wasps (Hymentoptera: Vespidae). *Zoology, 105*, 225–237.

Frantsevich, L., & Gorb, S. N. (2004). Structure and mechanics of the tarsal chain in the hornet, *Vespa crabro* (Hymenoptera: Vespidae): Implications on the attachment mechanism. *Arthropod Structure and Development, 33*, 77–89.

Frantsevich, L., Ji, A., Dai, Z., Wang, J., Fransevich, L., & Gorb, S. N. (2008). Adhesive properties of the arolium of a lantern-fly, *Lycorma delicatula* (Auchenorrhyncha, Fulgoridae). *Journal of Insect Physiology, 54*, 818–827.

Friedemann, K., Schneeberg, K., & Beutel, R. G. (2014a). Fly on the wall-attachment structures in lower Diptera. *Systematic Entomology, 39*, 460–473.

Friedemann, K., Spangenberg, R., Yoshizawa, K., & Beutel, R. G. (2014b). Evolution of attachment structures in the highly diverse Acercaria (Hexapoda). *Cladistics, 30*, 170–201.

Friedemann, K., Kunert, G., Gorb, E., Gorb, S. N., & Beutel, R. G. (2015). Attachment forces of pea aphids (*Acyrthosiphon pisum*) on different legume species. *Ecological Entomology, 40*, 732–740.

Full, R., & Tu, M. (1991). Mechanics of a rapid running insect: Two-, four-, and six-legged locomotion. *The Journal of Experimental Biology, 156*, 215–231.

Gao, H., Wang, X., Yao, H., Gorb, S., & Arzt, E. (2005). Mechanics of hierarchical adhesion structures of geckos. *Mechanics of Materials, 37*, 275.

Ghasi-Bayat, A., & Hasenfuss, I. (1980a). Die Oberflächenstrukturen des Prätarsus von *Elasmucha ferrugata* (Fabricius) (Acanthosomatidae, Heteroptera). *Zoologischer Anzeiger, 205*, 76–80.

Ghasi-Bayat, A., & Hasenfuss, I. (1980b). Über den Transportweg der Haftflüssigkeit der Pulvilli bei *Coptosoma scutellatum* (Geoffr.) (Plataspididae, Heteroptera). *Nachrichtenblatt der Bayerischen Entomologen, 1980*, 58–58.

Gillett, J. D., & Wigglesworth, V. B. (1932). The climbing organ of an insect, *Rhodnius prolixus* (Hemiptera; Reduviidae). *Proceedings of the Royal Society B, 111*, 364–376.
Gillot, C. (2005). *Entomology* (3rd ed.). Springer.
Gnaspini, P., Antunes-Carvalho, C., Newton, A. F., & Leschen, R. A. (2017). Show me your tenent setae and I tell you who you are - Telling the story of a neglected character complex with phylogenetic signals using Leiodidae (Coleoptera) as a case study. *Arthropod Structure and Development, 46*, 662–685.
Goetzke, H. H., Pattrick, J. G., & Federle, W. (2019). Froghoppers jump from smooth plant surfaces by piercing them with sharp spines. *Proceedings. National Academy of Sciences. United States of America, 116*, 3012–3017.
Goldberg, J., Bresseel, J., Constant, J., Kneubühler, B., Leubner, F., Michalik, P., & Bradler, S. (2015). Extreme convergence in egg-laying strategy across insect orders. *Scientific Reports, 5*, 7825.
Gorb, S. N. (1998). The design of the fly adhesive pad: Distal tenent setae are adapted to the delivery of an adhesive secretion. *Proceedings of the Royal Society B, 265*, 747–752.
Gorb, S. N. (2001). *Attachment devices of insect cuticle*. Springer.
Gorb, S. N. (2008a). Smooth attachment devices in insects. In J. Casas & S. J. Simpson (Eds.), *Advances in insect physiology: Insect mechanics and control* (Vol. 34, pp. 81–116). Elsevier Ltd, Academic Press.
Gorb, S. N. (2008b). Biological attachment devices: Exploring nature's diversity for biomimetics. *Philosophical Transactions of the Royal Society of London, Series A, 366*, 1557–1574.
Gorb, S. N. (2011) Biological fibrillar adhesives: Functional principles and biomimetic applications. In: *Handbook of adhesion technology* (pp. 1410–1436). Springer.
Gorb, S. N., & Beutel, R. G. (2001). Evolution of locomotory attachment pads of hexapods. *Naturwissenschaften, 88*, 530–534.
Gorb, E. V., & Gorb, S. N. (2002). Attachment ability of the beetle *Chrysolina fastuosa* on various plant surfaces. *Entomologia Experimentalis et Applicata, 105*, 13–28.
Gorb, S. N., & Gorb, E. V. (2004). Ontogenesis of the attachment ability in the bug *Coreus marginatus* (Heteroptera, Insecta). *The Journal of Experimental Biology, 207*, 2917–2924.
Gorb, S. N., & Scherge, M. (2000). Biological microtribology: Anisotropy in frictional forces of orthopteran attachment pads reflects the ultrastructure of a highly deformable material. *Proceedings of the Royal Society B, 267*, 1239–1244.
Gorb, S., Jiao, Y., & Scherge, M. (2000). Ultrastructural architecture and mechanical properties of attachment pads in *Tettigonia viridissima* (Orthoptera Tettigoniidae). *Journal of Comparative Physiology. A, 186*, 821–831.
Gorb, S. N., Gorb, E. V., & Kastner, V. (2001). Scale effects on the attachment pads and friction forces in syrphid flies (Diptera, Syrphidae). *The Journal of Experimental Biology, 204*, 1421–1431.
Gorb, S. N., Beutel, R., Gorb, E. V., Jiao, Y., Kastner, V., Niederegger, S., Popov, V. L., Scherge, M., Schwarz, U., & Vötsch, W. (2002). Structural design and biomechanics of friction-based releasable attachment devices in insects. *Integrative and Comparative Biology, 42*(6), 1127–1139.
Gorb, E. V., Kastner, V., Peressadko, A., Arzt, E., Gaume, L., Rowe, N., & Gorb, S. N. (2004). Structure and properties of the glandular surface in the digestive zone of the pitcher in the carnivorous plant *Nepenthes ventrata* and its role in insect trapping and retention. *The Journal of Experimental Biology, 207*, 2947–2963.
Gorb, S. N., Niederegger, S., Hayashi, C. Y., Summers, A., & V. & Walther, P. (2006). Silk-like secretion from tarantula feet. *Nature, 443*, 407.
Gorb, S. N., Sinha, M., Peressadko, A., Daltorio, K. A., & Quinn, R. D. (2007). Insects did it first: A micropatterned adhesive tape for robotic applications. *Bioinspiration & Biomimetics, 2*, S117–S125.
Gottardo, M. (2011). A new genus and new species of Philippine stick insects (Insecta: Phasmatodea) and phylogenetic considerations. *The Comptes Rendus Biologies, 334*, 555–563.

Gottardo, M., & Heller, P. (2012). An enigmatic new stick insect from the Philippine Islands (Insecta: Phasmatodea). *Comptes Rendus Biologies, 335*, 594–601.
Gottardo, M., Vallotto, D., & Beutel, R. G. (2015). Giant stick insects reveal unique ontogenetic changes in biological attachment devices. *Arthropod Structure and Development, 44*, 195–199.
Green, D. M., & Alberch, P. (1981). Interdigital webbing and skin morphology in the neotropical salamander genus *Bolitoglossa* (Amphibia; Plethodonitidae). *Journal of Morphology, 170*, 273–282.
Grimaldi, D., & Engel, M. S. (2005). *Evolution of the insects*. Cambridge University Press.
Grohmann, C., Henze, M. J., Norgaard, T., & Gorb, S. N. (2015). Two functional types of attachment pads on a single foot in the Namibia bush cricket *Acanthoproctus diadematus* (Orthoptera: Tettigoniidae). *Proceedings of the Royal Society B, 282*, 20142976.
Gunderson, S., & Schiavone, R. (1989). The insect exoskeleton: A natural structural composite. *JOM, 41*, 60–63.
Günther, K. (1953). Über die taxonomische Gliederung und die geographische Verbreitung der Insektenordnung der Phasmatodea. *Beiträge zur Entomologie, 3*, 541–563.
Haas, F., & Gorb, S. (2004). Evolution of locomotory attachment pads in the Dermaptera (Insecta). *Arthropod Structure & Development, 33*, 45–66.
Hanna, G., & Barnes, W. J. P. (1990). Adhesion and detachment of the toe pads of tree frogs. *The Journal of Experimental Biology, 155*, 103.
Hasenfuss, I. (1977). Die Herkunft der Adhäsionsflüssigkeit bei Insekten. *Zoomorphology, 87*, 51–64.
Hasenfuss, I. (1978). Über das Haften von Insekten an glatten Flächen - Herkunft der Adhäsionsflüssigkeit. *Zoologische Jahrbücher Abteilung, 99*, 115–116.
Heming, B. S. (1971). Functional morphology of the thysanopteran pretarsus. *Canadian Journal of Zoology, 49*, 91–108.
Heming, B. S. (1972). Functional morphology of the pretarsus in larval Thysanoptera. *Canadian Journal of Zoology, 50*, 751–766.
Heming, B. S. (1973). Metamorphosis of the pretarsus in *Frankliniella fusca* (Hinds) (Thripidae) and *Haplothrips verbasci* (Osborn) (Phlaeothripidae) (Thysanoptera). *Canadian Journal of Zoology, 51*, 1211–1234.
Henning, W., (1973). Diptera (Zweiflügler). In *Handbuch der Zoologie* (Vol. 4). Walter de Gruyter.
Henning, B. (1974). Morphologie und Histologie der Tarsen von *Tettigonia viridissima* L. (Orthoptera, Ensifera). *Zoomorphology, 79*, 323–342.
Hilker, M., & Fatouros, N. E. (2015). Plant responses to insect egg deposition. *Annual Review of Entomology, 60*, 493–515.
Hiller, U. (1968). Untersuchungen zum Feinbau und zur Funktion der Haftborsten von Reptilien. *Zeitschrift für Morphologie der Tiere, 62*, 307.
Holway, R. (1935). Preliminary note on the structure of the pretarsus and its possible phylogenetic significance. *Psyche, 42*, 1–24.
Homann, H. (1957). Haften Spinnen an einer Wasserhaut? *Naturwissenschaften, 44*(11), 318–319.
Hoyle, G. (1976). Arthropod walking. In R. M. Herman, S. Grillner, P. S. G. Stein, & D. G. Stuart (Eds.), *Neural control of locomotion* (pp. 137–179). Springer.
Hu, D. L., Chan, B., & Bush, J. W. (2003). The hydrodynamics of water strider locomotion. *Nature, 424*, 663–666.
Huber, G., Gorb, S. N., Spolenak, R., & Arzt, E. (2005a). Resolving the nanoscale adhesion of individual gecko spatulae by atomic force microscopy. *Biology Letters, 1*, 2.
Huber, G., Mantz, H., Spolenak, R., Mecke, K., Jacobs, K., Gorb, S. N., & Arzt, E. (2005b). Evidence for capillarity contributions to gecko adhesion from single spatula nanomechanical measurements. *Proceedings of the National Academy of Sciences of the United Science of Americal, 102*(45), 16293.
Ignasov, J., Kapilavai, A., Filonenko, K., Larsen, J. C., Baird, E., Hallam, J., Büsse, S., Kovalev, A. E., Gorb, S. N., Duggen, L., & Manoonpong, P. (2018). Bio-inspired design and movement generation of dung beetle-like legs. *Artificial Life and Robotics, 23*, 555–563.

Ishii, S. (1987). Adhesion of a leaf-feeding ladybird *Epilachna vigintioctomaculata* (Coleoptera, Coccinellidae) on a vertically smooth surface. *Applied Entomology and Zoology, 22*, 222–228.
Jagota, A., & Bennison, S. J. (2002). Mechanics of adhesion through a fibrillar microstructure. *Integrative and Comparative Biology, 42*, 1140–1145.
Jayaram, K., & Full, R. J. (2016). Cockroaches traverse crevices, crawl rapidly in confined spaces, and inspire a soft, legged robot. *Proceedings. National Academy of Sciences. United States of America, 113*, E950–E957.
Jiao, Y., Gorb, S. N., & Scherge, M. (2000). Adhesion measured on the attachment pads of *Tettigonia viridissima* (Orthoptera, Insecta). *The Journal of Experimental Biology, 203*, 1887–1895.
Johnson, K. L., Kendall, K., & Roberts, A. D. (1971). Surface energy and the contact of elastic solids. *Proceedings of the Royal Society of London A, 324*(1558), 301.
Kendall, M. (1970). The anatomy of the tarsi of *Schistocerca gregaria* Forskal. *Cell and Tissue Research, 109*, 112–137.
Kennedy, C. E. J. (1986). Attachment may be a basis for specialization in oak aphids. *Ecological Entomology, 11*, 291–300.
Kesel, A. B., Martin, A., & Seidl, T. (2003). Adhesion measurements on the attachment devices of the jumping spider *Evarcha arcuata*. *The Journal of Experimental Biology, 206*, 2733.
Kester, K. M., & Barbosa, P. (1994). Behavioral responses to host foodplants of two populations of the insect parasitoid *Cotesia congregata* (Say). *Oecologia, 99*, 151–157.
Kinzelbach, R. K. (1971). Morphologische Befunde an Fächerflüglern und ihre phylogenetische Bedeutung (Insecta: Strepsiptera). *Zoologica, 41*, 1–256.
Knoll, F. (1926). Die Arum-Blütenstände und ihre Besucher (Insekten und Blumen IV.). *Abhandlungen der Kaiserlich-Königlichen Zoologisch-Botanischen Gesellschaft in Wien, 12*, 379–481.
Knoll, F. (1930). Über Pollenkitt und Bestäubungsart. *Journal of Plant Physiology, 23*, 609–675.
Kölsch, G., & Betz, O. (1998). Ultrastructure and function of the adhesion-capture apparatus of *Stenus* species (Coleoptera, Staphylinidae). *Zoomorphology, 118*, 263–277.
Koppetsch, T., Böhme, W., Büsse, S., & Gorb, S. N. (2020). Comparative epidermal microstructure anatomy and limb and tail osteology of eyelid geckos (Squamata: Eublepharidae): Implications of ecomorphological adaptations. *Zoologischer Anzeiger, 287*, 45–60.
Kosaki, A., & Yamaoka, R. (1996). Chemical composition of footprints and cuticula lipids of three species of lady beetles. *The Journal Applied Entomology and Zoology, 40*, 47–53.
Kryuchkov, M., Lehmann, J., Schaab, J., Cherepanov, V., Blagodatski, A., Fiebig, M., & Katanaev, V. L. (2017a). Alternative moth-eye nanostructures: antireflective properties and composition of dimpled corneal nanocoatings in silk-moth ancestors. *Journal of Nanobiotechnology, 15*, 61.
Kryuchkov, M., Blagodatski, A., Cherepanov, V., & Katanaev, V. L. (2017b). Arthropod corneal nanocoatings: Diversity, mechanisms, and functions. In S. N. Gorb & E. V. Gorb (Eds.), *Functional surfaces in biology III* (pp. 29–52). Springer.
Labonte, D., Williams, J. A., & Federle, W. (2014). Surface contact and design of fibrillar 'friction pads' in stick insects (*Carausius morosus*): mechanisms for large friction coefficients and negligible adhesion. *Journal of the Royal Society Interface, 11*, 20140034.
Langer, M. G., Ruppersberg, J. P., & Gorb, S. N. (2004). Adhesion forces measured at the level of a terminal plate of the fly's seta. *Proceedings of the Royal Society of London B, 271*, 2209–2215.
Lankester, E. R. (1904). The structure and classification of the Arthropoda. *Journal of Cell Science, 47*, 523–582.
Lee, Y. I., Kogan, M., & Larsen, J. R., Jr. (1986). Attachment of the potato leafhopper to soybean plant surfaces as affected by morphology of the pretarsus. *Entomologia Experimentalis et Applicata, 42*, 101–107.
Lees, A., & Hardie, J. (1988). The organs of adhesion in the aphid *Megoura viciae*. *The Journal of Experimental Biology, 136*, 209–228.

Liu, Z., & Liang, A. P. (2016). Ultramorphology of the tarsal adhesive structures of eight leaf beetle species (Coleoptera: Chrysomelidae). *Journal of the Kansas Entomological Society, 89*, 215–230.

Magallón, S., & Castillo, A. (2009). Angiosperm diversification through time. *American Journal of Botany, 96*, 349–365.

Maki, T. (1936). Studies on the skeletal structure, musculature and nervous system of the Alder Fly *Chauliodes formosanus* Petersen. *Memoirs of the Faculty of Science and Agriculture; Taihoku Imperial University, 16*, 117–243.

Nelson, C. H. (2009). Surface ultrastructure and evolution of tarsal attachment structures in Plecoptera (Arthropoda: Hexapoda) *Aquatic Insects, 31*, 523–545.

Neville, A. C. (1993). *Biology of fibrous composites*. Cambridge University Press.

Niederegger, S., & Gorb, S. N. (2003). Tarsal movements in flies during leg attachment and detachment on a smooth substrate. *Journal of Insect Physiology, 49*, 611–620.

Niederegger, S., & Gorb, S. N. (2006). Friction and adhesion in the tarsal and metatarsal scopulae of spiders. *Journal of Comparative Physiology. A, 192*, 1223.

Niederegger, S., Gorb, S., & Jiao, Y. (2002). Contact behaviour of tenent setae in attachment pads of the blowfly *Calliphora vicina* (Diptera, Calliphoridae). *Journal of Comparative Physiology A, 187*, 961–970.

Nielsen, E. S., & Common, I. F. B. (1991). Lepidoptera. In CSIRO (Ed.), *The insects of Australia* (Vol. 2, pp. 817–915). Cornell University Press.

Orivel, J., Malherbe, M. C., & Dejean, A (2001). Relationships between pretarsus morphology and arboreal life in ponerine ants of the genus *Pachycondyla* (Formicidae: Ponerinae). *Annals of the Entomological Society of America, 94*, 449–456.

Oseto, C. Y., & Helms, T. J. (1976). Anatomy of the adults of *Loxagrotis albicosta*. *The University of Nebraska Studies, 52*, 136.

Pattrick, J. G., Labonte, D., & Federle, W. (2018). Scaling of claw sharpness: Mechanical constraints reduce attachment performance in larger insects. *The Journal of Experimental Biology, 221*, jeb188391.

Pearson, K. G., & Franklin, R. (1984). Characteristics of leg movements and patterns of coordination in locusts walking on rough terrain. *International Journal of Robotics Research, 3*, 101–112.

Peattie, A. M., & Full, R. J. (2007). Phylogenetic analysis of the scaling of wet and dry biological fibrillar adhesives. *Proceedings of the National Academy of Sciences of the United States of America, 104*(47), 18595–18600.

Peccoud, J., & Simon, J. C. (2010). The pea aphid complex as a model of ecological speciation. *Ecological Entomology, 35*, 119–130.

Peisker, H., & Gorb, S. N. (2012). Evaporation dynamics of tarsal liquid footprints in flies (*Calliphora vicina*) and beetles (*Coccinella septempunctata*). *The Journal of Experimental Biology, 215*, 1266–1271.

Perez-Goodwyn, P., Peressadko, A., Schwarz, H., Kastner, V., & Gorb, S. (2006). Material structure, stiffness, and adhesion: why attachment pads of the grasshopper (*Tettigonia viridissima*) adhere more strongly than those of the locust (*Locusta migratoria*)(Insecta: Orthoptera). *Journal of Comparative Physiology. A, 192*, 1233–1243.

Persson, B. N. J. (2003). On the mechanism of adhesion in biological systems. *The Journal of Chemical Physics, 118*, 7614.

Persson, B. N. J., & Gorb, S. N. (2003). The effect of surface roughness on the adhesion of elastic plates with application to biological systems. *The Journal of Chemical Physics, 119*, 11437963.

Petersen, D. S., Kreuter, N., Heepe, L., Büsse, S., Wellbrock, A. H., Witte, K., & Gorb, S. N. (2018). Holding tight to feathers-structural specializations and attachment properties of the avian ectoparasite *Crataerina pallida* (Diptera, Hippoboscidae). *The Journal of Experimental Biology, 221*, jeb179242.

Prete, F. R., & Hamilton, K. (1999). Prey capture. In F. R. Prete, H. Wells, P. H. Wells, & L. E. Hurd (Eds.), *The praying mantids* (pp. 194–223). Johns Hopkins University Press.

Rebora, M., Michels, J., Salerno, G., Heepe, L., Gorb, E. V., & Gorb, S. N. (2018). Tarsal attachment devices of the southern green stink bug *Nezara viridula* (Heteroptera: Pentatomidae). *Journal of Morphology, 279*, 660–672.

Rebora, M., Salerno, G., Piersanti, S., Gorb, E. V., & Gorb, S. N. (2020). Role of fruit epicuticular waxes in preventing *Bactrocera oleae* (Diptera: Tephritidae) attachment in different cultivars of *Olea europaea*. *Insects, 11*, 189.

Rees, B. E., & Ferris, G. F. (1939). The morphology of *Tipula reesi* Alex. *Microentomology, 4*, 143–178.

Richards, A. G., & Richards, P. A. (1979). The cuticular protuberances of insects. *International Journal of Insect Morphology and Embryology, 8*, 143–157.

Ritzmann, R. E., Quinn, R. D., & Fischer, M. S. (2004). Convergent evolution and locomotion through complex terrain by insects, vertebrates and robots. *Arthropod Structure and Development, 33*, 361–379.

Rizzo, N. W., Gardner, K. H., Walls, D. J., Keiper-Hrynko, N. M., Ganzke, T. S., & Hallahan, D. L. (2006). Characterization of the structure and composition of gecko adhesive setae. *Journal of the Royal Society Interface, 3*, 441.

Robertson, J. A., Bradler, S., & Whiting, M. F. (2018). Evolution of oviposition techniques in stick and leaf insects (Phasmatodea). *Frontiers in Ecology and Evolution, 6*, 216.

Röder, G. (1986). Zur Morphologie des Praetarsus der Diptera und Mecoptera. *Zoologische Jahrbücher. Abteilung für Anatomie und Ontogenie der Tiere, 114*, 465–502.

Roff, D. A. (1994). The evolution of flightlessness: Is history important? *Evolutionary Ecology, 8*, 639–657.

Roff, D. A., & Fairbairn, D. J. (2007). The evolution and genetics of migration in insects. *Bioscience, 57*, 155–164.

Rosenberg, H. I., & Rose, R. (1999). Volar adhesive pads of the feathertail glider, *Acrobates pygmaeus* (Marsupialia; Acrobatidae). *Canadian Journal of Zoology, 77*, 233–248.

Ross, E. S. (1991). Embioptera - Embiidina (embiids, web-spinners, foot-spinners). In I. D. Naumann, P. B. Carne, J. F. Lawrence, E. S. Nielsen, J. P. Spradbery, R. W. Taylor, M. J. Whitten, & M. J. Littlejohn (Eds.), *The insects of Australia* (pp. 405–409). Melbourne University Press.

Ross, E. S. (2000). Embia: Contributions to the biosystematics of the insect order Embiidina. Part 1, Origin, relationships and integumental anatomy of the insect order Embiidina. *Occasional Papers of the California Academy of Sciences, 149*, 1–53.

Roth, L., & Willis, E. (1952). Tarsal structure and climbing ability of cockroaches. *The Journal of Experimental Biology, 119*, 483–517.

Salerno, G., Rebora, M., Gorb, E. V., Kovalev, A. E., & Gorb, S. N. (2017). Attachment ability of the southern green stink bug *Nezara viridula* (Heteroptera: Pentatomidae). *Journal of Comparative Physiology. A, 203*, 601–611.

Salerno, G., Rebora, M., Gorb, E. V., & Gorb, S. N. (2018). Attachment ability of the polyphagous bug *Nezara viridula* (Heteroptera: Pentatomidae) to different host plant surfaces. *Scientific Reports, 8*, 1–14.

Schargott, M., Popov, V. L., & Gorb, S. N. (2006). Spring model of biological attachment pads. *Journal of Theoretical Biology, 243*, 48.

Schedl, W. (1991). Hymenoptera, Unterordnung Symphyta. In *Handbuch der Zoologie* (Vol. 4). Walter de Gruyter.

Scherge, M., & Gorb, S. N. (2001). *Biological micro- and nanotribology: Nature's solutions*. Springer.

Schliemann, H. (1970). Bau und Function der Haftorgane von *Thyroptera* und *Myzopoda* (Vespertilionoidea, Microchiroptera, Mammalia). *Zeitschrift für Wissenschaftliche Zoologie, 181*, 353–400.

Schliemann, H. (1971). Die Haftorgane von *Thyroptera* und *Myzopoda* (Microchiroptera, Mammalia) - Gedanken zu ihrer Entstehung als Parallelbildungen. *Journal of Zoological Systematics and Evolutionary Research, 9*, 61–80.

Schliemann, H. (1974). Haftorgane bei Fledermäusen. *Natur und Museum. 104*, 15–20.
Schliemann, H. (1975). Über die Entstehung von Haftorganen bei Chiropteren. *Mitteilungen aus dem Hamburgischen Zoologischen Museum und Institut, 72*, 249–259.
Schliemann, H. (1983). Haftorgane - Beispiele für gleichsinnige Anpassungen in der Evolution der Tiere. *Funktionelle Biologie und Medizin, 2*, 169–177.
Schliemann, H., & Rehn, C. (1980). Zur Kenntnis der Haftorgane von *Eudiscopus denticulus* (Osgood, 1932) (Mammalia, Microchiroptera, Vespertilionidae). *Zeitschrift für Säugetierkunde, 45*, 29–39.
Scholz, I., Baumgartner, W., & Federle, W. (2008). Micromechanics of smooth adhesive organs in stick insects: Pads are mechanically anisotropic and softer towards the adhesive surface. *Journal of Comparative Physiology. A, 194*, 373–384.
Schuh, R. T. (1976). Pretarsal structure in the Miridae (Hemiptera), with a cladistic analysis of relationships within the family. *American Museum Novitates, 2601*, 1–42.
Schulmeister, S. (2003). Morphology and evolution of the tarsal plantulae in Hymenoptera (Insecta), focussing on the basal lineages. *Zoologica Scripta, 32*, 153–172.
Sellick, J. T. (1997a). The range of egg capsule morphology within the Phasmatodea and its relevance to the taxonomy of the order. *The Italian Journal of Zoology, 64*, 97–104.
Sellick, J. T. (1997b). Descriptive terminology of the phasmid egg capsule, with an extended key to the phasmid genera based on egg structure. *Systematic Entomology, 22*, 97–122.
Simon, S., Letsch, H., Bank, S., Buckley, T. R., Donath, A., Liu, S., Machida, R., Meusemann, K., Misof, B., Podsiadlowski, L., Zhou, X., Wipfler, B., & Bradler, S. (2019). Old World and New World Phasmatodea: Phylogenomics resolve the evolutionary history of stick and leaf insects. *Frontiers in Ecology and Evolution, 7*, 345.
Singer, R. B. (2002). The pollination mechanism in *Trigonidium obtusum* Lindl. (Orchidaceae: Maxillariinae): Sexual mimicry and trapflowers. *Annals of Botany, 89*, 157–163.
Slifer, E. H. (1950). Vulnerable areas on the surface of the tarsus and pretarsus of the grasshopper (Acrididae, Orthoptera) with special reference to the arolium. *Annals of the Entomological Society of America, 43*, 173–188.
Snodgrass, R. (1956). *Anatomy of the Honey Bee*. Comstock Publishing Associates.
Song, Y., Dai, Z., Wang, Z., Ji, A., & Gorb, S. N. (2016). The synergy between the insect-inspired claws and adhesive pads increases the attachment ability on various rough surfaces. *Scientific Reports, 6*, 26219.
Spolenak, R., Gorb, S., & Arzt, E. (2005a). Adhesion design maps for bio-inspired attachment systems. *Acta Biomaterialia, 1*, 5.
Spolenak, R., Gorb, S., Gao, H., & Arzt, E. (2005b). Effects of contact shape on the scaling of biological attachments. *Proceedings of the Royal Society of London A, 461*(2054), 305.
Stork, N. E. (1980a). Role of waxblooms in preventing attachment to brassicas by the mustard beetle, *Phaedon cochleariae*. *Entomologia Experimentalis et Applicata, 28*, 100–107.
Stork, N. E. (1980b). Experimental analysis of adhesion of *Chrysolina polita* (Chrysomelidae: Coleoptera) on a variety of surfaces. *The Journal of Experimental Biology 88*, 91–108.
Stork, N. E. (1980c). A scanning electron microscope study of tarsal adhesive setae in the Coleoptera. *Zoological Journal of the Linnean Society, 68*, 173–306.
Stork, N. E. (1983a). A comparison of the adhesive setae on the feet of lizards and arthropods. *Journal of Natural History, 17*, 829–835.
Stork, N. E. (1983b). The adherence of beetle tarsal setae to glass. *Journal of Natural History, 17*, 583–597.
Theischinger, G. (1991). Megaloptera (Alderflies, dobsonflies). In CSIRO (Ed.), *The insects of Australia* (Vol. 1, pp. 521–524). Cornell University Press.
Thewissen, J. G. M., & Etnier, S. A. (1995). Adhesive devices on the thumb of vespertilionoid bats (Chiroptera). *Journal of Mammalogy, 76*, 925–936.
Thor, M., Strøm-Hansen, T., Larsen, L. B., Kovalev, A. E., Gorb, S. N., Baird, E., & Manoonpong, P. (2018). A dung beetle-inspired robotic model and its distributed sensor-driven control for walking and ball rolling. *Artificial Life and Robotics, 23*, 435–443.

Thorpe, W. H., & Caudle, H. B. (1938). A study of the olfactory responses of insect parasites to the food plant of their host. *Parasitology, 30*, 523–528.
Turing, A. M. (1952). The chemical theory of morphogenesis. *Philosophical Transactions of the Royal Society of London B, 237*, 37–72.
Vallotto, D., Bresseel, J., Constant, J., & Gottardo, M. (2016). Morphology of the terminalia of the stick insect *Dajaca napolovi* from Vietnam (Insecta: Phasmatodea). *Entomological Science, 19*, 376–382.
Varenberg, M., & Gorb, S. N. (2009). Hexagonal surface micropattern for dry and wet friction. *Advanced Materials, 21*, 483–486.
Vogel, S. (1965). Kesselfallen-Blumen. *Umschau in Wissenschaft und Technik, 65*, 12–17.
Voigt, D., & Gorb, S. (2010). Egg attachment of the asparagus beetle *Crioceris asparagi* to the crystalline waxy surface of *Asparagus officinalis*. *Proceedings of the Royal Society B, 277*, 895–903.
Voigt, D., Gorb, E. V., & Gorb, S. N. (2007). Plant surface-bug interactions: *Dicyphus errans* stalking along trichomes. *Arthropod-Plant Interact, 1*, 221–243.
Voigt, D., Schuppert, J. M., Dattinger, S., & Gorb, S. N. (2008). Sexual dimorphism in the attachment ability of the Colorado potato beetle *Leptinotarsa decemlineata* (Coleoptera: Chrysomelidae) to rough substrates. *Journal of Insect Physiology, 54*, 765–776.
Voigt, D., Karguth, A., & Gorb, S. N. (2012). Shoe soles for the gripping robot: Searching for polymer-based materials maximising friction. *Robotics and Autonomous Systems, 60*, 1046–1055.
Voigt, D., Tsipenyuk, A., & Varenberg, M. (2017a). How tight are beetle hugs? Attachment in mating leaf beetles. *Royal Society Open Science, 4*, 171108.
Voigt, D., Takanashi, T., Tsuchihara, K., Yazaki, K., Kuroda, K., Tsubaki, R., & Hosoda, N. (2017b). Strongest grip on the rod: tarsal morphology and attachment of Japanese pine sawyer beetles. *Zoological Letters, 3*, 16.
Voigt, D., Perez-Goodwyn, P., Sudo, M., Fujisaki, K., & Varenberg, M. (2019). Gripping ease in southern green stink bugs *Nezara viridula* L.(Heteroptera: Pentatomidae): Coping with geometry, orientation and surface wettability of substrate. *Entomological Science, 22*, 105–118.
Vötsch, W., Nicholson, G., Müller, R., Stierhof, Y.-D., Gorb, S., & Schwarz, U. (2002). Chemical composition of the attachment pad secretion of the locust *Locusta migratoria*. *Insect Biochemistry and Molecular Biology, 32*, 1605–1613.
Wagner, D. L., & Liebherr, J. K. (1992). Flightlessness in insects. *Trends in Ecology and Evolution, 7*, 216–220.
Walker, G., Yule, A. B., & Ratcliffe, J. (1985). The adhesive organ of the blowfly, *Calliphora vomitoria*: A functional approach (Diptera: Calliphoridae). *Journal of Zoology, 205*, 297–307.
Wang, J., & Liang, A. (2015). Ultrastructure of the fossula spongiosa and pretarsus in *Haematoloecha nigrorufa* (Stål) (Hemiptera: Heteroptera: Reduviidae: Ectrichodinae). *Zootaxa, 3963*, 230–239.
Wang, M., Béthoux, O., Bradler, S., Jacques, F. M. B., Cui, Y., & Ren, D. (2014). Under cover at pre-angiosperm times: A cloaked phasmatodean insect from the Early Cretaceous Jehol biota. *PLoS One, 9*, e91290.
Ward, L. K., & Spalding, D. F. (1993). Phytophagous British insects and mites and their food-plant families: Total numbers and polyphagy. *Biological Journal of the Linnean Society of London, 49*, 257–276.
Wedmann, S., Bradler, S., & Rust, J. (2007). The first fossil leaf insect: 47 million years of specialized cryptic morphology and behavior. *Proceedings. National Academy of Sciences. United States of America, 104*, 565–569.
Weirauch, C. (2005). Pretarsal structures in Reduviidae (Heteroptera, Insecta). *Acta Zoologica, 86*, 91–110.
Weirauch, C. (2007). Hairy attachment structures in Reduviidae (Cimicomorpha, Heteroptera), with observations on the fossula spongiosa in some other Cimicomorpha. *Zoologischer Anzeiger, 246*, 155–175.

Weygoldt, P. (1986). Arthropod interrelationships – The phylogenetic-systematic approach. *Journal of Zoological Systematics Evolutionsforsch, 24*, 19–35.
White, D., & Carver, M. (1971). Adhesive vesicles in some species of *Neophyllaphis* Takahashi, 1920 (Homoptera: Aphididae). *Austral Entomology, 10*, 281–284.
Whiting, M. F., Bradler, S., & Maxwell, T. (2003). Loss and recovery of wings in stick insects. *Nature, 421*, 264–267.
Wiebes, J. T. (1979). Co-evolution of figs and their insect pollinators. *Annual Review of Ecology and Systematics, 10*, 1–12.
Wieland, F. (2013). *The phylogenetic system of Mantodea (Insecta: Dictyoptera)*. Universitätsverlag Göttingen.
Wilson, E. O. (1969). Migration and dispersal of insects by flight. *Science, 166*, 592.
Wilson, E. O. (1990). *Success and dominance in ecosystems: The case of the social insects* (Vol. 2). Ecology Institute.
Wipfler, B., Letsch, H., Frandsen, P. B., Kapli, P., Mayer, C., Bartel, D., Buckley, T. R., Donath, A., Edgerly-Rooks, J. S., Fujita, M., Liu, S., Machida, R., Mashimo, Y., Misof, B., Niehuis, O., Peters, R. S., Petersen, M., Podsiadlowski, L., Schütte, K., Shimizu, S., Uchifune, T., Wilbrandt, J., Yan, E., Zhou, X., & Simon, S. (2019). Evolutionary history of Polyneoptera and its implications for our understanding of early winged insects. *Proceedings of the National Academy of Sciences of the United States of America, 116*, 3024–3029.
Wolff, J. O., & Gorb, S. N. (Eds.). (2016). *Attachment structures and adhesive secretions in arachnids* (p. 184). Berlin et al.
Wootton, R. J. (1992). Functional morphology of insect wings. *Annual Review of Entomology, 37*, 113–140.
Yang, H., Shi, C., Engel, M. S., Zhao, Z., Ren, D., & Gao, T. (2020). Early specializations for mimicry and defense in a Jurassic stick insect. *National Science Review, 8*, nwaa056.
Zeng, Y., O'Maley, C., Singhal, S., Rahim, F., Park, S., Chen, X., & Dudley, R. (2019). A tale of winglets: Evolution of flight morphology in stick insects. *Frontiers in Ecology and Evolution, 8*, 121.
Zurek, D., Voigt, D., & Gorb, S. N. (2007). The role of attachment organs in the larval locomotion of the beetle *Gastrophysa viridula*. *Comparative Biochemistry and Physiology Part A, 4*, S146.
Zurek, D. B., Gorb, S. N., & Voigt, D. (2015). Locomotion and attachment of leaf beetle larvae *Gastrophysa viridula* (Coleoptera, Chrysomelidae). *Interface Focus, 5*, 20140055.
Zurek, D. B., Gorb, S. N., & Voigt, D. (2017). Changes in tarsal morphology and attachment ability to rough surfaces during ontogenesis in the beetle *Gastrophysa viridula* (Coleoptera, Chrysomelidae). *Arthropod Structure and Development, 46*, 130–137.

Chapter 11
Convergence of Arboreal Locomotor Specialization: Morphological and Behavioral Solutions for Movement on Narrow and Compliant Supports

Jesse W. Young

Abstract Arboreal supports impose a set of locomotor challenges not typically encountered in other terrestrial ecosystems. Because all arboreal animals must negotiate this common set of physical challenges in an environment where clumsy mistakes can lead to tragedy (or at least to the increased energetic burden of having to fight gravity to regain a lost position), it is of little surprise that we see widespread convergence of locomotor morphology and behavior among arboreal amphibians, lizards, and mammals. In this chapter I consider the biomechanical challenges imposed by moving on narrow and compliant arboreal supports, and survey existing data on how arboreal amphibians, lizards, and mammals have arrived at morphological and behavioral solutions to these problems. I focus on the biomechanical problems of negotiating narrow and compliant supports given that these challenges are, to some degree, uniquely characteristic of the arboreal environment.

Narrow supports potentially compromise locomotor performance in two ways: (1) by increasing the probability that the animal may tangentially slip from the support and, (2) by challenging mediolateral (i.e., transverse/rolling plane) stability. Compliant supports, by contrast, have the potential to reduce locomotor performance by absorbing some of the mechanical energy that the animal could use to accelerate and redirect its center of mass, and then unpredictably returning this energy at random times and in random directions (at least with respect to the animal's desired movement dynamics). Widespread morphological solutions to the biomechanical problems of moving on narrow and compliant supports include small body size, appendicular joints with enhanced mobility, grasping extremities, and long tails. Convergent behavioral solutions for increasing stability on precarious arboreal supports include reducing speed, increased limb joint flexion, the use of "compliant" gait kinematics marked by elongated limb contact durations (i.e., duty factors), a switch to gaits that facilitate more continuous contact with the substrate (and fewer

J. W. Young (✉)
Department of Anatomy and Neurobiology, Northeast Ohio Medical University (NEOMED), Rootstown, OH, USA
e-mail: jwyoung@neomed.edu

V. L. Bels, A. P. Russell (eds.), *Convergent Evolution*, Fascinating Life Sciences,
https://doi.org/10.1007/978-3-031-11441-0_11

ballistic aerial phases), and a decrease overall limb stiffness typically accomplished via exaggerated limb joint excursions during the stance phase. Future research on arboreal locomotion in tetrapods should focus on integrating quantitative laboratory data on locomotor kinematics and kinetics with holistic ecological data on substrate use and support morphology gleaned in the field. Such integrated datasets will be critical for furthering our understanding of how locomotor anatomy and behavior are shaped by the rigors of the natural arboreal environment.

Keywords Adhesion · Center of mass · Duty factor · Force · Gait · Locomotor biomechanics · Perch · Prehensility · Stability · Torque

11.1 Introduction

Arboreal supports—used here to broadly include trees, shrubs, undergrowth, and other herbaceous vegetation—represent an extension of the terrestrial habitat, providing new foraging resources (Sussman & Raven, 1978; Sussman, 1991), expanded areas for nesting (Pruetz et al., 2008), and avenues for escape or camouflage from terrestrial predators (Dunbar, 1988; Isbell, 1994). Taken together, these factors are thought to reduce sources of "extrinsic mortality" in arboreal animals, increase lifespans and facilitate a slower pace of life history (van Schaik & Deaner, 2003; Shattuck & Williams, 2010; Healy et al., 2014). Given these benefits, it is not surprising that arboreal specialists are found in every major group of extant tetrapods, including mammals, birds, lizards, snakes, and amphibians–of the major tetrapod clades, only turtles have yet to invade the arboreal habitat (Hildebrand & Goslow, 2001). Moreover, morphometric signals of vertebrate arboreal adaptation can be found as early as the Late Permian (Fröbisch & Reisz, 2009), indicating that selection for arboreality has been present throughout much of tetrapod evolution

However, arboreality also introduces a set of challenges not typically encountered in other terrestrial ecosystems, chief among which are the biomechanical challenges of arboreal locomotion. Arboreal supports vary in diameter, angular orientation, compliance, and distribution in space (Fig. 11.1). Although potential locomotor substrates in ground-based terrestrial ecosystems can vary along similar dimensions—e.g., the narrow and distributed supports of a rocky outcrop, the steeply sloping sides of a mountain, or the damping compliance of a muddy field or sandy beach—it is rare to encounter locomotor substrates that are as widely variable as those of arboreal environments (Mattingly & Jayne, 2004). For instance, a recent survey of quadrupedal locomotor substrates used by wild platyrrhine monkeys found that support diameters ranged over two orders of magnitude (Fig. 11.2) (Dunham et al., 2018, 2019a, 2020). Detailed study of locomotor bouts in this sample found that individual monkeys in the sample could be required to negotiate as many as nine different supports in a single bout of locomotion, and up to four different substrates in a single stride, such that nearly every footfall contacts a different substrate, with each new support potentially varying from the last along multiple dimensions (McNamara et al., 2019). Moreover, locomotor errors in an

Fig. 11.1 Biomechanical relevance of functional variation in arboreal locomotor supports. (**a**) An Indian giant squirrel (*Ratufa indica*) foraging in the thin branch zone of the terminal canopy. (**b**) A blue-sided tree frog (*Agalychnis annae*) climbing down and an inclined leaf stalk. (**c**) A juvenile chameleon (*Chamaeleo chamaeleon*) perched on a compliant herbaceous stem. (**d**) An adult and a juvenile Sumatran orangutan (*Pongo abelii*) distributing their body weight among multiple disparate supports in the canopy. All images downloaded from Wikimedia Commons and reproduced under Creative Commons license CC BY-SA. 4.0 (A: photograph by Manoj Ashokkumar, link: https://commons.wikimedia.org/wiki/File:Ratufa_Indica_02.jpg; B: photograph by Charles J. Sharp, link: https://commons.wikimedia.org/wiki/File:Golden-eyed_tree_frog_(Agalychnis_annae)_1.jpg; C: photograph by Mehmet Karaca, link: https://commons.wikimedia.org/wiki/File:Baby_common_chameleon.jpg; D: photograph by Prayugo Utomo, link: https://commons.wikimedia.org/wiki/File:Orangutan_Tapanuli.jpg). No images have been altered from their original format

arboreal environment can impose a high cost (Wheatley et al., 2021). Studies of skeletal pathology indicate that injury from falling often accounts for the greatest incidence of long bone trauma in free-ranging primate populations (Schultz, 1944; Buikstra, 1975; Lovell, 1991; Jurmain, 1997; Carter et al., 2008).

All arboreal animals, regardless of phylogenetic history, must overcome the same biomechanical challenges presented by potentially narrow, steep, compliant, and distributed locomotor supports. It is thus reasonable to expect some degree of convergence in the morphological and behavioral strategies used to accommodate to these challenges. Alternatively, whether due to historical (phylogenetic) contingency or other stochastic processes, different lineages may have arrived at different "solutions" to the same problem or emphasized morphological versus behavioral strategies to varying degrees. Moreover, across all terrestrial animals, body size is a

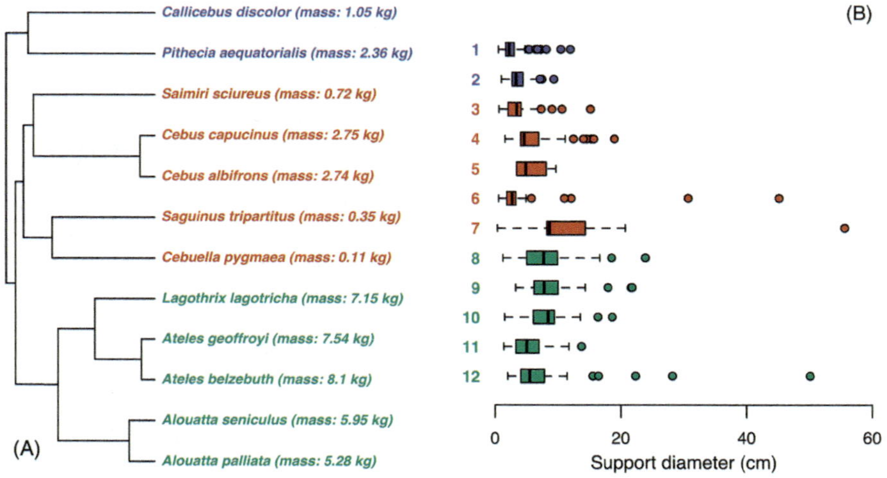

Fig. 11.2 Functional variation of locomotor substrates for free ranging platyrrhine monkeys. (**a**) Cladogram of platyrrhine species in the sample. Separate taxonomic families are distinguished by text color: purple = Pithecidae, orange = Cebidae, and green = Atelidae. Average masses for each species were taken from Smith and Jungers (1997) (reported mass for *Pithecia aequatorialis* is from *P. monachus* and reported mass for *Saguinus tripartitus* is from *S. fuscicollis*). Observational data were recorded opportunistically at Tiputini Biodiversity Station in Ecuador (August–October 2017) and La Suerte Biological Field Station in Costa Rica (June–July 2018) (Dunham et al., 2019a, 2020) (**b**) Box-and-whisker plots of variation in support diameters used during quadrupedal locomotion. Dark lines represent the median of each distribution, boxes extend across the interquartile range (IQR), and whiskers extend to ±150% of the IQR. Circles indicate outliers beyond this range. Species are arrayed in the same top-to-bottom order listed in the cladogram. Substrate diameters were recorded to within 1.5 mm of accuracy using remote photogrammetric methods. See Dunham et al. (2018) for more information

critical determinant of both the nature and magnitude of the challenges imposed by the arboreal habitat (Jenkins, 1974; Grand, 1984; Cant, 1992; Shapiro et al., 2014; Karantanis et al., 2015). The same terminal branch that is dangerously narrow and compliant for a large species could be comfortably broad for a smaller one.

Below, I separately consider the biomechanical challenges imposed by narrow and compliant arboreal supports, and survey existing data on how arboreal tetrapods have arrived at morphological and behavioral solutions to these problems. I focus on the biomechanical problems of negotiating narrow and compliant supports given that these challenges are, to some degree, uniquely characteristic of the arboreal environment. For recent reviews of the biomechanical difficulties of moving on inclined/declined perches and distributed supports—and of convergent morphological and behavioral solutions to these problems—see Birn-Jeffery and Higham (2014) and Graham and Socha (2020), respectively. Additionally, I focus on quadrupedal locomotion and leaping, as these are the predominant locomotor modes of most arboreal animals. For recent treatments of the biomechanics of arboreal locomotion in snakes and birds, see Jayne (2020) and Abourachid et al. (2017) respectively.

11.2 Biomechanical Challenges of Narrow Arboreal Supports

The challenges induced by moving on supports of finite diameter have received more attention than any other functional aspect of arboreal substrates. Support narrowness *per se* (i.e., divorced from support compliance, though narrowness and compliance are typically correlated in natural arboreal habitats; van Casteren et al., 2013; Dunham et al., 2018) potentially compromises locomotor performance in two ways, first by increasing the probability that the animal may slip from the support and, second, by challenging mediolateral (i.e., transverse/rolling plane) stability.

11.2.1 Narrow Supports Increase the Potential for Tangential Slipping

As support diameter decreases relative to the width of the animal's trunk, the tangential (i.e., shear) forces that are exerted along the edge of the branch gradually predominate relative to normal forces exerted towards the center of the branch (Fig. 11.3). In fact, a simple model of a standing quadruped with columnar, parasagittal limbs held alongside the trunk shows that tangential reaction forces will exceed normal reaction forces whenever the support diameter is $\leq$141% of trunk diameter. In these cases, preventing slippage from the support necessitates some form of enhanced grip onto the branch's surface, for example through surface friction, mechanical interlocking with claws, or employment of the dry/wet adhesion systems of arboreal lissamphibians and lizards (Cartmill, 1985). Note that, by this logic, any support with a diameter less than an animal's trunk width can properly be defined as a "narrow" support. Arboreal animals show several morphological and behavioral adaptations to reduce the potential for tangential slipping from the edge of narrow supports.

Body Size Reduced body size could be considered a morphological "solution" to the stability challenges of narrow supports (Jenkins, 1974; Gebo, 2004; Shapiro et al., 2014). Absolutely smaller body dimensions necessarily diminish the range of potential arboreal supports that can be considered narrow (see Fig 11.3), as well as reduce branch loading magnitudes, further promoting arboreal stability (particularly if supports are compliant; see Sect. 11.3). In general, arboreal lizards, carnivorans and primates are smaller than their terrestrial counterparts (Polk et al., 2000; Meiri, 2018). Several studies have also shown that overall body size influences preferred support diameters in wild populations of arboreal tetrapods. For instance, overall body mass is positively correlated with preferred perch diameters in Caribbean species of *Anolis* lizards (Irschick et al., 1997), and small overall body size is one of the defining morphological characteristics of the narrow-support specialized "twig" ecomorph of the Caribbean *Anolis* radiation (Huyghe et al., 2007; Losos,

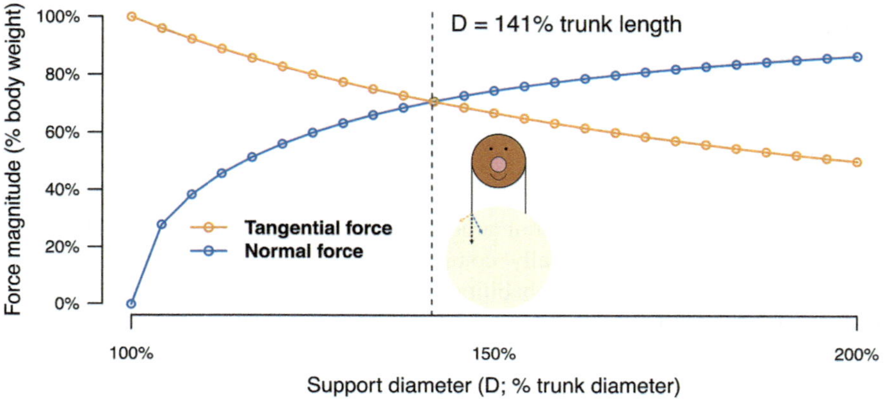

Fig. 11.3 Vertical forces exerted by an animal moving atop a branch (illustrated by the dashed black arrow in the inset figure) can be resolved into two components relative to the branch's cross-section: a normal component directed toward the center of the support (dashed blue arrow in inset) and a perpendicular tangential component directed along the edge of the support (dashed orange arrow in inset). As support diameter decreases relative to trunk diameter, the normal force component decreases as the tangential component increases, such that tangential forces predominate for any support diameter less than 141% of trunk diameter. For any support diameters less than or equal to trunk diameter, all force is tangential to the support. Decreasing support diameter therefore requires a mechanism to facilitate the substrate adhesion required to counteract tangential slipping (Cartmill, 1974, 1979, 1985). Note that this model assumes that the limbs are held in a neutral parasagittal position alongside the trunk. Limb adduction would necessarily decrease "effective trunk diameter" and allow the animal to move on narrower supports without increasing tangential force components (Higham & Jayne, 2004; Spezzano & Jayne, 2004)

2009). Similarly, field-based studies of sympatric primates have shown that both within and between species larger individuals preferentially travel and forage on larger supports (e.g., tree boughs), whereas smaller individuals use a more diverse array of supports (McGraw, 1998; Dunham et al., 2018; Williamson et al., 2021). Such correlations between body size and perch diameter are also present ontogenetically within species—Shapiro et al. (2014) showed that the small body size of juvenile short-tailed opossums (*Monodelphis domestica*) permits them to travel on cylindrical supports less than half as wide as the narrowest support traversable by adult conspecifics (e.g., Lemelin et al., 2003).

To test how arboreality potentially influences body size evolution, I used the PanTHERIA mammalian database (Jones et al., 2009) to examine the scaling of body mass versus body length in nine families of rodents (i.e., Capromyidae, Caviidae, Cricetidae, Echimyidae, Gliridae, Hystricidae, Muridae, Nesomyidae, Octodontidae, Sciuridae). I selected rodents overall, and these nine rodent families specifically, to provide a large sample of closely related species representing occupancy of both terrestrial and arboreal habitats. From the PanTHERIA database, I collated data on body mass (i.e., M_b), body length (i.e., L_b; the distance from the crown of the head to the base of the tail), and habitat preference (i.e., terrestrial *versus* arboreal, where "arboreal" was designated for any animal showing evidence

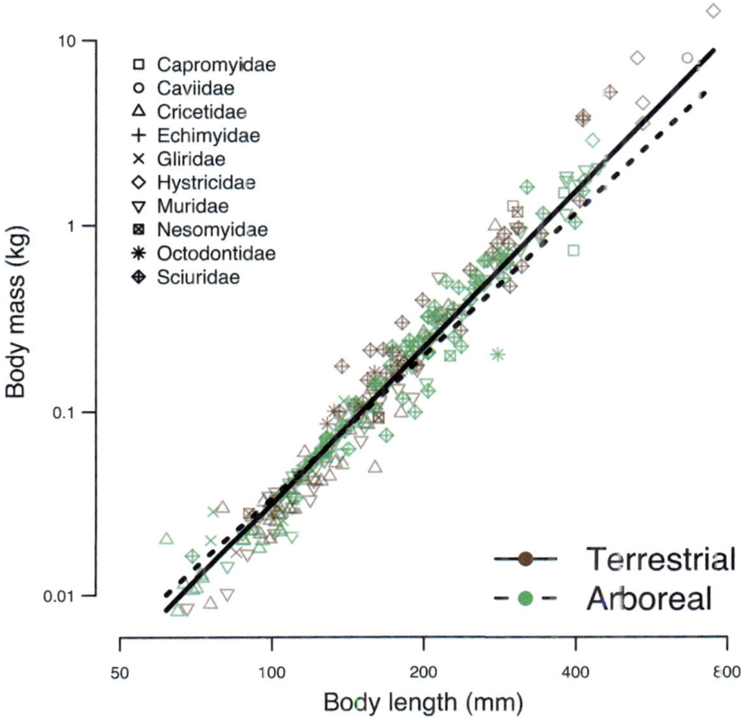

Fig. 11.4 Scaling of body mass to body length in arboreal and terrestrial rodents [morphological data from Jones et al. (2009)]. Trend lines represent phylogenetic least squares regression lines, grouped by habitat [phylogenetic relationships sampled from Bininda-Emonds et al. (2007)]. At longer body lengths, arboreal rodents have significantly lower body masses than closely related terrestrial species

of time spent on an above-ground support). Overall, the final dataset included 257 species of rodents, including 113 terrestrial species and 144 arboreal species. I used phylogenetic generalized least squares regression modeling to test the prediction that body mass versus length scales differently in terrestrial *versus* arboreal rodents, sampling phylogenetic relationships from the mammalian cladogram of Bininda-Emonds et al. (2007). Results supported the prediction that arboreality is associated with decreased body mass relative to body length, particularly as overall body size increases (Fig. 11.4). In terrestrial rodents, body mass scales to length at a value close to the isometric expectation of 3.0 (i.e., $M_b \propto L_b^{3.01}$), whereas in arboreal rodents, body mass scales to length with negative allometry (i.e., $M_b \propto L_b^{2.75}$). As a result, large-bodied arboreal rodents tend to be lighter than large-bodied terrestrial rodents of the same overall body length. At the upper end of the common body length distribution (i.e., a body length of 460 mm), the average body mass of an arboreal rodent is predicted to be only 74% that of a terrestrial rodent.

Body Shape Arboreal animals could reduce the propensity for slipping from narrow supports by altering other aspects of their body shape, independent of body size *per se*. For instance, all else being equal, a mediolaterally narrow trunk would have the effect of moving the limbs closer to the top of the support by reducing the mediolateral distance between the limbs, mitigating tangential forces, and promoting stability (or, conversely, decreasing the absolute diameter of the support the animal could safely walk on). This is perhaps the reason that chameleons, arguably the most committedly arboreal of all lizards (Fischer et al., 2010), are characterized by such mediolaterally compressed trunk shapes and more parasagittal limb orientations (Gans, 1967; Peterson, 1984). Similarly, within primates, arboreal monkeys - which most frequently travel and forage on top of arboreal supports - have mediolaterally narrow thoraces compared to the larger-bodied apes (who typically use suspensory postures to travel and forage beneath arboreal supports, mitigating stability challenges of narrow supports in a different way; Napier, 1967; Grand, 1972; Larson, 1998a). Even without changing trunk diameters, animals could, of course, position their extremities closer to the midline of the body, and closer to the top of the support, via increased limb adduction, flexion, and rotation (Fig. 11.1). For instance, Schmidt and Fischer (2010) showed that rats (*Rattus norvegicus*) decrease the horizontal distance between their hands and feet, respectively, by about 25–50% when moving on horizontal poles versus the flat ground, thereby shifting limb contacts much closer to the top of the support and reducing tangential forces. Similarly, lizards moving on narrow perches exhibit significantly greater limb flexion as support diameter decreases, primarily as a means of ensuring that the foot is placed closer to the top of the support despite their habitually sprawling limb posture (Higham & Jayne, 2004; Spezzano & Jayne, 2004). Overall, arboreal tetrapods typically have more mobile limb joints than their terrestrial counterparts (e.g., joints capable of greater angular displacements: Jenkins & Camazine, 1977; Peterson, 1984; Larson, 1998b; Larson et al., 2001; Sargis, 2002a, b; Fischer et al., 2010), enhancing their ability to reposition the limbs to take up more biomechanically advantageous positions during locomotion on narrow supports (Schmitt, 2003a). Coincident with such changes in limb position, previous studies have also shown that there is an increase in medially directed horizontal forces during locomotion on narrow supports versus on terrestrial ones, in accordance with increased limb adduction towards the midline of the support (Schmitt, 2003a; Carlson et al., 2005; Schmidt & Fischer, 2010; Krause & Fischer, 2013).

Barring morphological or behavioral adjustments to either reduce trunk diameter or adjust limb positioning, arboreal animals must necessarily develop mechanisms for increasing adhesion if they are to prevent tangential slippage from the sides of a narrow substrate. Increased adhesion can be accomplished in multiple ways. Non-mammalian arboreal taxa are often able to directly adhere to a support, either via wet adhesion (dependent on the formation of capillary bridges; seen in tree frogs) or dry adhesion (dependent on the formation of molecular bonds; seen in arboreal lizards) (Langowski et al., 2018; Chap. 9). Miller and Stroud (2021) recently showed that the presence of adhesive toepads in lizards is correlated with the evolution of arboreality, such that pad-bearing species more often transition to arboreality than padless species and are less likely to transition back to being terrestrial. In the absence of such direct adhesion mechanisms, protuberant volar pads (in many mammals) or specialized subdigital setae (in chameleons) can increase friction between the manus/pes and the surface of the support (Cartmill, 1974, 1979; Khannoon et al., 2014; Spinner et al., 2014; Maiolino et al., 2016), helping the animal produce reaction forces to counteract tangential sliding (Lammers & Biknevicius, 2004; Lammers, 2009b). Some studies have shown that arboreal carnivores and lizards tend to have more curved claws than terrestrial counterparts (Cartmill, 1985; Van Valkenburgh, 1987; Zani, 2000), permitting greater penetration into the substrate for increased mechanical interlocking and stability—though it is important to note that functional signals in claw curvature can be obscured by inadequate methods to quantify claw shape and confounding biases in how "functional groups" are delineated (Tinius & Russell, 2017). Lastly, strong, large grasping appendages permit arboreal animals to (1) increase normal forces independently of any forces induced by gravity and whole-body acceleration, thereby increasing friction and (2) reposition the center of pressure (i.e., the point on surface of the support through which resulting force vector is applied) to a more favorable location nearer the top of the branch (Cartmill, 1974, 1979, 1985; Lammers, 2009b). Several recent reviews have comprehensively considered specific morphological adaptations that increase grasping performance in tetrapods (Sustaita et al., 2013; Nyakatura, 2019; Pouydebat & Bardo, 2019; Chap. 12). Broadly, powerful arboreal grasping is facilitated by having large extremities, relatively long digital rays capable of increased abduction, flexion, and opposition, and well-developed and precise digital musculature (Manzano et al., 2008; Herrel et al., 2011, 2013a, b; Sustaita et al., 2013; Almécija et al., 2015; Young & Chadwell, 2020; Chap. 12).

11.2.2 Narrow Supports Challenge Mediolateral Stability

In addition to increasing the potential for tangential slipping from the sides of the support, narrow substrates also challenge mediolateral (i.e., transverse/rolling plane) stability. The mechanical rationale for this limitation is illustrated in Fig. 11.5. Translational (linear) vertical and mediolateral forces (i.e., F_V and F_{ML}), engendered by both gravity and muscle-powered acceleration of the body, have the potential to induce destabilizing torques (τ_{des}) in the transverse (rolling) plane. The magnitude of such torques depends on the magnitude of the forces and the distance of the animal's center of mass from the center of the support (illustrated by vectors *h* and *v* in Fig. 11.3). There are three fundamental strategies open to an animal to mitigate these destabilizing torques and maintain balance:

1. Reduce the net magnitude of the translational forces.
2. Reduce distance from the whole-body center of mass to the center of support.

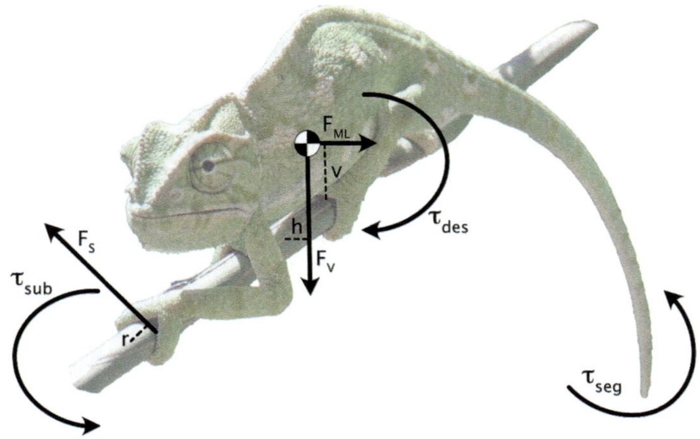

Fig. 11.5 Biomechanical determinants of mediolateral stability on narrow diameter supports, as represented by a free-body diagram of a common chameleon (*Chamaeleo chamaeleon*) balancing on a narrow stalk. Translational vertical (F_V) and mediolateral (F_{ML}) forces acting in the transverse/rolling plane engender a destabilizing torque (τ_{des}) which, unless counteracted, could topple the animal from the support. The magnitude of τ_{des} depends on four quantities: the magnitude of F_V, the magnitude of F_{ML}, and the perpendicular distances of each force vector from the center of the support (i.e., their moment arms, represented as h and v, respectively). To maintain stability, an animal moving on a narrow support must either reduce the magnitude of τ_{des} (by limiting the magnitudes of F_V, F_{ML}, h and/or v) or produce a countervailing stabilizing torque sufficient to mitigate τ_{des}. This stabilizing torque could arise from the animal's interactions with the substrate (i.e., τ_{sub}) or via angular acceleration of the animal's body segments relative to the whole-body center of mass (i.e., τ_{seg}). Chameleon photograph reproduced under a Public Domain CC0 license, downloaded from: https://pxhere.com/en/photo/817159)

3. Produce stabilizing torques to balance destabilizing torques. Such torque can come from interactions between the animal and the substrate (i.e., substrate reaction torques, τ_{sub}) or arise from dynamic movements of the animal's body segments relative to the whole body center of mass (i.e., segmental torques, τ_{seg}) (see Lammers & Zurcher, 2011a for further discussion).

Below, I consider several morphological and behavioral features of arboreal animals that permit them to exercise each of these strategies for increasing mediolateral stability on narrow supports.

11.2.2.1 Strategies that Promote Mediolateral Stability on Narrow Supports: 1. Reducing the Net Magnitude of Translational Forces

Several studies have demonstrated that animals can reduce the magnitude of translational forces imparted to the substrate when moving on narrow and confined supports (Schmitt, 2003b; Lammers & Biknevicius, 2004; Franz et al., 2005; Wallace & Demes, 2008; Young, 2009; Schmidt, 2011; Chadwell & Young, 2015;

Young et al., 2016; Young & Chadwell, 2020; Wölfer et al., 2021). Arboreal animals primarily reduce translational force magnitudes in two ways: by slowing down and by using compliant gait kinematics (also known as "grounded" locomotion).

Reducing Speed The first, and perhaps most straightforward strategy to reduce translational forces is by simply slowing down. Moving quickly requires increased forces to redirect and accelerate the center of mass (Farley, 1991; Weyand et al., 2000). Moving more slowly necessarily reduces the peak forces imparted to the support. Increased speed also increases the probability of improper limb placements and other locomotor "accidents" (Wheatley et al., 2015; Wynn et al., 2015; Amir Abdul Nasir et al., 2017). Accordingly, performance testing across a range of tetrapods (including anurans, squamates, marsupials, rodents, carnivores, and primates), has shown average speed is directly correlated with support diameter, such that the fastest speeds are observed on flat supports and speed progressively decreases with decreases in support size on cylindrical perches (Losos & Sinervo, 1989; Sinervo & Losos, 1991; Lammers & Biknevicius, 2004; Renous et al., 2010; Gálvez-López et al., 2011; Herrel et al., 2013a; Shapiro et al., 2014; Karantanis et al., 2015, 2017a, b, c, d; Clemente et al., 2019; Gaschk et al., 2019; Young & Chadwell, 2020; Granatosky et al., 2021; Wölfer et al., 2021). Although most of this research has been conducted in controlled laboratory environments, similar reductions in speed with decreasing support diameters have also been observed in field-based studies of *Anolis* lizards, tree squirrels, and primates (Mattingly & Jayne, 2004; Dunham et al., 2019b, 2020). Positive correlations between average speed and support diameter are particularly salient in locomotor generalists that are not habitually restricted to the arboreal habitat. For instance, studies of how perch diameter affects sprinting speeds in lizards have shown that more terrestrial species/populations decrease speed on narrower perches whereas more arboreal forms are able to maintain speed (Losos & Sinervo, 1989; Sinervo & Losos, 1991). Similar results have been obtained in studies comparing the locomotor kinematics of grasping versus non-grasping marsupials, rodents, and primates. Species able to grasp only poorly (e.g., *Monodelphis domestica*, *Sciurus carolinensis*, and *Callithrix jacchus*) reduce speed as support diameter narrows whereas size-matched species with more pronounced grasping abilities (e.g., *Petaurus breviceps* and *Saimiri boliviensis*) are able to maintain speed across changes in support diameter (Shapiro et al., 2014; Young & Chadwell, 2020).

Interestingly, some studies have also found that a few animals *increase* speed on narrower supports (Delciellos & Vieira, 2007; Gálvez-López et al., 2011; Camargo et al., 2016). In fact, many mammals predominantly use fast, asymmetrical gaits when moving on arboreal supports (Young, 2009; Schmidt, 2011; Schmidt & Fischer, 2011; Chadwell & Young, 2015; Young et al., 2016; Clemente et al., 2019; Dunham et al., 2019b, 2020; Gaschk et al., 2019; Young & Chadwell, 2020; Granatosky et al., 2021; Wölfer et al., 2021). This seemingly counterproductive approach could represent an alternative strategy for improving balance on narrow supports because moving faster can help animals maintain greater dynamic stability in the mediolateral plane (Bruijn et al., 2009; Young & Chadwell, 2020).

Table 11.1 Kinematic characteristics of compliant gaits

Feature	Definition
Increased duty factor	Quotient of stance phase duration and stride duration (Hildebrand, 1966).
Use of continuous contact gait patterns	Quadrupedal gaits are designated as symmetrical or asymmetrical (Dagg, 1973; Hildebrand, 1980; Abourachid, 2003; Hutchinson et al., 2019). In a symmetrical gait (e.g., walks, paces, and trots), paired movements occur between the fore- and hindlimbs on one side, and the movements left-side and right-side forelimb-hindlimb pairs mirror one another but are separated in time. In asymmetrical gaits (e.g., canters, gallops, and bounds), which are typically used at faster speeds, paired movements occur between limbs attached to the same girdle (i.e., between left/right forelimbs and left/right hindlimbs), and forelimb-hindlimb pairs do not mirror one another. Within symmetrical and asymmetrical gaits, certain footfall patterns emphasize quick, simultaneous limb contacts separated by whole-body aerial phases (e.g., symmetrical trots/paces or asymmetrical bounds), whereas others result in temporally distributed limb contacts with reduced aerial phases (e.g., symmetrical walks, or asymmetrical canters and gallops). Use of distributed limb contact gaits reduces peak force magnitudes and flattens the trajectory of the center of mass (Schmitt et al., 2006).
Increased stride length	Distance travelled by the center of mass during a complete locomotor stride (Alexander & Maloiy, 1984).
Increased limb angular excursion	Total angle swept by the forelimb/hindlimb during stance phase (Larson et al., 2001).
Increased mid-joint yield	Change in elbow/knee angle from the beginning to the middle of the stance phase (i.e., from limb touchdown to mid-stance).

Conversely, some highly arboreal animals, such as chameleons and lorises, appear to have become particularly specialized for slow-speed locomotion on narrow supports (Peterson, 1984; Abu-Ghalyun et al., 1988; Demes et al., 1990; Nekaris, 2005; Herrel et al., 2013b). In these cases, the advanced grasping abilities of these animals facilitate stability on even the narrowest perches, while slow speed can enhance crypsis during both predation and predator evasion.

Use of Compliant Gait Kinematics In addition to reducing speed, arboreal animals moving on narrow supports often adopt a coordinated suite of kinematic adjustments variously referred to as "compliant gait" (Alexander & Jayes, 1978; Schmitt, 1999), "grounded running" (Hutchinson et al., 2003; Rubenson et al., 2004; Biknevicius, 2006; Schmitt et al., 2006; Biknevicius et al., 2013), or "Groucho running" (in homage to the distinctive gait of Groucho Marx; McMahon et al., 1987; McMahon & Cheng, 1990). These adjustments include lengthened limb contact durations (i.e., duty factors), switching to gaits that facilitate more continuous contact with the substrate (and fewer whole-body aerial phases), and decreasing overall limb stiffness via increased limb joint excursions during stance phase (Table 11.1). Overall, compliant gait kinematics permit animals to smooth the path of the center of mass, reduce peak loading forces imparted to the substrate, reduce rapid redirections of the center of mass (i.e., "collisions"), and maintain substrate contact throughout the stride (McMahon, 1985; Ruina et al., 2005; O'Neill & Schmitt, 2012)—all of which should promote stability during locomotion on narrow and compliant supports, although at the possible cost of increased muscular work and energy expenditure (McMahon et al., 1987).

Several studies have indicated increased use of compliant gait kinematics on narrow supports in a wide variety of animals. For instance, studies of animals as diverse as lizards, rodents, marsupials, and primates broadly reveal that there is an inverse correlation between mean duty factor and support size, even after controlling for variation in speed (which is important, given the broadly negative correlation between speed and duty factor) (Lemelin & Cartmill, 2010; Schmidt & Fischer, 2010; Shapiro & Young, 2010, 2012; Foster & Higham, 2012; Shapiro et al., 2014, 2016b; Chadwell & Young, 2015; Karantanis et al., 2015, 2017a, b, c, d; Young et al., 2016; Clemente et al., 2019; Dunham et al., 2019b, 2020; Gaschk et al., 2019; Young & Chadwell, 2020; Wölfer et al., 2021). Typically, increases in duty factor are accompanied by changes in gait selection, shifting from footfall patterns characterized by simultaneous footfalls (i.e., symmetrical trots and asymmetrical bounds/half-bounds) to gaits with more distributed contacts (i.e., symmetrical walks or ambles and asymmetrical canters and gallops) (Young, 2009; Gálvez-López et al., 2011; Shapiro et al., 2016b; Young et al., 2016; Dunham et al., 2019b, 2020; Wölfer et al., 2021). Fewer studies have directly examined how limb joint kinematics change with support diameter but have generally found greater stance phase angular excursions with decreasing support diameter (Peterson, 1984; Schmitt, 1994, 1998, 1999, 2003b; Franz et al., 2005; Schmitt et al., 2010; O'Neill & Schmitt, 2012; Young, 2012). Moreover, broad surveys of limb kinematics in a variety of mammals have demonstrated that taxa that frequently travel arboreally (e.g., primates and marsupials) consistently move with greater stance phase limb joint excursions and use longer strides (hallmarks of compliant joint kinematics) relative to size matched terrestrial taxa, regardless of the type or diameter of locomotor support (Alexander & Maloiy, 1984; Reynolds, 1987; Larson et al., 2000, 2001; Larney & Larson, 2004). Similarly, chameleons (*Chamaeleo* sp.) move with exaggerated limb joint excursions and relatively longer strides when compared to less arboreal lizards, providing further support that arboreal adaptation leads to the consistent use of compliant joint kinematics during quadrupedal locomotion (Peterson, 1984; Fischer et al., 2010). Nevertheless, others have argued that, at least in the case of primates, what are referred as "compliant limb kinematics" are instead typical kinematic patterns for most small, non-cursorial mammals that only appear unique in comparison to the kinematic patterns of large mammalian cursors (Schmidt, 2005a). Further study of how primates and other arboreal animals adjust limb stiffness to support properties are needed to elucidate whether "compliant joint kinematics" may constitute a specific adaptive response to support precarity or may simply be a plesiomorphic movement pattern typical of non-cursorial animals.

11.2.2.2 Strategies that Promote Mediolateral Stability on Narrow Supports: 2. Reduction of the Distance of the from the Whole-Body Center of Mass to the Center of the Support

Animals travelling on narrow perches can also reduce the magnitude of destabilizing torques by limiting the distance between their center of mass and the center of the pole. The most straightforward means of doing so is reduce effective limb length (i.e., the distance from the hip/shoulder to the manus/pes), either morphologically by reducing anatomical limb length or behaviorally by increasing limb joint flexion. Previous studies of arboreal animals have found evidence for both solutions.

Shorter Limbs Species and populations of lizards that more commonly move on narrow arboreal supports frequently have relatively shorter limbs compared to groups that favor wider supports or terrestrial environments (Losos & Sinervo, 1989; Losos, 1990, 2009; Sinervo & Losos, 1991; Losos & Irschick, 1996; Hopkins & Tolley, 2011). Furthermore, Losos et al. (1997) found that populations of *Anolis sagrei* lizards experimentally introduced to narrower branch environments than evident in their native habitat evolved shorter limbs within a little as 10–14 years, indicating active directional selection on limb length in response to variation in perch diameter. Short-limbed narrow branch specialists are better able to maintain speed across decreases in perch diameter and suffer fewer falls when moving on narrow supports (Losos & Sinervo, 1989; Sinervo & Losos, 1991). Future research in this vein should also quantify how arboreality specifically affects the length of anatomical segments within the limb. For a sprawling animal, decreasing the length of the zeugopod (i.e., antebrachium/crus) would presumably have a greater effect on center of mass height than decreasing the length of the stylopod (brachium/thigh) or autopod (manus/pes).

Recently, Hagey et al. (2017) found evidence that among gekkonids narrow branch specialists tended to have relatively *longer* limbs than species that favored wider supports. They hypothesized that the greater clinging ability of gekkonid toepads might free them from the stability constraints seen in anoles and other previously investigated lizard species.

Increased Limb Flexion In contrast to most arboreal lizards, arboreal primates, carnivorans, and rodents tend to have relatively longer limbs for their size than terrestrial members of their orders (Polk et al., 2000). Although long limbs can be beneficial for other aspects of arboreal locomotion—such as bridging gaps in the arboreal canopy—long limbs also elevate the whole-body center of mass far above the support and potentially increase the magnitude of destabilizing torques. As such, many mammals (and chameleons) use more crouched postures with increased limb flexion when moving on narrow supports, thereby decreasing effective limb length (Napier, 1967; Taylor, 1970; Rose, 1973; Peterson, 1984; Cartmill, 1985; Walker, 1998; Schmitt, 2003b; Stevens, 2003; Spezzano & Jayne, 2004; Schmidt & Fischer, 2010). In this sense, the sprawling posture of arboreal lizards and amphibians may confer a preexisting advantage relative to relatively erect-limbed mammals, as a sprawling posture itself lowers center of mass height (Gatesy, 1991).

Additional Means of Managing Center of Mass Position Short limbs and crouched postures primarily reduce the average vertical distance between the center of mass and the support (i.e., position vector *v* in Fig. 11.5). Reducing the average mediolateral distance between the center of mass and the support (i.e., position vector *h* in Fig. 11.5) requires precise control of the path of the center of mass as the animal traverses the support. Studies of semicircular canal morphology in *Anolis* lizards, diprotodont marsupials, and rodents have suggested increased vestibular sensitivity (particularly in the mediolateral plane) in taxa that frequently move in complex arboreal environments with a high density of narrow supports (Schmelzle et al., 2007; Dickson et al., 2017; Bhagat et al., 2021), perhaps facilitating greater control over center of mass deviations during locomotion. Additionally, arboreal taxa frequently have more massive distal segments of their limbs (Preuschoft & Günther, 1994; Raichlen, 2005) and tails (Horner, 1954; Martin, 1968; Wilson, 1972; Siegel & van Meter, 1973; Grand, 1977; Irschick et al., 1997; Delciellos & Vieira, 2007; Hayssen, 2008; Russo & Shapiro, 2011; Sheehy III et al., 2015; Mincer & Russo, 2020), both of which could serve as counterweights to dampen center of mass movements, particularly during slow locomotion (Rose, 1974; Young et al., 2015).

11.2.2.3 Strategies for Promoting Mediolateral Stability on Narrow Supports: 3. Production of Stabilizing Torques to Balance Destabilizing Torques

The final strategy for mitigating destabilizing torques on narrow diameter supports is to produce countervailing stabilizing torques. Broadly, there are two means by which arboreal animals can produce such stabilizing torques (Lammers & Zurcher, 2011a): (1) by generating substrate reaction torques via direct interactions between the animal's extremities and the surface of the support (i.e., τ_{sub} in Fig. 11.5), or (2) by generating appropriately timed angular accelerations of the body's segments relative to the whole body center of mass (i.e., τ_{seg} in Fig. 11.5).

Substrate Reaction Torques Effectuating substrate reaction torques requires the production of shear-based reaction forces oriented tangentially to the surface of the support. Requisite shearing forces can be produced in two ways—as a byproduct of translational forces associated with supporting and moving body mass (see Fig. 11.3), or via the rotary action of muscular grasping limbs, acting independently of translational force production *per se*. Lammers and Gauntner (2008) refer to the resulting torques produced by these two means as, respectively, "substrate reaction force torque" (τ_{SRF}) and "muscular torque" (τ_{musc}) (see also Chadwell & Young, 2015). To produce muscular torques, an animal must exert opposing normal forces on either side of the support (Cartmill, 1974, 1985), either with a single grasping limb or a pair of left and right limbs (what Lammers and Gauntner (2008) refer to as the "opposing digits strategy" and the "opposing limbs strategy", respectively).

Only a few studies have directly measured rolling plane torque production during narrow support locomotion, likely because doing so requires specialized transducers (Lammers & Gauntner, 2008; Lammers, 2009a; Lammers & Zurcher, 2011a, b; Chadwell & Young, 2015; Young & Chadwell, 2015, 2020; Clemente et al., 2019). Previous studies have focused exclusively on relatively small mammals characterized by varying degrees of arboreality, including the gray short-tailed opossum (*Monodelphis domestica*) and quoll (*Dasyurus hallucatus*), both of which are marsupials, and the common marmoset (*Callithrix jacchus*) and squirrel monkey (*Saimiri boliviensis*), which are primates. Some commonalities have emerged from this research. For example, cumulative torque (i.e., total change in angular momentum) almost never equals zero over the duration of a single stride, suggesting that animals maintain stability across a bout of locomotion, rather than within strides *per se*. Additionally, previously-studied animals show a tendency to produce opposing τ_{SRF} with the left and right limbs, consistent with the placement of alternating limbs on either side of the support during locomotion (Fig. 11.6) (Lammers & Gauntner, 2008; Clemente et al., 2019). However, most studies have also found that total rolling plane torque cannot be predicted by τ_{SRF} alone, implying substantial use of τ_{musc} independent of τ_{SRF} (Fig. 11.6). Even "non-grasping" taxa (i.e., those possessing relatively short digits and small extremities) have been shown to exert substantial muscular torque during narrow support locomotion (Lammers & Gauntner, 2008; Chadwell & Young, 2015), likely indicating an "opposing limbs" strategy for torque production in lieu of the "opposing digits" strategy available to species with a more highly developed grasping apparatus. Many of the previously studied "non-grasping" mammals frequently use high-speed asymmetrical gallops, half-bounds, and bounds when moving on narrow supports (Lammers & Zurcher, 2011b; Chadwell & Young, 2015; Young & Chadwell, 2020; Granatosky et al., 2021; Wölfer et al., 2021). Chadwell and Young (2015) argued that asymmetrical gaits may permit such animals to exert opposing torques between the opposing left and right limbs of a girdle, creating an "effective grasping limb".

It has also been suggested that the atypical footfall sequences characteristic of the walking and running gaits of primates and other arboreal mammals may promote stability by facilitating the production of balanced torques from the contralateral fore- and hind- limbs (Schmidt, 2005b). Most quadrupedal tetrapods use "lateral sequence" gaits (adopting the terminology of Hildebrand, 1966, 1967), where the hindlimb contacts are followed by forelimb contacts on the same side of the body; e.g., a complete quadrupedal gait sequence could be right hindlimb → right forelimb → left hindlimb → left forelimb, such that the progression of hindlimb contacts to forelimb contacts occurs ipsilaterally underneath the body. In contrast, primates and several other arboreal mammals use what Hildebrand (1967) referred to as "diagonal sequence" gaits. Here, hindlimb contacts are followed by forelimb contacts on the contralateral side of the body; e.g., right hindlimb → left forelimb → left hindlimb → right forelimb, such that the progrssion of hindlimb contacts to forelimb contacts

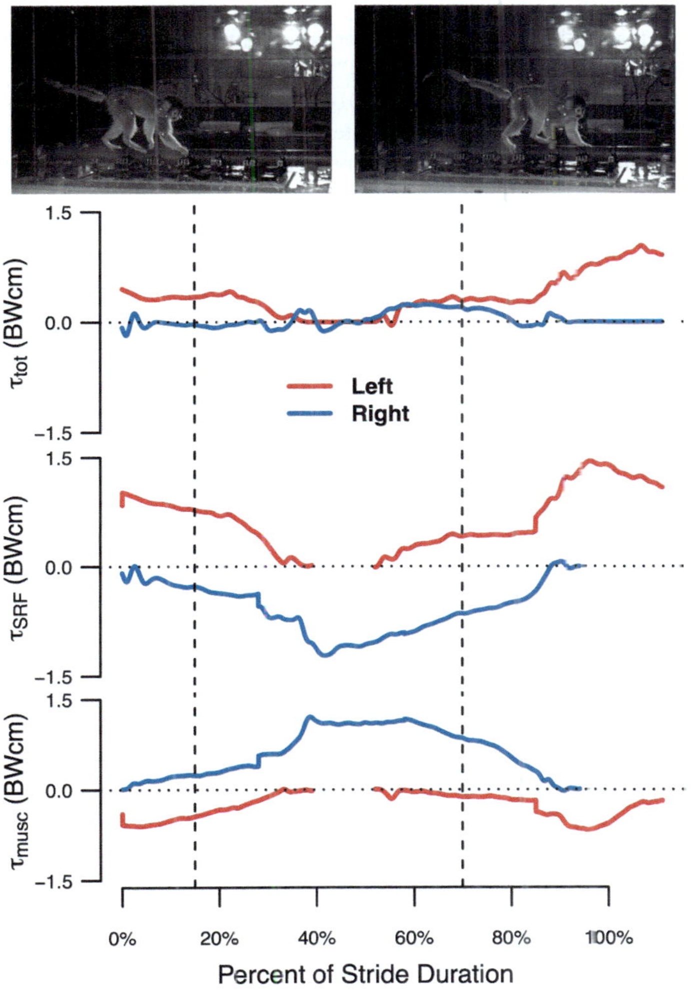

Fig. 11.6 Rolling plane torque production during narrow support locomotion. Plotted data are from an exemplar stride of a 526 g juvenile squirrel monkey (*Saimiri boliviensis*) walking across a 2.5 cm diameter instrumented pole (unpublished data from JW Young and BA Chadwell). From top to bottom, line series plots show total rolling torque (τ_{tot}) as well as its sub-components: substrate reaction force torque (τ_{SRF}) and muscular torque (τ_{musc}) (see Lammers and Gauntner (2008) and Chadwell and Young (2015) for explanations of how τ_{SRF} and τ_{musc} are calculated from τ_{tot}). Sample video frames indicate instances in which the animal was supported by a left-right pair of fore- and hindlimbs. Although left and right limbs often exert countervailing τ_{SRF}, commensurate with their placement on either side of the support, τ_{musc} typically acts in opposition to τ_{SRF}. As a result, left and right τ_{tot} occasionally move in opposition to one another, but also occasionally move in unison

proceeds diagonally underneath the body. As implied by their name, the diagonal sequence gaits used by primates and other arboreal mammals emphasize periods of support by contralateral (i.e., "diagonal") fore- and hindlimb couplets (Cartmill et al., 2002, 2007a; Shapiro et al., 2014), likely facilitating the production of countervailing τ_{SRF}. Although appropriately phased lateral sequence gaits can also emphasize periods of diagonal limb support, Cartmill et al. (2002, 2007a, b) have argued that primate-like diagonal sequence gaits alone facilitate diagonal bipedal support while simulataneously ensuring that a grasping hindlimb is securely planted near the whole-body center of mass at the moment of forelimb touchdown, permitting the animal to safely recover in the event of a misplaced forelimb step on an upcoming precarious support (see Usherwood & Smith, 2018 for a similiar argument based on strategies for mitigating pitching and rolling torques throughout the gait cycle). Though the specific mechanical predictions of the Cartmill et al. (2002, 2007a, b) hypothesis deserve further testing (Shapiro & Raichlen, 2005, 2007), the prediliction of arboreal mammals of favoring diagonal sequence gaits cannot be ignored. In a recent comprehensive phylogenetic analysis of quadrupedal gait sequencing in gnathostomes, Wimberly et al. (2021) demonstrated a strong association between arboreality and diagonal sequence gait usage across several mammalian clades, indicating multiple evolutionary convergences on this gait pattern associated with arboreality.

Finally, it is possible that the prehensile tails characteristic several arboreal animals could be used as an additional grasping limb capable of generating stabilizing torques on narrow arboreal supports. Prehensile tails have convergently evolved at least 14 times among arboreal mammals (Emmons & Gentry, 1983; Organ et al., 2009), as well in arboreal chameleons (Peterson, 1984; Boistel et al., 2010; Fischer et al., 2010; Herrel et al., 2011, 2013b)—strong evidence of their utility in a precarious arboreal environment. Although prehensile tails are most often considered as an adaptation relating to posture, particularly in a foraging context (Meldrum, 1998), some studies have also shown that prehensile tail use occurs during active arboreal locomotion, where the tail is used as a "fifth limb" that interacts with the support to produce balancing torques on precarious supports (Peterson, 1984; Bergeson, 1996; Garber & Rehg, 1999; Lawler & Stamps, 2002; Boistel et al., 2010).

Segmental Torques In addition to substrate reaction torques, arboreal animals could use segmental angular accelerations relative to the whole-body center of mass to counteract destabilizing torques and maintain stability on narrow supports. Because the fore- and hindlimbs are typically needed for direct substrate interactions, tails are often employed as a free segmental mass able to effect such torques about the whole-body center of mass (Schwaner et al., 2021). Arboreal taxa frequently have relatively longer and more massive tails than closely related terrestrial counterparts (Horner, 1954; Martin, 1968; Wilson, 1972; Siegel & van Meter, 1973; Grand, 1977; Irschick et al., 1997; Delciellos & Vieira, 2007; Hayssen, 2008; Russo & Shapiro, 2011; Sheehy III et al., 2016; Mincer & Russo, 2020). Long and massive tails promote stability in two ways. First, as mentioned above, they can serve as static counterweights to move the center of mass closer to the support (and even below it). However, tails can also be used dynamically to regulate whole-body angular momentum. Commensurate with this hypothesis, studies have shown that domestic cats, mouse lemurs, squirrel monkeys, and several species of cercopithecine monkeys all increase rotary movement of the tail as support dimensions narrow (Walker et al., 1998; Larson & Stern, 2006; Young et al., 2021). Similarly, experimental studies of lizards, mice, cats, and squirrel monkeys have shown that diminution or loss of the tail causes increased difficulty maintaining balance on narrow or mobile supports (Ballinger, 1973; Igarashi & Levy, 1981; Walker et al., 1998; Jusufi et al., 2008; Hsieh, 2016; Shapiro et al., 2016a).

11.3 Biomechanical Challenges Presented by Compliant Arboreal Supports

Mechanically, arboreal supports can generally be modeled as cantilevered beams (e.g., the projecting limbs, branches, and twigs of the tree crown) or suspended beams (e.g., lianas attached at either end to their host supports). In either case, bending predominates as the primary mode of loading (McMahon & Kronauer, 1976). Their mechanical stiffness (i.e., resistance to deflection) is thus proportional to the cross-sectional dimensions of support and the material properties of its constituent materials—in other words, narrow supports are typically bendy, but some can be more so than others. Because potential supports in an arboreal environment vary greatly in both their cross-sectional dimensions and their material properties (Fig. 11.2; Mattingly & Jayne, 2004), they are similarly variable in the degree to which they are compliant (Gilman & Irschick, 2013; van Casteren et al., 2013; Dunham et al., 2018). For instance, in surveys of representative locomotor supports used by orangutans in Southeast Asian, and platyrrhine monkeys in South American forests, van Casteren et al. (2013) and Dunham et al. (2018), respectively, found that support compliance varied over 2–4 orders of magnitude. Even more dramatically, Gilman and Irschick (2013) found that the compliance of arboreal locomotor supports used by *Anolis carolinensis* in Florida (Southeastern USA) ranged over six orders of magnitude. These studies have also shown that realized compliance of an arboreal support measured at a given distance along its length is proportional to (1) the diameter of the support, (2) the total length of the support (proportional to total support inertia), and (3) the distance of the load point from the proximal branching point (proportional to the load arm of the bending moment). Dunham et al. (2018) found that the combination of support diameter, total support length, and distance of the load point to the proximal attachment explained 80% of the variance in the branch compliance in their sample. Some of the unexplained variance is likely related to interspecific and ontogenetic variation in the material properties of the plants in the sample.

Compliant supports present two primary challenges to safe and efficient arboreal locomotion. Firstly, compliant supports absorb some of the mechanical energy that could otherwise be used to accelerate the center of mass during locomotion. Field studies of wild arboreal primates have noted that supports frequently move and sway during locomotion, indicative of energetic losses (Morbeck, 1977; Thorpe et al., 2007a, b, 2009; Young et al., 2019, 2021). Although it is theoretically possible to regain this energy when the support recoils, existing evidence suggests that doing so is rare for most arboreal animals (Alexander, 1991; Bonser, 1999). Specifically, the low resonant frequency of typical arboreal supports means that the period of recoil rarely coincides with critical periods of impulse generation during locomotion. As such, jumping animals will typically lose contact before the support has sufficiently recoiled to impart energy to the center of mass, compromising jumping performance as support compliance increases (Crompton et al., 1993; Demes et al., 1995; Warren & Crompton, 1997; Walker, 2005; Channon et al., 2011; Gilman et al., 2012; Gilman & Irschick, 2013; Reynaga et al., 2019). In an exception to this general rule, Astley et al. (2015) found that the exceedingly long limbs of jumping Cuban tree frogs can permit them to extend the period of impulse generation long enough to recover some of the lost energy when the support recoils.

Secondly, compliant supports complicate locomotor control by making the dynamics of substrate interactions during locomotion less predictable (MacLellan & Patla, 2006; Chang et al., 2010). Fundamentally, all locomotion depends on animals being able to harness Newton's Third Law of Motion: they impart forces to the environment and use the resulting reaction forces from this interaction to accelerate the center of mass. By absorbing some of the forces imparted by the animal, and then returning this energy at effectively random times and in random directions (at least with respect to animal's desired movement dynamics), compliant supports upset the predictability of this relationship. To put it simply, compliant supports are "unreliable partners" for locomotion.

11.3.1 Behavioral Strategies for Mitigating the Effects of Arboreal Support Compliance

Few studies have directly investigated how arboreal animals adapt their locomotion to increased support compliance. However, existing studies suggest two broad strategies for mitigating the energetic losses and instability induced by moving on compliant supports.

Firstly, field-based ecological studies of both lizards and mammals moving in their natural habitats have shown that arboreal animals tend to selectively avoid strongly compliant supports. For instance, Gilman and Irschick (2013) showed that wild green anole lizards (*Anolis carolinensis*) selectively jumped from perches that were significantly less compliant than the average perch available in the environment, and less compliant than the average perch they chose for other locomotor and postural activities (e.g., walking and sitting). Similarly, in a survey of perch use in a natural population of four sympatric species of *Anolis* lizards, Mattingly and Jayne (2004) showed that most species chose locomotor supports that were wider than the average diameter of potential supports in the environment, consistent with a strategy of selecting relatively stiff perches for locomotion. Leaping bushbabies (*Galago senegalensis*) and platyrrhine monkeys (*i.e., Chiropotes* and *Pithecia*) also choose launching substrates that are relatively broad, particularly in comparison to the substrates upon which they land. Preliminary data from my laboratory on support choice in seven sympatric monkey species at Tiputini Biodiversity Station in the Yasuní Biosphere Reserve in Ecuador show that large-bodied atelids (i.e., *Ateles, Alouatta*, and *Lagothrix*) selectively move on less compliant supports than the sympatric relatively small-bodied pithecids (i.e., *Callicebus* and *Pithecia*) and cebids (i.e., *Saguinus* and *Saimiri*) (Young et al., 2019). Finally, a recent semi-naturalistic study by Hunt et al. (2021) suggests that decisions made by leaping squirrels about *where* to launch from the length of a compliant perch are motivated by the need to mitigate energetic losses due to support displacement. Specifically, by varying both the compliance of the launch support and the distance to the landing target, Hunt et al. (2021) showed that squirrels consistently choose a launch point closer to the fixed end of the support, even if this means having to cover a greater distance in order to reach the target.

Secondly, a few laboratory and field studies have investigated how arboreal animals alter their locomotor kinematics to limit compliant support displacement, mitigating energy losses and promoting stability. Because support displacement is directly correlated with loading magnitudes, many of these adjustments are comparable those made to reduce translational forces on narrow supports (see Sect. 11.2.2.1). For example, in a laboratory study of how quadrupedal common marmosets (*Callithrix jacchus*) adjust their locomotor kinematics to variation in support diameter and compliance, Young et al. (2016) showed that speed, substrate contact duration, and total center of mass movement accounted for >60% of the variation in compliant support displacement, demonstrating that, just as with narrow supports, reducing speed and using "compliant" gait kinematics are effective strategies for promoting stability on compliant supports, likely by reducing the magnitude of forces imparted to the substrate. Similarly, jumping Cuban tree frogs (*Osteopilus septentrionalis*) reduce peak support loads during take-off by ~40% on compliant versus rigid substrates, despite the attendant reductions in jump power and consequent decrease in jump performance (Reynaga et al., 2019). At the other end of the body size spectrum of arboreal mammals, Thorpe and colleagues have shown that orangutans foraging in the terminal branch zone of the arboreal canopy—where most supports are highly compliant—distribute their body weight among a significantly greater number of supports and use more irregular gait sequences, both methods of effectively limiting compliant branch displacement (Thorpe et al., 2009; Myatt & Thorpe, 2011). Finally, large arboreal animals can effectively negotiate compliant supports by moving from above-branch positions to more stable suspensory postures (Grand, 1972; Bergeson, 1996; Guillot, 2011).

11.4 Conclusions

Movement on narrow and compliant arboreal supports presents a distinct set of challenges to locomotor efficiency and stability, challenges that are defined by a limited set of biomechanical relationships (Cartmill, 1985). In the context of the specific issues reviewed in this chapter, there are a limited number of ways to prevent tangential slipping and promote mediolateral stability on narrow supports, or to limit energy losses and mitigate support displacement on compliant ones. Because all arboreal animals must negotiate this common set of physical challenges, in an environment where clumsy mistakes can lead to tragedy (or at least to the increased energetic burden of having to fight gravity to regain a lost position in the canopy), it is not surprising that we see widespread convergence of locomotor morphology and behavior among arboreal tetrapods.

Common morphological adaptations to arboreal quadrupedalism include relatively small body size, well-developed prehensile extremities (both grasping hands/feet and prehensile tails), relatively long limbs (among some radiations), and long and heavy tails. To some degree, these morphological traits are common across several disparate arboreal radiations, though some patterns are more localized within specific clades. For instance, whereas arboreal mammals typically have relatively longer limbs than closely related terrestrial counterparts, arboreal lizards often have relatively shorter limbs than non-specialized taxa (but see Hagey et al., 2017). This discrepancy in arboreal adaptation between most lizards and mammals is likely due to differences in overall body mass. Although some arboreal lizards can be quite large—e.g., *Varanus salvadorii* can reach a snout-vent length (SVL) of nearly a meter and a body mass of 15 kg—most are quite small, with a median SVL of 80 mm and body mass of ~10 g (Meiri, 2018). Arboreal rodents, by contrast have a median body mass of ~100 g, and arboreal primates and carnivores are even larger (Jones et al., 2009). The small body size of arboreal lizards effectively limits the scope of their available habitat to supports in their immediate vicinity (Mattingly & Jayne, 2004)—though leaping between supports still provides one avenue of broadening the microhabitat (Toro et al., 2004). In such a situation, short limbs promote stability by limiting the magnitude of destabilizing torques, as described in Sect. 11.2.2.2. By contrast, the larger body size and longer limbs of arboreal mammals facilitates bridging across disparate arboreal supports, permitting greater access to all parts of the canopy—although at the potential cost of an increased distance of the center of mass from the support.

Existing data suggest that behavioral convergences among arboreal quadrupeds are likely to be even more widespread than morphological convergences. This pattern is not surprising, given the tendency for behavioral changes to often precede morphological adaptations, as well as the increased evolutionary lability of some (but not all) species-typical behaviors (Garland Jr & Losos, 1994; Rendall & Di Fiore, 2007; Higham et al., 2015). For instance, arboreal anurans, lizards, and mammals all respond to the increased stability challenges of precarious arboreal supports by reducing speed and increasing the use of compliant gait kinematics.

Such behavioral changes are typically more common and more pronounced in species of relatively large body size, although less so if such species possess additional morphological adaptations to increase stability (such as well-developed prehensile extremities) (Young & Chadwell, 2020; Schapker et al., 2022). There seem to be many ways to be a successful small arboreal quadruped, but relatively few ways to be a large one, at least without compromising safety and/or energetic burden (Jenkins, 1974; Shapiro et al., 2014).

Despite pervasive trends of morphological and behavioral convergence among arboreal quadrupeds, some broad performance-related differences still pertain between major groups. For instance, several studies have noted that arboreal lizards seem to be more challenged by decreases in perch diameter than other aspects of support variation (such as changes in steepness; Higham & Jayne, 2004; Spezzano & Jayne, 2004), whereas arboreal mammals show the opposite trend (Stevens, 2007; Nyakatura & Heymann, 2010; Shapiro & Young, 2010; Shapiro et al., 2014). This difference may ultimately stem from class-level differences in overall bauplan and basic locomotor mode. The sprawling posture of most non-mammalian tetrapods, which often correlates with increased lateral bending of the vertebral column during quadrupedalism (particularly when moving quickly) undoubtedly results in increased mediolateral movement of the center of mass, increasing the potential for destabilizing torques that could compromise mediolateral balance (see Fig. 11.5). By contrast, the comparatively parasagittal orientation of mammalian limbs, and the expanded gait repertoire of mammals, could help mitigate mediolateral balance disruptions on narrow perches (Fischer et al., 2002; Wimberly et al., 2021). Indeed, as part of their hyper-specialization for slow arboreal locomotion along discontinuous supports, chameleons' limb morphology and joint kinematics have evolved to become somewhat convergent on the parasagittal, erect-limb mammalian condition, a configuration that Peterson (1984) argued would beneficially permit them to limit mediolateral loading on unstable perches.

Two avenues for future research into arboreal locomotor adaptation would be particularly fruitful. First, studies integrating quantitative kinematic/kinetic data from the laboratory with holistic ecological data on substrate use and support morphology from the field will be critical for furthering our understanding of how locomotor anatomy and behavior are shaped by the rigors of the natural arboreal environment. Such studies are relatively depauperate in the literature on mammalian arboreal locomotion, and model studies on arboreal lizards could serve as templates for how to proceed (e.g., Garland Jr & Losos, 1994; Losos, 2009). Second, expanding available data on *in vivo* locomotor kinetics in non-mammalian arboreal quadrupeds (e.g., force and torque production during narrow support locomotion) could improve our understanding of how variation in fundamental bauplans and locomotor modes impacts stability and other aspects of performance in the arboreal milieu.

References

Abourachid, A. (2003). A new way of analysing symmetrical and asymmetrical gaits in quadrupeds. *Comptes Rendus Biologies, 326*(7), 625–630.

Abourachid, A., Fabre, A. C., Cornette, R., & Höfling, E. (2017). Foot shape in arboreal birds: Two morphological patterns for the same pincer-like tool. *Journal of Anatomy, 231*(1), 1–11.

Abu-Ghalyun, Y., Greenwald, L., Hetherington, T. E., & Gaunt, A. S. (1988). The physiological basis of slow locomotion in chamaeleons. *The Journal of Experimental Zoology, 245*(3), 225–231.

Alexander, R. M. (1991). Elastic mechanisms in primate locomotion. *Zeitschrift für Morphologie und Anthropologie, 78*, 315–320.

Alexander, R. M., & Jayes, A. (1978). Vertical movements in walking and running. *Journal of Zoology, 185*(1), 27–40.

Alexander, R. M., & Maloiy, G. M. O. (1984). Stride lengths and stride frequencies of primates. *Journal of Zoology London, 202*, 577–582.

Almécija, S., Smaers, J. B., & Jungers, W. L. (2015). The evolution of human and ape hand proportions. *Nature Communications, 6*, 7717. https://doi.org/10.1038/ncomms8717

Amir Abdul Nasir, A. F., Clemente, C. J., Wynn, M. L., & Wilson, R. S. (2017). Optimal running speeds when there is a trade-off between speed and the probability of mistakes. *Functional Ecology, 31*(10), 1941–1949.

Astley, H. C., Haruta, A., & Roberts, T. J. (2015). Robust jumping performance and elastic energy recovery from compliant perches in tree frogs. *The Journal of Experimental Biology, 218*(21), 3360–3363.

Ballinger, R. E. (1973). Experimental evidence of the tail as a balancing organ in the lizard, *Anolis carolinensis*. *Herpetologica, 29*(1), 65–66.

Bergeson, D. J. (1996). *The positional behavior and prehensile tail use of Alouatta palliata, Ateles geoffroyi, and Cebus capucinus*. Washington University.

Bhagat, R., Bertrand, O. C., & Silcox, M. T. (2021). Evolution of arboreality and fossoriality in squirrels and aplodontid rodents: Insights from the semicircular canals of fossil rodents. *Journal of Anatomy, 238*(1), 96–112.

Biknevicius, A. R. (2006). Locomotor mechanics of the tölt in Icelandic horses. *American Journal of Veterinary Research, 67*, 1505–1510.

Biknevicius, A. R., Reilly, S. M., McElroy, E. J., & Bennett, M. B. (2013). Symmetrical gaits and center of mass mechanics in small-bodied, primitive mammals. *Zoology, 116*(1), 67–74.

Bininda-Emonds, O. R. P., Cardillo, M., Jones, K. E., MacPhee, R. D. E., Beck, R. M. D., Grenyer, R., Price, S. A., Vos, R. A., Gittleman, J. L., & Purvis, A. (2007). The delayed rise of present-day mammals. *Nature, 446*(7135), 507–512.

Birn-Jeffery, A. V., & Higham, T. E. (2014). The scaling of uphill and downhill locomotion in legged animals. *Integrative and Comparative Biology, 54*(6), 1159–1172.

Boistel, R., Herrel, A., Daghfous, G., Libourel, P. A., Boller, E., Tafforeau, P., & Bels, V. (2010). Assisted walking in Malagasy dwarf chamaeleons. *Biology Letters*:1–5.

Bonser, R. H. (1999). Branching out in locomotion: The mechanics of perch use in birds and primates. *The Journal of Experimental Biology, 202*, 1459–1463.

Bruijn, S. M., van Dieën, J. H., Meijer, O. G., & Beek, P. J. (2009). Is slow walking more stable? *Journal of Biomechanics, 42*(10), 1506–1512.

Buikstra, J. E. (1975). Healed fractures in *Macaca mulatta*: Age, sex and symmetry. *Folia Primatol (Basel), 23*, 140–148.

Camargo, N. F., Sano, N. Y., Ribeiro, J. F., & Vieira, E. M. (2016). Contrasting the realized and fundamental niche of the arboreal walking performance of neotropical rodents. *Journal of Mammalogy, 97*(1), 155–166.

Cant, J. G. H. (1992). Positional behavior and body size of arboreal primates: A theoretical framework for field studies and an illustration of its application. *American Journal of Physical Anthropology, 88*(3), 273–283.

Carlson, K. J., Demes, B., & Franz, T. M. (2005). Mediolateral forces associated with quadrupedal gaits of lemurids. *Journal of Zoology, 266*(3), 261–273.
Carter, M. L., Pontzer, H., Wrangham, R., & Peterhans, J. K. (2008). Skeletal pathology in *Pan troglodytes schweinfurthii* in Kibale National Park, Uganda. *American Journal of Physical Anthropology, 135*, 389–403.
Cartmill, M. (1974). Pads and claws in arboreal locomotion. In F. A. Jenkins Jr. (Ed.), *Primate locomotion* (pp. 45–83). Academic.
Cartmill, M. (1979). The volar skin of primates: Its frictional characteristics and their functional significance. *American Journal of Physical Anthropology, 50*(4), 497–510.
Cartmill, M. (1985). Climbing. In M. Hildebrand, D. M. Bramble, K. F. Liem, & D. B. Wake (Eds.), *Functional vertebrate morphology* (pp. 73–88). Harvard University Press.
Cartmill, M., Lemelin, P., & Schmitt, D. (2002). Support polygons and symmetrical gaits in mammals. *Zoological Journal of the Linnean Society, 136*, 401–420.
Cartmill, M., Lemelin, P., & Schmitt, D. (2007a). Primate gaits and primate origins. In M. J. Ravosa & M. Dagosto (Eds.), *Primate origins: Adaptations and evolution* (pp. 403–435). Springer.
Cartmill, M., Lemelin, P., & Schmitt, D. (2007b). Understanding the adaptive value of diagonal-sequence gaits in primates: A comment on Shapiro and Raichlen, 2005. *American Journal of Physical Anthropology, 133*, 822–825.
Chadwell, B. A., & Young, J. W. (2015). Angular momentum and arboreal stability in common marmosets (*Callithrix jacchus*). *American Journal of Physical Anthropology, 156*, 565–576.
Chang, M. D., Sejdić, E., Wright, V., & Chau, T. (2010). Measures of dynamic stability: Detecting differences between walking overground and on a compliant surface. *Human Movement Science, 29*(6), 977–986.
Channon, A. J., Gunther, M. M., Crompton, R. H., D'Aout, K., Preuschoft, H., & Vereecke, E. E. (2011). The effect of substrate compliance on the biomechanics of gibbon leaps. *The Journal of Experimental Biology, 214*(4), 687–696.
Clemente, C. J., Dick, T. J., Wheatley, R., Gaschk, J., Nasir, A. F. A. A., Cameron, S. F., & Wilson, R. S. (2019). Moving in complex environments: A biomechanical analysis of locomotion on inclined and narrow substrates. *The Journal of Experimental Biology, 222*(6), jeb189654.
Crompton, R. H., Sellers, W. I., & Gunther, M. M. (1993). Energetic efficiency and ecology as selective factors in the saltatory adaptation of prosimian primates. *Proceedings of the Royal Society B: Biological Sciences, 254*(1339), 41–45.
Dagg, A. I. (1973). Gaits in mammals. *Mammal Review, 3*, 135–154.
Delciellos, A. C., & Vieira, M. V. (2007). Stride lengths and frequencies of arboreal walking in seven species of didelphid marsupials. *Acta Theriol (Warsz), 52*(1), 101–111.
Demes, B., Jungers, W. L., & Nieschalk, U. (1990). Size- and speed-related aspects of quadrupedal walking in slender and slow lorises. In F. K. Jouffroy, M. H. Stack, & C. Niemetz (Eds.), *Gravity, posture and locomotion in primates* (pp. 175–197). Florence.
Demes, B., Jungers, W. L., Gross, T., & Fleagle, J. (1995). Kinetics of leaping primates: Influence of substrate orientation and compliance. *American Journal of Physical Anthropology, 96*(4), 419–429.
Dickson, B. V., Sherratt, E., Losos, J. B., & Pierce, S. E. (2017). Semicircular canals in *Anolis* lizards: Ecomorphological convergence and ecomorph affinities of fossil species. *Royal Society Open Science, 4*(10), 170058.
Dunbar, R. I. M. (1988). *Primate social systems*. Croom Helm Ltd.
Dunham, N. T., McNamara, A., Shapiro, L. J., Hieronymus, T. L., & Young, J. W. (2018). A user's guide for the quantitative analysis of substrate characteristics and locomotor kinematics in free-ranging primates. *American Journal of Physical Anthropology, 167*, 569–584.
Dunham, N. T., McNamara, A., Shapiro, L. J., Hieronymus, T. L., Phelps, T., & Young, J. W. (2019a). Effects of substrate and phylogeny on quadrupedal gait in free-ranging platyrrhines. *American Journal of Physical Anthropology, 170*(4), 565–578.
Dunham, N. T., McNamara, A., Shapiro, L. J., Phelps, T., Wolfe, A. N., & Young, J. W. (2019b). Locomotor kinematics of tree squirrels (*Sciurus carolinensis*) in free-ranging and laboratory

environments: Implications for primate locomotion and evolution. *Journal of Experimental Zoology Part A Ecological Genetics and Physiology, 331*, 103–119.
Dunham, N. T., McNamara, A., Shapiro, L. J., Phelps, T., & Young, J. W. (2020). Asymmetrical gait kinematics of free-ranging callitrichines in response to changes in substrate diameter and orientation. *The Journal of Experimental Biology, 223*, jeb217562. https://doi.org/10.1242/jeb.217562
Emmons, L. H., & Gentry, A. H. (1983). Tropical forest structure and the distribution of gliding and prehensile-tailed vertebrates. *The American Naturalist, 121*(4), 513–524.
Farley, C. T. (1991). A mechanical trigger for the trot-gallop transition in horses. *Science, 253*, 306–308.
Fischer, M. S., Schilling, N., Schmidt, M., Harrhaus, D., & Witte, H. (2002). Basic limb kinematics in small therian mammals. *The Journal of Experimental Biology, 205*, 1315–1328.
Fischer, M. S., Krause, C., & Lilje, K. E. (2010). Evolution of chameleon locomotion, or how to become arboreal as a reptile. *Zoology, 113*, 67–74.
Foster, K. L., & Higham, T. E. (2012). How forelimb and hindlimb function changes with incline and perch diameter in the green anole, *Anolis carolinensis*. *The Journal of Experimental Biology, 215*(Pt 13), 2288–2300.
Franz, T. M., Demes, B., & Carlson, K. J. (2005). Gait mechanics of lemurid primates on terrestrial and arboreal substrates. *Journal of Human Evolution, 48*, 199–217.
Fröbisch, J., & Reisz, R. R. (2009). The Late Permian herbivore *Suminia* and the early evolution of arboreality in terrestrial vertebrate ecosystems. *Proceedings of the Royal Society B-Biological Sciences London, 276*(1673), 3611–3618.
Gálvez-López, E., Maes, L. D., & Abourachid, A. (2011). The search for stability on narrow supports: An experimental study in cats and dogs. *Zoology, 114*(4), 224–232.
Gans, C. (1967). The chameleon. *Natural History, 76*(4), 52–59.
Garber, P. A., & Rehg, J. A. (1999). The ecological role of the prehensile tail in white-faced capuchins (*Cebus capucinus*). *American Journal of Physical Anthropology, 110*(3), 325–339.
Garland, T., Jr., & Losos, J. (1994). Ecological morphology of locomotor performance in squamate reptiles. In P. C. Wainwright & S. Reilly (Eds.), *Ecological morphology: Integrative organismal biology* (pp. 60–98). University of Chicago Press.
Gaschk, J. L., Frère, C. H., & Clemente, C. J. (2019). Quantifying koala locomotion strategies: Implications for the evolution of arborealism in marsupials. *The Journal of Experimental Biology, 222*(24), jeb207506.
Gatesy, S. M. (1991). Hind limb movements of the American alligator (Alligator mississippiensis) and postural grades. *Journal of Zoology, 224*(4), 577–588.
Gebo, D. L. (2004). A shrew-sized origin for primates. *The Yearbook of Physical Anthropology, 47*, 40–62.
Gilman, C. A., & Irschick, D. J. (2013). Foils of flexion: The effects of perch compliance on lizard locomotion and perch choice in the wild. *Functional Ecology, 27*(2), 374–381.
Gilman, C. A., Bartlett, M. D., Gillis, G. B., & Irschick, D. J. (2012). Total recoil: Perch compliance alters jumping performance and kinematics in green anole lizards (*Anolis carolinensis*). *The Journal of Experimental Biology, 215*(Pt 2), 220–226.
Graham, M., & Socha, J. J. (2020). Going the distance: The biomechanics of gap-crossing behaviors. *Journal of Experimental Zoology Part A: Ecological and Integrative Physiology, 333*(1), 60–73.
Granatosky, M. C., Amanat, S., Panyutina, A. A., & Youlatos, D. (2021). Gait mechanics of a blind echolocating rodent: Implications for the locomotion of small arboreal mammals and proto-bats. *Journal of Experimental Zoology Part A: Ecological and Integrative Physiology, 335*(4), 436–453.
Grand, T. I. (1972). A mechanical interpretation of terminal branch feeding. *Journal of Mammalogy, 53*(1), 198–201.

Grand, T. I. (1977). Body weight: Its relation to tissue composition, segment distribution, and motor function. I. Interspecific comparisons. *American Journal of Physical Anthropology, 47*(2), 211–239.

Grand, T. I. (1984). Motion economy within the canopy: Four strategies for mobility. In P. S. Rodman & J. G. H. Cant (Eds.), *Adaptations for foraging in nonhuman primates: Contributions to an organismal biology of prosimians, monkeys, and apes* (pp. 54–72). Columbia University Press.

Guillot, D. (2011). Forelimb suspensory gait characteristics of wild *Lagothrix poeppigii* and *Ateles belzebuth*: Developing video-based methodologies in free-ranging primates. In K. D'Août & E. E. Vereecke (Eds.), *Primate locomotion: Linking field and laboratory research* (pp. 247–269). Springer.

Hagey, T. J., Harte, S., Vickers, M., Harmon, L. J., & Schwarzkopf, L. (2017). There's more than one way to climb a tree: Limb length and microhabitat use in lizards with toe pads. *PLoS One, 12*(9), e0184641.

Hayssen, V. (2008). Patterns of body and tail length and body mass in Sciuridae. *Journal of Mammalogy, 89*(4), 852–873.

Healy, K., Guillerme, T., Finlay, S., Kane, A., Kelly, S. B., McClean, D., Kelly, D. J., Donohue, I., Jackson, A. L., & Cooper, N. (2014). Ecology and mode-of-life explain lifespan variation in birds and mammals. *Proceedings of the Royal Society B: Biological Sciences, 281*(1784), 20140298.

Herrel, A., Measey, G. J., Vanhooydonck, B., & Tolley, K. A. (2011). Functional consequences of morphological differentiation between populations of the Cape Dwarf Chameleon (*Bradypodion pumilum*). *Biological Journal of the Linnean Society, 104*(3), 692–700.

Herrel, A., Perrenoud, M., Decamps, T., Abdala, V., Manzano, A., & Pouydebat, E. (2013a). The effect of substrate diameter and incline on locomotion in an arboreal frog. *The Journal of Experimental Biology, 216*(19), 3599–3605.

Herrel, A., Tolley, K. A., Measey, G. J., da Silva, J. M., Potgieter, D. F., Boller, E., Boistel, R., & Vanhooydonck, B. (2013b). Slow but tenacious: An analysis of running and gripping performance in chameleons. *The Journal of Experimental Biology, 216*(6), 1025–1030.

Higham, T. E., & Jayne, B. C. (2004). Locomotion of lizards on inclines and perches: Hindlimb kinematics of an arboreal specialist and a terrestrial generalist. *The Journal of Experimental Biology, 207*(2), 233–248.

Higham, T. E., Birn-Jeffery, A. V., Collins, C. E., Hulsey, C. D., & Russell, A. P. (2015). Adaptive simplification and the evolution of gecko locomotion: Morphological and biomechanical consequences of losing adhesion. *Proceedings of the National Academy of Sciences, 112*(3), 809–814.

Hildebrand, M. (1966). Analysis of the symmetrical gaits of tetrapods. *Folia Biotheoretica*, 1–22.

Hildebrand, M. (1967). Symmetrical gaits of primates. *American Journal of Physical Anthropology, 26*, 119–130.

Hildebrand, M. (1980). The adaptive significance of tetrapod gait selection. *American Zoologist, 20*, 255–267.

Hildebrand, M., & Goslow, G. E. (2001). *Analysis of vertebrate structure*. Wiley.

Hopkins, K., & Tolley, K. (2011). Ecomorphological variation in dwarf chameleons and the dynamics between natural and sexual selection. *Biological Journal of the Linnean Society, 102*, 878–888.

Horner, E. (1954). Arboreal adaptations of *Peromyscus*, with special reference to use of the tail. *Contributions from the Laboratory of Vertebrate Biology, University of Michigan, 61*, 1–84.

Hsieh, S.-T. T. (2016). Tail loss and narrow surfaces decrease locomotor stability in the arboreal green anole lizard (*Anolis carolinensis*). *The Journal of Experimental Biology, 219*, 364–373.

Hunt, N. H., Jinn, J., Jacobs, L. F., & Full, R. J. (2021). Acrobatic squirrels learn to leap and land on tree branches without falling. *Science, 373*(6555), 697–700.

Hutchinson, J. R., Famini, D., Lair, R., & Kram, R. (2003). Are fast-moving elephants really running? *Nature, 422*, 493–494.

Hutchinson, J. R., Felkler, D., Houston, K., Chang, Y.-M., Brueggen, J., Kledzik, D., & Vliet, K. A. (2019). Divergent evolution of terrestrial locomotor abilities in extant Crocodylia. *Scientific Reports, 9*(1), 1–11.

Huyghe, K., Herrel, A., Vanhooydonck, B., Meyers, J. J., & Irschick, D. J. (2007). Microhabitat use, diet, and performance data on the Hispaniolan twig anole, *Anolis sheplani*: Pushing the boundaries of morphospace. *Zoology, 110*(1), 2–8.

Igarashi, M., & Levy, J. K. (1981). Locomotor balance performance of short-tailed squirrel monkeys. *Journal of Medical Primatology, 10*, 136–140.

Irschick, D. J., Vitt, L. J., Zani, P. A., & Losos, J. B. (1997). A comparison of evolutionary radiations in mainland and Caribbean *Anolis* lizards. *Ecology, 78*(7), 2191–2203.

Isbell, L. (1994). Predation on primates: Ecological patterns and evolutionary consequences. *Evolutionary Anthropology: Issues, News, and Reviews, 3*, 61–71.

Jayne, B. C. (2020). What defines different modes of snake locomotion? *Integrative and Comparative Biology, 60*(1), 156–170.

Jenkins, F. A. (1974). Tree shrew locomotion and the origins of primate arborealism. In F. A. Jenkins (Ed.), *Primate locomotion* (pp. 85–115). Academic Press.

Jenkins, F. A., & Camazine, S. M. (1977). Hip structure and locomotion in ambulatory and cursorial carnivores. *Journal of Zoology, 181*(3), 351–370.

Jones, K. E., Bielby, J., Cardillo, M., Fritz, S. A., O'Dell, J., Orme, C. D. L., Safi, K., Sechrest, W., Boakes, E. H., & Carbone, C. (2009). PanTHERIA: A species-level database of life history, ecology, and geography of extant and recently extinct mammals: Ecological Archives E090-184. *Ecology, 90*(9), 2648–2648.

Jurmain, R. (1997). Skeletal evidence of trauma in African apes, with special reference to the Gombe chimpanzees. *Primates, 38*, 1–14.

Jusufi, A., Goldman, D. I., Revzen, S., & Full, R. J. (2008). Active tails enhance arboreal acrobatics in geckos. *Proceedings of the National Academy of Sciences, 105*(11), 4215–4219.

Karantanis, N.-E., Youlatos, D., & Rychlik, L. (2015). Diagonal gaits in the feathertail glider *Acrobates pygmaeus* (Acrobatidae, Diprotodontia): Insights for the evolution of primate quadrupedalism. *Journal of Human Evolution, 86*, 43–54.

Karantanis, N.-E., Rychlik, L., Herrel, A., & Youlatos, D. (2017a). Arboreal gaits in three sympatric rodents *Apodemus agrarius*, *Apodemus flavicollis* (Rodentia, Muridae) and *Myodes glareolus* (Rodentia, Cricetidae). *Mammalian Biology, 83*, 51–63.

Karantanis, N.-E., Rychlik, L., Herrel, A., & Youlatos, D. (2017b). Comparing the arboreal gaits of *Muscardinus avellanarius* and *Glis glis* (Gliridae, Rodentia): A first quantitative analysis. *Mammal Study, 42*(3), 161–173.

Karantanis, N. E., Rychlik, L., Herrel, A., & Youlatos, D. (2017c). Arboreal locomotion in Eurasian harvest mice *Micromys minutus* (Rodentia: Muridae): The gaits of small mammals. *Journal of Experimental Zoology Part A: Ecological and Integrative Physiology, 327*(1), 38–52.

Karantanis, N. E., Rychlik, L., Herrel, A., & Youlatos, D. (2017d). Arboreality in acacia rats (*Thallomys paedulcus*; Rodentia, Muridae): Gaits and gait metrics. *Journal of Zoology, 303*(2), 107–119.

Khannoon, E. R., Endlein, T., Russell, A. P., & Autumn, K. (2014). Experimental evidence for friction-enhancing integumentary modifications of chameleons and associated functional and evolutionary implications. *Proceedings of the Royal Society B: Biological Sciences, 281*(1775), 20132334.

Krause, C., & Fischer, M. S. (2013). Biodynamics of climbing: Effects of substrate orientation on the locomotion of a highly arboreal lizard (*Chamaeleo calyptratus*). *The Journal of Experimental Biology, 216*.

Lammers, A. R. (2009a). The effects of substrate texture on the mechanics of quadrupedal arboreal locomotion in the gray short-tailed opossum (*Monodelphis domestica*). *The Journal of Experimental Zoology, 311A*, 813–823.

Lammers, A. R. (2009b). Mechanics of generating friction during locomotion on rough and smooth arboreal trackways. *The Journal of Experimental Biology, 212*(Pt 8), 1163–1169.

Lammers, A. R., & Biknevicius, A. R. (2004). The biodynamics of arboreal loocmotion: The effects of substrate diameter on locomotor kinetics in the gray short-tailed opossum (*Monodelphis domestica*). *The Journal of Experimental Biology, 207*, 4325–4336.

Lammers, A. R., & Gauntner, T. (2008). Mechanics of torque generation during quadrupedal arboreal locomotion. *Journal of Biomechanics, 41*, 2388–2395.

Lammers, A. R., & Zurcher, U. (2011a). Stability during arboreal locomotion. In V. Klika (Ed.), *Theoretical biomechanics* (pp. 1–16).

Lammers, A. R., & Zurcher, U. (2011b). Torque around the center of mass: Dynamic stability during quadrupedal arboreal locomotion in the Siberian chipmunk (*Tamias sibiricus*). *Zoology, 114*(2), 95–103.

Langowski, J. K., Dodou, D., Kamperman, M., & van Leeuwen, J. L. (2018). Tree frog attachment Mechanisms, challenges, and perspectives. *Frontiers in Zoology, 15*(1), 1–21.

Larney, E., & Larson, S. G. (2004). Compliant walking in primates: Elbow and knee yield in primates compared to other mammals. *American Journal of Physical Anthropology, 125*, 42–50.

Larson, S. G. (1998a). Parallel evolution of the hominoid trunk and forelimb. *Evolutionary Anthroplogy*, 87–99.

Larson, S. G. (1998b). Unique aspects of quadrupedal locomotion in nonhuman primates. In E. Strasser, J. Fleagle, A. Rosenberger, & H. McHenry (Eds.), *Primate locomotion: Recent advances* (pp. 157–173). Plenum Press.

Larson, S. G., & Stern, J. T. (2006). Maintenance of above-branch balance during primate arboreal quadrupedalism: Coordinated use of forearm rotators and tail motion. *American Journal of Physical Anthropology, 129*, 71–81.

Larson, S. G., Schmitt, D., Lemelin, P., & Hamrick, M. (2000). Uniqueness of primate forelimb posture during quadrupedal locomotion. *American Journal of Physical Anthropology, 112*, 87–101.

Larson, S. G., Schmitt, D., Lemelin, P., & Hamrick, M. (2001). Limb excursion during quadrupedal walking: how do primates compare to other mammals? *Journal of Zoology London, 255*, 353–365.

Lawler, R. R., & Stamps, C. (2002). The relationship between tail use and positional behavior in *Alouatta palliata*. *Primates, 43*, 147–152.

Lemelin, P., & Cartmill, M. (2010). The effect of substrate size on the locomotion and gait patterns of the kinkajou (*Potos flavus*). *The Journal of Experimental Zoology, 313A*, 157–168.

Lemelin, P., Schmitt, D., & Cartmill, M. (2003). Footfall patterns and interlimb coordination in opossums (Family Didelphidae): Evidence for the evolution of diagonal sequence gaits in primates. *Journal of Zoology London, 260*, 423–429.

Losos, J. B. (1990). The evolution of form and function: Morphology and locomotor performance in West Indian *Anolis* lizards. *Evolution, 44*(5), 1189–1203.

Losos, J. (2009). *Lizards in an evolutionary tree*. University of California Press.

Losos, J. B., & Irschick, D. J. (1996). The effect of perch diameter on escape behaviour of *Anolis* lizards: Laboratory predictions and field tests. *Animal Behaviour, 51*(3), 593–602.

Losos, J. B., & Sinervo, B. (1989). The effects of morphology and perch diameter on sprint performance of *Anolis* lizards. *The Journal of Experimental Biology, 145*(1), 23–30.

Losos, J. B., Warheitt, K. I., & Schoener, T. W. (1997). Adaptive differentiation following experimental island colonization in *Anolis* lizards. *Nature, 387*(6628), 70–73.

Lovell, N. C. (1991). An evolutionary framework for assessing illness and injury in nonhuman primates. *The Yearbook of Physical Anthropology, 34*, 117–155.

MacLellan, M. J., & Patla, A. E. (2006). Adaptations of walking pattern on a compliant surface to regulate dynamic stability. *Experimental Brain Research, 173*(3), 521–530.

Maiolino, S. A., Kingston, A. K., & Lemelin, P. (2016). Comparative and functional morphology of the primate hand integument. In T. L. Kivell, P. Lemelin, B. G. Richmond, & D. Schmitt (Eds.), *The evolution of the primate hand* (pp. 195–224). Springer.

Manzano, A. S., Abdala, V., & Herrel, A. (2008). Morphology and function of the forelimb in arboreal frogs: Specializations for grasping ability? *Journal of Anatomy, 213*, 296–307.

Martin, R. D. (1968). Reproduction and ontogeny in tree-shrews (*Tupaia belangeri*), with reference to the general behaviour and taxonomic relationships. *Zeitschrift für Tierpsychologie, 25*(4).

Mattingly, W. B., & Jayne, B. C. (2004). Resource use in arboreal habitats: Structure affects locomotion of four ecomorphs of *Anolis* lizards. *Ecology, 85*(4), 1111–1124.

McGraw, W. S. (1998). Comparative locomotion and habitat use of six monkeys in the Tai Forest, Ivory Coast. *The American Journal of Physical Anthropologists, 105*(4), 493–510.

McMahon, T. A. (1985). The role of compliance in mammalian running. *The Journal of Experimental Biology, 115*, 263–282.

McMahon, T. A., & Cheng, G. C. (1990). The mechanics of running: How does stiffness couple with speed? *Journal of Biomechanics, 23*(Suppl 1), 65–78.

McMahon, T. A., & Kronauer, R. E. (1976). Tree structures: Deducing the principle of mechanical design. *Journal of Theoretical Biology, 59*(2), 443–466.

McMahon, T. A., Valiant, G., & Frederick, E. C. (1987). Groucho running. *Journal of Applied Physiology, 62*, 2326–2337.

McNamara, A., Dunham, N. T., Shapiro, L. J., & Young, J. W. (2019). The effects of natural substrate discontinuities on the quadrupedal gait kinematics of free-ranging *Saimiri sciureus*. *American Journal of Primatology, 81*, e23055.

Meiri, S. (2018). Traits of lizards of the world: Variation around a successful evolutionary design. *Global Ecology and Biogeography, 27*(10), 1168–1172.

Meldrum, D. J. (1998). *Tail-assisted hind limb suspension as a transitional behavior in the evolution of the platyrrhine prehensile tail. Primate locomotion* (pp. 145–156). Springer.

Miller, A. H., & Stroud, J. T. (2021). Novel tests of the key innovation hypothesis: Adhesive toepads in arboreal lizards. *Systematic Biology*.

Mincer, S. T., & Russo, G. A. (2020). Substrate use drives the macroevolution of mammalian tail length diversity. *Proceedings of the Royal Society B, 287*(1920), 20192885.

Morbeck, M. E. (1977). Positional behavior, selective use of habitat substrate and associated non-positional behavior in free-ranging *Colobus guereza* (Rüppel, 1835). *Primates, 18*, 35–58.

Myatt, J. P., & Thorpe, S. K. S. (2011). Postural strategies employed by orangutans (*Pongo abelii*) during feeding in the terminal branch niche. *American Journal of Physical Anthropology, 146*, 73–82.

Napier, J. R. (1967). Evolutionary aspects of primate locomotion. *American Journal of Physical Anthropology, 27*, 333–342.

Nekaris, K. A. I. (2005). Foraging behavior of the slender loris (*Loris lydekkerianus lydekkerianus*): Implications for theories of primate origins. *Journal of Human Evolution, 49*, 289–300.

Nyakatura, J. A. (2019). Early primate evolution: insights into the functional significance of grasping from motion analyses of extant mammals. *Biological Journal of the Linnean Society, 127*(3), 611–631.

Nyakatura, J. A., & Heymann, E. W. (2010). Effects of support size and orientation on symmetric gaits in free-ranging tamarins of Amazonian Peru: implications for the functional significance of primate gait sequence patterns. *Journal of Human Evolution, 58*(3), 242–251.

O'Neill, M. C., & Schmitt, D. (2012). The gaits of primates: Center of mass mechanics in walking, cantering and galloping ring-tailed lemurs, *Lemur catta*. *Journal of Experimental Biology, 215*(10), 1728–1739.

Organ, J. M., Teaford, M. F., & Taylor, A. B. (2009). Functional correlates of fiber architecture of the lateral caudal musculature in prehensile and nonprehensile tails of the Platyrrhini (Primates) and Procyonidae (Carnivora). *The Anatomical Record, 292*(6), 827–841.

Peterson, J. A. (1984). The locomotion of *Chamaeleo* (Reptilia: Sauria) with particular reference to the forelimb. *Journal of Zoology, 202*(1), 1–42.

Polk, J. D., Demes, B., Jungers, W. L., Biknevicius, A. R., Heinrich, R. E., & Runestad, J. A. (2000). A comparison of primate, carnivoran and rodent limb bone cross-sectional properties: Are primates really unique? *Journal of Human Evolution, 39*, 297–325.

Pouydebat, E., & Bardo, A. (2019). *An interdisciplinary approach to the evolution of grasping and manipulation*. Oxford University Press.

Preuschoft, H., & Günther, M. M. (1994). Biomechanics and body shape in primates compared with horses. *Zeitschrift für Morphologie und Anthropologie, 80*, 149–165.

Pruetz, J. D., Fulton, S., Marchant, L. F., McGrew, W. C., Schiel, M., & Waller, M. (2008). Arboreal nesting as anti-predator adaptation by savanna chimpanzees (*Pan troglodytes verus*) in southeastern Senegal. *American Journal of Primatology: Official Journal of the American Society of Primatologists, 70*(4), 393–401.

Raichlen, D. A. (2005). Effects of limb mass distribution on the ontogeny of quadrupedalism in infant baboons (*Papio cynocephalus*) and implications for the evolution of primate quadrupedalism. *Journal of Human Evolution, 49*, 415–431.

Rendall, D., & Di Fiore, A. (2007). Homoplasy, homology, and the perceived special status of behavior in evolution. *Journal of Human Evolution, 52*(5), 504–521.

Renous, S., Höfling, E., & Da Rocha, P. (2010). Effect of substrate on the locomotion behaviour of the South American iguanian lizard *Polychrus acutirostris*. *Italian Journal of Zoology, 77*(2), 216–226.

Reynaga, C. M., Eaton, C. E., Strong, G. A., & Azizi, E. (2019). Compliant substrates disrupt elastic energy storage in jumping tree frogs. *Integrative and Comparative Biology, 59*(6), 1535–1545

Reynolds, T. R. (1987). Stride length and its determinants in humans, early hominids, primates, and mammals. *American Journal of Physical Anthropology, 72*, 101–115.

Rose, M. D. (1973). Quadrupedalism in primates. *Primates, 14*, 337–357.

Rose, M. D. (1974). Postural adaptations in New and Old World monkeys. In F. A. Jenkins (Ed.), *Primate locomotion* (pp. 201–222). Academic Press.

Rubenson, J., Heliams, D. B., Lloyd, D. G., & Fournier, P. A. (2004). Gait selection in the ostrich: Mechanical and metabolic characteristics of walking and running with and without an aerial phase. *Proceedings of the Royal Society B-Biological Sciences, 271*(1543), 1091–1099.

Ruina, A., Bertram, J. E. A., & Srinivasan, M. (2005). A collisional model of the energetic cost of support work in walking and galloping, pseudo-elastic leg behavior in running and the walk-to-run transition. *Journal of Theoretical Biology, 237*, 170–192.

Russo, G. A., & Shapiro, L. J. (2011). Morphological correlates of tail length in the catarrhine sacrum. *Journal of Human Evolution, 61*(3), 223–232.

Sargis, E. J. (2002a). Functional morphology of the forelimb of tupaiids (Mammalia, Scandentia) and its phylogenetic implications. *Journal of Morphology, 253*(1), 10–42.

Sargis, E. J. (2002b). Functional morphology of the hindlimb of tupaiids (Mammalia, Scandentia) and its phylogenetic implications. *Journal of Morphology, 254*(2), 149–185.

Schapker, N. M., Chadwell, B. A., & Young, J. W. (2022). Robust locomotor performance of squirrel monkeys (*Saimiri boliviensis*) in response to simulated changes in support diameter and compliance. *Journal of Experimental Zoology Part A: Ecological and Integrative Physiology, 337*, 417–433.

Schmelzle, T., Sánchez-Villagra, M. R., & Maier, W. (2007). Vestibular labyrinth diversity in diprotodontian marsupial mammals. *Mammal Study, 32*(2), 83–97.

Schmidt, M. (2005a). Hind limb proportions and kinematics: Are small primates different from other small mammals? *The Journal of Experimental Biology, 208*(17), 3367–3383.

Schmidt, M. (2005b). Quadrupedal locomotion in squirrel monkeys (Cebidae: *Saimiri sciureus*): A cineradiographic study of limb kinematics and related substrate reaction forces. *American Journal of Physical Anthropology, 128*, 359–370.

Schmidt, A. (2011). Functional differentiation of trailing and leading forelimbs during locomotion on the ground and on a horizontal branch in the European red squirrel (*Sciurus vulgaris*, Rodentia). *Zoology (Jena, Germany), 114*(3), 155–164.

Schmidt, A., & Fischer, M. S. (2010). Arboreal locomotion in rats - The challenge of maintaining stability. *The Journal of Experimental Biology, 213*(21), 3615–3624.

Schmidt, A., & Fischer, M. S. (2011). The kinematic consequences of locomotion on sloped arboreal substrates in a generalized (*Rattus norvegicus*) and a specialized (*Sciurus vulgaris*) rodent. *The Journal of Experimental Biology, 214*(15), 2544–2559.

Schmitt, D. (1994). Forelimb mechanics as a function of substrate type during quadrupedalism in two anthropoid primates. *Journal of Human Evolution, 26*(441–458), 441–457.
Schmitt, D. (1998). Forelimb mechanics during arboreal and terrestrial quadrupedalism in Old World monkeys. In E. Strasser, J. Fleagle, A. Rosenberger, & H. McHenry (Eds.), *Primate locomotion: Recent advances* (pp. 175–200). Plenum Press.
Schmitt, D. (1999). Compliant walking in primates. *Journal of Zoology London, 248*, 149–160.
Schmitt, D. (2003a). Mediolateral reaction forces and forelimb anatomy in quadrupedal primates: Implications for interpreting locomotor behavior in fossil primates. *Journal of Human Evolution, 44*, 47–58.
Schmitt, D. (2003b). Substrate size and primate forelimb mechanics: Implications for understanding the evolution of primate locomotion. *International Journal of Primatology, 24*, 1023–1036.
Schmitt, D., Cartmill, M., Griffin, T. M., Hanna, J. B., & Lemelin, P. (2006). Adaptive value of ambling gaits in primates and other mammals. *The Journal of Experimental Biology, 209*, 2042–2049.
Schmitt, D., Gruss, L. T., & Lemelin, P. (2010). Brief communication: Forelimb compliance in arboreal and terrestrial opossums. *American Journal of Physical Anthropology, 141*(1), 142–146.
Schultz, A. H. (1944). Age changes and variability in gibbons. A morphological study on a populations sample of a man-like ape. *American Journal of Physical Anthropology, 2*, 1–129.
Schwaner, M., Hsieh, S., Swalla, B., & McGowan, C. (2021). An introduction to an evolutionary tail: EvoDevo, structure and function of post-anal appendages. *Integrative and Comparative Biology*.
Shapiro, L. J., & Raichlen, D. A. (2005). Lateral sequence walking in infant *Papio cynocephalus*: Implications for the evolution of diagonal sequence walking in primates. *American Journal of Physical Anthropology, 126*, 205–213.
Shapiro, L. J., & Raichlen, D. A. (2007). Primate gaits and arboreal stability: A response to Cartmill et al. *American Journal of Physical Anthropology, 133*, 825–827.
Shapiro, L. J., & Young, J. W. (2010). Is primate-like quadrupedalism necessary for fine-branch locomotion? A test using sugar gliders (*Petaurus breviceps*). *Journal of Human Evolution, 58*, 309–319.
Shapiro, L. J., & Young, J. W. (2012). Kinematics of quadrupedal locomotion in sugar gliders (*Petaurus breviceps*): Effects of age and substrate size. *The Journal of Experimental Biology, 215*(3), 480–496.
Shapiro, L. J., Young, J. W., & VandeBerg, J. L. (2014). Body size and the small branch niche: Using marsupial ontogeny to model primate locomotor evolution. *Journal of Human Evolution, 68*, 14–31.
Shapiro, L. J., Chadwell, B. A., & Young, J. W. (2016a). Tail kinematics during asymmetrical gaits in mouse lemurs (*Microcebus murinus*). *American Journal of Physical Anthropology, 159*(Suppl. 62), 38.
Shapiro, L. J., Kemp, A. D., & Young, J. W. (2016b). Effects of substrate size and orientation on quadrupedal gait kinematics in mouse lemurs (*Microcebus murinus*). *Journal of Experimental Zoology. Part A, Ecological Genetics and Physiology, 325*, 329–343.
Shattuck, M. R., & Williams, S. A. (2010). Arboreality has allowed for the evolution of increased longevity in mammals. *Proceedings of the National Academy of Sciences, 107*(10), 4635–4639.
Sheehy, C. M., III, Albert, J. S., & Lillywhite, H. B. (2016). The evolution of tail length in snakes associated with different gravitational environments. *Functional Ecology, 30*(2), 244–254.
Siegel, M. I., & van Meter, R. (1973). Skeletal correlates of ecological adaptation in two species of *Peromyscus*. *Journal of Mammalogy, 54*(1), 275–278.
Sinervo, B., & Losos, J. B. (1991). Walking the tight rope: Arboreal sprint performance among *Sceloporus occidentalis* lizard populations. *Ecology, 72*(4), 1225.
Smith, R. J., & Jungers, W. L. (1997). Body mass in comparative primatology. *Journal of Human Evolution, 32*, 523–559.

Spezzano, L. C., & Jayne, B. C. (2004). The effects of surface diameter and incline on the hindlimb kinematics of an arboreal lizard (*Anolis sagrei*). *The Journal of Experimental Biology, 207*(12), 2115–2131.
Spinner, M., Westhoff, G., & Gorb, S. N. (2014). Subdigital setae of chameleon feet: Friction-enhancing microstructures for a wide range of substrate roughness. *Scientific Reports, 4*(1), 1–9.
Stevens, N. J. (2003). *The influence of substrate size, orientation and compliance upon prosimian arboreal quadrupedalism* [Ph.D. Dissertation]. State University of New York at Stony Brook: Stony Brook University.
Stevens, N. J. (2007). The effect of branch diameter on primate gait sequence pattern. *American Journal of Primatology, 70*, 1–7.
Sussman, R. W. (1991). Primate origins and the evolution of angiosperms. *American Journal of Primatology, 23*, 209–223.
Sussman, R. W., & Raven, P. H. (1978). Pollination by lemurs and marsupials: An archaic coevolutionary system. *Science, 200*, 731–736.
Sustaita, D., Pouydebat, E., Manzano, A., Abdala, V., Hertel, F., & Herrel, A. (2013). Getting a grip on tetrapod grasping: Form, function, and evolution. *Biological Reviews, 88*(2), 380–405.
Taylor, M. E. (1970). Locomotion in some East African viverrids. *Journal of Mammalogy, 51*, 42–51.
Thorpe, S., Crompton, R., & Alexander, R. (2007a). Orangutans use compliant branches to lower the energetic cost of locomotion. *Biology Letters, 3*, 253–256.
Thorpe, S. K. S., Holder, R. L., & Crompton, R. H. (2007b). Origin of human bipedalism as an adaptation for locomotion on flexible branches. *Science, 316*(5829), 1328–1331.
Thorpe, S. K. S., Holder, R., & Crompton, R. H. (2009). Orangutans employ unique strategies to control branch flexibility. *Proceedings of the National Academy of Sciences, 106*(31), 12646.
Tinius, A., & Russell, A. P. (2017). Points on the curve: An analysis of methods for assessing the shape of vertebrate claws. *Journal of Morphology, 278*(2), 150–169.
Toro, E., Herrel, A., & Irschick, D. (2004). The evolution of jumping performance in Caribbean *Anolis* lizards: Solutions to biomechanical trade-offs. *The American Naturalist, 163*(6), 844–856.
Usherwood, J. R., & Smith, B. J. (2018). The grazing gait, and implications of toppling table geometry for primate footfall sequences. *Biology Letters, 14*(5), 20180137.
van Casteren, A., Sellers, W. I., Thorpe, S. K. S., Coward, S., Crompton, R. H., & Ennos, A. R. (2013). Factors affecting the compliance and sway properties of tree branches used by the Sumatran orangutan (*Pongo abelii*). *PLoS One, 8*(7), e67877–e67879.
van Schaik, C. P., & Deaner, R. O. (2003). Life history and cognitive evolution in primates. In F. B. M. de Waal & P. L. Tyack (Eds.), *Animal social complexity* (pp. 5–25). Harvard Unversity Press.
Van Valkenburgh, B. (1987). Skeletal indicators of locomotor behavior in living and extinct carnivores. *Journal of Vertebrate Paleontology, 7*(2), 162–182.
Walker, S. E. (1998). Fine-grained differences within positional categories: a case study of *Pithecia* and *Chiropotes*. In E. Strasser, J. Fleagle, A. Rosenberger, & H. McHenry (Eds.), *Primate locomotion: Recent advances* (pp. 31–43). Springer.
Walker, S. E. (2005). Leaping behavior of *Pithecia pithecia* and *Chiropotes satanas* in eastern Venezuela. *American Journal of Primatology, 66*(4), 369–387.
Walker, C., Vierck, C. J., & Ritz, L. A. (1998). Balance in the cat: Role of the tail and effects of sacrocaudal transection. *Behavioural Brain Research, 91*(1–2), 41–47.
Wallace, I. J., & Demes, B. (2008). Symmetrical gaits of *Cebus apella*: Implications for the functional significance of diagonal sequence gait in primates. *Journal of Human Evolution, 54*, 783–794.
Warren, R. D., & Crompton, R. H. (1997). Locomotor ecology of *Lepilemur edwardsi* and *Avahi occidentalis*. *American Journal of Physical Anthropology, 104*(4), 471–486

Weyand, P. G., Sternlight, D. B., Bellizzi, M. J., & Wright, S. (2000). Faster top running speeds are achieved with greater ground forces not more rapid leg movements. *Journal of Applied Physiology, 89*(5), 1991–1999.

Wheatley, R., Angilletta, M. J., Niehaus, A. C., & Wilson, R. S. (2015). How fast should an animal run when escaping? An optimality model based on the trade-off between speed and accuracy. *Integrative and Comparative Biology, 55*(6), 1166–1175.

Wheatley, R., Buettel, J. C., Brook, B. W., Johnson, C. N., & Wilson, R. P. (2021). Accidents alter animal fitness landscapes. *Ecology Letters, 24*(5), 920–934.

Williamson, R. E., Webb, S. E., Dubreuil, C., Lopez, R., Hernandez, S. C., Fedigan, L. M., & Melin, A. D. (2021). Sharing spaces: Niche differentiation in diet and substrate use among wild capuchin monkeys. *Animal Behaviour*.

Wilson, D. R. (1972). Tail reduction in *Macaca*. In R. H. Tuttle (Ed.), *The functional and evolutionary biology of primates* (pp. 241–261). Aldine-Atherton.

Wimberly, A. N., Slater, G. J., & Granatosky, M. C. (2021). Evolutionary history of quadrupedal walking gaits shows mammalian release from locomotor constraint. *Proceedings of the Royal Society B, 288*(1957), 20210937.

Wölfer, J., Aschenbach, T., Michel, J., & Nyakatura, J. (2021). *Mechanics of arboreal locomotion in Swinhoe's striped squirrels: A potential model for early Euarchontoglires.*

Wynn, M. L., Clemente, C., Nasir, A. F. A. A., & Wilson, R. S. (2015). Running faster causes disaster: Trade-offs between speed, manoeuvrability and motor control when running around corners in northern quolls (*Dasyurus hallucatus*). *The Journal of Experimental Biology, 218*(Pt 3), 433–439.

Young, J. W. (2009). Substrate determines asymmetrical gait dynamics in marmosets (*Callithrix jacchus*) and squirrel monkeys (*Saimiri boliviensis*). *American Journal of Physical Anthropology, 138*, 403–420.

Young, J. W. (2012). Ontogeny of limb force distribution in squirrel monkeys (*Saimiri boliviensis*): insights into the mechanical bases of primate hind limb dominance. *Journal of Human Evolution, 62*, 473–485.

Young, J. W., & Chadwell, B. A. (2015). The mechanics of arboreal stability in squirrel monkeys (*Saimiri boliviensis*). *The American Journal of Physical Anthropolology Supplement, 60*, 330.

Young, J. W., & Chadwell, B. A. (2020). Not all fine-branch locomotion is equal: Grasping morphology determines locomotor performance on narrow supports. *Journal of Human Evolution, 142*, 102767.

Young, J. W., Russo, G. A., Fellmann, C. D., Thatikunta, M. A., & Chadwell, B. A. (2015). Tail function during arboreal quadrupedalism in squirrel monkeys (*Saimiri boliviensis*) and tamarins (*Saguinus oedipus*). *Journal of Experimental Zoology. Part A, Ecological Genetics and Physiology, 323*(8), 556–566.

Young, J. W., Stricklen, B. M., & Chadwell, B. A. (2016). Effects of support diameter and compliance on common marmoset (*Callithrix jacchus*) gait kinematics. *The Journal of Experimental Biology, 219*, 2659–2672.

Young, J. W., McNamara, A., Dunham, N. T., & Shapiro, L.J. (2019). Influence of substrate compliance on wild primate gait kinematics. *88th Annual Meeting of the American Association of Physical Anthropologists*. Cleveland, OH.

Young, J. W., Chadwell, B. A., Dunham, N. T., McNamara, A., Phelps, T., Hieronymus, T., & Shapiro, L. J. (2021). The stabilizing function of the tail during arboreal quadrupedalism. *Integrative and Comparative Biology*. https://doi.org/10.1093/icb/icab096

Zani, P. (2000). The comparative evolution of lizard claw and toe morphology and clinging performance. *Journal of Evolutionary Biology, 13*(2), 316–325.

Chapter 12
Convergent Evolution of Manual and Pedal Grasping Capabilities in Tetrapods

Emmanuelle Pouydebat, Grégoire Boulinguez-Ambroise, Adriana Manzano, Virginia Abdala, and Diego Sustaita

Abstract Grasping behavior and manipulation using the hand and/or foot is widespread among tetrapods and can be used in various contexts in the daily life of many species. Activities such as feeding and movement through the environment may be assisted by grasping. Well-defined digits and digital musculature are synapomorphies of the tetrapod clade and from this foundation other features, such as opposable digits and tendon configurations, have evolved independently in many lineages. The evolutionary transitions leading to grasping and manipulative behaviors are complex and require better understanding. Here we survey the evolution of grasping autopodia and their forms and functions across four major tetrapod clades, revealing that the underlying morphological bases and ecological factors may differ among tetrapods. Further interdisciplinary approaches, including eco-ethology,

E. Pouydebat and G. Boulinguez-Ambroise contributed equally to the writing of the article.

E. Pouydebat (✉)
Departement d'Ecologie et de Gestion de la Biodiversité, UMR 7179 C.N.R.S/M.N.H.N, Paris, France

G. Boulinguez-Ambroise (✉)
Departement d'Ecologie et de Gestion de la Biodiversité, UMR 7179 C.N.R.S/M.N.H.N, Paris, France

Department of Anatomy and Neurobiology, Northeast Ohio Medical University (NEOMED), Rootstown, OH, USA

A. Manzano
CICyTTP-CONICET-UADER, Diamante, Entre Ríos, Argentina

Catedra de Embriologia y Anatomia animal, FCyT, Universidad Autonoma de Entre Ríos, Parana, Entre Ríos, Argentina

V. Abdala
Cátedra de Biologia General, Facultad de Ciencias Naturales, Instituto de Biodiversidad Neotropical, UNT-CONICET, Universidad Nacional de Tucumán, San Miguel de Tucumán, Tucumán, Argentina

D. Sustaita
Department of Biological Sciences, California State University San Marcos, San Marcos, CA, USA

V. L. Bels, A. P. Russell (eds.), *Convergent Evolution*, Fascinating Life Sciences,
https://doi.org/10.1007/978-3-031-11441-0_12

morphology, biomechanics, ontogeny, and even genetics, relating to grasping form and function within and among tetrapods must be developed for a better understanding of the role that object/substrate/food grasping abilities play in the evolutionary success of several tetrapod lineages.

Keywords Grasping · Manipulation · Tetrapods · Dexterity · Feeding · Locomotion

12.1 Introduction

The ability to grasp and manipulate substrates and/or objects is fundamental from an evolutionary point of view (Sustaita et al., 2013). Indeed, these actions are involved in locomotion, postural stabilization, food acquisition and processing, social interactions and have contributed to the evolutionary success of many groups of vertebrates. Thus, grasping is fundamental to the behavioral repertoires (i.e., locomotion, feeding, and reproduction) of many vertebrates, and has implications for fitness. Nevertheless, data relating to this are scarce other than those relating to human biomechanics, kinesiology, medicine and physical anthropology (e.g., Susman, 1998; Marzke & Marzke, 2000; Pouydebat et al., 2014; Feix et al., 2015). Many studies have analyzed the evolution of structural variation of the hominid hand with regard to prehensile capabilities, emphasizing how grasping capabilities were involved in the origins and use of tools (Napier, 1956; Marzke et al., 1992; Marzke, 1997; Susman, 1998; Kivell et al., 2011; Borel et al., 2016; Vigouroux et al., 2018; Bardo et al., 2020). Nevertheless, grasping behavior is much more widerspread. Defined as the application of functionally effective forces by an appendage to an object for a task, grasping can be accomplished by the limbs, the tail, the trunk, the tongue, the teeth, or other animal parts (Mackenzie & Iberall, 1994; Lefeuvre et al., 2020). When focusing on the autopodia (hands and feet), grasping involves orientating and positioning of the digits along with appropriate displacement of the limb to accomplish the correct location of the grasping structure in space (Mackenzie & Iberall, 1994). Gripping suggests a static posture, but grasping is achieved by the dynamic development of a posture (Malek, 1981). Napier (1956) described the power grasp that is used for stability and security. He defined it as a primary grasp that provides the ability to resist slipping. He distinguished it from the precision grasp that is used for dexterity and sensitivity, whereby the digits are able to sense and monitor small changes in force and position. Specialists in robotics have expanded these definitions and have distinguished several grasping types within the power vs. precision grasping dichotomy (Cutkosky & Wright, 1986; Cutkosky, 1989; Cutkosky & Howe, 1990). Thus, they have identified nine types of power grasp that are characterized by "large areas of contact between the grasped object and the surfaces of the fingers and palm and by little or no ability to impart motions with the fingers" (Cutkosky, 1989, p. 272). These nine types differ according to whether they result in wrapping (i.e., for a prismatic object), or employ radial symmetry (i.e., for a circular object), and include the lateral pinch (i.e., bringing into opposition the first digit [generally the thumb] and the other digits which act as one gripping surface) and the non-prehensile grasp (i.e., a flat platform holding the object). With regard to the precision grasp (the object being held with the tips of the digits and thumb), Cutkosky (1989) identified seven different types according to the pattern of radial symmetry adopted (i.e., disk, sphere, or tripod [only three digits participating in grasping]), or with opposition occurring between the thumb and other, more laterally-situated, digits (i.e., for prismatic objects). Even though these definitions and categorizations are based on human hands, they have been applied to other primates (Pouydebat et al., 2009, 2011; Bardo et al., 2016, 2017) and can be applied to other groups of tetrapods. Indeed, frogs and lizards have also evolved significant forelimb grasping capabilities (e.g., Manzano et al., 2008; Abdala et al., 2009; Anzeraey et al., 2017). Iwaniuk and Whishaw (2000) suggested that 'rudimentary skilled forelimb movements', including grasping and manipulating with the digits, likely originated at the base of the tetrapod clade. These types of movements are probably homologous in frogs and mammals, and various losses of these abilities across taxa may have occurred independently. The examination of the forelimb musculature of tetrapods that ultimately underlies these movements demonstrates a large number of homologies across clades (Abdala & Diogo, 2010; Kardong, 2011), revealing six relevant 'muscular complexes' of the hand and forearm (ulnar extensors/flexors, radial extensors/flexors, and digital extensors/flexors. Thus, skilled movement behaviors made by the hands and/or feet seem to be phylogenetically conserved in tetrapods, from tree frogs to the first stone tool users, but also exhibit strong selective versatility.

We already know that the ability to grasp with the hand is often presumed to result from selective pressures associated with arboreal locomotion (e.g., Grillner & Wallen, 1985; Bracha et al., 1990) and/or prey capture (Iwaniuk & Whishaw, 2000). It seems that both arboreal locomotion (e.g., Gebo, 1985; Feduccia, 1999; Youlatos, 2008) and food acquisition (e.g., Fowler et al., 2011) are also implicated in the evolution of grasping with the foot in tetrapods. However, compared to manual grasping behavior and its associated anatomy, data on pedal grasping abilities are scarce. Here, we review grasping behavior as the ability to grasp and manipulate objects or substrates through voluntary movements of the hand and/or foot by exerting force (Sustaita et al., 2013). Our objective is to explore the form, function, ecology, and evolution associated with autopodial grasping in the context of each major extant tetrapod clade: Lissamphibia, Lepidosauria, Aves, and Mammalia. We hope to improve our understanding of the phenotypic variation exhibited by grasping autopodia and the selective forces that have shaped the evolution of grasping ability by exploring contemporary approaches incorporating measurements of grasping performance (i.e., force and precision). This review presents the opportunity for clarifying both the functional and ecological consequences of variation in musculoskeletal morphology and behavioral patterns of grasping and highlighting patterns of convergence among disparate tetrapod clades.

12.2 Grasping in Lissamphibians

Anurans are characterized by the absence of discrete caudal vertebrae and a truncated axial skeleton. Girdles and paired appendages develop at larval stages and integrate with the axial skeleton simultaneously with tail regression (Rocǒková & Roček, 2005; Handrigan & Wassersug, 2007; Pugener & Maglia, 2009; Manzano et al., 2013; Fabrezi et al., 2014). With this derived morphology many locomotor modes develop, such as jumping, which is considered the primary locomotor activity from which the other modes of locomotion of anurans, such as hopping, walking, swimming and climbing (Emerson & Koehl, 1990; Gomes et al., 2009; Manzano et al., 2018), derive (Prĭkryl et al., 2009).

Swimming, walking, burrowing to construct refugia, building nests, spreading substances on their skin, or even grasping objects such as branches, each associated with a different behavior and ecological context, are reflected in anatomical adaptations (Robovska-Havelkova et al., 2014; Liao et al., 2015; Manzano et al., 2017; Hill et al., 2018; Frýdlová et al., 2019) (Fig. 12.1a–c). Those adaptations are mainly found in the autopodia (hands and feet), which exhibit extensive variation highlighting the ecological importance of the manus and pes (Duellman & Trueb, 1986; Irschick et al., 1996; Zaaf & Van Damme, 2001; Rothier et al., 2017). For example, anuran manūs bear four digits and their pedes carry five, and these digits may vary in

Fig. 12.1 Species of *Phyllomedusa* moving on different substrates. (**a**) *P. sauvagii* elevates its body during walking (modified from www.inaturalist.org/observations/22510236). Red arrow indicates body elevation during locomotion; (**b**) *P. sauvagii* adjusts its wrists and ankles to permit grasping during vertical locomotion on a narrow perch (modified from Manzano et al., 2017); (**c**) Grasping of *P. azurea* on an inclined, thin branch [modified from Herrel et al. (2013a)]. White arrows indicate the point of hand-perch contact during the grip on an inclined substrate

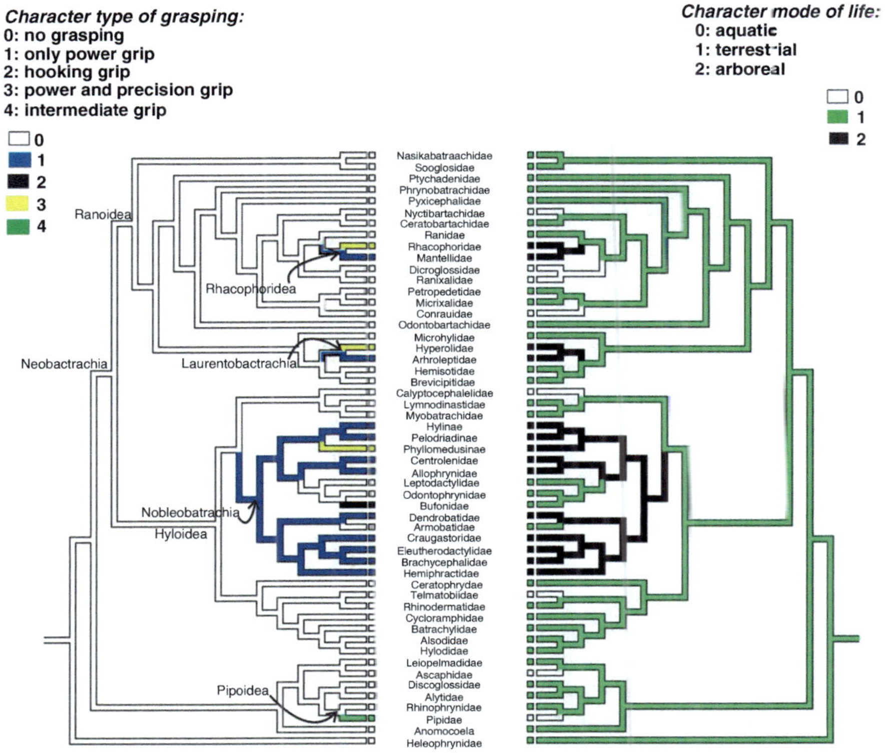

Fig. 12.2 Ancestral reconstruction of characters related to grasping capabilities in anuran amphibians using parsimony with Mesquite 2.7 (Maddison & Maddison, 2019), based on Jetz and Pyron (2018) and Feng et al. (2017). The comparison of grasping types (Table 12.1) and mode of life suggests that ancestral anurans lacked grasping abilities. The ability to grasp appeared at least three times within the Neobatrachia: Nobleobatrachia, Laurentobatrachia, and Rhacophoroidea. The precision grip specialization occurs within the groups Phyllomedusinae, Rhacophoridae, and Hyperolidae (in yellow). All frogs with a precision grip share an arboreal mode of life. The case of Pipidae could be interpreted as a novelty because they exhibit a different type of grip employed in its aquatic mode of life (Napier, 1956; Anzeraey et al., 2017). The hooking grip has been described only for Bufonidae and this manifests only as a behavior pattern

length due to reduction or loss of phalanges, or an acquisition of additional structures such as intercalary elements (additional connective tissue, bony or cartilaginous structures occurring between the penultimate and ultimate phalanx in the digits of many anurans; Duellman & Trueb, 1986; Manzano et al., 2007). Intercalary elements are considered to be an adaptation for climbing. Some frogs may also have sesamoid bones embedded in the flexor plate of hands and feet (Ponssa et al., 2010), the functional implications of which are unclear (Fig. 12.2a, b). All such adaptations of their limbs involve bone-muscular-ligament system modifications.

Table 12.1 Descriptions of the types of grip according to the position of digits and the forces exerted on an object

Grip type	Definition	Author
Power	The object is held in a clamp, involving the partially flexed fingers and the palmar and plantar surfaces. The arm and the leg exert the forces on the object.	Napier, (1956), Feix et al. (2016), Manzano et al. (2018)
Presicion	The objects are held with the tips of the fingers, which oppose each other. The opposing forces exerted between the fingers on the object are weak, but they provide dexterity to the hand or feet.	Napier (1956), Feix et al. (2016), Manzano et al. (2018)
Intermediate/scissor	The objects are held by the medial and lateral sides of two adjacent digits. Forces are intermediate between those of the above-described grips.	Napier (1956), Anzeraey et al. (2017), Vassallo et al. (2021)
Hooking	The distalmost phalanges of each digit hold the objects. Forces that are exerted do not require strong muscle contraction and are prolonged.	Napier (1956), Vassallo et al. (2021)

Despite adaptations for performing specific tasks some generalist frogs, such as *Rhinella marina* and *Rhinella arenarum*, also practice other skills, such as climbing, to escape when in danger or for exploring while foraging (Hudson et al., 2016; Vassallo et al., 2021) without any specific specializations. Their abilities involve strategies of behavior through the development of a hooking grip that allows them to achieve their objective. The hook-shaped terminal phalanges, combined with the action of the flexor tendons, enable them to climb occasionally to escape from an environment that presents obstacles (Vassallo et al., 2021), although—unlike tree frogs—, they cannot climb on smooth surfaces.

Climbing exclusively in arboreal environments has been considered to be the primary driver of the evolutionary development of skilled movements of the limbs, such as grasping (Manzano et al., 2008, 2018; Hildebrand, 1995; Gray et al., 1997; Cartmill, 1985). The limbs of many arboreal anuran species are relatively long, and intercalary skeletal elements and digital adhesive pads are often present on the hands and feet (Manzano et al., 2007), these being integrated with a muscle-ligament system to prevent the animal from falling (Hanna & Barnes, 1991) from smooth surfaces (Endlein et al., 2017). Additionally, in many arboreal frogs extensive divergence of the angles between the digits, such as those associated with opposability, are present on the hands or feet, or both (Fontanarrosa & Abdala, 2016; Manzano et al., 2018). In species of highly specialized arboreal frogs, such as *Phyllomedusa*, *Chiromantis*, and *Pseudis*, one or two digits have become rotated to lie opposite the others. These characteristics have been related to arboreality and, more specifically, to locomotion among thin branches in complex three-dimensional habitats (Herrel et al., 2013a).

Forelimbs are historically considered to be conserved among frogs and support the body during standing or walking, being *de facto* decoupled from a role in the generation of power for propulsion. Studies in this regard have focused mainly on the unique saltatory locomotion of anurans (Gans & Parsons, 1966; Lutz & Rome, 1994; Shubin & Jenkins, 1995). The forelimbs also, however, play an essential role in absorbing the impact of forces generated during landing (Nauwelaerts & Aerts, 2006; Akella & Gillis, 2011; but see Essner et al., 2010). Iwaniuk and Whishaw (2000) described specific forelimb movements of tetrapods as skilled movements associated with abilities to hold, reach, and manipulate objects, such as food.

During reproduction, anuran forelimbs play a fundamental role in amplexus: the male embraces the female while the eggs are deposited and fertilized. However, the hands are not particularly modified for the amplexus grasp, except for the presence of some epidermal calluses present on the ventral face of the hands of males. Sexual dimorphism is evident in the development of the muscles of the forearms and in the hands of the males, with the calluses (called nuptial pads) that are present, which help to hold the female during amplexus (Duellman & Trueb, 1986).

In general, the hands are not involved in feeding or even in the search for prey, except for some frogs with grasping hands that have been documented to hold the prey (Anzeraey et al., 2017; Manzano et al., 2018).

Historically, skilled forelimb movements were thought only to be encountered in the primate lineage (Napier, 1956, 1993; Landsmeer, 1962; Marzke et al., 1992; Susman, 1994), but it is now recognized that they are common among tetrapod taxa and probably share a common origin in early tetrapods. Skilled forelimb abilities in taxa other than hominids, primates, and mammals have been documented (Iwaniuk & Whishaw, 2000) and an increasing number of papers have noted the skilled limb abilities exhibited by anurans (e.g., Blaylock et al., 1976; Gray et al., 1997; Vaira, 2001; Sheil & Alamillo, 2005; Manzano et al., 2008; Herrel et al., 2013a; Anzeraey et al., 2017).

Most frogs with the ability to grasp with their hands exhibit similar pedal capabilities. However, studies of the feet of frogs are scarce and have focused mainly on toe pad anatomy and associated sticking abilities (Hanna & Barnes, 1991; Hill et al., 2018) and on the integrated modular system formed by intercalary elements and digital extensor muscles in relation to arboreal locomotion (Manzano et al., 2007). Arboreal walking is achieved using both the hands and feet to grasp branches, even those arrayed at different angles (Herrel et al., 2013a; Hill et al., 2018); see Fig. 12.1. Possible amphibian skills using manual and pedal movements, other than those related to locomotion, deserve more attention.

The hind feet can also display movements other than those associated with grasping during locomotion. The most complex limb movements involving the hands and feet occur during the "wiping behavior" observed in frogs with opposable digits, such as *Polypedates maculatus* and species of *Phyllomedusa* (Lillywhite et al., 1997; Barbeau & Lillywhite, 1999). During wiping frogs spread lipid substances all over their body using their hands and feet (Blaylock et al., 1976). Several arboreal frogs also use their hands and feet to build leaf nests into which their eggs are deposited (Kenny, 1966; Biju, 2009). These frogs also belong to arboreal groups that possess divergent opposable digits (Rhacophoridae, Hyperolidae, and Phyllomedusinae) (Fig. 12.2). Nevertheless, the opposability of digits is not always a characteristic of both the hands and feet.

Frogs capable of exercising skilled wrist movements that use their hands to reach for prey, grasp it and move it into or out of the mouth (Gray et al., 1997) do not, however, appear to use their feet to do this (except for *Phyllomedusa*; Manzano et al., 2018). Surprisingly, the grip used during locomotion in *Phyllomedusa bicolor* was described as being accomplished with a high level of dexterity (Manzano et al., 2008; Herrel et al., 2013a). It was also recorded that species of *Phyllomedusa* can perform power and precision grips (Table 12.1), taking prey by surrounding it with their hand (Manzano et al., 2018). Anzeraey et al. (2017) reported on an intermediate grip (Table 12.1) in the aquatic *Xenopus* (described as the 'scissor grip'), which is used to hold the prey item but that does not allow the hand to close around it. In addition, the aquatic *Pseudis*, a hylid frog genus with opposable digits on the hands, has fully webbed feet and limited digital movements. They use their hands mainly to float over the vegetation and no grasping has been reported; the feet are used for propulsion during swimming or jumping (Manzano & Barg, 2005).

12.2.1 Anatomical Bases of Grasping and the Precision Grip

As mentioned above, intercalary elements form an integral unit of the limbs of frogs that evolved independently of the phalanges and have been integrated into the developmental program of the forelimb and, in some groups, the hindlimb also (Manzano et al., 2007). The distalmost phalanges, intercalary elements, muscles, and digital adhesive pads act as integrated units to enhance climbing ability (Noble, 1931; Emerson & Diehl, 1980; Mcallister & Channing, 1983; Paukstis & Brown, 1987, 1991; Burton, 1996, 1998a, b). However, the presence of well-developed intercalary elements or digital pads is not always associated with the arboreal mode of life (Manzano et al., 2007). Regardless, they constitute parts of a successful device for preventing falls from slippery surfaces (Hill et al., 2018). Furthermore, the presence of widely divergent angles between digits, such as occurs with opposable digits or some degree of zygodactyly, is believed to reflect specialization for living in an arboreal environment, as can be seen in *Phyllomedusa*, *Chiromantis*, and some mantelids (Manzano et al., 2018). The opposability of one or two digits implies the ability to rotate them so that they face the other digits, with the possibility of their tips being able to touch each other, thereby exerting sufficient oppositional forces on the object to execute a precision grip (Table 12.1). Opposability of anuran digits has been reported (Sheil & Alamillo, 2005; Manzano et al., 2008; Sustaita et al., 2013), but the connection between opposing fingers and the capability of gripping are not always clear. An example of this is encountered among the members of the pseudine group (secondarily aquatic hylid frogs) that are unable to grasp objects with their hands or feet because their digits are practically immobile due to the presence of comple interdigital webbing. The immobile, cylindrical and mineralized intercalary elements of species of *Pseudis* (Hylidae) (Manzano et al., 2007) limit the mobility of the fingers and also their flexion, along with that of the palm, around objects. No reports mention their ability to grasp or climb. Selective pressure has seemingly been

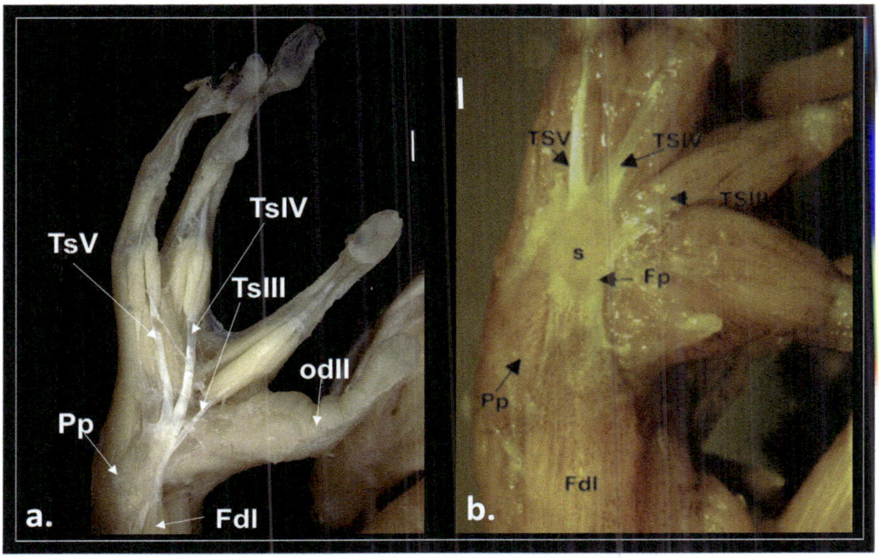

Fig. 12.3 Ventral view of the manus of (**a**) *Phyllomedusa iheringii* showing the m. palmaris profundus attached to the flexor tendons of digits IV and V. No flexor plate or sesamoid are present. (**b**) Ventral view of the manus of *Rhinella fernandezae* showing the sesamoid embedded within the flexor plate; the digital tendons arise from it. Abbreviations: *Pp* m. palmaris profundus ("m. flexor accesorius" according to Blotto et al., 2020); *TsII-V* superficial tendons of digits II, III, IV and V, respectively; *odII* opposable digit II; *Fdl* m. flexor digitorum longus; *s* sesamoid; *Fp* flexor plate. Scale: 1 mm

focused on the feet and their role in swimming, rather than on their hands which tend to remain immobile.

When present, grasping ability varies from taxon to taxon, and some specialists for walking on narrow branches demonstrate the most highly derived forelimb and hand movements (*sensu* Iwaniuk & Whishaw, 2000). Species of frogs with fully mobile, opposable fingers appear to have the finest motor control of movements of the hands and fingers. The presence of opposable digits has been associated with the ability for a precision grip (Table 12.1) (Napier, 1956) because as the contact of the tips increases so does the dexterity of finger movement. Even in those species reported to execute manual movements with a high degree of skill, such as *Phyllomedusa sauvagii*, the toe pads do not appear to be highly developed (Manzano et al., 2008) (Fig. 12.3a). They avoid slipping through the power and precision of their grip, with dexterity essentially being used during wiping behavior or other tasks, such as the grasping of narrow branches (Blaylock et al., 1976; Lillywhite et al., 1997).

When tree frogs move on narrow substrates they move their arms independently of one another (as opposed to simultaneous bilateral movements during landing or jumping), and also close their hands (i.e., execute a grip, *sensu* Napier, 1956) to resist rolling torques while walking on branches narrower than the width of their body (Hill et al., 2018). In the case of *Phyllomedusa* the body can also be raised during walking (Fig. 12.1a, c) and the wrist can be manipulated to ensure the grip (Fig. 12.1b) (Manzano et al., 2017). In species of this genus, the forearm muscles are highly differentiated and appear to be able to control each finger individually (Herrel et al., 2008a). The musculature of the hand of these frogs superficially resembles that of other tree frogs (there is no palmar sesamoid or aponeurosis) but seems to have more complex architecture (Fig. 12.3a). Manzano et al. (2008) mentioned a general increase in the length and cross-sectional area of the muscles, affecting the speed and force of contraction respectively. Also, the presence of strong and long tendons, such as those of the m. extensores breves and m. adductor indicis longus, reflect reduced compliance for greater control of the more distal elements as a result of increased tendon stress. Additionally, the main flexor tendons are independent, resulting in the ability of each finger to be able to be flexed independently (Fig. 12.3a). The presence of muscles with accessory branches (which result in additional insertion sites; Manzano & Lavilla, 1995) are some of the unique characteristics of *Phyllomedusa* that may be related to its greater manual dexterity (Manzano et al., 2008).

For example, there is a close anatomical and functional relationship between the m. palmaris profundus ("m. flexor accessorius" *sensu* Blotto et al., 2020) and the m. flexor digitorum longus as shown by stimulation experiments (Manzano et al., 2008) (Fig. 12.3a). Generally, in frogs (and also in other tree frogs, such as *Triprion petasatus*; see Blotto et al., 2020) the superficial tendons (the major flexor tendons) originate from the branches of the m. flexor digitorum longus or from a flexor plate and are united by a fascia to the m. palmaris profundus ("m. flexor accessorius" *sensu* Blotto et al., 2020) (Fig. 12.3a, b). In the genus *Phyllomedusa* the m. palmaris profundus attaches directly to the superficial tendon that arises from the medial branch of m. flexor digitorum longus, and when contracted it pulls that tendon laterally 2–3 mm (Manzano et al., 2008), thereby effectively increasing the moment arm of the latter. This actively assists in flexing the hand and wrist, ultimately allowing complete closure of the hand around a narrow perch (Manzano et al., 2008).

The ability to execute complex actions by the limbs, such as grasping, has been interpreted to be an exaptation of the specialization of the forelimbs and hindlimbs for arboreal locomotion (Manzano et al., 2008). However, Anzeraey et al. (2017) demonstrated that the aquatic frog *Xenopus laevis* can perform a complex repertoire of grasping and handling tasks, thus challenging perspectives on the ecological origin of grasping within anurans (Fig. 12.3). The hooking grip performed by the terrestrial generalist *Rhinella* shows unexpected functional capacities that could allow a species to colonize new niches (Vassallo et al., 2021).

12.2.2 Grasping Performance

In vivo measurements of grasping force and the results of muscle stimulation experiments suggest that arboreal frogs actively adjust the position of the hands during locomotion and include a grasping type of support (Fig. 12.1b) (Manzano et al., 2008, 2018; Herrel et al., 2013a). *Phyllomedusa bicolor* can generate greater grasping forces than the more generalized *Litoria caerulea*, which may assist in enhancing its stability and allow it to move more securely on narrow substrates. *Phyllomedusa bicolor* is also able to generate large forces through the abduction of digits II, IV, and V. Interestingly, the combined stimulation of the mm. flexor indicis superficialis proprius II and lumbricalis IV of *Phyllomedusa bicolor* produced pronounced adduction of digits II and IV, causing the extremities of the digits to touch one another, this being required for the generation of a precision grip (Napier, 1956; Feix et al., 2016). Species of *Phyllomedusa* can use both the hands and feet in the same skillful way.

Herrel et al. (2013a) demonstrated that in *Pithecopus azureus* (also a phyllomedusine), hand positions and grip types are highly dependent on the substrate. The substrate can vary in texture, size, diameter and inclination, interfering with the animal's stability during locomotion (Lammers & Zurcher, 2011). In these cases the animals change their grip to optimize interactions with the substrate (Fig. 12.1b, c). Primates can also vary their grip according to the substrate (Lemelin & Schmitt, 1998; Reghem et al., 2012). The effects of the diameter and inclination of the substrate on the grip type and kinematics, at least for primates and lizards, suggest that locomotor mechanics associated with movement on narrow substrates drive movement kinematics independently of morphology and phylogeny (Herrel et al., 2013a; Manzano et al., 2018).

12.2.3 Brain Correlates

Hand movements in humans and other primates involve complex neuronal patterns and functions in the fore- and hindbrain areas. The main center of movement coordination in tetrapods is the cerebellum, with organized layers of cells that regulate coorination of impulses, such as a granular layer of round and small cells and specialized Purkinje cell layers. Despite the conservative organization of the brain among tetrapods, neuroanatomical variation is evident among frogs (Ten Donkelaar, 1998; Manzano et al., 2017). A functionally-related trend towards increased cerebellum size is evident (Taylor et al., 1995). Indeed, Manzano et al. (2017) showed an increasingly complex network of Purkinje cells in the cerebellum of species of *Phyllomedusa* compared to other arboreal and terrestrial species of frogs. This was related to the generation of complex or subtle movements and their associated increased manual dexterity. Purkinje cells are inhibitory cells of the vestibular system that mature during frog metamorphosis and are involved in the cerebellum's sensory process (Gona & Uray, 1980; Llinàs et al., 1967; Ten Donkelaar, 1998).

Although Manzano et al. (2008) experimentally demonstrated the precision grip capacities of the hands and feet of these frogs, the coordination between the movement of the hands and the visual perception of the frogs seems to be limited. During locomotion, visual coordination for controlling landing is essential (Drew,

1991), as Cox et al. (2018) demonstrated. An experimental procedure of labyrinthectomy and ablation of the optic nerves, with the addition of bilateral transection of the sciatic and femoral nerves responsible for proprioception in cane toads, shows that vision is essential for fine-tuning this behavior (Cox et al., 2018). A combination of vision and proprioceptive postural behavior (Lambert & Straka, 2012) may be more highly developed in those frogs that must move in a three-dimensional environment, such as tree branches. In fact, in aquatic frogs, such as *Xenopus*, postural compensation and recovery after damage (the unilateral ablation of the endorgans of the vestibular system) are not possible whereas, for example, in terrestrial frogs they are (Lambert & Straka, 2012). Manipulation other than prehensility to avoid falling from a branch would arise as an exaptation from locomotion favoring increased size and complexity in those structures that allow arboreal locomotion. However, climber-walkers (see also Taylor et al., 1995), hopper-walkers and burrowing frogs have a large cerebellum, suggesting that these abilities, developed with the paired limbs in different locomotor contexts, would have implications for the evolution of the cerebellum in anurans, with the arboreal environment being a driver of more profound cerebellar modifications.

Given the complexity of limb movements observed in frogs and the fact that these evolved independently several times (Fig. 12.2), frogs provide an excellent taxon for better understanding the neurological context associated with the evolution of increased manual dexterity and grasping behavior.

12.3 Grasping in Non-avian Reptiles

The recorded non-avian reptilian species that exhibit manual or pedal grasping abilities are restricted to the lepidosaurs (Abdala et al., 2009; Herrel et al., 2011; Sustaita et al., 2013). The most recent synthesis of prehensility in lepidosaurs is that of Sustaita et al. (2013), wherein it was noted that most studies of limb function in lizards have focused on quadrupedal locomotion and running performance (e.g., Losos, 1990; Irschick & Garland, 2001) and, to a lesser extent, on clinging and climbing (e.g., Zani, 2000; Zaaf & Van Damme, 2001; Tulli et al., 2009, 2011). These studies highlight the ecological and functional diversity that lizards face in nature and the forces driving limb morphology evolution. Lizards use grasping mostly to accommodate locomotion in complex three-dimensional habitats that present discontinuities and gaps between perches. In general terms, feeding or mating behaviors play a lesser role in shaping the grasping skills of lizards than they do in other tetrapods, such as some anuran species (Anzeraey et al., 2017; Manzano et al., 2018).

The grip most commonly observed in lizards is that corresponding to a power grip as defined by Landsmeer (1962): "objects are held in a clamp formed by the partly flexed fingers and the palm, with counter pressure applied by the thumb lying more or less in the plane of the palm. In the power grip the combined fingers form one jaw of the clamp with the palm as the other jaw". Chameleons (one of the most

Fig. 12.4 Chameleon hand showing the "super-digits" generated by the syndactylous complexes formed by manual digits 1–3 and 4–5. Each complex constitutes a single functional unit (Molnar et al., 2017)

specialized arboreal groups of lizards) are considered the most adept graspers among lizards (Herrel et al., 2011). Chameleons exhibit zygodactylous manūs (Fig. 12.4) and pedes and a fully prehensile tail. Some species, such as *Chamaeleo vulgaris*, possess a carpus in which the centrale and distal bones are fused forming a single spherical element (Renous-Lécuru, 1973). However, in other congeneric species no fusion is evident (Herrel et al., 2013b). The role of this fusion of elements with regard to grasping remains unclear. Interesting data on the development of these autopodial specializations are provided by Diaz Jr. and Trainor (2015). They stressed that chameleons lack an astragalus-calcaneum complex typical of amniotes; additionally, phylogenetically derived chameleons exhibit an ankle structure convergent with that of amphibians (Diaz Jr. & Trainor, 2015). Remarkably, most of the muscles usually present in the hands and feet of chameleons are present in the same configuration as they are in other lizards (Mivart St., 1870; Ribbing, 1913; Gasc, 1963; Molnar et al., 2017). Some peculiarities are, among others, the broad, V-shaped plantar and palmar aponeuroses and the muscle orientation of the superficial short flexors originating from these aponeuroses, which contribute to the functioning of the "super digits" described for this group (Fig. 12.4, Molnar et al., 2017).

Until relatively recently it was thought that chameleons were the only lizards that exhibited autopodial prehensility, but the list has grown with more recent studies. Three other lizard lineages are capable of performing a power grip *sensu* Landsmeer (1962): geckos, *Polychrus*, and anolines (Abdala et al., 2009; Sustaita et al., 2013; Fontanarrosa & Abdala, 2014, 2016), although these have been examined much less extensively in this regard. Interestingly, prehensility in non-chameleon lizards is performed with a hand without the extreme modifications shown by chameleons. Recently, it has been demonstrated that grasping skills of the hands, feet, and tail seem to be affected by the amount of friction the animal can generate between its feet and the perch (Luger et al., 2020).

12.3.1 The Anatomy of the Hands and Feet of a Grasping Lizard

Several anatomical traits of the hand of lizards can be linked to grasping abilities (Abdala et al., 2009; Fontanarrosa & Abdala, 2014, 2016). These specializations are evident in relation to the tendons and bones of the lizard hand and the rather conservative intrinsic muscles of the hand (Russell & Bauer, 2008; Abdala et al., 2009; Abdala & Diogo, 2010; Diogo & Abdala, 2010). Three patterns of the tendons of the palm of the hand have been described: L, P and G (Fig. 12.5). These play a crucial role in the flexion at the metacarpo-phalangeal joints, which provides the main input for the power grip *sensu* Landsmeer (1962). The power grip allows the

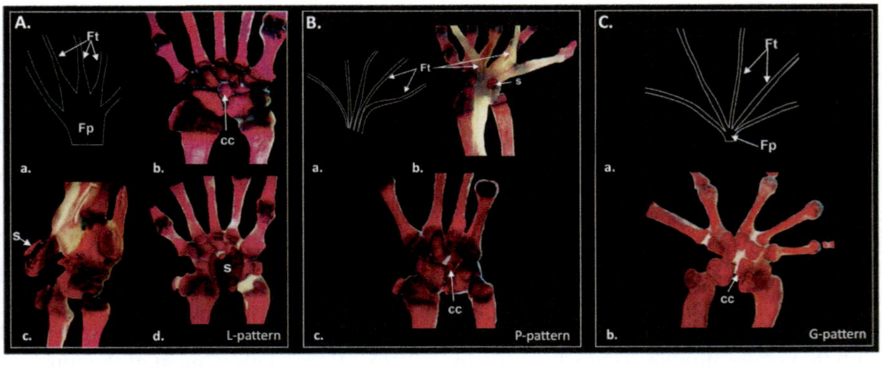

Fig. 12.5 (**A**) Hand of *Liolaemus cuyanus* showing the L-pattern of palmar tendons with (a) a schematic of the flexor plate and the digital flexor tendons; (b) the centrale located at the center of the palm of the hand; (c) lateral and (d) ventral view of the hand showing the location of the palmar sesamoid. (**B**) Hand of *Anolis cristatellus* showing the P-pattern of palmar tendons. (a) schematic of the independent flexor tendons to the digits and the reduced palmar sesamoid; (b) digital flexor tendons and the sesamoid embedded in a flexor plate; (c) elongated centrale. (**C**) Hand of *Homonota horrida* showing the G-pattern of palmar tendons. (a) schematic of the independent flexor tendons serving the digits and the flexor plate lacking the palmar sesamoid; (b) elongated centrale. *cc* centrale, *Fp* flexor plate, *Ft* digital flexor tendons, *s* sesamoid. Redrawn from Sustaita et al. (2013)

hand to close around a narrow perch or branch. The most widely distributed tendinous pattern within lizards is the L-pattern (Moro & Abdala, 2004; Abdala et al., 2009), consisting of a single tendinous plate that does not allow for extensive metacarpo-phalangeal flexion (Fig. 12.5A). The plate is associated with the m. flexor digitorum longus, the largest forearm muscle, and serves the digits through the digital flexor tendons. One or two sesamoids—the palmar sesamoids—are generally embedded in this 'flexor plate' (Haines, 1950; Abdala et al., 2009: Regnault et al., 2016). This single tendinous structure prevents independent movement of the digits and instead they move together as a single unit. However, the flexor plate may enhance the flexion of the distal phalanges thereby improving, for example, the grip provided by the claw (pers. obs. VA). Contrastingly, the P-pattern (Moro & Abdala, 2004; Abdala et al., 2009) has a small or no flexor plate and the m. flexor digitorum longus serves the digits with independent digital flexor tendons. Palmar sesamoids also tend to be small or absent (Fig. 12.5B). Most anolines and other lizards, such as *Polychrus*, exhibit the P-pattern. Finally, a third pattern, the G-pattern, is present in most geckos (Abdala et al., 2009), the flexor plate of which lacks embedded sesamoids (Fig. 12.5C). Experimental work was conducted showing that the different patterns (L, P, and G) correlate with hand movement capabilities and grasping performance (Abdala et al., 2009). Similar anatomical patterns are recognized in anurans, for example (see Fig. 12.3a, b), but this promising area of research has not been pursued recently. It would be interesting to evaluate experimentally, and with more ecological data from more tetrapod species, the consequences of having a flexor plate in relation to the possibilities for colonizing new niches in, for example, forest trees.

Lizards with a sesamoid embedded in the tendon of the m. flexor digitorum longus are not capable of flexing the metacarpophalangeal joints of the hand and therefore cannot execute a power grip. Although the function of sesamoids, even the palmar sesamoid, are still speculative, it can be inferred that the palmar sesamoid allows flexion of the distalmost phalanges through the tension transmitted by the digital flexor tendons, this allowing for a more accurate interaction with the perch (see, e.g., Vassallo et al., 2021). Absence a sesamoid or sesamoids in the palm of the hand appear to facilitate the flexion of the metacarpophalangeal joints, as exhibited by lizards with the P- and G- patterns. It can be inferred that lizards possessing a palmar sesamoid limit tendon movement, thereby resulting in incomplete flexion at the digital joints. In certain cases, however, the pressure of the palm of the hand against the perch combined with a gentle flexing of the terminal phalanges of the hand can generate a form of prensility in lizards with the L pattern of palmar tendons (Fig. 12.6).

Interestingly, a human clinical condition, "trigger-wrist," can provide clues about the impairment produced by the palmar sesamoid. Humans lack palmar sesamoids but in some circumstances a tumor or nodule occurs on the flexor tendon and/or tendon sheath. This passes through the carpal tunnel and may prevent the sliding of the tendons through this conduit (Förstner & Schaefer, 1998) (Fig. 12.7a). It may be that the tendon of the m. flexor digitorum longus, reinforced with a palmar sesamoid as in the L pattern, would have considerable difficulty sliding through the carpal

Fig. 12.6 A gentle flexure of the terminal phalanges of the hand produce a version of prensility in *Iguana* which has the L pattern of palmar tendons. Picture available in the royalty free photos released under public domain license site Pikist.com

tunnel (Fig. 12.7b). Moreover, in humans, under certain circumstances, proximal and distal interphalangeal (PIP and DIP, respectively) joint flexion precedes metacarpophalangeal (MCP) joint flexion, and substantially greater m. flexor digitorum profundus forces are required to effect similar flexion angles at the MCP joints, as is the case for the PIP and DIP (Nimbarte et al., 2008). Kamper et al. (2002) showed that contraction of the extrinsic flexor muscles simultaneously with flexion of all the digital joints generated substantially less flexion at the MCP. It can be thus deduced that a greater tendon excursion is required for complete flexion of the digital joints, especially the MCP joint, and the palmar sesamoid probably prevents this in lizards with an L pattern of palmar tendons.

Many studies have shown that the intrinsic hand muscles (those that originate and insert within the hand) do not exhibit particular innovations related to the ability to move the hand in lizards. The distal insertion of the forearm muscles in those tetrapods having particularly skilled hand movements seems to be of greater importance (Herrel et al., 2008b; Russell & Bauer, 2008; Abdala & Diogo, 2010; Diogo & Abdala, 2010; Sustaita et al., 2013).

Some studies have shown that the configuration of the wrist and hand bones of lizards also correlate with grasping ability. The osseous structures of the hand and their characteristics associated with grasping abilities have been analyzed by Fontanarrosa and Abdala (2014, 2016). Collectively these studies surveyed the anatomy of the carpus of 278 specimens distributed among 24 genera and 13 squamatan families qualitatively (Fontanarrosa & Abdala, 2014) and quantitatively (Fontanarrosa & Abdala, 2016). In addition to the potential for the palmar sesamoid to impede grasping they highlighted other characters, such as an elongated centrale (Figs. 12.5B, C) (as previously noted by Sustaita et al., 2013), which

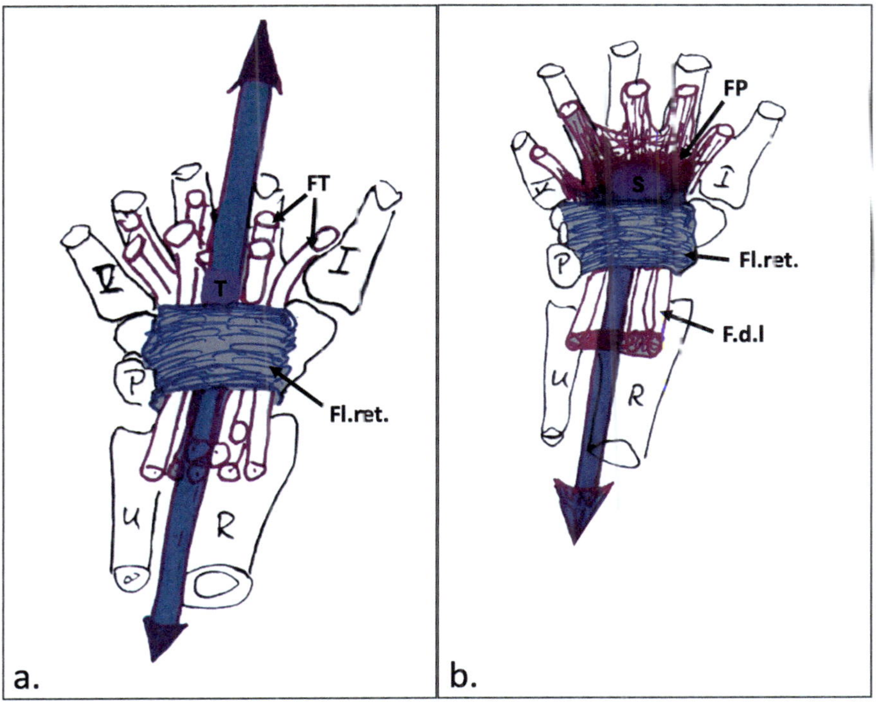

Fig. 12.7 (**a**) Anatomical structure of the human right hand in palmar view showing a tumor or nodule occurring on the flexor tendon and/or tendon sheath, preventing the slicing of the tendons in the carpal tunnel. (**b**) The same structures projected onto a "lizard hand in palmar view": the tendon of the m. flexor digitorum longus is reinforced with a palmar sesamoid, exhibiting the L pattern of palmar tendons. Here the tendons have considerable difficulty in sliding in the carpal tunnel and flexion of the wrist and fingers and their free extension are hampered. *U* ulna, *R* Radius, *P* pisiforme, *I–V* metacarpals, *FT* Flexor tendons, *Fl.ret.* flexor retinaculum, *F.d.l.* m. flexor digitorum longus, *S* sesamoid, *T* tumor or nodule, *FP* flexor plate. Drawings from H. Förstner

facilitate grasping. The centrale is the only element in the lacertilian middle carpal row (Russell & Bauer, 2008). In non-grasping lizards it is usually flanked by the radiale and ulnare (Fig. 12.5A). This pattern imposes restricted mobility because the close contact between the proximal portions of the radiale and ulnare prevents wrist movement. The palmar sesamoid(s) also prevents flexion of the digits. These one or two sesamoids lock the articulation between the first distal carpal and metacarpal I, and between the second distal carpal and metacarpal II. The proximal region of the hand becomes a rigid structure with the only possibilities of motion being flexion and extension of the distalmost phalanges. In those lizards capable of grasping (e.g., *Anolis* and *Polychrus*), the centrale has shifted positionally from the central row to the proximal one (Fig. 12.5B). Thus, it has become, functionally, a proximal carpal, acting as a pivot between the radiale and ulnare, and it is now more slender and elongate rather than being truncated and sub-spherical as it is in most other lizards.

Additionally, the proximal head of the first metacarpal is located in a space bounded medially by the radiale, proximally by the highly reduced distal carpal I, and laterally by the distomesial side of the centrale. In the hand of most lizards there is no differentiation into thenar and hypothenar regions. The altered shape and position of the centrale and the reduction of the palmar sesamoid may thus provide regional differentiation and mobility within the hand by allowing digit I to be located in a relatively more medial position. A quantitative analysis of the hand bones of lizards indicated that grasping is a functional consequence of the centrale's width and the proximodistal length of the palmar sesamoid (Fontanarrosa & Abdala, 2016). A similar displacement is present in relation to digit V in *Anolis* and *Polychrus*. This provides the hand with an entirely new dimension for movement about the long (proximo-distal) axis of the palm and may explain how these lizards are capable of grasping narrow branches. Moreover, a grasping hand exhibits a relatively narrower first metacarpal and a greater divergence angle between digits one and five (Fontanarrosa & Abdala, 2016) than is the case in lizards that lack grasping abilities. Finally, a grasping hand depends on the relative lengthening of its long bones, a feature shared by almost all arboreal tetrapods (Fontanarrosa & Abdala, 2016).

12.3.2 Pedal Grasping in Lizards

The morphology, kinematics, and ecomorphology of lizard hindlimbs and their role in locomotion have been intensively studied (Losos, 1990; Reilly & Delancey, 1997; Zaaf & Van Damme, 2001; Higham & Jayne, 2004; Kohlsdorf et al., 2001; Russell & Bauer, 2008). However, the topic of grasping feet in lizards has seldom been addressed (Brinkman, 1980; Russell & Bauer, 2008; Abdala et al., 2014). Remarkably, taxa that exhibit manual grasping abilities also show pedal grasping skills: this being evident in chameleons (Fischer et al., 2010), varanids (Mendyk & Horn, 2011), anoles, and geckos (Abdala et al., 2014). Contrary to what has been described in relation to manual grasping (e.g., Abdala et al., 2009), the few accounts of pedal tendon structure have revealed great homogeneity among most lizards (Russell, 1993; Russell & Bauer, 2008; Abdala et al., 2014), and no reports exist of differences in their complex plantar tendons that relate to particular functional abilities. The presence of a plantar sesamoid (Abdala et al., 2019) has seldom been reported (e.g., in the gecko *Ptenopus* spp. by Russell & Bauer, 2008). *Ptenopus* is secondarily terrestrial (Russell & Bauer, 2008), which seemingly suggests that the anatomical constraints acting on manual morphology may also play a role in driving pedal morphology. Overall, the morphology of the lizard foot is conserved (Russell & Bauer, 2008; Abdala et al., 2014). Some exceptions have been reported for the astragalocalcaneum of *Chamaleo*, which is different from all other lizard astragalocalacanea in being depressed and curved. *Varanus* presents an unusual structure of the mesotarsal joint through the elongation and orientation of the lateral process of the astragalocalcaneum and the structure of its distomesial border. The functional significance of these differences is, however, unknown (Russell & Bauer,

2008). In accordance with this overall structural similarity, lizard lineages generally fail to exhibit pedal grasping capabilities, except for those taxa mentioned above that exhibit manual grasping abilities. Pedal grasping in *Anolis* spp. appears to take place at the level of the distal interphalangeal joints (Robinson, 1975; V. Abdala, personal observations).

Abdala et al. (2014) analyzed the anatomy of the crus and pes of several lizard families in a phylogenetic context to relate them to grasping abilities. Once again, no particular anatomical trait was discovered that was found to be related to grasping. Most of the skeletal elements evaluated showed a strong phylogenetic signal. Even taxa such as *Polychrus* and *Anolis*, which can actively grasp using the pedal digits and curl them around narrow branches, lack any particular set of osteological attributes associated with this ability. Thus, phylogeny seems to be the best predictor of most osteological traits of the lizard foot, with ecological particularities playing a lesser role in shaping anatomy. Contrastingly, most of the K values, the metric indicating phylogenetic signal (Blomberg et al., 2003), for the variables based on muscle and tendon morphometric characters, indicated weak phylogenetic signal, suggesting that their variation cannot be explained by phylogeny alone. Perhaps it is in these soft tissues of the foot that the ability to adjust the grip resides.

12.3.3 Lizard Grasping Performance

Claws are a vital aspect of locomotion, with claw height contributing to clinging and climbing on rough surfaces (Zani, 2000). In most lizards, the hand forms an almost rigid plate, whatever flexibility there is seeming to occur primarily in the distal regions of the digits and at the claws (Zani, 2000; Tulli et al., 2009). Arboreal and saxicolous (moving on rocks) lizards use vertical substrata and tend to have shorter and significantly more highly curved claws. Contrastingly, species utilizing open terrestrial habitats have longer and relatively straighter claws (Tulli et al., 2009). A more extensive analysis of the major traits of claws in lizards has recently been published and assesses convergent evolution of these structures (Baeckens et al., 2020). Versatile claws allow lizards that cannot grasp to negotiate vertical substrates but do not facilitate movement on narrow branches. It should be noted, however, that some facility for grasping can be achieved by using flexure at the metacarpo-phalangeal (MP) joints, such as in the arboreal *Iguana* (Fig. 12.6), as explained above. The few studies of grasping performance of lizards reveal a tendency for grasping forces to differ among species (Abdala et al., 2009). The weakest grasping forces recorded are for *Pogona vitticeps*, which is unable to close its hands around narrow substrata (Abdala et al., 2009). Species able to grasp exhibited no significant difference in grasping force, although it was somewhat greater in *Anolis equestris* compared to *Gekko gecko* (Abdala et al., 2009). More data recorded from a wider variety of species are needed to enhance our understanding of the morpho-functional relationships among grasping species and provide insights into the advantages conferred by the different palmar tendinous patterns observed (Abdala et al.,

2009). A recent study (Feiner et al., 2020) analyzed the locomotor performance of some *Anolis* lizards and suggested that differences in structural habitats promote different styles of locomotion and perching behavior in this genus.

In summary it can be stated that lizards that can close the hand around branches of small diameter are able to do so mainly because their carpal joints are flexible, they lack large palmar sesamoids that would otherwise prevent them from closing the hand, and they exhibit flexor tendons emanating from the forearm muscles that serve each digit independently, enhancing their ability to move. These morphological traits are expected to facilitate the hand movements needed to exploit niches characterized by narrow branches.

Varanus beccarii is one of the few lizards reported to be able to perform extractive foraging through grasping movements (Mendyk & Horn, 2011) by using its hands to take food and push it into its mouth. Interestingly, *V. beccarii* is an arboreal lizard, further supporting the pervasive relationship between arboreality and skilled hand movements. As for some frogs, *V. beccarii* can free its hands from their role in locomotion and support in order to use them in a feeding context. It could be proposed that this decoupling is enabled by their grasping feet which ensure stability of contact with the substratum while performing manually-assisted foraging and feeding. In general, hands can only be employed in an entirely novel context, such as prey prehension, if they are able to be temporarily released from their roles in locomotion and substrate prehension. *V. beccarii* is the only lizard species for which pedal grasping that enables it to free its hands for use in a feeding context has been reported. Additionally, it should be noted that these lizards exhibit high levels of behavioral complexity (Horn & Visser, 1997; Sweet & Pianka, 2007). It is possible that the grasping hands and feet of this species, coupled with their cognitive abilities (Manrod et al., 2008; Cooper et al., 2020), account for their remarkably skilful activities.

12.3.4 What About Other Reptilian Groups: Turtles and Crocodiles?

The subject of the capability of grasping has been scarcely, if ever, investigated for turtles. Forelimb abilities of aquatic, fresh-water turtles have been reported by Manzano et al. (2015). Pleurodires are capable of complex and subtle hand movements that are associated with locomotion and certain grooming behaviors (Manzano et al., 2015). Several YouTube videos of copulating aquatic turtles reveal that the males of *Phrynops* hold the females by grasping their shells and curving the distalmost extremities of their digits around the border of the carapace. In sea turtles the male exerts pressure on the female's shell using the distal ends of its pectoral and pelvic flippers. Females of both *Phrynops* and sea turtles swim during the entire copulatory processes. Although the general anatomy of turtles is relatively conservative, some aquatic turtles exhibit specializations related to their grasping capabilities (Abdala et al., 2008), although there are no data about hand grasping abilities in any of the more than 400 species. Recently Fujii et al. (2018) documented evidence of marine turtles using both hands to catch prey, and suggested that such forelimb use could have originated in ancestral turtles approximately 70 million years ago (Fujii et al., 2018). As can be seen, much work remains to be conducted about turtles for this underrated issue.

Seemingly, the only record of skilled forelimb movements in crocodiles is that furnished by Iwaniuk and Whishaw (2000), although unfortunately it is based upon unpublished observations. Data on the wrist movements of alligators indicate that the crocodilian wrist mechanism functions to automatically lock their semi-pronated palms into a rigid column (Hutson & Hutson, 2014). Thus, it is possible that this inhibits the development of other skilled attributes of the hands of this group. Interestingly, YouTube videos show that crocodiles can use their almost rigid hands to assist females to receive the male during mating. Unfortunately, we were unable to locate any literature pertaining to the role of grasping during mating in turtles and crocodiles, indicating that much more work is required in the exploration of this subject. Crocodiles exhibit almost all locomotor modes present in quadrupedal mammals, with an impressive locomotor repertoire (Hutchinson et al., 2019). Thus, it can be inferred that locomotion was the primary function driving the evolution of the limbs in crocodiles.

12.4 Grasping in Birds

With the specialization of the forelimbs for flight throughout the evolutionary history of birds, the capacity for grasping, such as in perching, resides solely with the pes. The climbing of substrates, handling of food items and the manipulation of nesting materials, progressively became relegated to the hindlimbs. This poses distinct challenges for the development of grasping ability in birds because the feet are subject to a variety of functional demands beyond grasping, such as terrestrial and/or aquatic locomotion, preening, fighting, and thermoregulation (Lovette & Fitzpatrick, 2016). Despite the specializations exhibited by many groups of birds for one or few of these functions, many birds employ their feet, to varying extents, for several of these functions (Sustaita et al., 2013; Morrison 2018). Perhaps because of this, the avian foot has achieved a remarkable diversity of form and function despite having lost digit V (Bock & Miller, 1959). Possibly because of these potential constraints, most birds that perform pedal grasping are restricted to the execution of a 'power grasp' (as opposed to a 'precision grip'), as exemplified most dramatically by birds of prey (hawks, falcons, and owls) for seizing, and to some extent, killing, prey. Parrots, mousebirds, tits, and even crows (Melletti & Mirabile, 2010; Katzner, 2016) are able to hang upside down with acrobatic mastery. Nevertheless, some birds, such as parrots, have achieved comparable levels of digital dexterity to other tetrapods whose limbs are far more specialized for grasping (Sustaita et al., 2013). Thus, birds have independently converged on grasping form and function found in other groups of tetrapod vertebrates principally along four main avenues: (1) opposability of digits, (2) the presence of toe pads and claws, (3) modification of certain aspects of musculoskeletal morphology, and (4) through behavioral repertoires. Below, we consider each of these avenues in turn, highlighting how birds have attained similar grasping abilities to those of other tetrapods, albeit by different means. In doing so, we provide several new insights that have emerged since Sustaita et al.'s (2013) review of the topic.

12.4.1 Opposability of Digits

Digital opposability is a key feature of manual and pedal grasping among terapod lineages (Sustaita et al., 2013), particularly for arboreal lizards (e.g., chameleons) and mammals (e.g., primates). The ability to grasp arboreal perches is a hallmark of avian evolution (Sereno & Chenggang, 1992; Middleton, 2001). The reversal of the hallux has been associated with the ability to perch; as such, specifically *when* this ability evolved is subject to some debate, as the orientation of the hallux in the putative ancestor of modern birds, *Archaeopteryx*, is somewhat equivocal (Middleton, 2001, 2003; Fowler et al., 2011; Hattori, 2016). Conventional wisdom suggests that hallucial reversal evolved with arboreality in birds for grasping perches (Feduccia et al., 2007), but Fowler et al. (2011) suggest that reversal of the hallux might have been selected for predatory purposes in the terrestrial dromaeosaurid lineages leading to birds. Regardless, it is clear that the reversal and incumbency of the great toe (hallux) to form an opposable digit was a fundamental precursor of grasping (Sereno & Chenggang, 1992; Feduccia, 1999; Middleton, 2001; Fowler et al., 2011). Furthermore, we now have a better understanding of *how* digital opposability might have evolved in birds (Fig. 12.8). Pharmacological paralysis experiments performed on developing embryonic birds have shown that torsion of the cartilaginous immature first metatarsal, resulting from muscle activity, is the primary cause of hallucial reversal at an early developmental stage (Botelho et al., 2015a; Fig. 12.8). Degrees of this form of plasticity may have existed in ancestral lineages, leading to the various stages of hallucial reversal observed in fossil avialian taxa.

Digit opposability in birds occurs in different ways, such that digit I (hallux) opposes the other three, or different combinations of digits II, III, IV cluster with the hallux in opposition to the others (Fig. 12.9). Abourachid et al. (2017) highlighted how the "pincer-like" foot structure of arboreal birds, comprising different combinations of these forward- and rearward-facing toes, converges upon that of many other arboreal/climbing tetrapods, particularly chameleons and primates. Previously it was thought that the six typical toe arrangements found among avian taxa (zygodactyl, heterodactyl, syndactyl, anisodactyl, pamprodactyl, ectropodactyl) evolved from an ancestral anisodactyl (in which the caudally-directed hallux opposes toes II–IV) 'perching' foot (Bock & Miller, 1959). However, recent developmental studies suggest that the zygodactyl toe arrangement (in which the

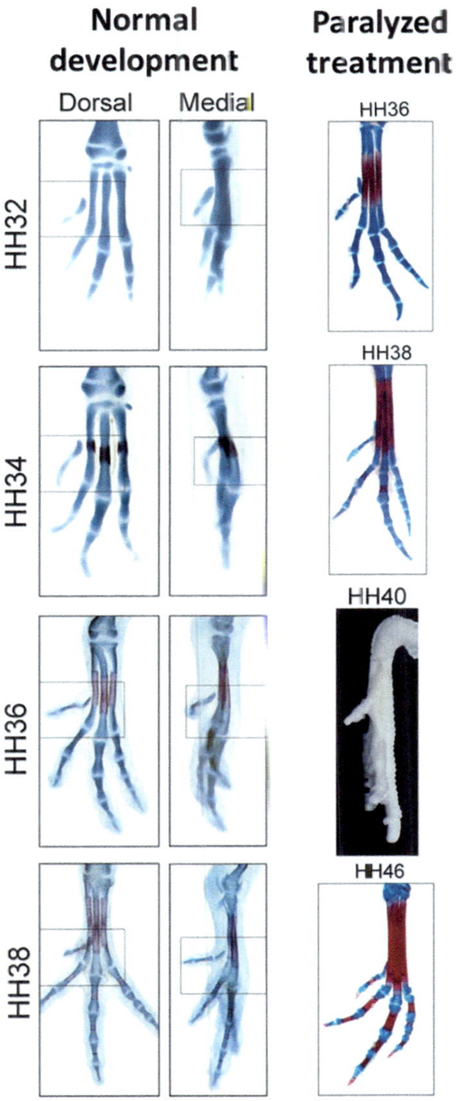

Fig. 12.8 Alcian/Blue Alizarin Red-stained feet of developing quail and chicken embryos at selected Hamburger and Hamilton (HH) developmental stages. HH32-HH38 for normally developing quail embryos show the progessive reorientation of the hallux. Lack of reorientation is evident in paralysed chicken embryos during stages HH36-HH46. Modified with permission from Botelho et al. (2015a), Scientific Reports, www.nature.com, Creative Commons CC-BY 4.0

caudally-directed toes I and IV oppose toes II and III), found in parrots, cuckoos, woodpeckers and allies, and facultatively in owls, some kites, and the osprey (Bock & Miller, 1959; Raikow, 1985; Tsang, 2012), may in fact represent the ancestral condition for crown-group birds (Botelho et al., 2014; Botelho et al., 2015b). Botelho et al. (2014) provided compelling evidence regarding the role of asymmetrical degeneration of the intrinsic muscles controlling digit IV for generating the zygodactyl configuration in budgerigars, similar to the process of hallucial reversal (Fig. 12.8). They further suggested that the loss of the abductor of digit IV in

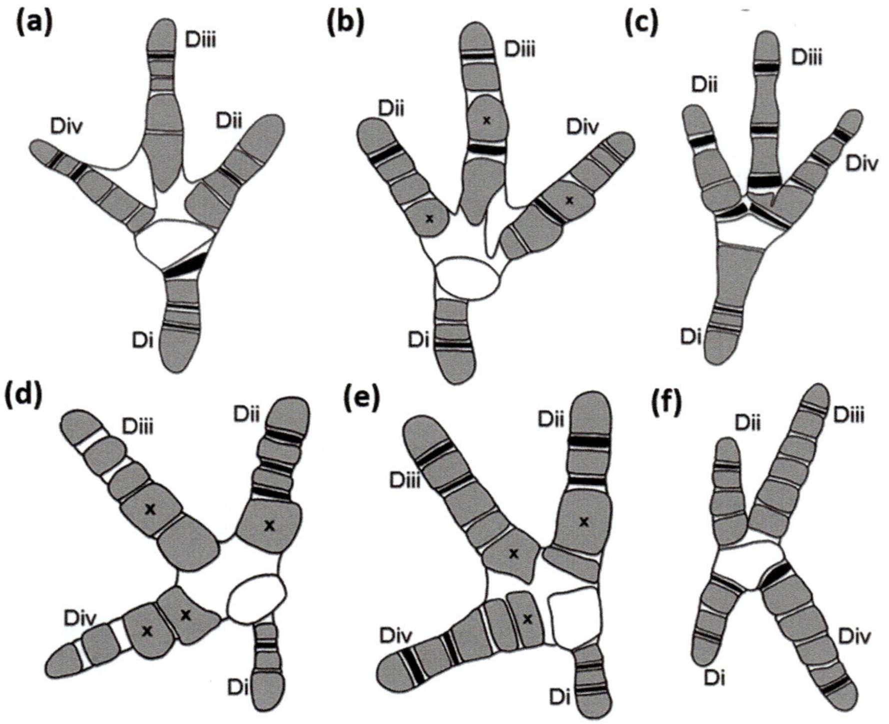

Fig. 12.9 Drawings of the plantar surface of the feet of various representatives of predatory and non-predatory avian taxa, left (L) or Right (R) as indicated: (**a**) *Aquila audax* (R); (**b**) *Falco cenchroides* (L); (**c**) *Corvus coronoides* (L); (**d**) *Ninox boobook* (R); (**e**) *Tyto alba* (R); (**f**) *Eolophus roseicapilla* (L). The drawings illustrate varying degrees of toe divarication among anisodactylous (**a**–**c**) and zygodactylous (**d**–**f**) taxa, as well as variation in the sizes, shapes, and distribution of toe pads (gray), furrows (black), and folds (white). Protrusional pads are marked with an "X." *Di* Digit I, *Dii* Digit II, *Diii* Digit III, *Div* Digit IV. Modified with permission from Tsang et al. (2019a), Journal of Morphology, Wiley

passeriforms resulted in their 'secondarily' anisodactyl configuration (Botelho et al., 2014, 2015b).

The functional significance of the various toe arrangements is not precisely clear. Many climbing specialists, such as woodpeckers, have zygodactyl (or "ectropodactyl" *sensu* Bock & Miller, 1959) feet. Parrots, which both climb and manipulate food and other objects with their toes, are strongly zygodactylous. Trogons are heterodactylous (with toes I and II opposing toes III and IV; Bock & Miller, 1959) and mostly use their feet for perching, whereas mousebirds that do a lot of climbing, hanging, and manipulation of food items, can assume toe configurations that range from anisodactyl, to zygodactyl, to pamprodactyl (Berman & Raikow, 1982). Perhaps these arrangements distribute the forces more evenly to enhance the grasping of vertical substrates (Bock & Miller, 1959). However, roadrunners are also zygodactylous, but are primarily cursorial. Moreover, owls and ospreys are semi-zygodactylous and use their feet for killing and grasping prey. The zygodactyl arrangement (Fig. 12.9) is thought to distribute the toes more symmetrically (Payne, 1962; Einoder & Richardson, 2007b; Tsang & McDonald, 2018), and the digital flexion forces more equitably (Ward et al., 2002), thereby enhancing prey-capture success. Indeed, based on an analysis of publicly available internet images and videos, Sustaita et al. (2019) found that semi-zygodactylous ospreys disproportionately use the zygodactylous toe arrangement when grasping (Fig. 12.12c). Facultative zygodactyly is characteristic of few, but phylogenetically disparate, taxa (Tsang, 2012; Botelho et al., 2015b). Tsang and McDonald (2018) showed that semi-zygodactylous taxa can assume a wide array of toe divarication angles (Fig. 12.9). They also showed how the raptorial foot morphotype is considerably more flexible than previously understood in its ability to assume a variety of toe divarication angles, particularly among prey generalists.

12.4.2 *Toe Pad and Claw Morphology*

Since Lennerstedt's (1974, 1975a, b) and Stettenheim's (2000) classic works on the topic, recent studies have reinvigorated the roles that keratin plays in avian foot form and function. Höfling and Abourachid (2020) recently described aspects of podothecal morphology that might also play an important role in grasping, such as the sharp, pointed and overlapping ventral scales of predatory or climbing species, thought to afford them greater traction. These authors also reported a greater prevalence of syndactyly (the partial fusion of certain foretoes) among bird taxa possessing each of the main toe configurations (anisodactyl, zygodactyl, and heterodactyl). They suggested that syndactyly may play roles in increasing the size of the sole for increasing friction with the substrate, and cite its potential advantage for perching by keeping the toes parallel and restricting their forces to those acting at a right angle to the branch (Höfling & Abourachid, 2020). Tsang et al. (2019a) found that the morphology of the ventral toe pads varied considerably among raptors, as well as between predatory and non-predatory taxa, and even among digits within individuals (Fig. 12.9). This variation in toe pad morphology is not unlike that displayed by the volar and plantar pads (or tubercles) on the paws of mammals, that are particularly well-developed in climbing and scansorial species (Cartmill, 1985; Barbera et al., 2019). They highlighted how, in predatory taxa, the toe pads are more pronounced and typically located at the interdigital joints, whereas in non-predatory taxa the folds between the pads are situated at the joints and the pads themselves are relatively smaller and flatter. Furthermore, toe pads are more protrusive in bird-eating raptors, presumably to aid in gaining purchase on more highly elusive prey, and more uniform, well-developed toe pads are characteristic of accipitrids that tend to pursue ground-dwelling prey (Tsang et al., 2019a).

Claw shape has obvious implications for grasping capability and many advancements have been made in our understanding of how claw shape varies among taxa and functional groups (Fig. 12.10a). Previous studies have reported significant

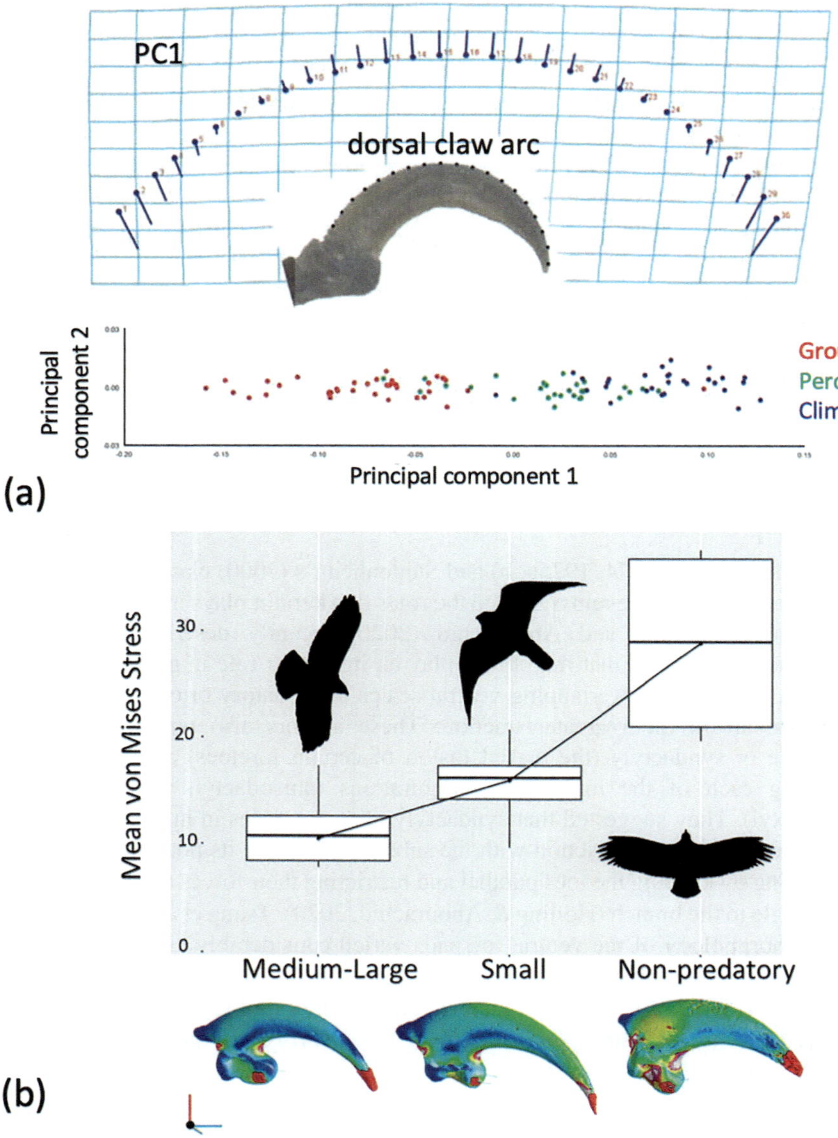

Fig. 12.10 (**a**) Variation in claw shape, measured using contemporary methods of geometric morphometrics, showing variation within and among "ground," "percher," and "climber" groups of taxa. Modified with permission from Tinius and Russell (2017), Journal of Morphology, Wiley. (**b**) Effects of claw shape on functional performance based on finite element analysis, among predators of medium-large and small prey, compared to non-predatory taxa. The warmer colors indicate regions of higher stress when subjected to external forces at the tips. Modified with permission from Tsang et al. (2019b), Scientific Reports, www.nature.com, Creative Commons CC-BY 4.0

differences in claw shape among functional groups (e.g., ground-dwellers, climbers, predatory birds), such that claw curvature is greater among arboreal perching/climbing and predatory than among ground-dwelling forms (Feduccia, 1993; Pike & Maitland, 2004). Glen and Bennett (2007) took their analysis in a different direction by testing for differences among foraging categories at the ordinal level, and found that claw curvature increases with degree of arboreal foraging. However, Birn-Jeffery et al. (2012) found that at the broadest taxonomic scales, including other tetrapods such as lizards, and after correcting for body size and phylogeny, claw shape differences only really separate ground-dwelling birds from other groups. These results were largely corroborated by recent studies based on geometric morphometric analyses of claw shape (Tinius & Russell, 2017; Hedrick et al., 2019) that failed to recover clear, discrete ecological groupings, but rather found claw shape to vary on a continuous scale and that this is confounded by body size (Fig. 12.10a). Nevertheless, taxa can be distinguished along various metrics of claw shape within functional groups. For example, among predatory birds Csermely and Rossi (2006) and Csermely et al. (2012) found clear differences in claw shape along phylogenetic lines. Furthermore, claw curvature (Fowler et al., 2009, 2011) and larger size (Einoder & Richardson, 2007a) have been associated with differences in prey immobilization technique (e.g., hawks and eagles use their highly curved talons to pin prey down during feeding) and prey-type specialization (e.g., piscivorous and mammal-eating raptors have long and robust talons, respectively). More recently the biomechanical consequences of claw shape and size have been explicitly tested with finite element modeling. Tsang et al. (2019b) argued that prey size profoundly influences claw shape and mechanical performance (Fig. 12.10b). They found that non-predatory species—and, to some extent, predatory species that pursue relatively small prey—have talons that are shorter, less curved, blunter, and experience higher von Mises stresses; the latter of which suggests a greater likelihood of structural failure (Fig. 12.10b). Conversely, the talons of predators that take relatively large prey are highly curved, with enlarged flexor tubercles, and experience lower von Mises stresses along their curvature (Tsang et al., 2019b).

Naturally the claws do not function in isolation from the rest of the foot and for scansorial/climbing species in particular, the arrangement of the toes (Bock & Miller, 1959), the hindlimb muscles and their moment arms, and the posture of birds play vital biomechanical roles in their abilities to cling, climb, and hang (Winkler & Bock, 1976; Norberg, 1979, 1986; Moreno & Carrascal, 1993; Zeffer & Norberg, 2003). Below we describe some additional musculoskeletal modifications related to grasping that birds share with other terapods.

12.4.3 Musculoskeletal Modifications for Grasping

Bird feet show a variety of adaptations for grasping deep to the skin and claws, exhibiting skeletal, muscular, and tendinous modifications. With regard to skeletal elements, Hopson (2001) demonstrated how the proportional lengths of the distal

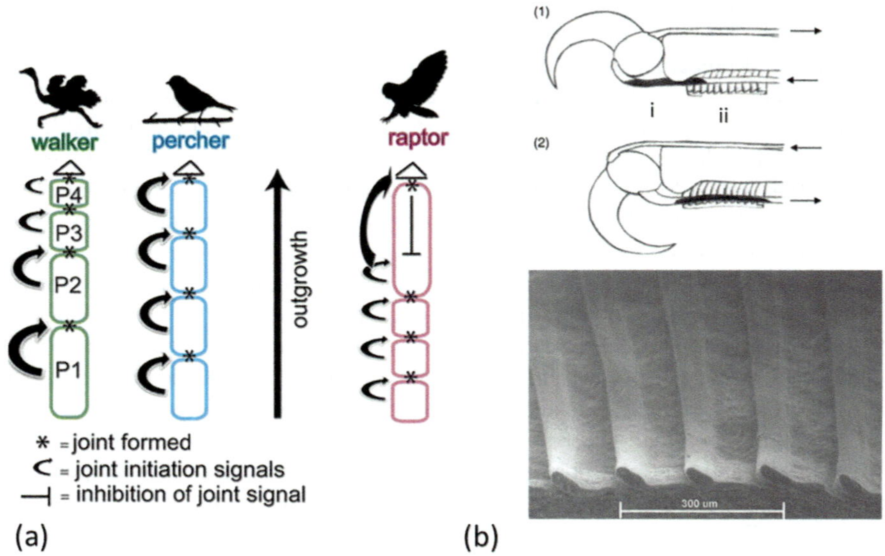

Fig. 12.11 (**a**) Variation in phalanx proportions among functional groups, showing how each phalanx signals the development of the next joint, such that in raptors the distal phalanx signal is inhibited, resulting in a long penultimate phalanx. Modified with permission from Kavanagh et al. (2013), Proceedings of the National Academy of Sciences, Highwire. (**b**) Raptor digit tendon-locking mechanism. Top panel shows engagement of the mechanisms (i = tubercle pad, ii = plicated sheath) from extended (1) to flexed (2). Modified with permission from Einoder and Richardson (2006), Ibis, British Ornitologists' Union. Lower panel shows a scanning electron micrograph of a longitudinally-sectioned tendon sheath of a barn owl, demonstrating the arrangement of plicae. Modified with permission from Einoder and Richardson (2007b), Emu, Royal Ornitologists Union, Csiro Publishing

phalanges of the third toe tend to be longer in arboreal climbing/ perching/predatory birds and shorter in terrestrial cursorial birds. This same pattern has more recently been confirmed by others (Kavanagh et al., 2013; Backus et al., 2015; Abourachid et al., 2017; Fig. 12.11a). Kavanagh et al. (2013) took the analysis a step further by uncovering the developmental basis of this pattern, thereby identifying a critical source of convergence among vertebrates. These authors indicated that the development of phalanges is modular, and this form of development restricts phalanx proportions in birds in the same way that it does in other vertebrate taxa. However, in birds the penultimate phalanx enjoys some developmental independence, and it is this phalanx that appears to show the greatest range of variation among grasping (relatively longer) and walking (relatively shorter) forms (Fig. 12.10a).

Interestingly, many arboreal and digging mammals demonstrate a similar pattern of increased distal phalangeal lengths (e.g., Ji et al., 2002), but seem to have arrived at this situation differently, through fusion of normally condensed phalanges and/or Fgf signaling (Kavanagh et al., 2013).

Other interesting osteological modifications in birds thought to foster grasping ability occur in parrots, such as a medially directed metatarsal I (which articulates with the hallux) and robust digits III and IV (Ksepka & Clarke, 2012). Furthermore, aspects of ungual phalanx morphology, reflective of the forces generated and incurred by the distal regions of the toes, vary in concert with substrate use and predatory behavior, such that the sizes of the articular surfaces and the digital flexor tubercles are relatively larger in arboreal and predatory birds (Mosto & Tambussi, 2014; Abourachid et al., 2017). The digital flexor musculature of birds is subdivided into a series of superficial flexors that insert on the proximal phalanges of toes II–IV, and two deep flexors that insert on the ungual phalanges; one onto those of toes II–IV, and another onto the ungual phalanx of the hallux (Hutchinson, 2002). The number and distribution of these flexor muscles, particularly in more derived passeriform taxa that have lost the intrinsic pedal muscles (Raikow, 1985), results in an 'underactuated mechanism' with fewer muscles and tendons relative to the degrees of freedom along the toe joints (Backus et al., 2015). Backus et al. (2015) performed a series of simulations, taking into consideration variation in phalangeal proportions and object sizes, to model the effects of multiple (superficial and deep) flexors vs. a single (just deep) flexor on grasping performance. Their results suggested that, hypothetically, a single deep flexor tendon serving the distal (ungual) phalanges is sufficient to oppose the (downward) weight of an object, and indeed, these deep digital flexors tend to be relatively larger than the superficial flexors in taxa that grasp objects. However, the addition of more proximally-inserted superficial flexors improves grasping performance with upwardly-directed (reaction) forces experienced during perching or walking. Here again, birds that tend to use their feet primarily for perching or walking have relatively more well-developed superficial flexors than deep flexors (Backus et al., 2015). Parrots and mousebirds enjoy a greater diversification and development of the intrinsic hindlimb digital muscles (e.g., m. extensor hallucis longus pars distalis and a branch of the m. extensor digitorum longus), which collectively provide for more 'delicate' control of the hallux (Berman & Raikow, 1982; Berman, 1984) for accessing and manipulating hanging food items (Harris, 1989). The neuromuscular coordination of grasping forces has not been extensively studied. Cutaneous (afferent) feedback from the digits can have profound implications for grasping performance (e.g., Shim et al., 2012). Lennerstedt (1975a, b) found Herbst corpuscles in the foot pad papillae of parrot feet, indicating a touch function of the papillae presumably associated with their climbing and pedal food handling habits.

With regard to tendon morphology, Raikow (1985) summarized the eight main types of digit flexor tendon arrangements found in birds. In the most common (in terms of the number of families represented) Type I configuration, the tendon of the m. flexor digitorum longus divides distally into three branches that insert onto toes II–IV, whereas that of the m. flexor hallucis longus inserts directly onto the hallux (Raikow, 1985). The tarsometatarsal portions of these deep flexor tendons are often connected together at some point along their lengths by a tendinous vinculum. Thus, in most plantar tendon arrangements the actions of the deep digital flexors are not independent, and contraction of the m. flexor hallucis longus assists in flexion of toes II–IV, but not vice versa (Raikow, 1985). Perhaps the most striking tendon modification is that of the digital tendon-locking mechanism (Quinn & Baumel, 1990), which is highly convergent with a similar mechanism in the toes of bats (Bennett, 1993; Quinn & Baumel, 1993; Simmons & Quinn, 1994). This mechanism works by virtue of the ratchet-like microstructure of the distal portions of the tendons and their associated sheaths (Quinn & Baumel, 1990; Einoder & Richardson, 2006; Fig. 12.11b). An additional mechanism has been proposed to work by way of flexion of the intertarsal joint, which places the digital flexor tendons that run caudad to it into tension (Ward et al., 2002; Einoder & Richardson, 2006). These two mechanisms presumably work in combination to maintain digital flexion forces during perching without the aid of continuous muscle contraction (Quinn & Baumel, 1990; Middleton, 2003; Einoder & Richardson, 2006). Incidentally however, Galton and Shepherd's (2012) surgical intervention experiments on European starlings (*Sturnus vulgaris*) demonstrated that they were still able to perch without these tendons being intact. Perhaps this mechanism is more important in forms of grasping other than perching.

12.4.4 Behavioral Repertoires

Birds perform grasping in a few major contexts: landing, perching, climbing, hanging, and handling/manipulating food and other objects (e.g., nesting materials). None of these tasks are particularly unique to birds and many other tetrapods regularly perform these forms of grasping, even with their feet as do birds. However, landing arguably presents different challenges from those experienced by most (non-flighted) tetrapods, and to some extent even bats [which land upside down! (Riskin et al., 2009)]. Bonser's (1999) extensive studies of the locomotor mechanics of perching revealed convergences in take-off and landing behavior between primates and birds. Specifically, landing forces are significantly lower than take-off forces in both, although interestingly perch compliance decreases landing forces in primates but not in birds. Provini et al. (2014) found that the hindlimbs of zebra finches and diamond doves reduced landing velocity by 60% and thereby contributed substantially to the absorption of kinetic energy during touchdown. Furthermore, they described how birds coordinate the use of the wings and hindlimbs to control landing speed, by producing higher wingbeat forces in the final stages prior to touchdown (Provini et al., 2014). Roderick et al. (2019) showed how foot, toe, and claw kinematics are also coordinated during landing in parrotletts (Fig. 12.12A). The foot 'pre-shapes' to the perching substrate ~30 milliseconds prior to making contact, which is thought to enable the grasp to be secured more quickly (Roderick et al., 2019; Fig. 12.12A). At the point of contact the claws curl around the perch until the toe pads and claw tips achieve the requisite amount of friction to prevent slippage, thereby minimizing the amount of squeezing that needs to be accomplished by the toes. This "overcompensation then relax" strategy is thought to balance safety and the reduction of energy expenditure during landing/grasping. Roderick et al. (2019)

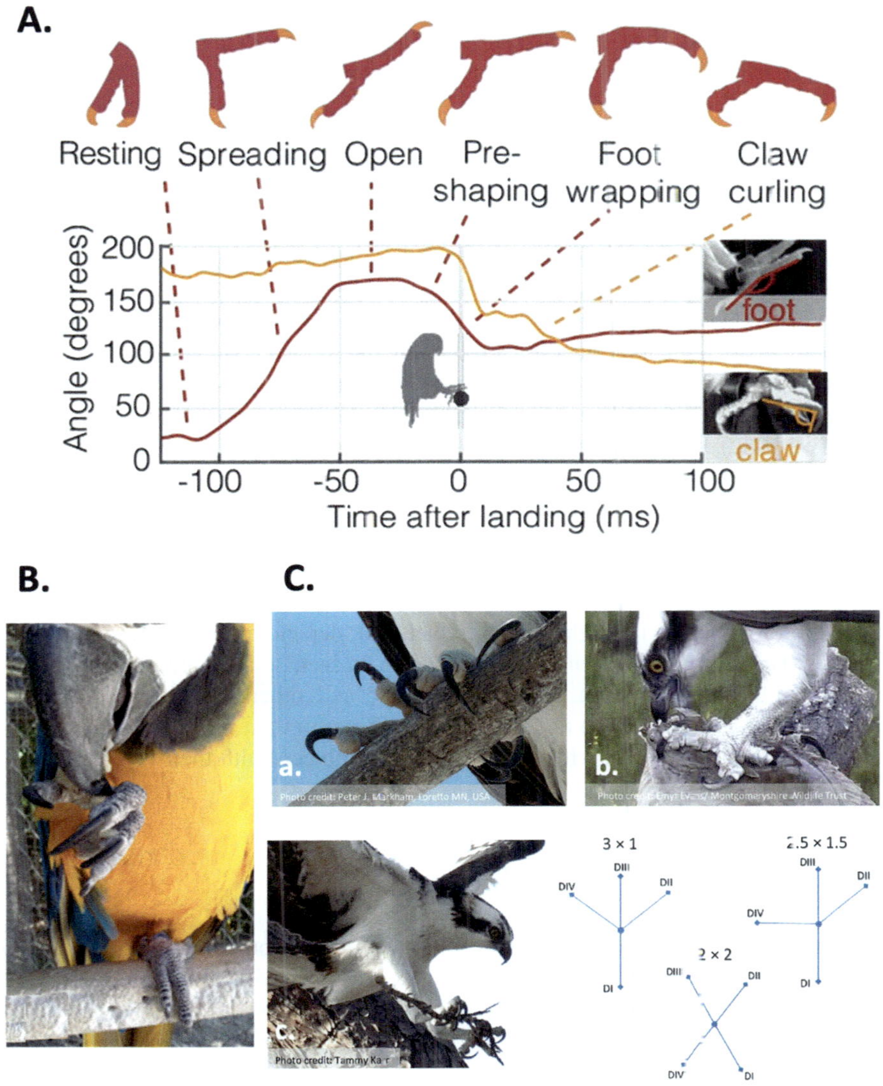

Fig. 12.12 (**A**) Changes in foot and claw angles before, during, and after landing in parrotlets. Modified with permission from Roderick et al. (2019), eLife, elifesciences.org, Creative Commons CC-BY 4.0. (**B**) Close-up of a blue and yellow macaw using its right foot to handle food during feeding (photo by D. Sustaita). (**C**) Different contexts of foot use in ospreys, showing the versatility in toe configuration, from anisodactyly (3 × 1; (a) left foot) to zygodactyly (2 × 2; (b) left foot and (c) left and right foot); a transitional configuration (2.5 × 1.5) being visible on the right foot in (a). *Di* Digit I, *Dii* Digit II, *Diii* Digit III, *Div* Digit IV. Modified with permission from Sustaita et al. (2019); PeerJ, Creative Commons CC-BY 4.0

indicated that while the coordinated wing, leg, and foot dynamics are largely stereotyped, the claw kinematics change with respect to perch diameter and substrate properties, and move remarkably rapidly to accommodate to the substrate after contact is made.

Pedal grasping in the context of handling and manipulating of objects, such as food (Clark, 1973; Fig. 12.12B) and/or nesting material (Sustaita et al., 2019; Fig. 12.12C), is less common among birds but enjoys a fairly broad phylogenetic distribution (Scooter, 1944; Smith, 1971; Tozer & Allen, 2004). Although feeding is likely to impose an important selective force for grasping capability in birds, on a larger evolutionary scale it appears to be confounded with arboreality (see Sustaita et al., 2013). Despite its broad phylogenetic representation, however, relatively few arboreal taxa regularly handle food with their feet, suggesting that such 'repurposing' of the grasping function may not necessarily be easy to accomplish. Raptors, parrots, tits (Moreno & Carrascal, 1993), and mousebirds (Berman & Raikow, 1982), for instance, probably represent extremes in their tendencies for pedal food manipulation. At one extreme raptors (hawks, falcons, and owls) are clearly adapted for generating high grasping forces (Goslow Jr., 1972; Csermely & Gaibani, 1998; Ward et al., 2002; Sustaita, 2008; Sustaita & Hertel, 2010).

At the other extreme parrots seem to exert finer control over their grasps for handling and manipulating food items (Berman, 1984; Fig. 12.12B), often with a predilection for using the right, left, or either foot, depending on the individual and species (Harris, 1989; Brown & Magat, 2011). In fact, some raptors have demonstrated similar lateralization (Csermely, 2004), and others, such as the African harrier-hawk (Burton, 1978) and the caracara (Biondi et al., 2010), are also particularly dexterous in reaching for and manipulating objects with their feet.

12.4.5 Summary and Prospects

The grasping behavior of birds is convergent with that of other tetrapods on several phenotypic levels. Birds share the propensity to grasp with other tetrapods for a variety of reasons; for maintaining stability and support on vertical and horizontal substrates, and for seizing, handling, and manipulating food items and other objects (Iwaniuk & Whishaw, 2000). As for other tetrapods, grasping in birds is affected by some combination of opposable digits and in climbing/scansorial species this conforms to similar morphological rules, such as the presence of relatively long distal phalanges. In addition, birds share other adaptations of the tendons, toe pads and claw morphology with several other scansorial/climbing tetrapod taxa. Nevertheless, there are unique aspects to their grasping capabilities that are explained by their commitment to a volant, and typically arboreal, lifestyle. The primary differences in grasping between birds and other tetrapods is that in birds grasping is restricted to the feet, albeit often with the aid of the bill. As a result, birds likely experience more conflicting demands on their foot form and function, since they cannot partition the roles of weight-bearing and object manipulation between the hind- and forelimbs as mammals do (Iwaniuk & Whishaw, 2000; Sustaita et al., 2013). In addition, birds experience the added complexity of landing, which we argue differs from the types of landing performed by gliding and other volant tetrapods, principally in the greater requirement for coordinating functions across wing/tail and hindlimb locomotor modules (Gatesy & Dial, 1996). Specifically, how this might constrain the evolution of bird feet is a subject of considerable interest. Identifying the trade-offs in foot form and function in light of competing demands is not only important for uncovering evolutionary pathways but is also of great utility for the bioinspired robotic design of grasping implements.

12.5 Grasping in Mammals

Mammals exhibit a great diversity of grasping forms and functions (see Fig. 12.13), which includes several key features described above. As soon as they are born, some young mammals actively grasp the parental fur when being carried, while others can grasp the same locomotor substrates that adults move on. Later during life, grasping, both manual and pedal, occurs extensively during food manipulation. Manual grasping is associated most prominently with feeding behavior, even if it is largely also involved in the grasping of arboreal substrates during locomotion. Pedal grasping is associated more with locomotor behaviors, even if the feet can also be used for grasping objects or food, according to species. Although many mammals

Fig. 12.13 Grasping in different contexts—locomotion, foraging and social interactions—in the young olive baboon (*Papio anubis*). (**a**) Gripping the fur when clinging to the mother. (**b**, **c**) Grasping during social interactions: grooming and play. (**d**) Fine precision grip of a small item between the thumb and the side of the index finger. (**e**) Bimanual grasping of a large food item. (**f**) Grasping of arboreal substrates during locomotion. Photograph credit: G. Boulinguez-Ambroise

have been studied with regard to their grasping abilities and its associated limb morphology, the literature on primates is by far the most extensive.

Given the number of studies on primates, we cannot be exhaustive here. On the other hand, we address questions that we consider the most relevant for understanding the ecology and evolution of grasping among mammals: what are the demands and potential trade-offs associated with food and substrate grasping? Can we trace the evolutionary origin and explore the selective pressures that underlie grasping evolution? What is the relationship between arboreality, complex manipulative skills and forelimb movements? To answer these questions, we first examine manual and pedal grasping abilities during arboreal locomotion and food acquisition and subsequently describe their underlying functional adaptations.

12.5.1 First and Early Grasping Experiences

Ultrasound scans have demonstrated that limb movements emerge during fetal life. From 14 weeks of gestation human fetuses already show exploratory hand movements such as pushing the uterine wall (also flattening and sliding the palm against it); they grasp and manipulate the umbilical cord, and even repeat hand-mouth contacts (Sparling et al., 1999). Fetal limb movements have also been observed in chimpanzees, with frequent forelimb contact with the head (Takeshita et al., 2006). After birth, in many primates—including strepsirrhine and haplorrhine species—juveniles are carried by the parents, usually the mother (see Fig. 12.13a). In some species, especially in New-World primates like titi monkeys (*Callicebus moloch*), the juvenile is almost exclusively carried by the father (Fragaszy et al., 1982; Mendoza & Mason, 1986). When clinging to the parental fur (using both the hands and the feet), young primates commonly press each finger toward the adjacent ones (i.e., involving a close contact between digits) while the fingertips are pressed toward the palm (Bishop, 1962; Peckre et al., 2016). This fur-grasping grip (see Fig. 12.13a) involves different hand surface areas and contacts than those recruited when grasping branches during arboreal locomotion (i.e., the whole palm and all palmar parts of the fingers, see Fig. 12.13f; Reghem et al., 2012; Peckre et al., 2016). Peckre et al. (2016) compared oral-carrying with fur-clinging strepsirrhine species and found that species that cling to parental fur have greater manual dexterity. The authors thus suggested, with regard to fur-grasping, that "such focus of control on the touch-pads is a likely forerunner of fine control of the hand" (Bishop, 1962, p. 329; Peckre et al., 2016). In olive baboons (*Papio anubis*) young individuals cling to their mother's fur using both the fore- and hind limbs, being almost exclusively cradled during the first 3 weeks (i.e., clinging to the belly) and are then carried dorsally for several months (Nash, 1978). At the juvenile stage, relatively wider and thicker manual and pedal phalanges (see Fig. 12.14)—compared to those of adults—allow young baboons to strongly grasp the maternal fur (Boulinguez-Ambroise et al., 2021) while the mother is free to walk, run, climb, or leap (i.e., exhibits the full locomotor repertoire). Well-developed grasping abilities thus appear to be

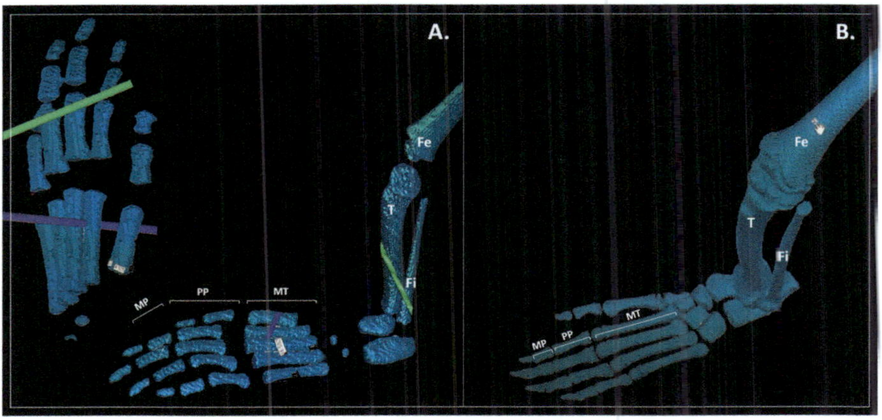

Fig. 12.14 Illustrations of hindlimb bone segments in juvenile and adult olive baboons (*Papio anubis*; 3D surface models segmented from CT-Scans; Photo courtesy of Gilles Berillon). (**a**) The juvenile morphology (with non-ossified epiphyseal plates) is best characterized by relatively wider phalanges and digit joints, compared to those of adults, while (**b**) length and thickness of the long bones and metapodia best characterize the adult morphology (Boulinguez-Ambroise et al., 2021). *MP* middle phalanges, *PP* proximal phalanges, *MT* metatarsals, *Fe* femur, *T* tibia, *Fi* fibula

fundamental to the survival of young baboons, whose grasping performance (first year of life) reaches 200% of the adult performance relative to body mass (Boulinguez-Ambroise et al., 2021).

Regarding marsupial neonates, the immature altricial young leaves the uterus and reaches the maternal pouch, where it will be carried and complete most of its development attached to the teat. Precociously developed forelimbs with separated digits and claws allow the tiny neonates to climb to the pouch at a stage in development in which the hindlimbs are still rudimentary buds (Lyne, 1964; Cooper & Steppan, 2010; Ashwell & Shulruf, 2014; Schneider & Gurovich, 2017). As in other mammals, marsupial neonates display forelimb movements even before birth; in the case of the tammar wallaby (*Macropus eugenii*) fetus, climbing movements have been reported about 3 days before birth, in preparation for those required for climbing to the pouch (Drews et al., 2013).

12.5.2 Manual and Pedal Substrate Grasping

In primates, a large number of studies have focused on hand use during food acquisition, and the grasping of objects or tools. However, an even greater number of studies have explored substrate grasping during locomotion. Indeed, the ability to grasp narrow branches safely and forcefully remains at the center of the debate on primate origins. Current hypotheses suggest that the use of narrow terminal branches to exploit fruits, flowers, insects and nectar may have been an important selective

Fig. 12.15 (**a**–**e**) Autopodial grasping configurations in young mouse lemurs (*Microcebus murinus*) on vertical and horizontal substrates. Configurations differ according to the position of the digits relative to the substrate: (**a**) Mesaxonic manual grasp with the axis running along digit 3; (**b**) Schizaxonic manual grasp between digits 2 and 3; (**c**) Pedal secure grasping; (**d**) Powerful telaxonic manual grasp: the thumb is fully opposed to the lateral digits; (**e**) Entaxonic manual grasp with the axis running along digit 2. Photograph credit: G. Boulinguez-Ambroise

pressure that led to the evolution of primate grasping (Cartmill, 1974; Godinot, 1991; Sussman, 1991). The evolution of specific hand and nail morphologies observed in primates might thus be linked to the use of thin terminal branches. Interspecific comparative studies in primates have been used to test this hypothesis. A study on the mouse lemur (*Microcebus murinus*) suggested that, moreso than an arboreal thin substratum, the frequent use of vertical supports may influence hand biomechanics toward ulnar deviation (see Fig. 12.15), as observed for lorisids and indriids (Reghem et al., 2012). The different types of grips employed in substrate grasping can be described according to the general posture of the hand/foot (midline) relative to the forearm's/leg's midline, the digits involved in substrate grasping (see Fig. 12.15), and the hand/foot areas that are in contact with the substrate (for a description, see Toussaint et al., 2020).

More generally, Lemelin and Schmitt (1998) observed that the use of ulnarly-deviated hand postures was associated with substrate preference in six haplorrhine species. The highly arboreal species displayed the most deviant manual postures both on poles and on the ground, whereas highly terrestrial species displayed only small deviations. The adaptations to the challenges of arboreal locomotion should be greater in very young arboreal primates, whose balance is not yet fully developed. Indeed, in young mouse lemurs the use of manual secure grasps—the most ulnarly deviant gripping postures (see Fig. 12.15d)—decrease during development, being greatly used shortly after birth on vertical and narrow substrates (Boulinguez-Ambroise et al., 2020a). Comparative studies of other terminal-branch specialists have tested the fine-branch hypothesis. Toussaint et al. (2020) compared 11 primate species (6 strepsirrhines and 5 platyrrhines) and 11 non-primate arboreal species (1 scandentian, 3 rodents, 3 carnivorans, and 4 marsupials) and found the possession of a grasping pollex and hallux to be crucial for climbing small vertical substrates. This study also revealed that carnivorans and rodents show a smaller repertoire of grasping postures than primates and marsupials, with primates having the greatest capability for postural adjustment (Toussaint et al., 2020). Tree shrews (Tupaiidae; Sargis, 2007) and some marsupials (Rasmussen, 1990; Rasmussen & Sussman, 2007) possess a hand and foot morphology that is functionally comparable to that of primates. In addition, the highly arboreal woolly opossum (*Caluromys* spp.) uses the terminal narrow branches of the canopy (Rasmussen, 1990; Grelle, 2003) and has developed relatively long digits and a long opposable nail-bearing hallux (Szalay, 1994; Lemelin, 1999; Argot, 2002). Such attributes provide the reasons for considering *Caluromys* the adaptive analog of a terminal-branch user capable of a powerful hallucal grasp, a key feature that is supposed to characterize primates (Hoffstetter, 1977; Youlatos, 2008). These results clearly suggest an evolutionary convergence of grasping in mammals. Besides being used by the woolly opossum (*Caluromys philander*) and the feathertail glider marsupial (*Acrobates pygmaeus*) (Youlatos, 2008; Youlatos et al., 2018), hallucal grasping is also used by small rodents moving on fine branches: namely, the harvest mouse (*Micromys minutus*), and domestic mice raised in a simulated fine-branch arboreal niche (Byron et al., 2011; Urbani & Youlatos, 2013). Hallucal grasping would ensure balance and safety for arboreal locomotor behaviors such as climbing and walking (Urbani & Youlatos, 2013). In fact, while such pedal grasping is maintained across the whole of mouse lemur development, manual postures become less ulnarly deviated, suggesting different functional roles between limbs and emphasizing the role of anchoring ensured by grasping feet (Boulinguez-Ambroise et al., 2020a). This more substantial role of the feet in primate locomotion is also suggested by other studies. Red ruffed lemurs (*Varecia rubra*) rely more on hind limb than on forelimb grasping during arboreal quadrupedal locomotion, with the toe flexors being activated more forcefully and for longer than the finger flexors (Patel et al., 2015). Alternatively, other species, like sciurid rodents, lack primate grasping adaptations but still move on fine terminal branches. Young and Chadwell (2020) compared the Eastern grey squirrel (*Sciurus carolinensis*) and two platyrrhine primates (*Callithrix jacchus* and *Saimiri boliviensis*); they found the primate grasping morphology to improve the locomotor performance on narrow supports compared to the squirrel limb morphology. So, according to the authors, such findings suggest that "basal primate morphological adaptations may have specifically facilitated improved locomotor performance in a fine-branch niche, rather than merely permitting access to the environment". Eastern grey squirrels and European red squirrels (*Sciurus vulgaris*), however, also feed and forage on terminal branches (Samaras & Youlatos, 2010; Orkin & Pontzer, 2011). Selection for locomotion on fine branches therefore seems to not be a sufficient condition for primate origins and this hypothesis may oversimplify the evolution of primates. Another specific feature of the primate grasping apparatus that has been related to locomotion on fine branches are their flattened nails (i.e., ungulae). These are homologous to, but different from, the claws (i.e., falculae) of non-primate mammals (for a description and review, see Maiolino et al., 2016). The form of the nail shows a high degree of variation among primates: flattened nails are not present on all digits in some primate species, such as some callitrichine monkeys that instead possess claw-like nails (i.e., tegulae) (Sussman & Kinzey, 1984; Ford, 1986). It has been suggested that non-primate mammal claws and primate tegulae facilitate the use of large vertical substrates (e.g., tree trunks that are too large to grip) when climbing up and down (Cartmill, 1974; Toussaint et al., 2020). In tree kangaroos and sloths, claws even take the form of large hooks to facilitate the grasping of arboreal substrates (Mendel, 1981; Iwaniuk et al., 1998; Warburton et al., 2011; Patiño et al., 2021). In parallel, it has been suggested that the possession of flattened nails and the lack of claws is related to increased grasping abilities, in that claws potentially hinder the grasping of thin and small items (Napier, 1993). However, in their study comparing primates and non-primate arboreal mammals, Toussaint et al. (2020) did not observe an advantage to possessing nails in moving on small substrates (while the possession of a grasping pollex and hallux was found to be crucial for negotiating small vertical substrates). As highlighted by Maiolino et al. (2016), the extensive diversity of nail morphology among primates suggests different roles for the nails of different species and more research is required to explain its functional significance. Primates possess many morphological features that did not evolve simultaneously, suggesting the operation of a wide range of selective pressures (Dagosto, 2007; Sargis et al., 2007).

12.5.3 Manual Food Grasping and Manipulation

The ability to grasp food or prey, or manipulate them, is often considered to be most advanced in mammals (Ivanco et al., 1996; Iwaniuk et al., 1999; Iwaniuk & Whishaw, 1999a, 2000; Endo et al., 2007; Sacrey et al., 2009). However, we still do not know whether the use of the hand for grasping represents convergent evolution, arising independently within each mammal lineage (Lassek, 1954; Bracha et al., 1990), or if this ability is plesiomorphic, having arisen early in mammalian evolution (Whishaw et al., 1992) before being lost or elaborated in different lineages. To answer this question, studies have explored hand grasping behavior in various taxa such as xenarthrans (sloths and anteaters; Taylor, 1985), pholidotes (pangolins; Grzimek, 1990), scandentians (tree shrews; Bishop, 1964), dermopterans (colugos; Macdonald, 1984), rodents (Whishaw, 1996; Whishaw et al., 1998), carnivorans (Boczek-Funcke et al., 1998; Iwaniuk et al., 1999), marsupials (Ivanco et al., 1996; Landy, 1997), and primates (Christel, 1993; Jones-Engel & Bard, 1996; Christel et al., 1998; Christel & Billard, 2002; Pouydebat et al., 2008, 2009, 2011; Pouydebat & Bardo, 2019). Mice and rats detect food using olfaction and typically adopt a 'sitting' posture on their haunches while grasping the food with the mouth and using their digits to manipulate it (Whishaw et al., 1998). Marsupials such as the northern quoll (*Dasyurus hallucatus*), opossums, and gliders, share some common characteristics with rodents (Schwensen, 1994; Ivanco et al., 1996; Landy, 1997): the food is detected using olfaction, or in some cases by proprioception, and is initially grasped by the mouth. On the contrary, carnivorous marsupial species (dasyurids and didelphids: Schwensen, 1994; Ivanco et al., 1996; phalangerids and petaurids: Landy, 1997) initially grasp some animal matter between digits II and III ('scissor grip') as well as with the entire hand (power grip) and ingest it (Landy, 1997). Most carnivorans use olfaction to detect food, but the marsh mongoose (*Atilax paludinosus*), small-clawed otters (*Amblonyx* spp.), and the raccoon (*Procyon lotor*) use both visual and tactile means. Raccoons essentially use their hands to grasp food, often using a bipedal posture (Iwaniuk & Whishaw, 1999b) and the scissor grip, as seen in carnivorous marsupials. They also often roll the food between the palms of both hands, as do otters when manipulating food or other objects such as stones during playing behaviors (personal observation). Other carnivorans are capable of fine manipulation of food, such as the giant panda (*Ailuropoda melanoleuca*), otters (*Aonyx* spp., *Lutrogale perspicillata*, and *Enhydra lutris*), the crab-eating and marsh mongooses (*Herpestes urva*, *Atilax paludinosus*; Ewer, 1973) and the African palm civet (*Nandinia binotata*; Estes, 1991). In contrast to raccoons, which seem to possess fine control of forepaw digits, the kinkajou (*Potos flavus*; Pocock, 1917) and the olingo (*Bassaricyon* spp.; Ewer, 1973) grasp objects with a power grip involving a single-handed converging grip (Rensch & Dücker, 1969; McClearn, 1992). The ringtail (*Bassariscus astutus*) and coatis (*Nasua* spp.) seem also to be less dexterous, having little fine control of digit movements (Ewer, 1973; McClearn, 1992). Interestingly, coatis are excellent diggers and shredders. Their lack of fine control of finger movements and the associated musculoskeletal parameters may not allow them to benefit from a complex grasping ability or to feed on the terminal branches as kinkajous, for example, can do (McClearn, 1992). Indeed, there is certainly a relationship between arboreality and skillful food grasping in carnivores (Fabre et al., 2013).

Manipulation of food items is not always linked to arboreality. For example, forepaw manipulation in semi-aquatic mustelids may result from their tendencies to handle a diversity of food items associated with their omnivorous diet (Hall & Schaller, 1964; Fujii et al., 2015; Zellmer et al., 2021). If we extend the reasoning further, the grasping abilities may not be linked to the hand at all! Indeed, in proboscideans, the trunk not only detects food (i.e., olfaction), but is also used as a prehensile organ to grasp it; the absence of bones in this organ allows for a high degree of freedom for movement in all directions. A study of African elephants (*Loxodonta africana*) revealed a plethora of manipulative strategies and grasping behaviors of the trunk during feeding activities (Lefeuvre et al., 2020). As food grasping strategies, the authors namely reported the catching of items between the trunk's fingers (i.e., finger-like distal projections), the wrapping of the trunk around bigger items, or the wrapping followed by torsion of the trunk allowing the breaking of the item (Lefeuvre et al., 2020). In primates, food detection occurs through a mixture of visual, olfactory and auditory stimuli in most strepsirrhines (lemuriforms and lorisiforms; Siemers et al., 2007; Piep et al., 2008; Perrenoud et al., 2015), and mainly visually in haplorhines (Martin, 1990). Strepsirrhines appear to grasp static foods first with the mouth and show no digital individualization (Petter, 1962; Bishop, 1964; Toussaint et al., 2013). In contrast, when grasping mobile prey, as well as during manipulation, cheirogaleids (*Microcebus murinus*; Reghem et al., 2011; Toussaint et al., 2015; Boulinguez-Ambroise et al., 2019) and several lorisiforms catch insects with one or both hands (Martin, 1972; Oates, 1984; Lemelin, 1996; Nekaris, 2005). Also, the specialized aye-aye (*Daubentonia madagascariensis*) is known to use the third finger to pry insects from holes in trees (Erickson, 1991; Milliken et al., 1991; Erickson et al., 1998). In contrast to strepsirrhines, haplorhines use their hands to grasp static foods and, in so doing, employ a great variety of hand and digit postures (Napier, 1956; Bishop, 1964; Christel, 1993; Spinozzi et al., 2004; Pouydebat et al., 2008). To a lesser extent food grasping in primates can involve social manipulation. It has been reported that orangutan mothers may manipulate their offspring as a social tool-use to retrieve food items, guiding their bodies and arms towards the food, and reorienting their hands so that they grab it (Völter et al., 2015). The various grip types used by great apes, namely chimpanzees (*Pan troglodytes*), when grasping static foods are highly comparable to those used by humans (Jones-Engel & Bard, 1996; Marzke & Wullstein, 1996; Byrne et al., 2001; Pouydebat et al., 2011). Several primates (i.e., great apes, capuchin monkeys [*Cebus* spp. and *Sapajus* spp.]) partially use the scissor grip, as do several marsupials and carnivorans (Pouydebat et al., 2009). The power grip may be one of the most commonly-used in mammals because it is employed by animals with opposable (great apes), pseudo-opposable (i.e., without a complete rotation of the first carpo-metacarpal joint as in platyrrhines), and non-opposable thumbs (carnivorans and marsupials). However, the hand of haplorrhines is capable of a diversity of movements, partly resulting from the ability for independent movement of the digits.

12.5.4 Functional Adaptations and Ecological Consequences

The mammalian hand possesses a morphology that perfectly reflects its adaptations for the prehension of arboreal substrates and the strategies for food acquisition with variable properties (e.g., static, mobile, various sizes and structures, etc.) (Hamrick, 2003). A large number of studies have explored the potential adaptive variation of the mammalian hand skeleton and the integumentary structures in relation to foraging behaviors and locomotion (e.g., Jouffroy & Lessertisseur, 1979; Van Valkenburgh, 1987; Norberg, 1994; Szalay, 1994; Thewissen & Etnier, 1995; Hamrick et al., 1999; Rosenberg & Rose, 1999; Lemelin, 1999, 2000; Hamrick, 2001a, b, c).

Interestingly, climbing mammals such as primates, tree shrews, and burramyid marsupials have developed papillary ridges on their fingers that improve their ability to grasp arboreal substrates (Whipple, 1904; Le Gros Clark, 1936; Cartmill, 1974, 1985; Hamrick, 1998; Rosenberg & Rose, 1999; Lemelin, 2000). When comparing two small-bodied marsupials, one arboreal (*Petaurus breviceps*) and the other mainly terrestrial (*Monodelphis domestica*), kinematic divergences were highlighted between the two species: in contrast to *Petaurus, Monodelphis* has relatively shorter digits that are associated with reduced grasping ability and a greater need for stabilizing mechanisms on narrow substrates (Shapiro et al., 2014). The gliding phalangers (*Petaurus breviceps*), as for other marsupials such as koalas (*Phascolarctos cinereus*), can cling to large-diameter substrata using their sharp and keeled claws, whereas pygmy possums (Burramyidae) are better adapted for climbing on narrow branches with their thin and flat nails (Iwaniuk & Whishaw, 2000). Counter to what was previsously thought (Napier, 1993), claws are not incompatible with prehensile hands and do not always obstruct the grasping process. The claws of tree kangaroos (*Dendrolagus* spp.) help facilitate unimanual grasping of food objects (Iwaniuk et al., 1998) and clawed rodents are able to grasp food with one hand (Whishaw et al., 1998), as can arboreal tupaiids (*Ptilocercus lowii* and *Tupaia minor*; Sargis, 2001). Similarly, even though colugos (Dermoptera) lack epidermal ridges on their fingers, they are still able to move on arboreal substrates (Lemelin, 2000). Improving friction with smooth arboreal substrates is probably easier for some bats (e.g., *Thyroptera tricolor*) which have adhesive pads on their hands and feet (Wimsatt & Villa, 1970; Thewissen & Etnier, 1995). Anatomical exploration has revealed that many adaptations exist. In order to maintain flexion forces with no additional muscular effort, bats (e.g., Quinn & Baumel, 1993), dermopterans (Simmons & Quinn, 1994), and some climbing rodents (Haffner, 1996) possess intrinsic digital tendon-locking mechanisms that differ among groups in form and function. Interestingly, adaptations may differ at different taxonomic levels and between taxa. For example, bats, birds and rodents have locking tendons that operate using different mechanisms. Bats and birds have a micro-anatomical ratchet mechanism composed of tendon tubercles and tendon sheath plicae (Quinn & Baumel, 1990, 1993). On the other hand, rodents have ventral tendon thickenings that offer better resistance during digital flexion (Haffner, 1996). It is quite possible that adaptive modifications of the integumentary structures of the hand, at least among mammals, have played a major role in the diversification of foraging strategies. In addition to the study of tendons, comparative studies of hand proportions have been very informative and have shown a link between the evolution of these anatomical data and ecological niche (Hamrick, 2001c). For example, primates and marsupials that feed on fruit and insects by foraging on thin branches have relatively short palms, long fingers and small claws (Jouffroy et al., 1991; Hamrick et al., 1999; Lemelin, 1999).

As has been shown for lizards, mammals possess pedal grasping adaptations that may have contributed to the evolution of manual grasping abilities by using the feet to grasp the substratum for support, thereby liberating the hands for other functions (e.g., Mac Neilage et al., 1987). The arboreal marsupials *Caluromys* spp. have long digits, a widely divergent hallux, and a developed hallucal eminence and pad (Argot, 2002; Lemelin et al., 2003), and are capable of 'powerful' pedal grasping (Sargis et al., 2007). An opposable hallux, allowing the foot to grasp, occurs in the molossid bat *Cheiromeles* spp. (Vaughan et al., 2011) and is a shared derived feature of non-human primates (Cartmill, 1972; Martin, 1990). This morphology and ability may have preceded the evolution of manual grasping (Byron et al., 2011 and references therein). Lemurid strepsirrhines can grasp the arboreal substrate between the first and second digits by hallucial grasping (Cartmill, 1985; Szalay & Dagosto, 1988; Gebo, 1993; Lemelin, 1999; Boyer et al., 2007) due to relatively large m. adductor hallucis, and an active m. peroneus longus (Kingston et al., 2010). Finally, pedal phalangeal curvature in great apes may be indicative of increased grasping during suspensory and climbing behaviors, in addition to their well-developed hallux (Nakatsukasa et al., 2002; see Congdon, 2012 for review). Besides the opposable hallux, the opposable thumb—allowing numerous manipulative and grasping behaviors (i.e., for food or locomotor substrate grasping or social interactions)—is present in most primate species and shows morphological variability.

In haplorrhines, the saddle-shaped carpo–metacarpal joint of the thumb allows the thumb to oppose the other digits (Napier & Napier, 1985) and to produce, in some cases, a precision grip involving contact between the distal tips of the thumb and other fingers (Marzke, 1997). It appears that chimpanzees have relatively shorter (average) thumb muscle moment arm lengths than humans, allowing them to apply an amount of force lower than that generated by humans when deploying the precision grip (Marzke et al., 1999). These morphological differences might explain why chimpanzees much more frequently use the lateral grip, involving the tip of their thumb and the lateral side of their index finger (Pouydebat et al., 2011). The forces produced by great apes during grasping tasks remain to be tested as we have so far only quantified pulling strength for some primate species, such as olive baboons and mouse lemurs. The measure of pulling strength allows us to assess how well a subject can grasp and hold onto a substrate with its forelimbs or hind limbs. Whereas a rat can pull only 7% of its body weight (40 g; Clark et al., 2004), adult mouse lemurs can pull over 100 times their own body weight (91g; Thomas et al., 2016), just like other specialized narrow branch walkers such as chameleons (Herrel et al., 2013b), suggesting that there has been strong selection for increased grasping strength in arboreal species. Interestingly, as mentioned above, 1-year old olive baboons demonstrate very high pulling strength (i.e., 200% of the adult performance, relative to body mass), this being consistent with the presence of relatively wider phalanges and digit joints in juveniles (Boulinguez-Ambroise et al., 2021). As baby baboons actively cling onto the mother's fur during their first months of life, the effect of infant-carrying should be considered when discussing the origins of grasping in primates.

Comparing the hands of primates overall it is evident that the human hand possesses many derived musculoskeletal traits (Lewis, 1989; Tocheri et al., 2008). Human hands exhibit long, robust thumbs, relatively larger joint surfaces, and hypertrophic thenar muscles, derived traits that are sometimes associated with the origins of making and using stone tools (Marzke, 1997; Susman, 1998). Longer fingers necessitate that relatively less muscle force is needed to stabilize digital joints, and the joints are subjected to relatively lower joint contact stresses during stone tool use. This is reflected in the increased robusticity of metacarpals and phalanges (i.e., robusticity can be assessed as the ratio of the bone midschaft circumference to the bone length (Bass, 1971; Cope et al., 2005), or as the head breadth of the bone relative to the bone length (Richmond et al., 2016) in humans relative to chimpanzees (Rolian et al., 2011). However, manual forces and pressures acting on the hand during the supposed first stone tool production (by the nearly two-million-year-old hominin fossil populations from Olduvai Tanzania) showed that peak normal force, pressure, and impulse, are significantly lower for the thumb than for digits II and/or III (Williams et al., 2012).

Revelations from experimental studies showing that some great apes with small thumbs (e.g., bonobos) are able to make stone tools similar to Oldowan tools (Toth et al., 1993; Schick et al., 1993), suggest that the dependence on a derived thumb in the evolution of stone tool use should be re-evaluated. The assumptions linking modern human thumb robusticity specifically to load resistance during stone tool production should be tested again with new methods and approaches, as has recently been done for Neanderthals (Bardo et al., 2020). Finally, the use of the hand among haplorhines (especially in apes and capuchin monkeys) seems to be the most complex among mammals and subject to great variability, but whether this complexity and variability of the grasping and manipulative behaviors are linked to their arboreal origins remains to be examined in a phylogenetic framework (e.g., Fig. 12.16).

12.5.5 Hand Preference, Social Interaction and Emotion in Primate Grasping Behavior

For humans hand preference has been well assessed for both unimanual and asymmetric bimanual grasping tasks, with a population-level bias towards the right hand approximating 90% (for a meta-analysis, see Papadatou-Pastou et al., 2020). Hand preference at both individual- and population- levels has been demonstrated for several non-human primate species as well (chimpanzees: Hopkins, 1996; New-World primates: Hook-Costigan & Rogers, 1997; Olive baboons: Molesti et al., 2016). Evidently handedness for unimanual grasping in these primates seems to be weaker than for bimanual grasping (Fagard & Marks, 2000; Meguerditchian et al., 2015; Molesti et al., 2016) and less sensitive than bimanual manipulations for detecting population-level bias (McGrew & Marchant, 1992; Vauclair et al., 2005). Several studies reported an effect of the mobility, the position, or the size of the item that was being grasped (Lehman, 1993; Meunier et al., 2011; Toussaint et al., 2013; Pouydebat et al., 2014). Such asymmetric use of the hands for bimanual grasping has been correlated with contralateral brain structural asymmetries in a section of the central sulcus related to the motor hand area, suggesting hemispheric specialization of the motor system for the limbs: in chimpanzees (*Pan troglodytes*; Hopkins & Cantalupo, 2004) and Olive baboons (*Papio anubis*; Margiotoudi et al., 2019), as is the situation for manipulative tasks in humans (Hammond, 2002). Interestingly, at the population level it has been reported for

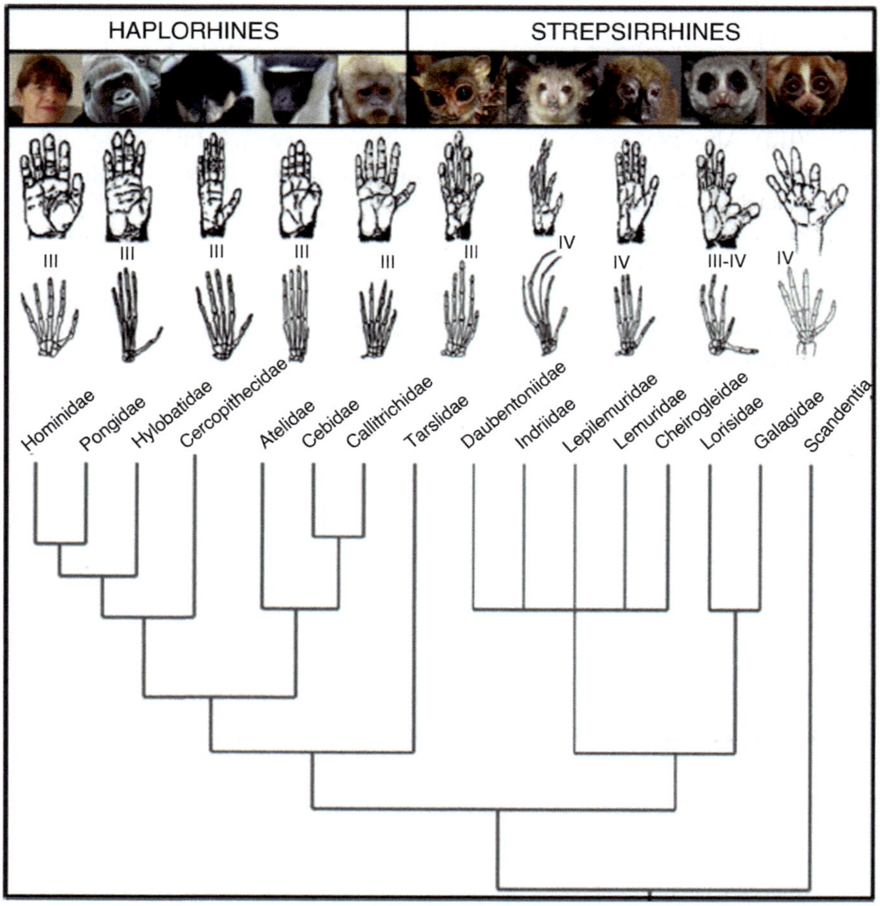

Fig. 12.16 Phylogenetic branching patterns showing morphological and functional differences of the hand of primates. Modified and adapted from Schultz (1972; hand skeleton figures), Hershkovitz (1977; hand skeleton figures), and Chiu and Hamrick (2002; phylogenetic topology). Photos courtesy of A. Bardo (Hylobatidae and Cercopithecidae), D. Haring (Tarsiidae, Daubentoniidae, Cheirogaleidae, Galagidae); all others by E. Pouydebat

chimpanzees and gorillas (*Gorilla gorilla*) that, while the right hand (i.e., left hemisphere activation) is more involved in unimanual grasping actions directed toward inanimate targets (e.g., objects, food items), no hand preference is reported for animate targets (e.g., conspecifics) (Forrester et al., 2011, 2012). In both humans and great apes, the use of the left hand for self-touching of the face is greater than the use of the right hand (Dimond & Harries, 1984), which may be explained by the fact that grasping actions towards animate targets (i.e., social manipulation) likely imply emotional involvement, and there is a general dominance of the right hemisphere for all kinds of emotions (Gainotti, 2019). In fact, there might be a neural distinction between targets requiring functional and social manipulation (Baldachini et al. 2021).

The mechanisms that may influence the development of handedness are widely debated on both theoretical and empirical grounds (Boulinguez-Ambroise et al., 2022a, b). Empirical lines of evidence highlight that, besides genetic factors (McManus & Bryden, 1992; Yeo & Gangestad, 1993; Laland et al., 1995), other nongenetic factors associated with the early developmental environment, likely play a role in the development of handedness (Hopkins & Ronnqvist, 1998; Hopkins, 2004; Fagard, 2013). A key factor of the developmental environment is the actions of the mother on the immediate environment of the fetus and then the infant. In humans and non-human primates such as chimpanzees, gorillas and baboons, maternal cradling of newborns is lateralized at the individual level and shows a left-side bias at the population-level, which means the use of left arm is favored over the right arm for cradling the infant in the majority of individuals (Manning et al., 1994; Boulinguez-Ambroise et al., 2020b). For Olive baboons, if the infant is cradled on the left the infant embraces and holds onto the left side of the mother with its right arm, the left hand being free, and vice versa. The hand that is not recruited for clinging to the fur is free to be able to reach and for fine manipulative grasping actions, thereby providing greater motor and neurological stimulation than for the other hand. In fact, in this species (*Papio anubis*), early postnatal individual hand preference for unimanual grasping within the first months of life positively correlates with maternal cradling lateralization (Boulinguez-Ambroise et al., 2021). As maternal left-cradling bias likely reflects brain right hemisphere specialization for emotion (see Manning & Chamberlain, 1991; Malatesta et al., 2019; Forrester et al., 2019), the early emergence of handedness in baboons might be indirectly related to emotional processing. However, very little comparative research on manual lateralization has been conducted for non-primate mammals (Ströckens et al., 2013; Versace & Vallortigara, 2015; Boulinguez-Ambroise et al., 2022b). Interestingly, Giljov et al. (2015) assessed handedness in marsupial species, one of the other large mammalian lineages. The authors reported a population-level manual preference for multiple behaviors (e.g., unimanual feeding, grooming) in red (*Macropus rufus*) and grey (*Macropus giganteus*) kangaroos, which mainly employ a bipedal gait, thereby freeing the hands for performing other tasks. By comparing mainly bipedal and quadrupedal marsupial species Giljov et al. (2015) highlighted the crucial role that postural characteristics (e.g., bipedality), rather than phylogeny, may play in the origin of handedness in mammals. These works on lateralization open many perspectives for comparison within tetrapods: the questions addressed and new ones that arise can be applied to other species outside primates (Karenina et al., 2017).

12.5.6 Concluding Remarks About Grasping in Mammals

To conclude, although many tetrapods grasp substrates for moving and manipulating food, most mammals demonstrate advanced finger mobility. Skilled reaching movements in rodents and primates are, in part, similar (Bishop, 1964; Jeannerod, 1988; Whishaw, 1996), suggesting an ancestral origin (homology) of skilled forelimb movements (Sacrey et al., 2009). However, grasping ability with the feet and hands can no longer be used to set primates apart from other mammals, or even tetrapods (Le Gros Clark, 1959; Martin, 1990). The great variability of postures and complex in-hand movements and repositioning abilities quantified for several primate species (Crast et al., 2009; Bardo et al., 2016, 2017) suggest that additional data for non-primate species are needed to determine whether grasping behavior is homologous or homoplasious. The high diversity of grasping patterns among primates demonstrates the necessity for developing integrative approaches (e.g., eco-ethology, biomechanics, physiology, morphology) to further our understanding of the complexity and the evolution of this function (Pouydebat & Bardo, 2019).

12.6 Conclusions

Grasping behavior plays an essential role in various contexts including locomotion, feeding, and social interactions in a great diversity of tetrapod vertebrates, but has received relatively little attention beyond the anthropological and biomedical literature. Although the ability to reach for food or prey, to hold it in a forepaw, or manipulate it with the digits exists in most tree-dwelling frogs, it is often considered to be most highly developed in mammals. Furthermore, although birds are limited to grasping with the hindlimbs, many species demonstrate comparable levels of grasping force and digital dexterity as those evident in mammals. Grasping modalities may differ from group-to-group, but they share common musculoskeletal bases and have been molded by similar selective pressures.

Among lissamphibians, anurans demonstrate the greatest complexity of forelimb movements, and such abilities have evolved several times independently. Features such as relatively long forelimbs, intercalary elements in the skeleton of the hand, adhesive sub-digital pads, and opposable digits facilitate their abilities to perform both power and precision gripping that ultimately enhance their arboreal locomotion and feeding abilities. Species of *Phyllomedusa* are described as having the most refined examples of manual dexterity among anurans and can generate greater forces and have specific modifications of their hand musculature compared to other tree frogs. These anatomical modifications include highly differentiated forearm muscles that appear to be able to control each finger individually, allowing complete closure of the hand around narrow substrates. Just as in primates, some frog species can adopt various hand positions, allowing them to change the form of their grip to optimize interactions with the features of the substrate they are interacting with (e.g., size, texture, inclination), assuring stability. Beside the various manual gripping configurations, grasping with the feet has also been reported for multiple behaviors and should receive more attention in future research.

Grasping in lizards appears to be driven largely by selection for locomotor attributes associated with navigating in complex three-dimensional habitats, and plays relatively less of a role in other behaviors, such as feeding. The most commonly observed grip employed by lizards is that corresponding to a power grip. With regard to this, the tendinous pattern of the palm of the hand plays a key role allowing flexion at the metacarpo-phalangeal joints. Complete power grasping abilities are restricted to lizards exhibiting specific tendon structures. Additionally, the configuration of the wrist and hand bones also appears to correlate with grasping ability. Those taxa, such as chameleons (the most specialized arboreal group of lizards), but also other lineages including anoles, geckos, and varanids that exhibit varying degrees of manual grasping abilities, also exhibit pedal grasping. However, the mechanisms behind lizard pedal grasping abilities are still poorly understood and require further investigation. Although phylogeny seems to best explain osteological traits of the lizard foot, it explains little with regard to the variation reported for muscle and tendon anatomy, and it is these that may better predict the ability of the pes to adjust so as to achieve a grip.

With the evolution of flight in birds the capacity for manipulating objects progressively became relegated to the hindlimbs, resulting in enhanced pedal grasping abilities relative to other tetrapods. An important precursor to the evolution of grasping in birds was the reversal and incumbency of the hind toe (hallux) to form an opposable digit. Digital flexor muscle size and complexity and tendon-locking mechanisms likely play important roles in generating and maintaining gripping forces. It is to be noted that although most birds able to perform pedal grasping are restricted to the execution of a power grasp, some birds have achieved high levels of digital dexterity, similar to those of other tetrapods whose limbs are far more specialized for grasping. In this regard, further investigation is required to elucidate the functional significance of several features of the avian foot, such as the various toe and digital flexor tendon configurations, proportional phalanx lengths, and claw size and shape. Such features are likely involved in enhancing grasping capability. Although feeding has been a major selective force driving grasping capability, on a broader evolutionary scale the ability to grasp is confounded because of its association with the adoption of an arboreal existence.

In mammals, grasping occurs extensively during food manipulation, namely while moving or standing on arboreal supports, but is also employed in various social interactions. The vast majority of work in this vein pertains to primates. Current hypotheses propose that the use of fine terminal branches (specifically vertical ones) for exploiting fruits, flowers, insects, and nectar as food resources may have constituted an important selective pressure driving the evolution of primate grasping abilities. However, further investigation is required to enable better explanation of variations observed about hand dexterity among species, specifically by examining early grasping experiences in juveniles, such as infants clinging to their mothers. While the power grip may be one of the most commonly employed in mammals (because it is used by animals with opposable, pseudo-opposable, and non-opposable thumbs) many primate species display a great variety of hand and digit postures that permit a fine precision grip. Arboreal mammals, including primates but also marsupials and rodents, show hallucial grasping and seem to rely more on hind limb than on forelimb grasping during quadrupedal locomotion. This is suggestive of a more substantial role of the feet in locomotion that may have contributed to the evolution of manual dexterity by freeing the hands for other functions (e.g., feeding, social behavior). Adaptive changes in the integumentary structures of the hand in mammals might have been essential in the diversification of foraging strategies. Interestingly, according to the idea of a lateralized "social brain", there might be a neural distinction in grasping between targets requiring functional and social manipulation and this should be further considered in future research.

Grasping ability and its underlying forearm musculature is fairly well conserved among most tetrapod clades. The presence of opposable digits and the relative development of the digital extensor and flexor muscles appear to underlie the grasping abilities of most tetrapods, although these are not necessarily prerequisites for successful grasping. Digital muscle and tendon complexity may limit, or enhance, digital independence, which, in turn, could have important implications for gripping force production and/or digital dexterity. Neither within nor among the major tetrapod clades is there a one-to-one mapping of grasping form to function, but the extent to which different configurations yield similar levels of performance awaits quantification. Despite the role of feeding in selection for grasping performance, the crux of tetrapod fore- and hindlimb prehension appears to be the arboreal context within which more complex forms of grasping are presumed to have arisen. A potentially important corollary to the selective context of grasping among tetrapods is the decoupling of fore- and hindlimb apparatuses from one another, and from the task of locomotion. Grasping performance might play a more critical role in tetrapod evolution than currently understood. However, more comprehensive data on grasping behavior and functional morphology, involving a greater diversity of taxa, are required to allow this hypothesis to be tested in a rigorous phylogenetic framework.

Acknowledgments We warmly thank Vincent Bels for inviting us to participate in this book project. We appreciate the generosity of Hartmut Förstner for providing us with the schematics shown in Fig. 12.7, and of Gilles Berillon for providing the 3D models shown in Fig. 12.14. Part of this work was funded by PICT 2016-2772, 2018-0832 and PIP 0389, and supported by a SU emergence funding "HUMDEXT". Finally, we would like to warmly thank those experts who helped us to improve this article, in particular Christine Böhmer and Anthony Russell.

References

Abdala, V., & Diogo, R. (2010). Comparative anatomy, homologies and evolution of the pectoral and forelimb musculature of tetrapods with special attention to extant limbed amphibians and reptiles. *Journal of Anatomy, 217*, 536–573.

Abdala, V., Manzano, A. S., & Herrel, A. (2008). The distal forelimb musculature in aquatic and terrestrial turtles: Phylogeny or environmental constraints? *Journal of Anatomy, 213*, 159–172.

Abdala, V., Manzano, A. S., Tulli, M. J., & Herrel, A. (2009). The tendinous patterns in the palmar surface of the lizard manus: Functional consequences for grasping ability. *The Anatomical Record: Advances in Integrative Anatomy and Evolutionary Biology, 292*, 842–853.

Abdala, V., Tulli, M. J., Russell, A. P., Powell, G. L., & Cruz, F. B. (2014). Anatomy of the crus and pes of neotropical iguanian lizards in relation to habitat use and digitally based grasping capabilities. *The Anatomical Record, 297*, 397–409. https://doi.org/10.1002/ar.22851

Abdala, V., Vera, M. C., Amador, L. I., Fontanarrosa, G., Fratani, J., & Ponssa, M. L. (2019). Sesamoids in tetrapods: the origin of new skeletal morphologies. *Biological Reviews of the Cambridge Philosophical Society, 94*(6), 2011–2032. https://doi.org/10.1111/brv.12546

Abourachid, A., Fabre, A. C., Cornette, R., & Hofling, E. (2017). Foot shape in arboreal birds: Two morphological patterns for the same pincer-like tool. *Journal of Anatomy, 231*(1), 1–11. https://doi.org/10.1111/joa.12614

Akella, T., & Gillis, G. B. (2011). Hopping isn't always about the legs: Forelimb muscle activity patterns during toad locomotion. *The Journal of Experimental Zoology, 315*, 1–11.

Anzeraey, A., Herrel, H., Aumont, M., Descamp, T., & Pouydebat, E. (2017). Effect of food properties on grasping abilities in *Xenopus laevis*. *Journal of Experimental Biology, 220*(Pt 23), 4486–4491.

Argot, C. (2002). Functional-adaptive analysis of the hind limb anatomy of extant marsupials and the paleobiology of the Paleocene marsupials *Mayulestes ferox* and *Pucadelphys andinus*. *Journal of Morphology, 253*, 76–108.

Ashwell, K. W. S., & Shulruf, B. (2014). Vestibular development in marsupials and monotremes *Journal of Anatomy, 224*, 447–458.

Backus, S. B., Sustaita, D., Odhner, L. U., & Dollar, A. M. (2015). Mechanical analysis of avian feet: Multiarticular muscles in grasping and perching. *Royal Society Open Science, 2*, 140350. https://doi.org/10.1098/rsos.140350

Baeckens, S., Goeyers, C., & Van Damme, R. (2020). Convergent evolution of claw shape in a transcontinental lizard radiation. *Integrative and Comparative Biology, 60*(1), 10–23. https://doi.org/10.1093/icb/icz151

Baldachini, M., Regaiolli, B., Llorente, M., Riba, D., & Spiezio, C. (2021). The influence of target animacy and social rank on hand preference in barbary macaques (*Macaca sylvanus*). *International Journal of Primatology*. https://doi.org/10.1007/s10764-020-00193-0

Barbeau, T. R., & Lillywhite, H. B. (1999). Comparative morphology and histochemistry of lipid-containing glands in the skin of treefrogs. *The American Zoologist, 39*, 115A.

Barbera, A. M., Delaunay, M. G., Dougill, G., & Grant, R. A. (2019). Paw morphology in the domestic guinea pig (*Cavia porcellus*) and brown rat (*Rattus norvegicus*). *Anat Rec (Hoboken), 302*(12), 2300–2310. https://doi.org/10.1002/ar.24271

Bardo, A., Borel, A., Meunier, H., Guéry, J.-P., & Pouydebat, E. (2016). Manual abilities in great apes during a tool use task. *American Journal of Physical Anthropology. doi, 10*, 1002.

Bardo, A., Cornette, R., Borel, A., & Pouydebat, E. (2017). Manual function and performance in humans, gorillas and orangutans during the same tool use task. *American Journal of Physical Anthropology*. https://doi.org/10.1002/ajpa.2332

Bardo, A., Moncel, M. H., Dunmore, C. J., et al. (2020). The implications of thumb movements for Neanderthal and modern human manipulation. *Scientific Reports, 10*, 19323.

Bass, W. M. (1971). *Human osteology: A laboratory and field manual of the human skeleton*. Missouri Archaeological Society.

Bennett, M. (1993). Structural modifications involved in the fore and hindlimb grip of some flying foxes. *Journal of Zoology, 229*, 237–248.

Berman, S. L. (1984). The hindlimb musculature of the white-fronted Amazon (*Amazonia albifrons*, Psittaciformes). *Auk, 101*, 74–92.

Berman, S. L., & Raikow, R. J. (1982). The hindlimb musculature of the mousebirds (Coliiformes). *Auk, 99*, 41–57.

Biju, S. D. (2009). A novel nesting behaviour of a treefrog, *Rhacophorus lateralis* in the Western Ghats, India. *Current Science, 97*, 433–437.

Biondi, L. M., Garcia, G. O., Bo, M. S., & Vassallo, A. I. (2010). Social learning in the Caracara Chimango, *Milvago chimango* (Aves: Falconiformes): an age comparison. *Ethology, 116*(8), 722–735. https://doi.org/10.1111/j.1439-0310.2010.01794.x

Birn-Jeffery, A. V., Miller, C. E., Naish, D., Rayfield, E. J., & Hone, D. W. (2012). Pedal claw curvature in birds, lizards and mesozoic dinosaurs--complicated categories and compensating for mass-specific and phylogenetic control. *PLoS One, 7*(12), e50555. https://doi.org/10.1371/journal.pone.0050555

Bishop, A. (1962). Control of the hand in lower primates. *Annals of the New York Academy of Sciences, 102*, 316–337.

Bishop, A. (1964). Use of the hand in lower primates. In J. Buettner-Janusch (Ed.), *Evolutionary and genetic biology of primates* (pp. 133–223). Academic.

Blaylock, L., Ruibal, R., & Platt-Aloia, K. (1976). Skin structure and wiping behavior of Phyllomedusinae frogs. *Copeia, 2*, 283–295.

Blomberg, S. P., Garland, T., & Ives, A. R. (2003). Testing for phylogenetic signal in comparative data: Behavioral traits are more labile. *Evolution, 57*, 717–745.

Blotto, B., Pereyra, M., Grant, T., & Faibovich, J. (2020). Hand and foot musculature of anura: Structure, homology, terminology, and synapomorphies for major clades. *Bulletin of the American Museum of Natural History* Number 443, 155 p., 14 plates, 1 table Issued November 6.

Bock, W. J., & Miller, W. D. (1959). The scansorial foot of the woodpeckers, with comments on the evolution of perching and climbing feet in birds. *American Museum Novitates, 1931*, 1–45.

Boczek-Funcke, A., Kuhtz-Buschbeck, J. P., Raethjen, J., Paschmeyer, B., & Illert, M. (1998). Shaping of the cat paw for food taking and object manipulation: An X-ray analysis. *European Journal of Neuroscience, 10*, 3885–3897.

Bonser, R. H. (1999). Branching out in locomotion: The mechanics of perch use in birds and primates. *The Journal of Experimental Biology, 202*(Pt 11), 1459–1463.

Borel, A., Chèze, L., & Pouydebat, E. (2016). Sequence analysis of grip and manipulation during tool using tasks: A new method to analyze hand use strategies and examine human specificities. *Journal of Archaeological Method and Theory, 24*(3), 751–775.

Botelho, J. F., Smith-Paredes, D., Nunez-Leon, D., Soto-Acuna, S., & Vargas, A. O. (2014). The developmental origin of zygodactyl feet and its possible loss in the evolution of Passeriformes. *Proceedings of the Biological Sciences, 281*(1788), 20140765. https://doi.org/10.1098/rspb.2014.0765

Botelho, J. F., Smith-Paredes, D., Soto-Acuna, S., Mpodozis, J., Palma, V., & Vargas, A. O. (2015a). Skeletal plasticity in response to embryonic muscular activity underlies the development and evolution of the perching digit of birds. *Scientific Reports, 5*, 9840. https://doi.org/10.1038/srep09840

Botelho, J. F., Smith-Paredes, D., & Vargas, A. O. (2015b). Altriciality and the evolution of toe orientation in birds. *Evolutionary Biology, 42*, 502–510.

Boulinguez-Ambroise, G., Zablocki-Thomas, P., Aujard, F., Herrel, A., & Pouydebat, E. (2019). Ontogeny of food grasping in mouse lemurs: Behavior, morphology and performance. *Journal of Zoology, 308*, 1–8.

Boulinguez-Ambroise, G., Herrel, A., & Pouydebat, E. (2020a). Ontogeny of locomotion in mouse lemurs: Implications for primate evolution. *Journal of Human Evolution, 142*, 102732.

Boulinguez-Ambroise, G., Pouydebat, E., Disarbois, E., & Meguerditchian, A. (2020b). Human-like maternal left-cradling bias in monkeys is altered by social pressure. *Scientific Reports, 10*, 11036.

Boulinguez-Ambroise, G., Herrel, A., Berillon, G., Young, J. W., Cornette, R., Meguerditchian, A., Cazeau, C., Bellaiche, L., & Pouydebat, E. (2021). Increased performance in juvenile baboons is consistent with ontogenetic changes in morphology. *American Journal of Physical Anthropology, 175*, 546–558. https://doi.org/10.1002/ajpa.24235

Boulinguez-Ambroise, G., Pouydebat, E., Disarbois, E., & Meguerditchian, A. (2022a). Maternal cradling bias in baboons: The first environmental factor affecting early infant handedness development? *Developmental Science, 25*, e13179. https://doi.org/10.1111/desc.13179

Boulinguez-Ambroise, G., Aychet, J., & Pouydebat, E. (2022b). Limb preference in animals: New insights into the evolution of manual laterality in hominids. *Symmetry, 14*(1), 96.

Boyer, D. M., Patel, B. A., Larson, S. G., & Sternjr, J. T. (2007). Telemetered electromyography of peroneus longus in *Varecia variegata* and *Eulemur rubriventer*: Implications for the functional significance of an enlarged peroneal process. *Journal of Human Evolution, 53*, 119–134.
Bracha, V., Zhuravin, I. A., & Burges, J. (1990). The reaching reaction in the rat: A part of the digging pattern? *Behavioural Brain Research, 36*, 53–64.
Brinkman, D. (1980). Structural correlates of tarsal and metatarsal functioning in *Iguana* (Lacertilia; Iguanidae) and other lizards. *Canadian Journal of Zoology, 58*, 277–289.
Brown, C., & Magat, M. (2011). The evolution of lateralized foot use in parrots: A phylogenetic approach. *Behavioral Ecology, 22*(6), 1201–1208. https://doi.org/10.1093/beheco/arr114
Burton, P. J. K. (1978). The intertarsal joint of the harrier-hawks *Polyboroides* spp. and the crane hawk *Geranospiza caerulescens*. *Ibis, 120*, 171–177.
Burton, T. C. (1996). Adaptation and evolution in the hand muscles of Australo-Papuan hylid frogs (Anura: Hylidae: Pelodryadinae). *Australian Journal of Zoology, 44*, 611–623.
Burton, T. C. (1998a). Are the distal extensor muscles of the fingers of Anurans an adaptation to arboreality? *Journal of Herpetology, 32*, 611–617.
Burton, T. C. (1998b). Variation in the hand and superficial throat musculature of Neotropical leptodactylid frogs. *Herpetologica, 54*, 53–72.
Byrne, R. W., Corp, N., & Byrne, J. M. (2001). Manual dexterity in the gorilla: Bimanual and digit role differentiation in a natural task. *Animal Cognition, 4*, 347–361.
Byron, C., Kunz, H., Matuszek, H., Lewis, S., & Van Valkinburgh, D. (2011). Rudimentary pedal grasping in mice and implications for terminal branch arboreal quadrupedalism. *Journal of Morphology, 272*, 230–240.
Cartmill, M. (1972). Arboreal adaptations and the origin of the order of primates. In R. H. Tuttle (Ed.), *The functional and evolutionary biology of primates* (pp. 97–122). Aldine Atherton.
Cartmill, M. (1974). Rethinking primate origins. *Science, 184*(4135), 436–443.
Cartmill, M. (1985). Climbing. In M. Hildebrand, D. Bramble, K. Liem, & D. Wake (Eds.), *Functional vertebrate morphology* (pp. 73–88). Harvard University Press.
Chiu, C. H., & Hamrick, M. W. (2002). Evolution and development of the primate limb skeleton. *Evolutionary Anthropology, 11*, 94–107.
Christel, M. (1993). Grasping techniques and hand preferences in Hominoidea. In H. Preuschoft & D. J. Chivers (Eds.), *Hands of primates* (pp. 91–108). Springer.
Christel, M. I., & Billard, A. (2002). Comparison between macaques' and humans' kinematics of prehension: The role of morphological differences and control mechanisms. *Behavioural Brain Research, 131*, 169–184.
Christel, M. I., Weiss, P., & Bavar, S. (1998). The temporal pattern and the asymmetry of performances in precise reach-to-grasp movements-a comparison between *H. sapiens* and *M. nemestrina.* In R. Blickhan, A. Wisser, & W. Nachtigall (Eds.), *Biona report 13. Motion systems.* Akad. d. Wissensch, Mainz, G. Fischer Stuttgart. Proc. Jena, 1997 (pp. 120–121).
Clark, G. A. J. (1973). Holding food with the feet in passerines. *Bird Banding 44*(2), 91–99.
Clark, B. D., Al-Shatti, T. A., Barr, A. E., Amin, M., & Barbe, M. F. (2004). Performance of a high-repetition, high-force task induces carpal tunnel syndrome in rats. *The Journal of Orthopaedic and Sports Physical Therapy, 34*, 244–253.
Congdon, K. (2012). Interspecific and ontogenetic variation in proximal pedal phalangeal curvature of Great Apes (*Gorilla gorilla*, *Pan troglodytes*, and *Pongo pygmaeus*). *International Journal of Primatology, 33*, 418–427.
Cooper, W. J., & Steppan, S. J. (2010). Developmental constraint on the evolution of marsupial forelimb morphology. *Australian Journal of Zoology, 58*, 1–15.
Cooper, T. L., Zabinski, C. L., Adams, E. J., Berry, S. M., Pardo-Sanchez, J. Reinhardt, E. M., Roberts, K. M., Watzek, J., Brosnan, S. F., Hill, R. L., Weigel, E. G., & Mendelson, J. R. (2020). Long-term memory of a complex foraging task in monitor lizards (Reptilia: Squamata: Varanidae). *Journal of Herpetology, 54*(3), 378–383.

Cope, J. M., Berryman, A. C., Martin, D. L., & Potts, D. D. (2005). Robusticity and osteoarthritis at the trapeziometacarpal joint in a Bronze Age population from Tell Abraq, United Arab Emirates. *American Journal of Physical Anthropology, 126*(4), 391–400.

Cox, S., Ekstrom, L. J., & Gillis, G. B. (2018). The influence of visual, vestibular, and hindlimb proprioceptive ablations on landing preparation in cane toads. *Integrative and Comparative Biology, 58*(5), 894–905.

Crast, J., Fragaszy, D., Hayashi, M., & Matsuzawa, T. (2009). Dynamic in-hand movements in adult and young juvenile chimpanzees (*Pan troglodytes*). *American Journal of Physical Anthropology, 138*, 274–285.

Csermely, D. (2004). Lateralisation in birds of prey: Adaptive and phylogenetic considerations. *Behavioural Processes, 67*(3), 511–520. https://doi.org/10.1016/j.beproc.2004.08.008

Csermely, D., & Gaibani, G. (1998). Is foot squeezing pressure by two raptor species sufficient to subdue their prey? *Condor, 100*, 757–763.

Csermely, D., & Rossi, O. (2006). Bird claws and bird of prey talons: Where is the difference? *Italian Journal of Zoology, 73*(1), 43–53.

Csermely, D., Rossi, O., & Nasi, F. (2012). Comparison of claw geometrical characteristics among birds of prey and non-raptorial birds. *Italian Journal of Zoology, 79*(3), 410–433.

Cutkosky, M. R. (1989). On grasp choice, grasp models, and the design of hands for manufacturing tasks. *IEEE Transactions on Robotics and Automation, 5*, 269–279.

Cutkosky, M. R., & Howe, R. D. (1990). Human grasp choice and robotic grasp analysis. In S. T. Venkatataman & T. Iberall (Eds.), *Dextrous robot hands* (pp. 5–31). Springer.

Cutkosky, M. R., & Wright, P. K. (1986). Friction, stability and the design of robotic fingers. *International Journal of Robotics Research, 5*, 20–37.

Dagosto, M. (2007). The postcranial morphotype of primates. In M. J. Ravosa & M. Dagosto (Eds.), *PRIMATE ORIGINS: Adaptations and evolution*. Springer.

Diaz, R. E., Jr., & Trainor, P. A. (2015). Hand/foot splitting and the 're-evolution' of mesopodial skeletal elements during the evolution and radiation of chameleons. *BMC Evolutionary Biology, 15*, 184.

Dimond, S., & Harries, R. (1984). Face touching in monkeys, apes and man evolutionary origins and cerebral asymmetry. *Neuropsychologia, 22*(2), 227–233.

Diogo, R., & Abdala, V. (2010). The head muscles of dipnoans - a review on the homologies and evolution of these muscles within Vertebrates. In J. M. Jorgensen & J. Joss (Eds.), *Biology of lungfishes* (pp. 169–218). Science Publishers and Taylor & Francis.

Drew, T. (1991). Visuomotor coordination in locomotion. *Current Opinion in Neurobiology, 1*, 652–657.

Drews, B., et al. (2013). Ultrasonography of wallaby prenatal development shows that the climb to the pouch begins in utero. *Scientific Reports, 3*, 1458.

Duellman, W. E., & Trueb, L. (1986). *Biology of amphibians*. McGraw-Hill.

Einoder, L., & Richardson, A. (2006). An ecomorphological study of the raptorial digital tendon locking mechanism. *Ibis, 148*, 515–525. https://doi.org/10.1111/j.1474-919X.2006.00541.x

Einoder, L. D., & Richardson, A. M. M. (2007a). Aspects of the hindlimb morphology of some Australian birds of prey: A comparative and quantitative study. *Auk, 124*(3), 773–788.

Einoder, L., & Richardson, A. (2007b). The digital tendon locking mechanism of owls: Variation in the structure and arrangement of the mechanism and functional implications. *Emu, 107*, 223–230. https://doi.org/10.1071/MU06019

Emerson, S. B., & Diehl, D. (1980). Toe pad morphology and mechanisms of sticking in frogs. *Biological Journal of the Linnean Society, 13*, 199–216.

Emerson, S. B., & Koehl, M. A. R. (1990). The interaction of behavioral and morphological change in the evolution of a novel locomotor type: 'Flying' frogs. *Evolution, 44*, 1931–1946.

Endlein, T., Ji, A., Yuan, S., Hill, I., Wang, H., Barnes, W., Jon, P., Dai, Z., & Sitti, M. (2017). The use of clamping grips and friction pads by tree frogs for climbing curved surfaces. *Proceedings of the Royal Society B, 284*, 20162867.

Endo, H., Hama, N., Niizawa, N., Kimura, J., Itou, T., Koie, H., & Sakai, T. (2007). Three-dimensional analysis of the manipulation system in the lesser panda. *Mammal Study, 32*, 99–103.

Erickson, C. J. (1991). Percussive foraging in the aye-aye, *Daubentonia madagascariensis*. *Animal Behaviour, 41*(5), 793–801.

Erickson, C. J., Nowicki, S., Dollar, L., et al. (1998). Percussive foraging: Stimuli for prey location by aye-ayes (*Daubentonia madagascariensis*). *International Journal of Primatology, 19*, 111–122.

Essner, R. L., Suffian, D. J., Bishop, P. J., & Reilly, S. M. (2010). Landing in basal frogs: Evidence of saltational patterns in the evolution of anuran locomotion. *Naturwissenschaften, 97*, 935–939.

Estes, R. D. (1991). *The behavior guide to African mammals*. University of California Press.

Ewer, R. F. (1973). *The carnivores*. Cornell Univ. Press.

Fabre, A. C., Cornette, R., Slater, G., Argot, C., Peigné, S., Goswami, A., & Pouydebat, E. (2013). Getting a grip on the evolution of grasping in carnivores: A three-dimensional analysis of forelimb shape. *Journal of Evolutionary Biology, 26*(7), 1521–1535.

Fabrezi, M., Manzano, A. S., Abdala, V., & Lobo, F. (2014). Anuran locomotion: Ontogeny and morphological variation of a distinctive set of muscles. *Evolutionary Biology, 41*, 308–326.

Fagard, J. (2013). The nature and nurture of human infant hand preference. *Annals of the New York Academy of Sciences, 1288*, 114–123.

Fagard, J., & Marks, A. (2000). Unimanual and bimanual tasks and the assessment of handedness in toddlers. *Developmental Science, 3*, 137–147.

Feduccia, A. (1993). Evidence from claw geometry indicating arboreal habits of *Archaeopteryx*. *Science, 259*(5096), 790–793. https://doi.org/10.1126/science.259.5096.790

Feduccia, A. (1999). *The origin and evolution of birds* (2nd ed.). Yale University Press

Feduccia, A., Martin, L., & Tarsitano, S. (2007). *Archaeopteryx* 2007: Quo vadis? *Auk, 124*, 373–380.

Feiner, N., Jackson, I. S. C., Munch, L. K., Radersma, R., & Uller, T. (2020). Plasticity and evolutionary convergence in the locomotor skeleton of Greater Antillean *Anolis* lizards. *eLife, 9*, e57468.

Feix, T., Kivell, T. L., Pouydebat, E., & Dollar, A. M. (2015). Estimating thumb-index finger precision grip and manipulation potential in extant and fossil primates. *Journal of Royal Society Interface, 12*(106).

Feix, T., Romero, J., Schmiedmayer, H. B., Dollar, A. M., & Kragic, D. (2016). The grasp taxonomy of human grasp types. *IEEE Transactions of Human-Machine Systems, 46*, 66–77.

Feng, Y. J., Blackburn, D. C., Liang, D., Hillis, D. M., Wake, D. B., Cannatella, D. C., & Zhang, P. (2017). Phylogenomics reveals rapid, simultaneous diversification of three major clades of Gondwanan frogs at the Cretaceous–Paleogene boundary. *Proceedings of the national Academy of Sciences USA, 114*, E5864–E5870.

Fischer, M. S., Krause, C., & Lilje, K. E. (2010). Evolution of chameleon locomotion, or how to become arboreal as a reptile. *Zoology, 113*, 67–74.

Fontanarrosa, G., & Abdala, V. (2014). Anatomical analysis of the lizard carpal bones in the terms of skilled manual abilities. *Acta Zoologica, 95*, 249–263.

Fontanarrosa, G., & Abdala, V. (2016). Bone indicators of grasping hands in lizards. *PeerJ, 4*, e1978. https://doi.org/10.7717/peerj.1978

Ford, S. (1986). Systematics of the New World monkeys. In D. R. Swindler & J. Erwin (Eds.), *Comparative primate biology* (Vol. 1, pp. 73–135). Liss.

Forrester, G. S., Leavens, D. A., Quaresmini, C., & Vallortigara, G. (2011). Target animacy influences gorilla handedness. *Animal Cognition, 14*, 903–907.

Forrester, G. S., Quaresmini, C., Leavens, D. A., Spiezio, C., & Vallortigara G. (2012). Target animacy influences chimpanzee handedness. *Animal Cognition, 15*, 1121–1127.

Forrester, G. S., Davis, R., Mareschal, D., Malatesta, G., & Todd, B. K. (2019). The left cradling bias: An evolutionary facilitator of social cognition? *Cortex, 118*, 116–131.

Förstner, H., & Schaefer, H. E. (1998). Schnellendes Handgelenk--Trigger Wrist. Eine Fallbeschreibung [Trigger wrist. A case report]. *Handchir Mikrochir Plast Chir, 20*(1), 33–35.

Fowler, D. W., Freedman, E. A., & Scannella, J. B. (2009). Predatory functional morphology in raptors: Interdigital variation in talon size is related to prey restraint and immobilisation technique. *PLoS One, 4*(11), e7999.

Fowler, D. W., Freedman, E. A., Scannella, J. B., & Kambic, R. E. (2011). The predatory ecology of *Deinonychus* and the origin of flapping in birds. *PLoS One, 6*(12), e28964. https://doi.org/10.1371/journal.pone.0028964

Fragaszy, D. M., Schwarz, S., & Shimosaka, D. (1982). Longitudinal observations of care and development of infant titi monkeys (*Callicebus moloch*). *American Journal of Primatology, 2*, 191–200. https://doi.org/10.1002/ajp.1350020207

Frýdlová, P., Sedláčková, K., Žampachová, B., Kurali, A., Hýbl, J., Škoda, D., Kutılek, P., Landová, E., Cerný, R., & Frynta, D. (2019). A gyroscopic advantage: Phylogenetic patterns of compensatory movements in frogs. *Journal of Experimental Biology, 222*, jeb186544.

Fujii, J. A., Ralls, K., & Tinker, M. T. (2015). Ecological drivers of variation in tool-use frequency across sea otter populations. *Behavioral Ecology, 26*, 519–526.

Fujii, J. A., McLeish, D., Brooks, A. J., Gaskell, J., & Van Houtan, K. S. (2018). Limb-use by foraging marine turtles, an evolutionary perspective. *PeerJ, 6*, e4565. https://doi.org/10.7717/peerj.4565

Gainotti, G. (2019). Emotions and the right hemisphere: Can new data clarify old models? *The Neuroscientist, 25*, 258–270.

Galton, P. M., & Shepherd, J. D. (2012). Experimental analysis of perching in the European starling (*Sturnus vulgaris*: Passeriformes; Passeres), and the automatic perching mechanism of birds. *Journal of Experimental Zoology. Part A, Ecological Genetics and Physiology, 317*(4), 205–215. https://doi.org/10.1002/jez.1714

Gans, C., & Parsons, T. S. (1966). On the origin of the jumping mechanism in frogs. *Evolution, 20*, 92–99.

Gasc, J.-P. (1963). Adaptation à la marche arboricole chez le cameleon. Archive d'Anatomie, d'Histologie et d'Embryologie Normales et Expérimentales, 46, 81–115.

Gatesy, S. M., & Dial, K. P. (1996). Locomotor modules and the evolution of avian flight. *Evolution, 50*(1), 331–340.

Gebo, D. L. (1985). The nature of the primate grasping foot. *American Journal of Physical Anthropology, 67*, 269–277.

Gebo, D. L. (1993). Functional morphology of the foot in primates. In D. L. Gebo (Ed.), *Postcranial adaptation in nonhuman primates* (pp. 175–196). Northern Illinois University Press.

Giljov, A., Karina, K., Ingram, J., & Malashichev, Y. (2015). Parallel emergence of true handedness in the evolution of marsupials and placentals. *Current Biology, 25*(14), 1878–1884.

Glen, C. L., & Bennett, M. B. (2007). Foraging modes of Mesozoic birds and non-avian theropods. *Current Biology, 17*(21), R911–R912. https://doi.org/10.1016/j.cub.2007.09.026

Godinot, M. (1991). Approches fonctionnelles des mains de primates paléogènes. *Geobios, M.S., 13*, 161–173.

Gomes, F. R., Rezende, E. L., Grizante, M. B., & Navas, C. A. (2009). The evolution of jumping performance in anurans: morphological correlates and ecological implications. *Journal of Evolutionary Biology, 22*, 1088–1097.

Gona, A. G., & Uray, N. J. (1980). Ultrastructural studies on Purkinje cells of the frog tadpole cerebellum. *Brain, Behavior and Evolution, 17*(3), 241–254.

Goslow, G. E., Jr. (1972). Adaptive mechanisms of the raptor pelvic limb. *Auk, 89*, 47–64.

Gray, L. A., O'Reilly, J. C., & Nishikawa, K. C. (1997). Evolution of forelimb movement patterns for prey manipulation in anurans. *The Journal of Experimental Zoology, 277*, 417–424.

Grelle, C. (2003). Forest structure and vertical stratification of small mammals in a secondary Atlantic forest, southeastern Brazil. *Studies on Neotropical Fauna and Environment, 32*(2), 81–85.

Grillner, S., & Wallen, P. (1985). Central pattern generators for locomotion, with special reference to vertebrates. *Annual Review of Neuroscience, 8*, 233–261.

Grzimek, B. (Ed.). (1990). *Grzimek's encyclopedia of mammals*. McGraw-Hill.

Haffner, M. (1996). A tendon-locking mechanism in two climbing rodents, *Muscardinus avellanarius* and *Micromys minutus* (Mammalia, Rodentia). *Journal of Morphology, 229*, 219–227.

Haines, R. W. (1950). The flexor muscles of the forearm and hand in lizards and mammals. *Journal of Anatomy, 84*, 13–29.

Hall, K. R. L., & Schaller, G. B. (1964). Tool-using behavior of the California Sea Otter. *Journal of Mammalogy, 45*, 287–298.

Hammond, G. (2002). Correlates of human handedness in primary motor cortex: A review and hypothesis. *Neuroscience and Biobehavioral Reviews, 26*, 285–292.

Hamrick, M. W. (1998). Functional and adaptive significance of primate pads and claws: Evidence from New World anthropoids. *American Journal of Physical Anthropology, 106*, 113–127.

Hamrick, M. W., Rosenman, B. A., & Brush, J. A. (1999). Phalangeal morphology of the Paromomyidae (Primates, Plesiadapiformes): The evidence for gliding behavior reconsidered. *American Journal of Physical Anthropology, 109*(3), 397–413.

Hamrick, M. W. (2001a). Primate origins: Evolutionary change in digital ray patterning and segmentation. *Journal of Human Evolution, 40*(4), 339–351.

Hamrick, M. W. (2001b). Morphological diversity in digital skin microstructure of didelphid marsupials. *Journal of Anatomy, 198*, 683–688.

Hamrick, M. W. (2001c). Development and evolution of the mammalian limb: Adaptive diversification of nails, hooves, and claws. *Evolution & Development, 3*(5), 355–363.

Hamrick, M. W. (2003). Evolution and development of mammalian limb integumentary structures. *Journal of Experimental Zoology. Part B, Molecular and Developmental Evolution, 298*(1), 152–163.

Handrigan, G. R., & Wassersug, R. J. (2007). The anuran Bauplan: A review of the adaptive, developmental, and genetic underpinnings of frog and tadpole morphology. *Biological Reviews, 82*, 1–25.

Hanna, G., & Barnes, W. J. P. (1991). Adhesion and detachment of the toe pads of tree frogs. *Journal of Experimental Biology, 155*, 103–125.

Harris, L. J. (1989). Footedness in parrots: Three centuries of research, theory, and mere surmise. *Canadian Journal of Psychology, 43*(3), 369–396.

Hattori, S. (2016). Evolution of the hallux in non-avian theropod dinosaurs. *Journal of Vertebrate Paleontology, 36*(4), e1116995. https://doi.org/10.1080/02724634.02722016.01116995

Hedrick, B. P., Cordero, S. A., & Zanno, L. E. (2019). Quantifying shape and ecology in avian pedal claws: The relationship between the bony core and keratinous sheath. *Ecology and Evolution, 9*, 11545–11556. https://doi.org/10.1002/ece3.5507

Herrel, A., Schaerlaeken, V., Ross, C., Meyers, J., Nishikawa, K., Abdala, V., Manzano, A., & Aerts, P. (2008a). Electromyography and the evolution of motor control: Limitations and insights. *Integrative and Comparative Biology, 48*, 261–271.

Herrel, A., Vanhooydonck, B., Porck, J., & Irschick, D. J. (2008b). Anatomical basis of differences in locomotor behavior in *Anolis* lizards: A comparison between two ecomorphs. *Bulletin of the Museum of Comparative Zoology, 159*, 213–238.

Herrel, A., Measey, G. J., Vanhooydonck, B., & Tolley, K. A. (2011). Functional consequences of morphological differentiation between populations of the Cape Dwarf Chameleon (*Bradypodion pumilum*). *Biological Journal of the Linnean Society, 104*, 692–700.

Herrel, A., Perrenoud, M., Decamps, T., Abdala, V., Manzano, A., & Pouydebat, E. (2013a). The effect of substrate diameter and incline on locomotion in an arboreal frog. *Journal of Experimental Biology, 216*, 3599–3605.

Herrel, A., Tolley, K. A., Measey, G. J., da Silva, J. M., Potgieter, D. F., Boller, E., Boistel, R., & Vanhooydonck, B. (2013b). Slow but tenacious: An analysis of running and gripping performance in chameleons. *The Journal of Experimental Biology, 216*, 1025–1030.

Hershkovitz, P. (1977). *Living new world monkeys (Platyrrhini), Volume 1: With an introduction to primates*. University of Chicago Press.
Higham, T. E., & Jayne, B. C. (2004). Locomotion of lizards on inclines and perches: Hindlimb kinematics of an arboreal specialist and a terrestrial generalist. *Journal of Experimental Biology, 207*, 233–248.
Hildebrand, M. (1995). *Analysis of vertebrate structure*. Wiley.
Hill, I. D., Benzheng, D., Barnes, J. P., Aihong, J., & Endlein, T. (2018). The biomechanics of tree frogs climbing curved surfaces: A gripping problem. *Journal of Experimental Biology, 221*, jeb168179. https://doi.org/10.1242/jeb.168179
Hoffstetter, R. (1977). Phylogénie des primates. Confrontation des résultats obtenus par les diverses voies d'approche du problème. *Bulletins et Memoires de la Societe d'Anthropologie de Paris, 4*, 327–346.
Höfling, E., & Abourachid, A. (2020). The skin of birds' feet: Morphological adaptations of the plantar surface. *Journal of Morphology*. https://doi.org/10.1002/jmor.21284
Hook-Costigan, M. A., & Rogers, L. J. (1997). Hand preferences in New World primates. *International Journal of Comparative Psychology, 9*, 173–207.
Hopkins, W. D. (1996). Chimpanzee handedness revisited: 54 years since Finch (1941). *Psychonomic Bulletein and Review, 3*, 449–457.
Hopkins, B., & Ronnqvist, L. (1998). Human handedness: Developmental and evolutionary perspectives. In F. Simon & G. Butterworth (Eds.), *The development of sensory, motor and cognitive capacities of early infancy: From perception to cognition* (pp. 191–236). Psychology Press.
Hopkins, W. D. (2004). Laterality in maternal cradling and infant positional biases: Implications for the development and evolution of hand preferences in nonhuman primates. *International Journal of Primatology, 25*, 1243–1265.
Hopkins, W. D., & Cantalupo, C. (2004). Handedness in chimpanzees (*Pan troglodytes*) is associated with asymmetries of the primary motor cortex but not with homologous language areas. *Behavioral Neuroscience, 118*(6), 1176.
Hopson, J. A. (2001). Ecomorphology of avian and nonavian theropod phalangeal proportions: Implications for the arboreal versus terrestrial origin of bird flight. In J. Gauthier & L. F. Gall (Eds.), *New perspectives on the origin and early evolution of birds: Proceedings of the international symposium in honor of John H. Ostrom* (pp. 211–235). Yale University.
Horn, H.-G., & Visser, G. J. (1997). Review of reproduction of monitor lizards *Varanus* spp. in captivity 2. *International Zoo Yearbook, 35*, 227–246.
Hudson, C. M., Gregory, P. B., & Shine, R. (2016). Athletic anurans: The impact of morphology, ecology and evolution on climbing ability in invasive cane toads. *Biological Journal of the Linnean Society, 119*(4), 992–999.
Hutchinson, J. R. (2002). The evolution of hindlimb tendons and muscles on the line to crown-group birds. *Comparative Biochemistry and Physiology a-Molecular and Integrative Physiology, 133*(4), 1051–1086.
Hutchinson, J. R., Felkler, D., Houston, K., et al. (2019). Divergent evolution of terrestrial locomotor abilities in extant Crocodylia. *Scientific Reports, 9*, 19302. https://doi.org/10.1038/s41598-019-55768-6
Hutson, J. D., & Hutson, K. N. (2014). A repeated-measures analysis of the effects of soft tissues on wrist range of motion in the extant phylogenetic bracket of dinosaurs: Implications for the functional origins of an automatic wrist folding mechanism in Crocodilia. *The Anatomical Record, 297*, 1228–1249. https://doi.org/10.1002/ar.22903
Irschick, D. J., Austin, C. C., Petren, K., Fisher, R. N., Losos, J. B., & Ellers, O. (1996). A comparative analysis of clinging ability among pad–bearing lizards. *Biological Journal of the Linnean Society, 59*, 21–35.
Irschick, D. J., & Garland, T. J. (2001). Integrating function and ecology in studies of adaptation: Investigations of locomotor capacity as a model system. *Annual Review of Ecology and Systematics, 32*, 367–396.

Ivanco, T. L., Pellis, S. M., & Whishaw, I. Q. (1996). Skilled forelimb movements in prey catching and in reaching by rats (*Rattus norvegicus*) and opossums (*Monodelphis domestica*): Relations to anatomical differences in motor systems. *Behavioural Brain Research, 79*, 163–181.

Iwaniuk, A. N., Nelson, J. E., Ivanco, T. L., Pellis, S. M., & Whishaw, I. Q. (1998). Reaching, grasping and manipulation of food objects by two tree kangaroo species, *Dendrolagus lumholtzi* and *Dendrolagus matschiei*. *Australian Journal of Zoology, 46*, 235–243.

Iwaniuk, A. N., & Whishaw, I. Q. (1999a). Brain size is not correlated with forelimb dexterity in fissiped carnivores (Carnivora): A comparative test of the principle of proper mass. *Brain, Behavior and Evolution, 54*, 167–180.

Iwaniuk, A. N., & Whishaw, I. Q. (1999b). How skilled are the skilled limb movements of the raccoon (*Procyon lotor*)? *Behavioural Brain Research, 99*(1), 35–44.

Iwaniuk, A. N., Nelson, J. E., & Whishaw, I. Q. (1999). The relationships between brain regions and forelimb dexterity in marsupials (*Marsupialia*): A comparative test of the principle of proper mass. *Australian Journal of Zoology, 48*, 99–110.

Iwaniuk, A. N., & Whishaw, I. Q. (2000). On the origins of skilled forelimb movements. *Trends in Neurosciences, 23*, 372–376.

Jeannerod, M. (1988). *The neural and behavioural organization of goal-directed movements*. Clarendon Press.

Jetz, W., & Pyron, R. A. (2018). The interplay of past diversification and evolutionary isolation with present imperilment across the amphibian tree of life. *Nature Ecology & Evolution, 2*, 850–858. https://doi.org/10.1038/s41559-018-0515-5

Ji, Q., Luo, Z. X., Yuan, C. X., Wible, J. R., Zhang, J. P., & Georgi, J. A. (2002). The earliest known eutherian mammal. *Nature, 416*(6883), 816–822. https://doi.org/10.1038/416816a

Jones-Engel, L., & Bard, K. (1996). Precision grips in young chimpanzees. *American Journal of Primatology, 39*, 1–15.

Jouffroy, F. K., & Lessertisseur, J. (1979). Relationships between limb morphology and locomotor adaptations among prosimians: an osteometric study. In M. E. Morbeck, H. Preuschoft, & N. Gomberg (Eds.), *Environment, behavior, and morphology: Dynamic interactions in primates* (pp. 143–181). Gustav Fischer.

Jouffroy, F. K., Godinot, M., & Nakano, Y. (1991). Biometrical characteristics of primate hands. *Human Evolution, 6*, 269–306.

Kamper, D. G., Hornby, T. G., & Rymer, W. Z. (2002). Extrinsic flexor muscles generate concurrent flexion of all three finger joints. *Journal of Biomechanics, 35*, 1581–1589.

Kardong, K. V. (2011). *Vertebrates: Comparative anatomy, function, evolution* (6th ed.). McGraw-Hill.

Karenina, K., Giljov, A., Ingram, J., et al. (2017). Lateralization of mother–infant interactions in a diverse range of mammal species. *Nature Ecology and Evolution, 1*, 0030.

Katzner, T. (2016). Hanging out at the Airport: Unusual upside-down perching behavior by Eurasian Jackdaws (*Corvus monedula*) in a human-dominated environment. *The Wilson Journal of Ornithology, 128*(4), 926–930.

Kavanagh, K. D., Shoval, O., Winslow, B. B., Alon, U., Leary, B. P., Kan, A., & Tabin, C. J. (2013). Developmental bias in the evolution of phalanges. *Proceedings of the National Academy of Sciences of the United States of America, 110*(45), 18190–18195 https://doi.org/10.1073/pnas.1315213110

Kenny, J. S. (1966). Nest building in *Phyllomedusa trinitatis* Mertens. *Caribbean Journal of Science, 6*, 15–22.

Kingston, A. K., Boyer, D. M., Patel, B. A., Larson, S. G., & Stern, J. T., Jr. (2010). Hallucal grasping in *Nycticebus coucang*: further implications for the functional significance of a large peroneal process. *Journal of Human Evolution, 58*, 33–42.

Kivell, T. L., Kibii, J. M., Churchill, S. E., Schmid, P., & Berger, L. R. (2011). *Australopithecus sediba* hand demonstrates mosaic evolution of locomotor and manipulative abilities. *Science, 333*, 1411–1417.

Kohlsdorf, T., Garland, T., Jr., & Navas, C. A. (2001). Limb morphology in relation to substrate usage in *Tropidurus* lizards. *Journal of Morphology, 248*, 151–164.

Ksepka, D. T., & Clarke, J. A. (2012). A new stem parrot from the green river formation and the complex evolution of the grasping foot in Pan-Psittaciformes. *Journal of Vertebrate Paleontology, 32*(2), 395–406. https://doi.org/10.1080/02724634.2012.641704

Laland, K. N., Kumm, J., Van Horn, J. D., & Feldman, M. W. (1995). A gene-culture model of human handedness. *Behavior Genetics, 25*, 433–445.

Lambert, F., & Straka, H. (2012). The frog vestibular system as a model for lesion-induced plasticity: Basic neural principles and implications for posture control. *Frontiers in Neurology, 3*, 42. https://doi.org/10.3389/fneur.2012.00042

Lammers, A. R., & Zurcher, U. (2011). Stability during arboreal locomotion. In K. Vaclav (Ed.), *Theoretical biomechanics* (pp. 319–334). InTechOpen Limited.

Landsmeer, J. M. (1962). Power grip and precision handling. *Annals of the Rheumatic Diseases, 21*, 164–170.

Landy, M. (1997). An analysis of skilled forelimb movements during feeding in possums and gliders. Unpublished Honours Thesis. Monash University.

Lassek, A. M. (1954). Motor deficits produced by posterior rhizotomy versus section of the dorsal funiculus. *Neurology, 4*, 120–123.

Lefeuvre, M., Gouat, P., Mulot, B., Cornette, R., & Pouydebat, E. (2020). Behavioural variability among captive African elephants in the use of the trunk while feeding. *PeerJ, 8*, e9678.

Le Gros Clark, W. E. (1936). The problem of the claw in the Primates. *Proceedings of the Zoological Society of London*, 1–24.

Le Gros Clark, W. E. (1959). *The antecedents of man*. Edinburgh University Press.

Lehman, R. A. W. (1993). Manual preference in prosimians, monkeys, and apes. In J. P. Ward & W. D. Hopkins (Eds.), *Primate laterality: Current behavioral evidence of primate asymmetries* (pp. 107–124). Springer.

Lemelin, P. (1996). Relationships between hand morphology and feeding strategies in small-bodied prosimians. *American Journal of Physical Anthropology, 22*(Suppl), 148.

Lemelin, P., & Schmitt, D. (1998). The relation between hand morphology and quadrupedalism in primates. *American Journal of Physical Anthropology, 105*, 185–197.

Lemelin, P. (1999). Morphological correlates of substrate use in didelphid marsupials: implications for primate origins. *Journal of Zoology, 247*, 165–175.

Lemelin, P. (2000). Micro-anatomy of the volar skin and interordinal relationships of primates. *Journal of Human Evolution, 38*, 257–267.

Lemelin, P., Schmitt, D., & Cartmill, M. (2003). Footfall patterns and interlimb co-ordination in opossums (Family Didelphidae): Evidence for the evolution of diagonal-sequence walking gaits in primates. *Journal of Zoology (London), 260*, 423–429.

Lennerstedt, I. (1974). Pads and papillae on the feet of nine passerine species. *Ornis Scandinavica, 5*(2), 103–111.

Lennerstedt, I. (1975a). A functional study of papillae and pads in the foot of passerines, parrots and owls. *Zoologica Scripta, 4*(2–3), 111–123.

Lennerstedt, I. (1975b). Pattern of pads and folds in the foot of European Oscines (Aves, Passeriformes). *Zoologica Scripta, 4*, 101–109.

Lewis, O. J. (1989). *Functional morphology of the evolving hand and foot*. Clarendon Press.

Liao, W. B., Lou, S. L., Zeng, Y., & Merilä, J. (2015). Evolution of anuran brains: Disentangling ecological and phylogenetic sources of variation. *Journal of Evolutionary Biology, 28*, 1986–1996.

Lillywhite, H. B., Mittal, A. K., Garg, T. K., & Agrawal, N. (1997). Integumentary structure and its relationship to wiping behaviour in the common Indian tree frog, *Polypedates maculatus*. *Journal of Zoology, 243*, 675–687.

Llinàs, R., Precht, W., & Kitai, S. T. (1967). Cerebellar purkinje cell projection to the peripheral vestibular organ in the frog. *Science, New Series, 158*(3806), 1328–1330.

Losos, J. B. (1990). The evolution of form and function: morphology and locomotion performance in West Indian *Anolis* lizards. *Evolution, 44*, 1189–1203.

Lovette, I., & Fitzpatrick, J. (2016). *Cornell Lab of Ornithology's handbook of bird biology* (3rd ed.). Wiley.

Luger, A., Vermeylen, V., Herrel, A., & Adriaens, D. (2020). Do substrate type and gap distance impact gap-bridging strategies in arboreal chameleons? *bioRxiv*. https://doi.org/10 1101/2020.08.21.260596

Lutz, G. J., & Rome, L. C. (1994). Built for jumping – The design of the frog muscular system. *Science, 263*, 370–372.

Lyne, A. G. (1964). Observations on the breeding and growth of the marsupial *Perameles nasuta* Geoffroy, with notes on other bandicoots. *Australian Journal of Zoology, 12*, 322–339.

Macdonald, D. (1984). *The encyclopedia of mammals*. Facts on File Publications.

Mackenzie, C. L., & Iberall, T. (1994). *The grasping hand: Advances in psychology* (Vol. 104). North-Holland.

Mac Neilage, P. F., Studdert-Kennedy, M. G., & Lindblom, B. (1987). Primate handedness reconsidered. *Behavioral and Brain Sciences, 10*, 247–303.

Maddison, W. P., & Maddison, D. R. (2019). Mesquite: A modular system for evolutionary analysis. *Version, 3*, 61.

Maiolino, S. A., Kingston, A. K., & Lemelin, P. (2016). Comparative and functional morphology of the primate hand integument. In T. Kivell, P. Lemelin, B. Richmond, & D. Schmitt (Eds.), *The evolution of the primate hand. Developments in primatology: Progress and prospects*. Springer.

Malatesta, G., Marzoli, D., Rapino, M., & Tommasi, L. (2019). The left-cradling bias and its relationship with empathy and depression. *Scientific Reports, 9*, 6141.

Malek, R. (1981). The grip and its modalities. In R. Tubiana (Ed.), *The hand* (Vol. 1, pp. 469–476). W. B. Saunders and Company.

Manning, J. T., & Chamberlain, A. T. (1991). Left-side cradling and brain lateralization *Ethology and Sociobiology, 12*, 237–244.

Manning, J. T., Heaton, R., & Chamberlain, A. T. (1994). Left-side cradling: Similarities and differences between apes and humans. *Journal of Human Evolution, 26*, 77–83.

Manrod, J. D., Hartdegen, R., & Burghardt, G. M. (2008). Rapid solving of a problem apparatus by juvenile black-throated monitor lizards (*Varanus albigularis albigularis*). *Animal Cognition, 11*, 267–273.

Manzano, A. S., & Barg, M. (2005). The iliosacral articulation in Pseudinae (Anura: Hylidae). *Herpetologica, 61*, 259–267.

Manzano, A. S., & Lavilla, E. O. (1995). Notas sobre la miología apendicular de *Phyllomedusa hypocondrialis* (Anura, Hylidae). *Alytes, 12*, 169–174.

Manzano, A. S., Fabrezi, M., & Vences, M. (2007). Intercalary elements, treefrogs, and the early differentiation of a complex system in the Neobatrachia. *Anatomical Record, 290*, 1551–1567.

Manzano, A. S., Abdala, V., & Herrel, A. (2008). Morphology and function of the forelimb in arboreal frogs: Specializations for grasping ability? *Journal of Anatomy, 213*, 296–307.

Manzano, A., Abdala, V., Ponssa, M. L., & Soliz, M. (2013). Ontogeny and tissue differentiation of the pelvic girdle and hind limbs of anurans. *Acta Zoologica (Stockholm), 94*, 420–436.

Manzano, A., Fontanarrosa, G., Prieto, Y., & Abdala, V. (2015) La prensilidad en anfibios y reptiles: Perspectivas evolutivas basadas en la anatomia y la función. *En Morfología de Vertebrados* (pp. 59–82). EUDEM.

Manzano, A. S., Herrel, A., Fabre, A. C., & Abdala, V. (2017). Variation in brain anatomy in frogs and its possible bearing on their locomotor ecology. *Journal of Anatomy, 231*, 38–58.

Manzano, A. S., Fontanarrosa, G., & Abdala, V. (2018). Manual and pedal grasping among anurans: a review of relevant concepts with empirical approaches. *Biological Journal of the Linnean Society, 127*(3), 598–610.

Margiotoudi, K., Marie, D., Claidière, N., Coulon, O., Roth, M., Nazarian, B., Lacoste, R., Hopkins, W. D., Molesti, S., Fresnais, P., Anton, J.-L., & Meguerditchian, A. (2019).

Handedness in monkeys reflects hemispheric specialization within the central sulcus. An in vivo MRI study in right- and left-handed olive baboons. *Cortex, 118*, 203–211.
Martin, R. D. (1972). Adaptive radiation and behaviour of the Malagasy lemurs. *Philosophical Transactions of the Royal Society London B, 264*, 295–352.
Martin, R. D. (1990). *Primate origins and evolution*. Princeton University Press.
Marzke, M. W., Wullstein, K. L., & Viegas, S. F. (1992). Evolution of the power ('squeeze') grip and its morphological correlates in hominids. *American Journal of Physical Anthropology, 89*, 283–298.
Marzke, M. W., & Wullstein, K. W. (1996). Chimpanzee and human grips: A new classification with a focus on evolutionary morphology. *International Journal of Primatology, 17*, 117–139.
Marzke, M. W. (1997). Precision grips, hand morphology, and tools. *American Journal of Physical Anthropology, 102*, 91–110.
Marzke, M. W., Marzke, R. F., Linscheid, R. L., Smutz, P., Steinberg, B., Reece, S., & An, K. N. (1999). Chimpanzee thumb muscle cross sections, moment arms and potential torques, and comparisons with humans. *American Journal of Physical Anthropology, 110*, 163–178.
Marzke, M. W., & Marzke, R. F. (2000). Evolution of the human hand: approaches to acquiring, analysing and interpreting the anatomical evidence. *Journal of Anatomy, 197*, 121–140.
Mcallister, W., & Channing, A. (1983). Comparison of toe pads of some southern African climbing frogs. *South African Journal of Zoology/Suid Afrikaanse Tydskrif vir Dierkunde, 18*, 110–114.
McClearn, D. (1992). Locomotion, posture, and feeding behavior of kinkajous, coatis, and raccoons. *Journal of Mammalogy, 73*(2), 245–261.
McGrew, W. C., & Marchant, L. F. (1992). Are gorillas right- handed or not? *Human Evolution, 1*, 17–23.
McManus, I. C., & Bryden, M. P. (1992). The genetics of handedness, cerebral dominance and lateralization. In I. Rapin & S. J. Segalowitz (Eds.), *Handbook of neuropsychology* (Developmental neuropsychology, Part 1) (Vol. 6, pp. 115–144). Elsevier.
Meguerditchian, A., Phillips, K. A., Chapelain, A., Mahovetz, L. M., Milne, S., Stoinski, T., Bania, A., Lonsdorf, E., Schaeffer, J., Russell, J., & Hopkins, W. D. (2015). Handedness for unimanual grasping in 564 great apes: The effect on grip morphology and a comparison with hand use for a bimanual coordinated task. *Frontiers in Psychology, 6*, 1794.
Melletti, M., & Mirabile, M. (2010). Hanging behavior of the hooded crow (*Corvus cornix*). *The Wilson Journal of Ornithology, 122*(1), 183–185. https://doi.org/10.1676/09-041.1
Mendel, F. C. (1981). The hand of two-toed sloths (*Choloepus*): Its anatomy and potential uses relative to size of support. *Journal of Morphology, 169*, 1–19.
Mendyk, R. W., & Horn, H.-G. (2011). Skilled forelimb movements and extractive foraging in the arboreal monitor lizard *Varanus beccarii* (Doria, 1874). *Herpetological Review, 42*, 343–349.
Mendoza, S. P., & Mason, W. A. (1986). Parental division of labour and differentiation of attachments in a monogamous primate (*Callicebus moloch*). *Animal Behaviour, 34*(5), 1336–1347.
Meunier, H., Blois-Heulin, C., & Vauclair, J. (2011). A new tool for measuring hand preference in non-human primates: Adaptation of Bishop's Quantifying Hand Preference task for olive baboons. *Behavioural Brain Research, 218*, 1–7.
Middleton, K. M. (2001). The morphological basis of hallucal orientation in extant birds. *Journal of Morphology, 250*(1), 51–60.
Middleton, K. M. (2003). *Morphology, evolution, and function of the avian hallux*. Brown University.
Milliken, G. W., Ward, J. P., & Erickson, C. J. (1991). Independent digit control in foraging by the aye-aye (*Daubentonia madagascariensis*). *Folia Primatol (Basel), 56*(4), 219–224.
Mivart St., G. (1870). On the myology of *Chamaeleon parsonii*. *Proceedings of the Zoological Society (Calcutta), London*, 850–890.
Molesti, S., Vauclair, J., & Meguerditchian, A. (2016). Hand preferences for unimanual and bimanual coordinated actions in olive baboons (*Papio anubis*): Consistency over time and across populations. *Journal of Comparative Psychology, 130*(4), 341–350.

Molnar, J., Diaz Jr, R. E., Skorka, T., Dagliyan, D., & Diogo, R. (2017). Comparative musculoskeletal anatomy of chameleon limbs, with implications for the evolution of arboreal locomotion in lizards and for teratology. *Journal of Morphology., 2017*, 1–21.

Moreno, E., & Carrascal, L. M. (1993). Ecomorphological patterns of aerial feeding in oscines (Passeriformes: Passeri). *Biological Journal of the Linnean Society, 50*, 147–165.

Moro, S., & Abdala, V. (2004). Analisis descriptivo de la miología flexora y extensora del miembro anterior de *Polychrus acutirostris* (Squamata, Polychrotidae). *Papeis Avulsos de Zoologia (Sao Paulo), 44*, 81–89.

Morrison, M. L. (2018). *Ornithology: Foundation, analysis, and application*. Johns Hopkins University Press.

Mosto, C. M., & Tambussi, P. C. (2014). Qualitative and quantitative analysis of talons of diurnal bird of prey. *Anatomia, Histologia, Embryologia, 43*(1), 6–15. https://doi.org/10.1111/ahe.12041

Nakatsukasa, M., Kunimatsu, Y., Nakano, Y., & Ishida, H. (2002). Morphology of the hallucial phalanges in extant anthropoids and fossil hominoids. *Zeitschrift für Morphologie und Anthropologie, 83*, 361–372.

Napier, J. R. (1956). The prehensile movements of the human hand. *Journal of Bone and Joint Surgery. British Volume (London), 38B*, 902–913.

Napier, J. R., & Napier, P. H. (1985). *The natural history of the primates*. The MIT Press.

Napier, J. R. (1993). In *Hands* (Rev. by R. H. Tuttle), p. 180. Princeton University Press.

Nash, L. T. (1978). The development of the mother-infant relationship in wild baboons (*Papio anubis*). *Animal Behaviour, 26*(3), 746–759.

Nekaris, K. A. I. (2005). Foraging behaviour of the slender loris (*Loris lydekkerianus lydekkerianus*): Implications for theories of primate origins. *Journal of Human Evolution, 49*, 289–300.

Nauwelaerts, S., & Aerts, P. (2006). Take-off and landing forces in jumping frogs. *Journal of Experimental Biology, 209*, 66–77.

Nimbarte, A. D., Kaz, R., & Li, Z. M. (2008). Finger joint motion generated by individual extrinsic muscles: A cadaveric study. *Journal of Orthopaedic Surgery and Research, 3*, 27.

Noble, G. K. (1931). *The biology of the amphibians*. McGraw-Hill.

Norberg, U. M. (1979). Morphology of the wings, legs and tail of 3 coniferous forest tits, the goldcrest, and the treecreeper in relation to locomotor pattern and feeding station selection. *Philosophical Transactions of the Royal Society of London Series B-Biological Sciences, 287*(1019), 131–165. https://doi.org/10.1098/rstb.1979.0054

Norberg, R. Å. (1986). Treecreeper climbing; mechanics, energetics, and structural adaptations. *Ornis Scandinavica, 17*, 191–209.

Norberg, U. M. (1994). Wing design, flight performance and habit use in bats. In P. C. Wainwright & S. M. Reilly (Eds.), *Ecological morphology: Integrative organismal biology* (pp. 205–239). University of Chicago Press.

Oates, J. F. (1984). The niche of the potto, *Perodicticus potto*. *International Journal of Primatology, 5*, 51–61.

Orkin, J. D., & Pontzer, H. (2011). The Narrow Niche hypothesis: Gray squirrels shed new light on primate origins. *American Journal of Physical Anthropology, 144*(4), 617–624.

Papadatou-Pastou, M., Ntolka, E., Schmitz, J., Martin, M., Munafò, M. R., Ocklenburg, S., & Paracchini, S. (2020). Human handedness: A meta-analysis. *Psychological Bulletin, 146*(6), 481–524.

Patel, B. A., Wallace, I. J., Boyer, D. M., Granatosky, M. C., Larson, S. G., & Stern, J. T., Jr. (2015). Distinct functional roles of primate grasping hands and feet during arboreal quadrupedal locomotion. *Journal of Human Evolution, 88*, 79–84.

Patiño, S., Pérez Zerpa, J., & Fariña, R. A. (2021). Finite element and morphological analysis in extant mammals' claws and quaternary sloths' ungual phalanges. *Historical Biology, 33*(6), 857–867.

Paukstis, G. L., & Brown, L. E. (1987). Evolution of the intercalary cartilage in chorus frogs, genus *Pseudacris* (Salientia: Hylidae). *Brimleyana, 13*, 55–61.

Paukstis, G. L., & Brown, L. E. (1991). Evolutionary trends in the morphology of the intercalary phalanx of anuran amphibians. *Canadian Journal of Zoology, 69*, 1297–1301.

Payne, R. S. (1962). How the Barn Owl locates prey by hearing. *Living Bird, 1*, 151–159.

Peckre, L., Fabre, A. C., Wall, C. E., Brewer, D., Ehmke, E., Haring, D., Shaw, E., Welser, K., & Pouydebat, E. (2016). Holding-on: Co-evolution between infant carrying and grasping behaviour in strepsirrhines. *Scientific Reports, 6*, 37729.

Perrenoud, M., Herrel, A., Borel, A., & Pouydebat, E. (2015). Strategies of food detection in a captive cathemeral lemur, *Eulemur rubriventer. Belgium J. Zool., 145*(1), 69–75.

Petter, J. J. (1962). Recherches dur l'ecologie et l'ethologie des lemuriens malagashes. *Mémoires Museum National Histoire Naturelle, Paris, 27*, 1–146.

Piep, M., Radespiel, U., Zimmermann, E., Schmid, S., & Siemers, B. M. (2008). The sensory basis of prey detection in captive-born grey mouse lemurs, *Microcebus murinus. Animal Behaviour, 75*, 871–878.

Pike, A. V. L., & Maitland, D. P. (2004). Scaling of bird claws. *Journal of Zoology, 262*, 73–81.

Pocock, R. I. (1917). *The classification of existing* Felidæ. *Annals and Magazine of Natural History, 20*, 329–350.

Ponssa, M. L., Goldberg, J., & Abdala, V. (2010). Sesamoids in anurans: New data, old issues. *The Anatomical Record, 293*, 1646–1668.

Pouydebat, E., Laurin, M., Gorce, P., & Bels, V. (2008). Evolution of grasping among anthropoids. *Journal of Evolutionary Biology, 21*, 1732–1743.

Pouydebat, E., Gorce, P., & Bels, V. (2009). Biomechanical study of grasping according to the volume of the object: Human versus non-human primates. *Journal of Biomechanics, 42*(3–9), 266–272.

Pouydebat, E., Reghem, E., Borel, A., & Gorce, P. (2011). Diversity of grip in adults and young humans and chimpanzees (*Pan troglodytes*). *Behavioural Brain Research, 218*(1–17), 21–28.

Pouydebat, E., Fragaszy, D., & Kivell, T. (2014). Grasping in primates: To feed, to move and human specificities. *Bulletins et Mémoires de la Société d'Anthropologie de Paris, 26*, 129–133.

Pouydebat, E., & Bardo, A. (2019). An interdisciplinary approach of the evolution of grasping and manipulation? *Biological Journal of the Linnean Society*. https://doi.org/10.1093/biolinnean/blz058

Přikryl, T., Aerts, P., Havelková, P., Herrel, A., & Roček, Z. (2009). Pelvic and thigh musculature in frogs (Anura) and origin of anuran jumping locomotion. *Journal of Anatomy, 214*, 100–139.

Provini, P., Tobalske, B. W., Crandell, K. E., & Abourachid, A. (2014). Transition from wing to leg forces during landing in birds. *The Journal of Experimental Biology, 217*(Pt 15), 2659–2666. https://doi.org/10.1242/jeb.104588

Pugener, L. A., & Maglia, A. M. (2009). Developmental evolution of the anuran sacro-urostylic complex. *South American Journal of Herpetology, 4*, 193–209.

Quinn, T. H., & Baumel, J. J. (1990). The digital tendon locking mechanism of the avian foot. *Zoomorphology, 109*, 281–293.

Quinn, T. H., & Baumel, J. J. (1993). Chiropteran tendon locking mechanism. *Journal of Morphology, 216*(2), 197–208. https://doi.org/10.1002/jmor.1052160207

Raikow, R. J. (1985). The locomotor system. In A. S. King & J. McLelland (Eds.), *Form and function in birds* (Vol. 3, pp. 57–147). Academic Press.

Rasmussen, D. T. (1990). Primate origins: Lessons from a Neotropical marsupial. *American Journal of Primatology, 22*, 263–277.

Rasmussen, D. T., & Sussman, R. W. (2007). Parallelisms among possums and primates. In M. Ravosa & M. Dagosta (Eds.), *Primate origins: Adaptation and evolution* (pp. 39–380). Springer.

Reghem, E., Tia, B., Bels, V., & Pouydebat, E. (2011). Food prehension and manipulation in *Microcebus murinus* (Prosimii, Cheirogaleidae). *Folia Primatologica, 82*, 177–188.

Reghem, E., Byron, C., Bels, V., & Pouydebat, E. (2012). Hand posture in the grey mouse lemur during arboreal locomotion on narrow branches. *Journal of Zoology, 288*, 76–81.
Regnault, S., Hutchinson, J. R., & Jones, M. E. (2016). Sesamoid bones in tuatara (*Sphenodon punctatus*) investigated with X-ray microtomography, and implications for sesamoid evolution in Lepidosauria. *Journal of Morphology, 278*, 62–72.
Reilly, S. M., & Delancey, M. J. (1997). Sprawling locomotion in the lizard *Sceloporus clarkii*: The effects of speed on gait, hindlimb kinematics, and axial bending during walking. *Journal of Zoology, 243*, 417–433.
Renous-Lécuru, S. (1973). Morphologie comparée du carpe chez les Lepidosauriens actuels (Rhynchocéphales, Lacertiliens, Amphisbéniens). *Gegenbaurs Morphologisches Jahrbuch Leipzig, 119*, 727–766.
Rensch, B., & Dücker, G. (1969). Manipulierfähigkeit eines Wickelbären bei längeren Handlungsketten. *Zeitschrift für Tierpsychologie, 26*, 104–112.
Ribbing, L. (1913). Kleinere Muskelstudien V. Die distale Extremitätenmuskulatur von *Chamaeleon vulgaris*. *Lunds Universitets Årsskrift N. F. Afd., 2*(8), 3–15.
Richmond, B. G., Roach, N. T., & Ostrofsky, K. R. (2016). Evolution of the early hominin hand. In T. Kivell, P. Lemelin, B. Richmond, & D. Schmitt (Eds.), *The evolution of the primate hand. Developments in primatology: Progress and prospects*. Springer.
Riskin, D. K., Bahlman, J. W., Hubel, T. Y., Ratcliffe, J. M., Kunz, T. H., & Swartz, S. M. (2009). Bats go head-under-heels: The biomechanics of landing on a ceiling. *The Journal of Experimental Biology, 212*(Pt 7), 945–953. https://doi.org/10.1242/jeb.026161
Robinson, P. L. (1975). The functions of the fifth metatarsal in lepidosaurian reptiles. *Colloques Internationaux du Centre National de la Recherche Scientifique, 218*, 451–483.
Robovska-Havelkova, P., Aerts, P., Rocek, Z., Prikryl, T., Fabre, A. C., & Herrel, A. (2014). Do all frogs swim alike? The effect of ecological specialization on swimming kinematics in frogs. *The Journal of Experimental Biology, 217*, 3637–3644.
Rocǒková, H., & Roček, Z. (2005). Development of the pelvis and posterior part of the vertebral column in the Anura. *Journal of Anatomy, 206*, 17–35.
Roderick, W. R., Chin, D. D., Cutkosky, M. R., & Lentink, D. (2019). Birds land reliably on complex surfaces by adapting their foot-surface interactions upon contact. *eLife, 8*. https://doi.org/10.7554/eLife.46415
Rolian, C., Lieberman, D. E., & Zermeno, J. P. (2011). Hand biomechanics during simulated stone tool use. *Journal of Human Evolution, 61*, 26–41.
Rosenberg, H. I., & Rose, R. (1999). Volar adhesive pads of the feathertail glider, *Acrobates pygmaeus* (Marsupialia; Acrobatidae). *Canadian Journal of Zoology, 77*(2), 233–248.
Rothier, P. S., Brandt, R., & Kohlsdorf, T. (2017). Ecological associations of autopodial osteology in Neotropical geckos. *Journal of Morphology, 278*, 290–299.
Russell, A. P. (1993). The aponeuroses of the lacertilian ankle. *Journal of Morphology, 218*, 65–84.
Russell, A. P., & Bauer, A. M. (2008). The appendicular locomotor apparatus of *Sphenodon* and normal-limbed squamates. In C. Gans, A. S. Gaunt, & K. Adler (Eds.), *Biology of Reptilia, Volume 21: The skull and appendicular locomotor apparatus of Lepidosauria, contributions to herpetology 24* (pp. 1–465). Society for the Study of Reptiles & Amphibians.
Sacrey, L. A., Alaverdashvili, M., & Whishaw, I. Q. (2009). Similar hand shaping in reaching-for-food (skilled reaching) in rats and humans provides evidence of homology in release, collection, and manipulation movements. *Behavioural Brain Research, 204*(1), 153–161.
Samaras, A., & Youlatos, D. (2010). Use of forest canopy by European red squirrels *Sciurus vulgaris* in Northern Greece: Claws and the small branch niche. *Acta Theriologica, 55*, 351–360.
Sargis, E. J. (2001). The grasping behavior, locomotion, and substrate use of tree shrews *Tupaia minor* and *T. tana* (Mammalia, Scandentia). *Journal of Zoology, 253*, 485–490.
Sargis, E. J. (2007). The postcranial morphology of *Ptilocercus lowii* (Scandentia, Tupaiidae) and its implications for primate supraordinal relationships. In M. J. Ravosa & M. Dagosto (Eds.), *Primate Origins: Adaptations and evolution*. Springer.

Sargis, E. J., Boyer, D. M., Bloch, J. I., & Silcox, M. T. (2007). Evolution of pedal grasping in Primates. *Journal of Human Evolution, 53*, 103–107.

Schick, K., Toth, N., Garufi, G., Rumbaugh, E., Rumbaugh, D., & Sevcik, R. (1993). *Pan* the tool-maker: Investigations into the stone tool-making and tool-using capabilities of a bonobo (*Pan paniscus*). *Journal of Archaeological Science, 20*, 821–832. https://doi.org/10.1006/jasc.1998.0350

Schneider, N. Y., & Gurovich, Y. (2017). Morphology and evolution of the oral shield in marsupial neonates including the newborn monito del monte (*Dromiciops gliroides*, Marsupialia Microbiotheria) pouch young. *Journal of Anatomy, 231*(1), 59–83.

Schultz, A. H. (1972). Les primates. In *La Grande Encyclopédie de la Nature* (Volume, 17 ed. Bordas), p. 383. Editions Rencontre, Lausanne.

Schwensen, K. (1994). The importance of the somatosensory system in skilled limb-use by the northernquoll, *Dasyurus hallucatus*. B.Sc. (Honours) Thesis, Monash University, Melbourne.

Scooter, C. A. (1944). Methods of grasping and carrying prey. *Condor, 46*(2), 88.

Sereno, P. C., & Chenggang, R. (1992). Early evolution of avian flight and perching: New evidence from the lower cretaceous of china. *Science, 255*(5046), 845–848. https://doi.org/10.1126/science.255.5046.845

Shapiro, L. J., Young, J. W., & VandeBerg, J. L. (2014). Body size and the small branch niche: Using marsupial ontogeny to model primate locomotor evolution. *Journal of Human Evolution, 68*, 14–31.

Sheil, C. A., & Alamillo, H. (2005). Osteology and skeletal development of *Phyllomedusa vaillanti* (Anura: Hylidae: Phyllomedusinae) and a comparison of this arboreal species with a terrestrial member of the genus. *Journal of Morphology, 265*, 343–368.

Shim, J. K., Karol, S., Kim, Y. S., Seo, N. J., Kim, Y. H., Kim, Y., & Yoon, B. C. (2012). Tactile feedback plays a critical role in maximum finger force production. *Journal of Biomechanics, 45*(3), 415–420. https://doi.org/10.1016/j.jbiomech.2011.12.001

Shubin, N. H., & Jenkins, F. A. (1995). An early jurassic jumping frog. *Nature, 377*, 49–52.

Simmons, N., & Quinn, T. (1994). Evolution of the digital tendon locking mechanism in bats and dermopterans: A phylogenetic perspective. *Journal of Mammalian Evolution, 2*, 231–251.

Siemers, B. M., Goerlitz, H. R., Robsomanitrandrasana, E., Piep, M., Ramanamanjato, J. B., Rakotondravony, D. R., Ramilijaona, O., & Ganzhorn, J. U. (2007). Sensory basis of food detection in wild *Microcebus murinus*. *International Journal of Primatology, 28*, 291–304.

Smith, G. A. (1971). The use of the foot in feeding, with especial reference to parrots. *Avicultural Magazine, vol 77*.

Sparling, J. W., Van Tol, J., & Chescheir, N. C. (1999). Fetal and neonatal hand movement. *Physical Therapy, 79*, 24–39.

Spinozzi, G., Truppa, V., & Lagana, T. (2004). Grasping behavior in tufted capuchin monkeys (*Cebus apella*): Grip types and manual laterality for picking up a small food item. *American Journal of Physical Anthropology, 125*, 30–41.

Stettenheim, P. R. (2000). The integumentary morphology of modern birds - An overview. *American Zoologist, 40*, 461–477.

Ströckens, F., Güntürkün, O., & Ocklenburg, S. (2013). Limb preferences in non-human vertebrates. *Laterality, 18*, 536–575.

Sussman, R. W., & Kinzey, W. G. (1984). The ecological role of the Callitrichidae: A review. *American Journal of Physical Anthropology, 64*, 419–449.

Susman, R. L. (1994). Fossil evidence for early hominid tool use. *Science, 265*, 1570–1573.

Susman, R. L. (1998). Hand function and tool behavior in early hominids. *Journal of Human Evolution, 35*, 23–46.

Sussman, R. W. (1991). Primate origins and the evolution of angiosperms. *American Journal of Primatology, 23*, 209–223.

Sustaita, D. (2008). Musculoskeletal underpinnings to differences in killing behavior between North American accipiters (Falconiformes: Accipitridae) and falcons (Falconidae). *Journal of Morphology, 269*, 283–301.

Sustaita, D., & Hertel, F. (2010). In vivo bite and grip forces, morphology and prey-killing behavior of North American accipiters (Accipitridae) and falcons (Falconidae). *Journal of Experimental Biology, 213*(15), 2617–2628.

Sustaita, D., Pouydebat, E., Manzano, A., Abdala, V., Hertel, F., & Herrel, A. (2013). Getting a grip on tetrapod grasping: Form, function, and evolution. *Biological Reviews of the Cambridge Philosophical Society, 88*(2), 380–405. https://doi.org/10.1111/brv.12010

Sustaita, D., Gloumakov, Y., Tsang, L. R., & Dollar, A. M. (2019). Behavioral correlates of semi-zygodactyly in Ospreys (*Pandion haliaetus*) based on analysis of internet images. *PeerJ, 7*, e6243. https://doi.org/10.7717/peerj.6243

Sweet, S. S., & Pianka, E. R. (2007). Monitors, mammals and Wallace's Line. Third multidisciplinary world conference on monitor lizards. *Mertensiella, 16*, 79–99.

Szalay, F. S. (1994). *Evolutionary history of the marsupials and an analysis of osteological characters*. Cambridge University Press.

Szalay, F. S., & Dagosto, M. (1988). Evolution of hallucial grasping in the primates. *Journal of Human Evolution, 17*, 1–33.

Taylor, B. K. (1985). Functional anatomy of the forelimb in vermilinguas (anteaters). In G. G. Montgomery (Ed.), *The evolution and ecology of armadillos, sloths, and vermilinguas* (pp. 151–171). Smithsonian Institution Press.

Taylor, G. M., Nol, E., & Boire, D. (1995). Brain regions and encephalization in anurans: Adaptation or stability? *Brain, Behavior and Evolution, 45*, 96–109.

Takeshita, H., Myowa-Yamakoshi, M., & Hirata, S. (2006). A new comparative perspective on prenatal motor behaviors: Preliminary research with four-dimensional ultrasonography. In T. Matsuzawa, M. Tomonaga, & M. Tanaka (Eds.), *Cognitive development in chimpanzees*. Springer.

Ten Donkelaar, H. J. (1998). Anurans. In R. Nieuwenhuys, H. J. Ten Donkelaar, & C. Nicholson (Eds.), *The central nervous system of vertebrates* (pp. 1151–1314). Springer.

Thewissen, J. G. M., & Etnier, S. A. (1995). Adhesive devices on the thumb of vespertilionoid bats (Chiroptera). *Journal of Mammalogy, 76*(3), 925–936.

Thomas, P., Pouydebat, E., Le Brazidec, M., Aujard, F., & Herrel, A. (2016). Determinants of pull strength in captive grey mouse lemurs (*Microcebus murinus*). *Journal of Zoology, 298*, 77–81.

Tinius, A., & Patrick Russell, A. (2017). Points on the curve: An analysis of methods for assessing the shape of vertebrate claws. *Journal of Morphology, 278*(2), 150–169. https://doi.org/10.1002/jmor.20625

Tocheri, M. W., Orr, C. M., Jacofsky, M. C., & Marzke, M. W. (2008). The evolutionary history of the hominin hand since the last common ancestor of *Pan* and *Homo*. *Journal of Anatomy, 212*, 544–562.

Toth, N., Schick, K., Savage-Rumbaugh, E. S., Sevcik, R. A., & Rumbaugh, D. M. (1993). *Pan* the tool-maker: Investigations into the stone tool-making and tool-using capabilities of a bonobo (*Pan paniscus*). *Journal of Archaeological Science, 20*, 81–91.

Toussaint, S., Reghem, E., Chotard, H., Herrel, A., Ross, C. F., & Pouydebat, E. (2013). Food acquisition on arboreal substrates by the grey mouse lemur: implication for primate grasping evolution. *Journal of Zoology, 291*, 235–242.

Toussaint, S., Herrel, A., Ross, C. F., Aujard, F., & Pouydebat, E. (2015). Substrate diameter and orientation in the context of food type in the gray mouse lemur, *Microcebus murinus*: Implications for the origins of grasping in primates. *International Journal of Primatology, 36*(3), 583–604.

Toussaint, S., Llamosi, A., Morino, L., & Youlatos, D. (2020). The central role of small vertical substrates for the origin of grasping in early primates. *Current Biology, 30*, 1600–1613.

Tozer, D. C., & Allen, M. L. (2004). Adult gray jay captures an adult black-capped chickadee. *Wilson Bulletin, 116*(4), 357–359.

Tsang, L. R. (2012). Facultative zygodactyly in the black-shouldered kite *Elanus axillaris*. *Australian Field Ornithology, 29*, 89–92.

Tsang, L. R., & McDonald, P. G. (2018). A comparative study of avian pes morphotypes, and the functional implications of Australian raptor pedal flexibility. *Emu - Austral Ornithology*. https://doi.org/10.1080/01584197.2018.1483203

Tsang, L. R., Wilson, L. A. B., & McDonald, P. G. (2019a). Comparing the toepads of Australian diurnal and nocturnal raptors with nonpredatory taxa: Insights into functional morphology. *Journal of Morphology, 280*(11), 1682–1692. https://doi.org/10.1002/jmor.21057

Tsang, L. R., Wilson, L. A. B., Ledogar, J., Wroe, S., Attard, M., & Sansalone, G. (2019b). Raptor talon shape and biomechanical performance are controlled by relative prey size but not by allometry. *Scientific Reports, 9*(1), 7076. https://doi.org/10.1038/s41598-019-43654-0

Tulli, M. J., Cruz, F. B., Herrel, A., Vanhooydonck, B., & Abdala, V. (2009). The interplay between claw morphology and microhabitat use in Neotropical iguanian lizards. *Zoology, 112*, 379–392.

Tulli, M. J., Abdala, V., & Cruz, F. B. (2011). Relationships among morphology, clinging performance and habitat use in Liolaemini lizards. *Journal of Evolutionary Biology, 24*, 843–855.

Urbani, B., & Youlatos, D. (2013). Positional behavior and substrate use of *Micromys minutus* (Rodentia: Muridae): Insights for understanding primate origins. *Journal of Human Evolution, 64*, 130–136.

Vaira, M. (2001). Breeding biology of the leaf frog, *Phyllomedusa boliviana* (Anura, Hylidae). *Amphibia-Reptilia, 22*, 421–429.

Van Valkenburgh, B. (1987). Skeletal indicators of locomotor behavior in living and extinct carnivores. *Journal of Vertebrate Paleontology, 7*(2), 162–182.

Vassallo, A. I., Manzano, A. S., Abdala, V., & Muzio, R. (2021). Can Anyone Climb? The skills of a non-specialized toad and its bearing on the evolution of new niches. *Evolutionary Biology*. https://doi.org/10.1007/s11692-021-09539-9

Vauclair, J., Meguerditchian, A., & Hopkins, W. D. (2005). Hand preferences for unimanual and coordinated bimanual tasks in baboons (*Papio anubis*). *Cognitive Brain Research, 25*, 210–216.

Vaughan, T., Ryan, J., & Czaplewski, N. (2011). *Mammalogy* (5th ed.). Jones and Bartlett Publishers, Greater Sudbury.

Versace, E., & Vallortigara, G. (2015). Forelimb preferences in human beings and other species: Multiple models for testing hypotheses on lateralization. *Frontiers in Psychology, 6*, 233.

Vigouroux, L., Bardo, A., Kivell, T. L., & Pouydebat, E. (2018). Biomechanical analyses of the consequences of different hand proportion on tool grasp abilities using musculo-skeletal simulation: A preliminary study. *Journal of Human Evolution, 125*, 106–121.

Völter, C. J., Rossano, F., & Call, J. (2015). From exploitation to cooperation: Social tool use in Orang-Utan mother– offspring dyads. *Animal Behaviour, 100*, 126–134.

Warburton, N. M., Harvey, K. J., Prideaux, G. J., & O'Shea, J. E. (2011). Functional morphology of the forelimb of living and extinct tree-kangaroos (Marsupialia: Macropodidae). *Journal of Morphology, 272*(10), 1230–1244.

Ward, A. B., Weigl, P. D., & Conroy, R. M. (2002). Functional morphology of raptor hindlimbs: Implications for resource partitioning. *Auk, 119*(4), 1052–1063.

Whipple, I. L. (1904). The ventral surface of the mammalian cheiridium, with special reference to conditions found in man. *Zeitschrift für Morphologie und Anthropologie, 7*, 261–368.

Whishaw, I. Q., Pellis, S. M., & Gorny, B. P. (1992). Skilled reaching in rats and humans: Evidence for parallel development or homology. *Behavioural Brain Research, 47*(1), 59–70.

Whishaw, I. Q. (1996). An endpoint, descriptive, and kinematic comparison of skilled reaching in mice (*Mus musculus*) with rats (*Rattus novergicus*). *Behavioural Brain Research, 78*, 101–111.

Whishaw, I. Q., Sarna, J. R., & Pellis, S. M. (1998). Rodent-typical and species-specific limb use in eating: Evidence for specialized paw use from a comparative analysis of ten species. *Behavioural Brain Research, 96*, 79–91.

Williams, E. M., Gordon, A. D., & Richmond, B. G. (2012). Hand pressure distribution during Oldowan stone tool production. *Journal of Human Evolution, 62*, 520–532.

Wimsatt, W. A., & Villa, B. (1970). Locomotor adaptations of the disc-winged bat *Thyroptera tricolor*. I. Functional organization of the adhesive discs. *The American Journal of Anatomy 129*, 89–120.

Winkler, H., & Bock, W. (1976). Analyse der Kräfteverhaltnisse bei Klettervögeln. *Journal für Ornithologie, 117*, 397–418.

Yeo, R., & Gangestad, S. W. (1993). Developmental origins of variation in human hand preference. *Genetica, 89*, 281–296.

Youlatos, D. (2008). Hallucal grasping behavior in *Caluromys* (Didelphimorphia: Didelphidae): Implications for primate pedal grasping. *Journal of Human Evolution, 55*, 1096e1101.

Youlatos, D., Moussa, D., Karantanis, N.-E., & Rychlik, L. (2018). Locomotion, postures, substrate use, and foot grasping in the marsupial feathertail glider *Acrobates pygmaeus* (Diprotodontia: Acrobatidae): insights into early euprimate evolution. *Journal of Human Evolution, 123*, 148–159.

Young, J. W., & Chadwell, B. A. (2020). Not all fine-branch locomotion is equal: Grasping morphology determines locomotor performance on narrow supports. *Journal of Human Evolution, 142*, 102767.

Zaaf, A., & Van Damme, R. (2001). Limb proportions in climbing and ground-dwelling geckos (Lepidosauria, Gekkonidae): A phylogenetically informed analysis. *Zoomorphology, 121*, 45–53.

Zani, P. A. (2000). The comparative evolution of lizard claw and toe morphology and clinging performance. *Journal of Evolutionary Biology, 13*, 316–325.

Zeffer, A., & Norberg, U. M. L. (2003). Leg morphology and locomotion in birds: Requirements for force and speed during ankle flexion. *Journal of Experimental Biology, 206*, 1085–1097.

Zellmer, N. T., Timm-Davis, L. L., & Davis, R. W. (2021). Sea otter behavior: Morphologic, physiologic, and sensory adaptations In R. W. Davis & A. M. Pagano (Eds.), *Ethology and behavioral ecology of sea otters and polar bears. Ethology and behavioral ecology of marine mammals*. Springer.

Chapter 13
Convergence in Gliding Animals: Morphology, Behavior, and Mechanics

Pranav C. Khandelwal, Shane D. Ross, Haibo Dong, and John J. Socha

Abstract Gliding locomotion has convergently evolved in multiple vertebrate and invertebrate taxa, spanning terrestrial and aquatic animals. The selective pressures attributed to the evolution of gliding include the topography of the environment as well as the capabilities for rapidly escaping predation, foraging over larger spatial areas, and landing safely after falling. Although gliding locomotion has likely evolved in response to these multiple factors in diverse lineages, extant taxa exhibit convergent morphologies and behaviors related to gliding. Understanding the relevance of specific gliding features is informed by the laws of physics: to successfully execute a glide, the animal must use a combination of body shape/size changes (morphology) along with attaining and modulating a favorable body posture (behavior) to generate sufficient aerodynamic forces to slow and control the descent. Gliding animals employ a diverse range of aerodynamic surfaces to generate lift and drag forces, from membrane wings in mammals, *Draco* lizards, fish, and squid, to smaller structures including skin flaps, flattened bodies, and appendages in geckos, snakes, frogs, spiders, and ants. These force-generating surfaces vary in their shape, size, and anatomical structure, but serve a common function of increasing the total body surface area of the animal compared to their non-gliding relatives, enabling them to produce significantly higher aerodynamic forces. Convergence is also observed in takeoff, gliding, and landing behaviors, necessary for the animal to execute a successful glide trajectory. Takeoff behaviors vary from jumping from vertical or horizontal substrates in terrestrial gliders, to launching from below or on top of the water surface in fish and squid. Once airborne, gliding animals produce

P. C. Khandelwal
Max Planck Institute for Intelligent Systems, Stuttgart, Germany

S. D. Ross
Department of Aerospace and Ocean Engineering, Virginia Tech, Blacksburg, VA, USA

H. Dong
Department of Mechanical Engineering, University of Virginia, Charlottesville, VA, USA

J. J. Socha (✉)
Department of Biomedical Engineering and Mechanics, Virginia Tech, Blacksburg, VA, USA
e-mail: jjsocha@vt.edu

V. L. Bels, A. P. Russell (eds.), *Convergent Evolution*, Fascinating Life Sciences,
https://doi.org/10.1007/978-3-031-11441-0_13

and modulate aerodynamic forces of lift and drag through adjustments in their body-airfoil or posture, and/or interactive combinations of both. In some taxa, modulation of aerodynamic forces enables the animal to undertake aerial maneuvers to navigate spatially complex habitats and to land. The evolution of dedicated primary wings in mammalian gliders and *Draco* flying lizards allows them to substantially slow their descent and transition into a more upright position to land, mostly on vertical substrates. Gliders that lack wings, including snakes, geckos, ants, and spiders, use a landing strategy involving impact with the substrate without a significant reduction in speed, using a combination of the body and appendages to land. Flying fish and squid attain a more streamlined posture by tucking their fins to reduce drag while entering the water surface. In this chapter, we provide a broad overview of gliding in diverse lineages, highlighting the ecological and physical pressures that have shaped this form of aerial locomotion in the animal kingdom.

13.1 Introduction

The capability of moving through air is not common in the animal kingdom. Those that can fly reap tremendous ecological benefits - enjoying fast travel, with some migrating over long distances (including over the highest mountains and largest oceans), access to largely inaccessible resources, and means for foraging or avoiding predators. Muscle-powered flapping flight has convergently evolved only four times in the history of life, in birds, bats, insects, and pterosaurs (Rayner, 1988; Wigglesworth, 1973), but each of these groups has benefitted from access to vertical airspace, as reflected in the high diversity of their extant species.

Gliding (defined below) is also a form of flight, but an unpowered one in which the animal falls through the air in a controlled fashion, trading the potential energy of height for kinetic energy that drives horizontal travel (Dudley et al., 2007; Socha et al., 2015). Gliding is far less physically and biomechanically demanding than flapping flight, but entails its own set of challenges. Against the pull of gravity, the gliding animal must create aerodynamic forces of lift and drag that both help to support its weight and direct it horizontally, as well as distribute those forces in a way that keeps the body stable. Without the additional input of power, the frictional losses of drag dictate that the glider can only produce a net path that is downward, yet angled, in still air. With this capability, gliding animals can use rapid aerial locomotion to evade predators or hunt prey, forage, and engage with other conspecifics via mate-seeking or territoriality behaviors, each such behavior being essential for their survival. Because their source of energy is limited by their starting height, gliders are constrained to relatively short flights in comparison to flapping flyers, often balancing the requirement to cover horizonal distance with maneuvering around obstacles mid-air in often spatially cluttered environments (Khandelwal & Hedrick, 2020). Nonetheless, these short flights also enable them to cross otherwise limiting gaps in the environment (Graham & Socha, 2020), expanding the scope of available resources compared to their non-flying relatives. Overall, with fewer

Fig. 13.1 Animal gliders of the world. Terrestrial gliders exclusively inhabit arboreal environments, including mammals (squirrels, sugar gliders, and colugos), lizards including *Draco* and the gecko *Hemidactylus*, snakes, frogs, spiders, and ants. The only known gliders that inhabit oceanic environments are fishes and squid. The animals in this figure are original illustrations by Dr. Mary Salcedo, inspired by images from multiple sources. Not to scale

physical and physiological requirements, gliding flight has convergently evolved far more often than has flapping flight, in groups that include mammals, reptiles, amphibians, insects, arachnids, fishes, and cephalopods (Fig. 13.1), with body sizes spanning over 4 orders of magnitude (from ~0.05 g in *Cephalotes atratus* (Yanoviak et al., 2005) ants to ~3.2 kg in the red giant flying squirrel (Jackson, 2000), *Petaurista petaurista*).

In this chapter, we discuss the morphological and behavioral convergences that underlie the ability of such a great diversity of animals to glide successfully. Perhaps the most common theme related to convergence in gliding is an increase of body surface area, usually through some dynamic action such as spreading of skin to form a wing, defined here as a morphological structure that functions to generate substantial aerodynamic forces in flight. But there is more to an airfoil than what might be seen in traditional bilaterally symmetrical wings: animal gliders also take advantage of morphological shape differences in flattened body structures relative to non-gliding relatives. Consequently, differences in morphology lead to convergence in specific aerial behaviors that the animals must employ on a day-to-day basis to become airborne, generate aerodynamic forces, and land successfully. We discuss such features in the context of the full repertoire of gliding, from takeoff to landing, informed by the physics that strongly guide and constrain such behaviors. For a historical perspective, we also point the reader to previous reviews of gliding, including those that focus on adaptations (Rayner, 1981), evolution (Dudley et al., 2007), and morphology and trajectory (Socha et al., 2015).

13.1.1 What Is Gliding?

We consider gliding, following the conceptual framework of recent authors (Dudley et al., 2007; Socha et al., 2015), to be any aerial behavior in which the animal actively produces aerodynamic forces to effectuate horizontal travel in a controlled fashion. In doing so, animals must be able of at least partially supporting their body weight (thereby also slowing their descent), and also controlling their trajectory either by modulating aerodynamic forces or using inertial mechanisms. The underlying physical mechanism of aerodynamic force production is described by the equations for lift and drag:

$$L = \frac{1}{2} C_L \rho S v^2 \text{ and } D = \frac{1}{2} C_D \rho S v^2.$$

where L and D are the lift and drag forces, C_L is the coefficient of lift, C_D is the coefficient of drag, ρ is the fluid (air) density, S is typically the effective surface area of the wing/airfoil, and v is the airspeed of the animal. Most gliders reconfigure a part of their body to form an airfoil that produces lift, but all morphological structures produce drag, important for the stability characteristics of the glider. The farther the center of aerodynamic pressure is from the center of mass of the glider, the greater the resulting rotational moment, meaning that even small amounts of drag can serve to topple the glider from a stable position, or be put to positive use for balance and to induce aerial maneuvers.

13.1.2 Driving Forces of Convergence in Gliders

A common benefit for all gliding animals is the ability to rapidly cover distance by moving through air instead of moving on the ground, climbing, or even swimming. Consequently, rapid aerial movement can help the gliding animal evade predators, forage, and cover a larger area of the habitat to perform behaviors, including the seeking of mates and establishment of territory. The evolution of gliding has been attributed to multiple factors including habitat structure, predator avoidance, foraging, injury avoidance, and energetic expenditure during locomotion (Fig. 13.2). It is highly likely that many of these factors collectively influenced the evolution of gliding across different taxa. We begin our discussion by examining terrestrial gliding animals with regard to each of the factors listed above and then extend it

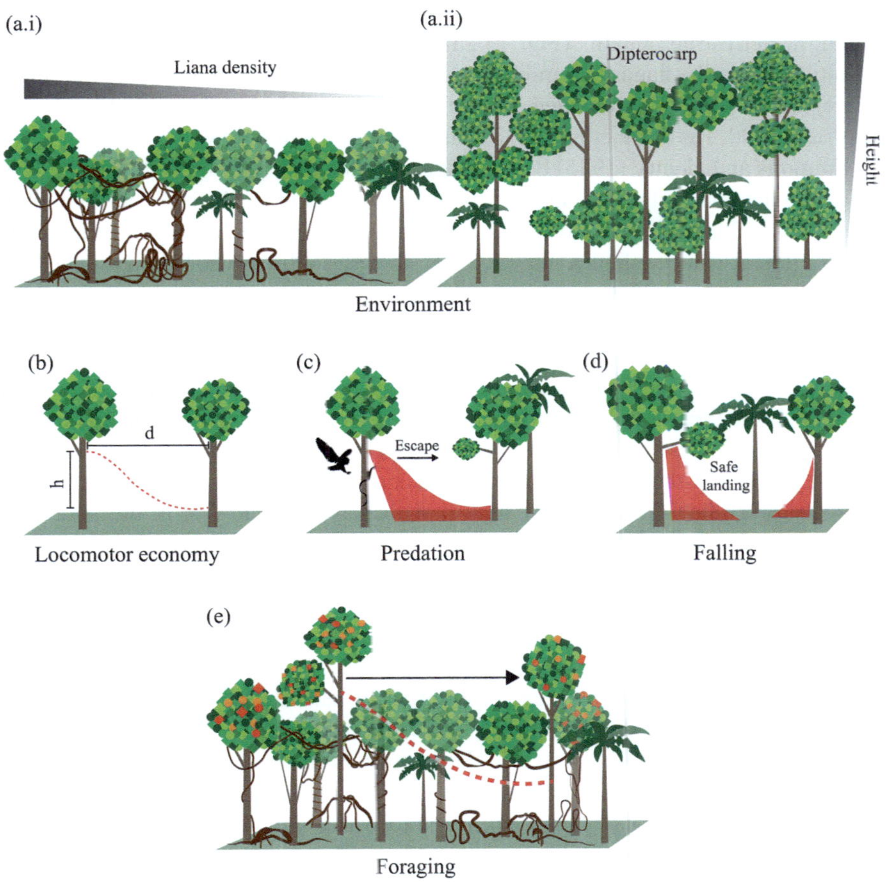

Fig. 13.2 Possible selective pressures that drove the evolution of gliding in arboreal taxa. Hypotheses include environmental conditions related to forest structure (**a**.i: density of vine-like lianas, and **a**.ii: tree height); energetics (**b**: locomotor economy); and behavior (**c**: predation, **d**: safe landing after falling, and **e**: foraging efficacy)

to the less frequently discussed selective pressures acting on aquatic gliders, including flying fish and squid.

13.1.3 Environment

Empirical evidence suggests that different forest structures across the world are correlated with different modes of locomotion by arboreal animals (Emmons & Gentry, 1983). Therefore, it follows that the variability in the structure of forests would have played a crucial role in the evolution of gliding in animals. Along these lines, there have been two prominent hypotheses put forth to explain the evolution of gliding in terrestrial animals (Fig. 13.2a). The first deals with the distribution and density of vine-like lianas (Emmons & Gentry, 1983), and second with the presence of tropical rainforests dominated by members of the Dipterocarpaceae (Dudley & DeVries, 1990, Heinicke et al., 2012), discussed in detail below.

Emmons and Gentry (1983) hypothesized that the evolution of gliding flight was related to the distribution of lianas in forest canopies (Fig. 13.2a.i). Lianas are long, woody vines that climb trees to gain access to sunlight. By extensively covering trees and the forest canopy, lianas can provide arboreal pathways for canopy-dwelling animals. Emmons and Gentry surveyed liana density in Africa (Afrotropical realm), South America (Neotropical realm), and Asia (Indo-Malayan realm) and showed that Africa has the highest liana density, followed by South America, with the lowest density in Asia, leading to their hypothesis that the scarcity of lianas in tropical Asia favors the relatively long-distance strategy of gliding. Furthermore, the increase in liana density from Asia to America should also correspond to an increase in glider diversity. Indeed, this is the case, with Asian tropical forests home to a greater diversity of gliding animals compared to Africa, with the lowest in the Americas (Corlett, 2007). Even within the Americas, the southern flying squirrel, *Glaucomys volans*, is restricted to high-elevation deciduous forests consisting of oak and oak-pine, and is not found in liana-rich lowland Neotropical forests (Lambert & Halsey, 2015, Jackson & Schouten, 2012). Although the hypothesis of variation in liana density seems promising, it has faced criticism based on contrary results. Dudley and DeVries (1990) argued that the mean liana density is only 30% higher in Africa compared to the Neotropics, yet there are relatively few gliding animals found in either geographical region (Jackson & Schouten, 2012; McCay, 2003). Furthermore, Appanah et al. (1993) have shown that Asian forests are not poor in diversity and abundance of lianas compared to Africa and South America.

Dudley and DeVries (1990) extended the discussion of the evolutionary influence of the environment on gliding to include the height of the forests (Fig. 13.2a.ii). In particular, trees of the Dipterocarpaceae reach heights upwards of 60 m in the Indo-Malayan forests compared to the average forest height of 30–40 m in Africa and America. The presence of taller trees could provide gliders with opportunities to cover larger horizontal distances, reduce their energetic cost of transport, and provide vertical habitat stratification, which should promote diversification of gliding taxa. Furthermore, the presence of taller trees could favor the evolution of larger body size in gliding animals, with increased takeoff height enabling longer glide distances and lowering their cost of transport. Dudley and DeVries (1990) also hypothesized that gliding animals would climb as high as possible to increase their horizontal gliding distance and reduce their energetic costs. However, recent work by Khandelwal and Hedrick (2020) has shown that glide distances are independent of the takeoff height in glides performed by *Draco* flying lizards in their natural environment. A complementary evolutionary pressure arising from the ecology of dipterocarp forests is the prevalence of food patchiness and the supra-annual cycles of flowering and fruiting, which could promote spatially larger foraging bouts and gliding as an efficient energetic means of doing so (Corlett, 2007; Heinicke et al.,

2012). Further supporting the influence of the Dipterocarpaceae on the evolution of gliding, Heinicke et al. (2012) demonstrated a strong temporal overlap between the emergence of dipterocarp forests and the origin of gliding in at least six lineages.

Convergence of phylogenetically distant groups of animals upon similar gliding behaviors, coupled with the structural difference in glider habitats, highlights the evolutionary pressure imposed by the environment on extant gliders worldwide. Both hypotheses, that based on liana density and that based upon the Dipterocarpaceae, likely contribute towards our understanding of the evolution of gliding in animals. However, neither hypothesis can stand on its own, falling short of explaining how intermediate forms of gliding evolved into the variety of gliding lineages that we see today (Lambert & Halsey, 2015).

13.1.4 Locomotor Economy

The locomotor economy hypothesis (Fig. 13.2b) states that the energetic cost per unit distance is less for gliding in comparison to various forms of terrestrial or aquatic locomotion. Studies testing this hypothesis have been carried out primarily for mammalian gliders by comparing the cost of climbing and gliding a particular distance to the cost of quadrupedal locomotion for traveling the same glide distance horizontally (Byrnes et al., 2011; Dial, 2003; Flaherty et al., 2010; Scheibe & Robins, 1998; Scheibe et al., 2006).

Dial (2003) used a simplistic model derived from equilibrium gliding mechanics to show that an intermediate body size, in the range of 200–400 g, would favor gliding as an energetically efficient means of transport compared to animals of small or large body size. Dial (2003) further argued that the energetic advantage for mid-size animals would be evident in their home range size, with intermediate-sized gliding mammals having larger home ranges and exploiting calorie-poor diets compared to their non-gliding counterparts, as opposed to similar home range size for small and large-sized gliding mammals. Overall, the energetic calculations presented by Dial (2003) support gliding being favorable in open habitats, thereby corroborating the liana density hypothesis.

A true test of the locomotor economy hypothesis would require experimental measurement of the cost of locomotion and its comparison with the overall energy budget of the animal over a prolonged period of time. To this end, Flaherty et al. (2010) used an open-flow respirometry technique to calculate the resting and running energetics of the northern flying squirrel (*Glaucomys sabrinus*) and estimated that 64% energetic savings were accrued by gliding. Flaherty et al. (2010) showed that the net cost of quadrupedal locomotion for flying squirrels was significantly greater than the allometric estimates predicted, with the metabolic rate for horizontal travel increasing exponentially with velocity, and that extended bouts of quadrupedal locomotion in flying squirrels were energetically expensive compared to gliding. Studying a gliding mammal from a different order, Byrnes et al. (2011)

tackled the locomotor economy hypothesis by tracking gliding behavior in wild colugos using on-body sensors, providing an indirect estimate of overall energy budget during extended durations of time. Comparing bouts of climbing versus gliding showed that the energy required to climb was ~1.5 times greater than moving an equivalent glide distance using quadrupedal locomotion. The study further argued that the locomotor ecology of mammalian gliders shows that less than 10% of the animal's daily energy expenditure is expended in climbing, reducing its selective influence on the evolution of gliding in animals.

An alternative approach in support of the locomotor economy hypothesis and the liana density hypothesis is presented by Lambert and Halsey (2015), who argue that because gliding has evolved in an arboreal context, the added energetic cost of climbing has already been evolutionarily 'paid' and should not be considered in testing the locomotor economy hypothesis. Instead, if the energetics of an animal climbing and traversing horizontally (either through the treetops or by descending to the ground) are compared with the energetic cost of gliding the same horizontal distance, gliding is in fact an energetically efficient means of locomotion, especially in relatively open, liana-free environments.

Although there are opposing lines of evidence for the locomotor economy hypothesis, it remains unclear whether gliding was sufficiently efficient as a means of locomotion to act as a selective pressure for evolutionarily intermediate forms of gliders. Lambert and Halsey (2015) present an example wherein the benefits of gliding energetics are gained only after the animal is able to glide; in other words, the animals representing the intermediate phases that led to gliding animals would still drop to the forest floor and not have a selective energetic advantage compared to their conspecifics.

13.1.5 Predation

Emmons and Gentry (1983) proposed that predation is likely to be a selective pressure relating to the evolution of gliding (Fig. 13.2c). The three-dimensional structure of arboreal habitats provides the possibility of escaping in a variety of directions, potentially adding an element of uncertainty for the predator (Byrnes & Spence, 2011). Furthermore, falling followed by gliding can be a faster escape response compared to jumping (Socha, 2006) and/or sprinting. Because arboreal predators have different hunting strategies and may use different sensory cues, the escape benefits of jumping/escaping and gliding would largely be determined by the composition of predators present when gliding evolved (Lambert & Halsey, 2015). Gliding also allows the animal to avoid the forest floor or small water bodies and, in turn, the majority of non-volant predators. The rapid nature of a glide minimizes travel time, reducing the time gliders are exposed to potential predators (Byrnes & Spence, 2011; Dudley et al., 2007). In support of the predation avoidance hypothesis, flying animals have been shown to have lower mortality rates than non-volant animals (Pomeroy, 1990), and arboreality has been linked to greater longevity of animals (except for marsupials) compared to their terrestrial relatives (Shattuck & Williams, 2010). Holmes and Austad (1994) compared the lifespans of gliding mammals with those of non-volant eutherians and showed that gliding mammals have significantly greater longevity. However, Stapp (1994) has urged caution with regard to the approach used by Holmes and Austad (1994) in comparing life spans of gliding and non-gliding animals, noting that a nocturnal and gliding lifestyle may not offer additional protection from predators relative to that already achieved by being arboreal and cavity-nesting (Fan & Jiang, 2009; Kavanagh, 1988 Stapp, 1994). For example, predation has been reported on giant flying squirrels by black-crested gibbons (Fan & Jiang, 2009) and on marsupial greater gliders by the powerful owl *Ninox strenua* (Kavanagh, 1988). The latter study suggested a staggering decline of ~80% in the marsupial greater glider population over a period of 46 months due to predation (Kavanagh, 1988; Fan & Jiang, 2009).

13.1.6 Foraging

The foraging efficiency hypothesis (Fig. 13.2e) has been addressed mainly through studies of mammalian gliders. The ability to glide allows animals to access patchily-distributed food and to forage over a larger area in a smaller amount of time, thereby potentially providing a selective advantage to gliding behavior compared to non-gliding arboreal animals. Byrnes and Spence (2011) showed that gliding originated independently in six lineages of mammals. Gliding evolved twice each in ancestors with folivorous, frugivorous, and exudivorous diets. Of these types, there were significant associations between changes in binary character (gliding versus non-gliding) for both exudivory and folivory. No significant association has been found between gliding and frugivory, despite gliding having evolved twice in ancestors exhibiting frugivory (Byrnes & Spence, 2011).

13.1.7 Falling

An open-space environment, gliding efficiency, predatory pressure, and foraging: all of these are possible selective factors relating to the evolution of gliding. However, it is unclear how each of these factors would have benefited the intermediate stages of gliding that preceded extant gliders. For example, for traversing between spatially distant trees efficiently, the lack of a well-evolved gliding apparatus and physiology could result in the organism falling to the ground in an uncontrolled fashion, making it prone to injuries and exposure to predators. Furthermore, the inability to glide long distances would prevent the organism from accessing sparsely distributed food resources.

Paskins et al. (2007) and Lambert and Halsey (2015) argue that controlling accidental falls in an arboreal habitat (Fig. 13.2d) would have provided a strong selective pressure for the evolution of gliding. Organisms with the capability of safely landing after slipping in an arboreal environment would have a survival advantage compared to those unable to do so. Furthermore, what might be considered as an only partial or small degree of morphological specialization for gliding can still augment drag forces that slow down the falling animal, as has been observed for extant gliders with only small skin folds (Russell, 1979). Controlling accidental falls so as to land safely and reduce injury is likely of high relevance to intermediate forms of gliders. Caple et al. (1983) showed that an increase of up to 5% in the total lift force (relative to body weight) would allow significant control of the body around the pitch and roll axes, both necessary for correcting body orientation during a fall. Furthermore, it is plausible that intermediate gliding forms were able to generate up to 5% of body weight as lift force and thereby benefit from controlling body pitch and roll during a fall. Paskins et al. (2007) used force measurements at takeoff and landing for flying squirrels to demonstrate that, in the absence of aerial control or the ability to slow down, flying squirrels would sustain impact forces up to 28 times their body weight. A study on flying frogs also emphasizes the importance of body control and suggests that aerial maneuverability, rather than horizontal glide distance, could be an important driving factor in the evolution of flying frogs (Emerson & Koehl, 1990).

13.1.8 A Note on the Evolutionary Arguments for Gliding

As previously mentioned, it is highly likely that more than one selective pressure played a crucial role in the convergence of gliding-related morphologies and behaviors that we see today. Furthermore, the majority of arguments put forth in support of one or more of the hypotheses about the evolution of gliding are informed by studies and/or observations conducted largely on mammalian gliders (but see Dudley & Yanoviak, 2011; McGuire, 2003; McGuire & Dudley, 2005). Few studies describe or provide anecdotal evidence in support of the evolution of gliding hypotheses in gliding animals apart from mammalian gliders, highlighting our lack of understanding of the ecological and biomechanical context of the majority of extant gliding animals (Dudley et al., 2007; Emerson & Koehl, 1990).

Aquatic aerial gliders (flying fish and flying squid) have received even less attention. Like all of the other non-mammalian gliders, studies have been limited in number due to constraints relating to experimental methodology and data collection, which present particular difficulties for these open-water oceanic groups. Davenport (1994) described the evolution of gliding in flying fish in regard to the combination of enlarged pectoral fins and an asymmetric caudal fin. Together, these adaptations allowed the fish to swim near the surface in the high-drag zone where it could extend its pectoral fins to lift the front part of its body above the water surface, further reducing drag and, in conjunction with the ground effect, enable it to enact high-speed 'skittering' on the water surface. Taxiing on the water surface was thus a possible precursor to gliding flight seen in present-day flying fish. Here, we briefly mention two hypotheses that have been put forth for the evolution of gliding in flying fish and are likely applicable to flying squid.

The locomotor economy and predation avoidance hypotheses are generally used to explain the evolution of gliding flight in flying fish and squid (Davenport, 1994; Xu et al., 2013; Rayner, 1986). Explanations that undergird the predation avoidance hypothesis include the ability of the animal to exit the water and pass into the air, thereby resulting in the predator losing a clear line of sight and being faced with uncertainty about the point at which the potential prey might re-enter the water. However, observations of flying fish exiting the water surface in the absence of obvious sub-surface predators (Davenport, 1994), along with video recordings of both fish and birds catching flying fish mid-air and tracking them from beneath the water (Davenport, 1994; BBC footage: https://www.youtube.com/watch?v=bk7McNUjWgw), suggest that there could be multiple ecological reasons for the fish taking to the air. Rayner (1986) suggested that periodic flights of flying fish could be an energy-saving strategy, especially over long distances of travel. However, Davenport (1994) argues that the fish's muscle that likely contributes to the high-speed takeoff from the water surface is energetically expensive, and its use is not economical for frequent short-duration aerial glides.

13.2 Morphology

The vast majority of gliding animals have evolved morphologies that increase their total body surface area compared to their non-gliding counterparts, perhaps the largest convergence demonstrated amongst these flyers. The combined effect of change in total body surface area and its interaction with the surrounding air during descent allows the animal to generate lift and drag forces to control its glide trajectory. Gliding-specific morphologies range from dedicated and novel primary wing structures in mammalian gliders and flying lizards, to modifications of pre-existing body parts in ants, spiders, and flying snakes. Some taxa exhibit multiple specializations for gliding, combining flattened appendages and a primary wing membrane, as is the case in *Draco* flying lizards, further enhancing their gliding capabilities. To a lesser extent, flying frogs and geckos employ skin flaps and webbed feet that function to increase the total body surface area.

13.2.1 Membrane Wings

Membrane wings (often referred to as 'patagia'; a single membrane is a patagium) have convergently evolved in mammals, reptiles, amphibians, fish and cephalopods. Among mammals, the membrane wing has evolved at least six times, perhaps suggesting that the genetic modifications required to evolve a patagium are not complex. The developmental and genetic underpinnings of wing membranes in mammals is a rich target for research, particularly given the possibility that similar changes may underlie the evolution of flapping flight in bats (Bishop, 2008).

Membrane wings in reptiles are found only in *Draco* flying lizards, and in invertebrates, only in flying squid. Across these taxonomic groups, the membrane wings are extended/stretched once airborne to generate aerodynamic forces of lift and drag, and begin to retract just before landing, followed by complete retraction after the end of the glide. Mechanically, there is a distinction between the supported mammalian membrane wing and that of most other gliding wings, distinguished by the inclusion of some stiffened strut. In *Draco* lizards, frogs, and fish, the struts are bony elements (ribs or toes), whereas in squid they are tentacles. These struts provide additional mechanical integrity, but they also form small ridges on the surface that possibly influence the wings' aerodynamics. Unlike all the rest, the flying geckos (genus *Gekko*, previously *Ptychozoon*) possess lateral skin flaps on the trunk (Russell et al., 2001), which can be considered as unsupported membrane wings of low aspect ratio (the ratio of the wing span to the wing chord).

Mammalian gliders have a propatagium between the head and the forelimbs, a patagium between their forelimb and hindlimb, and variably a uropatagium extending from the hindlimbs to the tail (Panyutina et al., 2015). In colugos, an extensive uropatagium extends from the hindlimbs and incorporates the entire tail, contributing to the colugo's relatively larger aerodynamic surface compared to that of all other gliding animals. Anatomically, the membrane wings are only supported by distal attachments to the animal's body. For example, the patagium is supported at the leading edge, trailing edge, and the base of the wing where it joins to the body. Because there are no intrinsic muscles in the membrane (such as the plagiopatagiales proprii muscles in bats; Cheney et al., 2014), changing the shape of the membrane wing requires the animal to reposition its limbs and/or other body parts that connect to the membrane.

Squid possess two separate flight surfaces that help them glide once they jet beyond the water surface. The first is a pair of wings formed by the fins positioned towards the anterior of the body, and the second is formed by the spreading of tentacles along the posterior part of the body, with some sort of membrane between them. Azuma (2006) suggested that the tentacle/membrane wing provides greater lift force than does the fins. Squid also stiffen their mantle using muscle contraction during jetting, facilitating pitch-up control of the body. When re-entering the water, squid fold their fins and tentacles, which likely serves to streamline the body and reduce drag while entering the water surface.

For *Draco* flying lizards, the primary wing consists of a patagium supported by elongated thoracic ribs operated by specialized musculature (Colbert, 1967; Russell & Dijkstra, 2001). The opening and closing of the patagium are controlled by the combination of highly modified external and internal intercostal muscles, external oblique muscles, and a system of ligaments spanning the individual elongated thoracic ribs. The muscle actuation to rotate the ribs laterally is greatest for the first two anterior ribs. The large actuation to pull the first rib forward is achieved by the iliocostalis muscle extending far anteriorly to its origin on the anterior ribs and vertebrae. The great length of the muscle is necessary to allow a degree of contraction sufficient to rotate the ribs far out to completely stretch the patagium. The anteriorly located second rib has a similar but less extensive development of the intercostal muscle. The remaining ribs are pulled anteriorly more by the ligaments that interconnect the ribs and less by the musculature, thereby completing the anterior extension of the ribs that consequently stretches the patagium open.

Mammalian gliders and *Draco* flying lizards are both proficient gliders that possess a specialized primary wing membrane to glide. However, the difference in the patagial anatomy between mammalian gliders and *Draco* flying lizards might also have an influence on the climbing speed and gait on trees in their natural habitat. The lack of connection between the patagium and the fore and hind limbs in *Draco* flying lizards might provide a higher degree of unconstrained motion compared to mammalian gliders.

The fully webbed feet of flying frogs and geckos also form membranous, bone-supported wings, significantly increasing the total lifting surface of the animal. The fully webbed feet of these gliding animals act as control surfaces that can easily be repositioned during descent to control the glide trajectory (Emerson & Koehl, 1990).

13.2.2 Skin Flaps

Gliding animals including flying frogs and flying geckos have evolved additional skin flaps instead of a primary wing membrane, along with fully webbed hands and feet. For example, flying frogs (genus *Rhacophorus*) have additional skin flaps on their legs compared to non-flying frogs (Inger, 1966). Flying geckos (*Gekko*) have cutaneous flaps extending laterally on either side of the body, and smaller ones around the hind limbs (Young et al., 2002).

In addition to the main membrane-strut wing, *Draco* flying lizards also have an additional set of lappets, small triangular structures on the lateral margins of the head that also extend immediately after takeoff and are retracted after landing. Because the surface area of these lappets is much smaller than the primary wing, their relative aerodynamic role is likely much smaller. However, their location far forward of the centers of mass and aerodynamic pressure mean that they could be influential in pitch control. The structure and function of lappets is potentially similar to the canards found in present-day aircraft in providing stability during gliding, but specific work is needed to understand this unique aspect of the lizard's morphology, which may also function to delay stall at high angles of attack (Wibowo et al., 2018).

13.2.3 Flattened Body

For some gliders, the dorsoventrally flattened body itself provides the main aerodynamic force-generating surface. A flattened body employed in gliding has convergently evolved in flying snakes (Socha, 2011), the lizards *Leiolepis belliana* (Losos et al., 1989) and *Holaspsis guentheri* (Arnold, 2002; Mertens, 1960; Schiøtz & Volsøe, 1959; Vanhooydonck et al., 2009), and the gliding spider *Selenops*

(Yanoviak et al., 2015). These taxa either employ behavioral flattening upon becoming airborne (discussed in detail in sect. 13.3.3) or are morphologically flattened at rest. *Draco* flying lizards may also exhibit a small degree of morphological flattening at rest, but this feature has not been well characterized within the genus. Recently, wandering salamanders (*Aneides vagrans*) have been observed to slow their aerial descent and change directions in a wind tunnel (Brown et al., 2022). This work suggests that they might glide in the wild, an ability attributed to their relatively flattened body, long limbs, and large feet, which were held in a skydiving posture in this experiment.

The flattened body of gliding spiders of the genus *Selenops* is also characteristic of the entire family *Selenopidae*, known commonly either as 'flatties' or wall crab spiders. Based on estimates from published images (Crews, 2011; Yanoviak et al., 2015), their degree of flatness is ~3.4 (length/depth) and ~1.5 (width/depth), contributing to the spider's steep glide trajectories, with glide ratios (horizontal/vertical travel) ranging from ~0.2 to ~1. The *Selenopidae* includes nearly 300 species, most of which have not been investigated for glide performance. Given that the flattened morphology is widespread amongst the group, which includes both arboreal and terrestrial taxa, this group may be amenable for testing evolutionary hypotheses about the acquisition of features related to gliding.

13.2.4 Limbs

Although limbs can be thought of as simple cylinders that mostly induce drag, their orientation and location can influence the centers of mass and aerodynamic pressure, and thus the balance of forces and torques on the glider. A wide range of terrestrial gliders have limbs that are unattached to membranes, including *Draco* flying lizards, flying frogs, and arthropods that glide. The effect of leg posture has been explored for some species. Physical modeling of the flying frog *Rhacophorus nigropalmatum* demonstrated that the positioning of the legs and feet have a strong influence on glide performance (Emerson & Koehl, 1990; Emerson et al., 1990; McCay, 2001), but the specific aerodynamic role of the legs (or any other part of the anatomy, for that matter) has not been studied. The gliding ant *Cephalotes atratus* has front legs that are dorsoventrally flattened (Yanoviak et al., 2005), but as with frogs, the aerodynamic effects of the legs have not been studied.

13.2.5 Sensory Structures

There is a paucity of studies on the role of sensation and sensory feedback specific to gliding animals. Consider the wings of bats: the membrane is very highly mechanosensitive (Swartz & Konow, 2015), with an average sensitivity greater or equal to that of skin of the human fingertip (Chadha et al., 2011). Only recently in bats have small, sub-millimeter projections on the skin surface been identified to function as specialized airflow sensors, with directional asymmetry favoring sensitivity to flow from the leading edge to trailing edge (Sterbing-D'Angelo et al., 2011). There have been few reports of equivalent structures in gliding animals; the cutaneous mechanoreceptor sensilla found in the patagia of *Gekko kuhli* (Russell et al. 2001) may be one such candidate. However, gliding animals employ flight far less often than flapping flyers, suggesting that selection pressures that result in greater mechanosensory feedback are probably much lower in gliders. Alternatively, our lack of knowledge of gliding-specific sensory structures may reflect a dearth of focused studies. One starting point would be to examine the mechanical sensitivity of skin in gliders like lizards, snakes, and frogs to determine the spatial and velocity resolution of airflow that they can detect; such mechanosensitivity could be employed for feedback control.

Flying lizards and snakes are strictly diurnal compared to their mammalian counterparts. Therefore, it is likely that vision is a major sensory modality. For *Draco* flying lizards, Khandelwal and Hedrick (2020) modelled the glide trajectory based on a vision-based obstacle avoidance model and showed that real trajectories in the natural environment agreed well with the model's predictions. Their data also demonstrated that the initiation of landing behavior followed a relative retinal expansion velocity (RREV) model, previously seen in flapping flyers ranging from birds to insects (Lee & Reddish, 1981; Wagner, 1982). Furthermore, during landing, these flying lizards used a controlled collision approach to reach their desired target (Lee et al., 1993).

The vision of flying snakes has been preliminarily characterized in a recent study (Zamore et al., 2020). Using an optokinetic drum akin to that used in experiments conducted on flying insects (e.g., Mongeau et al., 2019), *Chrysopelea paradisi* was shown to possess a visual system that is broadly tuned and motion sensitive. These snakes also held an image fixed (i.e., maintained the image in view despite its movement), consistent with the observation that the head moves least of any body part during the snake's highly dynamic undulatory glide behavior (Socha et al., 2005, 2010; Yeaton et al., 2020). Three-dimensional head scanning and inverse perimetry (a method of ray tracing) suggest that the snake possesses a small region of binocular overlap in front of the snout, but that a much larger region of overlap extends above the head, which may be used for overhead predator detection (Socha & Sidor, 2005). Lastly, these snakes exhibit a sensitivity to light roughly equivalent to twilight, consistent with their ecology as diurnal gliders.

Although mammalian gliders are nocturnal, they too primarily rely on vision to glide in their natural habitat. Flying squirrels are known to lack the ability to form functional opsins in the eye, rendering them color-blind, but this visual trait may increase their spatial and temporal resolution for night-time vision (Carvalho et al., 2006; Niven, 2006). However, it is unclear if the trajectories of nocturnal gliders can be described using similar models as those employed for *Draco* flying lizards.

Because stability in the air is critical to successful gliding, the vestibular system has been recognized as a target of specialization (Boistel et al., 2011), capable of detecting accelerations in all three rotational axes (pitch, roll, and yaw). Boistel et al. (2011) used microtomography to study the three-dimensional morphology of the inner ear of many gliding and non-gliding lizards and snakes, and found differences between gliders and those species that have only a climbing or terrestrial lifestyle. However, these differences did not involve the size of the semi-circular canals, as might be expected, but rather the angles of the connections between parts of the vestibular system. The mechanistic linkage to stability remains a target for future examination.

13.3 Behavior

Although gliding animals employ a range of behaviors that allow them to become airborne, change trajectory in air, and land, these behaviors typically involve actively changing the body posture, shape, and size, as well as recruiting other body parts during the process. Convergence in behaviors likely results from the limited number of body configurations, within the constraints of the morphology, that can generate sufficient aerodynamic forces for gliding.

13.3.1 Takeoff Maneuvers

Most gliding animals begin a glide with a takeoff, using thrust produced by the animal. Land-based gliders produce thrust by pushing from the substrate using their limbs or their body (in the case of flying snakes). Flying squid and fish generate thrust by pushing against water using jetting and body undulation, respectively (Davenport, 1994; Muramatsu et al., 2013; O'Dor et al., 2013). For arboreal gliders, the takeoff maneuver provides multiple advantages over simply dropping into the air: it can determine the direction of the glide, it transitions the body into a posture favorable for generating sufficient aerodynamic forces, and it can influence the net dynamics of the glide (Khandelwal & Hedrick, 2020).

13.3.1.1 Jumping Takeoffs

For mammalian gliders and lizards, jumping involves using the limbs to push off from the substrate and reorienting the long axis of the body so that it is almost parallel to the ground. In an arboreal environment, takeoffs can start from a vertical position on a tree trunk (Dehling, 2017; Khandelwal & Hedrick, 2020) or a more horizontal position from branches (Essner, 2002), although vertical takeoffs seem to be the predominant mode used in the wild. *Draco* flying lizards have been shown to modulate the direction of their jump based upon the complexity of the habitat in both the horizontal and vertical plane, to accommodate for obstacles in their path and the location of their desired landing tree (Khandelwal & Hedrick, 2020).

A typical takeoff from a vertical position, frequently performed by mammalian gliders (Fig. 13.3a) and *Draco* flying lizards, includes aerial twisting (Dehling, 2017; Panyutina et al., 2015). For these animals, takeoff requires them to generate adequate

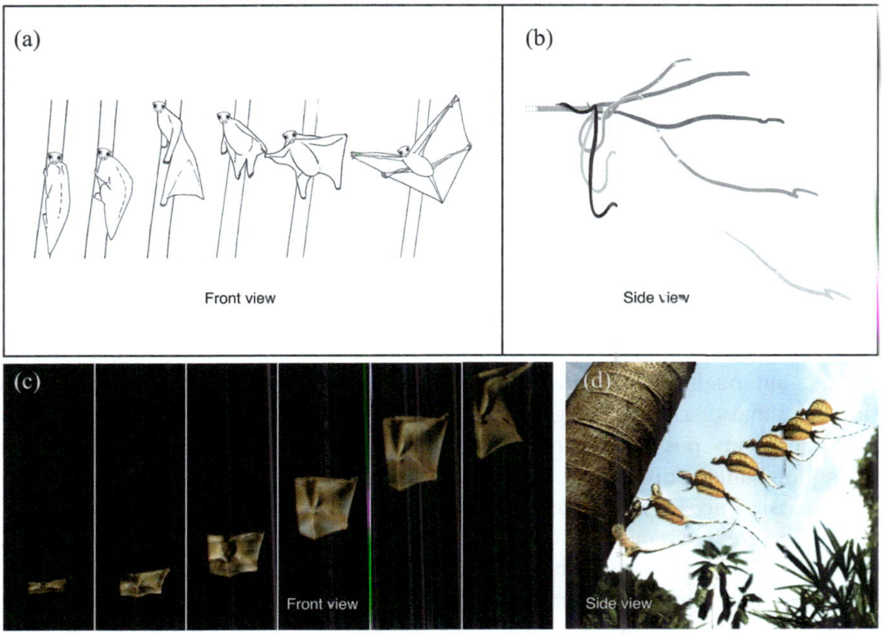

Fig. 13.3 Takeoff and landing in arboreal gliders. Mammals and *Draco* flying lizards employ a jump with a twist when taking off from a vertical tree trunk, illustrated in front view (**a**) in a colugo (modified with permission from Panyutina et al., 2015). The main takeoff used by flying snakes is the J-loop jump, shown in side view in (**b**) [modified with permission from Socha (2011)]. To land on a vertical substrate, mammals and *Draco* lizards enact a pitch-up maneuver that serves to reposition the legs for grasping while also avoiding direct contact of the head. The sequences in (**c**) are of a colugo (*Galeopterus variegatus*) and (**d**) a flying lizard (*Draco maculatus*) recorded by National Geographic television cameraman John Benam. Reproduced with permission from Socha et al. (2015)

thrust, deploy their wings, and correct their body orientation immediately after launching towards the landing target (Khandelwal & Hedrick, 2020). To accomplish this maneuver, the animal uses a combination of limb and tail movements along with dynamic wing and body morphing. *Draco* flying lizards rotate from a vertical to a horizontal orientation on the tree trunk and use their hind limbs to thrust themselves in the direction of the landing tree, accelerating at ~9 ms^{-2} and reaching a velocity of ~2.5 ms^{-1} by the time of complete wing deployment (Khandelwal et al., 2018). The main wing is extended and held in position independently of the limbs during takeoff. The forelimbs extend from the body and eventually are held parallel to the leading edge of the wing. Photographic evidence has been used to suggest that the hands grip the wing and aid in its deployment and modulation of aerodynamic force production (Dehling, 2017), but further experimental work is needed to test this hypothesis; the wrists may simply rest on the upper surface of the wings. Furthermore, pronounced tail movement along with changes in body roll, pitch, and yaw occur during takeoff, suggest that the tail may help control body orientation

(Khandelwal et al., 2018). *Draco* flying lizards complete their aerial twisting as they enter the ballistic dive phase of the glide.

The force generated via jumping has been recorded for flying squirrels (*Glaucomys sabrinus*) and the Malayan colugo (*Galeopterus variegatus*) (Paskins et al., 2007; Byrnes et al., 2008). Using on-body accelerometers, colugos were observed to generate takeoff forces ranging between 1 and 13 times their body weight (Byrnes et al., 2008). Flying squirrels generated takeoff forces ranging from 1.08 to 9.57 times body weight when measured from force sensors mounted on a horizontal platform (Paskins et al., 2007). In a semi-natural study of glide trajectories of flying squirrels (*Glaucomys sabrinus*), the launch velocity was observed to be 5.4 ± 1.0 m s^{-1} (Bahlman et al., 2013).

Detailed body kinematics during takeoff have not been studied in mammalian gliders, but qualitative video observations suggest execution of aerial twisting similar to that of *Draco* flying lizards, including reorientation from a vertical to a horizontal body position, followed by a jump and the spreading of the patagium during the ballistic dive phase of the takeoff.

Possessing no legs, the takeoffs of flying snakes are distinctly different (Socha, 2006). The predominant mode of beginning a glide is the J-loop jump (Fig. 13.3b), in which the snake hangs from a branch, accelerates upward with the anterior portion of the body while gripping the branch with the tail and hind end of the body, and then releases the branch as it arcs up and away from it. The anterior portion of the body that participates in the loop undergoes an axial twist, such that the ventral surface faces to the side, but the head is oriented in a more horizonal plane. As the snake enters the air it releases its grip on the branch and untwists, contorting the body to bring the ventral surface into a horizontal orientation as it begins to undulate. A variation of this takeoff is the sliding J-loop takeoff, which entails a smaller loop and no static grip on the branch; instead, the body is in continuous motion throughout the takeoff. Both of these dynamic takeoffs entail maximum jumping accelerations of ~20 ms^{-2}. In experimental glide trials, snakes were also observed to slide or deliberately fall off the branch (Socha, 2002; Socha et al., 2005, 2010; Yeaton et al., 2020). With little or zero preparatory movements or initial horizontal velocity, such takeoffs resulted in shorter glides than those initiated by jumps, and might be used as escape behaviors in the wild (Socha, 2006).

13.3.1.2 Launching from the Water

Flying fish and squid both shoot out of the water at an angle to begin their aerial trajectory (Davenport, 1994; Muramatsu et al., 2013). The many species of flying fish differ in their approach once emergent. Some fish gain height and begin gliding immediately, whereas others perform an intermediate tail-beating behavior, whereby the tail oscillates rapidly in the water with the body angled upward in the air. After a short tail-beating bout at the surface, the fish leaps into the air, deploys its fin-wings, and begins gliding. Unlike flying fish, which use undulatory propulsion to generate thrust and takeoff from the water surface, flying squid employ jet propulsion. Squid fill their mantle with water and can completely empty it in less than 3 s (O'Dor, 1988), ejecting it at velocities up to 3 ms^{-1} through their funnel (Alexander, 2003) as they exit the water at a low angle, and continue to rise in the air before gliding at a slight pitch-up orientation. (Muramatsu et al., 2013; O'Dor, 2013; O'Dor et al., 2013). The squid ejects water only during emergence and the initial phase of the aerial trajectory, which may serve both to provide an impulsive force and to empty the body of water, reducing mass and wing loading (the ratio of body weight to airfoil area) for its glide, and ultimately resulting in a more energetically favorable force-to-weight (power-to-weight) ratio.

Even though squid move less water than fish, they can achieve similar thrust by accelerating less water at much higher speeds. Furthermore, squid can exceed the swimming threshold relative to body length of fish (~25 body lengths s^{-1}) and achieve speeds up to ~49 body lengths s^{-1} (Muramatsu et al., 2013). While airborne, squid benefit from jetting water to generate propulsion in a medium that induces much less drag than in water.

13.3.2 Tail Movement

Mammalian gliders, *Draco* flying lizards, and flying snakes exhibit tail movements during some or all components of the glide (Socha et al., 2015; Khandelwal et al., 2018). As a long appendage, the tail may provide both inertial and aerodynamic functions that influence the glide. Tails essentially consist of a flexible vertebral column surrounded by muscle, providing the capability for active control. However, as a long thin appendage, the tail is also subject to passive fluid-structure interactions, and so it is unclear how tail kinematics reflect active versus passive effects for these gliders. From video recordings, the tails of flying snakes give the appearance of passive and somewhat haphazard movements that result from the body's undulation (see supplementary video, Socha et al., 2010), and mammals and *Draco* flying lizards appear to exhibit more deliberate control (unpublished data, Khandelwal, P. C.). However, no direct electromyographic evidence exists to quantify the role of muscle activation.

The tail might be used for controlling movements during all three major components of the glide: takeoff, gliding, and landing. In *Draco* flying lizards, rapid tail movement can be seen during takeoff while the animal reorients, such that its ventral side faces the ground. This reorientation is likely causally linked to the tail. Previous work on the gliding gecko *Hemidactylus platyurus* (previously, *Cosymbotus platyurus*) demonstrated the use of tail as an inertial appendage for reorientation while falling (Jusufi et al., 2008, 2010). *H. platyurus* rotates its tail during the fall and, taking advantage of the conservation of angular momentum reorients itself from ventral side up to the ventral side facing the direction of the fall.

Body reorientation may be a common feature of all gliders, part of an aerial righting reflex that adjusts the posture of the animal to one in which the dorsal side is uppermost and is advantageous for gliding or preventing injury during a fall (Jusufi et al., 2011). During a glide, small tail movements could be used to effect changes in roll, pitch, or yaw; however, their use to control stability or the direction of the glide trajectory remains largely unexplored. During the landing phase, the tail of *Draco* flying lizard is positioned closer to the dorsal side of the body, which might facilitate attaining an upright body pose to land on a vertical surface such that the forelimbs make contact with the tree first. Recently, it has also been hypothesized that the tail might be used to modulate the center of mass with respect to the center of aerodynamic pressure to initiate pitch control in gliding animals (Clark et al., 2021). In simulations that modulated tail position to maintain a fixed angle of attack (the angle of incidence between the wing and the oncoming air/wind), the *Draco* flying lizard glided nearly twice as far than without tail control.

Some tails move but are far more constrained anatomically. The tail of the colugo is connected to the hindlimb by an additional skin membrane, which drastically reduces the degrees of freedom available for tail motion. Nonetheless, high frequency oscillations of the tail of colugos associated with landing have been noted (Socha et al., 2015). A previous study of colugos quantified glide kinematics using an on-body sensor to measure acceleration, but it was not possible to measure postural adjustments performed during gliding (Byrnes et al., 2011). Future studies should use a combination of motion capture and inertial sensing to investigate and compare the use of the tail in colugos and flying squirrels/sugar gliders.

13.3.3 Body Flattening

As mentioned previously, some gliders flatten their body facultatively to create a greater surface area for generating aerodynamic forces. Flying snakes display both the greatest degree of dorsoventral flattening and reliance on this behavior to enable gliding. The snake transforms from a round to flattened cross-sectional shape as it jumps into the air (Socha, 2006). Flattening occurs sequentially from snout to vent within 0.5 s, resulting in a triangular cross-sectional shape with the apex oriented dorsally and a flat base ventrally, with two small ventral protrusions at the lateral edges of the base (Socha, 2011). Flattening occurs by rotation of the ribs both dorsally and anteriorly. The tail does not have ribs and hence does not flatten. The body is not uniformly wide, and as it narrows posteriorly toward the vent, the ventral protrusions may be less prominent or even absent. The cross-sectional shape of the paradise tree snake (*Chrysopelea paradisi*) has been determined qualitatively from stereo photos (Socha, 2011), but more rigorous imaging is required to determine its shape quantitatively. The net effect of dorsoventral flattening of the snake results in an increase of the width (up to 2×) and a decrease in the height of the body, creating a shape that can generate favorable lift-to-drag ratios (Holden et al., 2014; Jafari et al., 2021; Krishnan et al., 2014) within the aerodynamic regime of the gliding snake.

Aside from flying snakes, a few other reptiles use body flattening. However, the effect seems to be small in comparison. The butterfly lizard *Leiolepis belliana* flattens dorsoventrally, increasing its surface area by 9% (Losos et al., 1989). In experimental trials, its descent speed was found to be slowed by only 5%, a performance difference that suggests that while used when gliding, flattening did not evolve as an adaptation for gliding (Losos et al., 1989). In fact, this lizard may not even glide in the wild. Instead, flattening was suggested to function for thermoregulation, communication, or escape. The sawtail lizard *Holaspsis guentheri* may flatten more than the butterfly lizard, and also glide in the wild (Schiøtz & Volsøe, 1959), but its gliding has not been well studied; it appears to use flattening primarily to slow its descent (Vanhooydonck et al., 2009).

13.3.4 Airfoil Modification

Most gliding airfoils in the animal kingdom are soft, actuated, and compliant. These features allow changes in various airfoil properties including airfoil camber (curvature), surface area, and dihedral, the upward angle of the wing relative to the horizontal. Because a glider's airfoil typically offers fewer degrees of freedom than the flapping wings of birds and bats, airfoil modification is one of the primary ways by which the animal can modulate aerodynamic force generation during the glide. Such modulation is particularly crucial for gliding animals in spatially cluttered forest habitats for maneuvering around obstacles or performing evasive maneuvers in flight.

13.3.4.1 Camber Adjustment

The camber of a wing describes its curvature and shape, influencing the way it redirects airflow and therefore how it produces force. Typically, increasing camber increases force production, providing gliding animals with a mechanism for modulating flight performance (Pelletier & Mueller, 2000). Changes in camber during a glide have been observed for *Draco* flying lizards (Khandelwal, 2021; Khandelwal & Hedrick, 2022) and mammalian gliders (Bishop, 2006, 2007); however, it is unknown whether such changes are actively initiated by the animal or result passively from the material properties of the wing under aerodynamic load. The patagium of mammalian gliders and *Draco* flying lizards spans almost the entire length of the trunk and lacks muscles for actuation along the wing chord. Instead, it is possible that the animals use bending of the vertebral column to change the camber of the patagium.

Camber adjustment could be a way for gliding animals with low aspect ratio wings (<1.5) to increase aerodynamic force production at lower angles of attack, especially beneficial for animals inhabiting spatially cluttered environments (Pelletier & Mueller, 2000; Song et al., 2008). Wind tunnel tests have shown the advantage of a cambered airfoil for enhancing the force production ability of the airfoil; physical models of cambered airfoils of similar aspect ratio have shown changes of air flow on the curved surface, resulting in higher lift production at lower angles of attack (Shyy et al., 2008; Song et al., 2008). Furthermore, such a strategy can help enable a more robust gliding behavior through which the animal can actively adapt to changing environmental conditions. That is, the animal does not solely rely on the modulation of aerodynamic forces through changes in angle of attack, but can also tune force production by changing its airfoil camber (Khandelwal & Hedrick, 2022).

Draco flying lizards have been shown to actively modulate their camber during the mid-glide phase, positively and significantly affecting their coefficients of lift and drag (Khandelwal, 2021; Khandelwal & Hedrick, 2022). The advantage of changing body camber continues in the landing phase, allowing the animal to maintain sufficient lift while increasing the drag forces to slow down before touch-down on the desired target, potentially avoiding a full aerodynamic stall (Khandelwal, 2021; Khandelwal & Hedrick, 2022). Previous measurements made on flying squirrels, sugar gliders, and colugos have also shown their capability to modify camber in flight (Bishop, 2006, 2007; Panyutina et al., 2015; Stafford et al., 2002).

The flying snakes can also change the overall airfoil camber (Socha, 2011), but the morphing appears to be binary, from the rounded state pre-glide, to the flattened state as it enters the air (Socha, 2006). The flattened state is held until the snake lands. It is unknown if the snake is able to make small adjustments to its camber during flight, or if it returns to the rounded state pre-, during, or after landing.

13.3.4.2 Wing Area

The ability to change wing area can have benefits based on the desired aerial behavior, including a reduction in wing area. For example, reducing wing area can reduce drag and thereby increase airspeed.

In *Draco* flying lizards the patagium is not attached to the fore and hind limbs (Colbert, 1967), allowing lizards to potentially control wing opening during the takeoff phase of the glide. An advantage of such an approach could be to maintain a more streamlined body shape to increase airspeed and ultimately increase aerodynamic force when entering the mid-glide phase. This behavior can provide increased aerodynamic control earlier in the glide phase, which can be beneficial for maneuvering around obstacles in the glide path. Furthermore, anecdotal evidence also suggests the capability of these flying lizards to undertake asymmetrical wing opening during the takeoff phase (Khandelwal, P. C., unpublished data). Such control can lead to unbalanced aerodynamic force production, providing a mechanism for the animal to initiate roll response in the air.

Mammalian gliders have a patagium connected to the forelimb and the hindlimb (Jackson & Schouten, 2012). Here, wing area is modulated by changing the position of the fore- and hind- limbs; the patagium of gliding mammals is not supported by the ribs, in contrast to *Draco* flying lizards, nor does any other skeletal structure intrude into the web of the patagium. Furthermore, among mammalian gliders, colugos have an additional skin membrane connecting their hindlimbs with the tail

(Panyutina et al., 2015). The additional lifting surface may help to augment force production given the colugo's relatively large body mass, but the ability of the colugo to change this additional lifting surface is unknown.

13.3.4.3 Body Undulation

Flying snakes are the only gliders that employ large, full-body movements while gliding, an aerial undulation that gives the appearance of swimming through the air. Their net motion can be considered as the sum of two waves of motion, a large-amplitude horizontal wave and a much smaller vertical wave (Yeaton et al., 2020). In essence, the largest displacement occurs in the side-to-side direction, but vertical undulations also occur. Yeaton et al. (2020) first identified these small vertical movements, which are coupled at a 90° phase offset from the horizontal wave. Surprisingly, the vertical wave exhibits twice the frequency of the horizontal wave. Coupled horizontal and vertical waves of bending have only been identified in one other snake (sidewinders; Marvi et al., 2014), but at a 1-to-1 ratio of frequencies. The net effect of aerial undulation is to move the centers of aerodynamic pressure and mass in a way that increases the stability of the snake during gliding. In this way, the snake's aerial undulation functions in a unique fashion, both among gliding animals and all other locomoting animals.

13.3.5 Aerial Maneuvers

Maneuvering can be defined as any deliberate deviation from a constant straight path, representing any acceleration or directional change of the flying animal (Dudley, 2002). By this definition, it is likely that most gliders exhibit the ability to effectuate aerial maneuvers.

13.3.5.1 Pitch Control

The amount of aerodynamic force generated by a wing depends on the angle of attack, the angle of incidence between the wing and the oncoming air/wind. Most gliding animals have wings that are anatomically restricted and cannot be easily moved beyond the dorsal plane of the body (Endo et al., 1998; Gupta, 1966; Panyutina et al., 2015; Russell & Dijkstra, 2001; Russell et al., 2001). In such cases, varying the angle of the wing with respect to the oncoming airflow is linked to body orientation, specifically body pitch, which can be thought of as the nose-up or nose-down angle. Controlling body pitch thus plays a crucial role in allowing the animal to generate sufficient aerodynamic forces to perform a successful glide (Nave et al., 2019). Khandelwal and Hedrick (2022) showed *Draco* that flying lizards continuously change their body pitch to maintain an angle of attack of ~25° during the mid-glide, allowing them to operate close to their maximal lift-to-drag ratio. However, how gliders control pitch is not well understood. Some theoretical modeling has been conducted to understand pitch control in flying snakes (Jafari et al., 2014), and measurements of kinematics have demonstrated some correlates of pitch rotation with fine adjustments in sugar gliders. For example, movements of the forelegs away from the head were associated with nose-up rotations in sugar gliders (Bishop, 2007).

Flying squirrels, sugar gliders, and *Draco* flying lizards have a prominent tail that has been hypothesized to facilitate pitch control during gliding, whereby the tail is used as an inertial appendage to change the body pitch by changing its position in the sagittal plane of the body (Clark et al., 2021). Because aerodynamic forces are proportional to the square of the airspeed, the tail might play a larger aerodynamic role towards the latter part of the glide. However, the contribution of the inertial versus aerodynamic effects of the tail in pitch control remain untested.

Flying frogs and geckos (*Gekko*) possess enlarged webbed feet that can be used as aerodynamic control surfaces that can be moved independently to potentially change body pitch and consequently the body-airfoil angle of attack. In the absence of a tail, gliding ants, spiders, and frogs rely on drag-based limb movements to change the pitch of the body during gliding.

13.3.5.2 Rolls

Body roll—rotation around the transverse axis of the body—allows redirection of the aerodynamic lift force vector, enabling lateral movements in air. *Draco* flying lizards (Fig. 13.4a, b) and mammalian gliders have been observed to perform body rolls to change their glide direction (Khandelwal & Hedrick, 2020; Stafford et al., 2002). Stafford et al. (2002) observed that body rolls in the flying squirrel *Petaurista leucogenys* were induced by the adduction and ventral flexion of the forelimb, leading to asymmetrical left-right aerodynamic force production about the longitudinal axis of the body. Modelling of roll angle in *Draco* flying lizards has shown that they can maneuver around obstacles that lie directly in-line with their target by achieving a roll angle of approximately 21° while only losing roughly 7% of their total upward lift force (Khandelwal & Hedrick, 2020). However, the means by which they achieve body roll is unclear. It may be initiated by the use of the tail, or by the asymmetric extension of one side of the patagium relative to the other, or both.

Body rolls have also been investigated in the flying frog *Polypedates dennysi* using a wind tunnel (McCay, 2001). The tunnel was tilted to 45° to match the glide angle of this species, and the airspeed (12–14 ms^{-1}) was set so that the frog glided in place, enabling observation of body and limb orientation. Maneuvers were induced by revealing a plastic plant outside the tunnel that the frog would try to turn to reach. During rolls, the frog's feet were held at equal levels relative to the body, so it is unclear exactly how left-right force asymmetries were created. These frogs also exhibited a non-rolling maneuver called a crab turn, in which one rear foot is held higher than the other, inducing a yaw turn (a turn where the body rotates around the

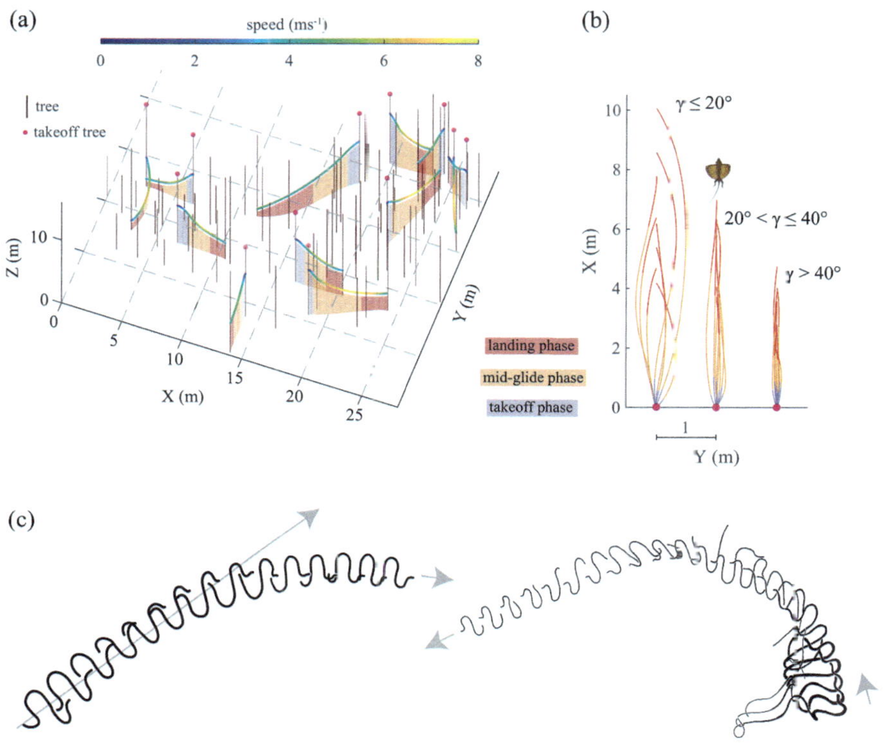

Fig. 13.4 Aerial maneuvering in *Draco* flying lizards and flying snakes. (**a**) Glides performed by the flying lizard *Draco dussumieri* in its natural environment. (**b**) Top view (X–Y plane) of glides grouped by the obstacle angles γ. The obstacle tree is defined as the tree closest to a straight glide path between the takeoff and the landing tree. The obstacle angle is the angle subtended by the obstacle tree on the takeoff tree in the X–Y plane. The figure highlights the maneuvering capabilities of the lizard around obstacles and the influence of obstacles on the glide path; the path curvature increases as the obstacle angle decreases. Note that the X and Y axes are not to scale in (**b**). Modified with permission from Khandelwal and Hedrick (2020). (**c**) Flying snakes appear to turn without banking. The sequences shown here were traced from video stills recorded at 60 Hz, depicting a gentle turn (left) and a sharper turn (right) that occurred just after takeoff. In both sequences, the snake appears to become smaller as it moves closer to the ground because recording took place using a top-view camera. Modified with permission from Socha (2011)

dorsoventral axis in the frontal plane). Both types of maneuvers are possible with only slight adjustments because the frog possesses only a small amount of positive stability in pitch and roll, and negative stability in yaw.

13.3.5.3 Drag-Inducing Limb Movement

All gliders experience drag forces while gliding, a consequence of frictional losses due to viscosity. Limbs can be considered as cylinders that create drag forces that influence the balance of forces on the animal. Changes in posture of the limbs can also be used to induce a moment that rotates the body in pitch, roll or yaw, thereby inducing an aerial maneuver. Gliders with free limbs include frogs, lizards, and arthropods. For frogs, the main gliding surface is the splayed webbed foot (including both fore- and hind- feet); therefore, postural changes of the leg also relocate the main aerodynamic surface, complicating analyses of the effects of leg movement. Gliding geckos (*Gekko*) present a similar situation, except that their foot surface is relatively smaller and therefore should have a weaker effect. Nonetheless, Young et al. (2002) have shown that bound feet in the flying gecko *Gekko kuhli* inhibited aerial maneuvers and prevented the gecko from entering the glide phase, demonstrating the importance of the feet as aerodynamic control surfaces. In contrast to the frogs and other gliding lizards, *Draco* flying lizard hangs its rear legs from the body; the front legs are held parallel to the leading edge of the wing. The relative role of the legs has not been rigorously studied in most gliding vertebrate taxa, including the movement of *Draco* flying lizards' front legs into the leading-edge position at the beginning of the trajectory.

For arthropods, legs represent a major aerodynamic surface. Asymmetrical leg movements in gliding ants result in the production of yaw turns, effectuated by a single hind leg raised dorsally on one side of the body (Yanoviak et al., 2010). Similar movements that lift the leg above the body axis are involved in aerial righting behaviors in ants (Yanoviak et al., 2010). In response to a threat, newly hatched stick insects (*Extatosoma tiaratum*) become airborne and tuck the legs against the body, which should reduce drag and increase speed through the air, helping to avoid predation (Zeng et al., 2020); these insects use a multitude of leg movements to change position in the air (Zeng et al., 2017). With eight independently actuated legs, gliding spiders possibly possess the greatest opportunity for aerial maneuvering using their jointed appendages. The gliding spider *Selenops* effects a stereotyped splayed posture in the air, with the two forelegs held laterally and anteriorly, and the remaining legs splayed posterolaterally (Yanoviak et al., 2015). Yanoviak et al. (2015) also report observing repeated twitching of the legs of *Selenops* while airborne, but the physical consequence of such vibration is unknown. They also note that these spiders steer with their forelegs; with the legs held anteriorly it is possible that such movements induce pitching moments as well. Overall, detailed kinematics and mechanical study are needed to understand how gliders use limb movements to manipulate drag forces.

13.3.5.4 Body Inertia

Despite the lack of obvious control surfaces, flying snakes are able to change direction in the air (Fig. 13.4c, d), an ability that likely varies with species. The paradise tree snake (*Chrysopelea paradisi*) has been observed, via kinematics studies, to turn while airborne (Socha, 2002; Socha et al., 2010; Yeaton et al., 2020); the golden tree snake (*Chrysopelea ornata*) has been reported to turn in one instance only, when takeoff was from a very high launch point (41 m; Heyer & Pongsapipatana, 1970), although no kinematic details were provided. Snakes appear to turn without banking. Although a rigorous mechanistic explanation of turning is still lacking, it is likely that snakes use an inertial mechanism, shifting body mass and taking advantage of principles of conservation of angular momentum. Yeaton et al. (2020) proposed that turns could be initiated using modifications to the large-amplitude horizontal waves during gliding, timed to occur when the inertial yaw moments are large.

13.3.5.5 Landing Maneuvers

Terrestrial gliding animals are strictly arboreal. Landing is potentially the most critical part of the glide and perhaps the most understudied aspect of gliding in animals. It involves the greatest risk of injury through impact with the arboreal or ground substrate, but it allows the animal to reach its target to perform various ecologically relevant behaviors. Generally, two types of landing are recognized: a pitch-up maneuver that orients the animal into a more vertical, head-up position prior to contact, or no maneuver at all (Socha, 2011), meaning that the animal essentially crash-lands with a direct impact.

Most glides of *Draco* flying lizards and mammals recorded in the wild have shown that they preferentially land on vertical tree trunks (Bahlman et al., 2013; Byrnes et al., 2008; Khandelwal & Hedrick, 2020; Krishna et al., 2016). To successfully land on a vertical substrate, these animals perform a landing maneuver involving large pitch-up rotation such that the ventral aspect of the body faces the vertical landing substrate (Fig. 13.3c, d), which serves both to move the legs and feet into a position favorable for gripping the substrate and also to avoid direct collision with the head. During a pitch-up rotation, the animal changes the angle of attack of the airfoil, causing a large increase in drag; if lift production ceases completely, aerodynamic stall occurs. This pitch-up maneuver allows the animal to expend kinetic energy in the form of drag, slow down, and perform a controlled landing. In *Draco* flying lizards, just before landing, the head is elevated toward the dorsal side and the forelimbs are extended forward, bringing them in contact with the landing substrate first (Khandelwal, 2021). Movement of the tail dorsally might also serve as a control movement. Overall, the mechanisms used by mammals and lizards for pitch-up landing maneuvers are not well understood.

In the absence of a primary enlarged wing surface, it appears that gliders including flying snakes and the gliding gecko *Hemidactylus platyurus* are unable to slow down sufficiently and/or rapidly to initiate a controlled landing maneuver (Siddall et al., 2021; Socha et al., 2005). Instead, they impact the landing substrate without significantly reducing their glide speed and use their body and/or tail to stick the landing. Recent work by Siddall et al. (2021) has shown that *H. platyurus* slams into the substrate and uses its tail to stabilize its landing. In experimental trials recorded in the forest, geckos approached the landing tree at speeds up to 7 ms^{-1} and crashed head-first with some part of the torso into the tree trunk. The high impact at landing led to a pitch-back movement with the torso bending dorsally, forcing the forelimbs to lose contact with the substrate. To prevent a fall, the gecko pushed its tail against the substrate to counteract the bending of the torso and stabilize the landing.

Landing in snakes has been described only qualitatively (Socha, 2011). When landing on the ground, they appear to contact the surface with the tail first and then 'roll' into the landing, with head contacting last. Such kinematics would serve to increase total impact time and reduce forces on the head. However, it is unclear if snakes preferentially use these landings when targeting the ground, or if there is a minimum trajectory distance needed to properly position the body for such a landing. Landing on arboreal substrates is even more enigmatic. Snakes appear to contact branches at a location somewhere along the body, and then the forward momentum of the remainder of the body results in a wrapping motion. It is not clear if or how flying snakes land on a vertical substrate.

13.4 Mechanics

Here, we provide an overview of the underlying physics that governs gliding in animals. Each animal has to counteract the pull of gravity by generating sufficient aerodynamic forces to slow and control its descent for the ecological task at hand. Specific aerodynamics result from a combination of body shape, size, and posture of the animal interacting with the surrounding air. Most or all of these factors can be actively modulated by the animal during the glide, which contributes to the greater complexity of analysis and prediction of mechanics compared to a rigid, fixed-wing glider of similar size moving through air. Convergent gliding morphologies and behaviors are likely dictated by these complex body-fluid interactions, with limited combinations of morphology and aerial behaviors leading to desirable aerodynamic output.

13.4.1 Aerodynamics

A large portion of the aerodynamic force produced by a gliding animal functions to support the animal's weight. As the animal glides, it experiences an airflow along the direction of the glide trajectory. The aerodynamic force is the resultant of the lift force, which acts perpendicularly to the glide path, and the drag force, a pressure- and friction-based force that acts to slow the glider along the glide path (Fig. 13.5). The cross-sectional shape and orientation of the airfoil plays a major role in determining the magnitude of lift and drag. The relatively large aerodynamic surfaces of mammalian gliders and *Draco* flying lizards allows them to generate significantly more lift than drag, resulting in relatively shallow glide trajectories (Fig. 13.6). However, smaller marsupial gliders, such as the feathertail gliders, have a much smaller aerodynamic surface along with a long feather-shaped tail, which must influence how they produce glide forces (Pridmore & Hoffmann, 2014).

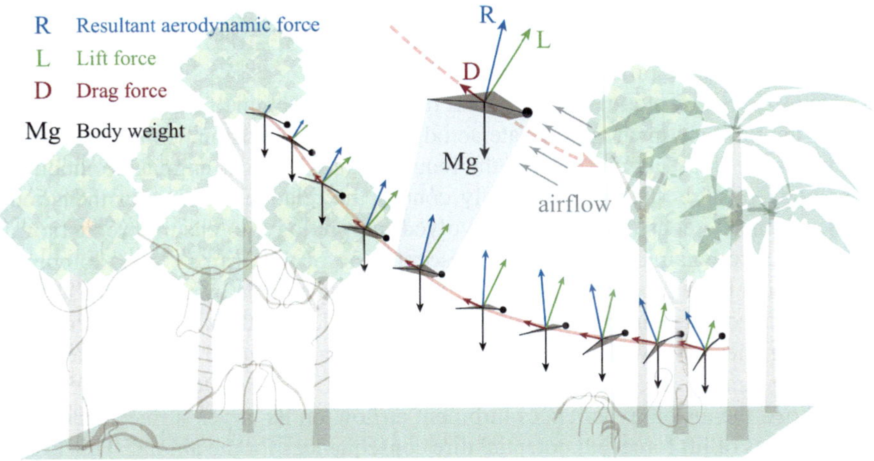

Fig. 13.5 Aerodynamic forces acting on a glider throughout its trajectory. The drag force (red) acts anti-parallel to the direction of travel, while the lift force (green) acts orthogonally. The relative magnitude of these two forces determines the resultant aerodynamic force (blue). The body weight of the glider determines the gravitational force (black), which acts downward. The changing direction and magnitude of the resultant aerodynamic force can result in a non-linear trajectory. Modified with permission from Khandelwal and Hedrick (2022)

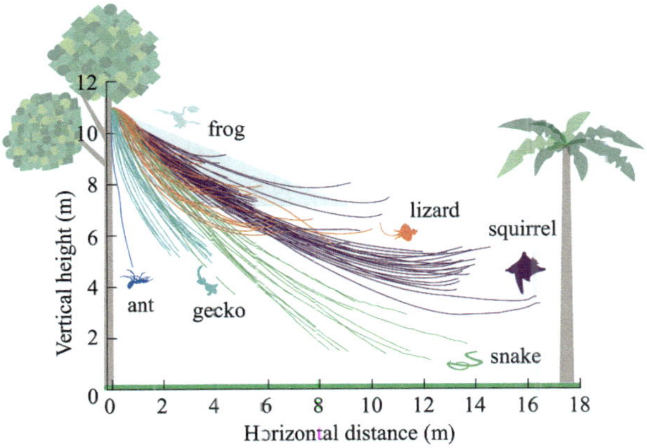

Fig. 13.6 Comparison of glide trajectories of some arboreal gliders. Trajectories shown are from experimental kinematic data from flying snakes (Socha et al., 2005), *Hemidactylus* geckos (Siddall et al., 2021), *Draco dussumieri* flying lizards (Khandelwal & Hedrick, 2020), *Cephalotes atratus* ants (Munk et al., 2015), *Glaucomys sabrinus* flying squirrels (Bahlman et al., 2013), and *Rhachophorus* frogs (Emerson & Koehl, 1990). Frog trajectories are depicted as minimum and maximum horizontal distances covered from a fixed height, as reported in the study

Conversely, gliding arthropods lack large flat surfaces and exhibit steep glide trajectories (Munk et al., 2015; Yanoviak et al., 2005, 2015), and so drag must dominate their force production profile.

Airfoil-like surface areas alone cannot result in production of aerodynamic forces if they are not held in an appropriate aerodynamic configuration. Convergent aerial behaviors are likely an outcome of the requirement of gliding animals to modulate their body posture and simultaneously control the relative position of their body parts, within the constraints of their evolved morphology, to ensure production of sufficient aerodynamic forces during descent. One of the primary ways to control aerodynamic force production is by changing the angle made by the animal's airfoil with the oncoming airflow, referred to as the angle of attack (AoA). Generally, increasing the AoA will increase the lift force on an airfoil at the expense of increasing drag force, up to certain point (Abbott & Von Doenhoff, 1959). At large AoA, the airfoil experiences turbulent airflow on its upper surface, leading to a drastic loss of lift force and potential dire consequences for the animal. This loss of lift is referred to as aerodynamic stall; many fixed rigid-wing airfoils experience it at AoA of less than 15° (Abbott & Von Doenhoff, 1959). Mammalian gliders and *Draco* flying lizards have been observed to operate at an AoA greater than 40° (Bishop, 2006, 2007; Khandelwal, 2021; Khandelwal & Hedrick, 2022), especially during the landing maneuver. At such high AoA, the animal is able to remain aloft without a significant loss in lift. Studies on gliding mammals have suggested that the ability to maintain lift generation at extreme AoA is likely due to camber (Khandelwal, 2021; Khandelwal & Hedrick, 2022). Field measurements of gliding behavior in *Draco* flying lizards have shown their ability to actively modulate camber during the glide and influence the production of lift and drag forces (Khandelwal, 2021; Khandelwal & Hedrick, 2022). These findings have been further corroborated through modelling studies, which have shown that cambered wings can delay stall (e.g., Song et al., 2008), and have also suggested the contribution of wing aspect ratio as a facilitator in lift production at high AoA (Torres & Mueller, 2004).

The magnitude of aerodynamic force production and how it varies with AoA also depends strongly on Reynolds number (Re), the non-dimensional number that indicates the relative role of inertial to viscous forces (Vogel, 1994). Reynolds numbers vary widely across gliders, from the small and slow arthropods to the large and fast mammals, ranging by three orders of magnitude from approximately a few thousand to more than 200,000 (Socha et al., 2015). The vast majority of gliders appear to glide at Re ~10,000–100,000, a fluidic regime that is understudied compared to high-Re fixed wings (Shyy et al., 2008), but is receiving increasing attention in recent years related to interest in engineering micro-aerial vehicles (MAVs). The flexibility of gliding surfaces in animals is also a topic of increasing interest (e.g., Shyy et al., 2010).

Clearly, the ability of animals to change their body posture, shape, and size during gliding has important consequences for their aerodynamic force production and the control of their glide trajectory. Moreover, modern data collection techniques, including on-body sensors, motion capture systems, and video, have shown that gliding animals do not often use equilibrium gliding (Bahlman et al., 2013; Byrnes et al., 2008; Khandelwal & Hedrick, 2020; McGuire & Dudley, 2005; Socha, 2002; Socha et al., 2005, 2010; Yeaton et al., 2020), a condition in which the aerodynamic forces exactly balance the weight during gliding, resulting in the animal descending at a constant velocity. Instead, gliders have been mostly observed to exhibit non-equilibrium dynamics (Yeaton et al., 2017; Khandelwal & Hedrick, 2020), whereby they modulate their aerodynamic forces during takeoff, mid-glide, and landing. During takeoff, the animal rapidly gains speed along with attaining a gliding pose to generate aerodynamic forces. Post-takeoff, the animal transitions into the mid-glide phase and may actively modulate the aerodynamic force production to navigate its spatially complex habitat and reach its desired target. Finally, the landing phase may involve slowing down and/or aligning the body with the landing target to safely land.

13.4.2 Inertial Mechanics

For gliding animals, maintaining an upright orientation during the glide is important due to the sensitivity of aerodynamic forces to orientation of the airfoil-body. Muscle activation is used in gliding animals to move their body parts as control surfaces, as well as to shift the mass distribution to maintain orientation and rotational stability. As the glider changes its shape during a glide, there is continuous redistribution of aerodynamic forces as well as of mass. The effect of shifting mass results in inertial effects that are not always intuitive, but play an important role in both rotational stability and maneuverability while airborne. Among gliders, the relative role of such effects is not well studied. For shifting mass, the effects could be considered to be very small or even negligible in gliders that maintain a relatively stable posture, to very large in flying snakes, which continuously move the body in three dimensions. Nonetheless, mass effects might be more important than previously considered, given that almost all gliders seem to make postural adjustments such as the previously discussed tail movements and changes in wing camber of *Draco* flying lizards and gliding mammals.

The strong coupling of control, force production, and body posture during gliding in animals requires a change in the conceptual framework to accurately understand a glide. The current framework is based upon fixed-wing aircraft theory for describing rotational stability, but given that gliding animals can be considered as living, morphing wings, a rigid-body approximation, common for aircraft, is inappropriate.

Consider instead a variable-geometry model (Garrido de Matos & Lino, 2013). In this approach, a gliding animal is considered not as a rigid body but as a "quasi-rigid body" (Goldreich & Toomre, 1969); that is, we can use some simplifying notation from the dynamics of rigid bodies, but consider that even in a frame co-moving with the animal's center of mass, the mass distribution is changing. This conceptual framework allows us to formulate general equations of motion for gliding animals that can be decomposed into inertial and aerodynamic components (Yeaton et al., 2020). The rotational dynamics are of special interest, as the inertial terms can be further decomposed into rigid-body terms and inertial moments due to the changing mass distribution,

$$\underbrace{\mathbf{M}_{\text{aero}}}_{\text{aerodynamic term}} = \underbrace{\mathbf{I} \cdot \dot{\omega} + \omega \times (\mathbf{I} \cdot \omega)}_{\text{rigid-body terms}} + \underbrace{\dot{\mathbf{I}} \cdot \omega + \dot{\gamma}}_{\text{variable-geometry terms}}$$

where M_{aero} is the aerodynamic moment, obtained from integrating over the body, I is the 3×3 inertia tensor, ω is the angular velocity, related to changing rates of yaw, pitch, and roll, and γ is an angular momentum-like term due to non-rigidity. Over-dots refer to time-rates of change. Because the mass distribution is changing in time, I is time-varying in a co-moving body frame due to the reconfiguration of the glider's shape, such as those brought about by relative body or tail movements. If the body were rigid, the mass distribution would not be changing and the inertial moments would vanish.

When starting from a low-speed jump, inertial moments always dominate at the beginning of a glide. Gliding animals can also maneuver in the air to achieve rotations and changes in direction of motion in any axis. Here we consider turns, which redirect the forward path in the lateral direction. We will refer broadly to two categories: low-speed turns, where aerodynamic effects are negligible due to the velocity squared dependence of lift and drag, and high-speed turns, where aerodynamic effects should dominate. Examples of low-speed turns are the aerial righting via tail movement seen in *Hemidactylus platyurus* geckos (Jusufi et al., 2011) and greater than 90° turns of flying snakes that occur in the space of a few body lengths (Socha, 2011). Such 'sharp' turns and reorientations likely occur when aerodynamic forces are negligible and are due primarily to internal moments, when the animal moves some part of its body (e.g., an appendage) relative to the main body, causing the body to rotate the other way (Jusufi et al., 2010). The rotation of the main body is independent of the speed of the appendage's rotation, and mathematically it is known as 'geometric phase' (Marsden et al., 1991), closely related to conservation of angular momentum (the rigid-body terms) because the aerodynamic moment is small. On the other hand, high-speed turns are those that depend largely on the aerodynamic moment, wherein the animal banks to redirect aerodynamic forces to achieve turns (Shin et al., 2019). The role of the variable-geometry moments, due to the time-varying mass distribution, is still relatively unknown and difficult to isolate, although they may be related to the vibration-induced and undulation-induced stability seen in insects (Taha et al., 2020) and flying snakes (Yeaton et al., 2020), respectively. Flying snakes may be a special case in which inertial reorientation is used more than, or in place of, banking (Yeaton et al., 2020), but a rigorous analysis of snake turning is needed to understand its mechanics.

13.4.3 Control

Some gliding animals control their glide primarily via aerodynamics by morphing their wing-body, although control authority may be enhanced via inertial moments like those seen in the tails of *Draco* flying lizards (Clark et al., 2021). Rapid tail movements have been observed during takeoff in flying mammals and lizards, though their contribution as an inertial appendage remains unclear. Control of inertial dynamics is significant for those animals that appear to have no large wing-like surfaces to create or direct flight forces, such as ants, spiders, or flying snakes, which can nonetheless reconfigure their bodies or appendages to redistribute aerodynamic forces or to take advantage of large inertial effects. For gliders that can use both aerodynamic-dominated and inertial-dominated control, it is possible that some degrees of freedom can be controlled with one, but not the other. For example, horizontal undulation alone may not be capable of providing any control of a flying snake's yaw degree of freedom (Jafari et al., 2017), which can only be effected indirectly, via aerodynamic moments influenced through pitch and roll degrees of freedom.

Gliding in many animals may be actively controlled; that is, a system of 'closed-loop' feedback incorporating sensory feedback to redirect aerodynamic forces to achieve desired objectives. Such feedback has not been rigorously investigated, but is evidenced in small control movements observed in kinematics studies (Bishop, 2006, 2007). But there is still a possible role for 'open-loop' control with no feedback, for example, via periodic changes in shape such as vibrations or undulations. For flying snakes, whose undulations have been well-documented (Socha, 2002; Socha et al., 2005, 2010; Yeaton et al., 2020), the closed-loop control system may need to work around a passively stable trajectory (Yeaton et al., 2017). Undulation may in fact lower the demand for a complex closed-loop control system in flying snakes (Jafari et al., 2017), but more dynamical modeling needs to be done. Overall, the use of an open- or closed-loop control system (or both) during gliding is largely understudied. Convergence in control strategies can be expected in gliding taxa with similar gliding morphologies and behaviors and/or based on the biological complexity of the animal.

References

Abbott, I. H., & Von Doenhoff, A. E. (1959). *Theory of wing sections*. Dover Publications.

Alexander, R. M. (2003). *Principles of animal locomotion*. Princeton University Press.

Appanah, S., Gentry, A., & LaFrankie, J. (1993). Liana diversity and species richness of Malaysian rain forests. *Journal of Tropical Forest Science*, 116–123.

Arnold, E. (2002). *Holaspis*, a lizard that glided by accident: mosaics of cooption and adaptation in a tropical forest lacertid (Reptilia, Lacertidae). *Bulletin-Natural History Museum Zoology Series, 68*(2), 155–163.

Azuma, A. (2006). *The Biokinetics of Flying and Swimming*. American Institute of Aeronautics and Astronautics.

Bahlman, J. W., Swartz, S. M., Riskin, D. K., & Breuer, K. S. (2013). Glide performance and aerodynamics of non-equilibrium glides in northern flying squirrels (*Glaucomys sabrinus*). *Journal of the Royal Society Interface, 10*(80), 20120794. https://doi.org/10.1098/rsif.2012.0794

Bishop, K. L. (2006). The relationship between 3-D kinematics and gliding performance in the southern flying squirrel, *Glaucomys volans*. *The Journal of Experimental Biology, 209*(4), 689–701.

Bishop, K. L. (2007). Aerodynamic force generation, performance and control of body orientation during gliding in sugar gliders (*Petaurus breviceps*). *The Journal of Experimental Biology, 210*(15), 2593–2606. https://doi.org/10.1242/jeb.002071

Bishop, K. L. (2008). The evolution of flight in bats: Narrowing the field of plausible hypotheses. *The Quarterly Review of Biology, 83*(2), 153–169.

Boistel, R., Herrel, A., Lebrun, R., Daghfous, G., Tafforeau, P., Losos, J. B., & Vanhooydonck, B. (2011). Shake rattle and roll: The bony labyrinth and aerial descent in squamates. *Integrative and Comparative Biology, 51*(6), 957–968. https://doi.org/10.1093/icb/icr034

Brown, C. E., Sathe, E. A., Dudley, R., & Deban, S. M. (2022). Gliding and parachuting by arboreal salamanders. *Current Biology, 32*, 441–456.

Byrnes, G., & Spence, A. J. (2011). Ecological and biomechanical insights into the evolution of gliding in mammals. *Integrative and Comparative Biology, 51*(6), 991–1001. https://doi.org/10.1093/icb/icr069

Byrnes, G., Lim, N. T. L., & Spence, A. J. (2008). Take-off and landing kinetics of a free-ranging gliding mammal, the Malayan colugo (*Galeopterus variegatus*). *Proceedings of the Royal Society B-Biological Sciences, 275*(1638), 1007–1013. https://doi.org/10.1098/rspb.2007.1684

Byrnes, G., Libby, T., Lim, N. T.-L., & Spence, A. J. (2011). Gliding saves time but not energy in Malayan colugos. *The Journal of Experimental Biology, 214*(16), 2690–2696.

Caple, G., Balda, R. P., & Willis, W. R. (1983). The physics of leaping animals and the evolution of preflight. *American Naturalist, 121*. https://www.jstor.org/stable/pdf/2460975.pdf?refreqid=excelsior%3Aef991e234ba0bf13fd4157c3de161468

Carvalho, L. D. S., Cowing, J. A., Wilkie, S. E., Bowmaker, J. K., & Hunt, D. M. (2006). Shortwave visual sensitivity in tree and flying squirrels reflects changes in lifestyle. *Current Biology, 16*(3), R81–R83.

Chadha, M., Moss, C., & Sterbing-D'Angelo, S. (2011). Organization of the primary somatosensory cortex and wing representation in the Big Brown Bat, *Eptesicus fuscus*. *Journal of Comparative Physiology A, 197*(1), 89–96.

Cheney, J. A., Konow, N., Middleton, K. M., Breuer, K. S., Roberts, T. J., Giblin, E. L., & Swartz, S. M. (2014). Membrane muscle function in the compliant wings of bats. *Bioinspiration & Biomimetics, 9*(2), 025007.

Clark, J., Clark, C., & Higham, T. E. (2021). Tail control enhances gliding in arboreal lizards: An integrative study using a 3D geometric model and numerical simulation. *Integrative and Comparative Biology, 61*(2), 579–588.

Colbert, E. H. (1967). Adaptations for gliding in the lizard *Draco*. *American Museum Novitates, 2283*, 1–20.

Corlett, R. T. (2007). What's so special about Asian tropical forests? *Current Science, 93*, 1551–1557.

Crews, S. C. (2011). A revision of the spider genus *Selenops* Latreille, 1819 (Arachnida, Araneae, Selenopidae) in North America, Central America and the Caribbean. *ZooKeys, 105*, 1.

Davenport, J. (1994). How and why do flying fish fly? *Reviews in Fish Biology and Fisheries, 4*(2), 184–214.

Garrido de Matos Lino, M.F. 2013. Design and Attitude Control of a Satellite with Variable Geometry. .

Dehling, J. M. (2017). How lizards fly: A novel type of wing in animals. *PLoS One, 12*(12), e0189573.

Dial, R. (2003). Energetic savings and the body size distributions of gliding mammals. *Evolutionary Ecology Research, 5*, 1–12.
Dudley, R. (2002). Mechanisms and implications of animal flight maneuverability. *Integrative and Comparative Biology, 42*(1), 135–140. https://doi.org/10.1093/icb/42.1 135
Dudley, R., & DeVries, P. (1990). Tropical rain forest structure and the geographical distribution of gliding vertebrates. *Biotropica, 22*, 432.
Dudley, R., & Yanoviak, S. P. (2011). Animal aloft: The origins of aerial behavior and flight. *Integrative and Comparative Biology, 51*(6), 926–936. https://doi.org/10.1093/icb/icr002
Dudley, R., Byrnes, G., Yanoviak, S. P., Borrell, B., Brown, R. M., & McGuire, J. A. (2007). Gliding and the functional origins of flight: Biomechanical novelty or necessity? *Annual Review of Ecology, Evolution, and Systematics, 38*(1), 179–201. https://doi.org/10.1146/annurev.ecolsys.37.091305.110014
Emerson, S. B., & Koehl, M. A. R. (1990). The interaction of behavioral and morphological change in the evolution of a novel locomotor type: "flying" frogs. *Evolution, 44*(8), 1931–1946.
Emerson, S. B., Travis, J., & Koehl, M. A. R. (1990). Functional complexes and additivity in performance: A test case with flying frogs. *Evolution, 44*(8), 2153–2157. https://doi.org/10.2307/2409624
Emmons, L. H., & Gentry, A. H. (1983). Tropical forest structure and the distribution of gliding and prehensile-tailed vertebrates. *The American Naturalist, 121*(4), 513–524
Endo, H., Yokokawa, K., Kurohmaru, M., & Hayashi, Y. (1998). Functional anatomy of gliding membrane muscles in the sugar glider (*Petaurus breviceps*). *Annals of Anatomy-Anatomischer Anzeiger, 180*(1), 93–96.
Essner, R. L. (2002). Three-dimensional launch kinematics in leaping, parachuting and gliding squirrels. *The Journal of Experimental Biology, 205*, 2469–2477.
Fan, P.-F., & Jiang, X.-L. (2009). Predation on giant flying squirrels (*Petaurista philippensis*) by black crested gibbons (*Nomascus concolor jingdongensis*) at Mt. Wuliang, Yunnan, China. *Primates, 50*(1), 45–49.
Flaherty, E. A., Ben-David, M., & Smith, W. P. (2010). Quadrupedal locomotor performance in two species of arboreal squirrels: Predicting energy savings of gliding. *Journal of Comparative Physiology B, 180*(7), 1067–1078.
Goldreich, P., & Toomre, A. (1969). Some remarks on polar wandering. *Journal of Geophysical Research, 74*(10), 2555–2567.
Graham, M., & Socha, J. J. (2020). Going the distance: The biomechanics of gap-crossing behaviors. *Journal of Experimental Zoology Part A: Ecological and Integrative Physiology, 333*(1), 60–73. https://doi.org/10.1002/jez.2266
Gupta, B. B. (1966). *Notes on the gliding mechanism in the flying squirrel*.
Heinicke, M. P., Greenbaum, E., Jackman, T. R., & Bauer, A. M. (2012). Evolution of gliding in Southeast Asian geckos and other vertebrates is temporally congruent with dipterocarp forest development. *Biology Letters, 8*(6), 994–997. https://doi.org/10.1098/rsbl.2012.0648
Heyer, W. R., & Pongsapipatana, S. (1970). Gliding speeds of *Ptychozoon lionatum* (Reptilia: Gekkonidae) and *Chrysopelea ornata* (Reptilia: Colubridae). *Herpetologica, 26*, 317–319.
Holden, D., Socha, J. J., Cardwell, N. D., & Vlachos, P. P. (2014). Aerodynamics of the flying snake *Chrysopelea paradisi*: How a bluff body cross-sectional shape contributes to gliding performance. *The Journal of Experimental Biology, 217*(3), 382–394. https://doi.org/10.1242/jeb.090902
Holmes, D. J., & Austad, S. N. (1994). Fly now, die later: Life-history correlates of gliding and flying in mammals. *Journal of Mammalogy, 75*(1), 224–226.
Inger, R. F. (1966). The systematics and zoogeography of the Amphibia of Borneo. *Fieldiana Zoology, 52*, 1–402.
Jackson, S. M. (2000). Glide angle in the genus petaurus and a review of gliding in mammals. *Mammal Review, 30*, 9–30.
Jackson, S., & Schouten, P. (2012). Gliding mammals of the World. *Gliding Mammals World*. https://doi.org/10.1071/9780643104051

Jafari, F., Ross, S. D., Vlachos, P. P., & Socha, J. J. (2014). A theoretical analysis of stability of gliding in flying snakes. *Bioinspiration & Biomimetics, 9*(2), 025014. https://doi.org/10.1088/1748-3182/9/2/025014

Jafari, F., Tahmasian, S., Ross, S. D., & Socha, J. J. (2017). Control of gliding in a flying snake-inspired n-chain model. *Bioinspiration & Biomimetics, 12*(6), 066002.

Jafari, F., Holden, D., LaFoy, R., Vlachos, P. P., & Socha, J. J. (2021). The aerodynamics of flying snake airfoils in tandem configuration. *The Journal of Experimental Biology, 224*(14). https://doi.org/10.1242/jeb.233635

Jusufi, A., Goldman, D. I., Revzen, S., & Full, R. J. (2008). Active tails enhance arboreal acrobatics in geckos. *Proceedings of the National Academy of Sciences of the United States of America, 105*(11), 4215–4219. https://doi.org/10.1073/pnas.0711944105

Jusufi, A., Kawano, D. T., Libby, T., & Full, R. J. (2010). Righting and turning in mid-air using appendage inertia: Reptile tails, analytical models and bio-inspired robots. *Bioinspiration and Biomimetics, 5*(4), 045001.

Jusufi, A., Zeng, Y., Full, R. J., & Dudley, R. (2011). Aerial righting reflexes in flightless animals. *Integrative and Comparative Biology, 51*(6), 937–943. https://doi.org/10.1093/icb/icr114

Kavanagh, R. P. (1988). The impact of predation by the powerful owl, *Ninox strenua*, on a population of the greater glider, *Petauroides volans*. *Australian Journal of Ecology, 13*(4), 445–450.

Khandelwal, P. (2021). How do animals glide in their natural habitat? *A holistic approach using the flying lizard Draco dussumieri* (p. 131).

Khandelwal, P. C., & Hedrick, T. L. (2020). How biomechanics, path planning and sensing enable gliding flight in a natural environment. *Proceedings of the Royal Society B: Biological Sciences, 287*, 20192888.

Khandelwal, P. C., & Hedrick, T. L. (2022). Combined effects of body posture and three-dimensional wing shape enable efficient gliding in flying lizards. *Scientific Reports, 12*, 1–11.

Khandelwal, P., Shankar, C., & Hedrick, T. (2018). Take-off biomechanics in gliding lizards. *Integrative and Comparative Biology, 1*, 38–41.

Krishna, M. C., Kumar, A., & Tripathi, O. (2016). Gliding performance of the red giant gliding squirrel *Petaurista petaurista* in the tropical rainforest of Indian eastern Himalaya. *Wildlife Biology, 22*(1), 7–12.

Krishnan, A., Socha, J. J., Vlachos, P. P., & Barba, L. A. (2014). Lift and wakes of flying snakes. *Physics of Fluids, 26*(3), 031901. https://doi.org/10.1063/1.4866444

Lambert, T. D., & Halsey, M. K. (2015). Relationship between lianas and arboreal mammals: Examining the Emmons–Gentry hypothesis. In *Ecology of Lianas* (pp. 398–406). Wiley.

Lee, D. N., & Reddish, P. E. (1981). Plummeting gannets: A paradigm of ecological optics. *Nature, 293*(5830), 293–294.

Lee, D. N., Davies, M. N., Green, P. R., & and. Van Der Weel, F. (1993). Visual control of velocity of approach by pigeons when landing. *The Journal of Experimental Biology, 180*(1), 85–104.

Losos, J. B., Papenfuss, T. J., & Macey, J. R. (1989). Correlates of sprinting, jumping, and parachuting performance in the butterfly lizard, *Leiolepis belliani*. *Journal of Zoology, London, 217*, 559–568.

Marsden, J., O'Reilly, O., Wicklin, F., & Zombros, B. (1991). Symmetry, stability, geometric phases, and mechanical integrators. *Nonlinear Science Today, 1*(1), 4–11.

Marvi, H., Gong, C., Gravish, N., Astley, H., Travers, M., Hatton, R. L., Mendelson, J. R., Choset, H., Hu, D. L., & Goldman, D. I. (2014). Sidewinding with minimal slip: Snake and robot ascent of sandy slopes. *Science, 346*(6206), 224–229.

McCay, M. G. (2001). Aerodynamic stability and maneuverability of the gliding frog *Polypedates dennysi*. *The Journal of Experimental Biology, 204*(16), 2817–2826.

McCay, M. G. (2003). Winds under the rain forest canopy: The aerodynamic environment of gliding tree frogs. *Biotropica, 35*(1), 94–102. https://doi.org/10.1111/j.1744-7429.2003.tb00266.x

McGuire, J. A. (2003). Allometric prediction of locomotor performance: An example from southeast Asian flying lizards. *The American Naturalist, 161*(2), 337–349.
McGuire, J. A., & Dudley, R. (2005). The cost of living large: Comparative gliding performance in flying lizards (Agamidae: *Draco*). *The American Naturalist, 166*(1), 93–106.
Mertens, R. (1960). Gliding and parachuting flight among the amphibians and reptiles. *The Bulletin of the Chicago Herpetological Society, 21*(1–2), 42–46.
Mongeau, J.-M., Cheng, K. Y., Aptekar, J., & Frye, M. A. (2019). Visuomotor strategies for object approach and aversion in *Drosophila melanogaster*. *The Journal of Experimental Biology, 222*(3), jeb193730.
Munk, Y., Yanoviak, S. P., Koehl, M. A. R., & Dudley, R. (2015). The descent of ant: field-measured performance of gliding ants. *The Journal of Experimental Biology*. https://doi.org/10.1242/jeb.106914
Muramatsu, K., Yamamoto, J., Abe, T., Sekiguchi, K., Hoshi, N., & Sakurai, Y. (2013). Oceanic squid do fly. *Marine Biology, 160*(5), 1171–1175. https://doi.org/10.1007/s00227-013-2169-9
Nave, G. K., & Ross, S. D. (2019). Global phase space structures in a model of passive descent. *Communications in Nonlinear Science and Numerical Simulation, 77*, 54–80. https://doi.org/10.1016/j.cnsns.2019.04.018
Niven, J. E. (2006). Colourful days, colourless nights. *The Journal of Experimental Biology, 209*(11), v–v.
O'Dor, R. (1988). The forces acting on swimming squid. *The Journal of Experimental Biology, 137*(1), 421–442.
O'Dor, R. K. (2013). How squid swim and fly. *Canadian Journal of Zoology, 91*(6), 413–419. https://doi.org/10.1139/cjz-2012-0273
O'Dor, R., Stewart, J., Gilly, W., Payne, J., Borges, T. C., & Thys, T. (2013). Squid rocket science: How squid launch into air. *Deep Sea Research Part II: Topical Studies in Oceanography, 95*, 113–118. https://doi.org/10.1016/j.dsr2.2012.07.002
Panyutina, A. A., Korzun, L. P., & Kuznetsov, A. N. (2015). Functional analysis of locomotor apparatus of colugos. In *Flight of mammals: From terrestrial limbs to wings* (pp. 205–225). Springer.
Paskins, K. E., Bowyer, A., Megill, W. M., & Scheibe, J. S. (2007). Take-off and landing forces and the evolution of controlled gliding in northern flying squirrels *Glaucomys sabrinus*. *The Journal of Experimental Biology, 210*(8), 1413–1423.
Pelletier, A., & Mueller, T. J. (2000). Low Reynolds number aerodynamics of low-aspect-ratio, thin/flat/cambered-plate wings. *Journal of Aircraft, 37*(5), 825–832.
Pomeroy, D. (1990). Why fly? The possible benefits for lower mortality. *Biological Journal of the Linnean Society, 40*(1), 53–65.
Pridmore, P. A., & Hoffmann, P. H. (2014). The aerodynamic performance of the feathertail glider, acrobates pygmaeus (Marsupialia: Acrobatidae). *Australian Journal of Zoology, 62*, 80–99.
Rayner, J. M. V. (1981). Flight adaptations in vertebrates. *Symposia of the Zoological Society of London, 48*, 137–172.
Rayner, J. M. V. (1986). Pleuston: Animals which move in water and air. *Endeavour, 10*, 58–64.
Rayner, J. M. V. (1988). The evolution of vertebrate flight. *Biological Journal of the Linnean Society, 34*, 269–287.
Russell, A. P. (1979). The origin of parachuting locomotion in gekkonid lizards (Reptilia, Gekkonidae). *Zoological Journal of the Linnean Society, 65*(3), 233–249. https://doi.org/10.1111/j.1096-3642.1979.tb01093.x
Russell, A. P., & Dijkstra, L. D. (2001). Patagial morphology of *Draco volans* (Reptilia: Agamidae) and the origin of glissant locomotion in flying dragons. *Journal of Zoology, 253*, 457–471. https://doi.org/10.1017/s0952836901000425
Russell, A. P., Dijkstra, L. D., & Powell, G. L. (2001). Structural characteristics of the patagium of *Ptychozoon kuhli* (Reptilia: Gekkonidae) in relation to parachuting locomotion. *Journal of Morphology, 247*(3), 252–263.

Scheibe, J. S., & Robins, J. H. (1998). Morphological and performance attributes of gliding mammals. In J. F. Merritt & D. A. Zegers (Eds.), *Ecology and Evolutionary Biology of Tree Squirrels* (pp. 131–144). Virginia Museum of Natural History.

Scheibe, J. S., Smith, W. P., Bassham, J., & Magness, D. (2006). Locomotor performance and cost of transport in the northern flying squirrel *Glaucomys sabrinus*. *Acta Theriologica, 51*(2), 169–178. https://doi.org/10.1007/bf03192668

Schiøtz, A., & Volsøe, H. (1959). The gliding flight of *Holapsis guentheri* Gray, a West-African lacertid. *Copeia, 1959*(3), 259–260.

Shattuck, M. R., & Williams, S. A. (2010). Arboreality has allowed for the evolution of increased longevity in mammals. *Proceedings of the National Academy of Sciences, 107*(10), 4635–4639.

Shin, W. D., Park, J., & Park, H.-W. (2019). Development and experiments of a bio-inspired robot with multi-mode in aerial and terrestrial locomotion. *Bioinspiration & Biomimetics, 14*(5), 056009.

Shyy, W., Lian, Y., Tang, J., Viieru, D., & Liu, H. (2008). *Aerodynamics of Low Reynolds Number Flyers*. Cambridge University Press.

Shyy, W., Lian, Y., Chimakurthi, S., Tang, J., Cesnik, C., Stanford, B., and Ifju, P. 2010. Flexible wings and fluid-structure interactions for micro-air vehicles. In *Flying insects and robots* (pp. 143–157). Springer.

Siddall, R., Byrnes, G., Full, R. J., & Jusufi, A. (2021). Tails stabilize landing of gliding geckos crashing head-first into tree trunks. *Communications Biology, 4*(1), 1–12.

Socha, J. J. (2002). Gliding flight in the paradise tree snake. *Nature, 418*, 603–604.

Socha, J. J. (2006). Becoming airborne without legs: The kinematics of take-off in a flying snake, *Chrysopelea paradisi*. *The Journal of Experimental Biology, 209*(17), 3358–3369.

Socha, J. J. (2011). Gliding flight in *Chrysopelea*: Turning a snake into a wing. *Integrative and Comparative Biology, 51*(6), 969–982. https://doi.org/10.1093/icb/icr092

Socha, J. J., & Sidor, C. A. (2005). *Chrysopelea ornata, C. paradisi* (Flying Snakes). Behavior. *Herpetological Review, 36*(2), 190–191.

Socha, J. J., O'Dempsey, T., & LaBarbera, M. (2005). A 3-D kinematic analysis of gliding in a flying snake, *Chrysopelea paradisi*. *The Journal of Experimental Biology, 208*(10), 1817–1833.

Socha, J. J., Miklasz, K., Jafari, F., & Vlachos, P. P. (2010). Non-equilibrium trajectory dynamics and the kinematics of gliding in a flying snake. *Bioinspiration & Biomimetics, 5*(4), 045002. https://doi.org/10.1088/1748-3182/5/4/045002

Socha, J. J., Jafari, F., Munk, Y., & Byrnes, G. (2015). How animals glide: From trajectory to morphology. *Canadian Journal of Zoology, 93*, 901–924. https://doi.org/10.1139/cjz-2014-0013

Song, A., Tian, X., Israeli, E., Galvao, R., Bishop, K., Swartz, S., & Breuer, K. (2008). Aeromechanics of membrane wings with implications for animal flight. *AIAA Journal, 46*(8), 2096–2106. https://doi.org/10.2514/1.36694

Stafford, B. J., Thorington, R. W., Jr., & Kawamichi, T. (2002). Gliding behavior of Japanese giant flying squirrels (*Petaurista leucogenys*). *Journal of Mammalogy, 83*(2), 553–562. https://doi.org/10.1644/1545-1542(2002)083<0553:gbojgf>2.0.co;2

Stapp, P. (1994). Can predation explain life-history strategies in mammalian gliders? *Journal of Mammalogy, 75*(1), 227–228.

Sterbing-D'Angelo, S., Chadha, M., Chiu, C., Falk, B., Xian, W., Barcelo, J., Zook, J. M., & Moss, C. F. (2011). Bat wing sensors support flight control. *Proceedings of the National Academy of Sciences, 108*(27), 11291–11296. https://doi.org/10.1073/pnas.1018740108

Swartz, S., & Konow, N. (2015). Advances in the study of bat flight: The wing and the wind. *Canadian Journal of Zoology, 93*(12), 977–990.

Taha, H. E., Kiani, M., Hedrick, T. L., & Greeter, J. S. (2020). Vibrational control: A hidden stabilization mechanism in insect flight. *Science robotics, 5*(46), eabb1502-eabb1502.

Torres, G. E., & Mueller, T. J. (2004). Low-aspect-ratio wing aerodynamics at low Reynolds numbers. *AIAA Journal, 42*(5), 865–873.

Vanhooydonck, B., Meulepas, G., Herrel, A., Boistel, R., Tafforeau, P., Fernandez, V., & Aerts, P. (2009). Ecomorphological analysis of aerial performance in a non-specialized lacertid lizard, *Holaspis guentheri*. *The Journal of Experimental Biology, 212*(15), 2475–2482. https://doi.org/10.1242/jeb.031856

Vogel, S. (1994). *Life in Moving Fluids* (2nd ed.). Princeton University Press.

Wagner, H. (1982). Flow-field variables trigger landing in flies. *Nature, 297*(5862), 147–148.

Wibowo, S. B., Sutrisno, S., & Rohmat, T.A. (2018). The influence of canard position on aerodynamic characteristics of aircraft in delaying stall conditions. In *AIP Conference Proceedings*. AIP Publishing LLC, p. 060028.

Wigglesworth, V. B. (1973). Evolution of insect wings and flight. *Nature, 246*, 127–129.

Xu, G.-H., Zhao, L.-J., Gao, K.-Q., & Wu, F.-X. (2013). A new stem-neopterygian fish from the Middle Triassic of China shows the earliest over-water gliding strategy of the vertebrates. *Proceedings of the Royal Society B: Biological Sciences, 280*(1750), 20122261.

Yanoviak, S. P., Dudley, R., & Kaspari, M. (2005). Directed aerial descent in canopy ants. *Nature, 433*, 624–626.

Yanoviak, S. P., Munk, Y., Kaspari, M., & Dudley, R. (2010). Aerial manoeuvrability in wingless gliding ants (*Cephalotes atratus*). *Proceedings of the Royal Society B: Biological Sciences, 277*(1691), 2199–2204. https://doi.org/10.1098/rspb.2010.0170

Yanoviak, S. P., Munk, Y., & Dudley, R. (2015). Arachnid aloft: Directed aerial descent in neotropical canopy spiders. *Journal of the Royal Society Interface, 12*(110), 20150534.

Yeaton, I. J., Socha, J. J., & Ross, S. D. (2017). Global dynamics of non-equilibrium gliding in animals. *Bioinspiration & Biomimetics, 12*(2), 026013. https://doi.org/10.1088/1748-3190

Yeaton, I. J., Ross, S. D., Baumgardner, G. A., & Socha, J. J. (2020). Undulation enables gliding in flying snakes. *Nature Physics, 16*(9), 974–982.

Young, B. A., Lee, C. E., & Daley, K. M. (2002). On a flap and a foot: Aerial locomotion in the "flying" gecko, *Pychozoon kuhli*. *Journal of Herpetology, 36*(3), 412–418.

Zamore, S. A., Araujo, N., & Socha, J. J (2020). Visual acuity in the flying snake, Chrysopelea paradisi. *Integrative Comparative Biology*. https://doi.org/10.1093/icb/icaa143

Zeng, Y., Lam, K., Chen, Y., Gong, M., Xu, Z., & Dudley, R. (2017). Biomechanics of aerial righting in wingless nymphal stick insects. *Interface Focus, 7*(1), 20160075.

Zeng, Y., Chang, S. W., Williams, J. Y., Nguyen, L. Y.-N., Tang, J., Naing, G., Kazi, C., & Dudley, R. (2020). Canopy parkour: movement ecology of post-hatch dispersal in a gliding nymphal stick insect, Extatosoma tiaratum. *The Journal of Experimental Biology, 223*(19), jeb226266.

Chapter 14
Convergence of Bipedal Locomotion: Why Walk or Run on Only Two Legs

François Druelle, Anick Abourachid, Menelia Vasilopoulou-Kampitsi, and Peter Aerts

Abstract The adaptive reason(s) as to why some tetrapods walk or run on only two legs is far from straightforward. Compared to a quadruped, maintaining balance is obviously challenging for a biped, since the number of ground contacts per cycle is reduced. Consequently, cyclic limb loading can also be expected to be higher. In association with these mechanical constraints, some species show clear adaptations that enable them to continuously walk and/or run bipedally. Others, however, can only perform bipedally for short bouts after which they proceed further on all fours. Apparently, in the latter case, an optimal functional anatomical context (the so-called '*evolutionary spandrel*') favours occasional bipedal performance. In this context, it is possible that the morphological features in the extant species that routinely practice bipedal walking and/or running (i.e., birds and humans) are convergent. Indeed, a similar (analogous) adaptive trait might have evolved independently in these 'bipedal' taxa as a result of similar selection pressures. Similarly, since occasional bipedalism should be a mandatory intermediate evolutionary stage leading to habitual, or obligate, bipedalism, the evolutionary pathways leading to bipedal adaptations can be explored in extant animals practicing different levels of bipedal performance.

F. Druelle (✉)
Histoire Naturelle de l'Homme Préhistorique, UMR 7194 CNRS, MNHN, UPVD, Paris, France

Functional Morphology Laboratory, University of Antwerp, Antwerp, Belgium
e-mail: francois.druelle@mnhn.fr

A. Abourachid
Mecadev, UMR 7179 CNRS, MNHN, Paris, France
e-mail: anick.abourachid@mnhn.fr

M. Vasilopoulou-Kampitsi
Functional Morphology Laboratory, University of Antwerp, Antwerp, Belgium

P. Aerts
Functional Morphology Laboratory, University of Antwerp, Antwerp, Belgium

Department of Movement and Sport Sciences, University of Ghent, Ghent, Belgium
e-mail: peter.aerts@uantwerpen.be

V. L. Bels, A. P. Russell (eds.), *Convergent Evolution*, Fascinating Life Sciences,
https://doi.org/10.1007/978-3-031-11441-0_14

In the present chapter, we discuss different functional and evolutionary contexts that have led species with different *Bauplans* to use bipedal walking and/or running (for short or longer behavioural bouts). Firstly, we consider the involuntary but dynamically imposed bipedal running observed in extant lizards. Secondly, the voluntary but anatomically constrained bipedal walking behaviour of extant non-human primates is discussed. Thirdly, the use of the bipedal posture in birds and humans is compared (i.e. for the species for which this mode of progression is anatomically imposed and constrained). Finally, in an attempt to link the mechanical constraints and the potential evolutionary pathways related to occasional, habitual and obligate bipedalism, we argue that bipedal tetrapods should converge upon the same functional anatomical outcomes. Indeed, based on a virtual limb that would connect the body's centre of mass to the foot, angular impulses about the body's centre of mass over a (half-) stance time of the (vertical) ground reaction forces should cancel each other out. This might leave only two adaptive pathways open that could lead to a more habitual use of bipedalism: (1) make the bipedal functional/anatomical limb coincide with the virtual limb, or (2) make the virtual limb coincide with the anatomical (quadrupedal) limb.

Keywords Acceleration · Angular momentum · Centre of mass · Hip joint · Kinematics · Knee joint · Limb posture · Locomotor repertoire · Manoeuvring · Trunk stabilization · Vertical ground reaction force

14.1 Introduction

In order to move around animals must interact with their environment. The associated propulsive reaction forces from the external world counter resistance and accelerate/decelerate the body. At the same time, animals must also maintain position, balance and posture. In most pelagic aquatic vertebrates, these propulsive (and drag) forces are spatially and temporally distributed over the entire body surface, and buoyancy (and potentially also dynamic lift) largely assist the maintenance of position and balance. With the evolutionary aquatic-terrestrial transition, however, the conditions of the interactions with the environment changed dramatically. Although the resistance imposed by the medium was reduced greatly (air is about 800 times less dense and 50 times less viscous than water), gravity was no longer countered by buoyancy and vertebrates evolved articulated limbs[1] to support the body and to move in a more or less efficient way.

This evolutionary transition came, however, with an important mechanical consequence. Propulsive reaction forces are no longer distributed evenly in time and

[1] Vertebrate 'articulated limbs' are convergent analogues, at least phenotypically, with the walking legs of arthropods. However, genetically, depending on the considered hierarchical level, it can be argued that these appendages are paralogues (i.e. results of parallel evolution; cf. Shubin et al., 1997).

spatially over the body surface in general but instead occur focused in time during stance located at the interface between the limbs and substrate. The result of this is impact-like loading coupled with high local stresses in the locomotor apparatus. Moreover, since buoyancy no longer cancels the effect of gravity, the means of maintaining position, balance and posture became important additional tasks for the locomotor apparatus. This led to a trading-off of propulsion generation with the potential loss of energy each time the limbs redirect body motion against the effects of gravity (collisional losses: Hobbs & Clayton, 2019; Lee et al., 2011; Ruina et al., 2005). These mechanical consequences impose constraints, but also offer opportunities for the functioning of the locomotor system. One example of a constraint is that animals cannot afford to fall over or collapse when manoeuvring in their environment while attempting to avoid a predator or catch prey. An example of an opportunity is the saving of energy by exploiting gravity when walking by the limb acting as an (spring-loaded) inverted pendulum or by converting the energy linked to the impact-like loading of the limbs into reusable elastic strain energy, primarily when using (faster) bouncing gaits.[2]

Moving on only two limbs (bipedally) undoubtedly amplifies the mechanical consequences linked to terrestrial locomotion. In order to illustrate this, we employ a simplified model of legged terrestrial locomotion. The essence of steady locomotion (i.e., moving the body centre of mass (BCoM) steadily from location A to B) resembles the progression of a rimless spoked wheel (e.g. Ruina et al., 2005) rolling down a gently sloped surface (i.e., gravity powers the motion, the propulsive component precisely countering resistance). Impact and loading are distributed over the number of spokes contacting the ground over a cycle. The more spokes the wheel has, the closer the resemblance to a rimmed wheel and the smoother the ride will be (Fig. 14.1a). However, perturbations, for instance as a result of irregularities in the terrain or external forces pushing on the wheel's BCoM, can occur. This implies undesired accelerations of the BCoM, away from the intended steady motion (Fig. 14.1b). To guarantee continued progression, balancing forces will not only be required to keep the BCoM on average directly above the supporting spokes,[3] but also to do work on the BCoM to compensate for the mechanical energy potentially lost (or gained) due to the perturbations. Therefore, the wheel needs instrumented 'magic' spokes that are able to adjust orientation and mechanical behaviour in response to perturbations (Fig. 14.1c). The greater the number of spokes the greater the number of possible corrective actions available per cycle for maintaining balance and posture and the safer the ride will be.

Even from this simplifying model that reduces a locomoting animal to a moving BCoM, it is obvious that maintaining balance (statically and dynamically) is a

[2]Inverted pendulum walking and bouncing gaits are functional convergences for tetrapods and arthropods (e.g. Blickhan & Full, 1987; Li et al., 2013).

[3]This view also 'reduces' balance maintenance to its essence. A detailed account on the mechanics of the control of (dynamic) balance (in bipeds), including the inertial effects of segmental movements, can be found in Hof (2008).

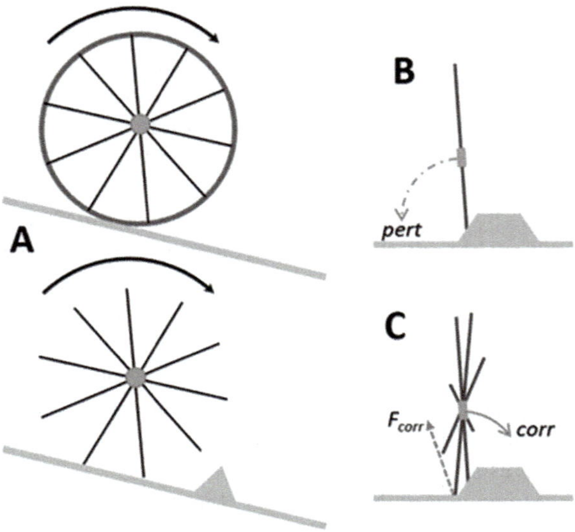

Fig. 14.1 (**a**) A wheel and a rimless wheel steadily rolling down a slope. (**b**) frontal view of the perturbed (*pert*) wheel; a spoke hits an irregularity in the terrain. (**c**) Frontal view of the 'magic wheel'; the spokes react (*Fcorr*) to correct (*corr*) the perturbation

challenge for a biped since only two limbs contact the ground per cycle. Moreover, for a given locomotor task, musculoskeletal stresses are probably higher for this biped as loading cannot be shared between limbs for most of the stance time (four beat walks or trotting gaits are preferable). Furthermore, collisional costs in locomotion may increase with a reduction in the number of footfalls per cycle (four beat patterns are preferable to two beat patterns; Ruina et al., 2005 but see Lee et al., 2011). It is therefore perhaps not too surprising that, within tetrapods, habitual or obligate bipedalism (see further) emerged only (with some exceptions) within the hominins and archosaurs (the latter with several independent transitions; cf. e.g. Grinham et al., 2019). However, the question is whether this 'locomotor behavioural bipedality as such' represents genuine evolutionary convergence in these cases: i.e., is it a similar (analogous[4]) adaptive trait that evolved independently in these taxa as the result of similar selection pressures?

If bipedal locomotion was the result of being selected for directly (i.e., being the adaptive trait), the drawbacks should be outweighed by the direct biological *fitness* gained. It seems, however, difficult to establish for what locomotor attributes (e.g., speed, manoeuvrability, metabolic costs, etc.) and under what selective circumstances bipeds would have gained this locomotor benefit and would have outperformed quadrupeds (cf. the adaptive scheme of Arnold, 1983; see also Aerts et al., 2000b). It is more likely that bipedal locomotion initially emerged and is retained (and eventually further refined by adaptive evolution) as the consequence of other adaptive features (behavioural, functional, etc.) with benefits that outweigh the evolutionary costs of bipedalism (e.g., Crompton et al., 2008; Harcourt-Smith, 2010;

[4]In theory, if not analogous (but homologous) one should consider the traits 'evolutionarily parallel'.

Rose, 1991; Thorpe et al., 2007). However, for whatever evolutionary reason, convergences in anatomy (e.g. limb morphology of megapods and rodents like jerboas or springhares), function (e.g. using bouncing gaits at higher speeds) or (motor) behaviour (e.g. limb postures and coordination) are expected because mechanical principles weigh heavily on selection (e.g., Taylor & Thomas, 2014). Nevertheless, the true nature of the apparent 'convergence' should always be questioned: are the observed trait similarities the result of (possibly constrained) direct adaptive selection (i.e. true evolutionary convergence), or are these similarities just the consequence of mechanical/constructional constraints without being the direct outcome of selection (eventually becoming evolutionary spandrels, cf. Gould & Lewontin, 1979)?

We discern three different kinds of bipedalism. We define habitual bipeds as those species that, under normal (yet sometimes specific) circumstances, move spontaneously and invariably on their hind limbs. For instance, the invariably used bipedal hopping gait of most macropod marsupials at high(er) speeds is habitual, despite the employment of a quadrupedal bounding gait when moving slowly. Also, humans are habitual bipeds, yet, for instance, spontaneously use all fours when clambering up steep slopes or trees (e.g. Grosprêtre & Lepers, 2016; Kraft et al., 2014). Often, habitual bipedality is associated with clear adaptations at the level of the hind limbs, associated with improving performance. Obligate bipeds are those species in which the forelimbs can no longer be used for terrestrial locomotion because they have become adapted for other functions. Birds are obligate bipeds, as were many dinosaurs. Finally, there is the category of taxa that are occasional or facultative bipeds. They normally move spontaneously and invariably on all fours but switch, under certain behavioural and ecological circumstances (sometimes imposed by mechanical constraints), to bipedal postures and locomotion. It has been argued that occasional bipedalism is an intermediate evolutionary stage that may lead to habitual or obligate bipedality (e.g. Harcourt-Smith, 2010; Persons & Currie, 2017; but see Grinham et al., 2019). From this point of view, better insights into the functional morphology and biomechanics of occasional bipedalism are essential for understanding the (mechanical) constraints, evolution and convergences evident in habitual and obligate bipeds.

The apparent lack of functional evidence directly related to locomotor performance that pertains to direct selection for bipedality in tetrapods serves as the starting point for what follows. We consider only time-symmetrical bipedal gaits (walk, run). In the above-mentioned context, we first focus on two cases of occasional bipedality, in lizards and non-human primates, and discuss, despite the great phylogenetic distance between them, the presence and nature of anatomical, functional and behavioural similarities (convergences) that they display. Next, we ask, and consider, the same questions in relation to birds, the only extant obligate bipeds. Finally, we consider possible convergences of, and transitions between, occasional, habitual and obligate bipedalism.

14.2 If Not Selected For, Why Walk or Run on Two Legs?

14.2.1 Occasional Bipedalism in Lizards: Involuntary But Dynamically Imposed

14.2.1.1 Setting the Mechanical Scene

Imagine the recording set of a, probably B-grade, movie. Two motor bikers are waiting, side by side, at a traffic light on a boulevard in a fashionable Mediterranean seaside resort. One rides a Monster 1200S, the legendary naked bike of Ducati. He sits a little lower and more at the front of his bike than his rival and leans over the handlebars. The other rides a Panigale 4V, the equally iconic racing machine of the same maker. He sits farther back and a bit higher. These machines develop the same maximum torque and have the same transmission. A few incidental passers-by watch them, initially carelessly. When the light switches to green, both open the throttle fully. They accelerate so powerfully that the bikes pitch upwards (enacting a "wheelie"). The Monster quickly goes onto two wheels again and takes a slight lead. The Panigale rider, however, continues on the rear wheel alone until they are out of sight and receives startled looks and shouts from the surprised bystanders.

What mechanically triggers this pitching and what causes the differences in behaviour and performance? The traction gained by the rear wheel of the bike (henceforth "bike" is used to refer to both the bike and its rider) results in a horizontal ground reaction force that primarily causes the bike to accelerate, but also balances both rolling and air resistance. Vertical ground reaction forces (at both or one wheel) balance the weight of the bike (and when present aerodynamic lift) and eventually accelerate its centre of mass (CoM) upwards. The moment of all external forces in play about a transverse axis at the rear axle (or any other parallel axis) must equal the rate of change of the angular momentum (or 'amount of rotation'[5]) of the bike. Above a threshold acceleration 'a_t' the counterclockwise and clockwise moments do not balance any longer and the 'bike' pitches upwards. This threshold 'a_t' can be deduced from the angular equation of motion (at 'a_t', there is no load on the front wheel, nor is there any pitching yet; see Fig. 14.2a and see Aerts et al. (2003) for a biological example). More simply:[6] when the forward acceleration 'a' is larger than minus 'g' (gravitational acceleration) times the horizontal distance 'd' between the rear axle and the CoM, divided by the CoM's height 'h' (i.e. $a > -g\ d/h$) the bikes

[5]In this simplified bike analogue, the angular momentum consists of the sum of (1) the bike's moment of inertia respective of the CoM times its angular velocity (i.e. the *local term*) and (2) the product of the bike's mass, the angular velocity of the CoM about the rear axle, and the distance between the CoM and axle squared (i.e. the so-called moment of the linear momentum or the *remote term*).

[6]The contribution of the rotating wheels to the angular momentum, which is small compared to that of the rest of the bike, and the moment imposed by aerodynamics (drag and lift) about the rear axle (probably close to zero as these forces are distributed over the entire surface) are not considered in this equation.

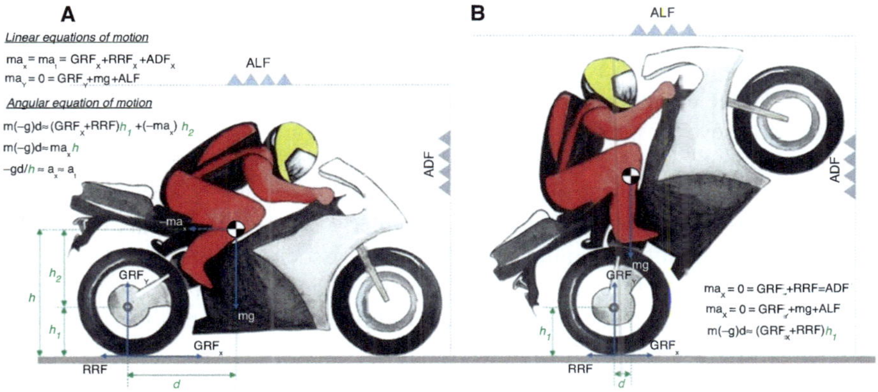

Fig. 14.2 (**a**) The bike at threshold acceleration a_t. No load is borne by the front wheel. The horizontal inertia (ma_t) equals the sum of all horizontal forces (GRF_x horizontal ground reaction force only acting at the rear wheel, RRF roll resistance force, ADF horizontal aerodynamic force distributed over the frontal surface). The vertical inertia (ma_y) is zero (i.e. vertical forces balance each other: GRF_y; mg = gravitational force acting at the bike's CoM ($g = -9.81\ m/s^2$); ALF = vertical aerodynamic force distributed over the upper surface). At a_t, angular acceleration of the bike is still zero and the moment of all forces about the rear axle, including the effect of the linear inertia, balance each other (GRF_y has no moment about the axle (see also footnote 3)). (**b**) The bike at zero forward acceleration proceeding in (unstable) pitched posture (also see text) [Illustrations: Menelia VK illustration]

move forward doing a wheelie.[7] As a result, some engine power is spent in this pitching motion and is thus not available to contribute to forward motion. Because of the 'anatomical' differences between the Monster and the Panigale, and primarily also because of the different positions and postures of the riders on their bikes, '*d*' is somewhat larger and 'h' somewhat smaller for the Monster, meaning that, when compared to the Panigale, the instantaneous acceleration drops below the pitching threshold sooner. At this point, gravity brings the front wheel back to the ground and all engine power is dedicated to forward motion. This explains why the Monster takes a slight lead. On the other hand, because the pitching of the Panigale proceeds for longer, it is easier for that biker to sense and assess the moment that acceleration ceases, the (intrinsically instable) posture at which the rate of change of the angular momentum equals zero,[8] and, in this way, to show off for the bystanders by driving further on the rear wheel only.

[7] An alternative explanatory view is that a torque, equal but opposite to the engine's torque about the axle, acts on the bike (minus the rear wheel) and will cause pitching when this can no longer be countered entirely by gravity. This is an internal torque, which should not be considered in the balance as described in the text.

[8] Upward and downward pitching moments are in equilibrium, i.e. moments about the axle of the gravitational force, the aerodynamical forces and the component of the propulsive force at the rear wheel that counters air resistance, cancel each other out, hence the CoM is essentially above the axle; cf. Fig. 14.2b.

14.2.1.2 Linking the Mechanical Framework to Lizard Locomotion

Only a few kilometers away from this (imaginary) Mediterranean urban scene, on a sunny day in the countryside of, for instance, the 'Alpes Maritimes' in France, it is easy to spot several specimens of small lacertid lizard species. The locomotor behaviour of these animals is very intermittent: fast sprints and stops, manoeuvring, sharp turns and jumps alternate with rest pauses (to enable proper respiration, cf. Carrier, 1987). Close observation also regularly reveals short bouts of bipedal running. This intermittent running style is probably associated with the lizard's small body size relative to the structural magnitude of the microhabitat: small terrain irregularities (stones, shrubs, crevices, etc.) readily present themselves as real obstacles that force the lizards to manoeuvre, stop, turn, and start sprinting again. Together with the relatively very high maximum speeds attained (up to 40 snout-vent lengths per second is not exceptional; cf. Van Damme & Vanhooydonck, 2001) such locomotor behaviour implies many accelerations per activity bout. For the south Mediterranean lacertid *Acanthodactylus boskianus,* for instance (snout-vent length of about 6 cm on average and a top speed of 3.6 m/s), accelerations up to 42 m/s^2 have been measured by Curtin Nancy et al. (2005). This by far exceeds the 'maximum acceleration performance' of the Panigale 4V (0–100 km/h in approximately 3 s, equal to approximately 9 m/s^2).

Is it conceivable that the observed short stretches of bipedal running in these lizards are, as in the bike-analogy above, simply the consequence of the accelerations inherent in the intermittent locomotor behaviour? Are the accelerations great enough to cross the thresholds (i.e. $a > -g\ d/h$; with 'd' the horizontal distance between the hip joint and the CoM)? If so, do 'Panigale's' also exist among lizards, exploiting this consequence in order to proceed bipedally when acceleration has ceased? Why would they do so and how should all this be framed in an evolutionarily adaptive and convergence-related context? These questions are now addressed.

14.2.1.3 Is Lizard-Bipedality Adaptive?: Part 1

Many lizards, scattered over the phylogenetic tree, show bipedal behaviour to various degrees (Aerts et al., 2003; Christian et al., 1994; Clemente, 2014; Clemente et al., 2018, 2008; Clemente & Wu, 2018; Irschick & Jayne, 1999; Snyder, 1949, 1952, 1954, 1962; Van Wassenbergh & Aerts, 2013). Some run only occasionally on their hindlimbs and for only short stretches at time (e.g. lizards like *Acanthodactylus boskianus*: Aerts et al., 2003; Curtin Nancy et al. 2005; Druelle et al., 2019b; *Pogona minor, Ctenophorus rubens*: Clemente et al., 2008; *Uma scoparia*: Irschick & Jayne, 1998, 1999). Others seem to excel at it and proceed bipedally for considerable distances (such as, for instance, *Lophognathus gilberti, Ctenophorus cristatus*: Clemente et al., 2008; *Callisaurus draconoides*: Irschick & Jayne, 1998, 1999; *Aspidoscelis sexlineata*: Olberding et al., 2012; *Liolaemus lutzae, Tropidurus torquatus*: Rocha-Barbosa et al., 2008). Invariably, however, they all initiate bipedal performance starting from a quadrupedal gait and posture (Clemente, 2014; Clemente & Wu, 2018). Moreover, and in accordance with the general consideration presented in the Introduction of this chapter, it seems very difficult to provide evidence for lizard bipedal running being an adaptation *per se* (Aerts et al., 2003; Clemente et al., 2008; Clemente & Wu, 2018). If so, selective benefits of bipedalism should exceed the involved evolutionary costs, but it is hard to identify any adaptive profits. It was argued that bipedal running enables lizards to run faster (Snyder, 1949, 1952, 1952, 1954) and morphological adaptations (e.g. long, robust hind limbs and muscles, narrow pelvis, short forelimbs, long tails) were linked to this (Snyder, 1954, 1962). Improved bipedal speed performance could, however, not be confirmed in experimental studies for a variety of species (Clemente et al., 2008; Irschick & Jayne, 1998, 1999) and the proposed morphological adaptations were instead argued to reflect selection for speed rather than bipedalism (Aerts et al., 2003; Clemente et al., 2008; Snyder, 1954, 1962). It was also suggested that lizards ran more economically when doing so bipedally (Snyder, 1949, 1952, 1954, 1962), but again this could not be confirmed experimentally (Clemente et al., 2008; note that bipedal and quadrupedal locomotor costs appear to be comparable for similar-sized animals in general: e.g. Fedak & Seeherman, 1979; Roberts et al , 1998; Taylor & Rowntree, 1973).

14.2.1.4 What Is the Probability of an Alternative Adaptive (and Convergent) Scenario?

In 2003, Aerts and colleagues proposed an alternative evolutionary scenario. They suggested that occasional bipedal bouts of many small, intermittently running, lizards (cf. above) emerged as a consequence of the combined need for manoeuvrability and speed to survive in a (for their size) structurally complex environment. In order to manoeuvre swiftly and rapidly, such lizards must reorient their body while heading in a new direction along a curvilinear path (Jindrich & Full, 1999). This body reorientation must be achieved by the front legs pushing sideways on the outside of the bend. This happens more easily when the whole body moment of inertia in the frontal plane about the body centre of mass (BCoM) is minimal (i.e. least resistance against reorientation) combined with a BCoM positioned far backwards (i.e. increasing the moment arm of the centripetal forces at the front leg; cf. Aerts et al., 2003). In this context, it is worthwhile noting that, for example, in *Acanthodactylus erythrurus,* the predicted optimal BCoM position and its measured position coincide very well, that is approximately 14 mm anterior to the pelvic girdle (Aerts et al., 2003). Mass distributions favouring such an optimal BCoM position may thus represent a convergent adaptation for improved manoeuvrability (with obvious *fitness* benefits) in swift lizard species (see below).

Such a morphological adaptation, however, carries with it important consequences, especially when improved manoeuvrability (the feature selected for) is coupled with intermittent locomotor behaviour characteristic of many lizards (cf. above). First, with a BCoM close to the pelvic girdle, most of the static and

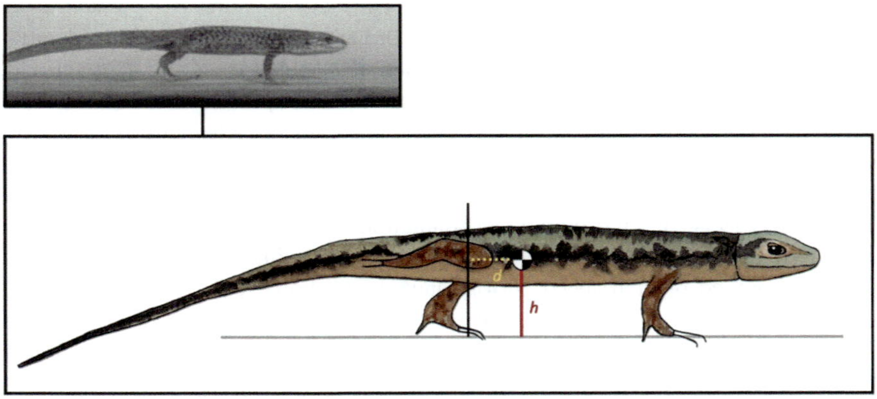

Fig. 14.3 Based on still from a high-speed video of a running *Podarcis melisellensis*. The black and white filled circle represents the position of the BCoM; 'd' is the horizontal distance from the hip joint to the BCoM, 'h' is the height of the BCoM (cf. text). Notice the difference in robustness and size between, for instance, the shank and the lower arm in the video-still [Illustration: Menelia VK illustration]

dynamic loading will be imposed upon the hind limbs (i.e. functionally close to bipedalism). Moreover, the posterior shift of the BCoM (combined with the low moment of inertia[9]) implies a small '*d*' (distance between BCoM and pelvis). This, in turn, reduces the threshold acceleration 'a_t' at which upward body pitching occurs. Even when below the threshold, acceleration will further lower the loading imposed on the front legs. It is probable that the latter not only explains the very conspicuous morphological difference in length and musculo-skeletal robustness between the front and hind limbs in many (primarily small) lizard species (the front limbs are almost redundant except in slow locomotion and when changing direction), but also the many observations of short stretches of bipedal running as a result of temporary pitching of the body.

The equation for estimating the magnitude of the threshold acceleration, $a_t = -g\ d/h$, reveals that the threshold is independent of size (in this case geometric similarity is assumed; but see below). Moreover, although admittedly very roughly, '*d*' and '*h*' will have about the same order of magnitude in lizards because of their sprawling limb posture. As a result, simplified whole body modelling predicts that the onset of bipedal running occurs when forward acceleration approaches '*-g*' (but see below). For *Podarcis melisellensis*, a small north-east Mediterranean lacertid, the ratio '*d/h*' equals approximately 0.75 (see Fig. 14.3). This results in an estimated threshold of approximately 7.4 m/s^2. This is far below the maximal acceleration (35 m/s^2) as measured for another *Podarcis* species that is very similar in size, ecology and behaviour (*Podarcis erhardii*; Vasilopoulou-Kampitisi, unpublished data).

[9]Notice that, because of the overall elongated and close to rotational-cylindrical body shape in lizards, the moments of inertia in the frontal and sagittal planes will be largely similar.

Furthermore, forward dynamic modelling of a sprint start of a generalized lacertid, based on the morphometrics and spatio-temporal running kinematics of *Acanthodactylus erythrurus* (Aerts et al., 2003), predicted up to 14 successive bipedal running steps that is, a clear and distinct bipedal running bout during the acceleration phase that precedes quadrupedal steady running at maximum speed. Simulating a slightly more strenuous sprint start resulted in the model toppling backwards.

The behaviour of the model was confirmed by tests on *Acanthodactylus boskianus* in which escape responses (that is, acceleration from stand still) were triggered by a sudden attempt to grasp the animals from the rear. Bipedal running bouts were frequently observed and, on a few occasions, the response was so vigorous that the specimen effectively made a back flip after a few steps (P. Aerts, unpublished observations). Furthermore, the proposed link between acceleration and bipedalism is supported by the experimental work performed by Clemente and colleagues (Clemente, 2014; Clemente et al., 2008; Clemente & Wu, 2018). They determined for many species the speed and acceleration (the latter assessed in multiple ways) at which running was quadrupedal, at which specimens performed bipedally and, if recorded, when the transition between both gait types occurred. These authors found no relationship between speed and bipedal behaviour. Bipedal running, however, did correlate with acceleration. Moreover, based on log-likelihood statistics they were able to estimate the threshold accelerations at which specimens would make the transition from quadrupedal to bipedal running. Generally, thresholds were lower than those predicted by the model (see below), but the ranking of the threshold magnitudes across the species was similar for the model and experiments. The same authors also determined the relationship between the frequency of bipedal performance and the position of the BCoM. Conforming to the theory, species with a BCoM closer to the hips were found to become bipedal more frequently.

Based on their modelling results, Aerts et al. (2003) argued that the occasional bipedalism of many (small) lizard species should be considered an 'evolutionary spandrel' [that is, a feature that '*owes its origin to a side consequence of another feature*' (after Gould, 1991, p. 53; see also Gould & Lewontin, 1979; Buss et al., 1998)]. From this point of view, occasional bipedal running in lizards is, at least in its initial emergence, a side effect of the requirement for manoeuvrability and does not represent an adaptation. In this way, the lizard's locomotor behaviour resembles that of the Monster 1200S in the bike analogy: the brief bipedal stretch is not really useful, but it is also almost harmless (it comes with negligible energetic cost associated with the brief upward pitching). The emergence of this 'spandrel' (that is, a pure side effect) is probably strongly size-dependent, despite the actual pitching threshold acceleration being argued to be largely size-independent. The reason for this is that the force *required* to accelerate the body is logically proportional to body mass (*i.e.*, F = ma). Available muscle force, however, is proportional to body mass$^{0.666}$ (on the premise of geometric similarity). Assuming that the acceleration of 35 m/s^2 reported for *Podarcis erhardii* concurs with the upper force limit of the species, and knowing that the specimens reaching these accelerations had, on average, a body mass of approximately 0.005 kg (Vasilopoulou-Kampitsi et al. in prep), it can be estimated that a geometrically similar 'monster' *Podarcis*' with a mass above 0.525 kg would be unable to reach the threshold acceleration of 7.4 m/s^2. The 'spandrel' would vanish! According to Pough (1980), 80% of all lizards have a body mass <0.02 kg, (far below the roughly estimated threshold size). Clearly, all these lizards do not have similar acceleration capacities, neither do they all exhibit a swift intermittent locomotor style, but all in all it is probably not too surprising that the bipedal spandrel is observed rather frequently in (small) lizards (cf. Aerts et al., 2003; Clemente, 2014).

14.2.1.5 Is Lizard Bipedality Adaptive?: Part 2

Nevertheless, Aerts et al. (2003) also argued that lizards, when confronted by the consequential pitching engendered by acceleration, may have begun to exploit this spandrel in order to proceed bipedally over longer distances. In this way, the behaviour resembles that of the Panigale 4V in the bike-analogy. Pitching goes on until the lizard has reached, and then maintains, the (unstable) equilibrium between the upward and downward pitching moments in order to run steadily on the hind legs alone. Theoretically, there are different ways for a lizard to achieve this and these options were explored 'in silico' by Van Wassenbergh and Aerts (2013). It is, however, thanks to the excellent experiments and analyses conducted by Clemente and co-workers that the presence of active spandrel-exploitation has been demonstrated and that several of the theoretical hypotheses have been experimentally tested (Clemente, 2014; Clemente et al., 2018, 2008; Clemente & Wu, 2018).

First, however, the evolutionary context should be considered. If lizards make use of the spandrel and continue in a bipedal running mode, it must be beneficial for them to do so. In the bike-analogy, 'signaling by impressing' is the only direct benefit the Panigale-rider gains and this must be traded-off against the following: the risks of crashing; the slightly increased energetic costs; and reduced performance (the latter two because of the greater air resistance encountered). The advantages for a lizard proceeding bipedally have not been compellingly demonstrated, but there are strong arguments for improved obstacle negotiation in terms of the potential for a more level trajectory of the BCoM (an energetic benefit), a better visual field for improved obstacle anticipation, and better limb clearance enabling the avoidance of stumbling (Clemente & Wu, 2018; Druelle et al., 2019b; Kohlsdorf & Biewener, 2006; Olberding et al., 2012; Tucker & McBrayer, 2012; Van Wassenbergh & Aerts, 2013). It is probable that such benefits can operate to trade-off the costs related to bipedal behaviour, as mentioned in the Introduction, and even more so if the capacity to exploit the spandrel is, over evolutionary time, further reinforced by natural selection. For lizard species such as, for instance, *Ctenophorus cristatus* (Clemente et al., 2008), *Callisaurus draconoides* (Irschick & Jayne, 1998, 1999) or *Aspidoscelis sexlineata* (Olberding et al., 2012), which excel at bipedal running, the adaptation initially favouring the optimal mass distribution for improved manoeuvrability could be considered to have served as an exaptation for bipedal running. This can be considered a true convergence since alternative behavioural strategies may have evolved independently to achieve the same goal: stable bipedal running. The question then remains, however, why do not all, or at least many more, (small) lizards excel at bipedalism?

14.2.1.6 Can Lizards Run Bipedally in a Controlled Way?

Finally, the mechanistic aspects of spandrel-exploitation must be evaluated. It was mentioned above that the experimentally determined threshold accelerations at which a quadrupedal run transitions into a bipedal one were generally (and mostly substantially) lower than the accelerations predicted by the model. As such, some such difference is not too surprising, because of the simplifications introduced into the model: to determine the threshold, Aerts et al. (2003) assumed the model-lizard to have head, trunk and tail extended, with its long axis horizontal to and parallel with the surface.[10] The point of application of the ground reaction force (at transition) falls directly below the hip and the height of the BCoM remains constant. At some junctures in a running bout this will be close to reality (cf. Fig. 14.3), but at others the body configuration may have changed considerably, almost inevitably resulting in lower thresholds (increase of '*d*' and '*h*'). However, the observed differences are often considerable and Clemente and colleagues rightly suggest that in these cases 'a deviation from accidental, morphologically based bipedalism toward dynamically controlled bipedalism' must be considered (cf. Clemente, 2014, p. 2178). Clemente (2014) also showed that the threshold accelerations decreased over evolutionary time, while the differences between model- and experimental threshold increased. Both are suggestive of selective improvement of dynamically controlled behaviour leading to more prolonged and sustained bouts of bipedalism. The role of behavioural control becomes even more pertinent when there is not only a reduction of the threshold, but also true 'Panigale'-like behaviour (bike analogy). Active dynamic control seems indispensable when, after upward pitching due to acceleration, the lizard's body must move further towards an (intrinsically unstable) bipedal posture that must be maintained while running long distances, often over uneven and complex terrain (Druelle et al., 2019b).

As long as the lizard's body continues to approach its 'steady' posture used during continued bipedal running, the total angular momentum (amount of rotation) will change and, as in the bike-analogy, the rate of change of this momentum must equal the moment of all external forces about a transverse axis at the level of the hip

[10] Note that for the forward dynamics simulations, the model consists of three segments (tail, trunk and head). Tail and head are constrained to a horizontal orientation and hip height remains constant. The point of application of the ground reaction forces moves symmetrically back and forth with respect to the hip (and shoulder when quadrupedal) during stance.

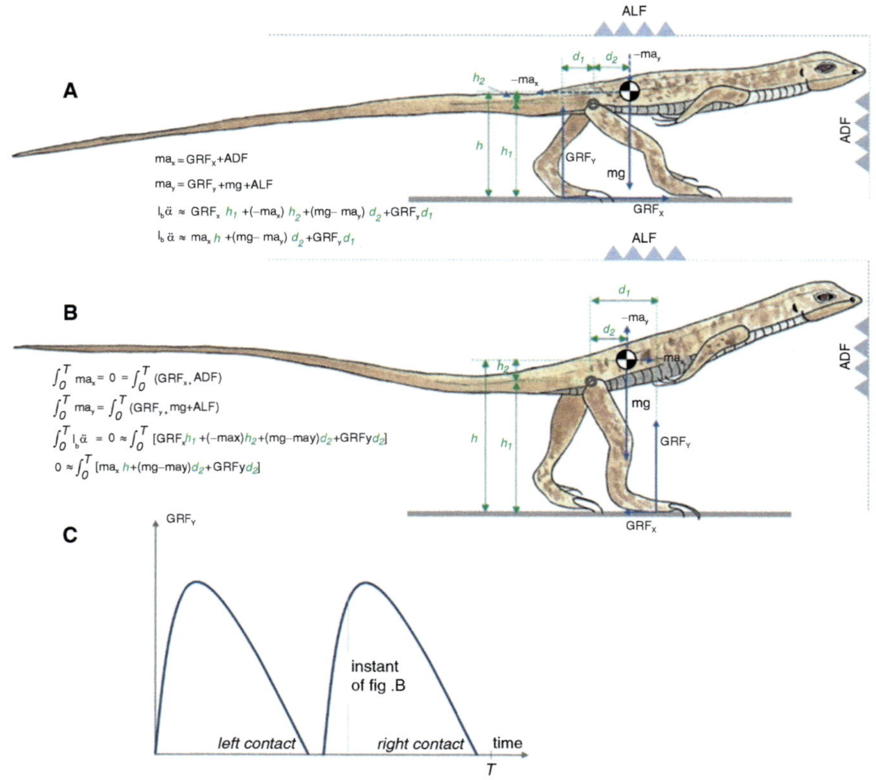

Fig. 14.4 (**a**) Generalized lacertid lizard, accelerating above 'threshold' acceleration a_t. The body pitches upward. At any instant, the horizontal inertia (ma_x) equals the sum of all instantaneous horizontal forces (GRF_x horizontal ground reaction force, ADF horizontal aerodynamic force distributed over the frontal surface). Similarly, instantaneous vertical inertia (ma_y) equals the sum of all instantaneous vertical forces (GRF_y; mg = gravitational force acting at the BCoM (*; $g = -9.81\ m/s^2$); ALF = vertical aerodynamic force distributed over the upper surface), and instantaneous angular inertia respective the hip ($I_b\ \ddot{\alpha}$, i.e. the body's moment of inertia respective of the hip times overall angular acceleration) equals the sum of the moments of all instantaneous forces about the hip, including the effect of the linear inertia (notice that the moment of the aerodynamic forces are not considered in the equation; cf. footnote[3] and text). (**b**) Generalized lacertid lizard, running steadily in a bipedal mode. This can only happen when, integrated over an entire stride (duration = T), accelerations cancel (no change of momentum measured over a stride). (**c**) Since the BCoM is always situated anterior to the hip, vertical ground reaction forces must be skewed early (see text). [Illustrations: Menelia VK illustration]

(or any other parallel axis; Fig. 14.4a).[11] At first glance, this may sound like a simple equation, but for the biological reality of a running lizard (even when reduced to a

---

[11] Note that each segment contributes a local (about its own CoM) and a remote (the moment of the segment's linear inertia about the hip) term (cf. footnote 3) to the total angular momentum.

sagittal plane phenomenon), disentangling the pure 'morphologically based' from the active 'dynamically controlled' pitching mechanics becomes quite complex. Active postural changes and kinematic adjustments (that is, those apparently independent of the acceleration-induced pitching) not only affect the change of total angular momentum (left hand side of the angular equation, Fig. 14.4a), but are also reflected in the external (ground reaction) force, as well as in the moment arms of the latter (right hand side of the angular equation, Fig. 14.4a). None the less, active tail lifting, extending the back and neck and altering the limb kinematics (arms and legs) can all influence the aforementioned torque balance[12] (Aerts et al., 2003; Clemente, 2014; Clemente et al., 2008; Clemente & Wu, 2018; Van Wassenbergh & Aerts, 2013).

Once running steadily (no overall acceleration), the situation changes. Ideally, measured over a cycle (or half a cycle, assuming time- and geometrically symmetrical stepping), there is no change in angular momentum since posture and instantaneous segmental velocities are identical at the start and the end of each (half-)cycle. Consequently, over the same interval, the summed time-integrals of the moments of all external forces (i.e. the angular impulses) about the hip must be zero. As a result, the angular impulses of the moments about the hip of the gravitational force (at the BCoM), the ground reaction force (at the interface between stance limb and substrate) and the aerodynamic forces (drag and potential lift; distributed over the surface) must cancel each other out (Fig. 14.4b). According to Van Wassenbergh and Aerts (2013), the contribution of aerodynamic forces to this is small and, to be effective, require extreme postures and speeds (based on computational fluid dynamics). With the impact of these forces being minimal, it appears that the 'zero-change in angular momentum over a (half-)cycle'-condition for steadily running lizards essentially reduces to the simple requirement of angular impulses of gravitational and vertical ground reaction forces about the hip summing to zero, since the angular impulse of the only remaining horizontal force (the horizontal ground reaction force) then also equals zero. Because the limb during stance oscillates about the hip, the moment of the vertical ground reaction forces about this joint fluctuate from counterclockwise to clockwise during stance, switching from one to the other when the force-application-point at the foot-substrate interface passes beneath the hip joint. The moment of the gravitational force about the hip is clockwise, since the BCoM is always situated anterior to the hip (Van Wassenbergh & Aerts, 2013). Consequently, the angular impulse of the vertical ground reaction force about the hip over a (half-) cycle must differ from zero and must be positive (counterclockwise) in order to cancel the negative (clockwise) angular impulse imposed by gravity.[13]

---

[12]Note that, employing a whole body approach, joint torques (being internal) do not appear directly in this moment balance (cf. footnote 4). As mentioned, these are reflected indirectly in the total angular momentum as well as in the external and inertial moments.

[13]Linear impulses of the gravitational and vertical ground reaction forces over the same time-interval [(half-) cycle] must also cancel each other.

This mechanical requirement constrains the potential locomotor behavioural solutions and thus determines the conditions for convergence. The requirement can only be met either by shifting the force application point anterior to the hip for the greater portion of stance (foot placement further in front of the hip; e.g. *A. boskianus* in Druelle et al., 2019c), by ensuring that the vertical ground reaction force profiles are asymmetrical over stance time (skewed), with the larger forces occurring early in the stance phase when acting in front of the hip, or by a combination of both (Clemente & Wu, 2018; Van Wassenbergh & Aerts, 2013; Fig. 14.4b). Although ground reaction forces are external and flow from the substrate to the lizard, they do result from segmental (hence musculo-skeletal) dynamics. Therefore, whatever the cause of the non-zero angular impulse of the vertical ground reaction (shifted or skewed force profiles), active adjustments of limb kinematics and joint torques are expected. Such a difference in kinematics between quadrupedal and bipedal performance was demonstrated for some species by Clemente and Wu (2018) and Irschick and Jayne (1999). Skewed ground reaction forces (with the larger forces occurring early in stance) were measured in bipedally-running lizards (Clemente et al., 2018; Clemente & Wu, 2018), but were also observed when the specimens ran quadrupedally (Clemente et al., 2018; Clemente & Wu, 2018; Sheffield et al., 2011). However, skewness tends to be greater during bipedal performance and is possibly common to all bipeds with the BCoM lying in front of the hip (Aminiaghdam et al., 2017; Clemente & Wu, 2018; see further). Such findings, together with the results of the phylogenetic analysis of Clemente (2014; cf. above) supports the idea of convergent locomotor behavioural evolution for sustained facultative bipedalism in lizards.

14.2.2 Occasional Bipedalism in Non-human Primates: Voluntary But Anatomically Constrained

The animals best-known for their ability to practice occasional bipedalism are definitely the non-human primates (NHPs). While visiting zoological parks, or hiking in the wild, many of us will have noticed the ease with which primates can suddenly stand on their hind legs and walk bipedally, as if to mimic humans and to provide an immediate reflection of our own condition. Many other aspects of NHP biology are obviously astonishing, but anthropologists and biomechanists have been greatly interested, for almost a century, in the capacity of NHPs to spontaneously stand and walk bipedally (Elftman, 1944; Elftman & Manter, 1935). Indeed, the study of our closest living relatives with regard to the way that they stand and walk on two legs, as well as in what ecological and behavioural contexts they do so, may provide important insights into how the first hominins began to walk bipedally (Foster et al., 2013; Hunt, 1994; Pontzer et al., 2014; Rose, 1976; Taylor & Rowntree, 1973; Thorpe et al., 2007; Wrangham, 1980).

Decades of research on the mechanics of bipedal walking in NHPs have led to the opinion that they are all morphologically able to stand and walk bipedally (e.g. Demes, 2011; Druelle & Berillon, 2014). In other words, their morphology is freed, to some extent, from the constraints related to strict quadrupedal locomotion and pronogrady. This allows them to make a transition, whenever they need, whenever they want, to adopting a bipedal posture or undertake a bipedal locomotor bout. It has been shown that NHPs use bipedalism in many specific behavioural contexts in their natural environments, such as when collecting fruit, carrying items, playing, displaying, during vigilance (Carvalho et al., 2012; Rose, 1976; Videan & McGrew, 2002; Wrangham, 1980), as well as for negotiating specific parts of their habitat in trees (Fleagle, 1976; Stanford, 2006; Thorpe et al., 2007) or on snowy ground (as seen in *Rhinopithecus roxellana*[14]). As a result, it is now clear that there are many behavioural and ecological opportunities for NHPs to stand and walk bipedally. However, although there are multiple reasons for using this mode, there is no obvious context that would require standing and walking on two legs for long periods of time. Also, it has been shown that the typical "bent-hip, bent-knee" posture of NHPs is energetically highly costly because it creates a high flexion moment about the hip and knee and thus requires strenuous muscular activity to support the body weight (Foster et al., 2013). For instance, compared to humans, the cost of transport during bipedal walking in NHPs is much higher (~0.2 J kg^{-1} m^{-1} in humans *versus* >0.55 J kg^{-1} m^{-1} in NHPs) (Demes et al., 2015).

14.2.2.1 Should We Infer that All NHPs Walk Bipedally in a Similar Way?

Overall, it is commonly claimed that the general pattern of bipedal gait in NHPs is similar across species (Demes, 2011; O'Neill et al., 2018; Pontzer et al., 2014; but see Ishida et al., 1974). Although primates are morphologically diverse (Fleagle, 2013), they indeed all appear to perform bipedally in a very similar, non-erratic, and stereotypical way. Basically, NHPs walk on two legs using the typical "bent-hip, bent-knee" (BHBK) posture coupled with a forward-leaning trunk, a limited stride length (mainly because of a limited ability to retract the hindlimbs; but see also the notion of the 'pelvic step' in Thompson et al., 2021), a high stride frequency relative to their quadrupedal locomotion, and knees that remain flexed during the stance phase. Compared to humans, extant NHPs do not exhibit any adaptations for bipedal walking [note that the indriids practice a sort of bipedal gallop when on the ground that is likely to be the result of specialisations for vertical clinging and leaping (Wunderlich & Schaum, 2007)]. Instead, they do what they can with the morphology that they have. Importantly, no structural specialisation is required for them to occasionally walk bipedally. The view that all NHPs walk bipedally in a similar

[14]"*Ce singe est le seul à savoir marcher sur la neige debout*" from National Geographic Wild France.

way (but see Ishida et al., 1974) is certainly reinforced by generally comparing their bipedal BHBK gait to the extended and very efficient one of humans (e.g. O'Neill et al., 2018; Thompson et al., 2021). Given the multiple locomotor and postural adaptations described for NHPs, including brachiation (Jungers & Stern, 1981; Michilsens et al., 2009, 2010), suspensory abilities (Myatt & Thorpe, 2011; Thorpe & Crompton, 2006; Zihlman et al., 2011), leaping (Aerts, 1998; Dunbar, 1988, 1994), climbing (DeSilva, 2009; Fleagle et al., 1981; Hanna et al., 2017; Hirasaki et al., 1995; Kozma et al., 2018; Thompson et al., 2018b), and quadrupedal knuckle-walking (Jenkins & Fleagle, 1975; Kivell & Schmitt, 2009; Tuttle, 1967), the bipedal pattern appears surprisingly consistent across NHP species. One wonders whether the general lens through which the bipedal gait in NHPs has been described is discriminating enough to identify subtle differences that might exist between species. It is possible, however, that some sort of mechanical constraint might "force" all NHP species to walk in a similar way, even if they are morphologically diverse.

14.2.2.2 The Challenge of Walking Bipedally in NHPs

As previously explained (see Introduction), maintaining balance (statically and dynamically) is a challenge during bipedalism because only two limbs contact the ground per cycle. There is thus a constant trade-off between forward propulsion of the body and the need to maintain stability. Bipedal equilibrium thus requires the upper body mass (Head + Arms + Trunk; HAT) to be balanced at the hip (Fig. 14.5) and the stabilization of the HAT segment is a fundamental aspect for efficiency (Ledebt & Bril, 2000; Thompson et al., 2018a; Witte et al., 1991). For example, in bipedally-trained macaques, trunk sway is very limited (~5°). Too many oscillations of a heavy HAT will obviously increase both mechanical work and the risk of losing balance (Winter, 1995). The challenges related to keeping a top-heavy HAT in balance over the hips are not entirely unequivocal. The more top-heavy, the larger the moment of inertia of the HAT and the greater will be the hip torques required to counter a given angular acceleration of that HAT. A larger moment of inertia of the HAT will concomitantly result in a smaller angular acceleration for any given perturbation torque. This, of course, does not affect the muscle torques needed to stabilize the HAT again (for whatever moment of inertia, the perturbation and recovery torque will be balanced). The benefit of the smaller angular acceleration as induced by the perturbation comes with the time available to initiate the proper (motor-)reaction.[15] In any case, the task of keeping the upper body "upright" is more challenging for the (hip extensor) muscles if the centre of mass of the HAT is high, as

[15] This can be compared to a heavy floor lamp accidently perturbed. When it is top-heavy it will initially topple slowly and there is sufficient time to grasp it. Once it has tilted too far, however, you lack the force to counter the large gravitational moment and save it. A lamp with a low moment of inertia will fall over very rapidly, and you will probably fail to grasp it in time. If you manage, however, you do not need too much force as the gravitational moment acting on it is smaller.

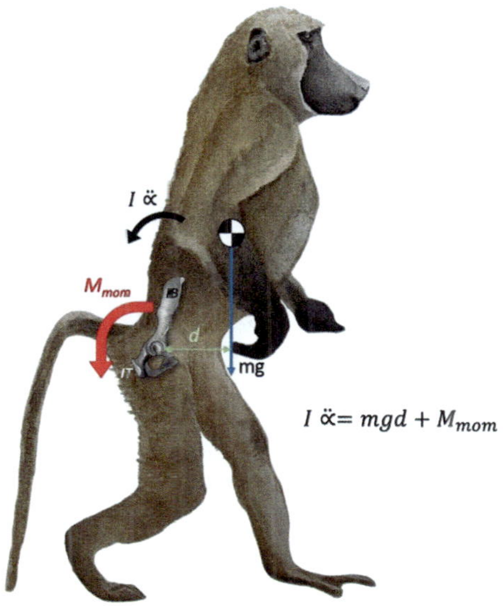

Fig. 14.5 The equilibrium of the HAT (Head + Arm + Trunk) during bipedal walking in the plane of progression is determined by the angular equation of motion (see also part 2.1 and footnote 3): I is the moment of inertia of the HAT, $\ddot{\alpha}$ is the angular acceleration of the HAT, m is the mass of the HAT, $g = -9.81\ m/s^2$ (gravitational acceleration), d is the moment arm of the gravitational force, M_{mom} is the net muscle torque at the hip. The pelvis is highlighted and allows depiction of the long and relatively flat iliac blade (IB), as well as the caudal orientation of the ischial tuberosity (IT). [Illustration: Menelia VK illustration]

it is the case for NHPs (compared to humans). Indeed, the centre of mass remains in front of the hip joints in the BHBK posture, thereby resulting in a high muscle torque at the hip.[16]

In this context, one can ponder why all NHPs walk with a BHBK posture. The vertebral column and the hip girdle are fundamental aspects of bipedal balance and efficiency and are obviously involved in compensating for this BHBK posture.

Specifically, given the caudal orientation of the ischial tuberosity in NHPs, it has been hypothesized that an extended posture (trunk and hind limbs) would greatly reduce the hip extensor (hamstring muscles) moment arm. The mechanical advantage, available with flexed hind limb postures, would indeed be completely lost when the hind limbs are extended (Aiello & Dean, 1990; Kozma et al., 2018; McHenry, 1975; but see Lewton & Scott, 2017). Hence, the BHBK posture enables the maintenance of the lever advantage of the hip extensors during occasional bipedal walking (Foster et al., 2013; Sockol et al., 2007). It has also been suggested that the BHBK posture results from the absence of mobility of the lumbar sector of

[16]EMG data on olive baboons walking bipedally show that, as soon as the bipedal posture is attained, the *latissimus dorsi* muscle is strongly and constantly activated (Druelle, *unpublished data*), but this is not the case during quadrupedal walking.

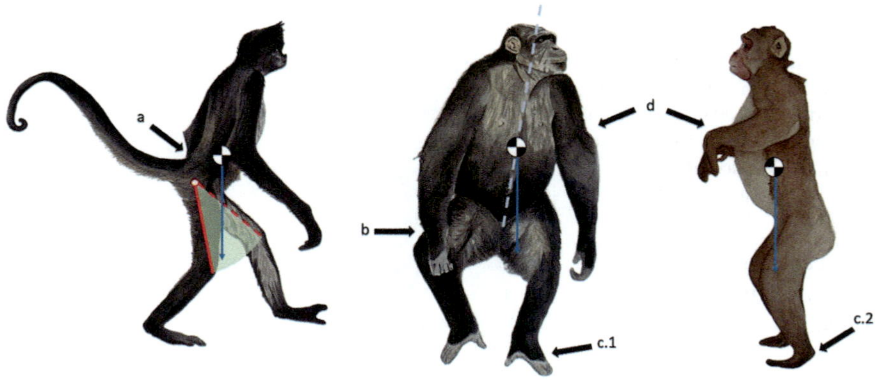

Fig. 14.6 Examples of the variability of postures adopted by non-human primates during bipedal walking and standing. Spider monkey (left); Chimpanzee (middle) and macaque (right). (a) the amount of lumbar entrapment varies between species, (b) the abduction of the thigh is important for chimpanzees, (c) the position of the foot varies between species, from plantigrady (c.1) to semi-plantigrady (c.2), (d) the movement of the forelimbs appears limited in cercopithecoids compared to hominoids. The angle indicated on the spider monkey shows the amplitude of the thigh movements during bipedal walking: the hindlimbs are mainly protracted during bipedalism thus producing the typical hip flexion. The blue dashed line on the chimpanzee shows the listing of the trunk in the frontal plane: the BCoM is shifted toward the supporting foot [Illustrations: Menelia VK illustration]

the vertebral column in NHPs, and more specifically in apes (Lovejoy, 2005; Lovejoy & McCollum, 2010; Machnicki et al., 2016). An "entrapment" of these vertebrae between the ilia would rigidify the lower back, thus impeding NHPs from positioning the upper body centre of mass inside the support polygon without flexing the hip and the knee (Fig. 14.6a). This feature also parallels the extensive protraction, but reduced retraction, of the hindlimbs when walking bipedally and reflects the need to position the foot under the centre of mass of the body. Finally, because the iliac blades are relatively "flat" and face mainly posteriorly in NHPs (this feature is even more accentuated in great apes), previous authors have pointed to a weak abductor mechanism at the hip. Indeed, the gluteal muscles (medius and minimus), which act mainly as hip rotators (in the transverse plane) in NHPs, are unlikely to be able to balance the hip adduction moment generated by a supporting limb in extension. Therefore, adopting a crouched posture allows the hip rotators to maintain the pelvic girdle in a stationary location in the frontal plane (Demes, 2011). Finally, note that some authors have suggested that anatomical limitations at the level of the muscles and fascia that connect the pelvic girdle and thigh in macaques restrict hip-joint extension (Ogihara et al., 2007). In addition, D'Août et al. (2002), by testing hindlimb movements and amplitudes of bonobo individuals under anaesthesia, observed that they are morphologically unable to extend both hip and knee synchronously.

The question remains as to whether all these features apply in the same way to all NHP species or whether each species solves the balance, propulsion, and energetic challenges (related to the important flexing moments) in its own way.

14.2.2.3 Bipedal Kinematics Among NHPs: Variability Within a Common BHBK Pattern

Given the morphology of NHPs, dealing with the moments acting on the upper body in a bipedal posture is feasible, but remains challenging. The data currently available provide the first resources for assessing whether NHPs have developed different strategies for maintaining their balance when walking on two legs. The chimpanzee, *Pan troglodytes,* is certainly the most studied NHP with regard to questions relating to human locomotor evolution (e.g. Demes et al., 2015; Kimura, 1991, 1996; Kimura & Yaguramaki, 2009; O'Neill et al., 2018; Pontzer et al., 2009; Sockol et al., 2007; Thompson et al., 2015), although its choice as a model is sometimes criticized because this species has certainly undergone important morphological modifications in relation to climbing and suspensory adaptations since branching from our last common ancestor (Lovejoy & McCollum, 2010). The chimpanzee, however, remains as a relevant model to study, although among many others (D'Août et al., 2014). The chimpanzee commonly walks bipedally using large step widths and flexed hindlimbs. With the thigh abducted (see Fig. 14.6b) the hip angle fluctuates around 110°, the knee around 120° and the ankle around 80°. The forward pitch of the trunk appears very variable (see Table 3 in Pontzer et al., 2014), but its average is around 30° (Kimura & Yaguramaki, 2009; Pontzer et al., 2014). The stance phase is relatively long (duty factor is well above 0.5) and chimpanzees increase speed by increasing both stride frequency and stride length through the use of a 'pelvic step' (Thompson et al., 2021). Dynamically speaking, and despite the BHBK posture, the oscillations of the centre of mass follow a pendular gait pattern (inverted pendulum model), that is, the maximum height of the centre of mass of the body (BCoM) generally occurs around the middle of the single support phase (Demes et al., 2015; Kimura, 1991; Pontzer et al., 2014). This is performed by shifting the BCoM toward the supporting foot, thus causing the pelvis and trunk to list toward the standing limb (i.e., depression of the pelvis) and the BCoM to rise (Fig. 14.6). This is also observed in the marked medio-lateral ground reaction force impulse during bipedal walking. In the chimpanzee a transfer between potential and kinetic energy is, therefore, possible but these kinematics also lead to considerable medio-lateral movements of the trunk (Demes et al., 2015; Thompson et al., 2018a). As a result, the amplitudes and phasic relationships between potential and kinetic energy remain too variable and dissimilar to be efficient and only low (2–45%) energy recovery rates are observed in this species, as well as high energetic costs (Demes et al., 2015; Pontzer et al., 2009; Sockol et al., 2007); note also that substantial variability exists between individuals.

The closest living relative of the chimpanzee is the bonobo, *Pan paniscus*, for which the bipedal pattern has also been extensively studied (e.g. Aerts et al., 2000a;

D'Août et al., 2002, 2001, 2004; Vereecke et al., 2003). Generally, chimpanzees and bonobos are very similar in the mechanics of their bipedal walking and the spatio-temporal parameters are alike. However, the knee and ankle of bonobos are more flexed, and the trunk is 10° less bent forward. It is uncertain, however, whether these (non-negligible) variations are due to morphological differences between the two species and they might instead result from the particular training regimes of the subjects: the bonobos studied by D'Août et al. (2004) were never trained and lived in their social group in a zoo, whereas the chimpanzees studied by Pontzer et al. (2014) received some training through their varied contact with humans. In support of this, it has been shown that trained Japanese macaques (*Macaca fuscata*), adopt more extended trunk and hindlimb postures during bipedal walking than untrained ones (Hirasaki et al., 2004; Ogihara et al., 2010). Moreover, early and regular training can also directly affect the musculoskeletal system of NHPs (Nakatsukasa et al., 1995; Preuschoft et al., 1988). Careful interpretation of these differences is therefore required. In contrast to chimpanzees, and by estimating functional leg length during a typical bipedal stride and observing the ground reaction force pattern, D'Août et al. (2002, 2004) suggested that there is no efficient inverted-pendulum mechanism in bonobos, although some energy transfer by a pendular mechanism remains possible, certainly at low speed, as is the case in chimpanzees and gibbons (Pontzer et al., 2014; Vereecke et al., 2006b). More generally, the foot mechanics of the African great apes are very similar and share common features. The foot is dorsiflexed during the swing phase and the heel and lateral midfoot commonly touch down simultaneously, rarely presenting a true heel-strike event (Vereecke et al., 2003). This "plantigrade" pattern is different from that of the lesser apes and cercopithecoids. Indeed, gibbons do not make contact with the heel at the beginning of the stance phase and foot contact is described as midfoot/heel-plantigrade (Vereecke et al., 2005). On the other hand, cercopithecoids commonly adopt a semi-plantigrade foot posture during bipedal walking and there is no contact of the heel with the ground, the foot being more plantarflexed (Fig. 14.6c; and see Berillon et al., 2010).

The third well-studied species is the Japanese macaque, *Macaca fuscata* (Blickhan et al., 2018; Hirasaki et al., 2004; Nakatsukasa et al., 2006; Ogihara et al., 2018, 2007). Two categories of macaque have been studied: animals well-trained for bipedal walking and others not specifically trained for bipedalism. Important differences are evident between the two types, with an obvious refinement of the bipedal gait in trained animals. Interestingly, by tracking the kinematics of the BCoM, researchers (Hirasaki et al., 2004; Ogihara et al., 2007) initially showed that trained macaques exhibit a pendular motion of the BCoM with some energy recovery (2–62%) being possible using this mechanism, mainly at low speed (Ogihara et al., 2010). However, the same authors recently reinterpreted their results based on vertical ground reaction force profiles, stating that there is no energy recovery using pendular mechanics in macaques (trained or not) because the double peak was never observed (see Ogihara et al., 2018). They concluded that the bipedal gait of macaques may exclusively rely on spring-mass mechanics, in which kinetic and potential energy fluctuate in-phase (Ogihara et al., 2018). Recent models, however, challenge this view because the ground reaction force profile may be inconsistent with the presence of a simple inverted-pendulum model (Demes et al., 2015; Geyer et al., 2006; Roberts & Azizi, 2011). As a result, it is probable that trained macaques do rely on a spring-loaded inverted pendulum, using their flexible and compliant legs (Schmitt, 1999). A recent study (O'Neill et al., 2018) compared the 3D kinematics of macaques and chimpanzees with those of humans and observed that macaques walk in a very similar way to chimpanzees, as far as the hindlimbs and pelvis movements are concerned. As in chimpanzees, and in contrast with humans, the pelvis rises on the swing limb side, creating an important elevation in the frontal plane (also known as hip hiking) and a rotation in the transverse plane (Kinoshita et al., 2021). The main differences between macaques and chimpanzees are evident in the amount of pelvis tilt, it being more anterior in macaques (by 9° on average) and the amount by which the hip is abducted (10° less) in macaques. The authors suggested that these differences support the idea that a lumbar region of the column freed from pelvic entrapment should allow more upright posture of the trunk (trained macaques maintain the trunk in a more upright posture than chimpanzees). Interestingly, the study of Machnicki et al. (2016) also corroborates this for spider monkeys, which exhibit pelvic and lumbosacral joint features (e.g. a short ilium, a large sacral alar width and a slightly invaginated lumbar column) that should enhance the capacity to carry the trunk more vertically (see Fig. 1 in Machnicki et al., 2016). As a result, hip extension angles are also much greater in spider monkeys than in chimpanzees, macaques, bonobos and baboons (Ishida et al., 1974). Interestingly, spider monkeys also exhibit the most caudal position of the BCoM among NHPs [that is, lying closer to the hip joint (Druelle et al., 2019a)].

Gibbons (*Hylobates lar*) exhibit peculiarities in their way of walking bipedally. Firstly, and in contrast to other NHPs, they increase speed mainly by increasing stride length instead of stride frequency (a pattern also observed in trained macaques) (Vereecke et al., 2006a, c). The forward pitch of the trunk is 15° in gibbons, less than in great apes and cercopithecoids, and the hindlimb joint angles fluctuate around 125° for the hip, 117° for the knee, and 100° for the ankle. It has also been suggested that, given the high percentage of congruity between the fluctuations of kinetic and potential energy, they specifically rely on a bouncing gait, or grounded run (Vereecke et al., 2006b), although at low speed the inverted pendular motion of the BCoM can also be observed (Vereecke et al., 2006b).

Capuchin monkeys, *Cebus apella*, have also been suggested to use a grounded run instead of a walking pattern (Demes, 2011; Demes & O'Neill, 2013). However, the mechanics of bipedal walking in capuchins is very different from that of gibbons. They use a high stride frequency and exhibit a trunk pitch of approximately 23° (trunk sway is 10°). The abduction of the thigh is less than it is in chimpanzees and falls within the range exhibited by macaques. The hip angle is 109°, the knee angle is 94° and the ankle is 105° on average.

Olive baboons (*Papio anubis*) exhibit similarities with macaques with regard to the foot and ankle region and the thigh undergoes little abduction (Berillon et al., 2011, 2010; Druelle et al., 2022). Otherwise, they use a long stance phase (duty factor ~0.7), as do bonobos and chimpanzees. The energy recovery rates (7–26%) related to the inverted pendulum mechanics are relatively low, as they are in capuchins, because of the in-phase fluctuations of the kinetic and potential energy (Druelle et al. *in preparation*). Nevertheless, it appears that there is a negative correlation between recovery rates and speed, thus suggesting that at low speed baboons would be able to use inverted pendulum mechanics.

Unfortunately, there are no bipedal kinematic data available for gorillas and orangutans, and very few for spider monkeys (Ishida et al., 1974). The large range of motion of the hip joints of orangutans appears to make them poor bipedal walkers on the ground, although they can use this posture efficiently in trees for negotiating flexible branches (Thorpe et al., 2007).

Although NHPs walk bipedally in terms of timing (duty factor > 0.5), it has been widely suggested that many species (e.g. capuchin monkeys, gibbons and macaques) actually use a bouncing gait, also called a "grounded run". This view is based on the dynamics of the BCoM, the kinetic and potential energy of which fluctuate in-phase as in the running gait of humans. On the other hand, chimpanzees and trained macaques (and maybe bonobos) are inclined to walk bipedally with an out-of-phase fluctuation of their BCoM. Furthermore, compared to humans, there is no strict transition in NHPs from pendular to mass-spring mechanics with speed.

Although NHPs do not make use of very efficient pendular mechanics, one can ask whether the "grounded run" described for NHPs is efficient. To be so, spring-mass mechanics require a passive absorption, storing and releasing energy (that is, external positive work is required somewhere in the body [hindlimbs]). In other words, it requires time-specific stretching and recoil of tendons and connective tissues in accordance with the fluctuations of the total power. This can occur in the foot (metatarsophalangeal joint, tarsometatarsal joint), at the ankle joint and/or at the knee. These aspects have been mainly studied for gibbons. Based on the kinematics of the joint angles during the bipedal cycle, Vereecke et al. (2006b) estimated that gibbons do not make use of efficient spring mechanics at the ankle joint but can instead store and release energy from the knee extensor muscle-tendon unit. This mechanism, however, is unlikely for capuchin monkeys, but pseudo-elastic springs (absorption phase only) are possible (Demes, 2011). More specifically, Vereecke and Aerts (2008) observed in gibbons that the long digital flexor tendons in the foot are stretched at touch down and elastic energy can be stored and released at the tarsometatarsal and talocrural joints. Later, Aerts et al. (2018) tested the efficiency of the well-developed Achilles tendon present in gibbons. They showed that a spring-like mechanism of the Achilles tendon is very unlikely in this species because only 7.5% of the energy is stored in this structure (compared to 35% in humans when running), and it is released at the wrong moment in the cycle.

14.2.2.4 BHBK Posture as a Case of Mechanical Convergence?

Although it has been increasingly claimed that all NHPs walk with a largely similar "bent-hip, bent-knee" posture, we can readily notice that non-subtle variations actually do exist between species (see above; Fig. 14.6). All NHPs are morphologically different, they all present specific locomotor adaptations (Fleagle, 2013), and they all walk bipedally in their own way. The BHBK posture, rather than reflecting a shared anatomical constraint, might instead (or in addition) reflect a mechanical constraint related to the relationship between the position of the BCoM and the functional leg (that is, the segment that connects the hip and the centre of pressure). It is interesting to observe that the spider monkey is the NHP species with the most caudally-situated BCoM and exhibits the most erect trunk, whereas the chimpanzee has one of the most cranially-situated BCoM and has the most forwardly-tilted trunk when walking bipedally. The typical BHBK strategy in NHPs may thus highlight a mechanical constraint that could induce an evolutionary convergence in species walking bipedally. The posture adopted by extant and extinct primates (including hominins) during bipedal walking might be driven by the position of the BCoM, with the hypothesis being that a more caudally-located position allows for a more erect bipedal posture (see above).

Walking bipedally with extended hindlimb joint postures (hip and knee), including an extension of the spine (lumbar lordosis), is clearly a human pattern (Hogervorst & Vereecke, 2014) and among primates only humans present a set of morphological adaptations for bipedal walking and running (Bramble & Lieberman, 2004; Hogervorst & Vereecke, 2015). With this morphological conformation, humans are able to be very stable on two legs and can maintain their BCoM, that is positioned low and close to the hip joints (see next section), within the support polygon between the feet.

14.2.3 *Extant Habitual and Obligate Bipedalism: Anatomically Imposed and Constrained*

Extant birds and humans can be defined as obligate and habitual bipeds, respectively (see Introduction), yet, their bipedal postures and kinematics are very different (Alexander, 2004). These are determined by very different *Bauplans* that result from different evolutionary histories. Yet, the capacity to effectively walk bipedally should be associated with clear adaptations at the level of the hind limbs that improve performance. Therefore, one can ask whether some morphological features related to bipedalism are convergent in birds and humans.

14.2.3.1 The Bauplans: From the Basal Tetrapod to Birds and Humans

In the basal tetrapod the trunk is supported by two pairs of limbs, with the hind limb articulating with the trunk via the pelvic girdle. The proximal part of the hindlimb, the thigh, is anchored to the trunk by the femoral head articulating with the pelvis at the acetabulum. The shank, situated between the knee and the ankle, has articulations of the tibia with the femur at the knee and with the tarsus at the ankle. The foot, or autopodium, forms the most distal part of the limb and is characterized by the proximally interlocking bones of the tarsus and metatarsus and the more mobile phalangeal bones forming the toes distally. Obviously, human and bird bipedalism are related to important modifications of the shape and positioning of the basic quadruped organizational plan, but these occur in very different contexts in these two groups and on different time-scales (see below).

In humans, the composition of the tetrapod limb is conserved with the same suite of bones. The limbs are parasagittaly oriented, as in all mammals, and the hind limbs are not markedly different from those of non-human primates that are only occasionally bipedal (see above). Habitual bipedalism did not require any major alteration of the limb itself. However, with regard to bipedal evolutionary refinements, the hind limb is relatively longer and has larger joints than other hominoids (see Table 5.1 in Harcourt-Smith, 2007, for a review; see also Bramble & Lieberman, 2004). It is worth noting that although there can be differences in size between human individuals, the proportions of the limbs are fairly consistent, with the tibia and femur varying by up to 6% of the total leg length among extant populations (Porter, 1999). Furthermore, in modern humans, the foot is highly derived with specific features such as the transversal and longitudinal arches that enhance stiffness, shock absorption, efficient push-off, storage and recoil of mechanical energy (e.g. Holowka & Lieberman, 2018; Ker et al., 1987; Venkadesan et al., 2020). The pelvic girdle has been extensively remodeled and "compacted" in humans, which allows the trunk to be held vertically in the so-called upright bipedal posture (e.g. Lovejoy, 2005; Lovejoy et al., 2009). The pelvic girdle takes the form of a basin that supports the contained organs by widening the sacrum, shortening the ilium (that is more laterally projected and flared) and tilting the pubis ventrally.

In birds, the limb is highly modified compared to the basal tetrapod *Bauplan*, not only through the parasagittal orientation of the hind limbs but also by reduction of the number and fusion of skeletal elements (see below). The trunk is greatly stiffened by an ankylosis of the dorsal vertebrae followed by a long series of sacral vertebrae fused into a synsacrum. The shape of the pelvic girdle is also substantially altered. The ilium is very elongated cranially and dorsal to the acetabulum, and is fused medially to the synsacrum; the ischium is fused to the ilium dorsally and forms a lateral part of the pelvic girdle, caudal to the acetabulum. The pubis, taking the form of a long and slender stick, lies ventral to the ischium. The hip joint of birds is always flexed, as is the knee. These joint configurations maintain the trunk and the femur in inclined positions compared to humans (Fig. 14.7). This is the first obvious difference between the hind limb structure of humans and birds. Whereas the trunk is rigid and is often in a forward inclined position in birds (and much more so than in non-human primates, see above), it is held vertically in humans and this is associated with fine adjustments of the spinal curvatures that balance the masses of the HAT (Head + Arms + Trunk) segment and thus stabilize it. Interestingly, stabilizing of the human trunk in a flexed, bent-forward, posture has been suggested to constrain leg function independently of its morphology, and a bird-like pattern of ground reaction force is observed (reduced braking and asymmetric, early skewed, vertical profile), as well as an increase in knee flexion throughout the stance phase, an increase in hip flexion and ankle dorsiflexion (Aminiaghdam et al., 2017).

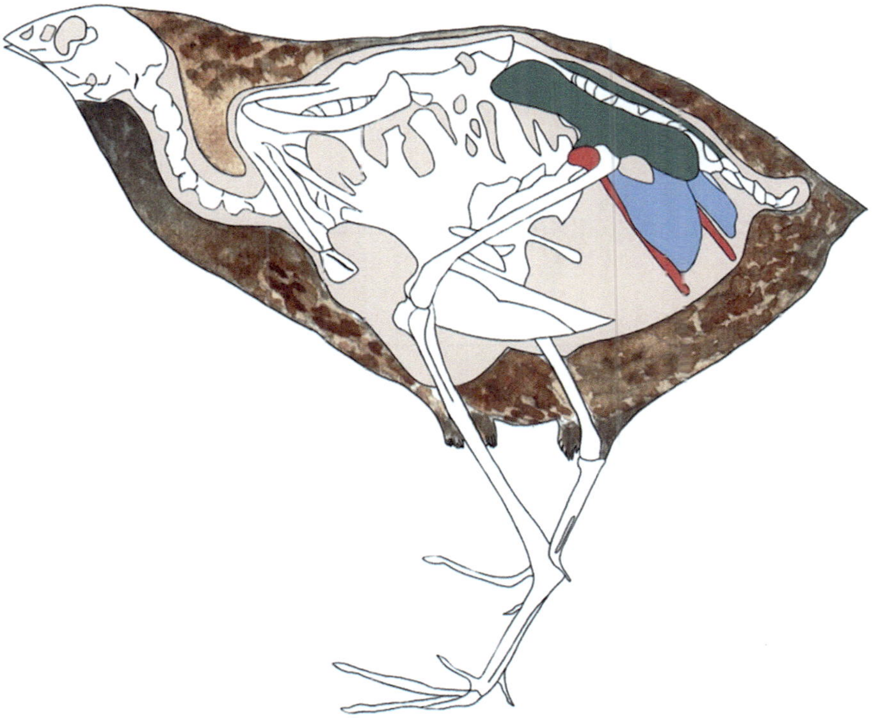

Fig. 14.7 Typical bipedal posture of a quail. The knee and hip joints are flexed so that the trunk and the femur are maintained in inclined positions. The pelvis is highlighted with three different colours and shows the ilium in green, the ischium in blue and the pubis in red. The hind limb consists of three long bones: the femur attached at the hip, then the tibiotarsus followed by the tarsometatarsus with which the four toes are articulated [Illustration: Menelia VK illustration]

A second fundamental difference between birds and humans is the number of limb bones they have. Birds are digitigrade while humans are plantigrade. Indeed, the distal hind limb bones are fused in birds: the tibia, fibula tarsals and metatarsals are distributed over two long bones, the tibiotarsus and the tarsometatarsus. In the standing bipedal position, support is thus provided by the phalanges of the toes that are situated directly below the knee. Therefore, topologically, the human ankle corresponds to the metatarsophalangeal joints of birds. More proximally in the limb, the tibia articulates with the femur in the human knee, and the tarsometatarsus of birds articulates with the tibiotarsus at the intertarsal joint. The femur articulates with the trunk at the hip in humans and the tibiotarsus articulates with the femur at the knee in birds (see Fig. 14.8). Therefore, although the body centre of mass (BCoM) is located above the hind limb, at the level of the iliac crests, in humans (~10 cm above the great trochanter; Palmer, 1944), the limb extends above the BCoM in birds. The avian thigh connects with the rigid and very obliquely-oriented

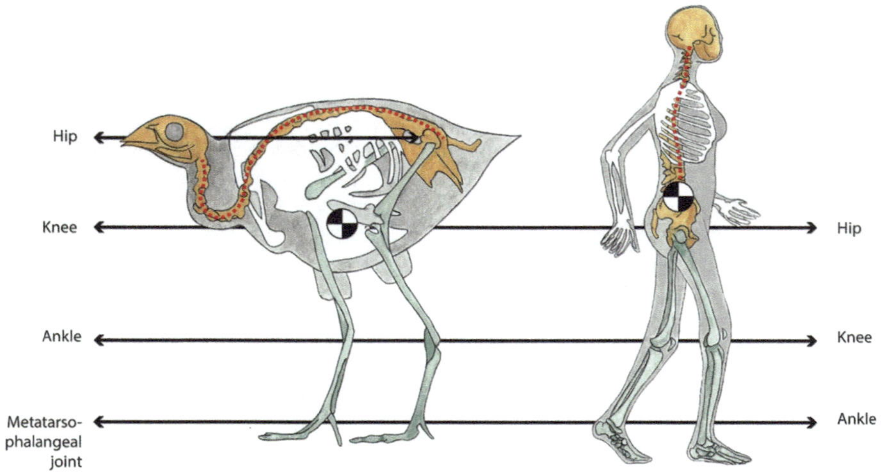

Fig. 14.8 Functional topology of the hindlimbs within the bird and human Bauplans. The human ankle corresponds to the metatarsophalangeal joint of birds, the human knee corresponds to the ankle joint of birds, and the hip in humans corresponds to the knee in birds. In birds, the hip is positioned above the BCoM (indicated by the black and white filled circle) [Illustrations: Menelia VK illustration]

trunk at the hip, which is located dorsally on the body. The enormous mass of the large pectoral flight muscles that attach ventrally to the sternum contribute to the low location of the BCoM, which is situated ventral to the vertebral axis at knee height (Fig. 14.8). As a result, the avian body mass is mainly located below the hips, while in the human system, more than two thirds (~68%) of the body mass is situated above the hips.

In the avian system, the hip moves the trunk, which can be held more or less vertical, depending on the species. In penguins for example, it is almost vertical although the BCoM is always located at knee height. The rigid thorax (Abourachid & Höfling, 2012), with a very large sternum, results in a rather stiff structure to which the wings are attached. Compared to mammals, the organization of the body of birds, except for the neck, is highly conserved in the entire clade. The neck of birds is long, encompassing from 10 to 26 vertebrae (Boas, 1929) and provides for the extensive mobility of the head relative to the (rigid) trunk. The adaptations of the hind limbs of birds are largely reflected in the width of the pelvis (Provini & Höfling, 2020) and the length of the segments incorporating the long bones (Gatesy & Middleton, 1997; Zeffer et al., 2003). The shape of the feet is also quite variable, although the number of toes (4) and the number of phalanges remains almost constant. The hallux is always oriented posteriorly, and toe III always anteriorly. In most cases (88%), toes II and IV are oriented anteriorly, but toe IV and, more rarely, toe II may be oriented posteriorly in perching birds. The hallux may be reduced in walking birds (Abourachid et al., 2017) and may even be absent. The skin of the feet also participates in their adaptations. For example, a web forms an oar in swimmers and skin sheaths connect the fingers, thus increasing the surface area of the foot, in perching birds (Höfling & Abourachid, 2021).

The hindlimb musculature is conserved throughout the avian phylogenetic tree, the fleshy part being mainly distributed in the thigh and shank, while it is mainly aponeurotic-tendinous in the tarsometatarsus and phalanges (that is, in the foot) (Abourachid & Höfling, 2012; Kardong, 2006). The cranio-caudal lengthening of the pelvic girdle provides long moment arms for the hip and knee muscles, whereas the long aponeurotic-tendinous features of the distal part of the limb participate in its lightening; such a limb morphology is commonly associated with cursorial animals (Carrano, 1999). In humans, the musculature of the thigh is voluminous proximally (hamstrings, quadriceps and gluteals), ensuring stability of the hip in extension and during the single stance phase (medio-lateral balance). Its proximal mass, at the level of its origin, is "globular" and forms the human buttocks. For instance, the gluteus maximus is greatly enlarged and its origin and insertion make it capable of generating great power. This also enables the maintenance of the hip in the extended position and the vertical carriage of the trunk (e.g. Hogervorst & Vereecke, 2015).

14.2.3.2 From Bauplans to Different Postural Stability Capacities

While standing bipedally humans and birds have their BCoM positioned above their support polygon, yet, the differences between the two biped *Bauplans* inherently affects their postural stability (Abourachid & Hugel, 2016). For instance, whereas humans lie down to sleep, the capacity to maintain balance while sleeping on one leg is an impressive behaviour in flamingos and is shared by many other birds (Chang & Ting, 2017). Human postural control can be regarded as a complex motor skill during both posture and movement (Pollock et al., 2000; Winter, 2009). Postural orientation involves the active alignment of the trunk and head with respect to gravity, support surfaces, the visual surroundings and internal references in order to control the position of the BCoM (e.g. Horak, 2006; Matthis et al., 2018). Postural stability involves the coordination of movement strategies that stabilize the BCoM with respect to the base of support that is delimited by the feet. To maintain balance in a standing position the ankle and/or hip is mobilized without moving the feet, the first for small amounts of sway on a firm surface, the second to quickly move the BCoM when standing on a narrow or compliant surface (Horak, 2006). Moving a foot is also often used to recover equilibrium. Far less is known about balance-keeping in birds. Observing the seemingly effortless single limb stance (even while sleeping) of many birds may suggest balance-keeping is less demanding (e.g. Chang & Ting, 2017). Similarly, when birds perch for sleeping they may just sit down without flexing the toes to grasp the substrate (Galton & Shepherd, 2012). However, balance control is definitely needed. Beyond the differences in the musculo-skeletal system (see above), the nervous system also participates in the differences in balance control between birds and other tetrapods. The lumbosacral organ is a structure of the lower spinal cord housed in the sacral vertebrae as part of the fused synsacrum of birds. This system acts as a mechanoreceptor that is sensitive to the movement of the trunk and participates in balancing the body (Necker, 2006; Stanchak et al., 2020). It is an analogue of the vestibular system, present in the inner ear of birds and other tetrapods. The presence of two mechanoreceptor systems might contribute to the marked dissociation of head and body movements in birds (Maurice et al., 2006) and play a role during locomotion (Necker, 2006; Stanchak et al., 2020).

14.2.3.3 Phylogenetic Histories and Movement Repertoires

Bipedal behaviour evolved in different contexts in birds and humans. The (habitual) bipedalism of humans is a feature of one genus, *Homo*, and bipedalism is a derived character, or an apomorphic trait, that is shared only, in some respects, with extinct hominins (since ~6 Ma; Harcourt-Smith, 2010). Birds originated from obligate bipedal dinosaurs, the theropods (Gatesy & Middleton, 1997), whose bipedalism was acquired from an archosaur ancestor in the Triassic (Gauthier, 1986). The obligate bipedalism of birds pertains to approximately ten thousand extant species over a large size range, from a few grams to several tens of kilograms (up to 150 kg for the largest ostriches; Abourachid & Höfling, 2012). In birds, flight is the apomorphic trait (that is, the evolutionary innovation shared by all members of the clade). Bipedalism is much older; it is a plesiomorphic, basal trait inherited from archosaurs.

Questions relating to the evolutionary transition toward habitual bipedalism in humans and obligate bipedalism in archosaurs remain intensely debated. With regard to hominin evolution, recent fossil discoveries and novel interpretations are sheding new light on bipedalism in hominoids, because Miocene apes might have used bipedal behaviours in the trees (Böhme et al., 2019, 2020; Hammond et al., 2020). Nevertheless, the current debate about the origin of habitual bipedalism in primates continues to be driven by the consideration of two competing concepts about the locomotor mode for the last common *Pan/Homo* ancestor. On the one hand, a chimpanzee-like repertoire that incorporates significant proportions of climbing behaviours and knuckle-walking on the ground (e.g. Richmond et al., 2001; Zihlman & Bolter, 2015), and on the other, a more generalist orangutan-like repertoire that includes careful suspensory and orthograde clambering locomotion (Crompton et al., 2010; Thorpe et al., 2007). Regardless, and given the potential mosaic of locomotor repertoires exhibited by Miocene apes (Almécija et al., 2021; Senut, 2007), the origin of bipedalism certainly arose from a "generalist" primate, or at least a primate largely freed from the constraints related to the quadrupedal (monkey-like) *Bauplan*, but not yet specialised in a *Pan*-like manner (e.g. *Oreopithecus bambolii*; Hammond et al., 2020). The dominant hypothesis about the emergence of hominoids advocates an increase in body size in a monkey-like arboreal quadruped (Cannon & Leighton, 1994; Hunt, 2016). This would have challenged their ability to effectively move in the trees, thus resulting in the evolution of new locomotor and behavioural strategies. In other words, new mechanical constraints related to an increasing body mass, either in stem hominoids such as in the proconsuloids (e.g. Rafferty et al., 1995; Ruff et al., 1989; Ward et al., 1991), or later in Miocene apes (Alba et al., 2015), that may have driven the evolutionary development of new locomotor strategies to enable them to continue to exploit the arboreal environment. A gradual increase in body size in Miocene apes might have opened the way towards new orthograde locomotor strategies such as brachiation, slow suspensory and clambering behaviours, as well as (arboreal) bipedalism. In this context, bipedal walking may have become the easiest way to move for some hypothetical extinct orthograde apes (well adapted for arboreal life) when "forced" to move on the ground, in a more open environment, for example (see Daver et al., 2022). Adopting an orthograde generalist repertoire in the trees may have resulted in a unique, bipedal solution when moving on the ground. In a changing environment that became gradually more open, bipedal walking would have been gradually refined (through initial pelvic adjustments) to effectively exploit this new ecological niche on the ground (e.g. Holowka & Lieberman, 2018; Lovejoy et al., 2009). In short, the evolutionary transition toward habitual bipedalism in hominins might be the result of initially a transition from a generalist, non-quadrupedal, hominoid to a more specialised one, later leading to the emergence of proto-hominins and bipedal walking (e.g. Lovejoy et al., 2009; Senut et al., 2001). A second stage may have refined the bipedal capacities toward endurance running in the genus *Homo* (Bramble & Lieberman, 2004).

In contrast to the human lineage, the origin of bipedalism in the archosaur lineage occurred long before birds arose. In Triassic archosaurs, bipedalism and cursoriality certainly coevolved; the relative length of the forelimb is negatively correlated with the relative length of metatarsal III (Kubo & Kubo, 2012; see also Persons & Currie, 2017). The behavioural constraints may have come from the necessity to escape or to catch prey, which require capacities for acceleration and manoeuvrability. High acceleration capacities may lead to involuntary bipedal running (see the first section on lizards). Even though a bipedal run generated by acceleration bouts would experience mass-related constraints for a lizard-like species weighing more than 0.5 kg (see Sect. 14.2.1.1), other *Bauplans* may have facilitated such an instantaneous transition. For example, the more dorsally-situated the CoM of the HAT (via more parasagittal leg postures; e.g. Carrano, 1999), the more rapidly the threshold acceleration is reached. Similarly, since habitual bipedal archosaurs present a specific *Bauplan* with a BCoM located closer to the hips (smaller '*d*', cf. Sect. 14.2.1.1) and longer relative hindlimbs (larger '*h*') compared to quadrupedal archosaurs (Bishop et al., 2020), a quadrupedal-bipedal transition resulting from the effects of acceleration seems plausible. From these bipedal ancestors, theropod dinosaurs acquired flying abilities associated with modification of the proportions of the body and legs and a reduction of the tail (Gatesy & Dial, 1996). However, in the extant birds (Neornithes) diversification and radiation occurred rapidly worldwide following the Cretaceous-Paleogene transition (~66 Mya; Claramunt & Cracraft, 2015). Extant avians represent an adaptive radiation that is related to flight capacities. The evolutionary success of modern birds is readily evident from the number of species, representing the greatest diversity among extant tetrapods, their wide size range, and their distribution across many environments. They have been able to colonize most of the Earth's biomes, from the poles to the equator, occupying terrestrial, aerial, and aquatic environments. Coupled with this is the great diversity in the way birds use their hindlimbs. In terrestrial environments they walk or hop on the ground, although running is not common because escape is mostly achieved through flight (but see below in our consideration of terrestrial birds). Birds are also commonly found in trees. The canopy is composed of a network of branches and leaves of complex geometry (the size and orientation of the substrates vary from branch to branch). Hence, the mechanical properties of the supports change from stable, for relatively large branches, to very flexible and mobile, for thin branches.[17] Most species have two main modes of locomotion, flying and walking (Earls, 2000; Provini & Abourachid, 2018; Provini et al., 2012; Tobalske & Dial, 2000), that rely upon a clear neuromotor decoupling of the fore- and hindlimbs. For example, when flying, only the wings are activated, the legs are at rest and stretched backwards (Gatesy & Dial, 1996). In addition to these modes, some species are also able to swim and dive (Baudinette & Gill, 1985). Interestingly, there is also a case of quadrupedal walking in one bird species: the nestling Hoatzin that climbs through the vegetation using alternating leg and wing movements (Abourachid et al., 2019). However, this is a unique case, in an immature bird, associated with an anatomical curiosity, the presence of functional claws on the wings, which disappear as soon as flight is acquired. Some species are secondarily flightless and can only move on the ground using their legs. In this context, the wings have lost their locomotor function. In contrast, the locomotor role of the legs never disappears, even in birds that hardly walk, as in the apodiforms (Zusi, 2013). The legs are also always employed in landing and take-off, and no bird species is strictly aerial, or aquatic, since their mode of reproduction requires them to come onto land to lay eggs in a nest and incubate them (Abourachid & Höfling, 2012). Therefore, the coordination of leg movements and their dynamics is highly diverse in birds, from alternating movements during walking, running and swimming (Abourachid et al., 2011, 2005; Abourachid & Renous, 2000; Provini et al., 2012) to more synchronous movements for landing, taking-off and hopping (Hayes & Alexander, 1983; Höfling et al., 2006; Provini & Abourachid, 2018; Verstappen et al., 2000).

Walking bipedally is the most efficient means of moving about for humans (~70% energy recovery using the pendulum principle). Bipedal walking by terrestrial birds is at least as efficient as the bipedal walking of humans (up to 80% of energy recovery in chickens; Muir et al., 1996). Running is used for increasing speed and leads to a clear change in dynamics in humans and birds (Cavagna et al., 1977). Obviously, other leg coordination modes are possible in humans, as in birds, such as climbing, clambering, swimming and jumping, using simultaneous or alternating (when running and jumping over obstacles) movements of the hind legs. This neuromotor flexibility results from their evolutionary origin and is firmly linked to

[17] Jumping is also a gait related to locomotion in this environment (Provini & Höfling, 2020). Some birds move easily vertically on tree trunks, up or down. Others, light enough not to break them, can perch on very thin and flexible supports, such as grass. Birds are capable of very acrobatic movements in trees when building nests (Hansell, 2000) or when reaching for succulent food, wherever it is located.

the arboreal environment (see above). The human locomotor repertoire is, therefore, not reduced to walking and running capacities, but the "natural locomotor repertoire" of humans has never been quantified and the other locomotor behaviours are generally considered occasional, yet are sometimes important for survival (see Kraft et al., 2014). Thus, in modern humans, walking and running are mainly performed on wide and relatively flat substrates (at a broad scale, because substrates may also be largely uneven) that are either level or inclined. During walking, the swinging of the arms, without being obligatory, balances the body by counterbalancing the rotation of the pelvis. Although many nestling birds use their wings to help their bipedal progression when escaping on inclined supports (the synchronous flapping of the wings helps the feet to adhere by pressing the bird against the support; Dial, 2003; Dial et al., 2008; Tobalske & Dial, 2007), the wings are not used when birds walk steadily on the ground. None the less, human bipedal locomotor capacities have allowed them to inhabit all kinds of terrestrial environments in all climates and human populations, therefore, can be found nearly everywhere.

14.2.3.4 Is the Foot of Birds and Humans Convergent?

The question can be posed as to whether avian and human bipedalism exhibit convergent anatomical features and whether they might be adapted to similar conditions. Considering the different environmental conditions of habitat and locomotor repertoire (see above) employed by each, the answer is not straightforward. Depending on the species, birds may be adapted to various ecological niches: arboreal, aerial, aquatic, and terrestrial. Overall, humans are clearly adapted to the terrestrial habitat. Thus, any potential for convergence should be explored by comparing terrestrial birds and humans. At first glance, in this regard, one may regard penguins, because of the upright carriage of their trunk, as an appropriate comparative model. It does not appear, however, that this is useful because the body shape of penguins is an adaptation to aquatic life and thermoregulation (Livezey, 1989). However, close consideration of the adaptations for arboreal and terrestrial locomotion reveals that the foot represents a fundamental limb segment that has been under strong selective pressure in both primates in general and birds. Indeed, the grasping autopod can certainly be considered as a convergent trait in the adaptation of primates and birds to arboreal life (Abourachid et al., 2017). The hallux, as opposed to the other toes, is used for gripping branches in primates (e.g. Toussaint et al., 2020). In birds, the hallux is relatively larger in arboreal species than in terrestrial ones (Abourachid et al., 2017). In primates, species living on thin flexible branches have a relatively larger hallux compared to that of primates that walk on the ground or on larger branches (Preuschoft et al., 1993). In humans, adapted for terrestrial environments, the hallux has lost its opposability and all the toes are oriented anteriorly. In terrestrial birds the hallux is relatively smaller than it is in arboreal birds (Abourachid et al., 2017) and birds that never perch. In the latter situation, it can be very small, as in the penguins (Sphenisciformes) or Kittiwake

(*Rissa tridactyla*) or even non-existent in the strictly terrestrial emu. In the ostrich, not only the hallux but also the second toe has disappeared.

Thus, convergence in human and bird locomotion is unlikely to be directly related to bipedalism as such, but rather the features of the segment that is at the interface between the substrate and the body, the foot, which may show some functional similarities. As in other convergences in tetrapods, for aquatic or aerial conditions, they mainly concern the autopodium, the most distal segment, through which forces are exchanged with the substrate. Hence, the autopodium of tetrapods appears to often be involved in cases of convergence. Adaptations for flapping, for example, in mammals and archosaurs fundamentally involve modifications of the hand through lengthening of the fingers in bats and their reduction in birds.

14.3 Mechanical Convergence in Extant Bipeds: Linking Occasional, Habitual and Obligate Bipedalism

Quadrupeds that are occasionally bipedal, as discussed in Sects. 14.2.1.1 and 14.2.1.2 (lizards and non-human primates) have in common the attribute that, while moving bipedally, the centre of mass remains located anterior to the hip (Figs. 14.4, 14.5, and 14.6). This implies that during steady bipedal walking/running the downward angular impulse of body weight about the hip, as measured over a (half-)stride (i.e. the time integral of the moment of body weight), must be balanced by an upward angular impulse of the vertical ground reaction force about the hip (cf. torque equation in Fig. 14.4b; see footnotes 3 and 7). To achieve this, a more anterior foot placement and/or greater vertical ground reaction forces early in stance (that is, early skewed force profiles; see Fig. 14.4c and Sect. 14.2.1.1 for more details) are essential. These functional requirements are linked to a suite of kinematic differences: a more horizontally-oriented thigh, flexed knee, longer functional limb length at foot placement than at lift-off, and shallower touch-down angle of the functional limb than the angle at lift-off (e.g. Andrada et al., 2014; Blickhan et al., 2015; and see, for instance, the spider monkey in Fig. 14.6). In a generalized habitual quadruped, hindlimbs can be assumed to have adapted to oscillate, more or less, symmetrically with respect to the hip. The above-mentioned kinematic and dynamic limb adjustments that this quadruped requires to enable it to proceed in a steady, occasional bipedal manner all represent asymmetries in the sagittal plane relative to the hip (e.g. Aminiaghdam et al., 2017; Andrada et al., 2014; Blickhan et al., 2015). These inevitably comply with considerably higher hip and knee extending torques than are normally required for moving on all fours. Indeed, because of the asymmetries linked to more anterior foot placement, thighs are continuously in an anteriorly-oriented, more horizontal position, and the knees are continuously flexed throughout stance (that is, held in an obligately 'bent-hip, bent-knee' posture) and increased extensor torques are essential at these joints to prevent collapse of the limb.

In turn, and because of the altered limb configuration (especially the more extended limb and shallower limb angle at foot placement), these increased joint torques also almost automatically lead to the early skewed vertical ground reaction forces. Higher torque requirements represent significantly higher locomotor costs (in a certain manner comparable to the additional costs related to accelerated running when torques must be higher to increase horizontal propulsive forces; this can be appreciated as soon as one experiences the exhausting nature of interval training). Moreover, as a result of the postural asymmetry of the limb with respect to the hip, the normal symmetrical limb flexion-extension cycle may be lost (e.g. the continuous greater knee flexion during stance in bipedally walking bonobos cf. D'Août et al., 2002) and elastic recoil mechanisms for energy recovery may become less efficient (or even absent). These mechanical constraints explain both the functional/constructional convergence (the "bent-hip, bent-knee" posture is dictated by the required balance of the angular impulses) and, when present, the 'occasional' nature of bipedalism (dictated by the high costs) in these quadrupeds.

As mentioned already in the Introduction of this chapter, it has been argued that occasional bipedalism is an intermediate evolutionary stage that may lead to habitual or obligate bipedality (e.g. Harcourt-Smith, 2010; Persons & Currie, 2017). Therefore, initially occasional bipedalism (either being a mechanical consequence or part of the postural-locomotor repertoire; cf. Sects. 14.2.1.1 and 14.2.1.2) should become favoured in an ongoing selective process because, for instance, enhanced bipedal behaviour became advantageous as such (cf. the evolutionary scenario outlined for the Miocene apes in Sect. 14.2.1.3), or because the forelimbs became too committed to functions other than locomotion and support (as probably happened in archosaurs, and definitely so in birds; cf. Sect. 14.2.1.3). Concomitantly, mechanical and constructional constraints that hampered the optimization of obligate or habitual bipedalism had then to be circumvented by further adaptive changes of the locomotor apparatus itself.

The above-mentioned functional/constructional convergence imposed by mechanical constraints in occasional bipeds (such as the "bent-hip, bent-knee" posture, with the thigh in a forward, more horizontal position to enable a more anterior touchdown of the foot, collectively resulting in asymmetrical limb oscillations relative to the hip) can conceptually be considered as a kind of virtual limb connecting the BCoM to the foot, symmetrically moving in front of and behind the BCoM (e.g. Andrada et al., 2014; Blickhan et al., 2015). According to this generalized simplified concept, angular impulses about the BCoM over a (half-)stance time of the (vertical) ground reaction forces cancel (= zero; notice that body weight itself logically exerts no moment about the BCoM) so that an upright posture can be maintained. However, the inherent 'high cost' problem remains and, in an evolutionary context, could only be solved in two basic ways: (a) make the bipedal functional/anatomical limb coincide with the virtual limb, or (b) make the virtual limb coincide with the anatomical (quadrupedal) limb (see also Blickhan et al., 2015).

The first option accords with the evolutionary solution seen in obligate bipedalism in birds (and bipedal archosaurs in general). In fact, birds retained the 'bent-hip, bent-knee' posture of the occasional biped. However, as mentioned in Sect. 14.2.1.3., the BCoM in birds is, throughout terrestrial locomotor cycles, situated more or less at the level of the knee. Making the functional/anatomical limb coincident with the virtual limb is thus simple: execute a stepping cycle primarily by oscillating the distal limb segments (tibiotarsus and tarsometatarsus) at the knee. With the musculature changed accordingly, the knee joint works as the 'functional hip' of other tetrapods (cf. Fig. 14.7). Angular movements at the anatomical hip, on the other hand, remain very small (e.g. Stoessel & Fischer, 2012; Verstappen et al., 2000). Therefore, work done at the anatomical hip remains limited (so that, to a great extent, the heavy HAT-segment can 'passively' hang from the hips; Blickhan et al., 2015). The ankle joint functions as a reversed classical tetrapod knee (cf. Fig. 14.7). In this way, the distal (functional) limb segments are able to work efficiently according the conservative spring-mass paradigm (e.g. Blickhan et al., 2015).

The second option represents the adaptations seen in human habitual bipedalism. The anatomical changes observed during human evolution (e.g. shape of the pelvis; lordosis of the spine; associated muscular changes; body proportions) resulted in a shift of the BCoM (on an evolutionary time-scale) backwards towards the vertically-oriented hip-level, necessitating that the oscillating virtual limb must move with it to keep bipedal performance possible. Since the moment arm of the body weight about the hip becomes gradually smaller during this process, the required additional extending torques (hence the costs) also gradually diminish together with the anatomical changes (thus providing a clear adaptive advantage). Ultimately, with the BCoM situated above the hip (on average throughout a stride), the anatomical and virtual limb quasi-coincide and the required torques become minimal.

The above mechanistic deduction thus effectively links the functional, mechanical and constructional/anatomical aspects of occasional, habitual and obligate bipedalism observed in tetrapods. The plausibility of the evolutionary scenario is, in fact, well supported, yet in a reversed way, by the elegant experiments carried out by Aminiaghdam et al. (2017). Testing people walking normally, as well as with the HAT (head-arms-trunk) in a more and more flexed position, revealed limb kinematics and dynamics that gradually change towards the predicted occasional (hence, conforming to the first evolutionary option, also bird-like obligate) bipedal behaviour: kinematic asymmetry of the oscillation regarding the hip and early skewed ground reaction forces.

References

Abourachid, A., & Höfling, E. (2012). The legs: A key to bird evolutionary success. *Journal of Ornithology, 153*, 193–198.

Abourachid, A. & Hugel, V. (2016). The natural bipeds, birds and humans: An inspiration for bipedal robots. In *Conference on biomimetic and biohybrid systems* (pp. 3–15). Springer.

Abourachid, A., & Renous, S. (2000). Bipedal locomotion in ratites (Paleognatiform): Examples of cursorial birds. *Ibis, 142*, 538–549.

Abourachid, A., Höfling, E., & Renous, S. (2005). Walking kinematics parameters in some paleognathous and neognathous neotropical birds. *Ornitologia Neotropical, 16*, 471–479.
Abourachid, A., Hackert, R., Herbin, M., Libourel, P. A., Lambert, F., Gioanni, H., Provini, P., Blazevic, P., & Hugel, V. (2011). Bird terrestrial locomotion as revealed by 3D kinematics *Zoology, 114*, 360–368.
Abourachid, A., Fabre, A.-C., Cornette, R., & Höfling, E. (2017). Foot shape in arboreal birds: Two morphological patterns for the same pincer-like tool. *Journal of Anatomy, 231*, 1–11.
Abourachid, A., Herrel, A., Decamps, T., Pages, F., Fabre, A.-C., Van Hoorebeke, L., Adriaens, D., & Amado, M. A. G. (2019). Hoatzin nestling locomotion: Acquisition of quadrupedal limb coordination in birds. *Science Advances, 5*, eaat0787.
Aerts, P. (1998). Vertical jumping in *Galago senegalensis*: The quest for an obligate mechanical power amplifier. *Philosophical Transactions of the Royal Society of London. Series B: Biological Sciences, 353*, 1607–1620.
Aerts, P., Van Damme, R., Van Elsacker, L., & Duchene, V. (2000a). Spatio-temporal gait characteristics of the hind-limb cycles during voluntary bipedal and quadrupedal walking in bonobos (*Pan paniscus*). *American Journal of Physical Anthropology, 111*, 503–517.
Aerts, P., Van Damme, R., Vanhooydonck, B., Zaaf, A., & Herrel, A. (2000b). Lizard locomotion: how morphology meets ecology. *Netherlands Journal of Zoology, 50*, 261–278.
Aerts, P., Van Damme, R., D'Août, K., & Van Hooydonck, B. (2003). Bipedalism in lizards: Whole-body modelling reveals a possible spandrel. *Philosophical Transactions of the Royal Society of London B: Biological Sciences, 358*, 1525–1533.
Aerts, P., D'Août, K., Thorpe, S., Berillon, G., & Vereecke, E. (2018). The gibbon's Achilles tendon revisited: Consequences for the evolution of the great apes? *Proceedings of the Royal Society B: Biological Sciences, 285*, 1880.
Aiello, L., & Dean, C. (1990). *An introduction to human evolutionary anatomy*. Academic Press.
Alba, D. M., Almécija, S., DeMiguel, D., Fortuny, J., de los Rìos, M. P., Pina, M., Robles, J. M., & Moyà-Solà, S. (2015). Miocene small-bodied ape from Eurasia sheds light on hominoid evolution. *Science, 350*.
Alexander, R. M. (2004). Bipedal animals, and their differences from humans. *Journal of Anatomy, 204*, 321–330.
Almécija, S., Hammond, A. S., Thompson, N. E., Pugh, K. D., Moyà-Solà, S., & Alba, D. M. (2021). Fossil apes and human evolution. *Science, 372*, 6542.
Aminiaghdam, S., Rode, C., Müller, R., & Blickhan, R. (2017). Increasing trunk flexion transforms human leg function into that of birds despite different leg morphology. *The Journal of Experimental Biology, 220*, 478–486.
Andrada, E., Rode, C., Sutedja, Y., Nyakatura, J. A., & Blickhan, R. (2014). Trunk orientation causes asymmetries in leg function in small bird terrestrial locomotion. *Proceedings of the Royal Society B: Biological Sciences, 281*, 20141405.
Arnold, S. J. (1983). Morphology, performance and fitness. *American Zoologist, 23*, 347–361.
Baudinette, R. V., & Gill, P. (1985). The energetics of 'flying'and 'paddling'in water: locomotion in penguins and ducks. *Journal of Comparative Physiology B, 155*, 373–380
Berillon, G., Daver, G., D'Août, K., Nicolas, G., de la Villetanet, B., Multon, F., Digrandi, G., & Dubreuil, G. (2010). Bipedal versus quadrupedal hind limb and foot kinematics in a captive sample of *Papio anubis*: Setup and preliminary results. *International Journal of Primatology, 31*, 159–180.
Berillon, G., D'Août, K., Daver, G., Dubreuil, G., Multon, F., Nicolas, G., Villetanet, B., & Vereecke, E. E. (2011). In what manner do quadrupedal primates walk on two legs? Preliminary results on Olive Baboons (*Papio anubis*). In L. Barrett (Ed.), *Primate locomotion* (pp 61–82). Springer.
Bishop, P. J., Bates, K. T., Allen, V. R., Henderson, D. M., Randau, M., & Hutchinson, J. R. (2020). Relationships of mass properties and body proportions to locomotor habit in terrestrial Archosauria. *Paleobiology, 46*, 550–568.

Blickhan, R., & Full, R. J. (1987). Locomotion energetics of the ghost crab: II. Mechanics of the centre of mass during walking and running. *Journal of Experimental Biology, 130*, 155–174.

Blickhan, R., Andrada, E., Müller, R., Rode, C., & Ogihara, N. (2015). Positioning the hip with respect to the COM: Consequences for leg operation. *Journal of Theoretical Biology, 382*, 187–197.

Blickhan, R., Andrada, E., Hirasaki, E., & Ogihara, N. (2018). Global dynamics of bipedal macaques during grounded and aerial running. *Journal of Experimental Biology, 221*(24), jeb178897.

Boas, J. E. V. (1929). *Biologisch-anatomische Studien über den Hals der Vögel. Det Kongelige Danske Videnskabernes Selskab Skrifter, Naturvidenskabelige og mathematiske Afdeling* (Vol. 9, pp. 101–222).

Böhme, M., Spassov, N., Fuss, J., Tröscher, A., Deane, A. S., Prieto, J., Kirscher, U., Lechner, T., & Begun, D. R. (2019). A new Miocene ape and locomotion in the ancestor of great apes and humans. *Nature, 575*, 489–493.

Böhme, M., Spassov, N., DeSilva, J. M., & Begun, D. R. (2020). Reply to: Reevaluating bipedalism in *Danuvius*. *Nature, 586*, E4–E5.

Bramble, D. M., & Lieberman, D. E. (2004). Endurance running and the evolution of *Homo*. *Nature, 432*, 345–352.

Buss, D. M., Haselton, M. G., Shackelford, T. K., Bleske, A. L., & Wakefield, J. C. (1998). Adaptations, exaptations, and spandrels. *American Psychologist, 53*, 533.

Cannon, C. H., & Leighton, M. (1994). Comparative locomotor ecology of gibbons and macaques: Selection of canopy elements for crossing gaps. *American Journal of Physical Anthropology, 93*, 505–524.

Carrano, M. (1999). What, if anything, is a cursor? Categories versus continua for determining locomotor habit in mammals and dinosaurs. *Journal of Zoology, 247*, 29–42.

Carrier, D. R. (1987). The evolution of locomotor stamina in tetrapods: Circumventing a mechanical constraint. *Paleobiology, 13*, 326–341.

Carvalho, S., Biro, D., Cunha, E., Hockings, K., McGrew, W. C., Richmond, B. G., & Matsuzawa, T. (2012). Chimpanzee carrying behaviour and the origins of human bipedality. *Current Biology, 22*, R180–R181.

Cavagna, G. A., Heglund, N. C., & Taylor, C. R. (1977). Mechanical work in terrestrial locomotion: Two basic mechanisms for minimizing energy expenditure. *American Journal of Physiology-Regulatory, Integrative and Comparative Physiology, 233*, R243–R261.

Chang, Y.-H., & Ting, L. H. (2017). Mechanical evidence that flamingos can support their body on one leg with little active muscular force. *Biology Letters, 13*, 20160948.

Christian, A., Horn, H., & Preuschoft, H. (1994). Bipedalie bei rezenten Reptilien. *Natur und Museum, 124*, 45–57.

Claramunt, S., & Cracraft, J. (2015). A new time tree reveals Earth history's imprint on the evolution of modern birds. *Science Advances, 1*, e1501005.

Clemente, C. J. (2014). The evolution of bipedal running in lizards suggests a consequential origin may be exploited in later lineages. *Evolution, 68*, 2171–2183.

Clemente, C. J., & Wu, N. C. (2018). Body and tail-assisted pitch control facilitates bipedal locomotion in Australian agamid lizards. *Journal of the Royal Society Interface, 15*, 20180276.

Clemente, C. J., Withers, P. C., Thompson, G., & Lloyd, D. (2008). Why go bipedal? Locomotion and morphology in Australian agamid lizards. *Journal of Experimental Biology, 211*, 2058–2065.

Clemente, C. J., Bishop, P., Newman, N., & Hocknull, S. (2018). Steady bipedal locomotion with a forward situated whole-body centre of mass: the potential importance of temporally asymmetric ground reaction forces. *Journal of Zoology, 304*, 193–201.

Crompton, R., Vereecke, E., & Thorpe, S. (2008). Locomotion and posture from the common hominoid ancestor to fully modern hominins, with special reference to the last common panin/hominin ancestor. *Journal of Anatomy, 212*, 501–543.

Crompton, R. H., Sellers, W. I., & Thorpe, S. K. (2010). Arboreality, terrestriality and bipedalism. *Philosophical transactions of the Royal Society of London. Series B, Biological sciences, 365*, 3301–3314.

Curtin Nancy, A., Woledge Roger, C., & Aerts, P. (2005). Muscle directly meets the vast power demands in agile lizards. *Proceedings of the Royal Society B: Biological Sciences, 272*, 581–584.

D'Août, K., Aerts, P., De Clercq, D., Schoonaert, K., Vereecke, E., & Van Elsacker, L. (2001). Studying bonobo (*Pan paniscus*) locomotion using an integrated setup in a zoo environment: preliminary results. *Primatologie, 4*, 191–206.

D'Août, K., Aerts, P., De Clercq, D., De Meester, K., & Van Elsacker, L. (2002). Segment and joint angles of hind limb during bipedal and quadrupedal walking of the bonobo (*Pan paniscus*). *American Journal of Physical Anthropology, 119*, 37–51.

D'Août, K., Vereecke, E., Schoonaert, K., De Clercq, D., Van Elsacker, L. & Aerts, P. (2004). Locomotion in bonobos (*Pan paniscus*): Differences and similarities between bipedal and quadrupedal terrestrial walking, and a comparison with other locomotor modes. *Journal of Anatomy, 204*, 353–361.

D'Août, K., Aerts, P., & Berillon, G. (2014). Using primate models to study the evolution of human locomotion: Concepts and cases. *Bulletins et Mémoires de la Société d'Anthropologie de Paris, 26*, 1–6.

Daver, G., Guy, F., Mackaye, H. T., Likius, A., Boisserie, J.-R., Moussa, A., Pallas, L., Vignaud, P., & Clarisse, N. D. (2022). Postcranial evidence of late Miocene hominin bipedalism in Chad. *Nature, 609*(7925), 94–100. https://doi.org/10.1038/s41586-022-04901-z

Demes, B. (2011). Three-dimensional kinematics of capuchin monkey bipedalism. *American Journal of Physical Anthropology, 145*, 147–155.

Demes, B., & O'Neill, M. C. (2013). Ground reaction forces and center of mass mechanics of bipedal capuchin monkeys: Implications for the evolution of human bipedalism. *American Journal of Physical Anthropology, 150*, 76–86.

Demes, B., Thompson, N. E., O'Neill, M. C., & Umberger, B. R. (2015). Center of mass mechanics of chimpanzee bipedal walking. *American Journal of Physical Anthropology, 156*, 422–433.

DeSilva, J. M. (2009). Functional morphology of the ankle and the likelihood of climbing in early hominins. *Proceedings of the National Academy of Sciences, 106*, 6567–6572.

Dial, K. P. (2003). Wing-assisted incline running and the evolution of flight. *Science, 299*, 402–404.

Dial, K. P., Jackson, B. E., & Segre, P. (2008). A fundamental avian wing-stroke provides a new perspective on the evolution of flight. *Nature, 451*, 985–989.

Druelle, F., & Berillon, G. (2014). Bipedalism in non-human primates: A comparative review of behavioural and experimental explorations on catarrhines. *BMSAP, 26*, 1–10.

Druelle, F., Berthet, M., & Quintard, B. (2019a). The body center of mass in primates: Is it more caudal than in other quadrupedal mammals? *American Journal of Physical Anthropology, 169*, 170–178.

Druelle, F., Goyens, J., Vasilopoulou-Kampitsi, M., & Aerts, P. (2019b). Compliant legs enable lizards to maintain high running speeds on complex terrains. *Journal of Experimental Biology, 222*, jeb195511.

Druelle, F., Goyens, J., Vasilopoulou-Kampitsi, M., & Aerts, P. (2019c). Small vertebrates running on uneven terrain: A biomechanical study of two differently specialised lacertid lizards. *Scientific Reports, 9*, 1–11.

Druelle, F., Özçelebi, J., Marchal, F., & Berillon, G. (2022). Development of bipedal walking in olive baboons, Papio anubis: A kinematic analysis. *American Journal of Biological Anthropology, 177*(4), 719–734. https://doi.org/10.1002/ajpa.24454

Dunbar, D. C. (1988). Aerial maneuvers of leaping lemurs: The physics of whole-body rotations while airborne. *American Journal of Primatology, 16*, 291–303.

Dunbar, D. C. (1994). The influence of segmental movements and design on whole-body rotations during the airborne phase of primate leaps. *Zeitschrift für Morphologie und Anthropologie, 89*, 109–124.

Earls, K. D. (2000). Kinematics and mechanics of ground take-off in the starling *Sturnis vulgaris* and the quail *Coturnix coturnix*. *Journal of Experimental Biology, 203*, 725–739.
Elftman, H. (1944). The bipedal walking of the chimpanzee. *Journal of Mammalogy, 25*, 67–71.
Elftman, H., & Manter, J. (1935). Chimpanzee and human feet in bipedal walking. *American Journal of Physical Anthropology, 20*, 69–79.
Fedak, M. A., & Seeherman, H. J. (1979). Reappraisal of energetics of locomotion shows identical cost in bipeds and quadrupeds including ostrich and horse. *Nature, 282*, 713.
Fleagle, J. G. (1976). Locomotion and Posture of the Malayan Siamang and Implications for Hominoid Evolution. *Folia Primatologica, 26*, 245–269.
Fleagle, J. G. (2013). *Primate adaptation and evolution* (3rd ed.). Academic.
Fleagle, J. G., Stern, J. T., Jungers, W. L., Susman, R. L., Vangor, A. K., & Wells, J. P. (1981). Climbing: A biomechanical link with brachiation and with bipedalism. *Symposium Zoological Society London, 48*, 359–375.
Foster, A. D., Raichlen, D. A., & Pontzer, H. (2013). Muscle force production during bent-knee, bent-hip walking in humans. *Journal of Human Evolution, 65*, 294–302.
Galton, P. M., & Shepherd, J. D. (2012). Experimental analysis of perching in the European starling (*Sturnus vulgaris*: Passeriformes; Passeres), and the automatic perching mechanism of birds. *Journal of Experimental Zoology Part A: Ecological Genetics and Physiology, 317*, 205–215.
Gatesy, S. M., & Dial, K. P. (1996). Locomotor modules and the evolution of avian flight. *Evolution, 50*, 331–340.
Gatesy, S. M., & Middleton, K. M. (1997). Bipedalism, flight, and the evolution of theropod locomotor diversity. *Journal of Vertebrate Paleontology, 17*, 308–329.
Gauthier, J. (1986). Saurischian monophyly and the origin of birds. *Memoirs of the California Academy of Sciences, 8*, 1–55.
Geyer, H., Seyfarth, A., & Blickhan, R. (2006). Compliant leg behaviour explains basic dynamics of walking and running. *Proceedings of the Royal Society B: Biological Sciences, 273*, 2861–2867.
Gould, S. J. (1991). Exaptation: A crucial tool for an evolutionary psychology. *Journal of Social Issues, 47*, 43–65.
Gould, S. J., & Lewontin, R. C. (1979). The spandrels of San Marco and the Panglossian paradigm: A critique of the adaptationist programme. *Proceedings of the Royal Society of London. Series B Biological Sciences, 205*, 581–598.
Grinham, L. R., VanBuren, C. S., & Norman, D. B. (2019). Testing for a facultative locomotor mode in the acquisition of archosaur bipedality. *Royal Society Open Science, 6*, 190569.
Grosprêtre, S., & Lepers, R. (2016). Performance characteristics of Parkour practitioners: Who are the traceurs? *European Journal of Sport Science, 16*, 526–535.
Hammond, A. S., Rook, L., Anaya, A. D., Cioppi, E., Costeur, L., Moyà-Solà, S., & Almécija, S. (2020). Insights into the lower torso in late Miocene hominoid *Oreopithecus bambolii*. *Proceedings of the National Academy of Sciences, 117*, 278–284.
Hanna, J. B., Granatosky, M. C., Rana, P., & Schmitt, D. (2017). The evolution of vertical climbing in primates: Evidence from reaction forces. *Journal of Experimental Biology, 220*, 3039–3052.
Hansell, M. H. (2000). *Bird nests and construction behaviour*. Cambridge University Press.
Harcourt-Smith, W. E. (2007). The origins of bipedal locomotion. *Handbook of Paleoanthropology, 3*, 1483–1518.
Harcourt-Smith, W. E. (2010). The first hominins and the origins of bipedalism. *Evolution: Education and Outreach, 3*, 333–340.
Hayes, G., & Alexander, R. M. (1983). The hopping gaits of crows (Corvidae) and other bipeds. *Journal of Zoology, 200*, 205–213.
Hirasaki, E., Kumakura, H., & Matano, S. (1995). Electromyography of 15 limb muscles in Japanese macaques (*Macaca fuscata*) during vertical climbing. *Folia Primatologica, 64*, 218–224.

Hirasaki, E., Ogihara, N., Hamada, Y., Kumakura, H., & Nakatsukasa, M. (2004). Do highly trained monkeys walk like humans? A kinematic study of bipedal locomotion in bipedally trained Japanese macaques. *Journal of Human Evolution, 46*, 739–750.

Hobbs, S. J., & Clayton, H. M. (2019). Collisional mechanics of the diagonal gaits of horses over a range of speeds. *PeerJ, 7*, e7689.

Hof, A. L. (2008). The 'extrapolated center of mass' concept suggests a simple control of balance in walking. *Human Movement Science, 27*(1), 112–125.

Höfling, E., & Abourachid, A. (2021). The skin of birds' feet: Morphological adaptations of the plantar surface. *Journal of Morphology, 282*(1), 88–97.

Höfling, E., Abourachid, A., & Renous, S. (2006). Locomotion behavior of the Lettered Aracari (*Pteroglossus inscriptus*) (Ramphastidae). *Ornithologia Neotropical, 17*, 363–371.

Hogervorst, T., & Vereecke, E. E. (2014). Evolution of the human hip. Part 1: The osseous framework. *Journal of Hip Preservation Surgery, 1*, 39–45.

Hogervorst, T., & Vereecke, E. E. (2015). Evolution of the human hip. Part 2: Muscling the double extension. *Journal of Hip Preservation Surgery, 2*, 3–14.

Holowka, N. B., & Lieberman, D. E. (2018). Rethinking the evolution of the human foot: Insights from experimental research. *Journal of Experimental Biology, 221*, jeb174425.

Horak, F. B. (2006). Postural orientation and equilibrium: what do we need to know about neural control of balance to prevent falls? *Age and Ageing, 35*, ii7-ii11.

Hunt, K. D. (1994). The evolution of human bipedality: ecology and functional morphology. *Journal of Human Evolution, 26*, 183–202.

Hunt, K. D. (2016). Why are there apes? Evidence for the co-evolution of ape and monkey ecomorphology. *Journal of Anatomy, 228*, 630–685.

Irschick, D. J., & Jayne, B. C. (1998). Effects of incline on speed, acceleration, body posture and hindlimb kinematics in two species of lizard *Callisaurus draconoides* and *Uma scoparia*. *Journal of Experimental Biology, 201*, 273–287.

Irschick, D. J., & Jayne, B. C. (1999). Comparative three-dimensional kinematics of the hindlimb for high-speed bipedal and quadrupedal locomotion of lizards. *Journal of Experimental Biology, 202*, 1047–1065.

Ishida, H., Kimura, T. & Okada, M. (1974). Patterns of bipedal walking in anthropoid primates. In *Proceedings of the 5th Congress of the International Primatological Society* (pp. 287–301). Japan Science Press.

Jenkins, F., & Fleagle, J. G. (1975). Knuckle-walking and the functional anatomy of the wrists in living apes. In R. H. Tuttle (Ed.), *Primate functional morphology and evolution* (pp. 213–227). Mouton.

Jindrich, D. L., & Full, R. J. (1999). Many-legged maneuverability: Dynamics of turning in hexapods. *Journal of Experimental Biology, 202*, 1603–1623.

Jungers, W. L., & Stern, J. T. (1981). Preliminary electromyographical analysis of brachiation in gibbon and spider monkey. *International Journal of Primatology, 2*, 19–33.

Kardong, K. V. (2006). *Vertebrates: Comparative anatomy, function, evolution*. McGraw-Hill.

Ker, R., Bennett, M., Bibby, S., Kester, R., & Alexander, R. M. (1987). The spring in the arch of the human foot. *Nature, 325*, 147–149.

Kimura, T. (1991). Voluntary bipedal walking of infant chimpanzees. *Human Evolution, 6*, 377–390.

Kimura, T. (1996). Centre of gravity of the body during the ontogeny of chimpanzee bipedal walking. *Folia Primatologica, 66*, 126–136.

Kimura, T., & Yaguramaki, N. (2009). Development of bipedal walking in humans and chimpanzees: A comparative study. *Folia Primatologica, 80*, 45–62.

Kinoshita, Y., Goto, R., Nakano, Y., & Hirasaki, E. (2021). A comparison of axial trunk rotation during bipedal walking between humans and Japanese macaques. *American Journal of Physical Anthropology, 174*(1), 66–75. https://doi.org/10.1002/ajpa.24136

Kivell, T. L., & Schmitt, D. (2009). Independent evolution of knuckle-walking in African apes shows that humans did not evolve from a knuckle-walking ancestor. *Proceedings of the National Academy of Sciences, 106*, 14241–14246.
Kohlsdorf, T., & Biewener, A. (2006). Negotiating obstacles: Running kinematics of the lizard *Sceloporus malachiticus*. *Journal of Zoology, 270*, 359–371.
Kozma, E. E., Webb, N. M., Harcourt-Smith, W. E., Raichlen, D. A., D'Août, K., Brown, M. H., Finestone, E. M., Ross, S. R., Aerts, P., & Pontzer, H. (2018). Hip extensor mechanics and the evolution of walking and climbing capabilities in humans, apes, and fossil hominins. *Proceedings of the National Academy of Sciences, 115*(16), 4134–4139.
Kraft, T. S., Venkataraman, V. V., & Dominy, N. J. (2014). A natural history of human tree climbing. *Journal of Human Evolution, 71*, 105–118.
Kubo, T., & Kubo, M. O. (2012). Associated evolution of bipedality and cursoriality among Triassic archosaurs: A phylogenetically controlled evaluation. *Paleobiology, 38*, 474–485.
Ledebt, A., & Bril, B. (2000). Acquisition of upper body stability during walking in toddlers. *Developmental Psychobiology, 36*, 311–324.
Lee, D. V., Bertram, J. E., Anttonen, J. T., Ros, I. G., Harris, S. L., & Biewener, A. A. (2011). A collisional perspective on quadrupedal gait dynamics. *Journal of the Royal Society Interface, 8*, 1480–1486.
Lewton, K. L., & Scott, J. E. (2017). Ischial form as an indicator of bipedal kinematics in early hominins: A test using extant anthropoids. *The Anatomical Record, 300*, 845–858.
Li, J., Zhang, X., Zou, M., Zhang, R., Chirende, B., Shi, R., & Wei, C. (2013). An experimental study on the gait patterns and kinematics of Chinese mitten crabs. *Journal of Bionic Engineering, 10*, 305–315.
Livezey, B. C. (1989). Morphometric patterns in Recent and fossil penguins (Aves, Sphenisciformes). *Journal of Zoology, 219*, 269–307.
Lovejoy, C. O. (2005). The natural history of human gait and posture: Part 1. Spine and pelvis. *Gait and Posture, 21*, 95–112.
Lovejoy, C. O., & McCollum, M. A. (2010). Spinopelvic pathways to bipedality: Why no hominids ever relied on a bent-hip-bent-knee gait. *Philosophical Transactions of the Royal Society B: Biological Sciences, 365*, 3289–3299.
Lovejoy, C. O., Suwa, G., Spurlock, L., Asfaw, B., & White, T. D. (2009). The pelvis and femur of *Ardipithecus ramidus*: The emergence of upright walking. *Science, 326*, 71-71e6.
Machnicki, A. L., Spurlock, L. B., Strier, K. B., Reno, P. L., & Lovejoy, C. O. (2016). First steps of bipedality in hominids: Evidence from the atelid and proconsulid pelvis. *PeerJ, 4*, e1521.
Matthis, J. S., Yates, J. L., & Hayhoe, M. M. (2018). Gaze and the control of foot placement when walking in natural terrain. *Current Biology, 28*, 1224–1233.e5.
Maurice, M., Gioanni, H., & Abourachid, A. (2006). Influence of the behavioural context on the optocollic reflex (OCR) in pigeons (*Columba livia*). *Journal of Experimental Biology, 209*, 292–301.
McHenry, H. M. (1975). The ischium and hip extensor mechanism in human evolution. *American Journal of Physical Anthropology, 43*, 39–46.
Michilsens, F., Vereecke, E. E., D'Août, K., & Aerts, P. (2009). Functional anatomy of the gibbon forelimb: Adaptations to a brachiating lifestyle. *Journal of Anatomy, 215*, 335–354.
Michilsens, F., Vereecke, E. E., D'Août, K., & Aerts, P. (2010). Muscle moment arms and function of the siamang forelimb during brachiation. *Journal of Anatomy, 217*, 521–535.
Muir, G. D., Gosline, J. M., & Steeves, J. D. (1996). Ontogeny of bipedal locomotion: Walking and running in the chick. *The Journal of Physiology, 439*, 589–601.
Myatt, J. P., & Thorpe, S. K. S. (2011). Postural strategies employed by orangutans (*Pongo abelii*) during feeding in the terminal branch niche. *American Journal of Physical Anthropology, 146*, 73–82.
Nakatsukasa, M., Hayama, S., & Preuschoft, H. (1995). Postcranial skeleton of a macaque trained for bipedal standing and walking and implications for functional adaptation. *Folia Primatologica, 64*, 1–29.

Nakatsukasa, M., Hirasaki, E., & Ogihara, N. (2006). Energy expenditure of bipedal walking is higher than that of quadrupedal walking in Japanese macaques. *American Journal of Physical Anthropology, 131*, 33–37.

Necker, R. (2006). Specializations in the lumbosacral vertebral canal and spinal cord of birds: evidence of a function as a sense organ which is involved in the control of walking. *Journal of Comparative Physiology A, 192*, 439.

O'Neill, M. C., Demes, B., Thompson, N. E., & Umberger, B. R. (2018). Three-dimensional kinematics and the origin of the hominin walking stride. *Journal of the Royal Society Interface, 15*, 20180205.

Ogihara, N., Hirasaki, E., Kumakura, H., & Nakatsukasa, M. (2007). Ground-reaction-force profiles of bipedal walking in bipedally trained Japanese monkeys. *Journal of Human Evolution, 53*, 302–308.

Ogihara, N., Makishima, H., & Nakatsukasa, M. (2010). Three-dimensional musculoskeletal kinematics during bipedal locomotion in the Japanese macaque, reconstructed based on an anatomical model-matching method. *Journal of Human Evolution, 58*, 252–261.

Ogihara, N., Hirasaki, E., Andrada, E., & Blickhan, R. (2018). Bipedal gait versatility in the Japanese macaque (*Macaca fuscata*). *Journal of Human Evolution, 125*, 2–14.

Olberding, J. P., McBrayer, L. D., & Higham, T. E. (2012). Performance and three-dimensional kinematics of bipedal lizards during obstacle negotiation. *Journal of Experimental Biology, 215*, 247–255.

Palmer, C. E. (1944). Studies of the center of gravity in the human body. *Child Development, 15*, 99–180.

Persons, W. S., & Currie, P. J. (2017). The functional origin of dinosaur bipedalism: Cumulative evidence from bipedally inclined reptiles and disinclined mammals. *Journal of Theoretical Biology, 420*, 1–7.

Pollock, A. S., Durward, B. R., Rowe, P. J., & Paul, J. P. (2000). What is balance? *Clinical Rehabilitation, 14*, 402–406.

Pontzer, H., Raichlen, D. A., & Sockol, M. D. (2009). The metabolic cost of walking in humans, chimpanzees, and early hominins. *Journal of Human Evolution, 56*, 43–54.

Pontzer, H., Raichlen, D. A., & Rodman, P. S. (2014). Bipedal and quadrupedal locomotion in chimpanzees. *Journal of Human Evolution, 66*, 64–82.

Porter, A. M. (1999). Modern human, early modern human and Neanderthal limb proportions. *International Journal of Osteoarchaeology, 9*, 54–67.

Pough, F. H. (1980). The advantages of ectothermy for tetrapods. *The American Naturalist, 115*, 92–112.

Preuschoft, H., Hayama, S., & Günther, M. (1988). Curvature of the lumbar spine as a consequence of mechanical necessities in Japanese macaques trained for bipedalism. *Folia Primatologica, 50*, 42–58.

Preuschoft, H., Godinot, M., Beard, C., Nieschalk, U. & Jouffroy, F. K. 1993 Biomechanical considerations to explain important morphological characters of primate hands. In *Hands of primates* (pp. 245–256). Springer.

Provini, P., & Abourachid, A. (2018). Whole-body 3D kinematics of bird take-off: key role of the legs to propel the trunk. *The Science of Nature, 105*, 12.

Provini, P., & Höfling, E. (2020). To hop or not to hop? The answer is in the bird trees. *Systematic Biology, 69*(5), 962–972.

Provini, P., Tobalske, B. W., Crandell, K. E., & Abourachid, A. (2012). Transition from leg to wing forces during take-off in birds. *Journal of Experimental Biology, 215*, 4115–4124.

Rafferty, K. L., Walker, A., Ruff, C. B., Rose, M. D., & Andrews, P. J. (1995). Postcranial estimates of body weight in Proconsul, with a note on a distal tibia of *P. major* from Napak, Uganda. *American Journal of Physical Anthropology, 97*, 391–402.

Richmond, B. G., Begun, D. R., & Strait, D. S. (2001). Origin of human bipedalism: The knuckle-walking hypothesis revisited. *American Journal of Physical Anthropology, 116*, 70–105.

Roberts, T. J., & Azizi, E. (2011). Flexible mechanisms: The diverse roles of biological springs in vertebrate movement. *Journal of Experimental Biology, 214*, 353–361.

Roberts, A., Soffe, S., Wolf, E., Yoshida, M., & Zhao, F. Y. (1998). Central circuits controlling locomotion in young frog tadpoles. *Annals of the New York Academy of Sciences, 860*, 19–34.

Rocha-Barbosa, O., Loguercio, M., Velloso, A., & Bonates, A. (2008). Bipedal locomotion in *Tropidurus torquatus* (Wied, 1820) and *Liolaemus lutzae* Mertens, 1938. *Brazilian Journal of Biology, 68*, 649–655.

Rose, M. (1976). Bipedal behavior of olive baboons (*Papio anubis*) and its relevance to an understanding of the evolution of human bipedalism. *American Journal of Physical Anthropology, 44*, 247–261.

Rose, M. (1991). The process of bipedalization in hominids. In Y. Coppens & B. Senut (Eds.), *Origine(s) de la bipédie chez les hominidés* (pp. 37–48). Editions du Centre National de la Recherche Scientifique.

Ruff, C. B., Walker, A., & Teaford, M. F. (1989). Body mass, sexual dimorphism and femoral proportions of Proconsul from Rusinga and Mfangano Islands, Kenya. *Journal of Human Evolution, 18*, 515–536.

Ruina, A., Bertram, J. E., & Srinivasan, M. (2005). A collisional model of the energetic cost of support work qualitatively explains leg sequencing in walking and galloping, pseudo-elastic leg behavior in running and the walk-to-run transition. *Journal of Theoretical Biology, 237*, 170–192.

Schmitt, D. (1999). Compliant walking in primates. *Journal of Zoology, 248*, 149–160.

Senut, B. (2007). The earliest putative hominids. In W. Henke & Y. Tattersall (Eds.), *Handbook of paleoanthropology* (pp. 1519–1538). Springer.

Senut, B., Pickford, M., Gommery, D., Mein, P., Cheboi, K., & Coppens, Y. (2001). First hominid from the Miocene (Lukeino Formation, Kenya). *Comptes Rendus de l'Académie des Sciences - Series IIA - Earth and Planetary Science, 332*, 137–144.

Sheffield, K. M., Butcher, M. T., Shugart, S. K., Gander, J. C., & Blob, R. W. (2011). Locomotor loading mechanics in the hindlimbs of tegu lizards (*Tupinambis merianae*): comparative and evolutionary implications. *Journal of Experimental Biology, 214*, 2616–2630.

Shubin, N., Tabin, C., & Carroll, S. (1997). Fossils, genes and the evolution of animal limbs. *Nature, 388*, 639–648.

Snyder, R. C. (1949). Bipedal locomotion of the lizard *Basiliscus basiliscus*. *Copeia, 1949*, 129–137.

Snyder, R. C. (1952). Quadrupedal and bipedal locomotion of lizards. *Copeia, 1952*, 64–70.

Snyder, R. C. (1954). The anatomy and function of the pelvic girdle and hindlimb in lizard locomotion. *American Journal of Anatomy, 95*, 1–45.

Snyder, R. C. (1962). Adaptations for bipedal locomotion of lizards. *American Zoologist, 2*, 191–203.

Sockol, M. D., Raichlen, D. A., & Pontzer, H. (2007). Chimpanzee locomotor energetics and the origin of human bipedalism. *Proceedings of the National Academy of Sciences, 104*, 12265–12269.

Stanchak, K. E., French, C., Perkel, D. J., & Brunton, B. W. (2020). The balance hypothesis for the avian lumbosacral organ and an exploration of its morphological variation. *Integrative Organismal Biology, 2*(1), obaa024.

Stanford, C. B. (2006). Arboreal bipedalism in wild chimpanzees: Implications for the evolution of hominid posture and locomotion. *American Journal of Physical Anthropology, 129*, 225–231.

Stoessel, A., & Fischer, M. S. (2012). Comparative intralimb coordination in avian bipedal locomotion. *Journal of Experimental Biology, 215*, 4055–4069.

Taylor, C. R., & Rowntree, V. J. (1973). Running on Two or on four legs: Which consumes more energy? *Science, 179*, 186–187.

Taylor, G., & Thomas, A. (2014). *Evolutionary biomechanics*. Oxford University Press.

Thompson, N. E., Demes, B., O'Neill, M. C., Holowka, N. B., & Larson, S. G. (2015). Surprising trunk rotational capabilities in chimpanzees and implications for bipedal walking proficiency in early hominins. *Nature Communications, 6*(1), 1–7.
Thompson, N. E., O'Neill, M. C., Holowka, N. B., & Demes, B. (2018a). Step width and frontal plane trunk motion in bipedal chimpanzee and human walking. *Journal of Human Evolution, 125*, 27–37.
Thompson, N. E., Rubinstein, D., & Larson, S. G. (2018b). Great ape thorax and shoulder configuration – An adaptation for arboreality or knuckle-walking? *Journal of Human Evolution, 125*, 15–26.
Thompson, N. E., Rubinstein, D., Parrella-O'Donnell, W., Brett, M. A., Demes, B., Larson, S. G., & O'Neill, M. C. (2021). The loss of the 'pelvic step' in human evolution. *Journal of Experimental Biology, 224*(16), jeb240440. https://doi.org/10.1242/jeb.240440
Thorpe, S. K. S., & Crompton, R. H. (2006). Orangutan positional behavior and the nature of arboreal locomotion in Homincidea. *American Journal of Physical Anthropology, 131*, 384–401.
Thorpe, S. K. S., Holder, R. L., & Crompton, R. H. (2007). Origin of human bipedalism as an adaptation for locomotion on flexible branches. *Science, 316*, 1328–1331.
Tobalske, B. W., & Dial, K. P. (2000). Effects of body size on take-off flight performance in the Phasianidae (Aves). *Journal of Experimental Biology, 203*, 3319–3332.
Tobalske, B. W., & Dial, K. P. (2007). Aerodynamics of wing-assisted incline running in birds. *Journal of Experimental Biology, 210*, 1742–1751.
Toussaint, S., Llamosi, A., Morino, L., & Youlatos, D. (2020). The central role of small vertical substrates for the origin of grasping in early primates. *Current Biology, 30*(9), 1600–1613.e3.
Tucker, D. B., & McBrayer, L. D. (2012). Overcoming obstacles: the effect of obstacles on locomotor performance and behaviour. *Biological Journal of the Linnean Society, 107*, 813–823.
Tuttle, R. H. (1967). Knuckle-walking and the evolution of hominoid hands. *American Journal of Physical Anthropology, 26*, 171–206.
Van Damme, R., & Vanhooydonck, B. (2001). Origins of interspecific variation in lizard sprint capacity. *Functional Ecology, 15*, 186–202.
Van Wassenbergh, S., & Aerts, P. (2013). In search of the pitching momentum that enables some lizards to sustain bipedal running at constant speeds. *Journal of the Royal Society Interface, 10*, 20130241.
Venkadesan, M., Yawar, A., Eng, C. M., Dias, M. A., Singh, D. K., Tommasini, S. M., Haims, A. H., Bandi, M. M., & Mandre, S. (2020). Stiffness of the human foot and evolution of the transverse arch. *Nature, 579*, 97–100.
Vereecke, E. E., & Aerts, P. (2008). The mechanics of the gibbon foot and its potential for elastic energy storage during bipedalism. *Journal of Experimental Biology, 211*, 3661–3670.
Vereecke, E., D'Août, K., De Clercq, D., Van Elsacker, L., & Aerts, P. (2003). Dynamic plantar pressure distribution during terrestrial locomotion of bonobos (*Pan paniscus*). *American Journal of Physical Anthropology, 120*, 373–383.
Vereecke, E., D'Août, K., Van Elsacker, L., De Clercq, D., & Aerts, P. (2005). Functional analysis of the gibbon foot during terrestrial bipedal walking: Plantar pressure distributions and three-dimensional ground reaction forces. *American Journal of Physical Anthropology, 128*, 659–669.
Vereecke, E., D'Août, K., & Aerts, P. (2006a). Locomotor versatility in the white-handed gibbon (Hylobates lar): A spatiotemporal analysis of the bipedal, tripedal, and quadrupedal gaits. *Journal of Human Evolution, 50*, 552–567.
Vereecke, E. E., D'Août, K., & Aerts, P. (2006b). The dynamics of hylobatid bipedalism: evidence for an energy-saving mechanism? *Journal of Experimental Biology, 209*, 2829–2838.
Vereecke, E. E., D'Août, K., & Aerts, P. (2006c). Speed modulation in hylobatid bipedalism: A kinematic analysis. *Journal of Human Evolution, 51*, 513–526.

Verstappen, M., Aerts, P., & Van Damme, R. (2000). Terrestrial locomotion in the black-billed magpie: Kinematic analysis of walking, running and out-of-phase hopping. *Journal of Experimental Biology, 203*, 2159–2170.

Videan, E. N., & McGrew, W. C. (2002). Bipedality in chimpanzee (*Pan troglodytes*) and bonobo (*Pan paniscus*): Testing hypotheses on the evolution of bipedalism. *American Journal of Physical Anthropology, 118*, 184–190.

Ward, C. V., Walker, A., & Teaford, M. F. (1991). Proconsul did not have a tail. *Journal of Human Evolution, 21*, 215–220.

Winter, D. A. (1995). Human balance and posture control during standing and walking. *Gait & Posture, 3*, 193–214.

Winter, D. A. (2009). *Biomechanics and motor control of human movement*. Wiley.

Witte, H., Preuschoft, H., & Recknagel, S. (1991). Human body proportions explained on the basis of biomechanical principles. *Zeitschrift für Morphologie und Anthropologie, 78*, 407–423.

Wrangham, R. W. (1980). Bipedal locomotion as a feeding adaptation in gelada baboons, and its implications for hominid evolution. *Journal of Human Evolution, 9*, 329–331.

Wunderlich, R., & Schaum, J. (2007). Kinematics of bipedalism in *Propithecus verreauxi*. *Journal of Zoology, 272*, 165–175.

Zeffer, A., Johansson, L. C., Marmebro, Ã., & s. (2003). Functional correlation between habitat use and leg morphology in birds (Aves). *Biological Journal of the Linnean Society, 79*, 461–484.

Zihlman, A. L., & Bolter, D. R. (2015). Body composition in *Pan paniscus* compared with *Homo sapiens* has implications for changes during human evolution. *Proceedings of the National Academy of Sciences, 112*(24), 7466–7471.

Zihlman, A. L., Mcfarland, R. K., & Underwood, C. E. (2011). functional anatomy and adaptation of male gorillas (*Gorilla gorilla gorilla*) with comparison to male orangutans (*Pongo pygmaeus*). *The Anatomical Record, 294*, 1842–1855.

Zusi, R. L. (2013). Introduction to the skeleton of hummingbirds (Aves: Apodiformes, Trochilidae) in functional and phylogenetic contexts. *Ornithological Monographs, 77*, 1–94.

Chapter 15
Aquatic Locomotion: Environmental Constraints That Drive Convergent Evolution

Frank E. Fish

Abstract The quintessential example of evolutionary convergence is that between the shark, ichthyosaur, and dolphin. Although not closely related, the three exemplar taxa have independently evolved adaptations in morphology, physiology, and behavior that result in concomitant levels of performance that meet the requirements associated with operating in a dense, viscous, and thermally conductive marine environment. These apex marine predators display a remarkable amount of homoplasy. All three taxa have developed streamlined fusiform bodies to reduce drag when swimming. The position, type, and morphology of the control surfaces (i.e., fins, flippers, flukes) are similar for the convergent taxa. The control surfaces have different internal support structures, but function similarly to generate lift forces for stability and maneuverability. The main departure in control surface design among the three taxa is that dolphins lack pelvic fins. For dolphins, the loss of pelvic appendages is directly related to the possession of horizontally oriented caudal flukes, which perform double duty as a propulsive device and posterior stabilizer for trim control. The flukes of dolphins and caudal fins of ichthyosaurs and sharks have a lunate shape that function as an oscillating wing to generate high efficiency, lift-based thrust for high-speed swimming. The three convergent taxa are homeothermic, with a body temperature above that of the water in which they live. The advantages of an elevated body temperature are the attainment of higher maximum swimming speeds, longer and faster sustained swimming speeds, improved digestion, brain heating, and enhanced visual acuity. The convergence of the shark, ichthyosaur, and dolphin with respect to morphology, physiology, and locomotor performance reflects similar selective pressures imposed by the physical fluid environment that have dictated the independent evolutionary trajectories of these high-performance marine predators.

Keywords Shark · Dolphin · Ichthyosaur · Thunniform · Homeotherm

F. E. Fish (✉)
Department of Biology, West Chester University, West Chester, PA, USA
e-mail: ffish@wcupa.edu

V. L. Bels, A. P. Russell (eds.), *Convergent Evolution*, Fascinating Life Sciences,
https://doi.org/10.1007/978-3-031-11441-0_15

15.1 Introduction

In terms of evolution, similarity is expressed in two different ways, resulting in homology (evolutionary similarity) and analogy (functional similarity). Homology results from similarity due to common descent and is expressed as derived traits within monophyletic groups that share common inheritance and phylogeny (Haas & Simpson, 1946; Chang & Kim, 1996; Wake et al., 2011). Homology is used to generate phylogenies to show the degree of relatedness among organisms or trace evolutionary pathways back through deep time. Analogy reflects similarity of function between different structures (Haas & Simpson, 1946). When analogy is combined with homoplasy, there is not only similarity in function but also similarity in appearance arrived at from multiple independent origins. Both analogy and homoplasy point to natural selection as the driver of the separate evolution of similar structures among lineages, but homology need not imply the operation of selective mechanisms (Gould, 2002).

Darwin (1859) considered analogy to be associated with "adaptive characters" that were beneficial to the individual that had been subjected to selection, but to be "valueless to the systematist." Much of evolutionary theory was driven by Darwin's idea of common descent traced through homologous characters, putting it at odds with the perceived importance and utility of analogous features, particularly where homoplasy was involved (Wake, 1991; Brooks, 1996). Homoplasy was viewed as false homology (Wake, 1991). Analogy and homoplasy were considered barriers to a full comprehension of evolutionary trajectories. However, homologous features and shared genetic attributes provide only an understanding of the historical connections among organisms and their adaptations but cannot be employed to address why particular adaptations have evolved. Analogy and homoplasy provide clues about factors leading to the origin of particular adaptations and their association with the ecology of the organisms under consideration.

The combination of similar functions (analogy) and designs (homoplasy) resulting from the independent evolution of unrelated lineages leads to convergence (Haas & Simpson, 1946; Wake, 1991; Wake et al., 2011; Stayton, 2015). In this case, convergence manifests as analogy and homoplasy without homology, and thus stems from different developmental-structural origins (Powell, 2007). The incorporation of shape and function differentiates convergence from mimicry (Haas & Simpson, 1946). Convergence differs from parallelism in that the latter entails analogy, homoplasy, and homology that depend upon the same developmental genetic mechanisms (Wake et al., 2011). Both convergence and parallelism yield similarity as a result of common selective pressures imposed by the interaction of organisms with their biotic and abiotic environments. Convergence and parallelism permit testing of adaptive hypotheses (Larson & Losos, 1996; Stayton, 2015). Comparative studies have used the concepts of convergence and parallelism to elucidate the mechanisms and constraints that underpin directed phylogenetic changes (Wake et al., 2011). The characters that evolve from unrelated or distantly related groups demonstrate similar adaptive solutions to common selective pressures imposed by similar relatively stable environmental constraints. Convergence is one of the most powerful lines of evidence for how and why natural selection drives form toward particular functional adaptations that are specific to similar ecological circumstances in varied localities in unrelated groups of organisms (Patterson, 1988; Wake, 1991).

It is not uncommon for multiple evolutionary pathways in unrelated groups to arrive independently at a common solution to an environmental challenge. Convergence not only demonstrates that there is more than one pathway to the same end point, but also demonstrates the power of natural selection to arrive at a favorable solution from a variety of starting points (Fish & Beneski, 2014). The organisms associated with any such evolutionary pathway are subject to the same physical laws of nature whether at the molecular, microscopic, or macroscopic level. The laws of motion and thermodynamics, in conjunction with the ecological niche and the available genetic variation within a lineage, constrain phylogenetic trajectories to a particular area of morphospace.

There are myriad ways of negotiating movement on or under land, in air, and through water. Locomotion in each of these environments imposes a specific set of selective pressures that influence the morphological and physiological evolution of animals to be able to operate with a positive energy budget. The majority of the surface of the earth is covered in expansive seas that harbor the majority of animal phyla. The physical properties of the aqueous medium place intense selective pressures on animals that attempt to move in it with speed based upon a high-energy economy. Design space is highly constrained in the aquatic realm.

To understand how and why convergence is driven to a similar endpoint, this chapter focuses on a textbook case associated with morphological and functional similarity in three highly derived types of aquatic vertebrates. Comparison of the shark, ichthyosaur, and dolphin body form represents a prominent example of the expression of convergent evolution (Fig. 15.1; Howell, 1930; Irving, 1966; Hildebrand, 1995; Liem et al., 2001; Thewissen & Nummela, 2008; Wicander & Monroe, 2012; Kardong, 2019). The three groups exhibit similarities in shape, physiology, and mechanics despite their phylogenetic separation (Fig. 15.2). This textbook example demonstrates how similar functional requirements have been met by different clades that have sharply focused on a nearly identical solution to the same environmental challenges. Such convergence, with its resulting analogy and homoplasy, is associated with similar environmental selection pressures imposed on these aquatic predators, enabling them to operate in the open ocean as fast, efficient swimmers that exploit similar trophic opportunities (Gans, 1974; Fish, 1996; Motani, 2000; Bernal et al., 2001a; Donley et al., 2004; Kelley & Motani, 2015; Fischer et al., 2016; Lingham-Soliar, 2016). If aquatic vertebrates are adapted to swim in a manner that minimizes energy expenditure, there should be distinct hydrodynamic advantages to morphological designs and propulsive modes employed by the most derived species in each group. Analysis of swimming mechanics and energetics may thus elucidate physical and biological constraints that may have influenced the convergence of these aquatic animals.

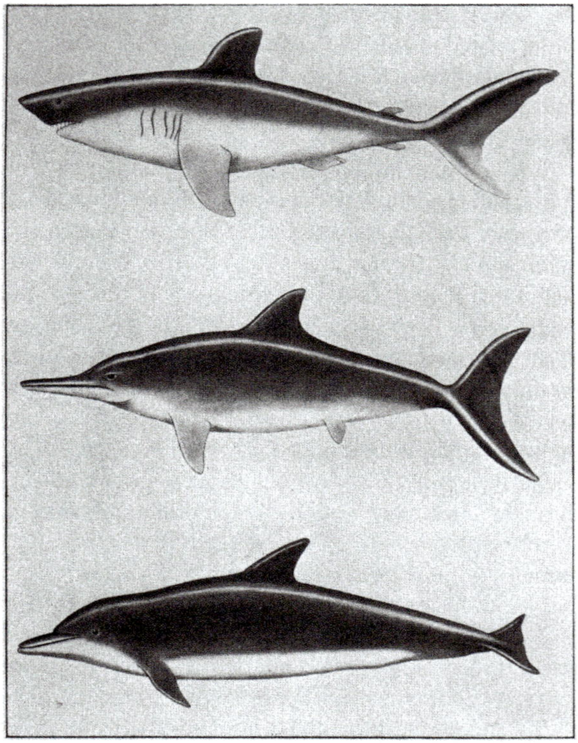

Fig. 15.1 Early illustration from Howell (1930) showing convergence in morphology of a shark, ichthyosaur, and dolphin for a fully aquatic life as fast-swimming predators. Figure from Howell, A. B., *Aquatic Mammals: Their Adaptations to Life in Water*, 1930. Courtesy of Charles C. Thomas, Publisher, Ltd, Springfield, Illinois

15.2 Physical Characteristics of Water That Affect Swimming Performance

To understand the limitations on locomoting through an aquatic medium that would lead to in convergence, a brief examination of the physical characteristics of water is required. The morphology and locomotor performance of animals in water are dictated by its physical properties (Daniel & Webb, 1987; Webb, 1988). The most pertinent of these physical properties with respect to movement in the aquatic environment, are density and viscosity (Webb, 1975; Daniel & Webb, 1987). Density is the mass per unit volume, whereas viscosity is the resistance to deformation (flow) by the fluid, when there is relative motion between different points in the fluid (Webb, 1975). Density affects inertial and pressure forces within a fluid, whereas viscosity is produced by friction within the fluid due to its "stickiness" resulting from cohesive forces between the water molecules and adhesion between the water and any solid surface (Webb, 1988; Denny, 1993). Pure water is 770 to 890 times denser than air at the same temperature, and at least 50 times more viscous. Thus, the resistance against moving through water is greater than that for air. Density

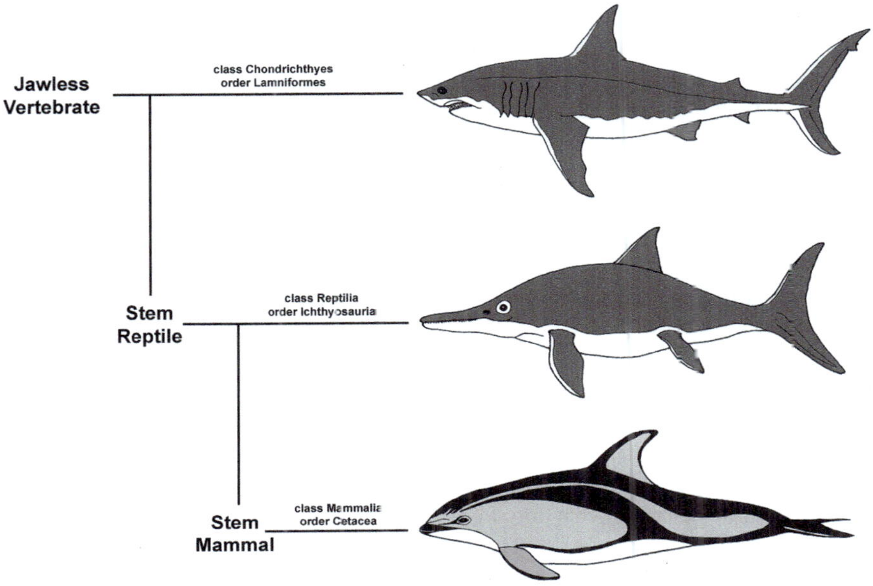

Fig. 15.2 Phylogenetic relationships of the morphologically convergent shark, ichthyosaur, and dolphin

and the pressure that it exerts is directed perpendicular to the surface of a submerged body, whereas viscosity is directed tangential to the surface of a submerged body.

As opposed to a solid, water as a fluid yields when pushed against (Lindsey, 1978). The distortion of the fluid can occur as a swimmer's body changes the direction and magnitude of the flow, known as vorticity, as it swims through stationary water, as moving water flows around a stationary body, or as the fluid is acted on due to the propulsive movement of the body and/or appendages. The increase in vorticity transfers the kinetic energy of the swimmer's movement to the water. Energy is lost to eddy formation and frictional forces in the water. As a consequence of the density and viscosity of water, movement through water imposes severe limitations on speed and energetic performance for swimming animals.

Water covers roughly 70% of the Earth's surface. Despite this enormous two-dimensional area, the oceans, seas, lakes and rivers have a third dimension of depth. As the density of water is related directly to its mass, water is affected by gravity so that the pressure experienced in a water column increases with depth. In seawater, the pressure increases by one atmosphere at sea level (760 mm of mercury, 14.7 pounds per square inch, 101 kPa) with every 10 meters of depth. Many of the tissues and total body composition of marine animals are close to the density of seawater, so they are not compressed with increased depth of submergence (Aleyev, 1977). However, air-filled spaces (e.g., swim bladder, lungs, sinuses, fur, feathers) are compressible (Brawn, 1962; Lovvorn et al., 1999; Fish et al., 2002; Moore et al.,

2011). This compression with depth reduces the volume of the air-filled spaces and consequently increases the animal's density and reduces its hydrostatic positive buoyancy (Kooyman, 1973; Ridgway & Howard, 1979; Moore et al., 2011).

Despite the limitations imposed on aquatic animals by the physical environment, there are advantages to moving in water. Aquatic locomotion can be the most economical form of transport (Tucker, 1970, 1975; Schmidt-Nielsen, 1972). As the density of the body tissues is close to that of water, a swimmer can be near neutrally buoyant, thereby negating the effect of gravity. In addition, both the near neutral buoyancy and viscosity of the water reduces sinking rates when air-filled spaces are compressed. Indeed, even with a loss of positive buoyancy, animals do not "sink like a stone" (Williams et al., 2000; Williams, 2001; Mitani et al., 2010). Unlike terrestrial and flying animals, swimmers do not have to expend energy to support the body against gravity during locomotion (Rayner, 1986; Withers, 1992). The power required for swimming is determined by speed and stress developed in muscles, but the effect of gravity is reduced (Pennycuick, 1992). This has allowed aquatic animals to attain huge sizes, ranging up to the 150-ton blue whale (*Balaenoptera musculus*) that are far beyond the size attainable by the largest terrestrial and flying animals (McClain et al., 2015; Goldbogen et al., 2019).

The near density match between marine animals and water that reduces the effect of gravity affects the energetics of swimming. Swimming fishes have a lower cost of transport (COT) compared to animals that fly and run (Tucker, 1975). COT is defined as the metabolic energy required to transport a unit mass a unit distance and is calculated by dividing the mass-specific metabolic rate by the swimming velocity (Fish, 1992). COT is inversely proportional to the efficiency of energy expenditure (Tucker, 1970). COT represents the energetic cost by which the metabolic power input is converted to thrust production (Tucker, 1970, 1975; Schmidt-Nielsen, 1972; Williams, 1987; Fish, 1992). The minimum COT is the most efficient and is considered to occur at the velocity which the animal can cover the greatest distance for the smallest energy cost.

Endothermic and homeothermic animals, which can include marine mammals and some species of tuna and lamnid (mackerel) shark, have elevated costs of transport compared to similarly sized ectothermic fishes (Williams, 1999; Fish, 2000; Watanabe et al., 2015). Williams (1999) asserted that the maintenance costs of endothermic tuna and aquatic mammals are higher than those of similarly-sized ectotherms due to the costs of the maintenance of an elevated body temperature and a high basal metabolism. When maintenance costs were omitted, yielding a net cost of transport, the endothermic swimmers were found to have similar locomotor costs, with the minimum COT, to ecothermic fishes. The maintenance costs of marine mammals are 22–77% of the gross COT (Williams, 1999; Fish, 2000). The similarity of locomotor cost of endothermic swimmers and ectothermic fishes, which have the lowest costs of transport among vertebrates, indicates that these endothermic swimmers have reached an optimum in terms of energetic performance (Williams, 1999). Swimming is also relatively economical because propulsive forces are easy to generate in water (Rayner, 1985, 1986).

As endothermy can affect swimming energetics, the thermal conductance and high heat capacity of water is an important physical property that challenges the retention of heat generated by the animal to maintain homeothermy. Water is about 23 times more thermally conductive than air (Denny, 1993). Thus, heat is transferred from an endotherm to the aquatic environment faster than in air, and this is exacerbated for endothermic aquatic animals that maintain an elevated homeothermic body temperature because the thermal gradient is increased. Furthermore, movement through water increases heat transfer by convective exchange. Heat transfer can be reduced with a body geometry that minimizes surface area, where exchange takes place, and the employment of specialized circulatory systems and insulative blubber.

15.3 Convergent Design

15.3.1 Body Streamlining

Design is acknowledged to have a major impact on the ecological performance of organisms (Liem, 1990). The similarity of body design to that of a shark is displayed by ichthyosaurs and cetaceans (Fig. 15.2). Sharks are one of the oldest vertebrate lineages to have become apex predators in the aquatic realm (Sternes & Shimada, 2020). Ichthyosaurs and dolphins have converged upon analogous biological roles to those of highly derived sharks. Ichthyosaurs lived in the oceans for 245 million years, whereas modern dolphins and whales first evolved around 34 million years ago (Motani, 2000; Thewissen et al., 2009). The similarity of the two tetrapods to sharks developed after their terrestrial ancestors returned to the sea (Howell, 1930; Motani, 2000; Thewissen & Bajpai, 2001; Caldwell, 2002; Gingerich, 2015). Such convergence is associated with constraints imposed on these swimmers by the physical environment and the similar biological roles that they fulfill. As apex predators, selection favored morphological, physiological, and behavioral adaptations for swimming performance that maximized speed and efficiency (Daniel & Webb, 1987; Blake, 1991; Fish, 1992; Motani, 2000; Donley et al., 2004; Lingham-Soliar, 2016). The evolution of highly derived convergent morphologies and swimming modes represents the culmination of a sequence of transitional stages displayed by chondrichthyans, reptiles, and mammals (Howell, 1930; Massare, 1988; Gingerich et al., 1990; Fish, 1992; Lingham-Soliar & Reif, 1998; Motani, 2000; Buchholtz, 2001a; Thewissen & Bajpai, 2001; Donley et al., 2004; Sternes & Shimada, 2020).

The most strikingly analogous features of the convergent shark, ichthyosaur, and dolphin are the shapes of their bodies and appendages (Fig. 15.2). Specifically, the highly derived group of sharks that can be considered the template for the convergent design are those in the family Lamnidae. Lamnid sharks include the great white shark (*Carcharodon carcharias*), mako shark (*Isurus oxyrinchus*), porbeagle shark (*Lamna nasus*), and salmon shark (*Lamna ditropis*). The most derived ichthyosaurs

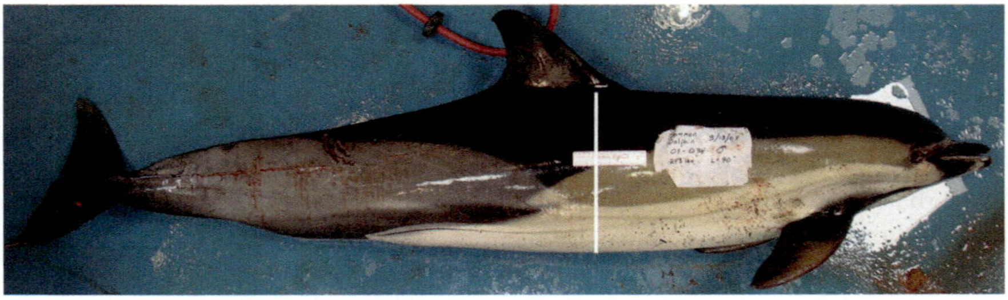

Fig. 15.3 Fusiform body shape of a common dolphin (*Delphinus delphis*). The white line indicates the position of the maximum thickness, called the shoulder position

are members of the clade Thunnosauria exemplified by the families Ichthyosauridae (e.g., *Ichthyosaurus* sp.), Ophthalmosauridae (e.g., *Ophthalmosaurus icenicus*), Stenopterygiidae (e.g., *Stenopterygius* sp.), and Temnodontosauridae (e.g., *Temnodontosaurus* sp.). The family of cetaceans that exemplifies convergence with lamnid sharks and ichthyosaurs is the Delphinidae (oceanic dolphins), which comprises about 32 species, including the ubiquitous bottlenose dolphin (*Tursiops truncatus*), common dolphin (*Delphinus delphis*), white-sided dolphins (*Lagenorhynchus* sp.), spotted dolphins (*Stenella* sp.), and the notorious killer whale (*Orcinus orca*). For the remainder of this discussion, unless otherwise specified, the highly derived exemplar species listed above are referred to broadly as sharks, ichthyosaurs, and dolphins.

Lamnid sharks, thunnosaurian ichthyosaurs and delphinids all exhibit similarities in body shape and appendage form that represent adaptations for optimizing swimming performance, maximizing energy efficiency, enhancing swimming speed, and maintaining stability in the open ocean. Overall body shape is the major determinant of resistance (i.e., drag) to movement through a fluid (Fish & Rohr, 1999: Gutarra et al., 2019). The body of the focal groups is streamlined and has smooth contours. Their streamlined bodies exhibit a fusiform shape similar to that of engineered high-performance hydrofoils (Hertel, 1966; Webb, 1975; Vogel, 1994; Fish, 1996, 2018). The fusiform shape resembles an elongate teardrop with a rounded leading edge that extends posteriorly to a maximum thickness and a gradually tapering tail (Fig. 15.3). This shape is accentuated by the anterior position of the bulk of the locomotor muscles (Pabst, 1990, 2000; Bernal et al., 2001a; Shadwick, 2005).

Streamlining minimizes the total drag on a body and reduces energy expenditure when swimming (Webb, 1975; Fish & Hui, 1991; Vogel, 1994; Fish et al., 2008; Gutarra et al., 2019). For a fully submerged, streamlined body, the total drag has two components: frictional or viscous drag and pressure or form drag (Webb, 1975; Fish, 1993a, b; Vogel, 1994). The frictional drag is due to the viscosity of the fluid, giving rise to tangential forces resulting from skin friction. Because the water immediately attached to the surface of a body does not move relative to the body (no slip condition; velocity = 0), a thin layer of water (boundary layer) encompasses a velocity gradient from the body surface to the free stream velocity outside of the boundary layer (i.e., outer flow). The velocity gradient generates shear (frictional) forces that consume kinetic energy from the movement of the body and transfers it to the water. The frictional drag is proportional to the wetted surface area of the body. The pressure drag component results from the distortion of the flow outside the boundary layer (i.e., deviation of the trajectory of the streamlines) and is dependent on pressure distribution as the body deflects the water. The pressure acts perpendicularly to the body surface.

The relationship between the pressure in a fluid and its velocity is expressed by the Bernoulli equation (Webb, 1975):

$$\frac{1}{2}\rho U - \rho g h + P = \text{constant}$$

where ρ is the density of the fluid, U is the velocity of the fluid along a streamline (i.e., line tangent to the direction of flow at every point in a flow field; Fox et al., 2009), *g* is the acceleration due to gravity, h is the depth of the streamline and P is the reference pressure. The first term is the dynamic pressure of the moving fluid, and the second term is the static pressure of the mass of fluid above the streamline. For streamlines of the same depth, the total pressure within the fluid is determined by the first term. The equation thus indicates that the pressure of the fluid is inversely proportional to the U^2, so a high velocity gives a low pressure and *vice versa*.

The pressure is highest at the rostrum or leading edge, where the flow stagnates. As the flow moves around the rostrum it accelerates and reaches a high velocity at the maximum thickness of the body. By Bernoulli's principle, the pressure at the rostrum is high because the velocity of the flow is zero, but as the velocity increases downstream, the pressure is reduced. This forebody region, therefore, has a favorable pressure gradient (i.e., water flow from high to low pressure). Downstream of the maximum thickness the flow starts to decelerate and pressure increases, but not to the extent displayed at the rostrum. This region has an adverse pressure gradient (i.e., water must move against an increasing pressure) and the boundary layer flow around the body must have enough energy to continue to remain alongside the body before separating into the wake. Separation of the boundary layer occurring at the trailing edge of the body will produce a narrow wake. If there is insufficient energy and momentum in the boundary layer flow, the increased pressure in the aft of the body will decelerate the flow. This reduction in flow velocity can lead to instabilities and flow reversal. Reversals in the form of eddies and vortices can interact within the boundary layer and cause it to prematurely detach from the body and interact with the outer flow. Such premature separation further increases the kinetic energy losses and associated drag on the body, which is manifested as a broad wake.

The flow within both the boundary layer and outer flow can be laminar, turbulent, or transitional (Webb, 1975). The type of flow is dependent on the Reynolds number (Re), which is the ratio of inertial to viscous forces in the fluid. Re is calculated as the product of the flow velocity, characteristic length, and density of the fluid divided by the viscosity. Typically, the flow will be laminar at $Re < 5 \times 10^5$, turbulent at $Re > 5 \times 10^6$, and transitional between these values (Webb, 1975; Fish, 1993a, b; Vogel, 1994). The shark, ichthyosaur, and dolphin swimming at cruising speeds will be in the turbulent regime with Re above 10^6. In a laminar flow, the virtual streamlines

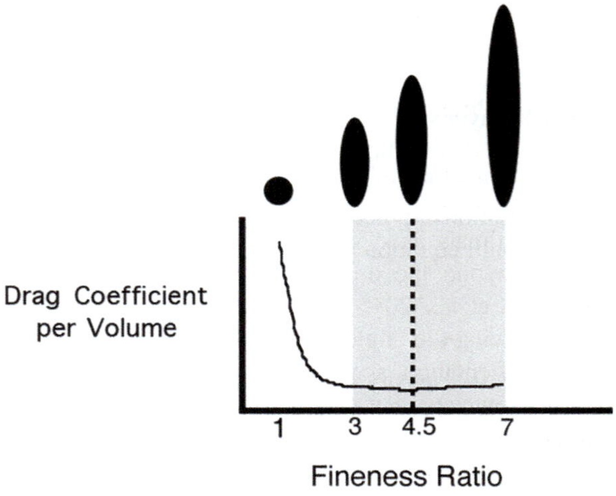

Fig. 15.4 Relationship of drag and Fineness ratio (black line) based on von Mises (1945). The dashed line shows the optimal Fineness ratio to be 4.5 for minimal drag based on airship design. The gray box illustrates the limits of Fineness ratio of 3–7 where drag increases by 10% from the optimal value

within the flow are parallel and orderly; whereas, in turbulent conditions, the flow is disordered and chaotic. There is increased momentum transfer and shear stresses in a turbulent flow. The ordered structure of laminar flow consumes less energy and has a lower drag than turbulent flow. However, laminar flow in the boundary layer is more susceptible to premature separation, with a concomitant increase in the pressure drag. A turbulent flow will energize and maintain attachment of the boundary layer and minimize the pressure drag at the expense of a slight increase in the frictional drag. Large animals swimming at high speeds, therefore, have a lower total drag with a turbulent boundary layer when compared to laminar conditions.

An indicator of the degree of streamlining is the fineness ratio (FR). FR is the ratio of the body length to the maximum diameter. An FR value of 1.0 would have a circular profile. The FR value of 4.5 is considered to induce the least drag and surface area for the maximum volume (Fig. 15.4; von Mises, 1945; Hertel, 1966; Webb, 1975; Ahlborn et al., 2009), although only a 10% increase in drag is realized in the range of FR of 3 to 7. The value of 4.5 was based on airship designs. Another study that examined the relationship of FR and drag on axisymmetrical torpedoes indicated that the optimal FR was 7 (Gertler, 1950). Depending on the contours of the body, animals should then have minimal drag in the FR range of 4.5 to 7.

FR for lamnid sharks, ichthyosaurs, and dolphins overlaps substantially (Fig. 15.5). FR for lamnid sharks is concentrated around the value of 4.5. However, FR for ichthyosaurs and dolphins has a large range spanning values between 4.5 and 7 (Fig. 15.5). Data for ichthyosaurs from Massare (1988) were based on a length that was measured from the anterior edge of the orbit on the skull to the tip of the terminal caudal vertebra. Inclusion of the elongate rostrum in the total length of ichthyosaurs would have skewed the data to higher values of FR. The data for the sharks and dolphins used the fork or notch length, respectively, which was measured as the length from the tip of the rostrum to the notch in the caudal fin or flukes. Despite the differences in measurement, the distribution of FR values for the three convergent

Fig. 15.5 Fineness ratio distribution for lamnid sharks, ichthyosaurs, and delphinid dolphins. The solid line shows the optimal Fineness ratio of 4.5 for minimal drag based on airship design (from von Mises, 1945) and the dashed line represents the optimal value for minimal drag based on axisymmetrical torpedo bodies (from Gertler, 1950)

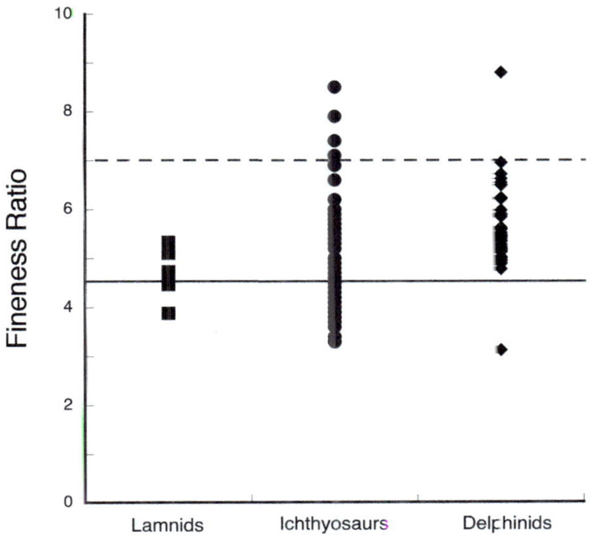

groups falls largely within the optimal range (4.5–7), indicating a low drag body form.

FR is a crude indicator of streamlining of the body because it does not provide information on changes in body contour. Body shapes can be compared to standardized two-dimensional airfoils that are classified by the United States National Advisory Committee for Aeronautics (NACA) (Abbott & von Doenhoff, 1959; Fish & Rohr, 1999). More recent data on engineered foil sections are available through Airfoil Tools (airfoiltools.com). The advantage of such comparisons is that the specific hydrodynamic characteristics have been measured for the foil sections. For example, a dolphin body has been compared to a NACA 66–018 foil (Hertel, 1966) and a killer whale (*Orcinus orca*) has a streamlined body shape similar to a NACA 66–026 (Pershin, 1983). Both these foil sections are designated as low drag laminar profiles (Fish & Rohr, 1999). However, the choice of foil design to represent body shape is subjective and overly simplistic in not accounting for body contours that do not match the idealized foil shape (Weber et al., 2009a; Fish, 2015). In addition, the animal's body is three-dimensional with attached appendages that are not accounted for in the two-dimensional foil design.

Another important parameter related to streamlining is the position of maximum thickness or shoulder position (SP) relative to the body length (BL). SP influences the hydrodynamic performance associated with lift and drag for a fusiform body profile (von Mises, 1945; Hoerner, 1965). SP can be associated with the point of transition from laminar to turbulent flow and boundary layer separation (Fish et al., 2008). A rearward displacement of SP allows for the maintenance of laminar flow over a larger portion of the body. However, a placement of SP too far aft on the body will trade off low drag laminar flow for the premature separation of the boundary layer and development of turbulence when not closely oriented to the oncoming flow, such as when maneuvering (Walters, 1962; Lang, 1963; Webb, 1975; Vogel, 1981). Hertel (1966) assigned a generic shark with a FR of 5.56 an SP of 0.44 BL. Kabasakal and Kabasakal (2013) measured an SP of 0.42 BL on a shortfin mako (*Isurus oxyrinchus*). SP for dolphins is 0.34–0.45 BL.

SP data have not been collected for ichthyosaurs. Massare (1988) modeled the bodies of ichthyosaurs and other Mesozoic aquatic reptiles as streamlined prolate spheroids and measured BL as the distance from the anterior edge of the orbit to the tip of the tail, which would complicate any SP measurement because BL includes the caudal fin and excludes the elongate rostrum. From Fig. 4 (Massare, 1988), based on the ichthyosaur *Ophthalmosaurus* (after Andrews, 1910), an SP of 0.31 BL is calculated. However, if the rostrum is included into BL, then the SP is 0.26 BL. If BL is measured at the angle in the tail, where the caudal fin may have originated, then SP is 0.31 BL. If the rostrum and caudal fin are excluded, then SP is 0.38 BL. This latter value is within the range of SP found in dolphins and close to the value for lamnid sharks, indicating convergence on a hydrodynamically-optimized shape.

Although the general body is fusiform for lamnid sharks, ichthyosaurs and delphinids, the rostrum is elongate in ichthyosaurs and to a lesser extent in some delphinids. The long, narrow rostrum is probably present to serve as a means of housing a battery of numerous teeth that enable feeding on small prey (e.g., fish, squid), despite potentially conflicting with the fusiform profile. However, the elongated rostrum of ichthyosaurs and some delphinids may act to reduce drag (Aleyev, 1977; Bandyopadhyay, 1989; Videler, 1993). The rostrum can potentially affect the pressure distribution over the forebody. The intersection of the rostrum and more posterior region of the head is marked by a concave region that facilitates transition from laminar to turbulent flow in the boundary layer (Aleyev, 1977). The turbulized boundary layer is less likely to separate and the pressure drag can be minimized over the aft of the body. In addition, the presence of an elongate rostrum reduces the pressure on the forebody, resulting in a reduction of the pressure differential over the body (Bandyopadhyay, 1989; Videler, 1993; Nesteruk, 2020).

15.3.2 Control Surfaces and Fin Shape

Aside from the analogous fusiform body shape exhibited by all three of the convergent taxa, it is the appendages, with their shape, position, and function that truly cement the idea of morphological convergence (Fish, 2004). The appendages are all considered to be highly streamlined (Fig. 15.6; Lang, 1966; Carey et al., 1971; Fish & Rohr, 1999; Motani, 2000). The fins, flippers, and flukes all function similarly as control surfaces. Control surfaces first evolved in aquatic organisms in association with their role in moderating stability and maneuvering (Nursall, 1962; Radinsky, 1987).

Control surfaces are structures that allow for adjustments to, and management of, the attitude of a body in a fluid (Fish & Lauder, 2017). When suspended within the

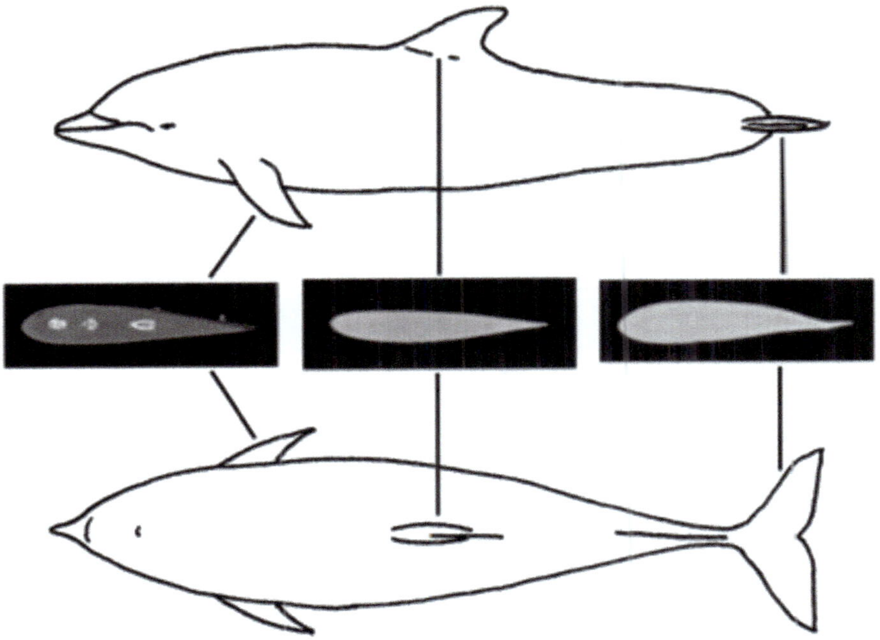

Fig. 15.6 Streamlining of body and control surfaces of a dolphin. Images of cross-sections of flipper, dorsal fin, and flukes were obtained from CT scans

water column, animals can move freely about three orthogonally arranged axes that intersect at the center of mass (CM) (Fig. 15.7). CM is the point where the weight of the animal is considered to be concentrated and acts as the 'balance point'. Movement about CM permits translation and rotation that give six degrees of freedom. The degrees of freedom of the translational movements are surge (anterior-posterior), heave (vertical displacement), and slip (lateral displacement), whereas rotation about the axes is termed roll for the longitudinal axis, pitch for the lateral axis, and yaw for the vertical axis (Fig. 15.7; Webb, 2004, 2006). Control about the roll axis governs lateral stability, about the yaw axis governs directional stability, and about the pitch axis imparts longitudinal stability. Longitudinal stability (i.e., horizontal orientation) is associated with 'trim', which relates to fore and aft balance. Trim is determined by the alignment of the centers of mass and buoyancy in the vertical axis, and the longitudinal axis of the horizontally oriented body (Burcher & Rydill, 1994).

Governance of each degree of freedom by various combinations of control surfaces maintains stability that combats internal and external perturbations (Fish & Lauder, 2017). Perturbations are forces and torques that cause undesired changes in attitude and trajectory (Webb, 2006). Internal perturbations are related to self-generated motions of the animal and changes in body density (e.g., muscle contraction, gas distribution). External perturbations are a function of forces impinging on

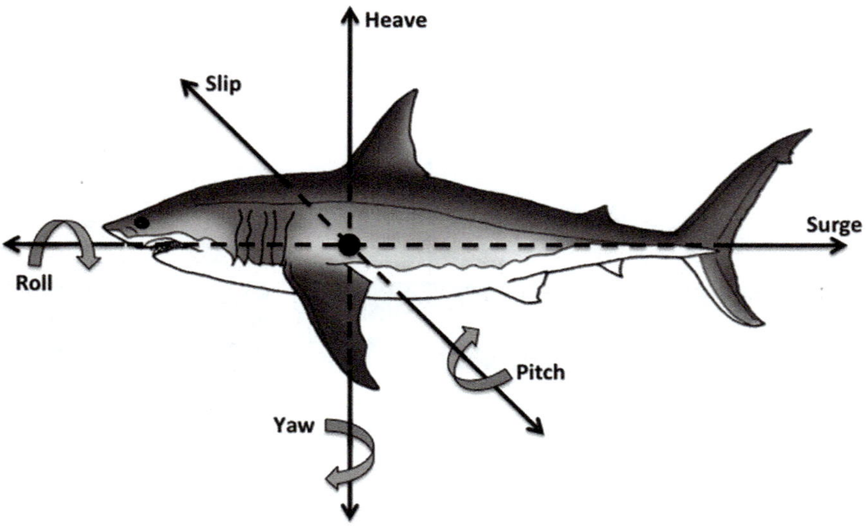

Fig. 15.7 The six degrees of freedom that specify movements of a free body. For the three orthogonal orientation axes running through the center of mass (black circle), there are three translational movements (Slip, Heave, Surge) and corresponding rotational movements (Pitch, Yaw, Roll)

the animal from the environment (e.g., waves, currents, vortices shed from structures in flow, interactions with other animals). The highly streamlined bodies of the shark, ichthyosaur, and dolphin should be unstable and necessitate the addition of control surfaces for stability (Triantafyllou, 2017; Fish & Lauder, 2017). The position, size, and geometry of the control surfaces help to maintain stability and suppress instabilities by generating forces that counter perturbations (Harris, 1936; Aleyev, 1977; Weihs, 1993; Bandyopadhyay et al., 1997; Fish, 2002, 2004; Fish & Lauder, 2017). Alternatively, when deployed asymmetrically these same stabilizing structures can initiate instabilities to assist maneuvering. It is this duality of function that makes control surfaces so important in the locomotor performance and is expressed in the similar design and position of the various appendages in the three convergent taxa.

Like the rudder and dive planes of a submarine (Burcher & Rydill, 1994; Gabler, 2000), the appendages have evolved into hydrofoils (i.e., wing-like planing surfaces) that hydrodynamically generate lift from the animal's movement dedicated to the control of stability and maneuverability (Harris, 1936, 1938; Lang, 1966; Fish & Shannahan, 2000; Fish, 2002, 2004; Webb, 2004, 2006; Cooper et al., 2008; Weber et al., 2009b, 2014; Fish & Lauder, 2017). The lift force can be vectored in a particular direction for stabilization or to generate instability for maneuvering. The lift force created by the control surface is a function of the surface area of its planform (i.e., shape) and Aspect Ratio (AR = fin span^2/planform area), where span is defined as the linear distance from base to tip of the control surface. Well-performing hydrofoils maximize the lift to drag ratio (L/D) (von Mises, 1945; Webb,

1975; Vogel, 1994; Weber et al., 2009b, 2014). Increased lift can be fostered by cambering (i.e., asymmetry between dorsal and ventral surfaces of a hydrofoil).

The control surfaces are modifications of the paired appendages (including the pectoral and pelvic limbs), the median dorsal fin, and the caudal fin or flukes (Harris, 1936, 1938; Riess, 1986; Fish, 2004; Lingham-Soliar, 2004; Fish & Lauder, 2017). Each of these control surfaces has an elongate teardrop design in cross-section (Lang, 1966; Fish, 2002). As with the fusiform body, the cross-sectional profile maintains low drag and reduces the energy lost by moving through the dense aquatic medium (Lang, 1966; Fish et al., 2007).

Although the caudal fin or flukes have a propulsive function (see below), these caudal extensions can do double duty as control surfaces. The caudal fins of the shark and ichthyosaur can be used as a rudder to stabilize the body in the yaw axis or to generate torques to effect a turn (Harris, 1936; Webb, 1975; Fish & Lauder, 2017); whereas, the horizontal orientation of the dolphin flukes permits control of pitch and the compressed peduncle can be laterally flexed to induce yawing motions (Fish, 2002).

All three of the convergent taxa possess pectoral appendages and caudal and dorsal "fins". Pelvic fins are present in both the shark and ichthyosaur, but not the dolphin. Another prominent difference is the orientation of the caudal fin. Both the shark and ichthyosaur have a vertically oriented caudal fin that can be displaced laterally, but the caudal flukes of the dolphin are oriented in the horizontal plane and move dorsoventrally. The dolphin's dorsoventral propulsive movements are a legacy of the change in posture to an up-right stance of mammals from a sprawling stance and an associated reorientation of the axial muscles (Fish, 2001). Another difference is that the shark has additional median fins, with a second, smaller dorsal fin and a ventrally located anal fin.

The paired (pectoral, pelvic) fins and the dorsal and anal fins of the shark are generally thin in cross-section. Flattened cartilaginous basal elements are situated at the base of the fin and numerous smaller radial cartilages extend distally from the basals. The majority of fin area is supported by a fan-like array of collagenous ceratotrichia (Kemp, 1977), which cannot be actively bent. However, it was proposed that an increase in the hydrostatic pressure within the body could be transferred through the cross-helical collagen fibers surrounding the body and into the dorsal fin of the white shark (*Carcharodon carcharias*) to stiffen the fin (Wainwright et al., 1978; Lingham-Soliar, 2005a). Pectoral and pelvic fin movements are somewhat limited in sharks (Harris, 1936, 1938; Wilga & Lauder, 2000, 2001; Fish & Shannahan, 2000).

The fore and hind limbs of ancestral tetrapods have been adapted for use in water as flippers in the aquatic tetrapods. These limbs enclose a bony skeleton homologous with the bones of terrestrial tetrapods (Williston, 1914; Howell, 1930; Caldwell, 2002; Fish, 2004; Cooper et al., 2007; Kelley & Pyenson, 2015; Massare & Lomax, 2019). Flippers represent modifications of the pectoral and pelvic limbs. There is limited mobility of the elements within these appendages. In particular, the digits are not separated, resulting in a loss of independent movement within the flippers. Flippers are used to control stability and maneuverability. Movement of the flippers

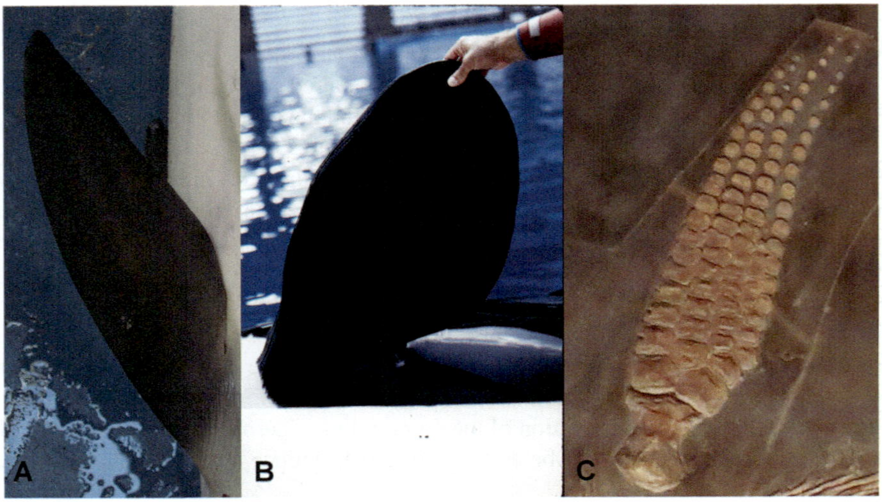

Fig. 15.8 Foreflipper design of tetrapods. (**a**) Pectoral flipper of the striped dolphin, *Stenella*, (**b**) pectoral flipper of the killer whale, *Orcinus*, and (**c**) pectoral flipper of the ichthyosaur, *Stenopterygius* (courtesy of Judy Massare)

is confined to the shoulder and hip joints (Williston, 1914; Howell, 1930; Felts, 1966; Fish, 2004; Cooper et al., 2007, 2008). The number of phalanges in each digit of the flipper is variable among species. Hyperphalangy is the condition found in cetaceans and ichthyosaurs in which the maximum number of phalanges in the digits often greatly exceeds the number in the ancestral state (Williston, 1914; Howell, 1930; Riess, 1986; McGowan, 1991; Fish & Battle, 1995; Caldwell, 2002; Cooper et al., 2007; Maxwell, 2012; Massare & Lomax, 2019). Polydactyly (i.e., extra digits) also occurs in some aquatic tetrapods (Fedak & Hall, 2004).

The effectiveness of the various control surfaces is dependent on their shape and *AR*. High *AR* indicates a long narrow control surface associated with high lift generation, whereas low *AR* indicates a broad surface area with a short span. High *AR* hydrofoils are characteristic of relatively fast swimmers and have a high lift to drag ratio (L/D). Highly effective fins maximize L/D (Webb, 1975; Weihs, 1989). The *AR* for the pectoral fins of lamnid sharks is about 3.0 (Hoffman et al., 2020). Dolphins, such as *Delphinus*, *Lagenorhynchus*, *Stenella*, and *Tursiops*, have pectoral flippers with a pointed tip and a range of *AR* of 3.5–6.3 (Fig. 15.8a), whereas *Orcinus*, with rounded flippers, has a lower *AR* of 1.9–2.9 (Fig. 15.8b; Fish unpubl. Data; Fish et al., 1988). Ichthyosaur specimens of *Ichthyosaurus* and *Stenopterygius* had pectoral flippers with *AR* of 4.5–6.0, within the range of dolphins with flippers of a similar planform (Fig. 15.8c).

It is curious that, unlike the shark and ichthyosaur, dolphins lack pelvic or hind flippers. Indeed, all modern cetacean and sirenian (manatee and dugong) species lack external hind flippers despite their quadrupedal ancestors having had hind legs (Gingerich et al., 1990, 1994, 2001, 2015; Thewissen & Fish, 1997; Domning,

2000, 2001; Bejder & Hall, 2002; Thewissen et al., 2007, 2009; Uhen, 2010; Díaz-Berenguer et al., 2018, 2019; Lambert et al., 2019). Internally the pelvis and hind limbs of cetaceans have become vestigial structures (Adam, 2009). The loss of the external hind limbs in modern species is associated with the absence of the genetically controlled signaling cascade for formation of the hind limb (Thewissen et al., 2006; Thewissen, 2018). However, hind limb buds do develop for a brief period before reabsorption in embryonic cetaceans, and occasionally atavistic rudimentary hind limbs appear in modern whales and dolphins (Struthers, 1893; Andrews, 1921; Ohsumi, 1965; Hall, 1984; Bejder & Hall, 2002; Thewissen et al., 2006; Thewissen, 2018). This atavism indicates that the genes controlling hind limb formation have not been entirely lost. The loss of the hind limbs, along with other morphological changes, in the transition from terrestrial to fully aquatic habits, has been heralded as one of the best characterized examples of macroevolution (Thewissen & Bajpai, 2001; Gingerich, 2015; Huelsmann et al., 2019).

Why were the hind limbs lost in cetaceans and not sharks and ichthyosaurs? As control surfaces, the possession of hind flippers in cetaceans would be redundant. A submarine has control surfaces with dive planes anterior to the center of mass either near the bow or on the conning tower and posteriorly in combination with the rudder. Submarines adjust their trim hydrostatically through buoyancy control with the ballast tanks or hydrodynamically with lift generated by the dive planes (Burcher & Rydill, 1994; Fish & Lauder, 2017). The fore and aft position of the two sets of dive planes balances the vessel to keep it in trim. The shark and ichthyosaur both move their caudal fins laterally for propulsion; therefore, to maintain trim, rely upon their pelvic fins/flippers, located posteriorly, working in concert with the anterior pectoral fins (Harris, 1936, 1938; Standen, 2008). However, cetaceans and sirenians have horizontally oriented flukes that act as a control surface at the posterior end of the animal in concert with the anterior pectoral flippers to maintain trim, while the flukes also are used for propulsion (Fish, 2002; Kojeszewski & Fish, 2007). Like the feathers of an arrow, the flukes located far posterior to the CM can generate large directionally-correcting torques because of their long lever arms (Harris, 1936; Wegner, 1991; Webb et al., 1996; Fish, 2002, 2004; Fish & Lauder, 2017). Thus, the posteriorly located flukes, in concert with the anterior pectoral flippers, can perform trim control, so that extra sets of horizontally-oriented control surface such as pelvic flippers are redundant.

In addition to the caudal flukes performing the task of trim control, loss of the pelvic flippers would be of benefit in reducing drag on cetaceans (Bejder & Hall, 2002). Possession of hind flippers with their additional surface area and concomitant increased drag would effectively limit swimming speed and increase energetic costs for locomotion. Possession of hind flippers would, therefore, be hydrodynamically disadvantageous. The lack of pelvic limbs in cetaceans is related to the orientation of the propulsive oscillatory movement of the spine. Dorsoventral movement of the spine is a legacy of the ancestral terrestrial mammals. These mammals stood on erect limbs and reorganized the axial muscles for dorsoventral bending of the spine from the reptilian lateral bending condition (Howell, 1930; Fish, 2001). The ichthyosaurs' reptilian ancestors retained the ancestral muscle architecture associated with the sprawling posture and employed lateral bending (undulation). Primitive ichthyosaurs had a more anguilliform swimming mode that was fostered by lateral undulations (Motani et al., 1996; Buchholtz, 2001a). This committed the ichthyosaurs to lateral oscillations of the tail, as performed by sharks. Because the tail beat laterally, there was need for posteriorly located pelvic fin control surfaces for trim control. Buchholtz (2001a), however, contended that *Stenopterygius quadriscissus* and *Ophthalmosaurus icenicus*, with small hind limbs, were in the process of reduction or loss of these trim stabilizers.

Irrespective of the paired appendages with their skeletal braces, other control surfaces lack internal skeletal support, such as the dorsal fin, flukes, and peduncle keels. In the shark, ichthyosaur, and dolphin, these features are reinforced by dense arrays of collagen fibers (Felts, 1966; Lingham-Soliar, 2001; Lingham-Soliar & Plodowski, 2007). The keels of the peduncle of dolphins are composed of blubber that contains structural collagenous and elastin fibers (Hamilton et al., 2004), which act as tensile stays. Collagen fibers with high tensile strength are found in the flukes of cetaceans (Sun et al., 2010a, b; Gough et al., 2018), and are arranged in a dense array of chordwise-oriented thin crossing fibers composing the inner core layer sandwiched between two layers of spanwise-oriented thick fibers composing the outer ligamentous layers (Felts, 1966; Sun et al., 2010a, b, 2011; Gough et al., 2018). This sandwich composite beam provides rigidity while allowing some bending (Sun et al., 2010b).

The presence of a relatively stiff, non-mobile triangular dorsal fin is a strong indicator of convergence upon a similar lifestyle of the three taxa (Lingham-Soliar & Plodowski, 2007). *AR* for the dorsal fins is generally less than 2. In all cases, the dorsal fin has an anterior insertion that is near the maximum diameter of the animals and coincides with the longitudinal position of CM (Fig. 15.7; Fish, 2002). The dorsal fin functions to resist roll and yaw (Fish, 2002; Lingham-Soliar, 2005a; Lingham-Soliar & Plodowski, 2007). Its position near CM allows the dorsal fin to perform as an anti-slip device to maintain the trajectory of rapid, small radius turns (Fish, 2002). The wing-like profile of the dorsal fin could also aid in generating lift that is oriented toward the center of rotation to supply the centripetal force for maneuvering.

Based on Fig. 1 of Lingham-Soliar (2016), comparison of the profiles of a great white shark (*Carcharodon carcharias*) and ichthyosaur (*Stenopterygius quadriscissus* SMF 457; Senckenberg Museum, Germany) permits the sweep of the dorsal fin to be measured. The sweep is measured as the angle between the line perpendicular to the longitudinal axis of the body and the one-quarter chord position on the fin (Fish & Rohr, 1999). The sweep was 30.0° and 31.3° for the ichthyosaur and shark, respectively. These values are within the range of sweep angles (29.7–57.5°) of the dorsal fins of dolphins (Fish unpubl. data).

The caudal fins/flukes of the three convergent taxa, the shark, ichthyosaur and dolphin, are extensions from the tail. The caudal fins/flukes are similar in their general planform shape (Fig. 15.9). These caudal extremities are relatively stiff, with a high *AR* lunate planform shape with tapering tips (Williston, 1914; Lighthill, 1969, 1970; Riess, 1986; Webb, 1975; Fish et al., 1988; Massare, 1988; Fish,

Fig. 15.9 Images of a shark (top), ichthyosaur (middle), and dolphin (bottom) showing the high aspect ratio, wing-like caudal fin and narrow peduncle. Shark image courtesy of Rodney Fox

1998a, b; Lingham-Soliar, 2005b, 2016; Lingham-Soliar & Plodowski, 2007). The cross-sectional profiles of these caudal fins/flukes have a streamlined fusiform shape similar to that of engineered hydrofoils, although the fossil remains of ichthyosaurs do not allow for an exact determination of the three-dimensional geometry of the caudal fin (Lang, 1966; Fish et al., 2007; Crofts et al., 2019).

The caudal fins and flukes are used primarily for propulsion (Fig. 15.9). They act as an oscillating wing or hydrofoil to generate a lift-based thrust (see below). The caudal fins and flukes are connected to the body by a narrow peduncle that can be flexed in the direction of oscillation. This 'narrow necking', along with keels oriented in the plane of oscillation, reduces the resistance of the peduncle to reciprocating propulsive oscillations (Lighthill, 1969, 1970; Zhang et al., 2020). In addition, the narrow peduncle, in concert with the large inertial mass in the anterior region of the body of the swimmer, will minimize kinetic energy losses due to recoil in the anterior body and reduce drag (Lighthill, 1969; Webb, 1975; Sfakiotakis et al., 1999; Ben-Zvi & Shadwick, 2013).

The caudal fins and flukes largely differ in their composition. The vertebral column extends into the dorsal lobe of the caudal fin of sharks, into the ventral lobe of the caudal fin of ichthyosaurs, and between the two caudal flukes of dolphins (Howell, 1930; Fish, 1998b; Fish et al., 2006; Crofts et al., 2019). For the shark, the blades of the caudal fin are supported by collagenous ceratotrichia (Kemp, 1977; Crofts et al., 2019). The caudal fin and flukes of the ichthyosaur and dolphin, respectively, are supported by compact arrays of collagen fibers (Felts, 1966; Lingham-Soliar, 2005b, 2016; Lingham-Soliar & Plodowski, 2007; Gough et al., 2018).

The flukes of dolphins are symmetrical and have a range of *AR* of 3.4–5.5. The sweep on the caudal flukes ranges from 25.0° to 41.3° (Fish unpubl. data; Fish et al., 1988). Based on the representative species mentioned above (Lingham-Soliar, 2016), *AR* for the caudal fins is 3.8 for the shark and 4.2 for the ichthyosaur. The dorsal (hypercaudal) lobe of the caudal fin and ventral (hypocaudal) lobe are asymmetrical for the shark and ichthyosaur. Comparatively, the planar area of the hypercaudal lobe is 34% greater than the hypocaudal lobe of the shark, whereas the planar area of the hypocaudal lobe is only 5% greater than the hypercaudal lobe for the ichthyosaur. The increased size of the respective lobes appears to be associated with the bending direction of the vertebral column.

The asymmetries of the caudal fins of the shark and ichthyosaur will affect the pitching moments of the animals. The heterocercal tail of the shark will induce an upward pitch at the tail and a corresponding downward pitch of the head (Harris, 1936; Alexander, 1965). The pectoral fins situated anterior of CM would then be used to generate a restoring lift force to maintain trim of the body (Harris, 1936; Fish & Shannahan, 2000). The caudal fin of the ichthyosaur is a reversed heterocercal or hypocercal tail, which would induce an opposite rotation of the body compared to that of the shark. Taylor (1987) considered that because of a near neutral buoyancy of the ichthyosaur, the hypocercal fin would induce an upward oriented thrust vector through the center of balance and negate any pitching, as indicated by the model by Thomson and Simanek (1977). However, Wilga and Lauder (2004), and Flammang et al. (2011) used digital particle image velocimetry of a swimming shark and found a strong posteroventrally directed jet flow from the motion of the heterocercal tail. This direct evidence validated the assertions of Harris (1936) and Alexander (1965), while falsifying the hypothesis of Thomson and Simanek (1977). This result would indicate an upward pitch of the heterocercal tail of the shark and conversely a downward pitch of the hypocercal tail for the ichthyosaur. As an obligate air-breather, downward pitch of the tail and upward pitching of the head would be advantageous for the ichthyosaur to facilitate surfacing for breathing, in contrast to the gill-breathing shark (Taylor, 1987; McGowan, 1992; Crofts et al., 2019).

15.3.3 Integument

The integument, or skin, is the one part of the body that is intimate contact with the environment and can function to enhance locomotion (Garten & Fish, 2020). The magnitude of the frictional drag component is a function of the texture of the skin The interaction of the surface of the body and appendages with the flow of water determines the development of the shear forces in the boundary layer.

Sharks have a rough skin, which appears counter-intuitive to the maintenance of an orderly, low-drag flow. Fast swimming sharks have scales that have flat crowns and sharp ridges oriented longitudinally, with rounded valleys between them (Pershin et al., 1976; Reif, 1978, 1985; Reif & Dinkelacker, 1982; Lang et al., 2008; Oeffner & Lauder, 2012). This scale morphology can potentially provide a 7–8% drag reduction by acting as riblets (Reidy, 1987; Walsh, 1990). Riblets are streamwise microgrooves that reduce drag by acting as small fences to break up spanwise vortices and reduce the surface shear stress and associated loss of momentum.

Small ridges, like riblets, on the epidermis of dolphins have been hypothesized to stabilize longitudinal vortices (Yurchenko & Babenko, 1980), but the geometry of the ridges, with rounded edges, does not suggest an effective analogy with riblets (Fish & Hui, 1991). The skin of cetaceans is generally described as being smooth (Shoemaker & Ridgway, 1991). Wainwright et al. (2019) validated that the skin is smooth, particularly when compared to other pelagic swimmers. It was shown that the size of the ridges on the body of cetaceans is small, or even absent, on skin of the control surfaces of most species.

Ridges were found on the fossilized skin of an ichthyosaur, although these ridges were interpreted to be due to post-mortem wrinkling (Delair, 1966; Lingham-Soliar, 1999). The skin exhibited orthogonally-oriented fibers arranged in dense pre-stressed layers (Lingham-Soliar, 1999). This arrangement was inferred to be for the prevention of wrinkling of the skin and bulging of the muscles in the living animal. Fibrous skin would maintain a smooth hydrodynamic surface (Lingham-Soliar, 1999, 2001). A layer of subdermal collagen fibers occurs in the body and tail of sharks, ichthyosaurs, and dolphins, which could be used for the transfer of muscle forces for swimming (Wainwright et al., 1978; Lingham-Soliar, 1999; Lingham-Soliar & Wesley-Smith, 2008; Flammang, 2010; Crofts et al., 2019).

15.4 Swimming Performance

15.4.1 Swimming Speed

The three convergent taxa are all marine predators and are considered to be capable of rapid and sustained swimming (Massare, 1988; Fish & Rohr, 1999; Buchholtz, 2001b). Lamnid sharks and oceanic dolphins undergo seasonal migrations maintaining a steady swimming speed over days (Miyazaki et al., 1974; Bonfil et al., 2005; Weng et al., 2007, 2008; Taylor et al., 2016). Satellite tag data for white sharks (*Carcharodon carcharias*) and salmon sharks (*Lamna ditropis*) indicate migratory speeds of 0.8 to 1.5 m/s (Weng et al., 2007, 2008; Bonfil et al., 2010). Dolphins routinely swim for prolonged periods at speeds of 0.4 to 3.7 m/s (Fish & Rohr, 1999). Based on the metabolically measured cost of transport, Williams et al. (1993) and Yazdi et al. (1999) each estimated that the optimal swimming speed for bottlenose dolphins (*Tursiops truncatus*) was 2.1 and 2.5 m/s, respectively. The optimal speed would coincide with the minimum cost of transport, which would have the highest efficiency.

Estimates of routine swimming speeds for ichthyosaurs were calculated by Massare (1988) and Motani (2002a, b). Both investigators used models based on metabolically and hydrodynamically derived energetics. Massare (1988) calculated speeds for ichthyosaurs ranging from 1.82 to 3.06 m/s. Initially, Motani (2002a) used a set of hydrodynamic equations to calculate the optimal swimming speed (the speed at which the energy consumption required to move a unit length is minimal) for *Stenopterygius*. Optimal swimming speed would correlate with the minimum cost of transport. The estimated optimal speeds ranged from 1.2 to 1.7 m/s. Motani (2002b) modified his model and calculated optimal swimming speeds based on the basal metabolic rates for each of three conditions (reptiles, tuna-leatherback turtle, and cetaceans-pinnipeds). His results produced speed ranges of 0.55 to 0.78 m/s (reptiles), 1.2 to 1.6 m/s (tuna-leatherback turtle), and 2.2 to 2.5 m/s (cetaceans-pinnipeds). Motani (2002b) considered the reptilian condition was probably not feasible, and he argued ichthyosaurs would have had a higher metabolism and an elevated body temperature due to their large body size. A cruising speed of at least 1 m/s would have been similar to the swimming performance of the blue marlin (*Makaira nigricans*) and yellowfin tuna (*Thunnus albacares*) that have a similar diet to *Stenopterygius* (Motani, 2002b).

While optimal swimming speed can be estimated to provide an indication of long-duration routine swimming speeds for activities such as cruising, searching, and migrating, burst swimming speeds cannot be predicted. Burst swimming represents an unsteady behavior of short-duration. High-speed swimming is most commonly associated with pursuit of prey and escape from predators (Webb, 1975). It is unknown what the maximum burst swimming speed was for ichthyosaurs, although given the similar morphology and swimming mode (see below), it can be assumed to be comparable to that attained by lamnid sharks and delphinid dolphins. Shortfin mako shark (*Isurus oxyrinchus*) is capable of a maximum speed of 19.44 m/s (Díez et al., 2015). Burst speeds for dolphins have been reported to range from 5.6 m/s up to 15.0 m/s (Fish & Rohr, 1999).

15.4.2 Swimming Mode

The shark, ichthyosaur, and dolphin share a thunniform (i.e., from *Thunnus*, the generic name for several species of tuna) bauplan with a stiff, streamlined fusiform body, lunate tail, dorsal fin, and extremely narrow necking of the peduncle (Lighthill, 1969; Webb, 1975; Lindsey, 1978; Motani, 2005; Shadwick, 2005; Lingham-Soliar, 2016; Gutarra et al., 2019). Aside from these morphological similarities, the designation of being thunniform (tuna-like) swimmers indicates similar kinematic patterns for highly efficient aquatic propulsion (Aleyev, 1977; Fish et al., 1988; Lindsey, 1978; Webb, 1975; Motani, 2005). Although there is no direct evidence for the swimming kinematics of ichthyosaurs, the morphological similarities with lamnid sharks and dolphins strongly indicate that these modern analogues can be used as a proxy for the swimming mode of the extinct taxon, particularly for highly derived ichthyosaurs (e.g., *Ichthyosaurus*, *Ophthalmosaurus*, *Stenopterygius*).

Thunniform swimming of the extant convergent species, as well as tuna, encompasses undulatory motions of one-half to one full wavelength within the body. Transmission of the undulations to the caudal-most portion of the body produces an oscillatory motion (Fierstine & Walters, 1968; Lighthill, 1969; Lindsey, 1978; Smits, 2019). Significant propulsive movements are confined to the peduncle and caudal fin (Webb, 1975; Lindsey, 1978; Fish, 1998a). However, the peak-to-peak amplitude (A) at the caudal fin is typically about 20% of body length during routine swimming (Webb, 1975), but can be greater than 30% of body length, particularly during rapid accelerations (Fierstine & Walters, 1968; Skrovan et al., 1999; Fish et al., 2014). The peak-to-peak amplitude of the caudal fin remains relatively constant over a range of swimming speeds. This constancy of the oscillatory amplitude at 20% of body length is tied to minimization of the energy required to swim (Saadat et al., 2017).

As the stroke amplitude remains constant, swimming speed (U) increases directly with the frequency of oscillations (f) of the tail (Fish, 1996; Rohr & Fish, 2004; Smits, 2019). Maximum propulsive efficiency is related to the non-dimensional Strouhal number, which combines the three kinematic parameters in the equation:

$$\mathrm{St} = f\,A/U.$$

The Strouhal number is a gauge of the effectiveness of flapping locomotion by indicating the distance a swimmer moves with each tail stroke (Saadat et al., 2017). The Strouhal number predicts that the maximum spatial amplification and optimal creation of thrust-producing jet vortices lies within a narrow range of 0.2–0.4 (Triantafyllou et al., 1993; Triantafyllou & Triantafyllou, 1995; Streitlien & Triantafyllou, 1998). The Strouhal numbers for thunniform swimmers occur within the optimal range and coincide with the maximal propulsive efficiency between 0.76 and 0.98 (Webb, 1975; Triantafyllou & Triantafyllou, 1995; Fish, 1998a; Rohr & Fish, 2004; Fish & Lauder, 2006).

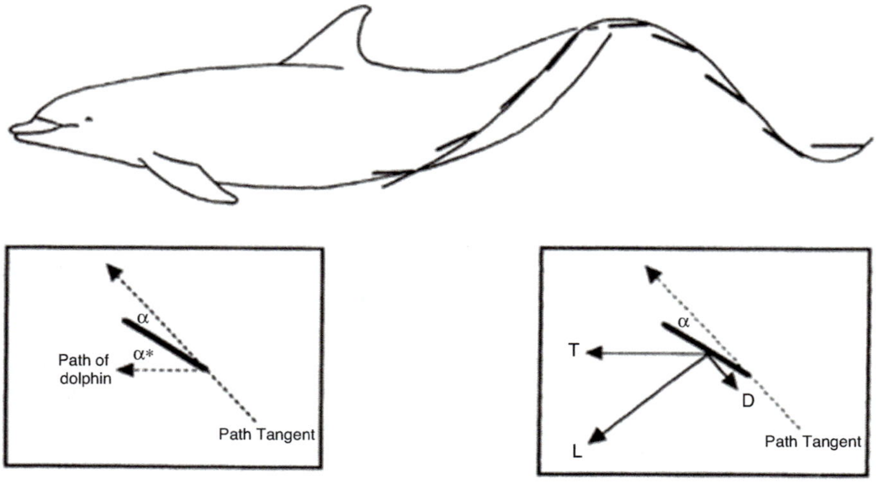

Fig. 15.10 Path of oscillating dolphin flukes through a stroke cycle. The tips of the flukes move along a sinusoidal path. Sequential fluke positions along the path are illustrated as straight lines. The box on the left shows the relationship between the tangent to the path of the flukes with the angle of attack, α, and the incident angle, α*. Angle of attack is the angle between the tangent of the fluke's path and the axis of the fluke's chord; pitch angle is the angle between the fluke axis and the translational movement of the animal. The box on the right shows the relationship between the major forces produced by the motion of the fluke. D is the drag, L is the lift, and T is the thrust resolved from L (from Fish, 1993b)

When swimming, the tip of the caudal fin traces out a sinusoidal trajectory for thunniform swimmers (Fig. 15.10). The thunniform tails of the shark and ichthyosaur are heaved (flexed) laterally at the peduncle (Shadwick, 2005), whereas the dolphin flukes are dorsoventrally heaved (Fish & Rohr, 1999). In addition, the base of the caudal fin is pitched so that the angle of inclination (i.e., angle of fin relative to longitudinal movement of swimmer) is maintained at a defined angle throughout the stroke cycle. Without this pitching motion, the fin would sweep through the stroke with a continuously changing angle of inclination. The combination of heave and pitch causes the trailing edge of the caudal fin to lag behind its leading edge, allowing the fin to slice through the water (Van Buren et al., 2020). The addition of heave to a pitching motion can dramatically increase the thrust and propulsive efficiency of an oscillating wing-like hydrofoil. Flexion at the peduncle and base of the caudal fin emulates a double-jointed system, which allows the angle of inclination (α*; Fig. 15.10) of the caudal fin to be adjusted throughout the stroke cycle, maintaining nearly continuous maximum thrust (Parry, 1949; Fierstine & Walters, 1968; Lindsey, 1978; Reif & Weishampel, 1986; Fish et al., 1988).

The thunniform namesake tuna possesses the double-jointed system, with joints at the bases of the anterior end of the peduncle and caudal fin. These two flexion points are separated by a few stiff, laterally keeled vertebrae, which act as pulleys for the tendons connecting with the large anterior muscle mass (Fierstine & Walters,

1968). Despite having keels on the peduncle (Lingham-Soliar & Reif, 1998; Bernal et al., 2001a; Hamilton et al., 2004), the lack of keels on the vertebrae and numbers of vertebrae in the peduncle of the three convergent taxa do not preclude an analogous double-jointed system with the tuna. Instead, the larger number of vertebrae with anteroposteriorly short centra in the peduncle permits increased flexibility for heaving motions (McGowan, 1992; Buchholtz, 2001a, 2001b, 2007; Lingham-Soliar, 2001; Buchholtz & Schur, 2004; Buchholtz et al., 2005; Kim et al., 2013). Flexural pitching of the caudal fin can be actuated by the insertions of long tendons from the anterior musculature (Fierstine & Walters, 1968; Pabst, 1990, 1996; Lingham-Soliar & Reif, 1998; Lingham-Soliar, 2001; Shadwick & Gemballa, 2005; Adams & Fish, 2019). Furthermore to prevent "bowstringing" of the tendons (i.e., lifting away from the joint), fibrous connective tissue in the skin would act like a retinaculum (Fierstine & Walters, 1968; Lingham-Soliar & Reif, 1998; Pabst, 2000). Maximum flexion occurs near the base of the flukes of dolphins at a vertebra known as the "ball vertebra" (Watson & Fordyce, 1993; Tsai, 1998; Fish et al., 2006). This vertebra has convex (rounded) anterior and posterior faces and differs from the flat (acoelous) faces of the other vertebrae. The ability to rotate the flukes about a pitching axis at the ball vertebra allows for control of the angle of the flukes when swimming (Long Jr. et al., 1997; Fish, 1998b). The abrupt dorsal or ventral tailbend of the vertebral column of the shark and ichthyosaur, respectively, indicate this position to be the flexion point for the pitch of the caudal fin.

Aquatic propulsion employing the thunniform mode results from the transfer of momentum from the animal to the water (Webb, 1988). The rate of momentum exchange between the propulsor and the water determines the amount of thrust generated (Daniel et al., 1992). Thrust is produced exclusively by the stiff, high-*AR*, lunate tail, which acts like an oscillating hydrofoil or wing (Williston, 1914; Lighthill, 1969; Riess, 1986; Webb, 1975; Fish et al., 1988; Massare, 1988; Fish & Hui, 1991; McGowan, 1992; Fish, 1998a, b; Lingham-Soliar, 2004; Shadwick, 2005). The momentum imparted to the water by the oscillating hydrofoil takes the form of a wake with two alternating rows of thrust-type vortices and a posteriorly directed jet stream (Weihs, 1972; Triantafyllou et al., 2000; Fish et al., 2014, 2018; Smits, 2019).

Thrust and efficiency are maximized by the use of lift (circulation)-based oscillating hydrofoils (Lighthill, 1969; Webb, 1975; Sfakiotakis et al., 1999; Fish & Lauder, 2006). Lift-based oscillatory swimming is associated with the radiation into pelagic habitats where steady swimming is required (Webb & de Buffrénil, 1990). Lift is generated as for an airplane wing, and results from differential flow between the surfaces of the hydrofoil (i.e., Bernoulli effect) because the foil is canted at an angle of attack (α; i.e., incident angle to on-coming flow). The heaving and pitching motions of the caudal fins of the thunniform swimmers are responsible for changes of angle of inclination and angle of attack throughout the stroke cycle (Fig. 15.10). As the caudal fin moves along a sinusoidal path through the water, the incident flow encountered by the fin is a combination of the forward movement of the swimmer and the heaving motion of the tail, where the pitch of the fin relative to its path is the angle of attack (Fig. 15.10). Lift is directed perpendicular to the pathway traversed by the caudal fin and can be resolved into an anteriorly directed thrust vector (Lighthill, 1969; Weihs & Webb, 1983; Fish, 1993b). Thrust is derived from a combination of the horizontal component of the lift force and leading-edge suction (Ahmadi & Widnall, 1986). Thrust derived from lift increases directly with increases in angle of attack. To maximize lift, the propulsor is maintained at an angle of attack (<30°) throughout the stroke cycle (Fish et al., 1988; Fish, 1993b). However, low angles of attack increase efficiency while reducing the probability of stalling (i.e., dramatic loss of lift) and decreased thrust production (Chopra, 1976; Van Buren et al., 2020). By restricting bending to the peduncle and base of the caudal fin, this permits rotational motion to maintain a positive angle of attack of the caudal fin to the oncoming flow (Webb, 1975). Thrust is thus generated continuously throughout a stroke cycle.

Some drag is produced by the oscillating fin as a result of skin friction and drag due to lift (i.e., induced drag). The induced drag is small compared to the lift. A high L/D ratio is a function of the high *AR* of the caudal fin (Bose & Lien, 1989; Liu & Bose, 1993; Fish, 1998a, b). Induced drag is also limited by sweepback of the caudal fin. Sweepback is the angle made by the leading edge with respect to the longitudinal axis of the body. Minimal induced drag is fostered by a swept wing planform with a triangular shape (Küchermann, 1953; Ashenberg & Weihs, 1984).

A tapered wing with sweepback or crescent design can reduce the induced drag by 8.8% compared with a wing with an elliptical planform (van Dam, 1987). The combination of low sweep with high *AR* allows for high efficiency rapid swimming (Azuma, 1983). Sweep angles of 30° and 40° produce more thrust than a 50° swept fin during the mid-stroke for the caudal fin, but as the fin reverses direction during the oscillation, the 50° sweep produces more thrust (Matta et al., 2019). However, a fin exceeding about 30° of sweep leads to a reduction in efficiency and large sweeps generate less lift (Chopra & Kambe, 1977; Matta et al., 2019). The sweep angle of the symmetrical caudal flukes of delphinids ranges from 4.4° to 41.3° and *AR* ranges from 3.0 to 5.5 (Fish, unpubl. data). Based on Fig. 1 from Lingham-Soliar (2016), the sweep angle for the hypercaudal and hypocaudal lobes of the caudal fin with *AR* of 3.8 of a great white shark (*Carcharodon carcharias*) is 37.8° and 51.0°, respectively. The sweep angle for the hypercaudal and hypocaudal lobes of the caudal fin with *AR* of 4.2 of the ichthyosaur (*Stenopterygius quadriscissus* SMF 457) is 33.0° and 50.9°, respectively.

The relationship between sweep and *AR* also indicates a structural limitation to the strength and stiffness of the propulsor (van Dam, 1987; Bose et al., 1990). The ability to sustain certain loads without breaking is considered a major constraint on increasing span and *AR* (Daniel, 1988).

15.5 Porpoising

Highly derived aquatic animals should avoid swimming close to the surface of the water. At the surface, kinetic energy from the motion of the animal is transferred as potential energy to the vertical displacement of water in the formation of waves (Denny, 1993; Vogel, 1994). This transfer increases the resistance on the swimming animals as wave drag. The additional drag on the body can be as much as five times the frictional drag on a fully submerged body (Hertel, 1966). The maximum increased drag occurs when the longitudinal axis of the body is at a depth of half a body diameter. Movement at or near the surface can thus substantially increase the energy expended and limit the speed of swimming (Hertel, 1966; Fish, 1996). For example, sea otters swimming submerged have an oxygen consumption 41% lower than when surface swimming and a maximum swimming speed when submerged that is 74% faster than at the surface (Williams, 1989). When towed at the surface, the drag on harbor seals (*Phoca vitulina*) was 2.5 times greater than when submerged at the identical velocity (Williams & Kooyman, 1985). Elimination of wave drag occurs at a submergence depth of ≥3 times the body diameter (Hertel, 1966), although wave drag was essentially zero below a depth of only two body diameters for a dolphin (*Lagenorhynchus*) at a high swimming speed (6.1–9.1 m/s) (Lang & Daybell, 1963). Thus, the locomotor strategy of submerged swimming can result in increased efficiency by the removal of wave drag.

For the shark, gas exchange via the gills means that it can remain submerged indefinitely and thus avoid wave drag, but the ichthyosaur and dolphin are obligate air-breathers and must return to the surface to ventilate the lungs. These aquatic tetrapods can utilize a strategy of swimming below the water surface for prolonged periods. The dolphin, like other marine mammals (cetaceans, pinnipeds), is adapted to extend periods of submergence because of its enhanced oxygen storage capacity resulting from increased blood volume and elevated levels of myoglobin in the muscles (Ridgway & Johnston, 1966; Kooyman, 1989; Elsner & Meiselman, 1995; Noren et al., 2002). Furthermore, the dolphin can use its oxygen reserves sparingly by reducing its metabolism utilizing diving bradycardia, peripheral vaso-constriction, and, if necessary, shifting to an anaerobic metabolism (Scholander, 1940; Kooyman, 1989; Williams et al., 1991, 2015). To prevent increased energy cost when coming to the surface to breathe, these animals limit such times and quickly ventilate the lungs before submerging. Dolphins can ventilate 90% of their lung volume in about 0.3 s (Irving et al., 1941; Ridgway et al., 1969; Kooyman & Cornell, 1981; Hui, 1989).

A variation on the strategy to limit increased drag at the water surface and prolong ventilation time is accomplished by porpoising (Hui, 1987, 1989; Williams, 1987; Fish & Hui, 1991). Porpoising consists of repetitive high-speed motions of rhythmic, long, ballistic jumps alternating with periods of submerged swimming close to the surface (Au et al., 1988; Fish & Hui, 1991; Weihs, 2002). Porpoising is conducted by fast swimmers, including dolphins, penguins, and pinnipeds. The energy required for leaping at low swimming speeds is greater than the energy expended in swimming just below the surface of the water. Porpoising becomes energetically efficient when the energy required to swim a given distance just below the water surface increases with swimming speed faster than the energy to leap that distance (Au & Weihs, 1980; Au et al., 1988; Blake, 1983). Above a critical speed, known as the crossover speed, where the energies of submerged swimming and leaping converge, there is an energetic advantage to swimming by porpoising. For dolphins, various models have indicated a crossover speed of at least 3.9 m/s for porpoising (Fish & Rohr, 1999).

Although energy could be conserved by porpoising, data on porpoising animals contradict the assertions of the models. The models have assumed an emergence angle of 45° to maximize the distance traveled through the air. Gordon (1980) indicated that to maintain forward speed an emergence angle of 30° would be desirable as a compromise between maximum distance and maximum forward speed of a leap. In reality, the emergence angle of the leaps is approximately 39°, with wide variation (Hui, 1989; Weihs, 2002), which lies between the optimal predicted angles. Also bringing into question the energy savings of porpoising are data indicating that the distances covered in the air are interspersed with swimming bouts of about twice the length of the leap. The assumptions of energy savings by porpoising predict that dolphins would spend more time leaping than swimming at speeds greater than crossover speed (Au et al., 1988).

Video data of free-ranging dolphins indicate a graded transition from minimal blowhole exposure at the surface at low swimming speeds to quasi-leaps, in which the dolphin is never completely out of the water at any instant, at medium swimming speeds, and complete porpoising leaps at the highest swimming speeds (Hui, 1989). These observations are consistent with maintaining a minimum blowhole exposure time for respiratory inhalation as swimming speed increases. Consequently, porpoising behavior may be energy conserving only in the sense that it is energetically the cheapest way to breathe, not energetically the cheapest way to swim (Fish & Hui, 1991).

The position of the external nares of ichthyosaurs might have made porpoising a more effective means of breathing while swimming. Unlike dolphins, ichthyosaurs had external nares situated at the end of the rostrum, anterior to the eyes, necessitating that the rostrum be lifted above the water surface to breathe. Although there is no direct evidence for porpoising, ichthyosaurs were envisioned to swim at high enough speeds to enable leaping out of the water and porpoising due to the convergence of design and swimming mode with dolphins. The artist Heinrich Harder (1858–1935) produced a rendering of a group of ichthyosaurs porpoising (Fig. 15.11). Similarly, a color plate in the book *Prehistoric Animals* by Augusta and Burian (1957) depicted a school of *Stenopterygius* porpoising like modern dolphins. Massare's (1988) and Motani's (2002a, b) calculated swimming speeds for ichthyosaurs were too low to allow porpoising, but their estimates were only for cruising speeds. Cowen (1996) considered that moderately-sized ichthyosaurs could swim at high enough speeds to porpoise, providing the necessary time to breathe while conserving energy. His assertion was even accompanied by a limerick,

Fig. 15.11 Illustration of porpoising ichthyosaurs by Heinrich Harder (1858–1935) from "Tiere der Urwelt" (Animals of the Prehistoric World) created in 1916. The work is in the United States Public Domain

Fast swimming air breathers are rare
 Some ichthyosaurs did it with flair
 They swam up in a leap
 (It's energetically cheap)
 And they took a deep breath in mid-air.

Although not known to porpoise, lamnid sharks and some other species of sharks have been observed to make spectacular leaps out of the water (Campana et al., 2005; Schwartz, 2013). Unlike porpoising, such leaping behaviors of sharks are not associated with energy conservation. The white shark will perform a "Polaris breach" to ambush pinnipeds and seabirds from beneath and launch the shark and prey completely out of the water (Martin et al., 2005; Hammerschlag et al., 2012). These vertical leaps require sufficient power to deliver the high escape velocity required to propel the whole body into the air.

15.6 Thermoregulation

As highly active predators, the three taxa have converged on physiological mechanisms to increase locomotor performance. The lamnid shark, ichthyosaur, and dolphin all are considered to be homeothermic (i.e., maintenance of constant body temperature) with a body temperature that is elevated above that of the environmental temperature. Muscles generally perform better at the elevated stable body temperatures of homeotherms compared to the lower body temperatures that is typical of poikilotherms (i.e., body temperature that is the same as the ambient temperature) (Irschick & Higham, 2016). Elevated body temperatures allow for higher maximum swimming speeds, longer and faster sustained swimming speeds, operation in cold water, migratory habits, enhanced digestion and assimilation rates, brain heating, and enhanced visual acuity (Block & Carey, 1985; Wolf et al., 1988; Bernal et al., 2001a, b; Watanabe et al., 2015).

The dolphin, as a mammal, is an endothermic homeotherm (Castellini & Mellish, 2016). Endothermic animals have an elevated metabolism that generates sufficient internal heat for homeothermy. The excess heat is derived from endogenous biochemical, energy-liberating reactions dedicated to the maintenance of vital functions along with increased muscular work. The body temperature of marine mammals, including dolphins, can be up to 35–40 °C above the temperature of polar waters and even 10 °C higher than tropical waters (Berta et al., 2006). The core body is insulated by the dolphin's large body size, vascular counter-current thermal exchangers, hypodermal layer of blubber, and low surface-to-volume ratios (Berta et al., 2006; Favilla & Costa, 2020). The FR near the optimal value of 4.5 (see above) maximizes the body volume and minimizes surface area for drag reduction while limiting heat loss (Fish, 1993a; Berta et al., 2006; Gutarra et al., 2019).

The body temperature mirrors the ambient water temperature in most fishes, which are ectothermic poikilotherms. Ectothermy refers to a metabolism that is determined by the external ambient temperature. The lack of any insulating mechanism to retain body heat means that any excess heat generated endogenously or from muscle contractions when swimming is rapidly transferred conductively and convectively to the water across the body surface area and from the blood through the gills (Bernal et al., 2001a, 2012). Lamnid sharks do not have any specific thermogenic tissues but are capable of maintaining an elevated body temperature (Bernal et al., 2001b). Sharks expend mechanical and metabolic energy to generate the propulsive power for swimming. Excess heat devoted to homeothermy in the shark is a by-product of the contraction of aerobic red muscle (RM) resulting from constant activation during swimming (Carey et al., 1971). To swim at high speeds, particularly in cold water, requires high muscle power outputs that are fostered by an elevated stable body temperature (Irschick & Higham, 2016).

Like some scombrid fishes (e.g., tuna), lamnid sharks possess a vascular arrangement, known as the *rete mirabile*, that acts to circumvent heat loss from the body (Carey, 1973; Carey et al., 1971, 1985). The rete is an array of small arteries and veins in juxtaposition that acts as a countercurrent heat exchanger. The rete acts to return heat back to the muscle. All lamnid sharks have retia for heat exchange to support RM endothermy and support maintenance of a body temperature above ambient (Carey & Teal, 1969a; Carey et al., 1985). Mako and porbeagle sharks maintain body temperature 7–10 °C above ambient (Carey & Teal, 1969a), although Bernal et al. (2001b) found only a 0.3–3.0 °C difference in RM temperature for the mako shark compared to the ambient water temperature. Carey et al. (1982) reported that the muscle temperature of the white shark was 5 °C warmer than the ambient water temperature. An increase of muscle temperature of about 5 °C would provide a three-fold increase in overall speed (Irschick & Higham, 2016). McCosker (1987) and Goldman (1997) measured the stomach temperature as a proxy for core body temperature and found it to be 7.4 °C to 14.3 °C above the water temperature for the white shark. Such temperature differentials indicate that lamnid sharks can inhabit cold water and still be active predators that forage for fast and agile prey (Goldman, 1997).

While temperature measurements and anatomical studies of the circulatory specializations can be performed on the extant shark and dolphin to determine their thermoregulatory capabilities, comparative analysis is not directly possible for the extinct ichthyosaur. The inability to take direct temperature and metabolic measurements has made the arguments about homeothermy versus poikilothermy and endothermy versus ectothermy for ichthyosaurs controversial (Motani, 2010). Given the phylogenetic relationship of ichthyosaurs with reptiles the null hypothesis would be that ichthyosaurs were ectothermic poikilotherms. However, the convergence of body plan with the shark and dolphin indicates a large, fast-swimming, highly active ocean predator that would have at least been homeothermic. The analogous ecology of ichthyosaurs, lamnid sharks and oceanic dolphins and the paleoclimatic distribution of ichthyosaurs showing them to have inhabited cold marine waters would indicate adaptations associated with homeothermy (Bernard et al., 2010). Whether this homeothermy was derived endogenously, as in the dolphin, or from RM endothermy, as in the lamnid shark, is subject to debate (Bernard et al., 2010; Motani, 2010).

Based on the bone histology of highly derived ichthyosaurs, de Buffrénil and Mazin (1990) considered that their pattern of bone remodeling was associated with rapid postnatal growth, as indicated in the skeleton of medium and large mammals. Remodeling of the cortical bone was found to be characterized by greater bone reabsorption than bone reconstruction, resulting in a relative loss of bone mass. Similar to remodeling of bone in the dolphin (*Delphinus delphis*), the compact cortical bone is secondarily reconstructed through ontogeny into cancellous bone (de Buffrénil & Schoevaert, 1988; de Buffrénil & Mazin, 1990). The pattern of bone growth and remodeling was argued to be associated with high endothermic metabolism, incipient endothermy, or gigantothermic metabolism (de Buffrénil & Mazin, 1990). Gigantothermy is defined as the maintenance of a constant high body temperature due to thermal inertia accompanying large body size (Houssaye, 2013) and would indicate homeothermy but not necessarily endothermy.

Evidence for homeothermy in ichthyosaurs was supported by Bernard et al. (2010). When comparing the oxygen isotope ($\delta^{18}O$) compositions of teeth of ichthyosaurs with coexisting fish (i.e., same sedimentary bed) Their results indicated that the body temperature of ichthyosaurs did not vary significantly with seawater temperature, even when the water temperature was as low as 12 °C. The body temperature of ichthyosaurs was considered to be as high as 35 °C. This high body temperature supported the idea that ichthyosaurs had some kind of endothermy. RM endothermy, if not a mammal-like endogenous metabolism, would have been sufficient for homeothermy if there was a large body size (i.e., gigantothermy) and heat conservation mechanisms (e.g., specialized circulatory patterns, blubber) (Motani, 2010). Based on hydrodynamically-derived estimates of optimal swimming speeds, Motani (2002a, b) estimated that the basal metabolic rate of *Stenopterygius* was between reptilian and mammalian levels (Motani, 2005). Gigantothermy associated with homeothermy is observed in the leatherback sea turtle (*Dermochelys coriacea*), and also tunas, which have specialized retia for temperature control and RM-generated endothermy (Carey & Teal, 1969b; Carey et al., 1971, 1984; Carey, 1973; Neill et al., 1974; Dizon & Brill, 1979; Paladino et al., 1990; Holland et al., 1992; Lutcavage et al., 1992; Dewar et al., 1994; Holland & Sibert, 1994; Bostrom & Jones, 2007; Casey et al., 2014). Tunas also possess a thick layer of adipose tissue just under the skin, which has been inferred to reduce conductive heat transfer (Carey et al., 1984). Similarly, ichthyosaurs had a layer of adipose tissue under the skin like the blubber layer of dolphins that would help retain body heat for homeothermy (Lindgren et al., 2018; Delsett et al., 2022).

15.7 Conclusions

As evolution is a probabilistic rather than random process that can be directed by similar selection pressures, there is a high probability that multiple lines of divergent taxa will arrive at common solutions to shared environmental challenges and converge on a distinct morphology. The physical nature of the aquatic environment places huge restrictions on the design and functioning of organisms that originally evolved to inhabit a particular niche space. Occupying the same niche space in the aquatic realm could only have occurred in independent clades at separate times or locations. Despite different phylogenies, the iconic example of lamnid sharks, post-Triassic ichthyosaurs, and oceanic dolphins all converged on a homoplasious design in concert with analogous kinematic and physiological mechanisms. The shape of the body and appendages when measured according to fineness ratio, shoulder position, and aspect ratio indicate that all three taxa evolved designs that are optimal for minimizing drag when swimming. This thunniform design, in concert with the use of caudal fin/fluke as an oscillating hydrofoil, allowed these swimmers to efficiently generate thrust for high-speed cruising.

The evolutionary trajectories for each of the three convergent taxa, with their distinct body plans and physiologies, originated for the sharks in the Paleozoic, ichthyosaurs in the Mesozoic, and dolphins in the Cenozoic. The temporal separation allowed each group to develop into highly mobile, fast swimming, epipelagic, apex predators, but the interaction with the physical environment dictated and constrained their eventual convergence. This convergence was not an exact duplication because each taxon independently derived its morphological design, construction, and physiology within the constraints imposed by their divergent phylogenetic trajectories. The necessity for high swimming speed with minimal energy expenditure in a dense, viscous, thermally conductive environment demanded a fusiform body shape, common stabilizing control surfaces, homeothermy, and a lift-based thunniform type of propulsion.

Other animals exhibit convergent designs and adaptations for high-speed swimming performance. Swordfish, sailfish, marlin, tuna and squid are among the fastest swimmers in the ocean which possess fusiform bodies with fineness ratios in the optimal region (Aleyev, 1977; Fish, 2020). The massive great whales (e.g., blue whale, fin whale, sperm whale) are not bulky, but possess highly streamlined bodies (Fish, 1993a). Likewise, other marine mammals like phocid seals, otariid sea lions, and even the slow swimming manatee have converged on the optimal body design (Webb, 1975; Aleyev, 1977; Fish et al., 1988; Fish, 1993a; Kojeszewski & Fish, 2007). Fast swimmers have adapted their caudal propulsor as a rigid lunate fin or flukes to generate lift-based thrust that functions with a high propulsive efficiency (Lighthill, 1969; Webb, 1975; Lindsey, 1978; Fish, 1993a, 1996, 1998a; Fish et al., 2021). The convergence on lift-based propulsion using wing-like structures occurs in sea lions, which use oscillations of their foreflippers for propulsion (Feldkamp, 1987). The speed and power generated by the flippers provides sea lions with the capability to porpoise (Leahy et al., 2021). Such cases of analogy and homoplasy associated with morphologies and mechanics reflective of adaptation to the marine environment give even greater primacy to the convergence of the shark, ichthyosaur, and dolphin as the quintessential example for understanding the relationship between physics and the environment in directing evolutionary change.

Acknowledgements I am grateful for the assistance of Danielle Adams, William Gough, Griffin Lewis, and Kelsey Tennett with research on the structure and swimming performance of highly-derived aquatic vertebrates. I also appreciate the fruitful discussions with George Lauder, Judy Massare, Anthony Nicastro, Hans Thewissen, Paul Webb, and Terrie Williams in formulating some of the ideas presented in this chapter. Judy Massare is also gratefully acknowledged for contributing images of ichthyosaurs, and Samrithaa Balakrishnan, Anthony Russell and Alex Werth for useful comments on the manuscript. The work was supported in part with a grant from the Office of Naval Research (N000141410533).

References

Abbott, I. H., & von Doenhoff, A. E. (1959). *Theory of wing sections*. Dover.

Adam, P. J. (2009). Hind limb anatomy. In W. F. Perrin, B. Würsig, & J. G. M. Thewissen (Eds.), *Encyclopedia of Marine Mammals* (pp. 562–565). Academic Press.

Adams, D. S., & Fish, F. E. (2019). Odontocete peduncle tendons for possible control of fluke orientation and flexibility. *Journal of Morphology, 280*(9), 1323–1331.

Ahlborn, B. K., Blake, R. W., & Chan, K. H. S. (2009). Optimal fineness ratio for minimum drag in large whales. *Canadian Journal of Zoology, 87*, 124–131.

Ahmadi, A. R., & Widnall, S. E. (1985). Energetics and optimum motion of oscillating lifting surfaces of finite span. *Journal of Fluid Mechanics, 162*, 261–282.

Alexander, R. M. N. (1965). The lift produced by the heterocercal tails of selachii. *The Journal of Experimental Biology, 43*, 131–138.

Aleyev, Y. G. (1977). *Nekton*. Junk.

Andrews, C. ddW. (1910). A descriptive catalogue of the marine reptiles of the Oxford Clay. British Museum (Natural History), London.

Andrews, R. C. (1921). A remarkable case of external hind limbs in a humpback whale. *American Museum Novitates, 9*, 1–6.
Ashenberg, J., & Weihs, D. (1984). Minimum induced drag of wings with curved planform. *Journal of Aircraft, 21*, 89–91.
Au, D., & Weihs, D. (1980). At high speeds dolphins save energy by leaping. *Nature, 284*(5756), 548–550.
Au, D., Scott, M. D., & Perryman, W. L. (1988). Leap-swim behavior of 'porpoising' dolphins. *Cetus, 8*(1), 7–10.
Augusta, J., & Burian, Z. (1957). *Prehistoric animals*. Spring Books.
Azuma, A. (1983). Biomechanical aspects of animal flying and swimming. In H. Matsui & K. Kobayashi (Eds.), *Biomechanics VIII-A: International series on biomechanics* (Vol. 4A, pp. 35–53). Human Kinetics Publishers.
Bandyopadhyay, P. R. (1989). Viscous drag reduction of a nose body. *AIAA Journal, 27*(3), 274–282.
Bandyopadhyay, P. R., Castano, J. M., Rice, J. Q., Philips, R. B., Nedderman, W. H., & Macy, W. K. (1997). Low-speed maneuvering hydrodynamics of fish and small underwater vehicles. *Transactions of the ASME, 119*, 136–144.
Bejder, L., & Hall, B. K. (2002). Limbs in whales and limblessness in other vertebrates: Mechanisms of evolutionary and developmental transformation and loss. *Evolution and Development, 4*, 445–458.
Ben-Zvi, M., & Shadwick, R. E. (2013). Exploring the mechanics of thunniform propulsion: A model study. *Canadian Journal of Zoology, 91*(10), 741–755.
Bernal, D., Dickson, K. A., Shadwick, R. E., & Graham, J. B. (2001a). Analysis of the evolutionary convergence for high performance swimming in lamnid sharks and tunas. *Comparative Biochemistry and Physiology Part A: Molecular & Integrative Physiology, 129*(2–3), 695–726.
Bernal, D., Sepulveda, C., & Graham, J. B. (2001b). Water-tunnel studies of heat balance in swimming mako sharks. *The Journal of Experimental Biology, 204*(23), 4043–4054.
Bernal, D., Carlson, J. K., Goldman, K. J., & Lowe, C. G. (2012). Energetics, metabolism, and endothermy in sharks and rays. In J. C. Carrier, J. A. Musick, & M. R. Heithaus (Eds.), *Biology of Sharks and Their Relatives* (2nd ed., pp. 211–237). CRC Press.
Bernard, A., Lécuyer, C., Vincent, P., Amiot, R., Bardet, N., Buffetaut, E., Cuny, G., Fourel, F., Martineau, F., Mazin, J. M., & Prieur, A. (2010). Regulation of body temperature by some Mesozoic marine reptiles. *Science, 328*(5984), 1379–1382.
Berta, A., Sumich, J. L., & Kovacs, K. M. (2006). *Marine mammals: Evolutionary biology*. Academic.
Blake, R. W. (1983). Energetics of leaping in dolphins and other aquatic animals. *Journal of the Marine Biological Association of the UK, 63*(1), 61–70.
Blake, R. W. (1991). *Efficiency and economy in animal physiology*. Cambridge University Press.
Block, B. A., & Carey, F. G. (1985). Warm brain and eye temperature in sharks. *Journal of Comparative Physiology. B, 156*, 229–236.
Bonfil, R., Meyër, M. A., Scholl, M. C., Johnson, R. L., O'Brian, S., Oosthuizen, W. H., Swanson, S., Kotze, D., & Patterson, M. (2005). Transoceanic migration, spatial dynamics, and population linkages of white sharks. *Science, 310*, 100–103.
Bonfil, R., Francis, M. P., Duffy, C., Manning, M. J., & O'Brien, S. (2010). Large-scale tropical movements and diving behavior of white sharks *Carcharodon carcharias* tagged off New Zealand. *Aquatic Biology, 8*(2), 115–123.
Bose, N., & Lien, J. (1989). Propulsion of a fin whale (*Balaenoptera physalus*): Why the fin whale is a fast swimmer. *Proceedings of the Royal Society of London B, 237*, 175–200.
Bose, N., Lien, J., & Ahia, J. (1990). Measurements of the bodies and flukes of several cetacean species. *Proceedings of the Royal Society of London B, 242*, 163–173.
Bostrom, B. L., & Jones, D. R. (2007). Exercise warms adult leatherback turtles. *Comparative Biochemistry and Physiology A, 147*(2), 323–331.

Brawn, V. M. (1962). Physical properties and hydrostatic function of the swimbladder of herring (*Clupea harengus* L.). *Journal of the Fisheries Research Board of Canada, 19*, 635–656.
Brooks, D. R. (1996). Explanations of homoplasy at different levels of biological organization. In M. J. Sanderson & L. Hufford (Eds.), *Homoplasy: The Recurrence of Similarity in Evolution* (pp. 3–36). Academic Press.
Buchholtz, E. A. (2001a). Swimming styles in Jurassic ichthyosaurs. *The Journal of Vertebrate Paleontology, 21*, 61–73.
Buchholtz, E. A. (2001b). Vertebral osteology and swimming style in living and fossil whales (Order: Cetacea). *Journal of Zoology, 253*(2), 175–190.
Buchholtz, E. A. (2007). Modular evolution of the cetacean vertebral column. *Evolution and Development, 9*(3), 278–289.
Buchholtz, E. A., & Schur, S. A. (2004) Vertebral osteology in Delphinidae (cetacea). *Zoological Journal of the Linnean Society, 140*(3), 383–401.
Buchholtz, E. A., Wolkovich, E. M., & Cleary, R. J. (2005). Vertebral osteology and complexity in *Lagenorhynchus acutus* (Delphinidae) with comparison to other delphinoid genera. *Marine Mammal Science, 21*(3), 411–428.
Burcher, R., & Rydill, L. (1994). *Concepts in submarine design*. Cambridge University Press.
Caldwell, M. W. (2002). From fins to limbs to fins: Limb evolution in fossil marine reptiles. *The American Journal of Medical Genetics, 112*, 236–249.
Campana, S. E., Marks, L., & Joyce, W. (2005). The biology and fishery of shortfin mako sharks (*Isurus oxyrinchus*) in Atlantic Canadian waters. *Fisheries Research, 73*, 341–352.
Carey, F. G. (1973). Fishes with warm bodies. *Scientific American, 288*(2), 36–44.
Carey, F. G., & Teal, J. M. (1969a). Mako and porbeagle: Warm–bodied sharks. *Comparative Biochemistry and Physiology, 28*, 199–204.
Carey, F. G., & Teal, J. M. (1969b). Regulation of body temperature by the bluefin tuna. *Comparative Biochemistry and Physiology, 28*(1), 205–213.
Carey, F. G., Teal, J. M., Kanwisher, J. W., Lawson, K. D., & Beckett, J. S. (1971). Warm-bodied fish. *American Zoologist, 11*, 137–143.
Carey, F. G., Kanwisher, J. W., Brazier, O., Gabrielson, G., Casey, J. G., & Pratt, H. L., Jr. (1982). Temperature and activities of a white shark, *Carcharodon carcharias*. *Copeia, 1982*, 254–260.
Carey, F. G., Kanwisher, J. W., & Stevens, E. D. (1984). Bluefin tuna warm their viscera during digestion. *The Journal of Experimental Biology, 109*, 1–20.
Carey, F. G., Casey, J. G., Pratt, H. L., Urquhart, D., & McCosker, J. E. (1985). Temperature, heat production and heat exchange in lamnid sharks. *Memoirs of the Southern California Academy of Sciences, 9*, 92–108.
Casey, J. P., James, M. C., & Williard, A. S. (2014). Behavioral and metabolic contributions to thermoregulation in freely swimming leatherback turtles at high latitudes. *The Journal of Experimental Biology, 217*(13), 2331–2337.
Castellini, M. A., & Mellish, J.-A. (2016). Thermoregulation. In M. A. Castellini & J.-A. Mellish (Eds.), *Marine mammal physiology: Requisites for ocean living* (pp. 193–216). CRC Press.
Chang, J. T., & Kim, J. (1996). The measurement of homoplasy: A stochastic view. In M. J. Sanderson & L. Hufford (Eds.), *Homoplasy: The recurrence of similarity in evolution* (pp. 189–203). Academic Press.
Chopra, M. G. (1976). Large amplitude lunate-tail theory of fish locomotion. *Journal of Fluid Mechanics, 74*, 161–182.
Chopra, M. G., & Kambe, T. (1977). Hydrodynamics of lunate-tail swimming propulsion. Part 2. *Journal of Fluid Mechanics, 79*, 49–69.
Cooper, L. N., Dawson, S. D., Reidenberg, J. S., & Berta, A. (2007). Neuromuscular anatomy and evolution of the cetacean forelimb. *The Anatomical Record, 290*, 1121–1137.
Cooper, L. N., Sedano, N., Johannson, S., May, B., Brown, J., Holliday, C., Kot, W., & Fish, F. E. (2008). Hydrodynamic performance of the minke whale (*Balaenoptera acutorostrata*) flipper. *The Journal of Experimental Biology, 211*, 1859–1867.

Cowen, R. (1996). Locomotion and respiration in marine air-breathing vertebrates. In D. Jablonski, D. H. Erwin, & J. H. Lipps (Eds.), *Evolutionary biology* (pp. 337–352). University of Chicago Press.
Crofts, S. B., Shehata, R., & Flammang, B. E. (2019). Flexibility of heterocercal tails: What can the functional morphology of shark tails tell us about ichthyosaur swimming? *Integrative Organismal Biology, 1*, obz002.
Daniel, T. (1988). Forward flapping flight from flexible fins. *Canadian Journal of Zoology, 66*, 630–638.
Daniel, T. L., & Webb, P. W. (1987). Physics, design, and locomotor performance. In P. Dejours, L. Bolis, C. R. Taylor, & E. R. Weibel (Eds.), *Comparative physiology: Life in water and on land* (pp. 343–369). Liviana Press, Springer.
Daniel, T. L., Jordon, C., & Grunbaum, D. (1992). Hydromechanics of swimming. In R. M. N. Alexander (Ed.), *Advances in comparative and environmental physiology* (Vol. 11, pp. 17–49). Springer.
Darwin, C. (1859). *On the origin of species by means of natural selection*. John Murray.
de Buffrénil, V., & Mazin, J. M. (1990). Bone histology of the Ichthyosaurs: Comparative data and functional interpretation. *Paleobiology, 16*(4), 435–447.
de Buffrénil, V., & Schoevaert, D. (1988). On how the periosteal bone of the delphinid humerus becomes cancellous; ontogeny of a histological specialization. *Journal of Morphology, 198*, 149–164.
Delair, J. B. (1966). Unusual preservation of fibrous elements in an ichthyosaur skull. *Nature, 212*(5062), 575–576.
Delsett, L. L., Friis, H., Kölbol-Ebert, M., & Hurum, J. H. (2022). The soft tissue and skeletal anatomy of two Late Jurassic ichthyosaur specimens from the Solnhofen archipelago. *PeerJ*. https://doi.org/10.7717/peerj.13173
Denny, M. W. (1993). *Air and water*. Princeton University Press.
Dewar, H., Graham, J., & Brill, R. (1994). Studies of tropical tuna swimming performance in a large water tunnel-thermoregulation. *The Journal of Experimental Biology, 192*(1), 33–44.
Díaz-Berenguer, E., Badiola, A., Moreno-Azanza, M., & Canudo, J. I. (2018). First adequately-known quadrupedal sirenian from Eurasia (Eocene, Bay of Biscay, Huesca, northeastern Spain). *Scientific Reports, 8*(1), 1–13.
Díaz-Berenguer, E., Houssaye, A., Badiola, A., & Canudo, J. I. (2019). The hind limbs of *Sobrarbesiren cardieli* (Eocene, Northeastern Spain) and new insights into the locomotion capabilities of the quadrupedal sirenians. *Journal of Mammalian Evolution*, 1–27.
Díez, G., Soto, M., & Blanco, J. M. (2015). Biological characterization of the skin of shortfin mako shark *Isurus oxyrinchus* and preliminary study of the hydrodynamic behaviour through computational fluid dynamics. *Journal of Fish Biology, 87*, 123–137.
Dizon, A. E., & Brill, R. W. (1979). Thermoregulation in tunas. *American Zoologist, 19*(1), 249–265.
Domning, D. P. (2000). The readaptation of Eocene sirenians to life in water. *Historical Biology, 14*, 115–119.
Domning, D. P. (2001). The earliest known fully quadrupedal sirenian. *Nature, 413*, 625–627.
Donley, J. M., Sepulveda, C. A., Konstantinidis, P., Gemballa, S., & Shadwick, R. E. (2004). Convergent evolution in mechanical design of lamnid sharks and tunas. *Nature, 429*, 61–65.
Elsner, R., & Meiselman, H. J. (1995). Splenic oxygen storage and blood viscosity in seals. *Marine Mammal Science, 11*, 93–95.
Favilla, A. B., & Costa, D. P. (2020). Thermoregulatory strategies of diving air-breathing marine vertebrates: A review. *Frontiers in Ecology and Evolution, 8*, 555509.
Fedak, T. J., & Hall, B. K. (2004). Perspectives on hyperphalangy: Patterns and processes. *Journal of Anatomy, 204*, 151–163.
Feldkamp, S. D. (1987). Foreflipper propulsion in the California sea lion *Zalophus californianus*. *Journal of Zoology (London), 212*, 43–57.

Felts, W. J. L. (1966). Some functional and structural characteristics of cetacean flippers and flukes. In K. S. Norris (Ed.), *Whales, dolphins, and porpoises* (pp. 255–276). University of California Press.
Fierstine, H. L., & Walters, V. (1968). Studies of locomotion and anatomy of scombrid fishes. *Memoirs of the Southern California Academy of Sciences, 6*, 1–31.
Fischer, V., Bardet, N., Benson, R. B. J., Arkhangelsky, M. S., & Friedman M. (2016). Extinction of fish-shaped marine reptiles associated with reduced evolutionary rates and global environmental volatility. *Nature Communications, 7*, 10825.
Fish, F. E. (1992). Aquatic locomotion. In T. Tomasi & T. Horton (Eds.), *Mammalian energetics: Interdisciplinary views of metabolism and reproduction* (pp. 34–64). Cornell University Press.
Fish, F. E. (1993a). Influence of hydrodynamic design and propulsive mode on mammalian swimming energetics. *Australian Journal of Zoology, 42*, 79–101.
Fish, F. E. (1993b). Power output and propulsive efficiency of swimming bottlenose dolphins (*Tursiops truncatus*). *The Journal of Experimental Biology, 185*, 179–193.
Fish, F. E. (1996). Transitions from drag-based to lift-based propulsion in mammalian aquatic swimming. *American Zoologist, 36*, 628–641.
Fish, F. E. (1998a). Comparative kinematics and hydrodynamics of odontocete cetaceans: Morphological and ecological correlates with swimming performance. *The Journal of Experimental Biology, 201*, 2867–2877.
Fish, F. E. (1998b). Biomechanical perspective on the origin of cetacean flukes. In J. G. M. Thewissen (Ed.), *The emergence of whales: Evolutionary patterns in the origin of Cetacea* (pp. 303–324). Plenum Press.
Fish, F. E. (2000). Biomechanics and energetics in aquatic and semiaquatic mammals: Platypus to whale. *Physiological and Biochemical Zoology, 73*(6), 683–698.
Fish, F. E. (2001). Mechanism for evolutionary transition in swimming mode by mammals. In J.-M. Mazin, P. Vignaud, & V. de Buffrénil (Eds.), *Secondary adaptation of tetrapods to life in water* (pp. 261–287). Verlag Dr. Friedrich Pfeil.
Fish, F. E. (2002). Balancing requirements for stability and maneuverability in cetaceans. *Integrative and Comparative Biology, 42*, 85–93.
Fish, F. E. (2004). Structure and mechanics of nonpiscine control surfaces. *IEEE Journal of Oceanic Engineering, 28*, 605–621.
Fish, F. E. (2015). Estimation of shape of the sea lion for hydrodynamic analysis. Response to 'Cambered profile of a California sea lion's body'. *The Journal of Experimental Biology, 218*, 1271–1272.
Fish, F. E. (2018). Streamlining. In W. F. Perrin, B. Würsig, & J. G. M. Thewissen (Eds.), *Encyclopedia of marine mammals* (3 edn, pp. 951–954). Academic.
Fish, F. E. (2020). Advantages of aquatic animals as models for bio-inspired drones over present AUV technology. *Bioinspiration & Biomimetics, 15*, 025001.
Fish, F. E., & Battle, J. M. (1995). Hydrodynamic design of the humpback whale flipper. *Journal of Morphology, 225*, 51–60.
Fish, F. E., & Beneski, J. T. (2014). Evolution and bio-inspired design: Natural limitations. In A. Goel, D. A. McAdams, & R. B. Stone (Eds.), *Biologically inspired design: Computational methods and tools* (pp. 287–312). Springer.
Fish, F. E., & Hui, C. A. (1991). Dolphin swimming: A review. *Mamm. Rev., 21*, 181–196.
Fish, F. E., & Lauder, G. V. (2006). Passive and active flow control by swimming fishes and mammals. *Annual Review of Fluid Mechanics, 38*, 193–224.
Fish, F. E., & Lauder, G. V. (2017). Control surfaces of aquatic vertebrates in relation to swimming modes. *The Journal of Experimental Biology, 220*, 4351–4363.
Fish, F. E., & Rohr, J. (1999). Review of dolphin hydrodynamics and swimming performance. *SPAWARS Systems Center Technical Repeort.* 1801, San Diego, CA.
Fish, F. E., & Shannahan, L. D. (2000). The role of the pectoral fins in body trim of sharks. *Journal of Fish Biology, 56*, 1062–1073.

Fish, F. E., Innes, S., & Ronald, K. (1988). Kinematics and estimated thrust production of swimming harp and ringed seals. *The Journal of Experimental Biology, 137*, 157–173.
Fish, F. E., Smelstoys, J., Baudinette, R. V., & Reynolds, P. S. (2002). Fur doesn't fly, it floats: Buoyancy of hair in semi-aquatic mammals. *Aquatic Mammals, 28*(2), 103–112.
Fish, F. E., Nusbaum, M. K., Beneski, J. T., & Ketten, D. R. (2006). Passive cambering and flexible propulsors: Cetacean flukes. *Bioinspiration & Biomimetics, 1*, S42–S48.
Fish, F. E., Beneski, J. T., & Ketten, D. R. (2007). Examination of the three-dimensional geometry of cetacean flukes using CT-scans: hydrodynamic implications. *The Anatomical Record, 290*, 614–623.
Fish, F. E., Howle, L. E., & Murray, M. M. (2008). Hydrodynamic flow control in marine mammals. *Integrative and Comparative Biology, 211*, 1859–1867.
Fish, F. E., Legac, P., Williams, T. M., & Wei, T. (2014). Measurement of hydrodynamic force generation by swimming dolphins using bubble DPIV. *The Journal of Experimental Biology, 217*(2), 252–260.
Fish, F. E., Williams, T. M., Sherman, E., Moon, Y. E., Wu, V., & Wei, T. (2018). Experimental measurement of dolphin thrust generated during a tail stand using DPIV. *Fluids, 2018*(3), 33.
Fish, F. E., Rybczynski, N., Lauder, G. V., & Duff, C. M. (2021). The evolution of the tail and lack thereof for aquatic propulsion. *Integrative and Comparative Biology, 64*, 398–413. https://doi.org/10.1093/icb/icab021
Flammang, B. E. (2010). Functional morphology of the radialis muscle in shark tails. *Journal of Morphology, 271*, 340–352.
Flammang, B. E., Lauder, G. V., Troolin, D. R., & Strand, T. (2011). Volumetric imaging of shark tail hydrodynamics reveals a three-dimensional dual-ring vortex wake structure. *Proceedings of the Royal Society B: Biological Sciences, 278*(1725), 3670–3678.
Fox, R. W., Prichard, P. J., & McDonald, A. T. (2009). *Introduction to fluid mechanics*. Wiley.
Gabler, U. (2000). *Submarine design*. Bernard & Graefe.
Gans, C. (1974). *Biomechanics: An approach to vertebrate biology*. Lippincott.
Garten, J. L., & Fish, F. E. (2020). Comparative histological examination of the integument of cetacean flukes. *Aquatic Mammals, 46*(4), 367–381.
Gertler, M. (1950). *Resistance experiments on a systematic series of streamlined bodies of revolution-for application to the design of high-speed submarines. Rep. C-297*. David W. Taylor Model Basin.
Gingerich, P. D. (2015). Evolution of whales from land to sea. In K. P. Dial, N. Shubin, & E. L. Brainerd (Eds.), *Great transformation in vertebrate evolution* (pp. 239–256). University of Chicago Press.
Gingerich, P. D., Smith, B. H., & Simons, E. L. (1990). Hind limbs of Eocene *Basilosaurus*: Evidence of feet in whales. *Science, 249*, 154–157.
Gingerich, P. D., Raza, S. M., Arif, M., Anwar, M., & Zhou, X. (1994). New whale from the Eocene of Pakistan and the origin of cetacean swimming. *Science, 368*, 844–847.
Gingerich, P. D., ul Haq, M., Zalmout, I. S., Khan, I. H., & Malkani, M. S. (2001). Origin of whales from early artiodactyls: hands and feet of Eocene Protocetidae from Pakistan. *Science, 293*, 2239–2242.
Goldbogen, J. A., Cade, D. E., Wisniewska, D. M., Potvin, J., Segre, P. S., Savoca, M. S., Hazen, E. L., Czapanskiy, M. F., Kahane-Rapport, S. R., DeRuiter, S. L., & Gero, S. (2019). Why whales are big but not bigger: Physiological drivers and ecological limits in the age of ocean giants. *Science, 366*, 1367–1372.
Goldman, K. J. (1997). Regulation of body temperature in the white shark, Carcharodon carcharias. *Journal of Comparative Physiology B, 167*(6), 423–429.
Gordon, C. N. (1980). Leaping dolphins. *Nature, 287*, 759.
Gough, W. T., Fish, F. E., Wainwright, D. K., & Bart-Smith, H. (2018). Morphology of the core fibrous layer of the cetacean tail fluke. *Journal of Morphology, 279*(6), 757–765.
Gould, S. J. (2002). *The structure of evolutionary theory*. Belknap Press.

Gutarra, S., Moon, B. C., Rahman, I. A., Palmer, C., Lautenschlager, S., Brimacombe, A. J., & Benton, M. J. (2019). Effects of body plan evolution on the hydrodynamic drag and energy requirements of swimming in ichthyosaurs. *Proceedings of the Royal Society B, 286*(1898), 20182786.

Haas, O., & Simpson, G. G. (1946). Analysis of some phylogenetic terms, with attempts at redefinition. *Proceedings. American Philosophical Society, 90*, 319–349.

Hall, B. K. (1984). Developmental mechanisms underlying the formation of atavisms. *Biological Reviews, 59*, 89–122.

Hamilton, J. L., Dillaman, R. M., McLellan, W. A., & Pabst, D. A. (2004). Structural fiber reinforcement of keel blubber in harbor porpoise (*Phocoena phocoena*). *Journal of Morphology, 261*, 105–117.

Hammerschlag, N., Martin, R. A., Fallows, C., Collier, R. S., & Lawrence, R. (2012). Investigatory behavior toward surface objects and nonconsumptive strikes on seabirds by white sharks, *Carcharodon carcharias*, at Seal Island, South Africa (1997–2010). In M. L. Domeier (Ed.), *Global perspectives on the biology and life history of the white shark* (pp. 91–103).

Harris, J. E. (1936). The role of the fins in the equilibrium of the swimming fish. I. Wind-tunnel tests on a model of *Mustelus canis* (Mitchill). *The Journal of Experimental Biology, 13*, 476–493.

Harris, J. E. (1938). The role of the fins in the equilibrium of the swimming fish. II. The role of the pelvic fins. *The Journal of Experimental Biology, 15*, 32–47.

Hertel, H. (1966). *Structure, form, movement*. Reinhold.

Hildebrand, M. (1995). *Analysis of vertebrate structure*. Wiley.

Hoerner, S. F. (1965). *Fluid-dynamic drag*. Published by Author.

Hoffman, S. L., Buser, T. J., & Porter, M. E. (2020). Comparative morphology of shark pectoral fins. *J. Morph., 2020*, 1–16.

Holland, K. N., & Sibert, J. R. (1994). Physiological thermoregulation in bigeye tuna, Thunnus obesus. *Environmental Biology of Fishes, 40*(3), 319–327.

Holland, K. N., Brill, R. W., Chang, R. K., Sibert, J. R., & Fournier, D. A. (1992). Physiological and behavioural thermoregulation in bigeye tuna (*Thunnus obesus*). *Nature, 358*(6385), 410–412.

Houssaye, A. (2013). Bone histology of aquatic reptiles: What does it tell us about secondary adaptation to an aquatic life? *Biological Journal of the Linnean Society, 108*(1), 3–21.

Howell, A. B. (1930). *Aquatic Mammals*. Charles C Thomas.

Huelsmann, M., Hecker, N., Springer, M. S., Gatesy, J., Sharma, V., & Hiller, M. (2019). Genes lost during the transition from land to water in cetaceans highlight genomic changes associated with aquatic adaptations. *Science Advances, 5*(9), eaaw6671.

Hui, C. A. (1987). The porpoising of penguins: An energy-conserving behavior for respiratory ventilation? *Canadian Journal of Zoology, 65*, 209–211.

Hui, C. A. (1989). Surfacing behavior and ventilation of free-ranging dolphins. *Journal of Mammalogy, 70*, 833–835.

Irving, L. (1966). Elective regulation of the circulation in diving animals. In K. S. Norris (Ed.), *Whales, dolphins, and porpoises* (pp. 381–395). University of California Press.

Irving, L., Scholander, P., & Grinnell, S. (1941). The respiration of the porpoise *Tursiops truncatus*. *Journal of Cellular and Comparative Physiology, 17*, 145–168.

Irschick, D. J., & Higham, T. E. (2016). *Animal athletes: An ecological and evolutionary approach*. Oxford University Press.

Kabasakal, H., & Kabasakal, Ö. (2013). First record of a shortfin mako shark, *Isurus oxyrinchus* Rafinesque, 1810 (Chondrichthyes: Lamnidae) from the Bay of Saroz (NE Aegean Sea). *Annales, Series Historia Naturalis, 23*(1), 27–32.

Kardong, K. V. (2019). *Vertebrates: Comparative anatomy, function, evolution*. New York.

Kelley, N. P., & Motani, R. (2015). Trophic convergence drives morphological convergence in marine tetrapods. *Biology Letters, 11*, 20140709.

Kelley, N. P., & Pyenson, N. D. (2015). Evolutionary innovation and ecology in marine tetrapods from Triassic to Anthropocene. *Science, 348*(6232), aaa3716.

Kemp, N. E. (1977). Banding pattern and fibrillogenesis of ceratotrichia in shark fins. *Journal of Morphology, 154*(2), 187–203.
Kim, S. H., Shimada, K., & Rigsby, C. K. (2013). Anatomy and evolution of heterocercal tail in lamniform sharks. *The Anatomical Record, 296*(3), 433–442.
Kojeszewski, T., & Fish, F. E. (2007). Swimming kinematics of the Florida manatee (*Trichechus manatus latirostris*): Hydrodynamic analysis of an undulatory mammalian swimmer. *The Journal of Experimental Biology, 210*, 2411–2418.
Kooyman, G. L. (1973). Respiratory adaptations in marine mammals. *American Zoologist, 13*, 457–468.
Kooyman, G. L. (1989). *Diverse divers: Physiology and behavior*. Springer.
Kooyman, G., & Cornell, L. (1981). Flow properties of expiration and inspiration in a trained bottle-nosed porpoise. *Physiological Zoology, 54*, 55–61.
Küchermann, D. (1953). The distribution of lift over the surface of swept wings. *Aeronautical Quarterly, 4*, 261–278.
Lambert, O., Bianucci, G., Salas-Gismondi, R., Di Celma, C., Steurbaut, E., Urbina, M., & de Muizon, C. (2019). An amphibious whale from the middle Eocene of Peru reveals early South Pacific dispersal of quadrupedal cetaceans. *Current Biology, 29*, 1352–1359.
Lang, T. G. (1963). Porpoise, whale, and fish: Comparison of predicted and observed speeds. *Naval Engineers Journal, 75*, 437–441.
Lang, T. G. (1966). Hydrodynamic analysis of dolphin fin profiles. *Nature, 209*, 1110–1111.
Lang, T. G., & Daybell, D. A. (1963). *Porpoise performance tests in a seawater tank*. Naval Ordnance Test Station Technical Report 3063.
Lang, A. W., Motta, P., Hidalgo, P., & Westcott, M. (2008). Bristled shark skin: A microgeometry for boundary layer control? *Bioinspiration & Biomimetics, 3*(4), 046005.
Larson, A., & Losos, J. B. (1996). Phylogenetic systematics of adaptation. In M. R. Rose & G. V. Lauder (Eds.), *Adaptation* (pp. 187–220). Academic Press.
Leahy, A. M., Fish, F. E., Kerr, S. J., Zeligs, J. A., Skrovan, S., Cardenas, K. L., Leftwich, M. C. (2021). The role of California sea lion (*Zalophus californianus*) hindflippers as aquatic control surfaces for maneuverability. J. Exp. Biol. JEXBIO/2021/242388 (In revision).
Liem, K. F. (1990). Aquatic versus terrestrial feeding modes: Possible impacts on the trophic ecology of vertebrates. *American Zoologist, 30*, 209–221.
Liem, K. F., Bemis, W. E., Walker, W. F., Jr., & Grande, L. (2001). *Functional anatomy of the vertebrates: An evolutionary perspective*. Harcourt.
Lighthill, J. (1969). Hydrodynamics of aquatic animal propulsion – A survey. *Annual Review of Fluid Mechanics, 1*, 413–446.
Lighthill, J. (1970). Aquatic animal propulsion of high hydromechanical efficiency. *Journal of Fluid Mechanics, 44*, 265–301.
Lindgren, J., Sjövall, P., Thiel, V., Zheng, W., Ito, S., Wakamatsu, K., Hauff, R., Kear, B. P., Engdahl, A., Alwmark, C., Eriksson, M. E., Jarenmark, M., Sachs, S., Ahlberg, P. E., Marone, F., Kuriyama, T., Gustafsson, O., Malmberg, P., Thomen, A., Rodríguez-Meizoso, I., Uvdal, P., Ojika, M., & Schweitzer, M. H. (2018). Soft-tissue evidence for homeothermy and crypsis in a Jurassic ichthyosaur. *Nature, 564*, 359–365.
Lindsey, C. C. (1978). Form, function, and locomotory habits in fish. In W. S. Hoar & D. J. Randall (Eds.), *Fish physiology: Locomotion* (Vol. 7, pp. 1–100). Academic Press.
Lingham-Soliar, T. (1999). Rare soft tissue preservation showing fibrous structures in an ichthyosaur from the Lower Lias (Jurassic) of England. *Proceedings Royal Society London Series B: Biological Science, 266*, 2367–2373.
Lingham-Soliar, T. (2001). The ichthyosaur integument: Skin fibers, a means for a strong, flexible and smooth skin. *Lethaia, 34*(4), 287–302.
Lingham-Soliar, T. (2004). Caudal fin allometry in the white shark *Carcharodon carcharias*: Implications for locomotory performance and ecology. *Naturwissenschaften, 92*, 231–236.
Lingham-Soliar, T. (2005a). Dorsal fin in the white shark, *Carcharodon carcharias*: A dynamic stabilizer for fast swimming. *Journal of Morphology, 260*, 1–11.

Lingham-Soliar, T. (2005b). Caudal fin in the white shark, *Carcharodon carcharias*: A dynamic propeller for fast, efficient swimming. *Journal of Morphology, 264*, 233–252.
Lingham-Soliar, T. (2016). Convergence in thunniform anatomy in lamnid sharks and Jurassic ichthyosaurs. *Integrative and Comparative Biology, 56*, 1323–1336.
Lingham-Soliar, T., & Plodowski, G. (2007). Taphonomic evidence for high-speed adapted fin in thunniform ichthyosaurs. *Naturwissenschaften, 94*, 65–70.
Lingham-Soliar, T., & Reif, W.-E. (1998). Taphonomic evidence for fast tuna-like swimming in Jurassic and Cretaceous ichthyosaurs. *Neues Jahrbuch für Geologie und Paläontologie Abh, 207*(2), 171–183.
Lingham-Soliar, T., & Wesley-Smith, J. (2008). First investigation of the collagen D-band ultrastructure in fossilized vertebrate integument. *Proceedings Royal Society London Series B: Biological Science, 275*(1648), 2207–2212.
Liu, P., & Bose, N. (1993). Propulsive performance of three naturally occurring oscillating propeller planforms. *Ocean Engineering, 20*, 57–75.
Long, J. H., Jr., Pabst, D. A., Shepherd, W. R., & McLellan, W. A. (1997). Locomotor design of dolphin vertebral columns: Bending mechanics and morphology of *Delphinus delphis*. *The Journal of Experimental Biology, 200*, 65–81.
Lovvorn, J. R., Croll, D. A., & Liggins, G. A. (1999). Mechanical versus physiological determinants of swimming speeds in diving Brunnich's guillemots. *The Journal of Experimental Biology, 202*, 1741–1752.
Lutcavage, M. E., Bushnell, P. G., & Jones, D. R. (1992). Oxygen stores and aerobic metabolism in the leatherback sea turtle. *Canadian Journal of Zoology, 70*(2), 348–351.
Martin, R. A., Hammerschlag, N., Collier, R. S., & Fallows, C. (2005). Predatory behaviour of white sharks (*Carcharodon carcharias*) at Seal Island, South Africa. *Journal of the Marine Biological Association of the UK, 85*, 1121–1136.
Massare, J. A. (1988). Swimming capabilities of Mesozoic marine reptiles: Implications for method of predation. *Paleobiology, 14*, 187–205.
Massare, J. A., & Lomax, D. R. (2019). Hindfins of *Ichthyosaurus*: Effects of large sample size on 'distinct' morphological characters. *Geological Magazine, 156*, 725–744.
Matta, A., Bayandor, J., Battaglia, F., & Pendar, H. (2019). Effects of fish caudal fin sweep angle and kinematics on thrust production during low-speed thunniform swimming. *Biology Open, 8*, bio040626.
Maxwell, E. E. (2012). Unraveling the influences of soft-tissue flipper development on skeletal variation using an extinct taxon. *The Journal of Experimental Zoology Part B, 318*, 545–554.
McClain, C. R., Balk, M. A., Benfield, M. C., Branch, T. A., Chen, C., Cosgrove, J., Dove, A. D. Gaskins, L. C., Helm, R. R., Hochberg, F. G., & Lee, F. B. (2015). Sizing ocean giants: Patterns of intraspecific size variation in marine megafauna. *PeerJ, 3*, e715.
McCosker, J. E. (1987). The white shark, *Carcharodon carcharias*, has a warm stomach. *Copeia, 1987*(1), 195–197.
McGowan, C. (1991). *Dinosaurs, Spitefires, & Sea Dragons*. Harvard University Press.
McGowan, C. (1992). The ichthyosaurian tail: Sharks do not provide an appropriate analogue. *Palaeontology, 35*, 555–570.
Mitani, Y., Andrews, R. D., Sato, K., Kato, A., Naito, Y., & Costa, D. P. (2010). Three-dimensional resting behaviour of northern elephant seals: Drifting like a falling leaf. *Biology Letters, 6*. 163–166.
Miyazaki, N., Kasuya, T., & Nishiwaki, M. (1974). Distribution and migration of two species of *Stenella* in the Pacific coast of Japan. *The Scientific Reports of the Whales Research Institute, Tokyo, 26*, 227–243.
Moore, M. J., Hammar, T., Arruda, J., Cramer, S., Dennison, S., Montie, E., & Fahlman, A. (2011). Hyperbaric computed tomographic measurement of lung compression in seals and dolphins. *The Journal of Experimental Biology, 214*, 2390–2397.
Motani, R. (2000). Rulers of the Jurassic seas. *Scientific American, 283*, 52–59.

Motani, R. (2002a). Scaling effects in caudal fin propulsion and the speed of ichthyosaurs. *Nature, 415*, 309–312.
Motani, R. (2002b). Swimming speed estimation of extinct marine reptiles: Energetic approach revisited. *Paleobiology, 28*(2), 251–262.
Motani, R. (2005). Evolution of fish-shaped reptiles (Reptilia: Ichthyopterygia) in their physical environments and constraints. *Annual Review of Earth and Planetary Sciences, 33*, 395–420.
Motani, R. (2010). Warm-blooded “sea dragons”? *Science, 328*(5984), 1361–1362.
Motani, R., You, H., & McGowan, C. (1996). Eel-like swimming in the earliest ichthyosaurs. *Nature, 382*(6589), 347–348.
Neill, W. H., Stevens, E. D., Carey, F. G., Lawson, K. D., Mrosovsky, N., & Frair, W. (1974). Thermal inertia versus thermoregulation in “warm” turtles and tunas. *Science, 184*(4140), 1008–1010.
Nesteruk, I. (2020). Fastest fish shapes and optimal supercavitating and hypersonic bodies of revolution. *Innovative Biosystems and Bioengineering, 4*(4), 169–178.
Noren, S. R., Lacave, G., Wells, R. S., & Williams, T. M. (2002). The development of blood oxygen stores in bottlenose dolphins (*Tursiops truncatus*): implications for diving capacity. *Journal of Zoology London, 258*, 105–113.
Nursall, J. R. (1962). Swimming and the origin of paired appendages. *American Zoologist, 2*, 127–141.
Oeffner, J., & Lauder, G. V. (2012). The hydrodynamic function of shark skin and two biomimetic applications. *The Journal of Experimental Biology, 215*(5), 785–795.
Ohsumi, S. (1965). A dolphin (*Stenella caeruleoalba*) with protruded rudimentary hind limbs. *The Scientific Reports of the Whales Research Institute, 19*, 135–136.
Pabst, D. A. (1990). Axial muscles and connective tissues of the bottlenose dolphin. In S. Leatherwood & R. R. Reeves (Eds.), *The Bottlenose dolphin* (pp. 51–67). Academic Press.
Pabst, D. A. (1996). Morphology of the subdermal connective tissue sheath of dolphins: A new fibre-wound, thin-walled, pressurized cylinder model for swimming vertebrates. *Journal of Zoology London, 238*, 35–52.
Pabst, D. A. (2000). To bend a dolphin: Convergence of force transmission designs in cetaceans and scombrid fishes. *American Zoologist, 40*, 146–155.
Paladino, F. V., O’Connor, M. P., & Spotila, J. R. (1990). Metabolism of leatherback turtles, gigantothermy, and thermoregulation of dinosaurs. *Nature, 344*(6269), 858–860.
Parry, D. A. (1949). The swimming of whales and a discussion of Gray’s paradox. *The Journal of Experimental Biology, 26*, 24–34.
Patterson, C. (1988). Homology in classical and molecular biology. *Molecular Biology and Evolution, 5*, 603–625.
Pennycuick, C. J. (1992). *Newton rules biology: A physical approach to biological problems*. Oxford University Press.
Pershin, S. V. (1983). Hydrobionic features for optimizing the exterior forms of the propeller system in cetaceans. *Bionika, 17*, 13–24. (translated from Russian).
Pershin, S. V., Chernyshov, L. F., Kozlov, L. F., Koval, A. P., & Zayets, V. A. (1976). Patterns in the integuments of fast-swimming fishes. *Bionika, 10*, 3–21.
Powell, R. (2007). Is convergence more than an analogy? Homoplasy and its implications for macroevolutionary predictability. *Biology and Philosophy, 22*, 565–578.
Radinsky, L. B. (1987). *The evolution of vertebrate design*. University of Chicago.
Rayner, J. M. V. (1985). Vorticity and propulsion mechanics in swimming and flying animals. In J. Riess & E. Frey (Eds.), *Konstruktionsprinzipen lebender und ausgestorbener Reptilien* (pp. 89–118). University of Tubingen.
Rayner, J. M. (1986). Pleuston: Animals which move in water and air. *Endeavour, 10*(2), 58–64.
Reidy, L. W. (1987). *Flat plate drag reduction in a water tunnel using riblets*. NOSC Technical Report 1169.
Reif, W. E. (1978). Protective and hydrodynamic function of the dermal skeleton of elasmobranchs. *Neues Jahrbuch für Geologie und Paläontologie, 157*, 133–141.

Reif, W. E. (1985). *Squamation and ecology of sharks*. Senckenbergische Naturforschende Gesellschaft.
Reif, W. E., & Dinkelacker, A. (1982). Hydrodynamics of the squamation in fast swimming sharks. *Neues Jahrbuch für Geologie und Paläontologie, 164*, 184–187.
Reif, W. E., & Weishampel, D. B. (1986). Anatomy and mechanics of the lunate tail in lamnid sharks. *Zoologische Jahrbücher Anatomie, 114*, 221–234.
Ridgway, S. H., & Howard, R. (1979). Dolphin lung collapse and intramuscular circulation during free diving: Evidence from nitrogen washout. *Science, 206*, 1182–1183.
Ridgway, S. H., & Johnston, D. G. (1966). Blood oxygen and ecology of porpoises of three genera. *Science, 151*, 456–457.
Ridgway, S. H., Scronce, B. L., & Kanwisher, J. (1969). Respiration and deep diving in the bottlenose porpoise. *Science, 166*, 1651–1654.
Riess, J. (1986). Fortbewegunsweise, Schwimmbiophysik und Phylogenie der Ichthyosaurier. *Palaeontographica A, 192*, 93–155.
Rohr, J. J., & Fish, F. E. (2004). Strouhal numbers and optimization of swimming by odontocete cetaceans. *The Journal of Experimental Biology, 207*(10), 1633–1642.
Saadat, M., Fish, F. E., Domel, A., Di Santo, V., Lauder, G. V., & Haj-Hariri, H. (2017). On the rules for aquatic locomotion. *Physical Review Fluids, 2*, 083102.
Schmidt-Nielsen, K. (1972). Locomotion: Energy cost of swimming, flying, and running. *Science, 177*, 222–228.
Scholander, P. F. (1940). Experimental investigations on the respiratory function in diving mammals and birds. *Hvalradets Skrifter, 22*, 1–131.
Schwartz, F. J. (2013). Jumping and spinning by carcharhinid sharks: Another view. *The Journal of the North Carolina Academy of Science, 129*(3), 107–110.
Sfakiotakis, M., Lane, D. M., & Davies, J. B. C. (1999). Review of fish swimming modes for aquatic locomotion. *IEEE Journal of Oceanic Engineering, 24*(2), 237–252.
Shadwick, R. E. (2005). How tuna and lamnid sharks swim: An evolutionary convergence. *American Scientist, 93*, 524–531.
Shadwick, R. E., & Gemballa, S. (2005). Structure, kinematics, and muscle dynamics in undulatory swimming. *Fish Physiology, 23*, 241–280.
Shoemaker, P. A., & Ridgway, S. H. (1991). Cutaneous ridges in odontocetes. *Marine Mammal Science, 7*(1), 66–74.
Skrovan, R. C., Williams, T. M., Berry, P. S., Moore, P. W., & Davis, R. W. (1999). The diving physiology of bottlenose dolphins (*Tursiops truncatus*). II. Biomechanics and changes in buoyancy at depth. *The Journal of Experimental Biology, 202*(20), 2749–2761.
Smits, A. J. (2019). Undulatory and oscillatory swimming. *Journal of Fluid Mechanics, 874*, P1. https://doi.org/10.1017/jfm.2019.284
Standen, E. M. (2008). Pelvic fin locomotor function in fishes: Three-dimensional kinematics in rainbow trout (*Oncorhynchus mykiss*). *The Journal of Experimental Biology, 211*, 2931–2942.
Stayton, C. T. (2015). The definition, recognition, and interpretation of convergent evolution, and two new measures for quantifying and assessing the significance of convergence. *Evolution, 69*(8), 2140–2153.
Sternes, P. C., & Shimada, K. (2020). Body forms in sharks (Chondrichthyes: Elasmobranchii) and their functional, ecological, and evolutionary implications. *Zoology, 140*, 125799.
Streitlien, K., & Triantafyllou, G. S. (1998). On thrust estimates for flapping foils. *The Journal of Fluids and Structures, 12*, 47–55.
Struthers, J. (1893). On the rudimentary hind-limb of a great fin-whale (*Balaenoptera musculus*) in comparison with those of the humpback whale and the Greenland right-whale. *Journal of Anatomy and Physiology, 27*(3), 291–335.
Sun, Q., Morikawa, H., Kobayashi, S., Ueda, K., Miyahara, H., & Nakashima, M. (2010a). Structure and mechanical properties on tail flukes of dolphin. *Journal of Aero Aqua Bio-Mechanics, 1*(1), 45–50.

Sun, Q., Morikawa, H., Kobayashi, S., Ueda, K., Miyahara, H., & Nakashima, M. (2010b). Structure and bending properties of central part of tail fin of dolphin. *Journal of Biomechanical Science and Engineering, 5*, 388–398.

Sun, Q., Morikawa, H., Ueda, K., Miyahara, H., & Nakashima, M. (2011). Bending properties of tail flukes of dolphin. *Journal of Biomechanical Science and Engineering, 6*(1), 15–25.

Taylor, M. A. (1987). A reinterpretation of ichthyosaur swimming and buoyancy. *Palaeontology, 30*, 531–535.

Taylor, A. R., Schacke, J. H., Speakman, T. R., Castleberry, S. B., & Chandler, R. B. (2016). Factors related to common bottlenose dolphin (*Tursiops truncatus*) seasonal migration along South Carolina and Georgia coasts, USA. *Animal Migration, 1*, 14–26.

Thewissen, J. G. M. (2018). Highlights of cetacean embryology. *Aquatic Mammals, 44*, 591–602.

Thewissen, J. G., & Bajpai, S. (2001). Whale origins as a poster child for macroevolution. *Bioscience, 51*, 1037–1049.

Thewissen, J. G. M., & Fish, F. E. (1997). Locomotor evolution in the earliest cetaceans: Functional model, modern analogues, and paleontological evidence. *Paleobiology, 23*(4), 482–490.

Thewissen, J. G. M., & Nummela, S. (2008). Introduction: On becoming aquatic. In J. G. M. Thewissen & S. Nummela (Eds.), *Sensory evolution on the threshold: Adaptations in secondarily aquatic vertebrates* (pp. 1–25). University of California Press.

Thewissen, J. G. M., Cohn, M. J., Stevens, L. S., Bajpai, S., Heyning, J., & Horton, W. E. (2006). Developmental basis for hind-limb loss in dolphins and origin of the cetacean bodyplan. *Proceedings of the National Academy of Sciences of the United States of America, 103*(22), 8414–8418.

Thewissen, J. G., Cooper, L. N., Clementz, M. T., Bajpai, S., & Tiwari, B. N. (2007). Whales originated from aquatic artiodactyls in the Eocene epoch of India. *Nature, 450*(7173), 1190–1194.

Thewissen, J. G. M., Cooper, L. N., George, J. C., & Bajpai, S. (2009). From land to water: The origin of whales, dolphins, and porpoises. *Evolution: Education and Outreach, 2*, 272–288.

Thomson, K. S., & Simanek, E. D. (1977). Body form and locomotion of sharks. *American Zoologist, 17*, 343–354.

Triantafyllou, M. S. (2017). Tuna fin hydraulics inspire aquatic robotics. *Science, 357*, 251–252.

Triantafyllou, G. S., & Triantafyllou, M. S. (1995). An efficient swimming machine. *Scientific American, 272*, 40–48.

Triantafyllou, G. S., Triantafyllou, M. S., & Grosenbaugh, M. A. (1993). Optimal thrust development in oscillating foils with applications to fish propulsion. *The Journal of Fluids and Structures, 7*, 205–224.

Triantafyllou, M. S., Triantafyllou, G. S., & Yue, D. K. P. (2000). Hydrodynamics of fishlike swimming. *Annual Review of Fluid Mechanics, 32*, 33–53.

Tsai, W-L. (1998). *Dorsoventral bending of the tail and functional morphology of the caudal vertebrae in the bottlenose dolphin, Tursiops truncatus*. Master's thesis. Oklahoma State University.

Tucker, V. A. (1970). Energetic cost of locomotion in animals. *Comparative Biochemistry and Physiology, 34*, 841–846.

Tucker, V. A. (1975). The energetic cost of moving about. *American Scientist, 63*, 413–419.

Uhen, M. D. (2010). The origin(s) of whales. *Annual Review of Earth and Planetary Sciences, 38*, 189–219.

Van Buren, T., Floryan, D., & Smits, A. J. (2020). Bioinspired underwater propulsion. In W. Soboyejo & L. Daniel (Eds.), *Bioinspired structures and design* (pp. 113–139). Cambridge University Press.

van Dam, C. P. (1987). Efficiency characteristics of crescent-shaped wings and caudal fins. *Nature, 325*, 435–437.

Videler, J. J. (1993). *Fish swimming*. Chapman & Hall.

Vogel, S. (1981). *Life in moving fluids*. Willard Grant Press.

Vogel, S. (1994). *Life in moving fluids*. Princeton University Press.

von Mises, R. (1945). *Theory of flight*. Dover.
Wainwright, S. A., Vosburgh, F., & Hebrank, J. H. (1978). Shark skin: Function in locomotion. *Science, 202*, 747–749.
Wainwright, D. K., Fish, F. E., Ingersoll, S., Williams, T. M., & St. Leger, J., Smits, A.J., Lauder, G.V. (2019). How smooth is a dolphin? The ridged skin of odontocetes. *Biology Letters, 15*, 20190103.
Wake, D. B. (1991). Homoplasy: The result of natural selection, or evidence of design limitations? *The American Naturalist, 138*(3), 543–567.
Wake, D. B., Wake, M. H., & Specht, C. D. (2011). Homoplasy: From detecting pattern to determining process and mechanism of evolution. *Science, 331*, 1032–1035.
Walsh, M. J. (1990). Riblets. *Progress in Astronautics and Aeronautics, 123*, 203–261.
Walters, V. (1962). Body form and swimming performance in scombrid fishes. *American Zoologist 2*, 143–149.
Watanabe, Y. Y., Goldman, K. J., Caselle, J. E., Chapman, D. D., & Papastamatiou, Y. P. (2015). Comparative analyses of animal-tracking data reveal ecological significance of endothermy in fishes. *Proceedings of the National Academy of Sciences of the United States of America, 112*, 6104–6109.
Watson, A. G., & Fordyce, R. E. (1993). Skeleton of two minke whales, *Balaenoptera acutorostrata*, stranded on the south-east coast of New Zealand. *New Zealand Natural Sciences, 20*, 1–14.
Webb, P. W. (1975). Hydrodynamics and energetics of fish propulsion. *Bulletin Fisheries Research Board of Canada, 190*, 1–158.
Webb, P. W. (1988). Simple physical principles and vertebrate aquatic locomotion. *American Zoologist, 28*, 709–725.
Webb, P. W. (2004). Maneuverability-general issues. *IEEE Journal of Ocean Engineering, 29*, 547–555.
Webb, P. W. (2006). Stability and maneuverability. In R. E. Shadwick & G. V. Lauder (Eds.), *Fish biomechanics* (pp. 281–332). Academic Press.
Webb, P. W., & de Buffrénil, V. (1990). Locomotion in the biology of large aquatic vertebrates. *Transactions of the American Fisheries Society, 119*, 629–641.
Webb, P. W., LaLiberte, G. D., & Schrank, A. J. (1996). Does body and fin form affect the maneuverability of fish traversing vertical and horizontal slits? *Environmental Biology of Fishes, 46*, 7–14.
Weber, P. W., Murray, M. M., Howle, L. E., & Fish, F. E. (2009a). Comparison of real and idealized cetacean flippers. *Bioinspiration & Biomimetics, 4*, 046001.
Weber, P. W., Howle, L. E., Murray, M. M., & Fish, F. E. (2009b). Lift and drag performance of odontocete cetacean flippers. *The Journal of Experimental Biology, 212*, 2149–2158.
Weber, P. W., Howle, L. E., Murray, M. M., Reidenberg, J. S., & Fish, F. E. (2014). Hydrodynamic performance of the flippers of large-bodied cetaceans in relation to locomotor ecology. *Marine Mammal Science, 30*, 413–432.
Wegner, P. P. (1991). *What makes airplanes fly?* Springer.
Weihs, D. (1972). Semi-infinite vortex trails, and their relation to oscillating airfoils. *Journal of Fluid Mechanics, 54*, 679–690.
Weihs, D. (1989). Design features and mechanics of axial locomotion in fish. *American Zoologist, 29*, 151–160.
Weihs, D. (1993). Stability of aquatic animal locomotion. *Contemporary Mathematics, 141*, 443–461.
Weihs, D. (2002). Dynamics of dolphin porpoising revisited. *Integrative and Comparative Biology, 42*(5), 1071–1078.
Weihs, D., & Webb, P. W. (1983). Optimization of locomotion. In P. W. Webb & D. Weihs (Eds.), *Fish biomechanics* (pp. 339–371). Praeger.

Weng, K. C., Boustany, A. M., Pyle, P., Anderson, S. D., Brown, A., & Block, B. A. (2007). Migration and habitat of white sharks (*Carcharodon carcharias*) in the eastern Pacific Ocean. *Marine Biology, 152*(4), 877–894.
Weng, K. C., Foley, D. G., Ganong, J. E., Perle, C., Shillinger, G. L., & Block, B. A. (2008). Migration of an upper trophic level predator, the salmon shark *Lamna ditropis*, between distant ecoregions. *Marine Ecology Progress Series, 372*, 253–264.
Wicander, R., & Monroe, J. S. (2012). *Historical geology: Evolution of earth and life through time*. Brooks/Cole.
Wilga, C. D., & Lauder, G. V. (2000). Three-dimensional kinematics and wake structure of the pectoral fins during locomotion in Leopard sharks, *Triakis semifasciata*. *The Journal of Experimental Biology, 203*, 2261–2278.
Wilga, C. D., & Lauder, G. V. (2001). Functional-morphology of the pectoral fins in bamboo sharks, *Chiloscyllium plagiosum*—benthic vs. pelagic station-holding. *Journal of Morphology, 249*, 195–209.
Wilga, C. D., & Lauder, G. V. (2004). Hydrodynamic function of the shark's tail. *Nature, 430*, 850.
Williams, T. M. (1987). Approaches for the study of exercise physiology and hydrodynamics in marine mammals. In A. C. Huntley, D. P. Costa, G. A. J. Worthy, & M. A. Castellini (Eds.), *Approaches to marine mammal energetics* (Vol. No. 1, pp. 127–145). Special Publication of the Society for Marine Mammalogy.
Williams, T. M. (1989). Swimming by sea otters: Adaptations for low energetic cost locomotion. *Journal of Comparative Physiology. A, 164*, 815–824.
Williams, T. M. (1999). The evolution of cost efficient swimming in marine mammals: Limits to energetic optimization. *Philosophical Transactions of the Royal Society London B, 353*, 1–9.
Williams, T. M. (2001). Intermittent swimming by mammals: A strategy for increasing energetic efficiency during diving. *American Zoologist, 41*, 166–176.
Williams, T. M., & Kooyman, G. L. (1985). Swimming performance and hydrodynamic characteristics of harbor seals *Phoca vitulina*. *Physiological Zoology, 58*, 576–589.
Williams, T. M., Kooyman, G. L., & Croll, D. A. (1991). The effect of submergence on heart rate and oxygen consumption of swimming seals and sea lions. *Journal of Comparative Physiology. B, 160*, 637–644.
Williams, T. M., Friedl, W. A., & Haun, J. E. (1993). The physiology of bottlenose dolphins (*Tursiops truncatus*): Heart rate, metabolic rate and plasma lactate concentration during exercise. *The Journal of Experimental Biology, 179*, 31–46.
Williams, T. M., Davis, R. W., Fuiman, L. A., Francis, J., Le, B. J., Horning, M., Calambokidis, J., & Croll, D. A. (2000). Sink or swim: Strategies for cost-efficient diving by marine mammals. *Science, 288*, 133–136.
Williams, T. M., Fuiman, L. A., Kendall, T., Berry, P., Richter, B., Noren, S. R., Thometz, N., Shattock, M. J., Farrell, E., Stamper, A. M., & Davis, R. W. (2015). Exercise at depth alters bradycardia and incidence of cardiac anomalies in deep-diving marine mammals. *Nature Communications, 6*(1), 1–9.
Williston, S. W. (1914). *Water reptiles of the past and present*. University of Chicago Press.
Withers, P. C. (1992). *Comparative animal physiology*. Saunders.
Wolf, N. G., Swift, P. R., & Carey, F. G. (1988). Swimming muscle helps warm the brain of lamnid sharks. *Journal of Comparative Physiology. B, 157*, 709–715.
Yazdi, P., Kilian, A., & Culik, B. (1999). Energy expenditure of swimming bottlenose dolphins (*Tursiops truncatus*). *Marine Biology, 134*, 601–607.
Yurchenko, N. F., & Babenko, V. V. (1980). Stabilization of the longitudinal vortices by skin integuments of dolphins. *Biophysics, 25*, 309–315.
Zhang, J.-D., Sung, H. J., & Huang, W.-X. (2020). Specialization of tuna: A numerical study on the function of caudal keels. *Physics of Fluids, 32*, 111902.

Chapter 16
Convergent Evolution of Attachment Mechanisms in Aquatic Animals

Jérôme Delroisse, Victor Kang, Anaïd Gouveneaux, Romana Santos, and Patrick Flammang

Abstract To resist hydrodynamic forces, two main underwater attachment strategies have evolved multiple times in aquatic animals: glue-like "bioadhesive secretions" and pressure-driven "suction attachment". In this chapter, we use a multi-level approach to highlight convergence in underwater attachment mechanisms across four different length-scales (organism, organ, microscopic and molecular). At the organism level, the ability to attach may serve a variety of functions, the most important being: (i) positional maintenance, (ii) locomotion, (iii) feeding, (iv) building, and (v) defense. Aquatic species that use bioadhesive secretions have been identified in 28 metazoan phyla out of the 34 currently described, while suction organs have a more restricted distribution and have been identified in five phyla. Although biological adhesives are highly diverse, it is possible to categorize them into four main types according to the time scale of operation: permanent, temporary, transitory, and instantaneous adhesion. At the organ level some common

---

Jérôme Delroisse and Victor Kang contributed equally to this work.

Supplementary Information The online version contains supplementary material available at [https://doi.org/10.1007/978-3-031-11441-0_16].

J. Delroisse · P. Flammang (✉)
Biology of Marine Organisms and Biomimetics Unit, Research Institute for Biosciences, University of Mons, Mons, Belgium
e-mail: jerome.delroisse@umons.ac.be; Patrick.Flammang@umons.ac.be

V. Kang
Department of Bioengineering, Imperial College, London, UK
e-mail: k.kang@imperial.ac.uk

A. Gouveneaux
EthoS (Éthologie animale et humaine) - CNRS UMR 6552, Université de Caen Normandie, Caen, France
e-mail: anaid.gouveneaux@unicaen.fr

R. Santos (✉)
Faculdade de Ciências, Departamento de Biologia Animal, Centro de Ciências do Mar e do Ambiente (MARE), Universidade de Lisboa, Lisbon, Portugal
e-mail: romana.santos@campus.ul.pt

V. L. Bels, A. P. Russell (eds.), *Convergent Evolution*, Fascinating Life Sciences,
https://doi.org/10.1007/978-3-031-11441-0_16

principles have independently evolved in different biological lineages: for example, animals with single-unit attachment organs can be distinguished from those with multi-unit organs. Fundamental design elements can also be recognized for both types of attachment mechanisms. Suction attachment systems comprise a circular or elliptical attachment disc, a sealing rim to prevent leakage and a mechanism to lower the internal pressure. Bioadhesive-producing organs, on the other hand, usually contain a glandular tissue associated with connective tissues or other types of load-bearing support structures and muscles that facilitate locomotion or mechanical detachment. At the microscopic level, similar designs and organizations appear once again to have emerged independently in different phylogenetic lineages. Independent of the taxon and type of adhesion, there are species in which the biosynthesis, packaging and release of adhesive secretions takes place at the level of a single type of secretory cell, whereas in others these secretions are produced by two or more secretory cell types. Duo-gland adhesive systems involved in temporary adhesion present an additional level of complexity as they also exhibit de-adhesive secretory cells. Yet, strikingly similar cellular organizations have been reported in highly disparate species. In the case of biological suction organs, regions of the organ that contact the substratum are highly textured with stiff microstructures. Although clearly non-homologous in different animals, these microstructures are thought to enhance friction on rough surfaces. At the molecular level, proteins are the main organic constituent of adhesive secretions in aquatic animals. We compared the global amino acid compositions of bioadhesives using principal component analysis to show that homologous adhesives from phylogenetically related species cluster together, and there is little overlap between taxonomic groups. However, several non-permanent adhesives are grouped together even though they belong to disparate phyla, indicating convergence in amino acid composition. We also investigated relatedness among individual adhesive proteins using a sequence similarity-based clustering analysis. While many proteins appear taxon-specific, some have clear sequence homologies based on shared protein domains between phylogenetically distant organisms. However, it is highly probable that these domains, which are also present in many non-adhesive proteins, were convergently acquired from ancestral proteins with unrelated general functions. We herein present morphological, structural, and molecular convergences between different attachment mechanisms in aquatic animals that likely arose in response to shared functional and selective pressures.

Keywords Metazoans · Adhesive organs · Suction organs · Functional morphology · Adhesive proteins

16.1 Introduction

The physical environment in our oceans and freshwater systems is drastically different from that on land (Ditsche & Summers, 2014). Terrestrial organisms must contend with gravity on a daily basis, and it is easy to feel the direct consequence of gravity when we lift heavy objects or do a pull-up for exercise. Conversely, due to the low density of our atmosphere, we are able to walk and run without undue effort, unless confronted by extreme conditions such as severe storms. The density of water, on the other hand, denies gravity the power to hold aquatic animals to the bottom, and buoyancy forces need to be balanced to control sinking or floating. In addition, in aquatic environments, forces exerted by flowing water (hydrodynamic forces) can be strong and directionally unpredictable, requiring specific mechanisms, sturdy in all directions, to counteract them. Indeed, many aquatic animals lack grasping limbs to grip onto solid objects. To survive in such conditions animals from multiple phyla have evolved the ability to attach to various substrates underwater, often using specialized appendages or regions of their bodies called attachment organs (Gorb, 2008; Ditsche & Summers, 2014). Such adaptations allow animals to stick to substrates in order to move (e.g., limpets, sea stars, and octopuses), to maintain position (e.g., barnacles, mussels, and remora fish), to feed (e.g., cephalopods and comb jellies), or to build shelter (e.g., sandcastle worms and caddisfly larvae).

Interestingly, despite the diversity in the morphology and function of metazoan attachment organs (Nachtigall, 1974; Gorb, 2008), aquatic animals from multiple phyla mainly rely on two strategies for underwater attachment: either glue-like 'bioadhesive secretions' or pressure-driven 'suction attachment.' Bioadhesive secretions are complex mixtures of proteins, sugars and lipids and are most often used for attaching an organism to a non-living surface, including dynamic attachment during locomotion and permanent fixation (Nachtigall, 1974; Hennebert et al., 2015; Davey et al., 2021). Some well-known examples of aquatic animals that use bioadhesive secretions are echinoderms (e.g., sea stars, sea urchins), barnacles, and mussels. Conversely, suction requires muscular contraction for the generation of the pressure difference required for attachment (although glandular secretions can help with sealing) and is strictly used for temporary attachment (Nachtigall, 1974). Animals that rely on suction include cephalopods (e.g., octopuses and squids), numerous fishes (e.g., remora fish, clingfish, gobies), and insects (e.g., net-winged midge larvae and diving beetles).

As with any effort to categorize and characterize form and function in biology, there will be exceptions that are not adequately captured by these two mechanisms. Moreover, it is at times difficult to distinguish the two strategies: sea star and sea urchin tube feet were long considered to be suction organs before it was shown that they rely solely on adhesive secretions for attachment (Hennebert et al., 2012). Finally, an organism may use both suction and adhesive attachments (e.g., lottiid limpets; see Sect. 16.3). Nevertheless, suction attachment and bioadhesive secretion represent the two most common approaches to biological attachment in wet environments (Ditsche & Summers, 2014). It is likely that the diverse aquatic animals that employ these approaches have repeatedly arrived at similar forms and strategies in response to overlapping physical conditions and demands. In other words, there appear to be many examples of convergence in the attachment strategies of aquatic organisms. How widespread is evolutionary convergence in suction attachment organs or bioadhesive-secreting organs? What additional insights can we gain from identifying characteristics that have repeatedly emerged in unrelated taxa? These are some of the questions that are explored in this chapter. As pointed out by Tyler (1988), convergence should be reflected in a lack of correspondence in the functional hierarchy of components. Hence, we use a multi-scale approach to highlight convergence in underwater attachment mechanisms at four different length-scales: (1) individual organisms, (2) organs, (3) cells and microscopic structures, (4) molecules (Fig. 16.1). At the largest length-scale it is expected that two morphologically similar structures, either homologous or analogous, are also functionally similar. However, differences between both structures in the way functions are performed at lesser length-scales are indications that the probability of convergence between them is high. The adhesive organs of many interstitial invertebrates, for example, have the same general function, namely temporary maintenance of position on sand grains, but the finer levels of the functional hierarchy are performed by different molecular components (Tyler, 1988).

16.2 An Organism-Level Approach to Attachment Mechanisms in Aquatic Organisms

An impressive diversity of aquatic organisms uses attachment mechanisms at one or more stages of their life cycle. Of the most current list of metazoan phyla (34 according to Giribet & Edgecombe, 2020), the vast majority contain species that attach using suction organs, bioadhesive secretions, or both (Fig. 16.2). Species that use bioadhesive secretions, or are strongly suspected to do so, have been identified in 29 phyla (28 if only aquatic organisms are considered), whereas suction organs have a more restricted distribution (at least if we are only considering this at the phylum level) and have been identified in five phyla (Craniata, Arthropoda, Platyhelminthes, Mollusca and Annelida). The widespread distribution of bioadhesive secretions within the animal kingdom may erroneously suggest that all metazoan adhesive organs are homologous; however, homology cannot be inferred from the simple presence/absence of an adhesive system. Instead, a detailed analysis of adhesive systems is needed to understand if and how they are interrelated. Moreover, the high proportion of taxa using bioadhesives at the phylum level can be explained, in part, by the fact that this attachment mechanism was considered to be present even when it has only been described for a very limited subset of species in the phylum. It may not be reflected, therefore, at lower taxonomic levels, indicating

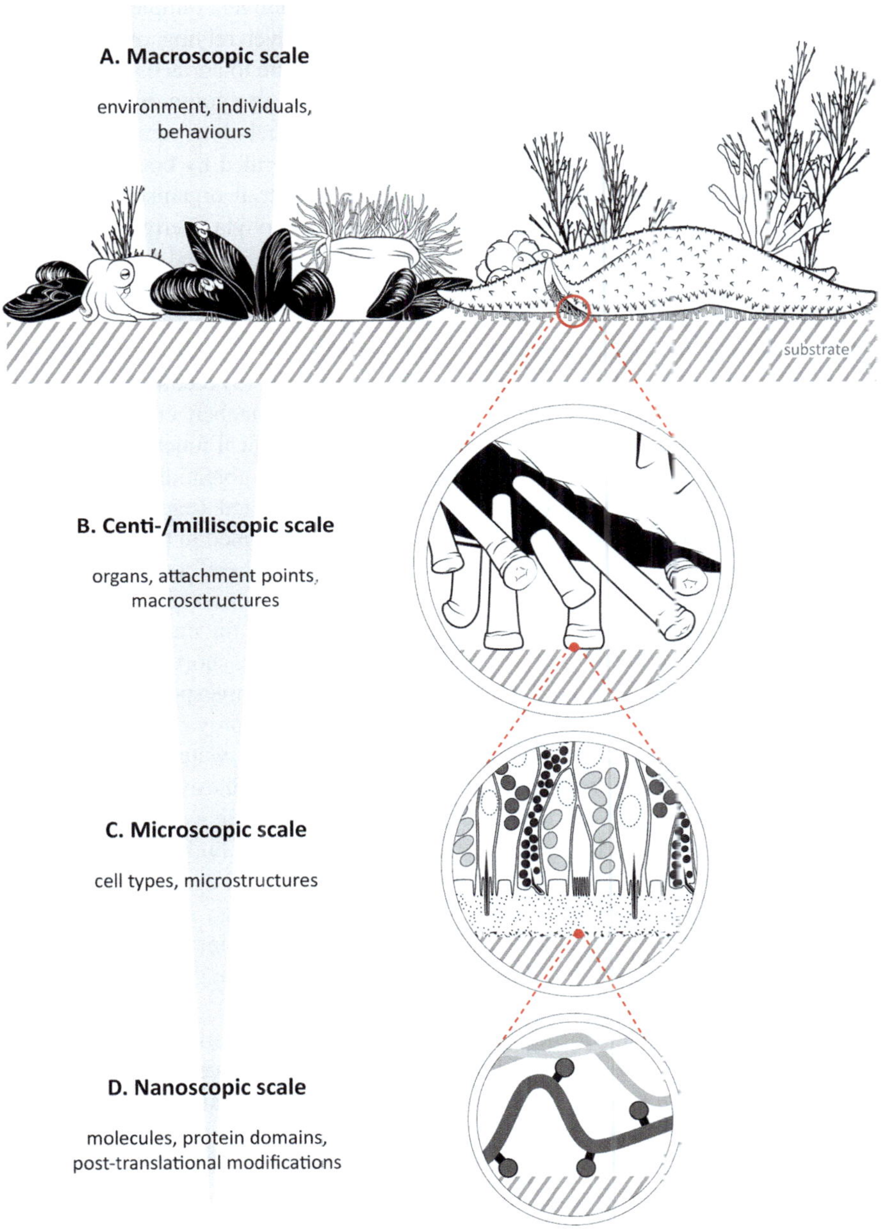

Fig. 16.1 Example of an intertidal community illustrating the multi-scale approach used in the present review. (**a**) At the macroscopic scale, many organisms inhabiting this hydrodynamically stressful environment rely on different attachment mechanisms for maintenance of position as well as for various other functions. Sea stars, for example, use temporary adhesion to the substratum for static sustained attachment to withstand the action of waves, for dynamic attachment during locomotion, or to grip and pry open the mussels on which they feed. (**b**) At the centi-/milliscopic scale their adhesion relies on a multitude of small appendages, the tube feet, each acting as an individual tether connecting the animal to the substratum. (continued overleaf)

several independent evolutionary events. As a representative example, most chaetognaths are pelagic (i.e., free-swimming organisms never relying on attachment mechanisms), but a few species are benthic and can adhere to surfaces (John, 1933; Feigenbaum, 1976). Moreover, although suction organs (also commonly referred to as suckers) are predominantly found in aquatic animals (including parasites living in the fluids of other animals), adhesive secretions are generated by both aquatic and terrestrial animals. For the sake of completeness, terrestrial organisms producing adhesive secretions were included in the total count of phyla. Terrestrial animals often use adhesives for prey capture or defense (e.g., spiders and velvet worms), although some use adhesive secretions for locomotion (e.g., snails and slugs) (Hennebert et al., 2015).

In metazoans, attachment mechanisms may serve a variety of functions, the most important being: (1) position maintenance, (2) locomotion, (3) feeding, (4) building shelter, and (5) defense (Fig. 16.2; Nachtigall, 1974; Hennebert et al., 2015). As expected, trying to define a strict terminology for the biological functions associated with attachment mechanisms is educatively useful but potentially problematic because these functions may be ecologically interconnected (e.g., a shelter can also be used to carry out feeding and for defense). As mentioned in the introduction, aquatic animals must resist hydrodynamic stresses and, therefore, many benthic species rely on bioadhesives or suction organs to attach to non-living surfaces or to other organisms to maintain their position. Depending on the biology of the species, this attachment may be long- or short-term. Dynamic, short-term attachment also allows for locomotion in turbulent environments. For some species, attachment mechanisms (i.e., bioadhesives or suction organs) also allow prey capture and, in the case of bioadhesives, the collection of food particles from the water column or from the bottom. Many filter feeding organisms that rely on adhesive mucus to trap particles therefore fall into this latter food collection category (e.g., some cnidarians, molluscs, annelids or brachiopods). The latter two functions, building and defense, only concern bioadhesives. Building involves the gluing of exogenous materials together for the construction of tubes, nests or burrows (e.g., sandcastle worms, caddisfly larvae, three-spined sticklebacks), and defense pertains to the release of a sticky material as a protective reaction against predators (e.g., sea cucumbers, centipedes, salamanders) (von Byern et al., 2017).

Although the diversity of biological adhesives is vast in terms of components, interactions and functions, some common principles have evolved independently in different biological lineages. In aquatic organisms, biological adhesives can be grouped together into four main types according to the time scale of operation: permanent, temporary, transitory and instantaneous adhesion (Fig. 16.2; Flammang,

---

Fig. 16.1 (continued) (**c**) At the microscopic scale the tube foot epidermis encloses a duo-gland adhesive system comprising two types of adhesive cells (in grey) that co-secrete the adhesive layer joining the tube foot to the substratum, and de-adhesive cells (in black) that produce a releasing secretion that allows detachment. (**d**) At the nanoscopic scale the adhesive material consists of a mixture of proteins and glycoproteins, some of which are involved in interfacial adhesive interactions and others in bulk cohesive function

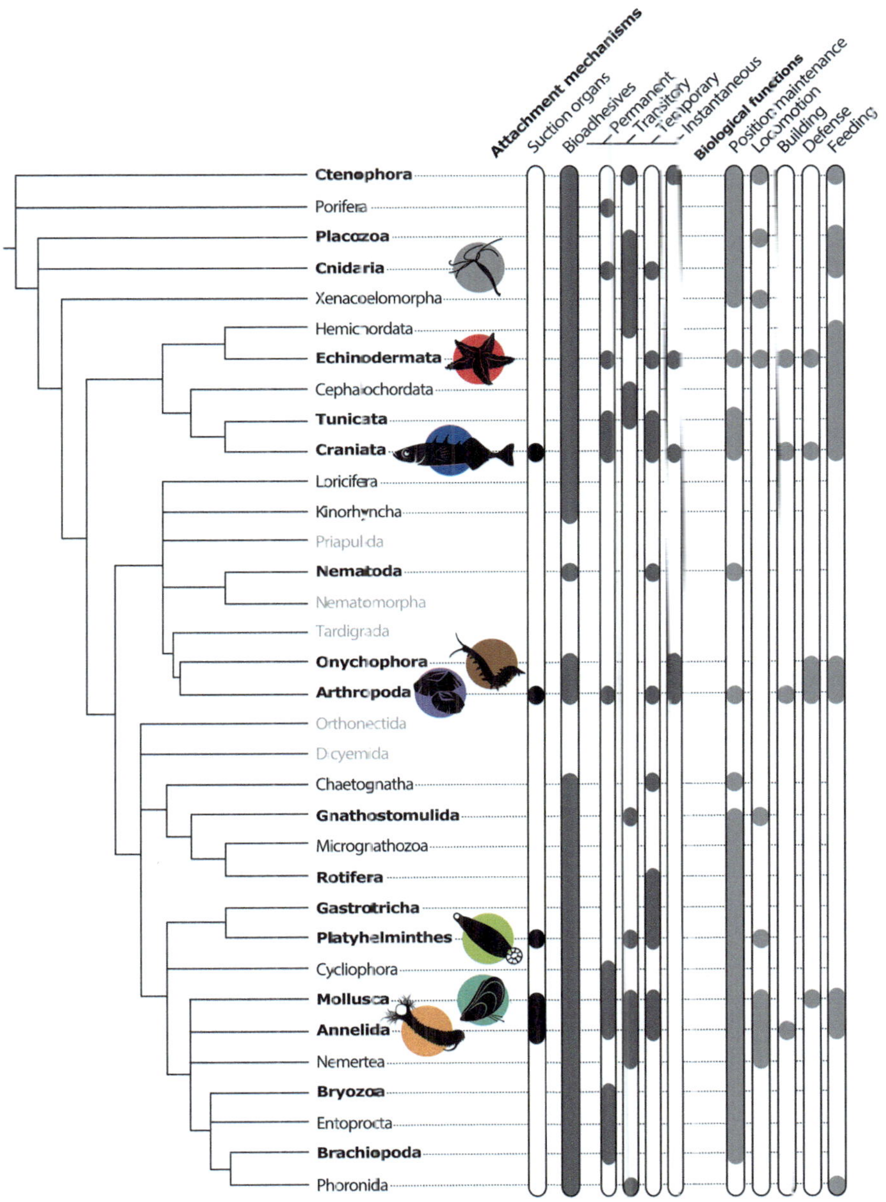

Fig. 16.2 Phylogenetic distribution of metazoans that use suction organs or bioadhesive secretions. The metazoan phylogenetic tree is based on Giribet and Edgecombe (2020). Biological functions associated with the attachment strategies are highlighted at the phylum-level. The names of taxa are shown in grey when no adhesion system is known; in black when adhesion systems are known and in bold when they have been studied to some extent (see Supplementary Table 16.1 for details). The species icons represent the taxa for which molecular data are available and which have been used in the molecular analyses that follow

1996; Whittington & Cribb, 2001; Flammang et al., 2005). *Permanent adhesion*, represented in 12 phyla, involves the secretion of a bioadhesive that hardens with time and forms a durable cement. As observed in barnacles, for example, this type of adhesion is seen in a number of phyla that include sessile benthic organisms that remain firmly fixed at the same place throughout their lifetime (e.g., Porifera, Cnidaria, Tunicata, Arthropoda, Mollusca, Bryozoa; Fig. 16.2). By definition, permanent adhesion is used exclusively for position maintenance or building. Temporary and transitory adhesion both correspond to a non-permanent type of adhesion and permit, for example, the combination of adhesion and locomotion at the same time, thus allowing adult organisms to graze, hunt or search for a mate, and larval forms to explore surfaces prior to metamorphosis. *Transitory adhesion* is used by many benthic and vagile organisms that creep on the substratum. It allows simultaneous adhesion and movement along a substratum, whereby the animals attach using a thin layer of secretion that is often left behind them as they move. This type of adhesion is characteristic of invertebrates that move along the substratum by ciliary gliding—mostly small soft-bodied invertebrates from the phyla Platyhelminthes, Nemertea, Gastrotricha and Annelida (Martin, 1978a, b). Larger animals, like sea anemones and gastropod molluscs, also use transitory adhesion, moving by means of waves of muscular contractions running along their attachment organ (the pedal disc in sea anemones, and the foot in gastropods) (Jones & Trueman, 1970; Edmunds et al., 1976). In limpets (Phylum Mollusca), the term transitory adhesion has recently been redefined to describe the regular switching between long-term and locomotory adhesion (Kang et al., 2020). By analogy to ciliary gliding, food collection using muco-ciliary systems is also classified as transitory adhesion. Considering both locomotion and feeding, transitory adhesion is represented in 13 phyla (Fig. 16.2). *Temporary adhesion*, on the other hand, is used by organisms such as sea stars and sea urchins (Echinodermata) that are able to adhere firmly yet temporarily to the substratum, allowing them to repeatedly attach and detach. This type of adhesion is also frequently found in small invertebrates that inhabit the interstitial environment, such as various species of Platyhelminthes, Gastrotricha, Nematoda and Annelida (Tyler, 1988; Lengerer & Ladurner, 2018). These animals use bioadhesives to temporarily secure themselves to the sand grains of marine or freshwater beaches to avoid dislodgement. Some echinoderms and mollusks also rely on temporary adhesion to capture their food and release it into the mouth. Overall, this type of adhesion occurs in 11 phyla (Fig. 16.2). *Instantaneous adhesion*, finally, describes a type of adhesion whereby the adhesive is rapidly discharged from single-use adhesive organs or glands and is immediately sticky. In aquatic animals, this type of adhesion is only seen in ctenophores during prey capture and in sea cucumbers through the release of Cuvierian tubules as a defense mechanism. Many bioadhesives produced by terrestrial animals for defense or prey capture satisfy the definition of instantaneous adhesives, even though they are not released by single-use organs or cells as seen in sea cucumbers. Examples of terrestrial species that use this type of adhesive can be found in four phyla (Fig. 16.2): Craniata (frogs and salamanders), Mollusca (slugs), Arthropoda (insects, centipedes, spiders), and Onychophora (velvet worms).

The fundamental design elements of a biological suction attachment system are a circular or elliptical attachment disc, a sealing rim to prevent leakage, and a mechanism to enable the lowering of the internal pressure. Using these parameters to identify suction organs it is clear that suction attachment has evolved independently in highly disparate branches of the tree of life (Fig. 16.2). It is important to bear in mind, however, that while morphological similarities are useful for an initial assessment of whether an attachment organ may be a suction organ, mechanistic studies are required to confirm that reduced pressure gives rise to the attachment force. It will become apparent that of the multiple species mentioned in this chapter as possessing suction attachment organs, only a few have fully satisfied this requirement, which highlights significant opportunities for future research. Examples of suction organs can be found in numerous species across five phyla (Annelida, Arthropoda, Craniata, Mollusca, and Platyhelminthes; Fig. 16.2 and Supplementary Table 16.1).

16.3 An Organ-Level Approach to Attachment Mechanisms in Aquatic Organisms (Macroscopic)

At the organ level, animals with single-unit attachment organs can be easily distinguished from those with multi-unit organs, irrespective of whether the organs rely on bioadhesives or suction attachment (Fig. 16.3). For example, whereas limpets and barnacles have evolved a single attachment pad, sea urchins, sea stars and mussels rely on multiple-point attachments. One might think these two distinct structural strategies are function-related (Nachtigall, 1974) but this is unlikely because both single- and multi-unit attachments may be used exclusively for anchoring (e.g., barnacle and mussel permanent adhesion) or cumulatively for anchoring, locomotion and feeding (e.g., limpet transitory adhesion and sea urchin/sea star temporary adhesion) (Fig. 16.3).

Limpets (Mollusca) are intertidal inhabitants that attach to the surface of rocks using a muscular pedal sole (Fig. 16.3a). The exact mechanism of attachment appears to differ between members of the families Patellidae (true limpets) and Lottiidae. Several studies have demonstrated that lottiid limpets alternate between suction attachment at high tide (when they are actively moving around to feed) and adhesive mucus secretion at low tide (when they are exposed to the environment and to predators and require more powerful, long-term attachment) (Smith, 1991a, 1992; Smith et al., 1993, 1999). Although patellid limpets also inhabit the intertidal zone and respond to the tide, they primarily rely on adhesive mucus secretions for attachment (Kang et al., 2020). Single attachment organs are also found in other common inhabitants of the intertidal zone, such as acorn and stalked barnacles (Arthropoda), which live their adult life permanently anchored to the substratum, and sea anemones (Cnidaria), some of which use transitory adhesion (Cowles, 1977; Young et al., 1988; Clarke et al., 2020).

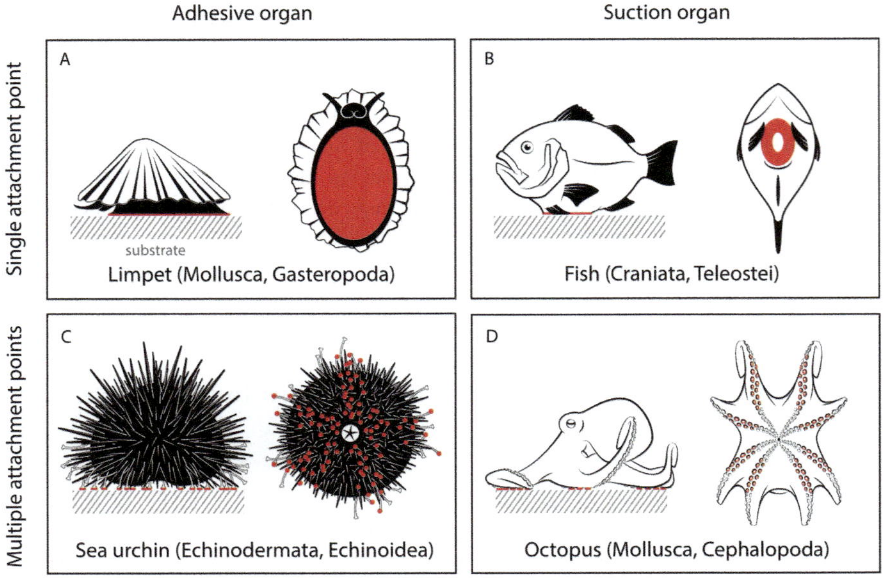

Fig. 16.3 Diversity of attachment organs in aquatic organisms. These organs may be distinguished based on the mechanism of attachment (columns: adhesive secretion *versus* suction) and the number of attachment points (rows: single *versus* multiple), as exemplified by four generalized organisms: (**a**) limpets, (**b**) fishes, (**c**) sea urchins, and (**d**) octopuses. In each case, a lateral view (left) and a ventral/oral view (right) are represented with the zone(s) of contact with the substratum highlighted in red

Sea urchins and sea stars (Echinodermata), meanwhile, attach using multiple specialized adhesive organs called tube feet (Figs. 16.1 and 16.3b) (Nichols, 1966). Tube foot attachment is temporary, allowing strong attachment to the substratum and easy detachment before the initiation of another attachment–detachment cycle (Thomas & Hermans, 1985; Flammang, 1996; Flammang et al., 2016; Federle & Labonte, 2019). Most tube feet consist of a basal hollow cylinder (the stem) and an enlarged and flattened apical extremity (the disc) that work together to make tube feet efficient and versatile, allowing echinoderms to resist hydrodynamic forces and to perform tasks such as climbing, righting, covering their bodies with objects, or opening mollusk shells (Lawrence, 1987; Flammang et al., 2016). Another example of multi-point attachment occurs in mussels (Mollusca) that permanently anchor to rocks using multiple thread-like tethers, collectively called byssus. Each thread contains three parts: a spatulate adhesive plaque, a stiff distal portion and a compliant proximal portion (Waite, 2017).

Within the organisms that possess multiple adhesive organs there is high variability in the number of organs and the adhesive contact areas of each organ. Even within the same taxonomic group (e.g., Echinoidea, the sea urchins) there are species that can increase their maximum adhesive surface area by increasing the number of adhesive organs (e.g., 0.8 tube feet/mm^2 of test area in *Colobocentrotus atratus* versus 0.2 tube feet/mm^2 in *Arbacia lixula*) or by increasing the contact area of each adhesive organ (e.g., 1.16 mm^2 tube foot disc area in *A. lixula* versus 0.81 mm^2 tube foot disc area in *C. atratus*) (Santos & Flammang, 2006, 2007, 2008). Moreover, unlike single adhesive organs, the strength of multi-component adhesive organs is the product of the number and mechanical properties of the individual tethers. This allows animals using multiple attachment points to adjust the number of tethers they use according to the environmental conditions. Sea urchins, for example, appear to respond to increased wave height by dedicating more tube feet to attachment, thereby increasing the overall attachment force (Santos & Flammang, 2007).

Bioadhesive-producing organs usually contain (or are associated with) connective tissues or other types of load-bearing support structures and muscles that facilitate locomotion or mechanical detachment. For example, support structures, such as ossicles and a circular plate of connective tissue within the adhesive discs of echinoderm tube feet, help to withstand the tensile forces that result from external loading. This connective tissue plate, at its proximal end, is continuous with the connective tissue sheath of the stem, and at its distal end divides into numerous branching connective tissue septa that attach apically to the support cells of the epidermis (Flammang et al., 2016). Additionally, these organs possess retractor muscles that might facilitate detachment, thereby complementing the action of a de-adhesive secretion (discussed in Sect. 16.4) (Lengerer & Ladurner, 2018). Because a chain is only as strong as its weakest link, there should be a good balance between the adhesive strength developed by the secretion and the mechanical properties of the load-bearing parts of the adhesive organ. However, other factors may play a role. In sea urchins, the force needed to break the stem (the proximal part of the tube foot linked to the animal) is greater than that needed to detach the distal disc (the distal part attached to the substratum). This can be explained by the fact that if the disc detaches from the substratum it can easily re-attach as re-attachment requires only a fresh adhesive secretion; if the stem breaks, however, the tube foot must be completely regenerated (Santos & Flammang, 2005, 2006, 2008). In mussels, byssal threads converge to a structure, also called the stem, which is contiguous with the byssal retractor muscles (within the body of the mussel) used to control thread tension (Waite, 1992; Sagert et al., 2006). The weakest link of the byssus is typically the proximal region of the thread (the part linked to the animal) or the adhesive plaque (the part attached to the substrate) (Bell & Gosline, 1996; Carrington et al., 2015). Therefore, multi-component adhesive organs or structures, although clearly not homologous, might be similarly designed so as to balance energy costs against over-engineered material properties.

Muscles and structural parts can also play a significant role in single adhesive organs, regardless of the type of adhesion involved. In barnacles (Arthropoda, Crustacea) the retractor muscle pulls the peripheral shell plate downward at the time the permanent cement is secreted, thereby improving adhesion (Kamino, 2016). In limpets, the contraction of the powerful foot muscle clamps the shell against the substratum, playing an important role in the adhesion mechanism because friction generated by this behavior resists dislodgement by shear forces (Ellem et al., 2002). Meanwhile, in reversibly attaching animals lacking a duo-gland system, detachment is mostly achieved through mechanical forces (Lengerer & Ladurner, 2018) (see also

Sect. 16.4). In *Hydra* and sea anemones (Cnidaria) release is induced by muscular contractions in the basal disc (Rodrigues et al., 2016a). Some cephalopods (Mollusca), such as *Idiosepius*, *Euprymna* and *Sepia*, seem to detach as a result of dermal muscle contraction (von Byern & Klepal, 2006).

If we shift our focus to biological suction attachments, it is also evident that suction organs have evolved multiple times. This speaks to the utility of the organ for carrying out a variety of biological functions, from maintaining a position against strong hydrodynamic forces to facilitating locomotion, feeding, and reproduction. Whenever animals with suction organs are discussed, perhaps the most recognizable, and one of the most well-studied examples is the octopus (Fig. 16.3d). Octopuses use numerous suckers on their arms to catch prey, manipulate objects, locomote, and maintain position. (Although octopod suckers also serve as mechano- and chemosensors, we here focus on their role in attachment.) Decapods, such as squids and cuttlefish, are related to octopuses and they also possess suckers that serve similar functions as octopod suckers; hence, we refer to them collectively as coleoid suckers. Suckers present on the arms of octopods and decapods are superficially similar, whereas some tentacular suckers of decapods may also possess large hooks and spines for piercing prey (Nixon & Dilly, 1977). In general, coleoid suckers are circular in ventral view, with a rim for sealing, a central opening, and musculature that helps lower the pressure within the cavity enclosed between the sucker and the substrate (Kier & Smith, 1990, 2002; Smith, 1991b). (Microstructures present on the sucker surface are explored in Sect. 16.4.)

Among annelids, leeches (e.g., *Placobdella parasitica* and *Hirudo verbana*) use suction organs (one at the anterior and another at the posterior end) for locomotion and maintenance of position. Leech suction organs have muscles for raising the central region of the attachment disc and both *in vivo* pressure recordings and attachment performance measurements have confirmed that both reduced pressures and proper sealing are important for attachment (Gradwell, 1972a; Kampowski et al., 2016). In arthropods, suction organs are found in two disparate families: net-winged midges (Blephariceridae) and diving beetles (Dytiscidae). Blephariceridae larvae are found in fast-flowing alpine streams and each larva uses six specialized suction organs to attach to rock surfaces (Rietschel, 1961; Kang et al., 2019, 2021). These suction organs bear a striking resemblance to coleoid suckers, with a circular attachment disc, a sealing rim, and a central piston controlled by muscles that lower the pressure upon retraction. In dytiscid beetles the males alone carry numerous suckers on their prolegs and these are primarily used for holding onto females during courtship and copulation (Aiken and Khan 1992; Karlsson et al., 2013; Chen et al., 2014). There are no muscles within the individual suckers and it is thought that suction attachment is afforded passively through a combination of stored elastic energy and larger movements of the leg and body.

An impressive variety of suction organs has evolved in the Craniata. Many fishes and amphibian larvae use suction attachment for locomotion and maintenance of position. In fish, ventral suckers have been developed through modifications of the pelvic fins and pelvic girdle, pectoral fins and pectoral girdle, or the periphery of the mouth (Arita, 1967; Lujan & Conway, 2015). These analogous structures thus appear to be derived from different organs illustrating multiple evolutionary convergences. Clingfishes (Gobiesocidae), lumpsuckers (Cyclopteridae; Fig. 16.3c), and snailfishes (Liparidae) use their suckers to maintain position, either against strong currents or crashing tidal waves (Arita, 1967; Budney & Hall, 2010). Clingfishes are one of the model species for the study of biological suction attachment, and several detailed investigations of their attachment performance to various substrates are available (Wainwright et al., 2013; Ditsche et al., 2014, 2017). Gobies (Gobiidae) have independently evolved ventral suction organs that are also derived from their pelvic fins (Budney & Hall, 2010). Some gobies use suction attachment (employing both oral and posterior suction organs) to climb waterfalls, which is a well-documented behavior that clearly demonstrates the adhesive power that can be generated by suckers (Schoenfuss & Blob, 2003; Maie et al., 2012). The remoras (Echeneidae) are another well-studied group of fishes that have a single large elliptical suction pad derived from a highly modified dorsal fin. The remoras use their suction pad to attach to many different hosts, including turtles, sharks, dolphins, whales, and other fishes. The morphology and function of these organs are explored in more detail in Sect. 16.4.

Although suction feeding is a common feeding strategy in fishes, many species living in fast-flowing waterways have modified the periphery of their mouthparts to facilitate suction attachment (Lujan & Conway, 2015). Species of the genus *Garra* (Cyprinidae), for example, inhabit sub-Himalayan mountain streams and use their suction organs for maintaining position (Das & Nag, 2006, 2009). Their suction organs are derived from modified lips and encircle the mouth, the lower lip being further modified into a structure called the callous pad, which appears to have retractor muscles that can reduce the pressure during attachments (Saxena & Chandy, 1966). It would be remiss to mention oral suckers without acknowledging lampreys (Petromyzontidae). Although there are no detailed studies of the mechanism of attachment, lampreys are capable of generating significantly reduced pressures within the oral hood (Gradwell, 1972b). While it is unclear whether lampreys possess a specialized sealing rim, the margin of their mouth is free of teeth and could function as a soft sealing rim. In addition, their numerous teeth may provide additional friction by piercing the skin of the host and anchoring the lamprey.

Before proceeding to the next phylum, there is one more aquatic taxon in the Craniata that uses suction attachment organs: frogs. The larvae of several families of frogs (Bufonidae, Ranidae, and Hylidae) possess oral and abdominal suction organs and are collectively referred to as gastromyzophorous tadpoles. While some species inhabit fast-flowing streams (e.g., *Rhinella quechua, Huia cavitympanum, Atelopus* sp.), hyliid tadpoles (e.g., *Phyllodytes gyrinaethes*) develop in bromeliads (Kaplan, 1997; Aguayo et al., 2009; Haad et al., 2014; Gan et al., 2016; Vera Candioti et al., 2017). The musculature beneath abdominal suckers suggests that the suckers can actively reduce the internal pressure, although further functional studies are required for verification. Species of *Atelopus* also possess protuberances on the posterior part of the abdominal suckers that may increase friction during attachment (Kaplan, 1997). Several other anuran genera have enlarged oral suckers (e.g., *Litoria,*

Mixophyes, Ascaphus) that resemble the specialised oral suckers of fishes and can actively reduce the pressure during attachments (Gradwell, 1971, 1975).

In Mollusca, besides coleoids, lottiid limpets are capable of reducing the pressure beneath their muscular feet during attachment (Smith, 1991a). In contrast, patellid limpets do not produce as low sub-pedal pressures as lottiid limpets (approximately −0.6 kPa compared to −20 kPa, relative to ambient) (Jones & Trueman, 1970; Smith, 1991a; Kang et al., 2020). As mentioned in the introductory text, it can be difficult to clearly delineate between attachments that rely on suction and adhesive secretions, and thorough investigations using pressure recordings and molecular biological techniques are necessary for a more complete understanding of the underlying mechanism(s).

Although numerous tapeworms (Platyhelminthes) possess circular attachment structures that resemble suckers, additional studies are needed to verify whether they are able to reduce the pressure within the cavity. The anterior part of the tapeworm, the scolex, bears remarkably diverse structures ranging from suckers and hooks to hair-like structures called microtriches (de Chambrier & Scholz, 2008). Some authors refer to sucker-like structures as bothridia, but the distinction between suckers and bothridia is unclear. Suckers have longitudinal and radial muscles (Pospekhova & Bondarenko, 2014), and bothridia contain radial muscles and a single retractor muscle (Jones, 2000), but these differences in musculature do not appear to be used for categorisation. While both structures have been imaged with plugs of tissue within their cavities, it is currently unknown if the organs act as mechanical clamps or as suction organs by contracting their muscles to create pressure difference-based attachment (Andersen & Lysfjord, 1982; Borucinska & Caira, 1993; Ibraheem, 1998). Since we only have morphological data relating to tapeworm attachment, further work is needed to verify that tapeworm "suckers" or bothridia can indeed function as suction organs.

It is interesting to note that, like adhesive secretion organs, suction organs can be found as a single relatively large attachment unit (*e.g.*, in remora fish, lottiid limpets, gastromyzophorous tadpoles, lampreys, and clingfish) or as a group of many relatively small suckers (*e.g.*, in coleoids, blephariceridid larvae, diving beetles, and tapeworms). There are advantages and disadvantages for both strategies: in terms of benefits, having a single large attachment organ means that the same suction attachment force can be generated with a lower internal pressure, which demands less work from the muscles. On the other hand, a larger contact area increases the probability of encountering a random topography that interrupts the seal, thereby weakening the suction attachment or causing failure. In contrast, an attachment organ comprising many smaller suckers has the advantage that each unit is less likely to come into contact with a challenging surface feature, and even if one fails, there are numerous others to provide attachment. The disadvantage of multiple suckers is that if the total contact area is less than that of a single large organ (*e.g.*, if the boundary is constrained to a circle, even the most optimal packing arrangement of smaller circles will result in ~20% loss of area), then each sucker must work harder to attain the same amount of total attachment force. Based on our current understanding, there does not appear to be a strong determinant for whether an organism uses a single or multiple suction attachments.

16.4 A Cell/Microstructure-Level Approach to Attachment Mechanisms in Aquatic Organisms (Microscopic)

At the microscopic level, similar designs and organizations appear to have emerged independently in different phylogenetic lineages for both adhesive and suction organs.

The biosynthesis, packaging and release of adhesive secretions take place at the level of specialized secretory cells. In some rare cases (e.g., the cement glands of barnacles; Liang et al., 2019), these secretory cells are associated with collecting ducts to form complex glands. In most cases, however, each secretory cell delivers its products directly at the epithelial surface of the body area where adhesion takes place. These secretory cells can, however, be aggregated to form large secretory structures which are also often named glands in the literature—this is the case, for example, for the cement glands of annelid tubeworms (Becker et al., 2012) or the byssal glands of mussels (Waite, 2017). Alternatively, secretory cells may be homogeneously distributed among other cell types, as in the pedal sole epidermis of gastropod molluscs such as limpets (Grenon & Walker, 1978; Kang et al., 2020). Independent of the taxon and type of adhesion, there are animals in which the adhesive material is produced by a single type of secretory cell (e.g., barnacle cement cells (Liang et al., 2019), platyhelminth rhabdite-secreting cells (Martin, 1978c), sea urchin tube feet adhesive cells (Flammang et al., 2016), and ctenophore collocytes (von Byern et al., 2010) for permanent, transitory, temporary and instantaneous adhesion, respectively) and others in which this material is made up by the blending of molecules produced by two or more secretory cell types (e.g., polychaete tubeworm cement cells (Becker et al., 2012), limpet pedal glands (Kang et al., 2020), and sea star tube foot adhesive cells (Flammang et al., 2016) for permanent, transitory and temporary adhesion, respectively).

Duo-gland adhesive systems involved in temporary adhesion present an additional level of complexity as, in addition to adhesive secretory cells, they also incorporate de-adhesive secretory cells, hence their name (Fig. 16.4). De-adhesive cells release a second type of secretion, poorly characterized to date, that allows the detachment of the adhesive organ from the substratum (Lengerer & Ladurner, 2018). Indeed, temporary adhesion can be defined as a reversible attachment process in which strong adhesion is followed, after a certain interval, by voluntary detachment leading to a loss of contact between the adhesive organ and the surface (Lengerer & Ladurner, 2018). Duo-gland adhesive structures are found in many unrelated taxa. They were originally described for small invertebrates inhabiting the interstitial environment (Boaden, 1968; Tyler, 1976). In these meiofaunal organisms, belonging to the phyla Platyhelminthes (Tyler, 1976; Lengerer et al., 2014), Gastrotricha

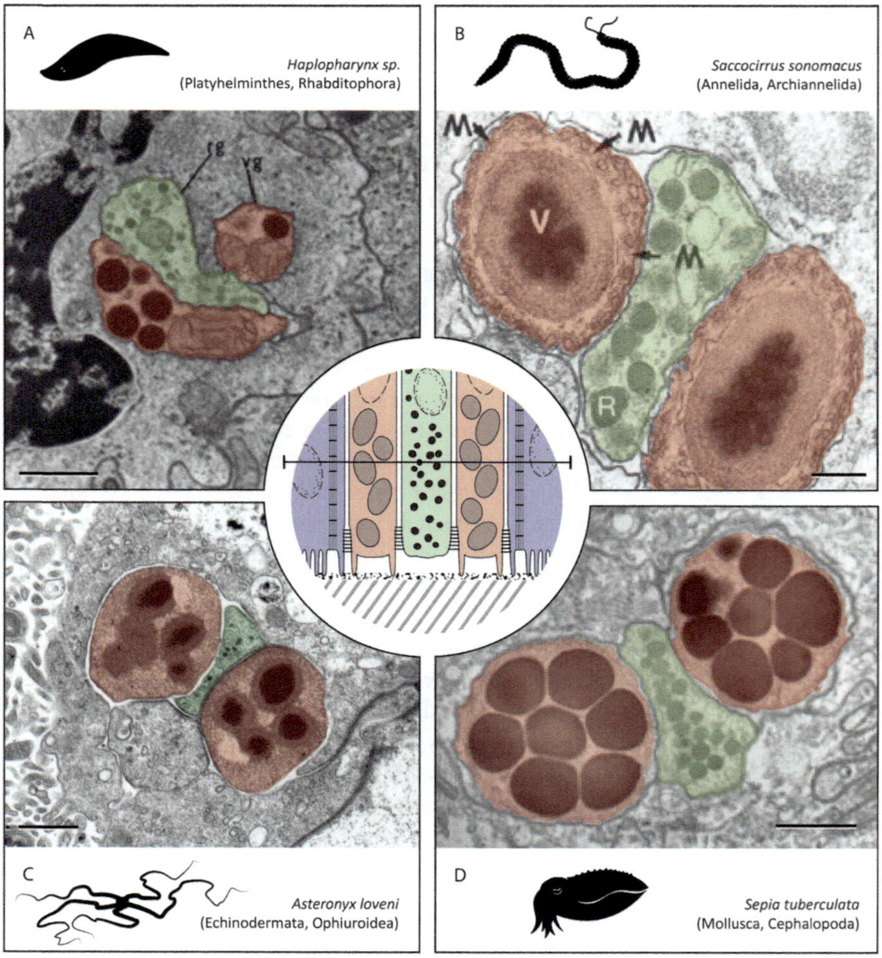

Fig. 16.4 Convergent cellular organisation of duo-gland secretory complexes in different metazoans. Transmission electron microscopic images of transverse sections through the adhesive epidermis of (**a**) a turbellarian flatworm body wall (adapted from Tyler (1976) with permission from Springer Nature), (**b**) a polychaete worm pygidium (adapted from Martin (1978a) with permission from Springer Nature), (**c**) a brittle star tube foot (original), and (**d**) a cuttlefish ventral mantle (adapted from von Byern et al. (2011) with permission from Wiley). The center of the figure shows a generalized drawing of a longitudinal section through such a secretory complex with the horizontal line showing the plane of section for images **a** to **d** (original drawing). Adhesive gland cells are indicated in red and de-adhesive gland cells in green. Scale bars: 1 μm. *M* microtubule, *R* releasing (de-adhesive) granule, *rg* releasing (de-adhesive) gland, *V* viscid (adhesive) granule, *vg* viscid (adhesive) gland

(Tyler & Rieger, 1980), Nematoda (Adams & Tyler, 1980), and Annelida (Martin, 1978a), they are involved in maintaining position. Duo-gland adhesive systems have also been described for echinoderm tube feet (Santos et al., 2009b; Flammang et al., 2016). Tube feet can be involved in position maintenance and locomotion (sea stars, sea urchins, and sea cucumbers), feeding (sea cucumbers, brittle stars, and feather stars), or shelter building (burrowing sea urchins). A duo-gland adhesive system has also been suggested to be present in the captacula (i.e., the food-collecting tentacles) of scaphopod molluscs (Shimek, 1988; Byrum & Ruppert, 1994), further widening the distribution range of this adhesive system in aquatic invertebrates.

Despite the more important morphological and functional complexity of duo-gland adhesive systems, strikingly similar cellular organizations have been reported for distantly related animals. In every species studied, the adhesive structures contain two types of closely associated secretory cells (Fig. 16.4). Adhesive cells are specialized epidermal cells, morphologically similar to the secretory cells involved in the other types of adhesion. They are filled with secretory granules which can vary greatly in shape, size, and contents. De-adhesive cells are thought to be derived from nerve cells in different taxa (Tyler, 1976; Flammang, 1996). They generally enclose small spherical, electron-dense secretory granules. The simplest organization of a duo-gland adhesive system consists of one adhesive cell with one de-adhesive cell, as seen in the flatworm *Macrostomum lignano* and in the sea urchin *Echinocardium cordatum* (Flammang et al., 1991; Lengerer et al., 2014). In the former, these two secretory cells are associated with one epidermal anchor cell, and the set of three cells has been named the duo-gland adhesive organ. In the latter, the adhesive and de-adhesive cells are associated with two sensory cells and the resulting structures have been called sensory-secretory complexes. There are also slightly more complex systems made up of the association of two adhesive cells flanking one de-adhesive cell (Fig. 16.4). This organization has been described for groups as diverse as flatworms (Tyler, 1976), annelids (Martin, 1978a), brittle stars (Flammang, 1996), and cuttlefishes (von Byern et al., 2011). For this last-mentioned cephalopod, *Sepia tuberculata*, it was proposed that detachment results from muscular contraction (von Byern et al., 2011). However, the close morphological convergence with other duo-gland adhesive systems suggests that de-adhesive secretions could help mechanical detachment.

As emphasized by the mechanism of detachment in *Sepia*, duo-gland adhesive systems do not seem to be the only adhesive systems involved in temporary adhesion. In a few taxa, structures possessing only one type of secretory cell attach and detach quickly. Such adhesive systems occur in some turbellarians (Tyler, 1976), gastrotrichs (Tyler & Rieger, 1980), and nematodes (Lippens, 1974). These structures were also described for cnidarians: the medusae of several species of hydrozoan possess adhesive tentacles that can attach and detach repeatedly (Honegger, 1984). Finally, barnacle larvae also fit into this category (Raine et al , 2020). In all these organisms the detachment process is purely mechanical (Lengerer & Ladurner, 2018).

In artificial suction cups (e.g., rubber suction cups used to attach mobile devices to glass), the disc wall—the side that attaches to the surface—is smooth. This is rarely the case in biological suction organs, where regions of the organ that contacts the attachment surface are highly textured. This texturing may arise from stiff microstructures (e.g., remora suction pads and net-winged midge larvae suction discs), dense arrays of cilia or microvilli (e.g., clingfish, lumpfish, limpets), or networks of channels and polygonal microstructures (e.g., coleoid suckers and clingfish). We provide an overview of the morphology and function of stiff microstructures below.

Spine-like microstructures called spinules are found within the suction pad of remoras (Fig. 16.5a–c). Spinules are mineralized projections that are approximately 500 μm in length and are found on top of lamellae (Fulcher & Motta, 2006; Beckert et al., 2015; Wang et al., 2017). The lamellae can be erected so that the spinules come into contact with the host surface. Several studies have demonstrated that the stiff spinules enhance friction on rough surfaces, thereby increasing drag resistance underwater (Beckert et al., 2015; Wang et al., 2017; Gamel et al., 2019). It is important that the spinules are sufficiently stiff and strong so that they retain their structural integrity when in contact with rough surfaces—otherwise they might either buckle or break under high loads. Interestingly, the orientation of the spinules with respect to drag forces may facilitate passive engagement. The spinules are posterior-facing, the drag force on hitch-hiking remoras (which act in the anterior-to-posterior direction) automatically promotes interlocking against surface asperities (Fulcher & Motta, 2006; Beckert et al., 2015). In addition, the soft fleshy rim around the suction disc also plays an important role during attachment as it creates a seal and contributes to friction on smooth surfaces (Fulcher & Motta, 2006; Wang et al., 2017).

Surprisingly, spine-like microstructures similar to remora spinules are found on suction organs of a family of insects (Blephariceridae; Fig. 16.5d–f). As mentioned previously, blephariceríd larvae are found in fast-flowing alpine water systems, where they use their suction organs to attach to and move on rocks. The spine-like microstructures on their suction organs are called microtrichia, and evidence suggests that they are stiff cuticular structures capable of interlocking with surface asperities (Rietschel, 1961; Kang et al., 2019, 2021). Since microtrichia tips are oriented towards the center of the suction organ, inward sliding of the organ likely results in passive engagement with the surface to increase friction, in a similar fashion to the posterior-facing spinules in remoras (Kang et al., 2019).

Many fish species that live in fast-flowing waters (rheophilic fish) have microstructures called unculi within their suction organs (note that unculi can also be present on other regions of their bodies) (Roberts, 1982). Unculi are keratinized outgrowths of single epithelial cells of approximately 10 to 20 μm in length, and in *Garra* sp. they are found atop tubercles within the oral sucker (Fig. 16.5g–i) (Saxena, 1959; Roberts, 1982; Teimori et al., 2011; Hussain & Bordoloi, 2018). It is possible that some fishes use their unculi to help scrape food from the substrate; however, unculi are also present on the frictional pads of the pectoral fins of non-suctorial fishes (Conway et al., 2012; De Meyer & Geerinckx, 2014), which

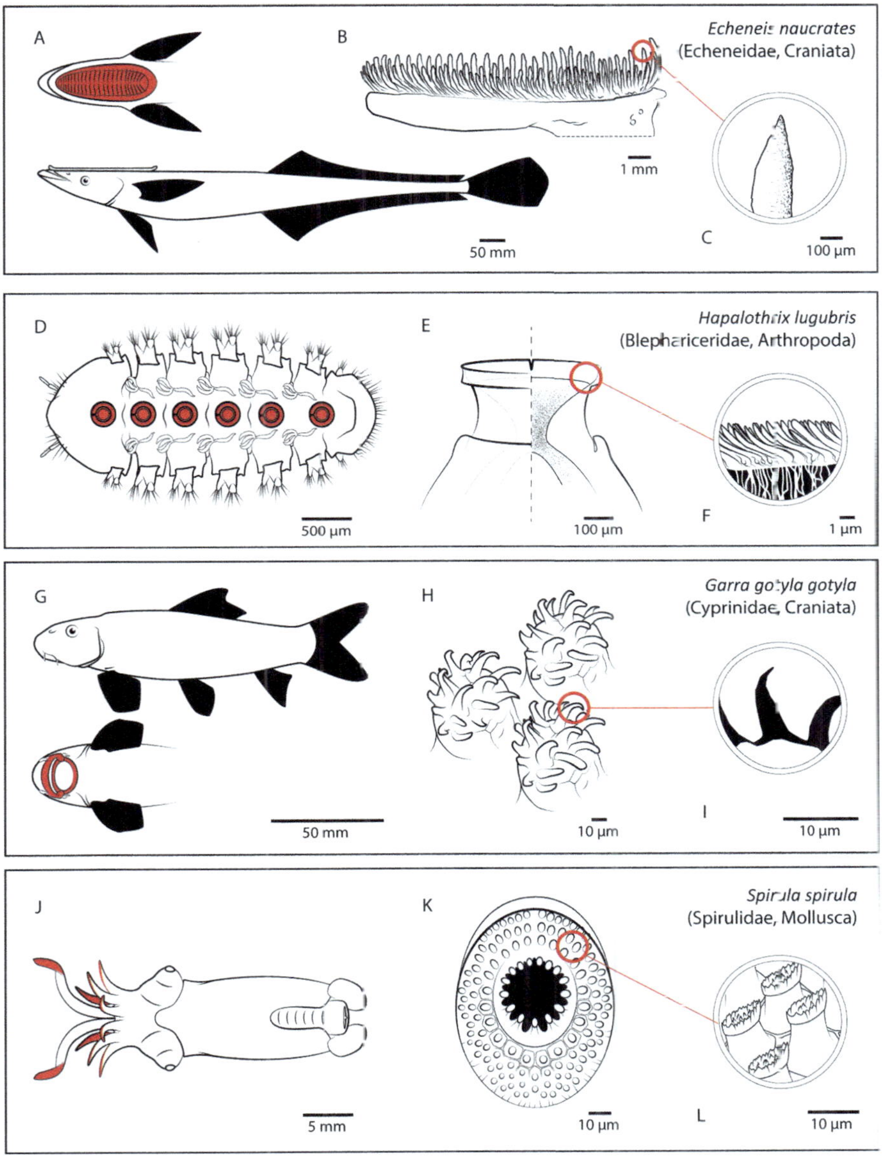

Fig. 16.5 Stiff microstructures in suction organs of disparate metazoans. Unlike synthetic suction cups, the surfaces of suction organs that contact the substrate are often highly textured. Stiff microstructures found on the suction discs of remora fish (spinules, **a–c**) and blepharicerid larvae (microtrichia, **d–f**) interlock with surface asperities to increase friction and help resist strong drag forces. Keratinized protuberances on the suction organ of cyprinid fish (e.g., in *Garra gotyla gotyla*, **g–i**) and cuticular pegs found on coleoid suckers (e.g., *Spirula spirula*, **j–l**) are also thought to increase friction during suction attachment

suggests a role in friction enhancement. Indeed, a functional study of attachment performance in *Hypostomus* sp. demonstrated that both oral suction organs and frictional pads contribute towards improved resistance against high flow rates (Gerstner, 2007).

Lastly, returning to the group of animals that symbolises biological suction attachment, many coleoid species possess microstructures called cuticular pegs on their suckers (Fig. 16.5j–l) (Nixon & Dilly, 1977; Salcedo-Vargas, 1995; Schmidtberg, 1999; Minnocci et al., 2015). Cuticular pegs, similarly to the arthropod cuticle, are made of chitin-protein complexes and may also be stiff structures (Hunt & Nixon, 1981; Accogli et al., 2017). Morphological studies have shown that cuticular pegs are found on a region of the suction organ called the infundibulum, which is firmly pressed against the surface during attachment (Nixon & Dilly, 1977; Kier & Smith, 1990; Salcedo-Vargas, 1995). Despite coleoid suckers being one of the most studied biological suction systems, not much is known about the function of these cuticular pegs. Researchers have hypothesized that they may increase friction when in contact with the surface and could also help to maximize attachment strength by transmitting low internal pressures throughout the disc (Kier & Smith, 1990). High mechanical wear from contacting and increasing friction may be why the lining of the infundibulum is periodically shed (Kier & Smith, 1990; Minnocci et al., 2015). Further research is needed to reveal the material properties of these cuticular pegs and how they generate additional friction during coleoid suction attachment.

16.5 A Molecule-Level Approach to Attachment Mechanisms in Aquatic Organisms (Nanoscopic)

In aquatic animals, the biochemical composition of adhesive secretions varies greatly from one taxonomic group to another (Tyler, 1988; Whittington & Cribb, 2001; Flammang et al., 2005, 2016). As a general rule, permanent adhesives consist almost exclusively of proteins. On the other hand, non-permanent adhesives (transitory as well as temporary) are made up of an association of proteins and carbohydrates, the latter being represented mostly in the form of acidic and sulfated glycans conjugated or associated to proteins (Hennebert et al., 2018). There is typically more protein than carbohydrate, usually in a ratio of approximately 2:1 (Flammang et al., 2016), but there may be substantial variation on this. The composition of instantaneous adhesives has only been investigated for sea cucumber Cuvierian tubules. Their adhesive is reminiscent of non-permanent adhesives through its constitution of proteins and carbohydrate in a 3:2 ratio (De Moor et al., 2003). However, it differs from them in that the carbohydrate fraction is in the form of neutral rather than acidic sugars. In all aquatic metazoans, therefore, adhesive secretions are predominantly made up of proteins (Hennebert et al., 2015). It is now well-established that the common properties of aquatic bioadhesives (e.g., the ability to displace water from the substratum, to spread and rapidly form strong adhesive bonds with the surface, and to cure) are related to the physico-chemical characteristics of their constituent proteins, including their post-translational modifications such as hydroxylation, phosphorylation and glycosylation (Stewart et al., 2011; Petrone, 2013; Davey et al., 2021). Thus far, hydroxylation and phosphorylation are the most thoroughly investigated modifications (Davey et al., 2021). Studies on mussel and tubeworm adhesive composition have revealed a high content of 3,4-dihydroxy-L-phenylalanine (DOPA), which is formed by post-translational hydroxylation of tyrosine, and of phosphoserine, which results from the phosphorylation of serine residues. These modified amino acids play important interfacial and cross-linking roles in aquatic adhesive secretions (Sagert et al., 2006), and in the case of mussel and tubeworm permanent adhesives, they are thought to be the result of convergent evolution (Kamino, 2010).

As far as the amino acid composition of the protein fraction is concerned, aquatic adhesives also vary considerably from one species to another. We used the method of Rocha et al. (2019) to quantify the level of relatedness among proteins. We performed a principal component analysis (PCA) of the amino acid compositions of bulk adhesive secretions that are usually mixtures of different proteins. The PCA, based on a variance-covariance matrix, was performed using the PAST 4.02 software (Hammer et al., 2001) on the relative amino acid content of whole adhesive secretions from 34 species belonging to seven phyla, including some terrestrial glues. An average protein (based on UniProtKB/SwissProt databases) and an average human secreted protein amino acid compositions were included for comparison (see Supplementary Table 16.2). Some post-translationally modified amino acids (*i.e.*, half-cystine and DOPA) were included in the analysis because they are important constituents of some aquatic adhesives (Kamino, 2010; Hennebert et al., 2015; Davey et al., 2021), but phosphoserine residues were not considered as they are dephosphorylated into serine residues during the acid hydrolysis step of the amino acid analysis (Stewart et al., 2004). Similarly, aspartic acid and asparagine, and glutamic acid and glutamine were grouped as Asx and Glx respectively since the acid hydrolysis induces a deamidation of Asn and Gln. Two principal factors extracted from the PCA, PC1 and PC2, accounted for 64.2% of the cumulative variance. Figure 16.6 shows that the adhesives of phylogenetically related species using the same type of adhesion generally cluster together, which suggests they are homologous. For example, the permanent cements of both acorn and goose barnacles, the transitory adhesives of limpets, and the instantaneous adhesives of sea cucumber Cuvierian tubules form tight, taxon-specific clusters. For some other taxa, however, the species are more distantly spaced but are still clustered together: for example, the temporary adhesives of echinoderm tube feet, the permanent adhesives of mussels and of tubeworms, and the slimes of velvet worms. It should be noted, however, that some of these taxa are represented by only two species. A notable exception is the loose cluster comprising the temporary adhesives of monogenean flatworms, for which divergence between species is more pronounced. Terrestrial glues are intermixed with aquatic adhesives, although they tend to cluster in the lower right part of the PCA plot.

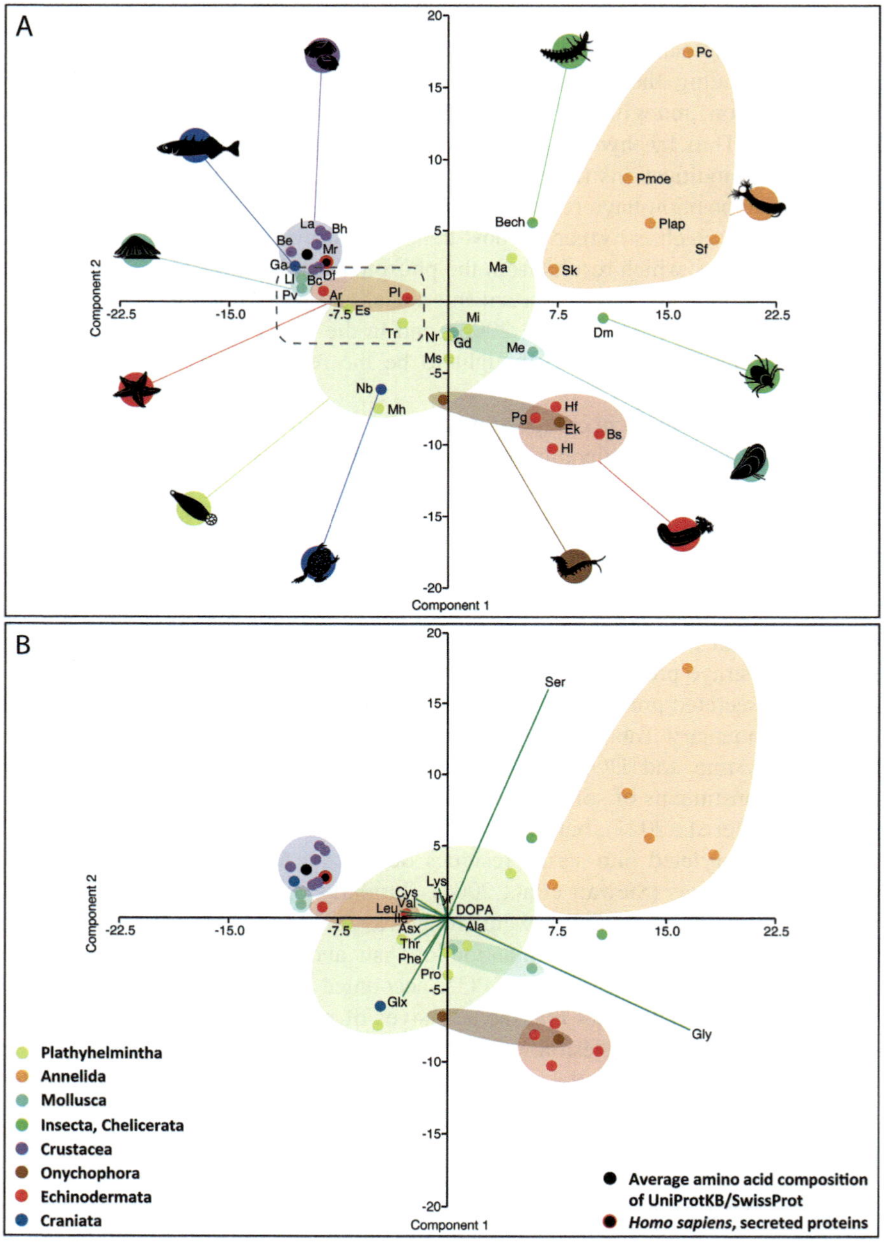

Fig. 16.6 Comparative amino acid composition of bioadhesives (PCA). Scatter plot of principal component axis 1 (PC1) and axis 2 (PC2) based on the relative amino acid composition of the secreted adhesives from various animal species, where the two first principal components (1 × 2) account for 64.2% of the cumulative variance. (**a**) Projection of metazoan bioadhesives (each dot represents one species; n = 34) on the factor plane showing clusters based on phylogenetic and functional aspects. (**b**) Projection of amino acid (n = 18) levels on the factor plane, showing amino acids that contribute the most to the characterisation of each group of bioadhesives.

Glycine and serine are over-represented in the adhesives of almost half of the species included in the analysis, and the first component of the PCA separates adhesives with a bias towards these amino acids (right part of Fig. 16.6b) from adhesives with a more average composition (left part of Fig. 16.6b). In most cases, there is little overlap between taxonomic groups, but several non-permanent adhesives from a number of species are grouped together even though they belong to disparate phyla (i.e., platyhelminths (Hamwood et al., 2002), mollusks (Grenon & Walker, 1980; Smith et al., 1999), and echinoderms (Flammang et al., 1998; Santos et al., 2009a) (Fig. 16.6a, grey dotted frame). This relationship might indicate convergence in amino acid composition driven by shared function and selective pressures. A similarity between transitory and temporary adhesives was already evident in terms of glycan composition (Hennebert et al., 2018; Kang et al., 2020) (see also above). In contrast, no such compositional convergence is observed for adhesives from sessile species using permanent adhesion. Indeed, the adhesives of mussels, tubeworms and barnacles differ greatly from each other (Fig. 16.6). In their composition, the protein fractions of mussel byssal plaques and polychaete cement have the presence of DOPA in common in their composition (Benedict & Waite, 1986; Jensen & Morse, 1988; Waite et al., 1989). However, the tubeworm adhesives are separated by their high content of phosphoserine (Mitterer, 1971; Stewart et al., 2004), which is a characteristic they share with the adhesive silk of caddisfly larvae (Stewart & Wang, 2010), a permanent adhesive used in building shelters. Barnacle cements, on the other hand, contain neither DOPA nor phosphoserine, and seem to have more in common with non-permanent adhesives, in which disulfide bonds serve an important function (Fig. 16.6b) (Walker, 1972; Kamino et al., 1996, Naldrett & Kaplan, 1997; Engel et al., 2021). As for the instantaneous adhesives of holothuroid Cuvierian tubules, they differ from all other aquatic bioadhesives because they are particularly rich in glycine (De Moor et al., 2003; Flammang et al., 2005), and instead share resemblance to the defensive onychophoran slimes (Röper, 1977; Benkendorff et al., 1999).

For all investigated species, adhesive secretions consist of at least two or more proteins. According to their sequence and structure, these proteins may achieve

Fig. 16.6 (continued) Ar Asterias *rubens, Bc Balanus crenatus, Be Balanus eburneus, Bech Brachycentrus echo, Bh Balanus hameri, Bs Bohadschia subrubra, Df Dosima fascicularis, Dm Dermacentor marginatus, Ek Euperipatoides kanangrensis, Es Entobdella soleae, Ga Gasterosteus aculeatus, Gd Geukensia demissa, Hf Holothuria forskali, Hl Holothuria leucospilota, La Lepas anatifera, Ll Lottia limatula, Ma Merizocotyle australensis, Me Mytilus edulis, Mi Merizocotyle icopae, Mh Monocotyle helicophallus, Mr Megabalanus rosa, Ms Monocotyle spiremae, Nb Notaden bennetti, Nr Neoheterocotyle rhinobatidis, Pc Phragmatopoma californica, Pg Pearsonothuria graeffei, Pl Paracentrotus lividus, Plap Phragmatopoma lapidosa, Pm Peripatopsis moseleyi, Pmoe Phragmatopoma moerchi, Pv Patella vulgata, Sf Sabellaria floridensis, Sk Sabellaria kaiparaensis, Tr Troglocephalus rhinobatidis*. The average amino acid composition of proteins from the UniProtKB/SwissProt database and of human secreted proteins are also included as comparison points (black dots). Amino acid compositions and references can be found in Supplementary Table 16.2

various sub-functions within the secreted adhesive (e.g., interfacial adhesive or bulk cohesive interactions). This means that bioadhesives are usually composed of a variety of different proteins. Thus, although the amino acid composition of barnacle cement resembles that of an average secreted mammalian protein (Fig. 16.6), these cements are in fact made up of several proteins of very different compositions and sequences (Rocha et al., 2019) (Fig. 16.7).

The evolutionary origins of metazoan adhesive proteins remain largely enigmatic. While some authors have proposed a complete independent evolution of bioadhesive proteins (Kamino, 2010), more recent works—driven by omics approaches—suggest some evolutionary-related sequence similarities and, more specifically, the presence of common protein domains between different bioadhesive proteins (Davey et al., 2021). Indeed, although some adhesive protein sequences are short and intrinsically disordered, others are long or very long, comprising multiple domains involved in various subtasks important for their adhesion and/or cohesive functions. Protein domains are "high-level parts of proteins that either occur alone or together with partner domains on the same protein chain" (Forslund & Sonnhammer, 2012). Many protein domains can perform a particular function or contribute in a specific way to the function of the overall protein. Most domains correspond to tertiary structural elements and are able to fold independently.

We also investigated adhesive proteins secreted by a wide variety of aquatic animals (i.e., Cnidaria, Annelida, Mollusca, Platyhelminthes, Echinodermata, Craniata) using a sequence similarity-based clustering analysis to highlight potential similarities between these bioadhesives. Sequence similarity searches (often performed using BLAST) can identify "homologous" proteins by detecting excess similarity corresponding to the statistically significant similarity that reflects common ancestry (Pearson, 2013). Adhesive protein sequences were retrieved from publicly-accessible databases or from previous studies (Rodrigues et al., 2016b). The sequence similarity-based clustering was performed using CLANS (Frickey & Lupas, 2004). An all-against-all BLASTp was conducted using the scoring matrix BLOSUM62 and linkage clustering was performed with an E-value of 1E-10 to identify coherent clusters. The clustering was first performed in 3-dimensions and then collapsed into 2D in order to generate the plot shown in Fig. 16.7a (see Supplementary Table 16.3 for the list of adhesive proteins). The connections between the dots indicate clear similarity and highlight potential homology between the proteins. Our analyses only included protein sequences that have been confirmed to be part of bioadhesive secretions. Many candidates that did not meet our rigorous criteria could be included in the future as new evidence becomes available.

While many proteins appear to be specific to the investigated organisms (represented as isolated dots or clusters of dots of the same color in our analysis; Fig. 16.7a), some exhibit clear sequence homologies between phylogenetically distant organisms (shown as connections between dots of different colors). At least four clusters of adhesive proteins from phylogenetically distant organisms have been identified. Our protein domain analyses showed that the similarity between all of these adhesive proteins is specifically associated with similar (and likely homologous) protein domains: lectin domains, epidermal growth factor-like (EGF)

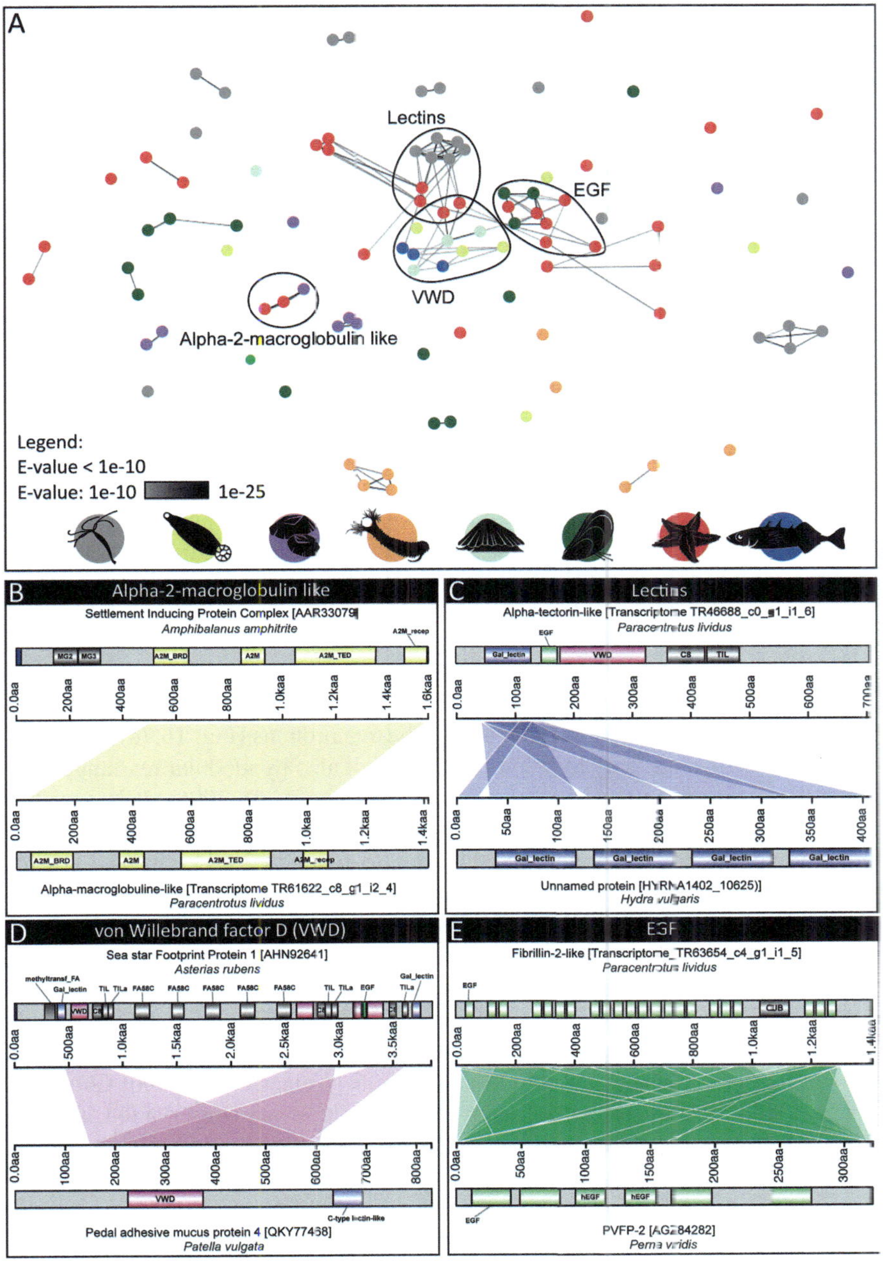

Fig. 16.7 Cluster analysis of adhesive protein sequences and identification of shared protein motifs. (**a**) CLANS analysis of selected adhesive proteins using an E-value threshold of 1E-10. (In BLAST analyses, the E-value is defined as the probability, due to chance, that there is another alignment with a similarity greater than the obtained score). Only proteins presenting a similarity above the threshold are connected by lines. The lines are color-coded according to their E-values. (**b**) Alpha-macroglobulin domains observed in barnacle settlement-inducing protein complex and echinoderm adhesive proteins. (**c**) Lectin domains observed in echinoderm (continued overleaf)

domains, alpha-2-macroglobulin-like (A2M) domains, and von Willebrand factor type D (VWD) domains. These domains are known to bind to other proteins and sugar groups, forming oligomers and adsorbing onto substrates—functions that are particularly relevant for adhesive proteins (Davey et al., 2021). A2M domains are specifically shared by two echinoderm proteins (found in the sea star *Asterias rubens* and in the sea urchin *Paracentrotus lividus*) and one barnacle protein (Settlement Inducing Protein Complex or SIPC of *Amphibalanus amphitrite*) (Fig. 16.7a). Comparison of the protein domains highlighted a general similarity of the two proteins that share a rather long alpha-2-macroglobulin-like multi-subdomain set of around 800 amino acids (Fig. 16.7b). Galactose/rhamnose binding lectin domains are observed in multiple adhesive proteins of *Hydra* (Rodrigues et al., 2016b) and are also present in various echinoderm adhesive proteins (i.e., from both sea stars and sea urchins) (Fig. 16.7c). VWD domains have also been found in various adhesive proteins from fish (*Gasterosteus*), flatworms, limpets, and echinoderms. As illustrated in the protein domain prediction, the Sea star Footprint Protein 1 of *Asterias rubens* contains numerous domains including three VWD domains. This domain is also found in one of the adhesive proteins isolated from the limpet *Patella vulgata* (*P-vulgata_4*), although only in one "copy". EGF domains have been detected in various adhesive proteins, including proteins from mussels and echinoderms (Fig. 16.7d). This domain also occurs in adhesive proteins from limpets and flatworms, but it appears that, with our stringent threshold, connections between these proteins and those of mussels and echinoderms are not visible on the CLANS analyses (Fig. 16.7a). It is noteworthy, however, that in most of the cases EGF domains are present in multiple copies in adhesive proteins (Fig. 16.7d).

Proteins evolve not only by point mutations but also by modular rearrangements generally occurring at the level of domains (Weiner et al., 2006). It is generally accepted that the vast majority of proteins have domain architectures that emerged through evolutionary descent rather than due to functional necessity and convergence (Gough, 2005). Many biological processes involved in the evolutionary emergence of domain architectures have been studied to date, including: gene fusion by a mobile element (such as a retrotransposon), gene fusion by loss of a stop signal or deletion of much of the intergenic region, domain insertion through recombination, gene fission by the introduction of transcription stop and start codons, and domain loss by the introduction of a stop codon with subsequent degeneration of the now untranslated domain (Björklund et al., 2005; Weiner et al., 2006; Chothia & Gough, 2009). Because protein domains exhibit evolutionary conservation, adhesive proteins from phylogenetically distant organisms undoubtedly share related features. However, it is highly probable that these domains, which are also present in a variety of non-adhesive proteins, were convergently acquired from ancestral proteins with unrelated general functions (even though the general domain subfunctions could be

---

Fig. 16.7 (continued) and *Hydra* adhesive proteins. (**d**) VWD domains observed in echinoderm and limpet adhesive proteins. (**e**) EGF domains found in echinoderm and mussel adhesive proteins. The list of the adhesive proteins used in the CLANS analysis can be found in Supplementary Table 16.3

similar or identical). Thus, it seems that there is no common ancestral bioadhesive protein; instead, evolutionarily related protein domains were likely repurposed to achieve similar functions in different bioadhesives.

16.6 Conclusion and Outlook

Investigating how multiple evolutionary scenarios converge on functionally similar traits is important for understanding the evolution of complex biological processes. Many aquatic animals, whether they are sessile or mobile, marine or freshwater, require strategies to allow them to attach to substrates in wet environments. We have explored the metazoan phylogeny and identified the two main mechanisms of aquatic attachment: bioadhesive secretions and suction attachment. Based on our survey, most of the recognized extant metazoan phyla contain at least one species that uses bioadhesives or suction organs, and numerous cases of convergent evolution can be identified that span the length-scales from molecules to organisms. We have shown that attachment systems are complex traits with similar functions that have emerged repeatedly during evolution. From the molecular point of view, it is likely that homologous features (i.e., protein domains) were independently requisitioned in different lineages. There remain, however, many gaps in our knowledge of biological attachment strategies and their evolution. For instance, although a growing number of studies have isolated and characterized proteins and sugars from adhesive secretions, functional studies of the individual components are scarce. Likewise, while it is relatively easy to classify an organ as a suction attachment organ, it is much more challenging to convincingly prove that the animal indeed generates pressure differences for attachment. Future studies that successfully explore these aspects in detail will be of great value to the bioadhesive community.

Our review demonstrates the utility of a multi-level approach in exploring the evolution of biological attachment strategies in aquatic metazoans. We show that convergence can be identified at many different organizational levels, which means that studies focusing solely on one level (e.g., adhesive proteins) can miss insights into other important components of adhesive systems (e.g., the glandular system that delivers the proteins to the substrate). Due to a combination of the breadth of our taxonomic coverage and the lack of studies that quantify convergence of specific traits of adhesive systems, our work is light on detailed discussions. We believe that there are ample opportunities for both continuing to explore the tree of life for strategies of adhesion as well as delving deeper into identified species to better understand the mechanism of action. Furthermore, if our multi-level approach is adopted in future studies, we expect a more holistic understanding of attachment strategies within and across different species to emerge. Such endeavors will undoubtedly uncover new and exciting examples of adhesion and will help to enrich our understanding of the role of convergent evolution in the development of complex biological traits.

Acknowledgments This work was supported by the ARC project "PROTEST" (Production and testing of recombinant sea star adhesive proteins; "Communauté française de Belgique—Actions de Recherche Concertées", ARC-17/21 UMONS 3), by the Fund for Scientific Research of Belgium (F.R.S.-FNRS) under grant "Projet de Recherche" n° T.0088.20, by the Fundação para a Ciência e Tecnologia through Project grants (IF/00006/2015/CP1276/CT0001; UIDP/04292/2020), and by the European Cooperation in Science and Technology (COST) Action CA15216. JD and PF are respectively Postdoctoral Researcher and Research Director of the F.R.S.-FNRS. This paper is a contribution to the Centre Interuniversitaire de Biologie Marine (CIBIM).

References

Accogli, G., Scillitani, G., Mentino, D., & De Santis, S. (2017). Characterization of the skin mucus in the common octopus *Octopus vulgaris* (Cuvier). Reared paralarvae. *European Journal of Histochemistry, 61*, 2815.

Adams, P. J., & Tyler, S. (1980). Hopping locomotion in a nematode: Functional anatomy of the caudal gland apparatus of *Theristus caudasaliens* sp. n. *Journal of Morphology, 164*, 265–285.

Aguayo, R., Lavilla, E. O., Vera Candioti, M. F., & Camacho, T. (2009). Living in fast-flowing water: Morphology of the gastromyzophorous tadpole of the bufonid *Rhinella quechua* (*R. veraguensis* group). *Journal of Morphology, 270*, 1431–1442.

Aiken, R. B., & Khan, A. (1992). The adhesive strength of the palettes of males of a boreal water beetle, *Dytiscus alaskanus* J. Balfour Browne (Coleoptera: Dytiscidae). *Canadian Journal of Zoology, 70*, 1321–1324.

Andersen, K. I., & Lysfjord, S. (1982). The functional morphology of the scolex of two *Tetrabothrius* Rudolphi 1819 species (Cestoda; Tetrabothriidae) from penguins. *Parasitology Research, 67*, 299–307.

Arita, G. S. (1967). *A comparative study of the structure and function of the adhesive apparatus of the Cyclopteridae and Gobiesocidae*. Doctoral dissertation, The University of British Columbia.

Becker, P. T., Lambert, A., Lejeune, A., Lanterbecq, D., & Flammang, P. (2012). Identification, characterization, and expression levels of putative adhesive proteins from the tube-dwelling polychaete *Sabellaria alveolata. The Biological Bulletin, 223*, 217–225.

Beckert, M., Flammang, B. E., & Nadler, J. H. (2015). Remora fish suction pad attachment is enhanced by spinule friction. *Journal of Experimental Biology, 218*, 3551–3558.

Bell, E., & Gosline, J. (1996). Mechanical design of mussel byssus: Material yield enhances attachment strength. *Journal of Experimental Biology, 199*, 1005–1017.

Benedict, C. V., & Waite, J. H. (1986). Ultrastructure and composition of the byssus of *Mytilus edulis* L. *Journal of Morphology, 189*, 261–270.

Benkendorff, K., Beardmore, K., Gooley, A. A., Packer, N. H., & Tait, N. N. (1999). Characterisation of the slime gland secretion from the peripatus, *Euperipatoides kanangrensis* (Onychophora: Peripatopsidae). *Comparative Biochemistry and Physiology Part B: Biochemistry and Molecular Biology, 124*, 457–465.

Björklund, A. K., Ekman, D., Light, S., Frey-Skött, J., & Elofsson, A. (2005). Domain rearrangements in protein evolution. *Journal of Molecular Biology, 353*, 911–923.

Boaden, P. J. S. (1968). Water movement—a dominant factor in interstitial ecology. *Sarsia, 34*, 125–136.

Borucinska, J., & Caira, J. N. (1993). A comparison of mode of attachment and histopathogenicity of four tapeworm species representing two orders infecting the spiral intestine of the nurse shark, *Ginglymostoma cirratum. The Journal of Parasitology, 79*, 238–246.

Budney, L. A., & Hall, B. K. (2010). Comparative morphology and osteology of pelvic fin-derived midline suckers in lumpfishes, snailfishes and gobies. *Journal of Applied Ichthyology, 26*, 167–175.

von Byern, J., & Klepal, W. (2006). Adhesive mechanisms in cephalopods: a review. *Biofouling, 22*, 329–338.
von Byern, J., Mills, C. E., & Flammang, P. (2010). Bonding tactics in ctenophores — morphology and function of the colloblast system. In J. von Byern & I. Grunwald (Eds.), *Biological adhesive systems* (pp. 29–40). Springer Vienna.
von Byern, J., Müller, C., Voigtländer, K., Dorrer, V., Marchetti-Deschmann, M., Flammang, P., & Mayer, G. (2017). Examples of bioadhesives for defence and predation. In S. N. Gorb & E. V. Gorb (Eds.), *Functional surfaces in biology III, biologically-inspired systems* (pp. 141–191). Springer International Publishing.
von Byern, J., Scott, R., Griffiths, C., Micossi, A., Grunwald, I., & Cyran, N. (2011). Characterization of the adhesive areas in *Sepia tuberculata* (Mollusca, Cephalopoda). *Journal of Morphology, 272*, 1245–1258.
Byrum, C. A., & Ruppert, E. E. (1994). The ultrastructure and functional morphology of a captaculum in *Graptacme calamus* (Mollusca, Scaphopoda). *Acta Zoologica, 75*, 37–46.
Carrington, E., Waite, J. H., Sarà, G., & Sebens, K. P. (2015). Mussels as a model system for integrative ecomechanics. *Annual Review of Marine Science, 7*, 443–469.
de Chambrier, A., & Scholz, T. (2008). Tapeworms (Cestoda: Proteocephalidea) of firewood catfish *Sorubimichthys planiceps* (Siluriformes: Pimelodidae) from the Amazon River. *Folia Parasitologica, 55*, 17–28.
Chen, Y., Shih, M.-C., Wu, M.-H., Yang, E.-C., & Chi, K.-J. (2014). Underwater attachment using hairs: The functioning of spatula and sucker setae from male diving beetles. *Journal of the Royal Society, Interface, 11*, 20140273.
Chothia, C., & Gough, J. (2009). Genomic and structural aspects of protein evolution. *The Biochemical Journal, 419*, 15–28.
Clarke, J. L., Davey, P. A., & Aldred, N. (2020). Sea anemones (*Exaiptasia pallida*) use a secreted adhesive and complex pedal disc morphology for surface attachment. *BMC Zoology, 5*, 5.
Conway, K. W., Lujan, N. K., Lundberg, J. G., Mayden, R. L., & Siegel, D. S. (2012). Microanatomy of the paired-fin pads of ostariophysan fishes (Teleostei: Ostariophysi). *Journal of Morphology, 273*, 1127–1149.
Cowles, D. (1977). Locomotion by *Epiactis prolifera* (Coelenterata: Actiniaria). *Marine Biology, 39*, 67–70.
Das, D., & Nag, T. C. (2006). Fine structure of the organ of attachment of the teleost, *Garra gotyla gotyla* (Ham). *Zoology, 109*, 300–309.
Das, D., & Nag, T. C. (2009). Organs of adhesion in some mountain-stream teleosts of India: Structure-function relationship. In S. N. Gorb (Ed.), *Functional surfaces in biology* (pp. 105–122). Springer.
Davey, P. A., Power, A. M., Santos, R., Bertemes, P., Ladurner, P., Palmowski, P., et al. (2021). Omics-based molecular analyses of adhesion by aquatic invertebrates. *Biological Reviews, 96*, 1051–1075.
De Meyer, J., & Geerinckx, T. (2014). Using the whole body as a sucker: Combining respiration and feeding with an attached lifestyle in hill stream loaches (Balitoridae, Cypriniformes). *Journal of Morphology, 275*, 1066–1079.
De Moor, S., Waite, J. H., Jangoux, M., & Flammang, P. (2003). Characterization of the adhesive from Cuvierian tubules of the sea cucumber *Holothuria forskali* (Echinodermata, Holothuroidea). *Marine Biotechnology, 5*, 45–57.
Ditsche, P., Hicks, M., Truong, L., Linkem, C., & Summers, A. (2017). From smooth to rough, from water to air: The intertidal habitat of Northern clingfish (*Gobiesox maeandricus*). *Naturwissenschaften, 104*, 33.
Ditsche, P., & Summers, A. P. (2014). Aquatic versus terrestrial attachment: Water makes a difference. *Beilstein Journal of Nanotechnology, 5*, 2424–2439.
Ditsche, P., Wainwright, D. K., & Summers, A. P. (2014). Attachment to challenging substrates--fouling, roughness and limits of adhesion in the northern clingfish (*Gobiesox maeandricus*). *Journal of Experimental Biology, 217*, 2548–2554.

Edmunds, M., Potts, G. W., Swinfen, R. C., & Waters, V. L. (1976). Defensive behaviour of sea anemones in response to predation by the opisthobranch mollusc *Aeolidia papillosa* (L.). *Journal of the Marine Biological Association of the United Kingdom, 56*, 65–83.
Ellem, G. K., Furst, J. E., & Zimmerman, K. D. (2002). Shell clamping behaviour in the limpet *Cellana tramoserica*. *Journal of Experimental Biology, 205*, 539–547.
Engel, B., Suppan, J., Nürnberger, S., Power, A. M., & Marchetti-Deschmann, M. (2021). Revisiting amino acid analyses for bioadhesives including a direct comparison of tick attachment cement (*Dermacentor marginatus*) and barnacle cement (*Lepas anatifera*). *International Journal of Adhesion and Adhesives, 105*, 102798.
Federle, W., & Labonte, D. (2019). Dynamic biological adhesion: Mechanisms for controlling attachment during locomotion. *Philosophical Transactions of the Royal Society of London Series B, Biological Sciences, 374*, 20190199.
Feigenbaum, D. L. (1976). Development of the adhesive organ in *Spadella schizoptera* (Chaetognatha) with comments on growth and pigmentation. *Bulletin of Marine Science, 26*, 600–603.
Flammang, P. (1996). Adhesion in echinoderms. In *Echinoderm Studies* (Vol. 5, pp. 1–60). Balkema.
Flammang, P., Demeuldre, M., Hennebert, E., & Santos, R. (2016). Adhesive secretions in echinoderms: A review. In A. M. Smith (Ed.), *Biological adhesives* (pp. 193–222). Springer International Publishing.
Flammang, P., De Ridder, C., & Jangoux, M. (1991). Ultrastructure of the penicillate podia of the spatangoid echinoid *Echinocardium cordatum* (Echinodermata) with special emphasis on the epidermal sensory-secretory complexes. *Acta Zoologica, 72*, 151–158.
Flammang, P., Michel, A., Cauwenberge, A. V., Alexandre, H., & Jangoux, M. (1998). A study of the temporary adhesion of the podia in the sea star *Asterias rubens* (Echinodermata, Asteroidea) through their footprints. *Journal of Experimental Biology, 201*(Pt 16), 2383–2395.
Flammang, P., Santos, R., & Haesaerts, D. (2005). Echinoderm adhesive secretions: From experimental characterization to biotechnological applications. *Progress in Molecular and Subcellular Biology, 39*, 201–220.
Forslund, K., & Sonnhammer, E. L. L. (2012). Evolution of protein domain architectures. *Methods in Molecular Biology, 856*, 187–216.
Frickey, T., & Lupas, A. (2004). CLANS: a Java application for visualizing protein families based on pairwise similarity. *Bioinformatics, 20*, 3702–3704.
Fulcher, B. A., & Motta, P. J. (2006). Suction disk performance of echeneid fishes. *Canadian Journal of Zoology, 84*, 42–50.
Gamel, K. M., Garner, A. M., & Flammang, B. E. (2019). Bioinspired remora adhesive disc offers insight into evolution. *Bioinspiration & Biomimetics, 14*, 056014.
Gan, L. L., Hertwig, S. T., Das, I., & Haas, A. (2016). The anatomy and structural connectivity of the abdominal sucker in the tadpoles of *Huia cavitympanum*, with comparisons to *Meristogenys jerboa* (Lissamphibia: Anura: Ranidae). *Journal of Zoological Systematics and Evolutionary Research, 54*, 46–59.
Gerstner, C. L. (2007). Effect of oral suction and other friction-enhancing behaviors on the station-holding performance of suckermouth catfish (*Hypostomus* spp.). *Canadian Journal of Zoology, 85*, 133–140.
Giribet, G., & Edgecombe, G. D. (2020). *The invertebrate tree of life*. Princeton University Press.
Gorb, S. N. (2008). Biological attachment devices: Exploring nature's diversity for biomimetics. *Philosophical Transactions Series A, Mathematical, Physical, and Engineering Sciences, 366*, 1557–1574.
Gough, J. (2005). Convergent evolution of domain architectures (is rare). *Bioinformatics, 21*, 1464–1471.
Gradwell, N. (1971). *Ascaphus* tadpole: Experiments on the suction and gill irrigation mechanisms. *Canadian Journal of Zoology, 49*, 307–332.

Gradwell, N. (1972a). Behaviors of the leech, *Placobdella*, and transducer recordings of suctorial pressures. *Canadian Journal of Zoology, 50*, 1325–1332.

Gradwell, N. (1972b). Hydrostatic pressures and movements of the lamprey, *Petromyzon*, during suction, olfaction, and gill ventilation. *Canadian Journal of Zoology, 50*, 1215–1223.

Gradwell, N. (1975). Experiments on oral suction and gill breathing in five species of Australian tadpole (Anura: Hylidae and Leptodactylidae). *Journal of Zoology, 177*, 81–98.

Grenon, J. F., & Walker, G. (1978). The histology and histochemistry of the pedal glandular system of two limpets, *Patella vulgata* and *Acmaea tessulata* (Gastropoda: Prosobranchia). *Journal of the Marine Biological Association of the UK, 58*, 803–816.

Grenon, J. F., & Walker, G. (1980). Biochemical and rheological properties of the pedal mucus of the limpet, *Patella vulgata* L. *Comparative Biochemistry and Physiology, 66B*, 451–458.

Haad, M. B., Candioti, F. V., & Baldo, D. (2014). The stream tadpoles of *Rhinella rumbolli* (Anura: Bufonidae). *Herpetologica, 70*, 184–197.

Hammer, Ø., Harper, D. A. T., & Ryan, P. D. (2001). PAST: Paleontological statistics software package for education and data analysis. *Palaeontological Association, 4*, 1–9.

Hamwood, T. E., Cribb, B. W., Halliday, J. A., Kearn, G. C., & Whittington, I. D. (2002). Preliminary characterisation and extraction of anterior adhesive secretion in monogenean (platyhelminth) parasites. *Folia Parasitologica, 49*, 39–49.

Hennebert, E., Gregorowicz, E., & Flammang, P. (2018). Involvement of sulfated biopolymers in adhesive secretions produced by marine invertebrates. *Biology Open, 7*.

Hennebert, E., Maldonado, B., Ladurner, P., Flammang, P., & Santos, R. (2015). Experimental strategies for the identification and characterization of adhesive proteins in animals: A review. *Interface Focus, 5*, 20140064.

Hennebert, E., Santos, R., & Flammang, P. (2012). Echinoderms don't suck: Evidence against the involvement of suction in tube foot attachment. *Zoosymposia, 7*, 25–32

Honegger, T. G. (1984). Ultrastructure of the adhesive tentacles of the limnomedusa *Vallentinia gabriella* (Hydrozoa, Olindiadidae). *Zoomorphology, 104*, 26–32.

Hunt, S., & Nixon, M. (1981). A comparative study of protein composition in the chitin-protein complexes of the beak, pen, sucker disc, radula and oesophageal cuticle of cephalopods. *Comparative Biochemistry and Physiology Part B: Comparative Biochemistry, 68*, 535–546.

Hussain, J. F., & Bordoloi, S. (2018). Adaptive modifications in four fish species of the genus *Garra* (Teleostei; Cyprinidae) in Basistha river, Assam, India. *Microscopy and Microanalysis, 24*, 310–317.

Ibraheem, M. H. (1998). The mode of attachment of two cestode species infesting the Nile electric catfish *Malapterurus electricus* (Froskal), with special reference to the functional morphology of the scolex and histopathology. *Arab Gulf Journal of Scientific Research, 16*, 159–172.

Jensen, R. A., & Morse, D. E. (1988). The bioadhesive of *Phragmatopoma californica* tubes: A silk-like cement containing L-DOPA. *Journal of Comparative Physiology B, 158*, 317–324.

John, C. C. (1933). Habits, structure, and development of *Spadella cephaloptera*. *Journal of Cell Science, 2*, 625–696.

Jones, H. D., & Trueman, E. R. (1970). Locomotion of the limpet, *Patella vulgata* L. *Journal of Experimental Biology, 52*, 201–216.

Jones, M. K. (2000). Ultrastructure of the scolex, rhyncheal system and bothridial pits of *Otobothrium mugilis* (Cestoda: Trypanorhyncha). *Folia Parasitologica, 47*, 29–38.

Kamino, K. (2010). Molecular design of barnacle cement in comparison with those of mussel and tubeworm. *The Journal of Adhesion, 86*, 96–110.

Kamino, K. (2016). Barnacle underwater attachment. In A. M. Smith (Ed.), *Biological adhesives* (pp. 153–176). Springer International Publishing.

Kamino, K., Odo, S., & Maruyama, T. (1996). Cement proteins of the acorn barnacle, *Megabalanus rosa*. *The Biological Bulletin, 190*, 403–409.

Kampowski, T., Eberhard, L., Gallenmüller, F., Speck, T., & Poppinga, S. (2016). Functional morphology of suction discs and attachment performance of the Mediterranean medicinal leech (*Hirudo verbana* Carena). *Journal of the Royal Society, Interface, 13*.

Kang, V., Johnston, R., van de Kamp, T., Faragó, T., & Federle, W. (2019). Morphology of powerful suction organs from blephariceridlarvae living in raging torrents. *BMC Zoology, 4*, 10.

Kang, V., Lengerer, B., Wattiez, R., & Flammang, P. (2020). Molecular insights into the powerful mucus-based adhesion of limpets (*Patella vulgata* L.). *Open Biology, 10*, 200019.

Kang, V., White, R. T., Chen, S., & Federle, W. (2021). Extreme suction attachment performance from specialised insects living in mountain streams (Diptera: Blephariceridae). *eLife, 10*, e63250.

Kaplan, M. (1997). Internal and external anatomy of the abdominal disc of *Atelopus* (Bufonidae) larvae. *Caldasia, 19*, 61–69.

Karlsson Green, K., Kovalev, A., Svensson, E. I., & Gorb, S. N. (2013). Male clasping ability, female polymorphism and sexual conflict: fine-scale elytral morphology as a sexually antagonistic adaptation in female diving beetles. *Journal of the Royal Society, Interface, 10*, 20130409.

Kier, W. M., & Smith, A. M. (1990). The morphology and mechanics of octopus suckers. *The Biological Bulletin, 178*, 126–136.

Kier, W. M., & Smith, A. M. (2002). The structure and adhesive mechanism of octopus suckers. *Integrative and Comparative Biology, 42*, 1146–1153.

Lawrence, J. M. (1987). *A functional biology of echinoderms*. Croom Helm.

Lengerer, B., & Ladurner, P. (2018). Properties of temporary adhesion systems of marine and freshwater organisms. *Journal of Experimental Biology, 221*, jeb182717.

Lengerer, B., Pjeta, R., Wunderer, J., Rodrigues, M., Arbore, R., Schärer, L., Berezikov, E., Hess, M. W., Pfaller, K., Egger, B., Obwegeser, S., Salvenmoser, W., & Ladurner, P. (2014). Biological adhesion of the flatworm *Macrostomum lignano* relies on a duo-gland system and is mediated by a cell type-specific intermediate filament protein. *Frontiers in Zoology, 11*, 12.

Liang, C., Strickland, J., Ye, Z., Wu, W., Hu, B., & Rittschof, D. (2019). Biochemistry of barnacle adhesion: An updated review. *Frontiers in Marine Science, 6*.

Lippens, P. L. (1974). Ultrastructure of a marine nematode, *Chromadorina germanica* (Buetschli, 1874). *Zoomorphology, 78*, 181–192.

Lujan, N. K., & Conway, K. W. (2015). Life in the fast lane: A review of rheophily in freshwater fishes. In R. Riesch, M. Tobler, & M. Plath (Eds.), *Extremophile fishes* (pp. 107–136). Springer International Publishing.

Maie, T., Schoenfuss, H. L., & Blob, R. W. (2012). Performance and scaling of a novel locomotor structure: Adhesive capacity of climbing gobiid fishes. *Journal of Experimental Biology, 215*, 3925–3936.

Martin, G. G. (1978a). The duo-gland adhesive system of the archiannelids *Protodrilus* and *Saccocirrus* and the turbellarian *Monocelis*. *Zoomorphology, 91*, 63–75.

Martin, G. G. (1978b). Ciliary gliding in lower invertebrates. *Zoomorphology, 91*, 249–261.

Martin, G. G. (1978c). A new function of rhabdites: Mucus production for ciliary gliding. *Zoomorphology, 91*, 235–248.

Minnocci, A., Cianchetti, M., Mazzolai, B., Sebastiani, L., & Laschi, C. (2015). Cryo-scanning electron microscopy investigation of the *Octopus vulgaris* arm structures for the design of an octopus-like arm artefact. *Microscopy Research and Technique, 78*, 1133–1145.

Mitterer, R. M. (1971). Comparative amino acid composition of calcified and non-calcified polychaete worm tubes. *Comparative Biochemistry and Physiology Part B: Comparative Biochemistry, 38*, 405–409.

Nachtigall, W. (1974). *Biological mechanisms of attachment: The comparative morphology and bioengineering of organs for linkage, suction, and adhesion*. Springer.

Naldrett, M. J., & Kaplan, D. L. (1997). Characterization of barnacle (*Balanus eburneus* and *B. cenatus*) adhesive proteins. *Marine Biology, 127*, 629–635.

Nichols, D. (1966). Functional morphology of the water-vascular system. *Physiology of Echinodermata, 219*, 240.

Nixon, M., & Dilly, P. (1977). Sucker surfaces and prey capture. *Symposia of the Zoological Society of London, 38*, 447–511.

Pearson, W. R. (2013). An introduction to sequence similarity ("homology") searching. *Current Protocols in Bioinformatics*, Chapter 3, Unit3.1.
Petrone, L. (2013). Molecular surface chemistry in marine bioadhesion. *Advances in Colloid and Interface Science, 195–196*, 1–18.
Pospekhova, N. A., & Bondarenko, S. K. (2014). Morpho-functional characteristics of the scolex of *Wardium chaunense* (Cestoda: Aploparaksidae). Penetrated into host intestine. *Parasitology Research, 113*, 131–137.
Raine, J. J., Aldred, N., & Clare, A. S. (2020). Anatomy and ultrastructure of the cyprid temporary adhesive system in two species of acorn barnacle. *Journal of Marine Science and Engineering, 8*, 968.
Rietschel, P. (1961). Bau, Funktion und Entwicklung der Haftorgane der Blepharoceridae. *Zeitschrift für Morphologie und Ökologie der Tiere, 50*, 239–265.
Roberts, T. R. (1982). Unculi (horny projections arising from single cells), an adaptive feature of the epidermis of ostariophysan fishes. *Zoologica Scripta, 11*, 55–76.
Rocha, M., Antas, P., Castro, L. F. C., Campos, A., Vasconcelos, V., Pereira, F., & Cunha, I. (2019). Comparative analysis of the adhesive proteins of the adult stalked goose barnacle *Pollicipes pollicipes* (Cirripedia: Pedunculata). *Marine Biotechnology, 21*, 38–51.
Rodrigues, M., Leclère, P., Flammang, P., Hess, M. W., Salvenmoser, W., Hobmayer, B., & Ladurner, P. (2016a). The cellular basis of bioadhesion of the freshwater polyp *Hydra*. *BMC Zoology, 1*.
Rodrigues, M., Ostermann, T., Kremeser, L., Lindner, H., Beisel, C., Berezikov, E., Hobmayer, B., & Ladurner, P. (2016b). Profiling of adhesive-related genes in the freshwater cnidarian *Hydra magnipapillata* by transcriptomics and proteomics. *Biofouling, 32*, 1115–1129.
Röper, H. (1977). Analytische Untersuchungen des Wehrsekretes von *Peripatopsis moseleyi* (Onychophora)/Analytical Investigations of the Defensive Secretion from *Peripatopsis moseleyi* (Onychophora). *Zeitschrift für Naturforschung C, 32*, 57-b.
Sagert, J., Sun, C., & Waite, J. H. (2006). Chemical subtleties of mussel and polychaete holdfasts. In A. M. Smith & J. A. Callow (Eds.), *Biological adhesives* (pp. 125–143). Springer.
Salcedo-Vargas, M. A. (1995). Systematic value of the ultrastructure of the sucker surface in the squid family Mastigoteuthidae (Mollusca: Cephalopoda). *Contributions to Zoology, 65*, 65–77.
Santos, R., da Costa, G., Franco, C., Gomes-Alves, P., Flammang, P., & Coelho, A. V. (2009a). First insights into the biochemistry of tube foot adhesive from the sea urchin *Paracentrotus lividus* (Echinoidea, Echinodermata). *Marine Biotechnology, 11*, 686–698.
Santos, R., & Flammang, P. (2005). Morphometry and mechanical design of tube foot stems in sea urchins: A comparative study. *Journal of Experimental Marine Biology and Ecology, 315*, 211–223.
Santos, R., & Flammang, P. (2006). Morphology and tenacity of the tube foot disc of three common European sea urchin species: a comparative study. *Biofouling, 22*, 187–200.
Santos, R., & Flammang, P. (2007). Intra- and interspecific variation of attachment strength in sea urchins. *Marine Ecology Progress Series, 332*, 129–142.
Santos, R., & Flammang, P. (2008). Estimation of the attachment strength of the shingle sea urchin, *Colobocentrotus atratus*, and comparison with three sympatric echinoids. *Marine Biology, 154*, 37–49.
Santos, R., Hennebert, E., Coelho, A. V., & Flammang, P. (2009b). The echinoderm tube foot and its role in temporary underwater adhesion. In S. N. Gorb (Ed.), *Functional Surfaces in Biology* (pp. 9–41). Springer.
Saxena, S. C., & Chandy, M. (1966). Adhesive apparatus in certain Indian hill stream fishes. *Journal of Zoology, 148*, 315–340.
Saxena, S. C. (1959). Adhesive apparatus of a hill stream cyprinid fish *Garra mullya* (Sykes). *Proceedings of the National Institute of Sciences of India, 25*, 205–214.
Schmidtberg, H. (1999). Ultrastructural studies of the suckers of newly hatched *Eledone moschata* and *Octopus vulgaris* (Mollusca; Cephalopoda). In F. Olóriz & F. J. Rodríguez-Tovar (Eds.), *Advancing research on living and fossil cephalopods* (pp. 203–221). Springer US.

Schoenfuss, H. L., & Blob, R. W. (2003). Kinematics of waterfall climbing in Hawaiian freshwater fishes (Gobiidae): Vertical propulsion at the aquatic–terrestrial interface. *Journal of Zoology, 261*, 191–205.

Shimek, R. L. (1988). The functional morphology of scaphopod captacula. *The Veliger, 30*, 213–221.

Smith, A. M. (1991a). The role of suction in the adhesion of limpets. *Journal of Experimental Biology, 161*, 151–169.

Smith, A. M. (1991b). Negative pressure generated by octopus suckers: A study of the tensile strength of water in nature. *Journal of Experimental Biology, 157*, 257–271.

Smith, A. M. (1992). Alternation between attachment mechanisms by limpets in the field. *Journal of Experimental Marine Biology and Ecology, 160*, 205–220.

Smith, A. M., Kier, W. M., & Johnsen, S. (1993). The effect of depth on the attachment force of limpets. *The Biological Bulletin, 184*, 338–341.

Smith, A. M., Quick, T. J., & St Peter, R. L. (1999). Differences in the composition of adhesive and non-adhesive mucus from the limpet *Lottia limatula*. *The Biological Bulletin, 196*, 34–44.

Stewart, R. J., Ransom, T. C., & Hlady, V. (2011). Natural underwater adhesives. *Journal of Polymer Science Part B, Polymer Physics, 49*, 757–771.

Stewart, R. J., & Wang, C. S. (2010). Adaptation of caddisfly larval silks to aquatic habitats by phosphorylation of h-fibroin serines. *Biomacromolecules, 11*, 969–974.

Stewart, R. J., Weaver, J. C., Morse, D. E., & Waite, J. H. (2004). The tube cement of *Phragmatopoma californica*: A solid foam. *Journal of Experimental Biology, 207*, 4727–4734.

Teimori, A., Esmaeili, H., & Ansari, M. (2011). Micro-structure consideration of the adhesive organ in doctor fish, *Garra rufa* (Teleostei; Cyprinidae) from the Persian Gulf Basin. *Turkish Journal of Fisheries and Aquatic Sciences, 11*, 407–411.

Thomas, L. A., & Hermans, C. O. (1985). Adhesive interactions between the tube feet of a starfish, *Leptasterias hexactis*, and substrata. *The Biological Bulletin, 169*, 675–688.

Tyler, S. (1976). Comparative ultrastructure of adhesive systems in the Turbellaria. *Zoomorphology, 84*, 1–76.

Tyler, S. (1988). The role of function in determination of homology and convergence - Examples from invertebrates adhesive organs. *Fortschritte der Zoologie, 36*, 331–347.

Tyler, S., & Rieger, G. E. (1980). Adhesive organs of the Gastrotricha. *Zoomorphology, 95*, 1–15.

Vera Candioti, F., Haas, A., Altig, R., & Peixoto, O. (2017). Cranial anatomy of the amazing bromeliad tadpoles of *Phyllodytes gyrinaethes* (Hylidae: Lophyohylini), with comments about other gastromyzophorous larvae. *Zoomorphology, 136*, 61–73.

Wainwright, D. K., Kleinteich, T., Kleinteich, A., Gorb, S. N., & Summers, A. P. (2013). Stick tight: suction adhesion on irregular surfaces in the northern clingfish. *Biology Letters, 9*, 20130234.

Waite, J. H. (1992). The formation of mussel byssus: anatomy of a natural manufacturing process. *Results and Problems in Cell Differentiation, 19*, 27–54.

Waite, J. H. (2017). Mussel adhesion - Essential footwork. *Journal of Experimental Biology, 220*, 517–530.

Waite, J. H., Hansen, D. C., & Little, K. T. (1989). The glue protein of ribbed mussels (*Geukensia demissa*): a natural adhesive with some features of collagen. *Journal of Comparative Physiology B, 159*, 517–525.

Walker, G. (1972). The biochemical composition of the cement of two barnacle species, *Balanus hameri* and *Balanus crenatus*. *Journal of the Marine Biological Association of the United Kingdom, 52*, 429–435.

Wang, Y., Yang, X., Chen, Y., Wainwright, D. K., Kenaley, C. P., Gong, Z., Liu, Z., Liu, H., Guan, J., Want, T., Weaver, J. C., Wood, R. J., & Wen, L. (2017). A biorobotic adhesive disc for underwater hitchhiking inspired by the remora suckerfish. *Science Robotics, 2*, eaan8072.

Weiner, J., III, Beaussart, F., & Bornberg-Bauer, E. (2006). Domain deletions and substitutions in the modular protein evolution. *The FEBS Journal, 273*, 2037–2047.

Whittington, I. D., & Cribb, B. W. (2001). Adhesive secretions in the Platyhelminthes. *Advances in Parasitology, 48*, 101–224.

Young, G. A., Yule, A. B., & Walker, G. (1988). Adhesion in the sea anemones *Actinia equina* L. and *Metridium senile* (L.). *Biofouling, 1*, 137–146.

Chapter 17
Convergent Evolution: Theory and Practice for Bioinspiration

Chris Broeckhoven and Anton du Plessis

Abstract For many biological systems different strategies, morphologies and/or behaviours have evolved in response to similar functional demands (a concept known as convergent evolution). The biodiversity on Earth thus holds a wealth of natural strategies that may provide tailored solutions to the social, economic and environmental challenges the world faces—a practice often referred to as biomimicry, biomimetics or bioinspiration. Despite the great potential and increasing popularity of bioinspiration as a research approach, deciding which biological systems to explore remains a challenging and complex task. Not only does the incompleteness of the knowledge about biodiversity inhibit the identification of suitable biological strategies, but also practitioners in the field of bioinspiration often rely on the assumption that natural structures are the result of evolutionary processes that strive for optimization, thereby failing to acknowledge the processes that might constrain adaptive evolution. The purpose of this chapter is threefold. First, we shed light on the evolutionary constraints and limitations that pose potential pitfalls for using biodiversity as a source of inspiration for innovation. Second, we highlight the central role that the study of convergent evolution could and should play in addressing the current challenges to approaches to bioinspiration. Finally, we provide valuable insights into methodological trends that might facilitate the identification and experimental analysis of biological systems and thereby advance our understanding of biological structures in novel ways. By engaging with these three lines of thought, we present a perspective on future directions for bioinspiration, drawing attention to the opportunities for improving the translation of biological knowledge into innovative solutions.

C. Broeckhoven (✉)
Laboratory of Functional Morphology, Department of Biology, University of Antwerp, Wilrijk, Belgium
e-mail: chris.broeckhoven@uantwerpen.be

A. du Plessis
Research Group 3D Innovation, Department of Physics, Stellenbosch University, Stellenbosch, South Africa

Object Research Systems, Montréal, Canada

V. L. Bels, A. P. Russell (eds.), *Convergent Evolution*, Fascinating Life Sciences,
https://doi.org/10.1007/978-3-031-11441-0_17

Keywords Additive manufacturing · Adhesion · Bioinspiration · Biomimetics · Bionics · Computational modelling · Design · Engineering · Micro-computed tomography · Optimization · Trade-offs

17.1 Introduction

Bioinspiration is the scientific discipline that strives to learn from, or imitate, nature and transfer biological knowledge to the engineering domain. Over the last few decades several terms have been put forward to capture the idea of 'learning from nature', including biomimicry, biomimetics and bionics, which we here consider to be synonymous with bioinspiration (Box 17.1).

Box 17.1 Overview of Definitions (ISO/TC266, 2015)

Biomimetics: "Interdisciplinary cooperation of biology and technology or other fields of innovation with the goal of solving practical problems through the function (sic) analysis of biological systems, their abstraction into models, and the transfer into and application of these models to the solution."

Biomimicry or biomimetism: "Philosophy and interdisciplinary design approaches taking nature as a model to meet the challenges of sustainable development (social, environmental, and economic)."

Bio-inspired design or bioinspiration: "Creative approach based on the observation of biological systems."

Bionics: "Technical discipline that seeks to replicate, increase, or replace biological functions by their electronic and/or mechanical equivalents."

Despite differences in their definition and underlying philosophy (Wanieck & Beismann, 2021), these interpretations share the same final outcome: inspiration drawn from biological strategies. Although the definitions have only been established recently (Vincent et al., 2006), the idea of imitating nature has been around for centuries, with notable examples being the Chinese's attempt to produce artificial silk over 3000 years ago and Leonardo da Vinci's efforts to design bird-inspired ornithopters (Vincent et al., 2006). In modern times, the search for bioinspired applications has become a scientific field in its own right that has provided fertile ground for a growing number of product innovations. Indeed, many novel engineering applications aimed at addressing global challenges faced by humanity, or simply at improving the quality of daily life, have been inspired by the biodiversity on Earth. Some of the most well-known examples of bioinspiration include synthetic adhesives inspired by a gecko's ability to climb on walls and ceilings and the invention of Velcro® based on the observation that the burrs of the burdock plant became stuck to the fur of George de Mestral's dog while out for a walk (Fig. 17.1). The growing interest in bioinspiration is evident from the increasing number of studies (Snell-Rood, 2016), specialized journals (Lepora et al., 2013)

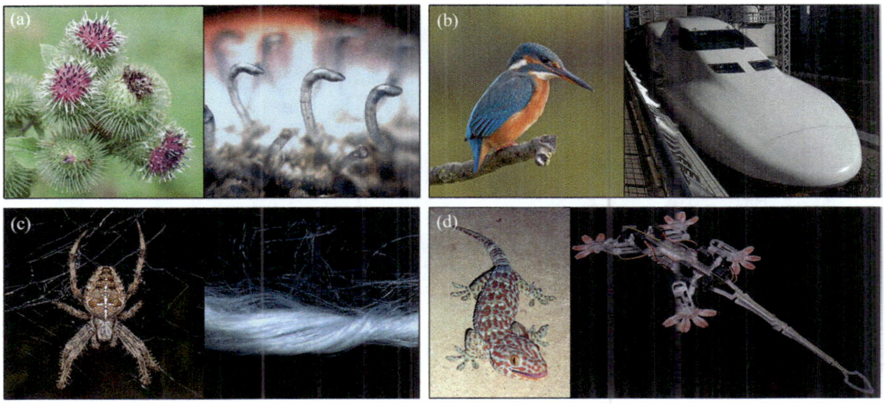

Fig. 17.1 Examples of bioinspired applications: (**a**) burr and hook Velcro® straps, (**b**) "nose" of a high-speed train inspired by the bill of the kingfisher, (**c**) spider-inspired artificial silk and (**d**) Stickybot inspired by the surface of a gecko's foot. Copyright images: Alex Greenhalgh, Spintex Engineering Ltd; Wikimedia Commons (Zeynel Cebeci, Olivepixel, Andreas Trepte, Vgenecr, Didier Descouens, Richard Ling, Dougleasy)

and international competence networks (von Gleich et al., 2010), online repositories with biological strategies such as AskNature (Deldin & Schuknecht, 2014) and the recent surge of patented inventions and spin-off companies (Farnsworth, 2020).

17.2 Current Approaches to Bioinspiration

Bioinspired innovations are by no means 'blueprints from nature', instead the process of bioinspiration is complex, consisting of numerous steps which involve close cooperation between biologists, engineers, designers and other scientists (Speck & Speck, 2008; Fayemi et al., 2017). Bioinspiration distinguishes between a bottom-up approach (also known as biology push or solution-based approach) and a top-down approach (also known as technology pull or problem-based approach), as shown in Fig. 17.2. The former approach starts with a question from biology or a biological discovery that leads to innovation, whereas the latter approach is triggered by a technical problem or the need for improvement of previously existing products or applications (Speck & Speck, 2008). Despite the different starting point and driving forces, both approaches converge at the identification of the study system, after which the relevant biological information is inferred (i.e., generation of the biological concept). The next steps involve the extraction of principles that underlie a specific function of interest, with these principles being separated from their biological model during the abstraction phases. The extraction and abstraction of principles are the most important, and, at the same time, most difficult, steps of the bioinspiration approach (Speck & Speck, 2008).

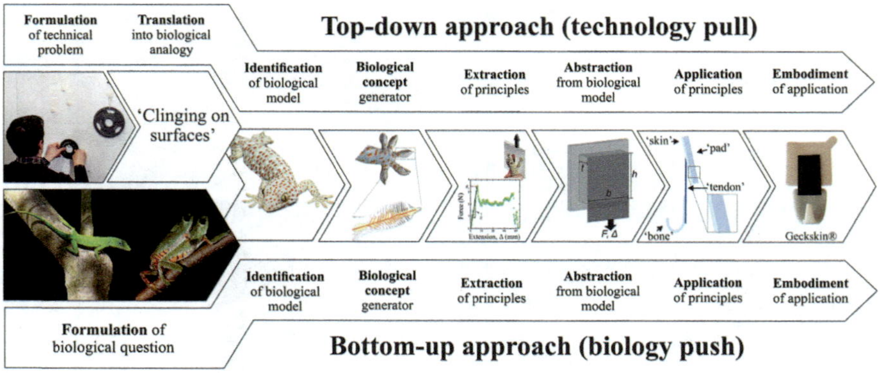

Fig. 17.2 Bioinspiration can be the result of a top-down or bottom-up approach, which stems from a technical problem or a biological question, respectively. After the identification of a suitable biological model or strategy, interdisciplinary knowledge transfer processes, including the extraction of principles and abstraction and application of principles, lead to the development of a bioinspired product of application, in this case the Geckskin® adhesive system. Diagram adapted from Speck et al. (2017) with inserts from Bartlett et al. (2012). Copyright images: Felsuma LLC, Geckskin® technology; Wikimedia Commons (José Roberto Leite, Filo gèn')

Most bioinspired applications to date are the result of bottom-up approaches based on biological discoveries. The development of the Geckskin® mechanical adhesive system, for instance, is a stunning achievement of such a bottom-up approach. Geckos are renowned for their exceptional ability to adhesively cling to smooth, vertical surfaces. Although past research has focussed predominantly on the small adhesive hair-like fibres, or setae, covering the pad of the gecko's foot, recent studies have shown that adhesion is accomplished by the interplay between the integument, muscles, tendons and vascular system (Russell, 2002; Bartlett et al., 2012; King et al., 2014). In geckos, the tendon is attached to the skin of the expanded scales that make up the toe pad at one end and to a muscle at the other end. As soon as a gecko places its toes on a surface, it is believed that the large blood sinuses within the foot pad swell up, pressing the integument against the surface (Russell, 2002). Simultaneously, the tendon pulls on the integument which causes a high level of stiffness in the direction of the applied force and results in the engagement of the adhesive setae. The interplay between the softness of the integument and the high level of stiffness obtained from the tendon helps geckos to hold onto vertical surfaces during climbing (Russell et al., 2019). These principles form the basis of the Geckskin® technology, which uses a soft rubber-like material to imitate the soft foot pad and a stiff fabric that provides the stiffness of the tendon. The Geckskin® adhesive system exemplifies the importance of correct translation of biological principles; earlier applications inspired by setae/spatulae do not scale up to allow large forces, whereas Geckskin® pads of 100 cm^2 can hold up to 300 kg (Bartlett et al., 2012; Patek, 2014).

17.3 Limitations of Bioinspiration

New insights into biological structures and functions gained during bottom-up approaches can often be applied to a large number of technical problems provided the principle extraction and abstraction steps are thoroughly carried out, hence this approach is often considered the most appropriate. However, the bottom-up approach can be extremely time-consuming, taking anywhere from 3 to 7 years for a biological structure with bioinspiration potential to result in the fabrication of a product or application (Speck & Speck, 2008). As a result, a recent shift, driven by industry, appears to have occurred in which the top-down approach is becoming the norm for bioinspiration. Despite the fact that the innovative leaps are expected to be smaller than those obtained through a bottom-up approach, the top-down approach can typically be completed in 6–18 months (Speck & Speck, 2008). In an attempt to save time, the task of biologists during a top-down process is often restricted to the identification of biological systems (Speck et al., 2017). Snell-Rood (2016) revealed that out of 300 biomimetic studies fewer than 10% included scientists working in the field of biology. In other words, biology is key to bioinspiration but, interestingly enough, biologists are more and more being considered as 'outsiders' (Graeff et al., 2019, 2021). Additionally, numerous tools have been developed over the last couple of decades, the purpose of which is the facilitation of the bioinspiration process (Fayemi et al., 2017; Wanieck et al., 2017). The majority of these tools, which paradoxically, are said to come from biology, have been designed by engineers to be used by engineers, not biologists (Graeff et al., 2019). Several challenges are likely to arise from an approach that fails to integrate biological knowledge. The aim of this chapter is to outline these challenges, demonstrate the urgency of addressing them and discuss approaches to solutions, particularly the role that the study of convergent evolution might play in resolving some of the issues of the current approach to bioinspiration.

17.3.1 Biodiversity

Lack of taxonomic knowledge of biodiversity might prevent the identification of suitable biological strategies (Müller et al., 2018). Only a small fraction of the world's biodiversity has been targeted by bioinspiration studies, limiting the number of matches that can be made between technical problems and potential biological solutions (Müller et al., 2018). Most researchers working in the field of bioinspiration consider only a single species and tend to focus on established study systems, including geckos, butterflies and lotus leaves, to name a few (Snell-Rood, 2016). Lotus leaves, for example, have since their original description (Barthlott & Neinhuis, 1997) served as inspiration for more than 1700 follow-up studies on their superhydrophobic properties (Müller et al., 2018) and the so-called *lotus effect* continues to play an important role in the development of bioinspired products (von Gleich et al., 2010). As a result, bioinspiration is characterized by bursts of innovation following biological discoveries, interspersed with long periods of minor incremental advances (Müller et al., 2018). Because the current pace of bioinspiration is not sufficient to serve our need for innovation, there have been substantial efforts to consolidate biological strategies into easily accessible repositories in order to bridge the gap between innovation and stagnation (Fayemi et al., 2017). Exploriation of a wide range of biological strategies will not only reduce the bottleneck and increase the pace of innovation, but will also provide practitioners of bioinspiration with the opportunity to learn from evolutionary patterns (Müller et al., 2018). One of the most comprehensive and well-known repositories is AskNature (Deldin & Schuknecht, 2014). By searching for the function "attach", for instance, 125 biological strategies, including the suction discs of remoras (fish), sticky toepads of tree frogs and adhesive tendrils of passion flowers (https://asknature.org), can be retrieved and used as a source of inspiration for adhesive systems. The advantage of such repositories is that the greater the number of biological strategies that can be identified, the likelier that the best fit can be found between a technical problem and a biological solution, thereby generating increased potential for innovation (Müller et al., 2018). There is, however, a catch in that biological strategies are subjectively selected based on the AskNature researcher's assessment of their bioinspiration potential and might, therefore, not be an accurate representation of the actual biodiversity (Deldin & Schuknecht, 2014). Furthermore, these repositories are often seen as tools to replace biological knowledge, yet descriptions of the biological strategies lack the necessary rigorous experimental data to serve as a suitable replacement for the biological concept generation steps.

17.3.2 Evolutionary Constraints and Limitations

Practitioners in the field of bioinspiration often rely on the assumption that natural structures are the result of evolutionary processes that strive for optimization (Adriaens, 2019). Although it has been widely accepted in the scientific community that evolutionary constraints yield suboptimal solutions that allow organisms to perform 'just good enough' within a specific environment (Fish & Beneski, 2014; Adriaens, 2019), the idea that biological systems are optimized by evolution remains rather persistent in the field of bioinspiration, This is likely due to the fact that optimization is the essence of engineering sciences. Among the main evolutionary constraints that can place limits on optimization, and thus create possible pitfalls for bioinspiration, are (1) phylogenetic constraints, (2) functional constraints and (3) environmental constraints.

Phylogenetic Constraint The results or components of the phylogenetic history of a lineage that limit the future evolutionary pathway of that lineage (McKitrick, 1993), known as phylogenetic constraints, play an important role in limiting adaptation. This implies that the extent to which evolutionary changes in an organism can

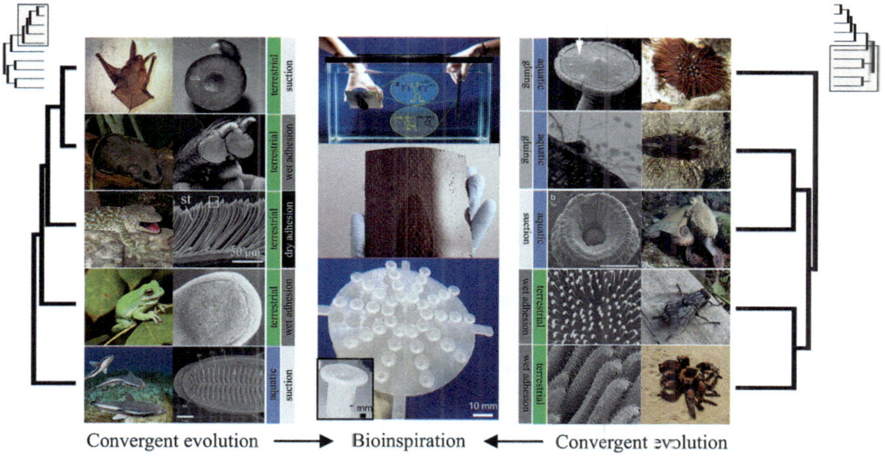

Fig. 17.3 Convergent evolution of adhesive organs in animals. Convergent origins are particularly useful for bioinspiration, not only because multiple biological solutions for a given problem can be explored, but also because they provide the set of environmental conditions within which they are most applicable. Adapted from Walker (1993); Rosenberg and Rose (1999); Riskin and Fenton (2001); Persson and Gorb (2003); Federle et al. (2006); Peattie et al. (2011); Bartlett et al. (2012); Waite et al. (2005); Flammang et al. (2016); Lee et al. (2016); Flammang and Kenaley (2017); North et al. (2017); Wu et al. (2022). Copyright images: Wikimedia Commons (Dan Riskin, Elias Neideck, Hinrich Kaiser, Bruce Dupree, Albert Kok, Jean-Marc Kuffer, Brocken Inaglory, Gilles Gonthier, George Chernilevsky)

occur depends on the pool of options available. To illustrate, there has been considerable convergent evolution upon efficient adhesion strategies in the animal kingdom, resulting in a huge diversity of adhesive organs in insects, arachnids, marine organisms such as echinoderms, cephalopods and molluscs, as well as numerous vertebrates such as fishes, frogs, lizards, bats and gliders (reviewed in Scherge et al., 2001; Fig. 17.3). Although adhesive organs likely evolved for a similar purpose—adhesion—each taxonomic group has its own unique structural, mechanical and chemical features. Furthermore, if the available options are not optimal for a particular solution, any evolutionary transition from the starting point is likely to remain suboptimal (Adriaens, 2019). This is somewhat in contrast to practitioners of bioinspiration who can start from any initial design and modify as needed. In lizards, for example, adhesive toepads evolved convergently in geckos (Gamble et al., 2012; Chap. 9), as well as anoles and a few species of skinks (Irschick et al., 1996). Studies investigating the clinging ability in the three clades have demonstrated that while the toe pads of anoles and geckos are largely functionally equivalent, those of the pad-bearing skinks are clearly functionally inferior (Irschick et al., 1996; Garner et al., 2019, 2021; Russell & Garner, 2021). Geckos and anoles converged upon the same functional endpoint via different evolutionary trajectories (Chap. 9), whereas in skinks, adaptations of the toe pads might be constrained as a result of phylogenetic conservatism, more specifically the ancestral and predominantly non-arboreal lifestyle of the members of the clade (Bars-Closel et al., 2017).

The diversity of adhesive organs in lizards holds significant potential for bioinspiration (Garner et al., 2019; Russell & Garner, 2021), as already demonstrated by the availability of gecko-inspired adhesive products. Although the majority of these studies have focused mainly on adhesion at the microscopic level, the functional control of adhesive organs, which require integration across multiple systems operating at different size scales, has received less attention (Autumn et al., 2002; Gamble et al., 2012). Repeated and convergent evolution of biological structures, such as adhesive organs, may provide a basis for understanding the minimal requirements for a biological system to be operable and functional, which, in turn, will allow researchers to eliminate some of the phylogenetic constraints and provide them with an understanding of the basic 'design rules' (Gamble et al., 2012; Russell & Garner, 2021). In line with the foregoing, rather than concentrating on a single species, "identifying distinct morphological modules from an array of separate evolutionary origins will permit a simpler and more directed approach to understanding how this functionally integrated complex operates" (Gamble et al., 2012).

Functional Constraints The structure-function relationships of the established bioinspiration examples appear to be well-resolved, yet most biological systems are studied with a specific relationship in mind, rather than taking the series of relationships that occur at the organismal level into account. While engineering applications are often designed with a single purpose in mind to maximize their efficiency, biological systems must balance a specific function with a number of competing functions that are vital for the organisms to survive (Fish & Beneski, 2014). Such competing demands on a specific trait that influence multiple aspects of performance are known to result in functional trade-offs (Holzman et al., 2011). While evolutionary changes in a trait might increase one aspect of performance, they might also decrease the performance of other traits. Trade-offs are thought to impose strong constraints on adaptive evolution and, more relevant to the field of bioinspiration, simultaneous optimization of multiple conflicting functions might yield suboptimal solutions (Gould & Lewontin, 1979). To illustrate, in addition to being adhesive, surface lipids and roughness imposed by the setae result in superhydrophobicity of gecko toe pads (Stark et al., 2016). Stark et al. (2016) found that adhesion to glass was greater after removal of the surface lipids, but no difference was present in adhesion to sandpaper or glass coated with hydrophobic octadecyltrichlorosilane self-assembled monolayers, suggesting that surface lipids might impair adhesion on hydrophilic surfaces. Hence, a clear trade-off is present between adhesion and the presence of surface lipids when adhering to hydrophilic surfaces (Stark et al., 2016). Identifying trade-offs is particularly challenging, because the performance costs resulting from competing demands on a specific trait might be mitigated by compensatory changes in another trait (Holzman et al., 2011). Geckos, like insects and spiders, possess claws in addition to their adhesive pads, which might be integral components of the complex attachment system (Naylor & Higham, 2019), and possibly mitigate some of the costs associated with a trade-off between adhesion and surface lipid presence. Identifying the functional trade-offs may therefore not always be trivial (or even possible) and require a proper understanding of the interactions between structures and functions, and those of the organisms in their environment (see also 'environmental constraints'). For example, extremely elaborate body armour evolved independently in two distantly-related species of girdled lizards, the armadillo lizard (*Ouroborus cataphractus*) and giant girdled lizard (*Smaug giganteus*), presumably in response to increased predation risk by mongooses (Broeckhoven et al., 2017). Although the dermal plates, or osteoderms, of both species are sufficiently strong to withstand mongoose bites, the puncture performance differs greatly between the species due to differences in microarchitecture of the osteoderms. In contrast to the rather solid osteoderms of *S. giganteus*, *O. cataphractus* evolved extensive internal vascularization of the osteoderms, which might provide thermoregulatory benefits in the hot environments inhabited by this species, whilst still providing sufficient protection against predators (Broeckhoven et al., 2017). Animals that possess some form of body armour have been the focus of a plethora of bioinspiration studies due to their impact-absorbing properties, yet the multifunctional nature of natural body armour and potential trade-offs, as well as their integration in an ecological context, has yet to receive any serious research attention.

Environmental Constraints One of the main objectives of bioinspiration is to understand how organisms are adapted to their environments and translate these adaptations into a technical context. Like functional trade-offs, unravelling the adaptive nature of organismal traits is not a straightforward process (Adriaens, 2019) and requires the contribution of different environmental pressures to be defined (Vincent, 2016). Importantly, the environment in which an organism evolves may fluctuate in both predictable and unpredictable ways and those changing conditions might influence the dynamics of the adaptive process itself (Boyer et al., 2021). To keep up with constant changes in a dynamic environment, organisms might benefit more from being plastic rather than adapting to a specific environment. By studying entire evolutionary lineages, a better understanding can be obtained of how morphology changes in response to a changing environment (Adriaens, 2019). The study of adaptive radiation—the divergence of species from a common ancestor along different ecological axes—and evolution of 'key innovations' that facilitate changes in these, often novel, environments, might be of great importance in this regard. The analogy to the field of bioinspiration lies in the fact that adaptive radiation provides solutions (morphological variation) to a range of problems (environmental variation). In other words, the biodiversity created through adaptive radiations could form a source of information for investigating how a single trait can be adapted for usage across a suite of related problems (Ma & Müller, 2011). Understanding the principle behind diversification in adaptive radiations would enable engineers to tailor customized solutions in a highly effective manner (Müller et al., 2018).

17.4 Convergent Evolution: An Opportunity to Integrate Evolutionary Constraints in Bioinspiration

The relevance of using biodiversity for bioinspiration goes beyond providing opportunities for individual case studies or bioinspired applications (Ma & Müller, 2011). One of the most powerful tools available to address the evolutionary constraints that limit the potential of bioinspiration and identify true adaptations, is the study of convergent evolution. Convergent evolution occurs when similar traits evolve independently across distantly related clades in response to similar selective pressures (McGhee, 2011). The study of convergent evolution could confer three major benefits for bioinspiration: (1) provide alternate solutions for a given technical problem, (2) provide a unified set of conditions underlying the convergence and thus determine the context within which the solution is most applicable and (3) indicate the minimal set of required attributes needed to solve the technical problem. Firstly, convergent evolution can be explored to determine whether multiple natural solutions exist for a specific biological problem or, conversely, whether there is a single universal solution across lineages to the biological problem. To demonstrate, a textbook example of convergent evolution is powered flight in insects, birds and bats (Taylor et al., 2012). Although the principle behind their flight mechanisms is similar, their basic Bauplan is noticeably different, which reflects differences in ancestry. In recent years, the wings of insects, birds and bats have served as bio-inspiration for the fabrication of unmanned-aerial-vehicles (UAV) including nano- and micro-aerial-vehicles, because their aerodynamic efficiency and high levels of manoeuvrability match the tasks that the UAVs must perform (McMichael & Francis, 1997). In this regard, much attention is devoted to understanding the unsteady flight mechanisms that govern flapping (Shyy et al., 2016). A review study by Abas et al. (2016) shows that, across flying animals, a significant relationship is present between the wing length of a species (i.e., bird, insect, bat) and the type of wing kinematics that is adopted for flying within a specific Reynolds number regime. Focussing on these interspecific differences in wing kinematics may be of particular importance for narrowing the specific needs of a UAV system. Secondly, comparing functional-morphological variation in an ecological context could not only be an additional approach to identify the natural solutions that evolved in response to specific biological problems (Adriaens, 2019), but also provide the set of environmental conditions within which it is most applicable. Referring back to an earlier example, suction attachment is common in aquatic organisms, having evolved convergently in several lineages of fish (e.g., lampreys, clingfish, remoras) and molluscs, amongst others (Fig. 17.3). In the field of bioinspiration, there has, for example, been a particular focus on the suckers of octopuses because of their capability of attaching to uneven surfaces and generating a very high negative pressure (Tramacere et al., 2015). In addition, the presence of a protrusion in the suction cup (i.e., the protuberance) which allows the octopus to remain firmly attached with minimal energy consumption has sparked much interest (Baik et al., 2017). While most research continues to explore the suction mechanisms in aquatic organisms, very little is known about the mechanisms used by terrestrial organisms to create and maintain suction and their associated morphological adaptations (Kuolt et al., 2021). Unlike in an aquatic environment, suction in a terrestrial environment places different mechanical demands on morphology, yet the latter have not been the topic of empirical research (Kier & Smith, 1990; Ditsche & Summers, 2014). Suction cups, for instance, evolved in the Spix's disk-winged bat (*Thyroptera tricolor*) to exploit the roosting opportunities presented by foliage (Riskin & Fenton, 2001) and are also present in leeches of the family Haemadipsidae which adopted a terrestrial lifestyle (Borda et al., 2008). Rather than continuing to focus on aquatic suction mechanisms, the study of convergent evolution could provide practitioners in the field of bioinspiration with insights into the principles of suction across the terrestrial-aquatic interface and be used for a diversity of applications.

Lastly, studying convergent evolution could provide opportunities for unravelling the principles behind a particular functionality across lineages to enable the generation of a set of 'design rules' that can be used to inspire a wider range of applications that might be suitable in a variety of conditions. Such an approach could help to tease apart the role of phylogenetic constraints versus ecological conditions on the evolution of a trait of interest. The importance of this approach was recently demonstrated by Russell and Garner (2021), who provided an overview of the research that examined the configuration of setae in adhesive toepad-bearing lizards, including the convergent evolution of the adhesive apparatus in geckos and anoles. To date, most bioinspired research aimed at designing and fabricating synthetic adhesives has focused on unravelling the complex adhesive apparatus of mainly one species of gecko, the Tokay (*Gekko gecko*). Russell and Garner (2021) show that there is no single morphology of setae for the toe pads of adhesive toepad-bearing lizards, but instead that the morphology (and size) of the setae varies considerably interspecifically and regionally. The authors point out how the study of convergent evolution of adhesive toepads in geckos and anoles has aided in determining the attributes of these complex structures (i.e., the 'design rules') that are necessary for an effective adhesive system capable of attaching to vertical surfaces and revealing the simplest configuration of components that is sufficient to fulfil this function. More specifically, Russell and Garner (2021) demonstrate that adhesive systems do not require high levels of morphological divergence from their ancestral phenotype to attain adhesive competence at whole-organismal levels. The availability of such information might not only simplify the, perhaps unnecessarily, complicated process of bioinspiration, but also could decrease the functional disparities between biological structures and their bioinspired analogues. Furthermore, the aforementioned example reiterates that the limitations of using a single species as a biological model must be recognized in the field of bioinspiration. Only after the evolutionary history of organisms is embraced can the effective translation from biology to the engineering domain be made and the study of convergent evolution might be vital for achieving this goal.

17.5 Tools and Technologies to Advance the Future of Bioinspiration

Taking the aforementioned factors into consideration, it becomes clear that a proper understanding of the evolution of organisms is pivotal for ensuring the success of bioinspired designs. Whereas genetic and genomic data have become increasingly affordable, available and abundant in recent times, the synthesis of accompanying phenotypic data has been lagging behind. Deriving function from biological structures using physical experimentation, for example, is an immensely costly and time-consuming process, especially when a large number of species are involved (Müller et al., 2018). We propose three ways which could increase the efficiency of the current bioinspiration approach: (1) the use of natural history collections and digitization of specimens, (2) the use of computer simulations to facilitate the extraction of form-function relations and (3) the use of additive manufacturing to overcome functional constraints and limitations.

17.5.1 Digitization of Natural History Collections

Practitioners in the discipline of bioinspiration often lack the taxonomic knowledge needed to identify species relevant to a specific technological problem, the evolutionary pressures acting on different features of the organism and the role of the organism within its ecological community (Green et al., 2019). Natural history collections provide a powerful resource for addressing these issues by granting scientists access to an impressive record of biodiversity on Earth. Although the main purpose of natural history collections is to preserve organisms for taxonomic classification, they provide a rich source of information for evolutionary and ecological studies (Holmes et al., 2016). Importantly, they allow researchers to investigate the response of organisms to changing environments, because specimens are often accumulated over time (Holmes et al., 2016; Green et al., 2019). Similarly, paleontological collections provide a window of opportunity for studying evolutionary change over time. The use of fossil specimens allows for the assessment of phenotypic change over extended periods of time in response to extreme environmental conditions. Additionally, paleontological collections could be studied to reveal morphological adaptations to selective pressures that might no longer be experienced by extant organisms. For example, glyptodonts and ankylosaurs convergently evolved extensive body armour, presumably to withstand the powerful tail club blows from conspecific individuals (Arbour & Zanno, 2020). Looking into the structure of osteoderms that comprise these armours, particularly in glyptodonts, could serve as bioinspiration for the development of lightweight impact-absorbing structures such as helmets (du Plessis et al., 2018). In addition to providing unique morphologies and functionalities that can serve as 'paleo-bioinspiration' for technological applications, paleontological collections offer an opportunity for increasing the number of convergent evolutionary events, which contributes additional knowledge to the data derived from extant organisms. To demonstrate, pterosaurs, a lineage of extinct flying reptiles, have been largely overlooked in the search for bioinspired solutions to UAVs (Chatterjee et al., 2013). Unlike birds and bats, pterosaurs supported a long and narrow wing membrane with their forelimb and a single elongated fourth finger. Furthermore, in contrast to the rather homogenous structure of bat wings, the skin membrane in pterosaurs had an elastic section in its proximal region and a stiff section more distally. The distal portions of the wings were strengthened by densely packed fibres, called actinofibrils, which allowed folding and unfolding of the wing membrane similar to the battens of a sail (Chatterjee et al., 2013). Looking at convergent evolution over the geological time-scale will undoubtedly open up new perspectives for the future development of bioinspired designs. Digitization of natural history collections will be pivotal in ensuring access to biodiversity and, over the years, much progress has been made in this regard (Hedrick et al., 2020). Of particular importance to the field of bioinspiration is the use of 3D digitization techniques such as micro-computed tomography, or micro-CT (du Plessis & Broeckhoven, 2019). Micro-CT allows for non-destructive imaging of both the internal and external anatomy of specimens at the macro- and micro-scale and, in the biological sciences, is widely used for taxonomic, morphological, evolutionary and ecological studies (Broeckhoven & du Plessis, 2018). This capability not only allows for high-resolution data to be extracted from a large number of specimens, thus increasing access to the huge diversity in biological systems (Müller et al., 2018), but also provides researchers with the possibility of looking into natural systems with an unprecedented level of detail, thereby improving our understanding of natural structures, all of which can be applied to technological or engineering problems (du Plessis & Broeckhoven, 2019; Fig. 17.4). Digitization of museum specimens can boost the impact of natural history collections by allowing access to highly accurate virtual representations of reference specimens, the so-called 'cybertypes' (Faulwetter et al., 2013) which can be shared by a broad community of actors working in the field of bioinspiration. Furthermore, unlike physical collections which are arranged according to taxonomic categories, it now becomes possible to group digitized collection data according to functionality whilst simultaneously including ecological, phylogenetic and life history information. Such an approach will allow for rapid detection of convergences in structure. Although the development of such repositories seems like a Herculean task, once a sufficient amount of digitized collection data is available, machine-learning algorithms can be explored to process this complex information (Muñoz & Price, 2019; Fig. 17.4).

17.5.2 Virtual Functional Morphology

An obvious limitation of using natural history collection specimens is the extent to which function can be inferred from morphological traits (Green et al., 2019).

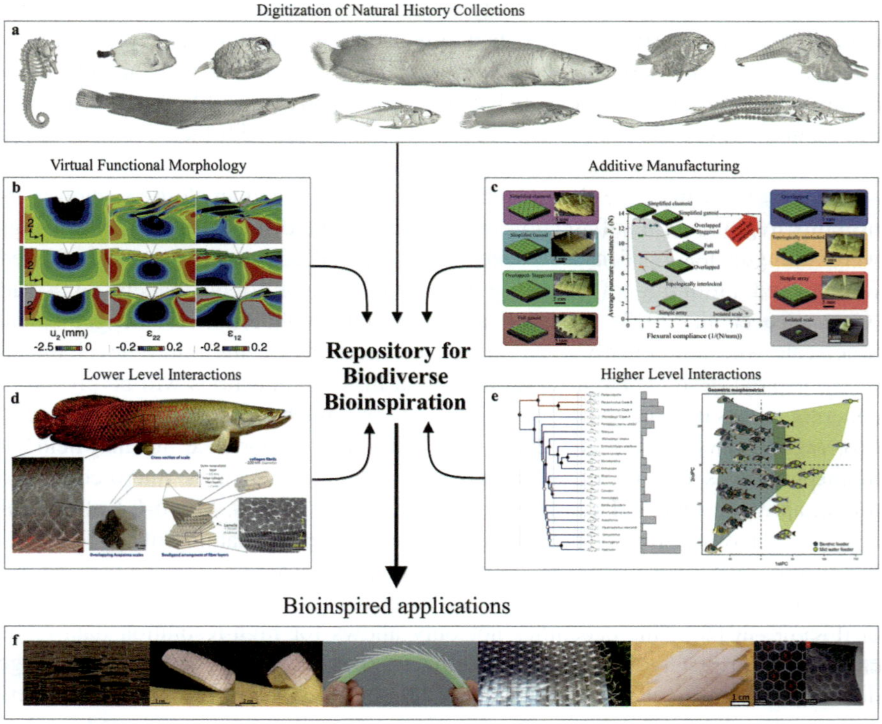

Fig. 17.4 Conceptual diagram illustrating how the digitization of museum collections, computational methods and additive manufacturing can be integrated in bioinspiration studies. In this example, museum specimens are first digitized and their dermal armour visualized by means of micro-computed tomography. Second, these 3D data are used as input for stress simulation analyses, validated by physical experimentation, and additively manufactured (3D-printed) models are created for further analyses, reducing the need to conduct destructive testing of natural specimens. Third, the data can be deposited into digital repositories along with additional information (e.g., life history, phylogeny). Lastly, these digital repositories can be consulted to explore biological strategies relevant to a particular technological problem and be used as bioinspiration for applications or products. In addition, these data open up new possibilities for studying evolutionary biology itself (i.e., 'reverse biomimetics'). From Broeckhoven and du Plessis, (2022)

Identifying the functional properties of biological structures of systems using physical experimentation is a costly, time-consuming endeavour and a (near) impossible task for fossil material and rare species or specimens. Fortunately, rapid advances in computational methods make it possible to accurately simulate the functional properties of biological structures (Rayfield, 2007). The analysis of the functional properties of structures by means of computer modelling and simulations goes hand in hand with computed tomography methods, mainly because these analytical techniques are highly dependent on accurate three-dimensional (3D) reconstructions. Engineering approaches such as finite element (FE) analysis, computational fluid dynamics (CFD) and multibody dynamics analysis, for example, have been successfully employed by applying them to digital models for testing various hypotheses regarding the functionality of biological structures, both for extant and extinct organisms (Cunningham et al., 2014). Most of these analytical techniques require the transformation of the digital model into an FE mesh—a painstaking task that is often prone to error. Recently, it has become possible to perform some of these simulations directly on voxel (i.e., 3D pixels) data derived from micro-CT scans without the need for meshing (Fig. 17.4). Such a method has been adopted, for example, for investigating whether the repeated evolution of venom-conducting fangs in snakes was associated with structural and mechanical changes (Broeckhoven & du Plessis, 2017). One of the main advantages is that biologists working with 3D specimen data are now being provided with an opportunity to explore functionality in a virtual way, without the need for a complete understanding of complex engineering approaches or software. Computational approaches using virtual data can make a significant contribution to the field of bioinspiration on various levels. First and foremost, computational methods play an important role in the convergence of the biological and engineering sciences. For instance, computational simulation methods used to test the mechanical performance of engineered structures can also be used to answer evolutionary questions (Adriaens, 2019) and are therefore pivotal for addressing one of the main hurdles of the current bioinspiration approach—the translation of biological function into engineering principles. Secondly, virtual phenotypes can be created through topology optimization methods and the performance thereof compared to existing phenotypes. This type of information can be used to examine to what extent biological structures are optimized and provide insights into evolutionary and functional constraints of suboptimal designs (Adriaens, 2019). For example, a study by Drol et al. (2019) on hedgehog spines used computer-aided design to create various three-dimensional models that show various levels of complexity, ranging from a simple hollow tube to the most realistic model with longitudinal stringers and transverse central plates. Finite-element analyses simulating flexural behaviour of hedgehog spines demonstrate that the most realistic model has the highest bending stiffness and appears to be the most optimized solution (Drol et al., 2019).

17.5.3 Additive Manufacturing

A critical factor for advancing bioinspiration will be to validate the results of computational modelling analyses by comparing them to measurements obtained from physical experiments conducted on live specimens. While this is a feasible approach for species that can be easily bred or collected, it is more difficult or impossible for the majority of taxa, particularly those that are extinct. An alternative approach is to create physical models of the specimens based on virtual data (e.g., micro-CT scans) by means of additive manufacturing (AM) or 3D-printing technologies (du Plessis & Broeckhoven, 2022; Fig. 17.4). These physical models can be easily and repeatedly subjected to mechanical testing or functional analyses (e.g., CFD) and the results can be compared to those obtained from the simulation analyses. The ability of AM to produce physical models not only reduces the need to remove specimens from their natural environments or to conduct destructive testing of natural specimens, but is also a powerful tool for the discovery of novel functionalities in rare or extinct taxa (Johnson & Carter, 2019). In addition to direct replication, AM offers the opportunity for creating hypothetical structures *de novo*, or with varying degrees of complexity, and opens the possibility for experimentally investigating the roles of evolutionary constraints, including functional trade-offs (Porter & Ravikumar, 2017). For example, using 3D-prints of sea horse tails, Porter et al. (2015) show that having a tail with a square prism cross-section is better for grasping and more resistant than one that is cylindrical. Moreover, by altering existing structures, the functional significance of biological structures can be explored by investigating the contribution of individual structural changes. To demonstrate, Martini et al. (2017) performed puncture and flexural tests on 3D-printed fish scales with increasingly complex geometries, ranging from isolated squares to complex scales with overlapping and interlocking features. Similarly to the study on hedgehog spines by Drol et al. (2019), the structures that offer the best combination of flexural compliance and puncture resistance have a similar geometry and arrangement to teleost and ganoid scales, suggesting a certain degree of shape optimization (Martini et al., 2017). Furthermore, AM can be used for isolating a particular structure from its organismal context. To illustrate, the hydrodynamic performance of sharks is not only determined by the drag-reducing properties of their skin but also depends on body shape and swimming performance (Domel et al., 2018). Additively manufactured shark skin will allow researchers to determine the exact contribution of the denticles (Wen et al., 2014, 2015) prior to their implementation in bioinspired designs. Last, but not least, by exploiting the diversity of materials that can be fabricated using additive manufacturing (du Plessis et al., 2019; Gibson et al., 2021), further insight can be provided into whether the functional performance of biological structures results from their actual form or shape, or the chemical composition of the material comprising the structures (Islam et al., 2021). This is particularly important because the materials and their properties of bioinspired applications will be different from those of the biological structures that served as inspiration. Ultimately, additive manufacturing has the latent ability to advance our understanding of evolutionary adaptations and provide the means of addressing the contribution of evolutionary constraints that limit our current approach to bioinspiration.

17.6 Conclusion and Future Perspectives

Bioinspiration is emerging as a highly promising interdisciplinary approach to innovation. Being an engineer-driven discipline, however, bioinspiration suffers from a number of conceptual shortcomings, particularly the lack of incorporation of evolutionary processes and principles. One of the key challenges outlined in this chapter is the need to address the evolutionary constraints that prevent biological structures from being optimized for a specific function. By considering these constraints as integral components of the bioinspiration process, opportunities can be created not only to improve the translation of biological knowledge into optimally functioning designs, but also to speed up the process of finding the best bioinspired solutions to technical problems. The study of biological systems that have converged towards the same adaptive solution might play a central role in this process. Specifically, it might provide a basis for understanding the minimal requirements for a biological system to be functional at the whole-organismal level and provide practitioners of bioinspiration with a set of 'design rules' free from evolutionary constraints. Although the extraction of accurate functional information from a large number of organisms might have been a limiting factor in the past, emerging new techniques such as micro-computed tomography, computational simulation analyses and additive manufacturing now allow for the collection of this information in a rapid and non-destructive manner. Additionally, these techniques allow researchers to extract functional information from fossil specimens, thereby greatly expanding the diversity of biological systems, and convergent events, that can be exploited for bioinspiration. Not only can the biological information obtained through these methods provide valuable insights for bioinspiration, they also open up new possibilities for studying evolutionary biology itself – a process referred to as 'reverse biomimetics' (Speck et al., 2017; Fig. 17.4).

References

Abas, M. F. B., Rafie, A. S. B. M., Yusoff, H. B., & Ahmad, K. A. B. (2016). Flapping wing micro-aerial-vehicle: kinematics, membranes, and flapping mechanisms of ornithopter and insect flight. *Chinese Journal of Aeronautics, 29*(5), 1159–1177.

Adriaens, D. (2019). Evomimetics: The biomimetic design thinking 2.0. In R. J. Martín-Palma, M. Knez, & A. Lakhtakia (Eds.), *Bioinspiration, biomimetics, and bioreplication IX* (p. 1096509). International Society for Optics and Photonics.

Arbour, V. M., & Zanno, L. E. (2020). Tail weaponry in ankylosaurs and glyptodonts: An example of a rare but strongly convergent phenotype. *The Anatomical Record, 303*(4), 988–998.

Autumn, K., Sitti, M., Liang, Y. A., Peattie, A. M., Hansen, W. R., Sponberg, S., Kenny, T. W., Fearing, R., Israelachvili, J. N., & Full, R. J. (2002). Evidence for van der Waals adhesion in gecko setae. *Proceedings of the National Academy of Sciences, 99*(19), 12252–12256.

Baik, S., Park, Y., Lee, T. J., Bhang, S. H., & Pang, C. (2017). A wet-tolerant adhesive patch inspired by protuberances in suction cups of octopi. *Nature, 546*(7658), 396–400.

Bars-Closel, M., Kohlsdorf, T., Moen, D. S., & Wiens, J. J. (2017). Diversification rates are more strongly related to microhabitat than climate in squamate reptiles (lizards and snakes). *Evolution, 71*(9), 2243–2261.

Barthlott, W., & Neinhuis, C. (1997). Purity of the sacred lotus, or escape from contamination in biological surfaces. *Planta, 202*(1), 1–8.

Bartlett, M. D., Croll, A. B., King, D. R., Paret, B. M., Irschick, D. J., & Crosby, A. J. (2012). Looking beyond fibrillar features to scale gecko-like adhesion. *Advanced Materials, 24*(8), 1078–1083.

Borda, E., Oceguera-Figueroa, A., & Siddall, M. E. (2008). On the classification, evolution and biogeography of terrestrial haemadipsoid leeches (Hirudinida: Arhynchobdellida: Hirudiniformes). *Molecular Phylogenetics and Evolution, 46*(1), 142–154.

Boyer, S., Hérissant, L., & Sherlock, G. (2021). Adaptation is influenced by the complexity of environmental change during evolution in a dynamic environment. *PLoS Genetics, 17*(1), e1009314.

Broeckhoven, C., & du Plessis, A. (2017). Has snake fang evolution lost its bite? New insights from a structural mechanics viewpoint. *Biology Letters, 13*(8), 20170293.

Broeckhoven, C., & du Plessis, A. (2018). X-ray microtomography in herpetological research: A review. *Amphibia-Reptilia, 39*(4), 377–401.

Broeckhoven, C., & du Plessis, A. (2022). Escaping the labyrinth of bioinspiration: Biodiversity as key to successful product innovation. *Advanced Functional Materials, 32*(18), 2110235.

Broeckhoven, C., du Plessis, A., & Hui, C. (2017). Functional trade-off between strength and thermal capacity of dermal armor: insights from girdled lizards. *Journal of the Mechanical Behavior of Biomedical Materials, 74*, 189–194.

Chatterjee, S., Lind, R., & Roberts, B. (2013). The novel characteristics of pterosaurs: Biological inspiration for robotic vehicles. *International Journal of Design & Nature and Ecodynamics, 8* (2), 113–143.

Cunningham, J. A., Rahman, I. A., Lautenschlager, S., Rayfield, E. J., & Donoghue, P. C. (2014). A virtual world of paleontology. *Trends in Ecology & Evolution, 29*(6), 347–357.

Deldin, J. M., & Schuknecht, M. (2014). The AskNature database: Enabling solutions in biomimetic design. In A. K. Goel, D. A. McAdams, & R. B. Stone (Eds.), *Biologically inspired design* (pp. 17–27). Springer.

Ditsche, P., & Summers, A. P. (2014). Aquatic versus terrestrial attachment: Water makes a difference. *Beilstein Journal of Nanotechnology, 5*(1), 2424–2439.

Domel, A. G., Domel, G., Weaver, J. C., Saadat, M., Bertoldi, K., & Lauder, G. V. (2018). Hydrodynamic properties of biomimetic shark skin: Effect of denticle size and swimming speed. *Bioinspiration & Biomimetics, 13*(5), 056014.

Drol, C. J., Kennedy, E. B., Hsiung, B. K., Swift, N. B., & Tan, K. T. (2019). Bioinspirational understanding of flexural performance in hedgehog spines. *Acta Biomaterialia, 94*, 553–564.

du Plessis, A., & Broeckhoven, C. (2019). Looking deep into nature: A review of micro-computed tomography in biomimicry. *Acta Biomaterialia, 85*, 27–40.

du Plessis, A., & Broeckhoven, C. (2022). Functional synergy of biomimicry and additive manufacturing: Toward a bio-enhanced engineering approach. In M. Eggermont, V. Shyam, & A. F. Hepp (Eds.), *Biomimicry for materials, design and habitats* (pp. 269–289). Elsevier.

du Plessis, A., Broeckhoven, C., Yadroitsev, I., Yadroitsava, I., & le Roux, S. G. (2018). Analyzing nature's protective design: The glyptodont body armor. *Journal of the Mechanical Behavior of Biomedical Materials, 82*, 218–223.

du Plessis, A., Broeckhoven, C., Yadroitsava, I., Yadroitsev, I., Hands, C. H., Kunju, R., & Bhate, D. (2019). Beautiful and functional: A review of biomimetic design in additive manufacturing. *Additive Manufacturing, 27*, 408–427.

Farnsworth, M. (2020). *Biomimicry and business: How companies are using nature's strategies to succeed.* Routledge.

Faulwetter, S., Vasileiadou, A., Kouratoras, M., Dailianis, T., & Arvanitidis, C. (2013). Micro-computed tomography: Introducing new dimensions to taxonomy. *ZooKeys, 263*, 1–15.

Fayemi, P. E., Wanieck, K., Zollfrank, C., Maranzana, N., & Aoussat, A. (2017). Biomimetics: Process, tools and practice. *Bioinspiration and Biomimetics, 12*(1), 011002.

Federle, W., Barnes, W. J. P., Baumgartner, W., Drechsler, P., & Smith, J. M. (2006). Wet but not slippery: Boundary friction in tree frog adhesive toe pads. *Journal of the Royal Society Interface, 3*(10), 689–697.

Fish, F. E., & Beneski, J. T. (2014). Evolution and bio-inspired design: Natural limitations. In A. K. Goel, D. A. McAdams, & R. B. Stone (Eds.), *Biologically inspired design* (pp. 287–312). Springer.

Flammang, B. E., & Kenaley, C. P. (2017). Remora cranial vein morphology and its functional implications for attachment. *Scientific Reports, 7*(1), 1–5.

Flammang, P., Demeuldre, M., Hennebert, E., & Santos, R. (2016). Adhesive secretions in echinoderms: A review. In A. M. Smith (Ed.), *Biological adhesives* (pp. 193–222). Springer.

Gamble, T., Greenbaum, E., Jackman, T. R., Russell, A. P., & Bauer, A. M. (2012). Repeated origin and loss of adhesive toepads in geckos. *PLoS One, 7*(6), e39429.

Garner, A. M., Wilson, M. C., Russell, A. P., Dhinojwala, A., & Niewiarowski, P. H. (2019). Going out on a limb: How investigation of the anoline adhesive system can enhance our understanding of fibrillar adhesion. *Integrative and Comparative Biology, 59*(1), 61–69.

Garner, A. M., Wilson, M. C., Wright, C., Russell, A. P., Niewiarowski, P. H., & Dhinojwala, A. (2021). The same but different: The setal arrays of anoles and geckos indicate alternative approaches to achieving similar adhesive effectiveness. *Journal of Anatomy, 238*(5), 1143–1155.

Gibson, I., Rosen, D., Stucker, B., & Khorasani, M. (2021). Materials for additive manufacturing. In I. Gibson, D. Rosen, B. Stucker, & M. Khorasani (Eds.), *Additive manufacturing technologies* (pp. 379–428). Springer.

Gleich, A., Pade, C., Petschow, U., & Pissarskoi, E. (2010). *Potentials and trends in biomimetics*. Springer.

Gould, S. J., & Lewontin, R. (1979). The spandrels of San Marco and the Panglossian paradigm: A critique of the adaptationist programme. *Proceedings of the Royal Society London B Biological Sciences, 205*, 581–598.

Graeff, E., Maranzana, N., & Aoussat, A. (2019). Biomimetics, where are the biologists? *Journal of Engineering Design, 30*(8-9), 289–310.

Graeff, E., Letard, A., Raskin, K., Maranzana, N., & Aoussat, A. (2021). Biomimetics from practical feedback to an interdisciplinary process. *Research in Engineering Design, 32*, 1–27.

Green, D. W., Watson, J. A., Jung, H. S., & Watson, G. S. (2019). Natural history collections as inspiration for technology. *BioEssays, 41*(2), 1700238.

Hedrick, B. P., Heberling, J. M., Meineke, E. K., Turner, K. G., Grassa, C. J., Park, D. S., Kennedy, J., Clarke, J. A., Cook, J. A., Blackburn, D. C., Edwards, S. V., & Davis, C. C. (2020). Digitization and the future of natural history collections. *Bioscience, 70*(3), 243–251.

Holmes, M. W., Hammond, T. T., Wogan, G. O., Walsh, R. E., LaBarbera, K., Wommack, E. A., Martins, F. M., Crawford, J. C., Mack, K. L., Bloch, L. M., & Nachman, M. W. (2016). Natural history collections as windows on evolutionary processes. *Molecular Ecology, 25*(4), 864–881.

Holzman, R., Collar, D. C., Mehta, R. S., & Wainwright, P. C. (2011). Functional complexity can mitigate performance trade-offs. *The American Naturalist, 177*(3), E69–E83.

Irschick, D. J., Austin, C. C., Petren, K., Fisher, R. N., Losos, J. B., & Ellers, O. (1996). A comparative analysis of clinging ability among pad-bearing lizards. *Biological Journal of the Linnean Society London, 59*(1), 21–35.

Islam, M. K., Hazell, P. J., Escobedo, J. P., & Wang, H. (2021). Biomimetic armour design strategies for additive manufacturing: A review. *Materials and Design, 205*, 109730.

ISO/TC266. (2015). *Biomimetics – Terminology, concepts and methodology*. Beuth. ISO 18458: 2015.

Johnson, E. H., & Carter, A. M. (2019). Defossilization: A review of 3D printing in experimental paleontology. *Frontiers in Ecology and Evolution, 7*, 430.

Kier, W. M., & Smith, A. M. (1990). The morphology and mechanics of octopus suckers. *The Biological Bulletin, 178*(2), 126–136.

King, D. R., Bartlett, M. D., Gilman, C. A., Irschick, D. J., & Crosby, A. J. (2014). Creating gecko-like adhesives for "real world" surfaces. *Advanced Materials, 26*(25), 4345–4351.

Kuolt, H., Kampowski, T., Poppinga, S., Speck, T., Tautenhahn, R., Moosavi, A., Weber, J., Gabriel, F., Pierri, E., & Dröder, K. (2021). Biomimetic suction cups for energy-efficient industrial applications. In K. Dröder & T. Vietor (Eds.), *Technologies for economic and functional lightweight design: Zukunftstechnologien für den multifunktionalen Leichtbau* (pp. 182–188). Springer.

Lee, H., Um, D. S., Lee, Y., Lim, S., Kim, H. J., & Ko, H. (2016). Octopus-inspired smart adhesive pads for transfer printing of semiconducting nanomembranes. *Advanced Materials, 28*(34), 7457–7465.

Lepora, N. F., Verschure, P., & Prescott, T. J. (2013). The state of the art in biomimetics. *Bioinspiration & Biomimetics, 8*(1), 013001.

Ma, J., & Müller, R. (2011). A method for characterizing the biodiversity in bat pinnae as a basis for engineering analysis. *Bioinspiration & Biomimetics, 6*(2), 026008.

Martini, R., Balit, Y., & Barthelat, F. (2017). A comparative study of bio-inspired protective scales using 3D printing and mechanical testing. *Acta Biomaterialia, 55*, 360–372.

McGhee, G. R. (2011). *Convergent evolution: Limited forms most beautiful*. MIT Press.

McKitrick, M. C. (1993). Phylogenetic constraint in evolutionary theory: Has it any explanatory power? *Annual Review of Ecology and Systematics, 24*(1), 307–330.

McMichael, J. M., & Francis, M. S. (1997). *Micro air vehicles – Toward a new dimension in flight*. Defense Advanced Research Projects Agency.

Müller, R., Abaid, N., Boreyko, J. B., Fowlkes, C., Goel, A. K., Grimm, C., Jung, S., Kennedy, B., Murphy, C., Cushing, N. D., & Han, J. P. (2018). Biodiversifying bioinspiration. *Bioinspiration & Biomimetics, 13*(5), 053001.

Muñoz, M. M., & Price, S. A. (2019). The future is bright for evolutionary morphology and biomechanics in the era of big data. *Integrative and Comparative Biology, 59*(3), 599–603.

Naylor, E. R., & Higham, T. E. (2019). Attachment beyond the adhesive system: The contribution of claws to gecko clinging and locomotion. *Integrative and Comparative Biology, 59*(1), 168–181.

North, M. A., Del Grosso, C. A., & Wilker, J. J. (2017). High strength underwater bonding with polymer mimics of mussel adhesive proteins. *ACS Applied Materials & Interfaces, 9*(8), 7866–7872.

Patek, S. N. (2014). Biomimetics and evolution. *Science, 345*(6203), 1448–1449.

Peattie, A. M., Dirks, J. H., Henriques, S., & Federle, W. (2011). Arachnids secrete a fluid over their adhesive pads. *PLoS One, 6*(5), e20485.

Persson, B. N. J., & Gorb, S. (2003). The effect of surface roughness on the adhesion of elastic plates with application to biological systems. *The Journal of Chemical Physics, 119*(21), 11437–11444.

Porter, M. M., & Ravikumar, N. (2017). 3D-printing a 'family' of biomimetic models to explain armored grasping in syngnathid fishes. *Bioinspiration & Biomimetics, 12*(6), 066007.

Porter, M. M., Adriaens, D., Hatton, R. L., Meyers, M. A., & McKittrick, J. (2015). Why the seahorse tail is square. *Science, 349*(6243), aaa6683.

Rayfield, E. J. (2007). Finite element analysis and understanding the biomechanics and evolution of living and fossil organisms. *Annual Review of Earth and Planetary Sciences, 35*, 541–576.

Riskin, D. K., & Fenton, M. B. (2001). Sticking ability in Spix's disk-winged bat, *Thyroptera tricolor* (Microchiroptera: Thyropteridae). *Canadian Journal of Zoology, 79*(12), 2261–2267.

Rosenberg, H. I., & Rose, R. (1999). Volar adhesive pads of the feathertail glider, *Acrobates pygmaeus* (Marsupialia; Acrobatidae). *Canadian Journal of Zoology, 77*(2), 233–248.

Russell, A. P. (2002). Integrative functional morphology of the gekkotan adhesive system (Reptilia: Gekkota). *Integrative and Comparative Biology, 42*(6), 1154–1163.

Russell, A. P., & Garner, A. M. (2021). Setal field transects, evolutionary transitions and gecko–anole convergence provide insights into the fundamentals of form and function of the digital adhesive system of lizards. *Frontiers of Mechanical Engineering, 6*, 1–17.

Russell, A. P., Stark, A. Y., & Higham, T. E. (2019). The integrative biology of gecko adhesion: Historical review, current understanding, and grand challenges. *Integrative and Comparative Biology, 59*(1), 101–116.

Scherge, M., Gorb, S. N., & Gorb, S. (2001). *Biological micro-and nanotribology*. Springer.

Shyy, W., Kang, C. K., Chirarattananon, P., Ravi, S., & Liu, H. (2016). Aerodynamics, sensing and control of insect-scale flapping-wing flight. *Proceedings of the Royal Society of London. Series A: Mathematical and Physical Sciences, 472*(2186), 20150712.

Snell-Rood, E. (2016). Interdisciplinarity: Bring biologists into biomimetics. *Nature News, 529*(7586), 277.

Speck, T., & Speck, O. (2008). Process sequences in biomimetic research. In C. A. Brebbia (Ed.), *Design & nature IV: Comparing design in nature with science and engineering* (pp. 3–11). WITPress.

Speck, O., Speck, D., Horn, R., Gantner, J., & Sedlbauer, K. P. (2017). Biomimetic bio-inspired biomorph sustainable? An attempt to classify and clarify biology-derived technical developments. *Bioinspiration & Biomimetics, 12*(1), 011004.

Stark, A. Y., Subarajan, S., Jain, D., Niewiarowski, P. H., & Dhinojwala, A. (2016). Superhydrophobicity of the gecko toe pad: Biological optimization versus laboratory maximization. *Philosophical Transactions of the Royal Society A, 374*, 20160184.

Taylor, G., Carruthers, A., Hubel, T., & Walker, S. (2012). Wing morphing in insects, birds and bats: Mechanism and function. In J. Valasek (Ed.), *Morphing aerospace vehicles and structures* (pp. 13–40). Wiley.

Tramacere, F., Follador, M., Pugno, N. M., & Mazzolai, B. (2015). Octopus-like suction cups: from natural to artificial solutions. *Bioinspiration and Biomimetics, 10*(3), 035004.

Vincent, J. F. (2016). The trade-off: A central concept for biomimetics. *Bioinspired Biomimetic and Nanobiomaterials, 6*(2), 67–76

Vincent, J. F., Bogatyreva, O. A., Bogatyrev, N. R., Bowyer, A., & Pahl, A. K. (2006). Biomimetics: Its practice and theory. *Journal of the Royal Society Interface, 3*(9), 471–482.

Waite, J. H., Andersen, N. H., Jewhurst, S., & Sun, C. (2005). Mussel adhesion: Finding the tricks worth mimicking. *The Journal of Adhesion, 81*(3-4), 297–317.

Walker, G. (1993). Adhesion to smooth surfaces by insects – a review. *International Journal of Adhesion and Adhesives, 13*(1), 3–7.

Wanieck, K., & Beismann, H. (2021). Perception and role of standards in the world of biomimetics. *Bioinspired, Biomimetic and Nanobiomaterials, 10*(1), 8–15.

Wanieck, K., Fayemi, P. E., Maranzana, N., Zollfrank, C., & Jacobs, S. (2017). Biomimetics and its tools. *Bioinspired, Biomimetic, and Nanobiomaterials, 6*(2), 53–66.

Wen, L., Weaver, J. C., & Lauder, G. V. (2014). Biomimetic shark skin: Design, fabrication and hydrodynamic function. *The Journal of Experimental Biology, 217*(10), 1656–1666.

Wen, L., Weaver, J. C., Thornycroft, P. J., & Lauder, G. V. (2015). Hydrodynamic function of biomimetic shark skin: Effect of denticle pattern and spacing. *Bioinspiration & Biomimetics, 10*(6), 066010.

Wu, M., Zheng, X., Liu, R., Hou, N., Afridi, W. H., Afridi, R. H., Guo, X., Wu, J., Wang, C., & Xie, G. (2022). Glowing sucker octopus (*Stauroteuthis syrtensis*)-inspired soft robotic gripper for underwater self-adaptive grasping and sensing. *Advanced Science, 9*(17), 2104382.

Chapter 18
Conclusion and Perspectives: What Convergent Evolution of Animal Forms and Functions Says About the Predictability of Evolution

Paul S. Katz

Abstract Investigations into the convergent evolution of form and function have led to the idea that evolution is, to some extent, predictable. Developmental and physical constraints limit the potential biological forms available for achieving particular functions. The more that forms and functions of animals are compared across the animal phylogeny, the closer we get to creating a mechanistic understanding of biological organization that allows us to make predictions about structure. This is also true for the nervous system, which has not been the subject of much phylogenetic study. Ideal solutions are not always feasible and must be taken into account when modeling neural circuits. For example, although mathematical theories predict that half-center oscillators consisting of two equal halves can produce stable oscillations of neural activity, symmetric half-center oscillators are biologically feasible only when identical contralateral neurons comprise the two halves. Dorsal-ventral rhythmic activity, including flexor-extensor alternation is not produced though a symmetric half-center oscillator because of developmental and physical constraints. It was thought that developmental constraints also limited gross changes to brain anatomy. However, studies of convergent evolution found that brains exhibit mosaic differences in the growth of areas under selection for particular functions. More common than mosaic growth are genomic and genetic convergences on protein expression in particular cells that create functions, as is seen in invertebrate central pattern generators, mammalian echolocation, and fish electrogenesis. Genomic convergence may allow researchers to predict functional convergence by searching for genetic signatures. By studying convergent evolution, we learn the rules of biological organization that allow us to predict form and function, leading to an understanding of the fundamental principles that apply to organisms under all conditions.

P. S. Katz (✉)
Department of Biology, University of Massachusetts Amherst, Amherst, MA, USA
e-mail: pkatz@umass.edu

V. L. Bels, A. P. Russell (eds.), *Convergent Evolution*, Fascinating Life Sciences,
https://doi.org/10.1007/978-3-031-11441-0_18

18.1 Introduction

When the architect, Louis Sullivan declared "*Form ever follows function*", he meant that the form of a modern building should be determined by its function (Sullivan, 1896). Modern biology follows a similar dictum that form follows function in convergent evolution. The chapters in this book profoundly illustrate the extent to which the animal kingdom is replete with examples of convergent evolution of both form and function. Without evolution as a context, finding two examples of the same form in different lineages of animals would be hailed as an amazing coincidence (Dobzhansky, 1973). But, evolutionary theory provides an explanation for the repeated appearance of similar forms as an indication of selective pressure to achieve a particular function and suggests predictability in evolution.

Darwin concludes the Origin of Species with this passage, "*. . . whilst this planet has gone cycling on according to the fixed law of gravity, from so simple a beginning endless forms most beautiful and most wonderful have been, and are being, evolved.*" (Darwin, 1859). He contrasts the consistent movement of the Earth through space, which obeys the fixed law of gravity, with the endless forms of life. Is he suggesting that life is not following a fixed law? More likely, he is subtly suggesting that a law of evolution by natural selection can account for endless forms. I would modify that proposition slightly; convergent evolution indicates that the forms are not endless, but bounded by constraints. The repeated evolution of the same form further suggests that, like the motion of this planet, evolution might also be predictable (Conway Morris, 2010; Stern, 2013; Pankey et al., 2014; Moen et al., 2016). By studying convergent evolution of structures we uncover the rules that govern form and function, leading to the creation of a mechanistic model of the natural world that has predictive value.

There are two types of constraints on evolution of form that lead to predictions: developmental and physical. For example, developmental constraints lead to the prediction that there will never be a Pegasus-type equine because there is no developmental program to allow wings to sprout from the back of a vertebrate as they do in insects. There are also physical constraints on form that lead us to predict that if powered flight ever evolved a fourth time in vertebrates, it would undoubtedly again involve modifications of the forelimbs to create a wing rather that the hind limbs because the physics of flight require balancing the center of gravity.

18.2 Evolution of Brains and Behaviors

Animal behavior is, in large part, governed by the nervous system. The brain is the most complex organ in the body; it has more cell types that display the most diverse morphologies, and it expresses a larger percentage of the genome than any other organ. This complexity causes it to be highly constrained; changing any part of it can have repercussions for other parts (Katz, 2011). Brain organization is highly phylotypic (Striedter & Northcutt, 2019), suggesting that selection does not generally act at the gross structural level. However, the structure of the nervous system is not just its visible morphology; structural elements that affect the function of the nervous system include cellular and subcellular composition such as the number of neurons, their synaptic connectivity, and their expression of genes. There is a revived interest in studying the evolution of neural structure and its relation to function (Barker, 2021; Burkhardt & Jékely, 2021; Iyer & Briggman, 2021; Oteiza & Baldwin, 2021; Tosches, 2021; Cisek & Hayden, 2022). As with gross morphological structures, convergent evolution of cellular and subcellular neural structures is informative of function (Nishikawa, 2002; Katz, 2011). It has been suggested that there could be genomic signatures for convergent behavioral phenotypes (O'Connell & Hofmann, 2012; Gallant & O'Connell, 2020; Barker, 2021; Jourjine & Hoekstra, 2021).

18.3 Convergence on Half-Center Organization Depends Upon Bilateral Symmetry

The simplest neural circuit imaginable is one with only two elements that mutually inhibit each other (Fig. 18.1a). This is the organization for a half-center oscillator first imagined by Brown (1914) to account for locomotion. The concept is quite simply that two identical halves when reciprocally inhibitory will alternate in their activity (Calabrese, 1995). The theory of operation for a half-center oscillator assumes that the two halves are identical (Wang & Rinzel, 1992). If one half were more excitable than the other or if the synapses were not of the same strength, then this simple circuit would not produce equal alternations and would, in many cases, become stuck on one side (Kristan & Katz, 2006).

Neural circuits in many species have convergently evolved this half-center circuit motif for generating alternating leftward and rightward movements. The natural symmetry of bilaterally symmetric animals provided an opportunity that evolution exploited several times to create a half-center oscillator composed of identical halves. In every case, where the circuitry is known, the neurons comprising the two halves of the oscillator are physically on the left and right sides of the nervous system. This is seen in the central pattern generator (CPG) circuits underlying rhythmic left-right body flexions of nudibranchs (Fig. 18.1a) (Sakurai et al., 2014; Sakurai & Katz, 2016), left-right alterations in leech heartbeat (Hill et al., 2001), and undulatory swimming in vertebrates such as lamprey, zebrafish, salamanders, and frog tadpoles (Ryczko et al., 2010). Thus, the inherent symmetry in nervous systems leads to the prediction that neural circuits for left-right alternation will have symmetrical halves.

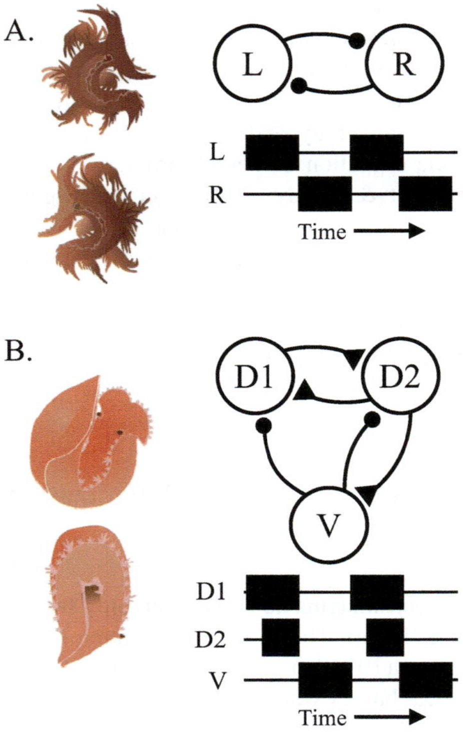

Fig. 18.1 The different neural circuit motifs for left-right and dorsal-ventral rhythmic motor patterns are exemplified in two nudibranchs. (**a**) The nudibranch *Dendronotus iris*, swims with alternating left-right body flexions. The central pattern generator (CPG) is a symmetric half-center oscillator that produces a symmetric left-right alternating motor pattern (bottom). (**b**) The nudibranch, *Tritonia exsulans*, swims with alternating dorsal-ventral body flexions. The CPG has two types of dorsal phase neurons (D1, D2) that alternate with a ventral phase neuron (V) to produce a rhythmic motor pattern (bottom). In the circuits filled circles represent inhibitory synapses and triangles represent excitatory synapses. Nudibranch drawings modified from Sakurai and Katz (2017)

18.4 Dorsal Ventral Asymmetry Constrains the Evolution of Half-Center Oscillators

Not all rhythmic movements involve left-right alternations; there are many examples of dorsally and ventrally directed rhythmic movements, which are synchronous across the midline rather than alternating. There is no natural symmetry in the dorsal-ventral axis for evolution to exploit to create a half-center oscillator. As a result, all of the CPG circuits that produce rhythmic motor patterns in the dorsal-ventral direction contain additional elements to compensate for the lack of symmetry. For example, the CPG for dorsal ventral body flexions in the nudibranch *Tritonia exsulans* (previously *T. diomedea*) has two dorsal elements and one ventral element in a three cell-type circuit (Fig. 18.1b) (Katz, 2009, 2018). Similarly, the segmental CPG underlying dorsal-ventral undulatory movements in leeches is also asymmetric (Kristan et al., 2005). The CPG underlying wing-flapping movements of the pelagic pteropod mollusc, *Clione limacina* also reflects this asymmetry; it has early and late phase dorsal interneurons and a single phase of ventral interneurons (Arshavsky et al., 1998). Thus, the absence of dorsal-ventral symmetry in the body plan is a developmental constraint that limits the evolution of CPG circuits, prohibiting the emergence of a symmetrical half-center oscillator in this plane. This leads to the prediction that the CPG circuits that involve dorsal-ventral oscillations will have asymmetric halves.

There is also a physical constraint that decreases the likelihood of ever evolving a half-center oscillator with dorsal-ventral symmetry, namely gravity. Unlike left-right movements, lift is generated differently by downwardly directed movements than upwardly directly movements. As would be predicted by developmental and physical constraints, the locust flight CPG is asymmetric in the control of elevator and depressor motor neurons (Mantziaris et al., 2020). Although the cellular composition of the central pattern generators underlying flight in birds and bats is not known, given the lack of dorsal ventral symmetry in the spinal cord and the physical constraints of flight, we can predict that the circuit will not be a half-center oscillator composed of identical units.

The vertebrate spinal cord is more complicated than invertebrate circuits. Models of "unit oscillators" that account for extensor - flexor alternation have been proposed. Nearly all of them make assumptions of dorsal – ventral symmetry (Juvin et al., 2007; Sherwood et al., 2011; Shevtsova et al., 2015). Although symmetry is the simplest assumption for a mathematical model, it does not take into account the developmental constraint of asymmetry in the dorsal-ventral axis. Therefore, I predict that these models will be shown to be incorrect in this detail. Work on the cell lineages of spinal neurons and their connectivity has led to a revision of one of those models to include dorsal-ventral asymmetries (Rybak et al., 2015; Ausborn et al., 2021). This illustrates how biological constraints have implications for theoretical predictions of form and function.

18.5 Cellular and Genetic Convergence in Oscillatory Circuits

Although the lack of dorsal-ventral symmetry creates a constraint on the evolution of half-center oscillators, other constraints and opportunities are present that lead to convergent evolution of CPG circuits. Dorsal-ventral swimming movements arose independently in *Tritonia* and another nudipleuran mollusc, *Pleurobranchaea californica* (Newcomb et al., 2012). Like many other protostomes, nudipleurans have neurons that are uniquely identifiable across individual animals within a species and across species regardless of the behaviors that the animals exhibit (Croll, 1987;

Katz & Quinlan, 2018). Two of the same (homologous) neurons were co-opted into the swim CPG circuits in *Tritonia* and *Pleurobranchaea*. Their independent incorporation into a CPG is probably not just a coincidence but rather reflects the fact that these neurons, unlike most others in the brain, are electrically coupled to their contralateral counterparts and synapse on efferent neurons that cause body flexion (Katz, 2009). Thus, they are needed for any whole-body, dorsal—ventral flexion behavior. Other nudipleuran species also have independently evolved this form of swimming (Newcomb et al., 2012). It could be predicted that the neural circuits underlying swimming in these species would incorporate the same set of neurons.

Not only is there a convergence in the incorporation of particular neurons in the CPG, there is also a molecular convergence in the expression of particular serotonin receptors in those homologous neurons. An important, perhaps necessary, feature of the swim CPG circuits in *Tritonia* and *Pleurobranchaea* is that they contain serotonergic neurons that enhance the synaptic strength of other neurons (Katz et al., 1994; Lillvis & Katz, 2013). The convergent evolution of this modulatory action includes convergence in the expression of the genes for two particular serotonin receptor subtypes in one of the identified neurons found in both CPGs (Tamvacakis et al., 2018). Thus, convergent evolution of a behavior controlled by a convergently-evolved CPG circuit composed of homologous neurons may have a genetic signature.

18.6 Genetic Convergence in the Evolution of Echolocation

Genetic signatures for convergent evolution can be seen in the convergent evolution of echolocation, which provides another example of the predictability of evolution. Bats and whales both use echolocation to find prey, yet they are phylogenetically distant, occupy different media (air and water), and are extremely different in size. Yet, despite the different forms of these animals, the different mechanisms they use for generating sounds, and even the difference in how they receive sound, there is a striking convergence of function (Madsen & Surlykke, 2013). There is, however, a convergence on form at the genomic and protein levels (Li et al., 2010; Parker et al., 2013; Lee et al., 2018; Marcovitz et al., 2019). In particular, the gene for the cellular motor protein, Prestin, which plays a role in the active movements of outer hair cells in the cochlea (He et al., 2014), has parallel changes to its amino acid sequence in echolocating bats and whales (Liu et al., 2014). Such a remarkable convergence points to the importance of this mechanism for echolocating mammals.

It was recently discovered that soft-furred tree mice (genus *Typhlomys*) form a third clade of mammals that can echolocate (Panyutina et al., 2017). The researchers predicted that Prestin, among other genes, would show convergence with that of bats and whales, which is what they found (He et al., 2021). Thus, all three echolocating clades of mammals (bats, whales, and tree mice) exploited an opportunity to refine the mechanism of hearing such that it could be sensitive enough to detect echoes of the animal's own vocalizations.

There is evidence for echolocation in other small mammals: shrews and tenrecs (Gould, 1965; Forsman & Malmquist, 1988). One might reasonably predict that these animals would also show a molecular convergence of Prestin sequence and in fact, that turned out to be the case for shrews (Chai et al., 2020). Moreover, a genomic search for rapid changes in Prestin sequences in mammals could be a way to identify additional echolocating clades that have not been behaviorally studied.

Echolocation also independently evolved in birds; oilbirds and swiftlets are both capable of echolocation (Konishi & Knudsen, 1979; Griffin & Thompson, 1982; Jordan Price et al., 2004; Brinkløv et al., 2013). Furthermore, birds also have the Prestin gene (He et al., 2014). However, there have not yet been bioinformatics studies to determine whether its sequence in oilbirds and swiftlets shows convergence with Prestin in mammalian echolocators. Binaural sound localization in the air arose independently the lineages leading to birds and mammals (Nothwang, 2016). The mechanical and neural mechanisms for sound localization have both similarities and differences (Schnupp & Carr, 2009). If echolocating birds do not show a convergence in Prestin gene, it might seem like a failure of the prediction, but as with any experiment that yields unpredicted results, this would better inform a model of the role of Prestin in echolocation and suggest that there may be a different solution to the same problem.

18.7 Genetic Convergence in the Evolution of Active Electrosensing in Fish

Echolocation involves precise sound localization. This ability has not arisen in fish, perhaps due to constraints in the auditory system. However, another ability that serves a similar function arose in fish, active electrosensing. Like echolocation, active electrosensing involves the generation of signal and the sensing of its distortion by prey. Also, like echolocation, active electrosensing allows fish to find prey that are much smaller than themselves under conditions in which visual pursuit would not be effective. Finally, like echolocation, active electrosensing evolved independently in two clades (Bullock et al., 1982; Crampton, 2019) and involves genetic convergence (Zakon et al., 2008; Gallant et al., 2014; Modrell et al., 2017).

There are three steps that needed to occur for the evolution of active electrosensing: (1) the sensory ability to detect electric fields, (2) the motor ability to control the generation of electric fields, and (3) the neural circuits in the brain able to perform the sensory motor integration. The first step, sensing electric fields is very common across fish lineages. Electric fields are readily transmitted in seawater. Hair cells of the lateral line, a mechanosensory organ common in fish, have become specialized to respond to electric fields by isolating them below the skin in ampullary organs. This is likely to be an ancestral trait that was lost in several lineages and regained independently in three teleost lineages (Bullock et al., 1983; Alves-Gomes, 2001). Although the development of electroreceptive organs has been well-studied

(Gibbs, 2004; Modrell et al., 2011), a genetic signature of the convergently-evolved electroreceptors in teleosts has not yet been found (Baker & Modrell, 2018).

The second step, generation of electric fields, evolved independently in at least six different fish clades. Each clade evolved a special type of cell, called an electrocyte, which is a modified muscle cell that has lost its contractile properties, changed its shape, and increased its expression of membrane cation channels. The genomic route to this functional convergence involved the same set of genes (Gallant et al., 2014). Furthermore, electrogenesis is energy intensive; in addition to a genomic signature for the electrocyte there is also strong convergence in the mitochondrial oxidative phosphorylation (OXPHOS) genes of electrogenic fish (Elbassiouny et al., 2020). Thus, electrogenesis in fish is somewhat predictable and has a genomic signature that could be searched in a database to possibly find other examples of electrogenic fish (Wang & Yang, 2021).

Self-generated electric fields for active sensing and communication arose independently in freshwater African mormyrid fish and South American gymnotids (Bullock et al., 1982). There are a variety of environmental factors that favor selection of this trait (Crampton, 2019). The species in these two lineages of weakly electric fish exhibit a great diversity in the duration of electric organ discharges, which derive from parallel changes in the expression and amino acid sequence of a particular sodium channel. Expression of one subtype of sodium channel gene is confined to the electrocytes and excluded from muscle cells. This permitted the sequence to be evolutionarily modified and expressed without affecting muscle function (Arnegard et al., 2010; Thompson et al., 2014). Parallel amino acid substitutions altered the gating kinetics to allow the electric organ discharge to vary in duration (Zakon et al., 2006). Thus, convergent evolution of signal diversity followed a gene duplication event that led to parallel neofunctionalization of that gene.

Active electrosensing is a computationally intense problem (Kawasaki & Guo, 1996; Chacron et al., 2011; McGillivray et al., 2012; Krahe & Maler, 2014). There are particular brain areas, such as the cerebellum, hindbrain, and torus semicircularis, that perform the sensory-motor transformations needed in active electrosensing (Heiligenberg, 1991; Metzner & Heiligenberg, 1991). These areas are highly enlarged in both the African and South American lineages of weakly electric fish (Sukhum et al., 2018; Schumacher & Carlson, 2021). Generally, brain areas are proportionally scaled to each other because of allometric developmental growth (Finlay & Darlington, 1995; Finlay et al., 2011). The enlargement of specific areas associated with active electrosensing occurring in lineages that independently evolved active electrosensing suggests that similar "mosaic" brain region enlargement may be a signature of convergent evolution of function, as has been suggested for mammals and birds (Barton & Harvey, 2000; De Winter & Oxnard, 2001; Moore & DeVoogd, 2017; Mantziaris et al., 2020). The gene regulatory network that leads to regional brain enlargement is not known, but future research may lead to genomic signatures of convergent brain mosaic evolution.

18.8 Summary and Conclusions

Convergent evolution is evidence of the predictability of evolution. There are developmental and physical constraints that have biased the direction of change. There are also genomic constraints and opportunities that shape the evolution of phenotypes. In addition to the examples discussed here, convergent evolution of social behaviors suggests genomic signatures for behavioral traits (Gallant & O'Connell, 2020; Nowicki et al., 2020). The expanding field of genomics and the increasing technological advances in single cell sequencing are likely to uncover new examples of genomic signatures for behavioral traits. An outcome of this could be that a future naturalist might discover behavioral abilities by searching through databases for genetic signatures.

The field of neuroscience has tended to focus on a small number of "model" organisms. Although this approach created tremendous advances in the mechanistic understanding of the neural basis of behavior in those species, it misses out on the opportunity to use convergent evolution as a tool for discovering general rules (Eisthen & Nishikawa, 2002; Nishikawa, 2002; Jourjine & Hoekstra, 2021). To have utility in predicting how brains work, models of neural function must incorporate biological constraints. The recognition of those constraints comes only from comparisons across species.

Although form follows function in biology, a greater challenge is the identification of functions that have converged independently of form. The conceptualization of the rules of function that are independent of form will result in principles that represent fundamental truths about biological organization applicable to all conditions.

References

Alves-Gomes, J. A. (2001). The evolution of electroreception and bioelectrogenesis in teleost fish: A phylogenetic perspective. *Journal of Fish Biology, 58*(6), 1489–1511.

Arnegard, M. E., Zwickl, D. J., Lu, Y., & Zakon, H. H. (2010). Old gene duplication facilitates origin and diversification of an innovative communication system--twice. *Proceedings of the National Academy of Sciences United States of America, 107*(51), 22172–22177.

Arshavsky, Y. I., Deliagina, T. G., Orlovsky, G. N., Panchin, Y. V., Popova, L. B., & Sadreyev, R. I. (1998). Analysis of the central pattern generator for swimming in the mollusk *Clione*. *Annals of the New York Academy of Sciences, 860*, 51–69.

Ausborn, J., Shevtsova, N. A., & Danner, S. M. (2021). Computational modeling of spinal locomotor circuitry in the age of molecular genetics. *International Journal of Molecular Sciences, 22*(13), 6835.

Baker, C. V., & Modrell, M. S. (2018). Insights into electroreceptor development and evolution from molecular comparisons with hair cells. *Integrative and Comparative Biology, 58*(2), 329–340.

Barker, A. J. (2021). Brains and speciation: Control of behavior. *Current Opinion in Neurobiology, 71*, 158–163.

Barton, R. A., & Harvey, P. H. (2000). Mosaic evolution of brain structure in mammals. *Nature, 405*(6790), 1055–1058.
Brinkløv, S., Fenton, M. B., & Ratcliffe, J. M. (2013). Echolocation in oilbirds and swiftlets. *Frontiers in Physiology, 4*, 123.
Brown, T. G. (1914). On the nature of the fundamental activity of the nervous centres; together with an analysis of the conditioning of rhythmic activity in progression, and a theory of the evolution of function in the nervous system. *The Journal of Physiology, 48*(1), 18–46.
Bullock, T. H., Northcutt, R. G., & Bodznick, D. A. (1982). Evolution of electroreception. *Trends in Neurosciences, 5*, 50–53.
Bullock, T. H., Bodznick, D. A., & Northcutt, R. G. (1983). The phylogenetic distribution of electroreception: Evidence for convergent evolution of a primitive vertebrate sense modality. *Brain Research, 287*(1), 25–46.
Burkhardt, P., & Jékely, G. (2021). Evolution of synapses and neurotransmitter systems: The divide-and-conquer model for early neural cell-type evolution. *Current Opinion in Neurobiology, 71*, 127–138.
Calabrese, R. L. (1995). Half-center oscillators underlying rhythmic movements. *Nature, 261*, 146–148.
Chacron, M. J., Longtin, A., & Maler, L. (2011). Efficient computation via sparse coding in electrosensory neural networks. *Current Opinion in Neurobiology, 21*(5), 752–760.
Chai, S., Tian, R., Rong, X., Li, G., Chen, B., Ren, W., Xu, S., & Yang, G. (2020). Evidence of echolocation in the common shrew from molecular convergence with other echolocating mammals. *Zoological Studies, 59*, e4.
Cisek, P., & Hayden, B. Y. (2022). *Neuroscience needs evolution*. The Royal Society.
Conway Morris, S. (2010). Evolution: like any other science it is predictable. *Philosophical Transactions of the Royal Society of London. Series B, Biological Sciences, 365*(1537), 133–145.
Crampton, W. G. (2019). Electroreception, electrogenesis and electric signal evolution. *Journal of Fish Biology, 95*(1), 92–134.
Croll, R. P. (1987). Identified neurons and cellular homologies. In M. A. Ali (Ed.), *Nervous systems in invertebrates*. Plenum Publishing Corporation.
Darwin, C. (1859). *On the origin of species by natural selection, or the preservation of favoured races in the struggle for life*. John Murray.
De Winter, W., & Oxnard, C. E. (2001). Evolutionary radiations and convergences in the structural organization of mammalian brains. *Nature, 409*(6821), 710–714.
Dobzhansky, T. (1973). Nothing in biology makes sense except in the light of evolution. *The American Biology Teacher, 35*, 125–129.
Eisthen, H. L., & Nishikawa, K. C. (2002). Convergence: Obstacle or opportunity? *Brain, Behavior and Evolution, 59*(5–6), 235–239.
Elbassiouny, A. A., Lovejoy, N. R., & Chang, B. S. (2020). Convergent patterns of evolution of mitochondrial oxidative phosphorylation (OXPHOS) genes in electric fishes. *Philosophical Transactions of the Royal Society B, 375*(1790), 20190179.
Finlay, B. L., & Darlington, R. B. (1995). Linked regularities in the development and evolution of mammalian brains. *Science, 268*, 1578–1584.
Finlay, B. L., Hinz, F., & Darlington, R. B. (2011). Mapping behavioural evolution onto brain evolution: The strategic roles of conserved organization in individuals and species. *Philosophical Transactions of the Royal Society of London. Series B, Biological Sciences, 366*(1574), 2111–2123.
Forsman, K., & Malmquist, M. (1988). Evidence for echolocation in the common shrew, Sorex araneus. *Journal of Zoology, 216*(4), 655–662.
Gallant, J. R., & O'Connell, L. A. (2020). Studying convergent evolution to relate genotype to behavioral phenotype. *The Journal of Experimental Biology, 223*(Pt Suppl 1), jeb208934.
Gallant, J. R., Traeger, L. L., Volkening, J. D., Moffett, H., Chen, P. H., Novina, C. D., Phillips, G. N., Jr., Anand, R., Wells, G. B., Pinch, M., Guth, R., Unguez, G. A., Albert, J. S., Zakon,

H. H., Samanta, M. P., & Sussman, M. R. (2014). Genomic basis for the convergent evolution of electric organs. *Science, 344*(6191), 1522–1525.

Gibbs, M. A. (2004). Lateral line receptors: where do they come from developmentally and where is our research going? *Brain, Behavior and Evolution, 64*(3), 163–181.

Gould, E. (1965). Evidence for echolocation in the Tenrecidae of Madagascar. *Proceedings of the American Philosophical Society, 109*(6), 352–360.

Griffin, D. R., & Thompson, D. (1982). Echolocation by cave swiftlets. *Behavioral Ecology and Sociobiology, 10*(2), 119–123.

He, D. Z., Lovas, S., Ai, Y., Li, Y., & Beisel, K. W. (2014). Prestin at year 14: Progress and prospect. *Hearing Research, 311*, 25–35.

He, K., Liu, Q., Xu, D.-M., Qi, F.-Y., Bai, J., He, S.-W., Chen, P., Zhou, X., Cai, W.-Z., Chen, Z.-Z., Liu, Z., Jiang, X.-L., & Shi, P. (2021). Echolocation in soft-furred tree mice. *Science, 372*(6548), eaay1513.

Heiligenberg, W. (1991). The neural basis of behavior: A neuroethological view. *Annual Review of Neuroscience, 14*, 247–267.

Hill, A. A., Lu, J., Masino, M., Olsen, O., & Calabrese, R. L. (2001). A model of a segmental oscillator in the leech heartbeat neuronal network. *Journal of Computational Neuroscience, 10*(3), 281–302.

Iyer, A. A., & Briggman, K. L. (2021). Amphibian behavioral diversity offers insights into evolutionary neurobiology. *Current Opinion in Neurobiology, 71*, 19–28.

Jordan Price, J., Johnson, P., Clayton, K. H., & D. (2004). The evolution of echolocation in swiftlets. *Journal of Avian Biology, 35*(2), 135–143.

Jourjine, N., & Hoekstra, H. E. (2021). Expanding evolutionary neuroscience: Insights from comparing variation in behavior. *Neuron, 109*(7), 1084–1099.

Juvin, L., Simmers, J., & Morin, D. (2007). Locomotor rhythmogenesis in the isolated rat spinal cord: a phase-coupled set of symmetrical flexion–extension oscillators. *The Journal of Physiology, 583*(1), 115–128.

Katz, P. S. (2009). *Tritonia* swim network. *Scholarpedia, 4*(5), 3638.

Katz, P. S. (2011). Neural mechanisms underlying the evolvability of behaviour. *Philosophical Transactions of the Royal Society of London. Series B, Biological Sciences, 366*(1574), 2086–2099.

Katz, P. S. (2018). The *Tritonia* swim central pattern generator. In G. Shepherd & S. Grillner (Eds.), *Handbook of microcircuits* (2nd ed.). Oxford University Press.

Katz, P. S., & Quinlan, P. D. (2018). The importance of identified neurons in gastropod molluscs to neuroscience. *Current Opinion in Neurobiology, 56*, 1–7.

Katz, P. S., Getting, P. A., & Frost, W. N. (1994). Dynamic neuromodulation of synaptic strength intrinsic to a central pattern generator circuit. *Nature, 367*, 729–731.

Kawasaki, M., & Guo, Y.-X. (1996). Neuronal circuitry for comparison of timing in the electrosensory lateral line lobe of the African wave-type electric fish Gymnarchus niloticus. *Journal of Neuroscience, 16*(1), 380–391.

Konishi, M., & Knudsen, E. I. (1979). The oilbird: Hearing and echolocation. *Science, 204*(4391), 425–427.

Krahe, R., & Maler, L. (2014). Neural maps in the electrosensory system of weakly electric fish. *Current Opinion in Neurobiology, 24*(1), 13–21.

Kristan, W. B., & Katz, P. (2006). Form and function in systems neuroscience. *Current Biology, 16*(19), R828–R831.

Kristan, W. B., Calabrese, R. L., & Friesen, W. O. (2005). Neuronal control of leech behavior. *Progress in Neurobiology, 76*(5), 279–327.

Lee, J. H., Lewis, K. M., Moural, T. W., Kirilenko, B., Borgonovo, B., Prange, G., Koessl, M., Huggenberger, S., Kang, C., & Hiller, M. (2018). Molecular parallelism in fast-twitch muscle proteins in echolocating mammals. *Science Advances, 4*(9), eaat9660.

Li, Y., Liu, Z., Shi, P., & Zhang, J. (2010). The hearing gene Prestin unites echolocating bats and whales. *Current Biology, 20*(2), R55–R56.

Lillvis, J. L., & Katz, P. S. (2013). Parallel evolution of serotonergic neuromodulation underlies independent evolution of rhythmic motor behavior. *The Journal of Neuroscience, 33*(6), 2709–2717.

Liu, Z., Qi, F. Y., Zhou, X., Ren, H. Q., & Shi, P. (2014). Parallel sites implicate functional convergence of the hearing gene prestin among echolocating mammals. *Molecular Biology and Evolution, 31*(9), 2415–2424.

Madsen, P. T., & Surlykke, A. (2013). Functional convergence in bat and toothed whale biosonars. *Physiology (Bethesda), 28*(5), 276–283.

Mantziaris, C., Bockemühl, T., & Büschges, A. (2020). Central pattern generating networks in insect locomotion. *Developmental Neurobiology, 80*(1–2), 16–30.

Marcovitz, A., Turakhia, Y., Chen, H. I., Gloudemans, M., Braun, B. A., Wang, H., & Bejerano, G. (2019). A functional enrichment test for molecular convergent evolution finds a clear protein-coding signal in echolocating bats and whales. *Proceedings of the National Academy of Sciences, 116*(42), 21094–21103.

McGillivray, P., Vonderschen, K., Fortune, E. S., & Chacron, M. J. (2012). Parallel coding of first- and second-order stimulus attributes by midbrain electrosensory neurons. *Journal of Neuroscience, 32*(16), 5510–5524.

Metzner, W., & Heiligenberg, W. (1991). The coding of signals in the electric communication of the gymnotiform fish Eigenmannia: From electroreceptors to neurons in the torus semicircularis of the midbrain. *Journal of Comparative Physiology A, 169*(2), 135–150.

Modrell, M. S., Bemis, W. E., Northcutt, R. G., Davis, M. C., & Baker, C. V. H. (2011). Electrosensory ampullary organs are derived from lateral line placodes in bony fishes. *Nature Communications, 2*(1), 496.

Modrell, M. S., Lyne, M., Carr, A. R., Zakon, H. H., Buckley, D., Campbell, A. S., Davis, M. C., Micklem, G., & Baker, C. V. H. (2017). Insights into electrosensory organ development, physiology and evolution from a lateral line-enriched transcriptome. *eLife, 6*, e24197.

Moen, D. S., Morlon, H., & Wiens, J. J. (2016). Testing convergence versus history: Convergence dominates phenotypic evolution for over 150 million years in frogs. *Systematic Biology, 65*(1), 146–160.

Moore, J. M., & DeVoogd, T. J. (2017). Concerted and mosaic evolution of functional modules in songbird brains. *Proceedings of the Royal Society B: Biological Sciences, 284*(1854), 20170469.

Newcomb, J. M., Sakurai, A., Lillvis, J. L., Gunaratne, C. A., & Katz, P. S. (2012). Homology and homoplasy of swimming behaviors and neural circuits in the Nudipleura (Mollusca, Gastropoda, Opisthobranchia). *Proceedings of the National Academy of Sciences of the United States of Americal, 109*(Suppl 1), 10669–10676.

Nishikawa, K. C. (2002). Evolutionary convergence in nervous systems: Insights from comparative phylogenetic studies. *Brain Behavior and Evolution, 59*(5–6), 240–249.

Nothwang, H. G. (2016). Evolution of mammalian sound localization circuits: A developmental perspective. *Progress in Neurobiology, 141*, 1–24.

Nowicki, J. P., Pratchett, M. S., Walker, S. P. W., Coker, D. J., & O'Connell, L. A. (2020). Gene expression correlates of social evolution in coral reef butterflyfishes. *Proceedings of the Royal Society B: Biological Sciences, 287*(1929), 20200239.

O'Connell, L. A., & Hofmann, H. A. (2012). Evolution of a vertebrate social decision-making network. *Science, 336*(6085), 1154–1157.

Oteiza, P., & Baldwin, M. W. (2021). Evolution of sensory systems. *Current Opinion in Neurobiology, 71*, 52–59.

Pankey, M. S., Minin, V. N., Imholte, G. C., Suchard, M. A., & Oakley, T. H. (2014). Predictable transcriptome evolution in the convergent and complex bioluminescent organs of squid. *Proceedings of the National Academy of Sciences of the United States of America, 111*(44), E4736–E4742.

Panyutina, A. A., Kuznetsov, A. N., Volodin, I. A., Abramov, A. V., & Soldatova, I. B. (2017). A blind climber: The first evidence of ultrasonic echolocation in arboreal mammals. *Integrative Zoology, 12*(2), 172–184.

Parker, J., Tsagkogeorga, G., Cotton, J. A., Liu, Y., Provero, P., Stupka, E., & Rossiter, S. J. (2013). Genome-wide signatures of convergent evolution in echolocating mammals. *Nature, 502*(7470), 228–231.

Rybak, I. A., Dougherty, K. J., & Shevtsova, N. A. (2015). Organization of the mammalian locomotor CPG: review of computational model and circuit architectures based on genetically identified spinal interneurons. *ENeuro, 2*(5).

Ryczko, D., Dubuc, R., & Cabelguen, J. M. (2010). Rhythmogenesis in axial locomotor networks: An interspecies comparison. *Progress in Brain Research, 187*, 189–211.

Sakurai, A., & Katz, P. S. (2016). The central pattern generator underlying swimming in *Dendronotus iris*: A simple half-center network oscillator with a twist. *Journal of Neurophysiology, 116*(4), 1728–1742.

Sakurai, A., & Katz, P. S. (2017). Artificial synaptic rewiring demonstrates that distinct neural circuit configurations underlie homologous behaviors. *Current Biology, 27*(12), 1721–1734. e1723.

Sakurai, A., Gunaratne, C. A., & Katz, P. S. (2014). Two interconnected kernels of reciprocally inhibitory interneurons underlie alternating left-right swim motor pattern generation in the mollusc *Melibe leonina*. *Journal of Neurophysiology, 112*(6), 1317–1328.

Schnupp, J. W., & Carr, C. E. (2009). On hearing with more than one ear: Lessons from evolution. *Nature Neuroscience, 12*(6), 692–697.

Schumacher, E. L., & Carlson, B. A. (2021). Convergent mosaic brain evolution is associated with the evolution of novel electrosensory systems in teleost fishes. *bioRxiv*.

Sherwood, W. E., Harris-Warrick, R., & Guckenheimer, J. (2011). Synaptic patterning of left-right alternation in a computational model of the rodent hindlimb central pattern generator. *Journal of Computational Neuroscience, 30*(2), 323–360.

Shevtsova, N. A., Talpalar, A. E., Markin, S. N., Harris-Warrick, R. M., Kiehn, O., & Rybak, I. A. (2015). Organization of left-right coordination of neuronal activity in the mammalian spinal cord: Insights from computational modelling. *The Journal of Physiology, 593*(11), 2403–2426.

Stern, D. L. (2013). The genetic causes of convergent evolution. *Nature Reviews. Genetics, 14*(11), 751–764.

Striedter, G. F., & Northcutt, R. G. (2019). *Brains through time: A natural history of vertebrates*. Oxford University Press.

Sukhum, K. V., Shen, J., & Carlson, B. A. (2018). Extreme enlargement of the cerebellum in a clade of teleost fishes that evolved a novel active sensory system. *Current Biology, 28*(23), 3857–3863.e3853.

Sullivan, L. H. (1896). The tall office building artistically considered. *Lippincott's Magazine*.

Tamvacakis, A. N., Senatore, A., & Katz, P. S. (2018). Single neuron serotonin receptor subtype gene expression correlates with behaviour within and across three molluscan species. *Proceedings of the Royal Society B: Biological Sciences, 285*, 20180791

Thompson, A., Vo, D., Comfort, C., & Zakon, H. H. (2014). Expression evolution facilitated the convergent neofunctionalization of a sodium channel gene. *Molecular Biology and Evolution, 31*(8), 1941–1955.

Tosches, M. A. (2021). From cell types to an integrated understanding of brain evolution: The case of the cerebral cortex. *Annual Review of Cell and Developmental Biology*.

Wang, X.-J., & Rinzel, J. (1992). Alternating and synchronous rhythms in reciprocally inhibitory model neurons. *Neural Computation, 4*(1), 84–97.
Wang, Y., & Yang, L. (2021). Genomic evidence for convergent molecular adaptation in electric fishes. *Genome Biology and Evolution, 13*(3), evab038.
Zakon, H. H., Lu, Y., Zwickl, D. J., & Hillis, D. M. (2006). Sodium channel genes and the evolution of diversity in communication signals of electric fishes: convergent molecular evolution. *Proceedings of the National Academy of Sciences of the United States of America, 103*(10), 3675–3680.
Zakon, H. H., Zwickl, D. J., Lu, Y., & Hillis, D. M. (2008). Molecular evolution of communication signals in electric fish. *The Journal of Experimental Biology, 211*(Pt 11), 1814–1818.